W9-CHD-988

ANNUAL REVIEW OF NUCLEAR AND PARTICLE SCIENCE

ANNUAL REVIEW OF NUCLEAR AND PARTICLE SCIENCE

VOLUME 42, 1992

J. D. JACKSON, *Editor*
University of California, Berkeley

HARRY E. GOVE, *Associate Editor*
University of Rochester

VERA LÜTH, *Associate Editor*
Superconducting Super Collider Laboratory, Dallas

ANNUAL REVIEWS INC 4139 EL CAMINO WAY P.O. BOX 10139 PALO ALTO, CALIFORNIA 94303-0897

ANNUAL REVIEWS INC.
Palo Alto, California, USA

International Standard Serial Number: 0163-8998
International Standard Book Number: 0-8243-1542-1
Library of Congress Catalog Card Number: 53-995

∞ The paper used in this publication meets the minimum requirements of American National Standards for Information Sciences—Permanence of Paper for Printed Library Materials, ANSI Z39.48-1984.

TYPESET BY MARYLAND COMPOSITION COMPANY, INC.
PRINTED AND BOUND IN THE UNITED STATES OF AMERICA

PREFACE

This volume is the forty second in the series that began as the *Annual Review of Nuclear Science* and the fifteenth with the augmented title, *Annual Review of Nuclear and Particle Science*. The change in title between 1977 and 1978 marked a change in Editorship and reflected the reality of related, but distinct, fields of nuclear and particle physics. In the interval since, the distinctions among nuclear physics, particle physics, and astrophysics have become, if anything, less sharp. Perhaps it is better to say that the fields each have retained a well-defined center, but have developed many regions of overlap and interdependence. That interdependence is most tellingly illustrated by the question of the number of light neutrinos with normal couplings to hadrons. Astrophysical arguments led to the belief that there could only be a "few," less than six or five or four depending on the arguer. Meticulous experiments at LEP, described in Volume 41 by Burkhardt and Steinberger, established the number as three, with very small errors. Meanwhile, some nuclear beta decay experiments were pointing to the emission with small probability of a different neutrino with a rest energy of 17 keV, light by the standards of LEP but disastrously heavy for cosmology. The 17-keV neutrino has not become established enough for review here, but such a possibility shows dramatically the coupling among seemingly disparate fields as we get ever closer to a coherent description of all subatomic phenomena.

In the past fifteen years, nuclear physics has increasingly exploited the opportunities provided by heavy-ion collisions at all energies to explore reaction dynamics, highly deformed nuclei, and the quest for the quark-gluon plasma, among other topics. The current volume illustrates the diversity, with reviews of excitation energy sharing in dissipative collisions (J. Tõke and W. U. Schröder), angular momentum distributions in subbarrier fusion reactions (R. Vandenbosch), and hadron interferometry in heavy-ion collisions (W. Bauer, C.-K. Gelbke, and S. Pratt). Other nuclear physics reviews in the current volume are on delayed fission (H. L. Hall and D. C. Hoffman) and nuclear structure at high excitations (J. J. Gaardhøje).

Relativistic heavy-ion collisions occupy increasing numbers of nuclear (and some particle) physicists at increasing energies. Fifteen years ago this series had an article on the high-energy interactions of nuclei by A. S. Goldhaber and H. H. Heckman, with a few GeV per mass unit the definition of high energy. Today we have a review of relativistic heavy-ion physics at CERN and BNL by J. Stachel and G. R. Young

(*continued*)

in which relativistic means 15 to several hundred GeV per mass unit. With multiparticle production, these collisions are different in degree, but not in kind, from so-called elementary particle collisions. The description of nucleons and mesons as collections of partons (see the current chapter on parton distribution functions of hadrons by J. F. Owens and W.-K. Tung) belies the adjective elementary. Indeed, existing and planned facilities provide beams of relativistic heavy ions as well as protons for interleaved experimental programs.

Nuclear theory has over the years addressed reaction dynamics, nuclear structure, and issues such as the conditions for, and signals of, the quark-gluon plasma with a variety of models, some based on QCD and others more empirical. In 1978 nuclear theory was represented by a review of self-consistent calculations of nuclear properties with phenomenological effective forces by P. Quentin and H. Flocard. In 1992 M. Oka and A. Hosaka review the theory of nucleons as Skyrmions, an approach that was one of the earliest uses of topological considerations in field theory, still with interest.

In 1978 particle physics was at the end of the "quark decade," in which the reality of quarks as the actual constituents of hadrons had gradually become established, first by the experiments on deep inelastic scattering of electrons and then of muons and neutrinos, and triumphantly in 1974–1976 by the discovery of the charm quark and soon after the b quark. In 1978, the general subject of quarks was reviewed by O. W. Greenberg; for heavy quarks T. Appelquist, M. Barnett, and K. Lane reviewed the progress and future prospects in Charm and Beyond. Today, charm and b quarks are still of interest, with a review of the hadroproduction of charm particles by J. A. Appel, while the prospects for detailed studies of *CP* violation in B physics are surveyed by Y. Nir and H. R. Quinn. Progress in accurate calculation of decay amplitudes for mesons consisting of a heavy and a light quark are reviewed by B. Grinstein. Together with the chapters in last year's volume of decays of B mesons by K. Berkelman and S. L. Stone and on charm baryons by J. G. Körner and H. W. Siebert, these reviews span much of current interest in heavy-quark systems. On the purely theoretical side, the chapter in the present volume on classical solutions in quantum field theories by E. J. Weinberg is an accessible review of a number of important concepts (solitons, instantons, etc) that find widespread application by theorists, but may be foreign to mere mortals.

The establishment of the reality of quarks has been accompanied by the building and verification of the Standard Model of electroweak interactions and QCD. Last year the impressive progress on the electroweak sector was summarized by H. Burkhardt and J. Steinberger in their review of work at LEP on electron-positron collisions in the vicinity of the Z resonance. This year S. Bethke and J. Pilcher review the LEP data on hadronic channels as tests of perturbative QCD. These two chapters present the considerable fruits of the initial program at LEP. They, together with the detailed comparison with theory last year by W. J. Marciano, show the Standard Model in excellent shape, with only the top quark and the Higgs particle missing from its completion (if that is what Mother Nature has in mind).

The nuclear and particle physics side of astrophysics and planetary physics has always found space in this series. Fifteen years ago C. Rolfs and H. P. Trautvetter reviewed experimental nuclear astrophysics and R. N. Clayton reviewed isotopic anomalies in the early solar system. In 1992 explosive hydrogen burning in stellar evolution is surveyed by A. E. Champagne and M. Wiescher.

The tools and techniques of our fields are worthy of review because of their intrinsic interest to practitioners in those fields and also because of their applications outside. In 1978 the uses of synchrotron radiation in condensed matter and biology was the subject of a chapter by H. Winick and A. Bienenstock, while the uses of muon spin rotation were reviewed by J. H. Brewer and K. M. Crowe. In the present volume this tradition is represented by the chapter on the nuclear microprobe by M. B. H. Breese, G. W. Grime, and F. Watt, with description of the tools and applications to fields as diverse as biology and microelectronics.

After fifteen volumes, Harry Gove, Roy Schwitters, and now Vera Lüth, and I (and all of you) mark an evolution toward higher energies in both nuclear and particle physics, with increasingly long intervals between new facilities. CEBAF and RHIC are still in the future for US nuclear physics. The electron-positron-proton collider facility HERA in Hamburg is just now coming into operation for physics, and the SSC and LHC are almost a decade away. In astrophysics, too, the time scale is lengthening, with satellite-borne facilities more and more dominating the frontier of observations. Are our fields slowing down and in danger because of cost and complexity? Certainly, facilities are growing larger and more expensive in material and human resources, but the two years of LEP running solved many problems that were the subject of investigations for decades. The discovery power of large detectors at the CERN proton-antiproton collider and now at the Fermilab Tevatron shows that large investment pays off in proportion. If

(*continued*)

the past is any guide, the next fifteen years will be as impressive as the last fifteen, as future experiment and theory address the difficult but fundamental issues of our fields. The *Annual Review of Nuclear and Particle Science* will, we hope, chronicle the progress.

J. D. JACKSON
EDITOR

Annual Review of Nuclear and Particle Science
Volume 42 (1992)

CONTENTS

SOME RELATED ARTICLES IN OTHER *ANNUAL REVIEWS*

From the Annual Review of Astronomy and Astrophysics, Volume 30 (1992):

Cosmological Applications of Gravitational Lensing, R. D. Blandford and R. Narayan

From Steam to Stars to the Early Universe, William A. Fowler

The Cosmological Constant, Sean M. Carroll, William H. Press, and Edwin L. Turner

Observations of the Isotropy of the Cosmic Microwave Background Radiation, Anthony C. S. Readhead and Charles R. Lawrence

The Origin of the X-Ray Background, A. C. Fabian and X. Barcons

From the *Annual Review of Earth and Planetary Sciences*, Volume 20 (1992):

Adventures in Lifelong Learning, Philip H. Abelson

From the *Annual Review of Energy and the Environment*, Volume 17 (1992):

Progress in Fusion Energy, U. Colombo and U. Farinelli

Radioactive-Waste Management in the United States: Evolving Policy Prospects and Dilemmas, John P. Holdren

From the *Annual Review of Fluid Mechanics*, Volume 24 (1992):

Dynamo Theory, P. H. Roberts and A. M. Soward

From the *Annual Review of Materials Science*, Volume 22 (1992):

X-Ray Tomographic Microscopy (XTM) Using Synchrotron Radiation, John H. Kinney and Monte C. Nichols

From the *Annual Review of Physical Chemistry*, Volume 43 (1992):

Atmospheric Ozone, Harold S. Johnston

ANNUAL REVIEWS INC. is a nonprofit scientific publisher established to promote the advancement of the sciences. Beginning in 1932 with the *Annual Review of Biochemistry*, the Company has pursued as its principal function the publication of high quality, reasonably priced *Annual Review* volumes. The volumes are organized by Editors and Editorial Committees who invite qualified authors to contribute critical articles reviewing significant developments within each major discipline. The Editor-in-Chief invites those interested in serving as future Editorial Committee members to communicate directly with him. Annual Reviews Inc. is administered by a Board of Directors, whose members serve without compensation.

ANNUAL REVIEWS OF
Anthropology
Astronomy and Astrophysics
Biochemistry
Biophysics and Biomolecular Structure
Cell Biology
Computer Science
Earth and Planetary Sciences
Ecology and Systematics
Energy and the Environment
Entomology
Fluid Mechanics
Genetics
Immunology
Materials Science
Medicine
Microbiology
Neuroscience
Nuclear and Particle Science
Nutrition
Pharmacology and Toxicology
Physical Chemistry
Physiology
Phytopathology
Plant Physiology and Plant Molecular Biology
Psychology
Public Health
Sociology

SPECIAL PUBLICATIONS

Excitement and Fascination of Science, Vols. 1, 2, and 3

Intelligence and Affectivity, by Jean Piaget

For the convenience of readers, a detachable order form/envelope is bound into the back of this volume.

Annu. Rev. Nucl. Part. Sci. 1992. 42:1–38

THE NUCLEAR MICROPROBE

M. B. H. Breese, G. W. Grime, and F. Watt

Nuclear Physics Laboratory, Keble Road, Oxford University, Oxford, OX1 3RH, United Kingdom

KEY WORDS: elemental analysis, quantitivity, imaging technique

---

CONTENTS

INTRODUCTION

The Nuclear Microprobe has continuously evolved over the last twenty years and has now reached a level of sophistication matching any analytical instrument available. The impetus for this development comes largely from the sensitivity and quantitivity of the analytical techniques

0163–8998/92/1201–0001$02.00

available, which can image trace element distributions, measure concentrations of parts per million, and determine the depth structure of samples.

In the Nuclear Microprobe a focused beam of MeV light ions is scanned across the sample surface. The most commonly used MeV ion is the proton, which is why the Nuclear Microprobe is also sometimes called the Scanning Proton Microprobe. However, other MeV light ions can generate the same analytical signals as protons, and are preferred for some of the analytical techniques described. The term Nuclear Microscopy is used in this review as a general title for measuring the analytical signals produced when a focused MeV light ion beam is incident on a sample. The basic operating principles of the Nuclear Microprobe are very similar to the scanning electron microscope, where the electrons are accelerated to keV energies and focused onto the sample surface. In a Nuclear Microprobe, a low energy ion beam, typically 20 keV, is injected into a megavolt electrostatic accelerator and emerges as an MeV ion beam. Because of the much higher energy and mass of MeV ions compared with keV electrons, conventional cylindrical electron focusing lenses are not strong enough to focus an MeV ion beam to a short enough working distance. Quadrupole lenses have a much stronger focusing action than cylindrical lenses because their major field component is perpendicular rather than parallel to the ion beam direction, and so most Nuclear Microprobes use a combination of magnetic quadrupoles as the final focusing element. The focused beam is scanned over the sample surface, and the strength of the relevant analytical signal is measured at each position in the scanned area to generate an image of the sample. Many of the analytical techniques available with the Nuclear Microprobe require at least 100 pA of beam current to generate enough signals to make a measurement in a reasonable time, and recent advances in quadrupole lens construction techniques have resulted in lenses capable of focusing 100 pA of 3 MeV protons into $0.3 \times 0.3\ \mu m^2$. This high resolution imaging capability is central to the success of the Nuclear Microprobe as an analytical facility.

This review discusses the analytical techniques available with the Nuclear Microprobe, and then its capabilities are assessed against other competing techniques to show its strengths and weaknesses. Problems in producing and focusing a beam of MeV light ions are then briefly discussed. Since the importance of the Nuclear Microprobe lies in its ability to give unique analytical information, the rest of this review is devoted to demonstrating the wide range of applications covered by the Nuclear Microprobe. Applications of the Nuclear Microprobe in microelectronics, crystallography, metallurgy, medicine, biology, geology, and archaeology are then described, and the advantages of Nuclear Microprobe analysis over other techniques are considered in each case.

1. ANALYTICAL TECHNIQUES

There are many different types of interaction that can occur when an MeV ion is incident on a sample, and each one forms the basis of an analytical technique. Many of these analytical techniques have widespread uses (1–4) even without using a focused ion beam but when they are used in conjunction with the high resolution imaging capability of the Nuclear Microprobe their range of applications greatly increases.

The MeV ion beam loses energy in the sample at a rate (dE/dx) that depends on the type of ion, its energy, and the sample atomic number; the rate is usually given in $(\text{mg cm}^{-2})^{-1}$. If the sample contains different elements, the compound energy loss (dE/dx) is

$$\left(\frac{dE}{dx}\right)_C = \sum p_Z \left(\frac{dE}{dx}\right)_Z, \qquad 1.$$

where p_Z and $(dE/dx)_Z$ are the weight fractions and energy loss due to each element present. The rate of energy loss of ions through matter has been extensively studied, and tabulated values accurate to within a few percent may be found in References 4–6. It is this accurate characterization of the energy loss and range of MeV ions in the sample that is crucial for the quantitivity of associated MeV analytical techniques. Only a small fraction of incident MeV ions approach close enough to interact with the sample atomic nuclei or inner electron shells, so many incident ions are needed to produce a measurable signal. An important criterion for each analytical technique is thus the probability, or cross section σ, for producing that analytical signal with a given number of incident ions. The higher the cross section the less beam current is needed to make a measurement in a given time. This is very important for Nuclear Microprobe analysis because the beam current is limited by the need to obtain a high spatial resolution focused probe. Nuclear Microprobe analysis is thus best suited to those analytical techniques with a large signal production cross section.

The statistical nature of the interaction of the MeV ions with the sample results in individual ions losing different amounts of energy after having traveled through the same sample thickness. This energy straggling limits the depth resolution attainable with MeV ion techniques (7). The interaction of the MeV ions with the sample also causes lateral straggle, which is a fundamental limitation on the spatial resolution attainable using the Nuclear Microprobe. Lateral straggle is, however, much less serious for MeV ions than for keV electrons. Thus, although the Nuclear Microprobe cannot focus an MeV ion beam into as small an area on the sample surface as an electron microscope can focus a keV electron beam, the spatial resolution attainable with the Nuclear Microprobe can exceed that of the electron microscope.

The most commonly used analytical techniques are outlined in the following sections.

1.1 *Particle-Induced X-Ray Emission (PIXE)*

A vacancy is created in the electron shell of a sample atom if energy greater than the binding energy is supplied. This principle is most commonly used with a keV electron beam in an electron microscope. A keV electron beam has a similar velocity to the inner shell electrons of the sample nuclei, which maximizes the vacancy production cross section and hence the x-ray yield. MeV light ions, typically 3-MeV protons, also have a high cross section for ejecting K- or L-shell electrons because their velocity is also comparable to the inner shell velocity (8, 9). The vacancy created in the inner shell is filled by an electron from an outer shell with subsequent emission of either a characteristic elemental x ray or an auger electron (10). PIXE analysis (8, 11–15) measures the x rays produced from the sample by an incident MeV ion beam, and because it has a high cross section compared to the other MeV techniques a beam current of about 100 pA is perfectly adequate for analysis.

The dominant source of background in the measured x-ray spectrum using electron probe microanalysis (EPMA) is the brehmsstrahlung radiation emitted by the incident keV electrons as they undergo many large angle scattering events with the sample electrons, which limits the elemental sensitivity. An incident MeV ion beam generates about a million times less brehmsstrahlung radiation (16) than a keV electron beam, making PIXE a much more sensitive technique for imaging trace element distributions. Figure 1 shows a comparison between the measured x-ray spectrum from the same area of a pollen tube using 2.5-MeV protons and 10-keV electrons (17). The high background under the EPMA x-ray spectrum results in the trace elements Fe, Cu, and Zn not being detected, whereas they are all visible in the PIXE spectrum. Legge et al give another example of the lower background in a PIXE spectrum compared with an EPMA spectrum (18).

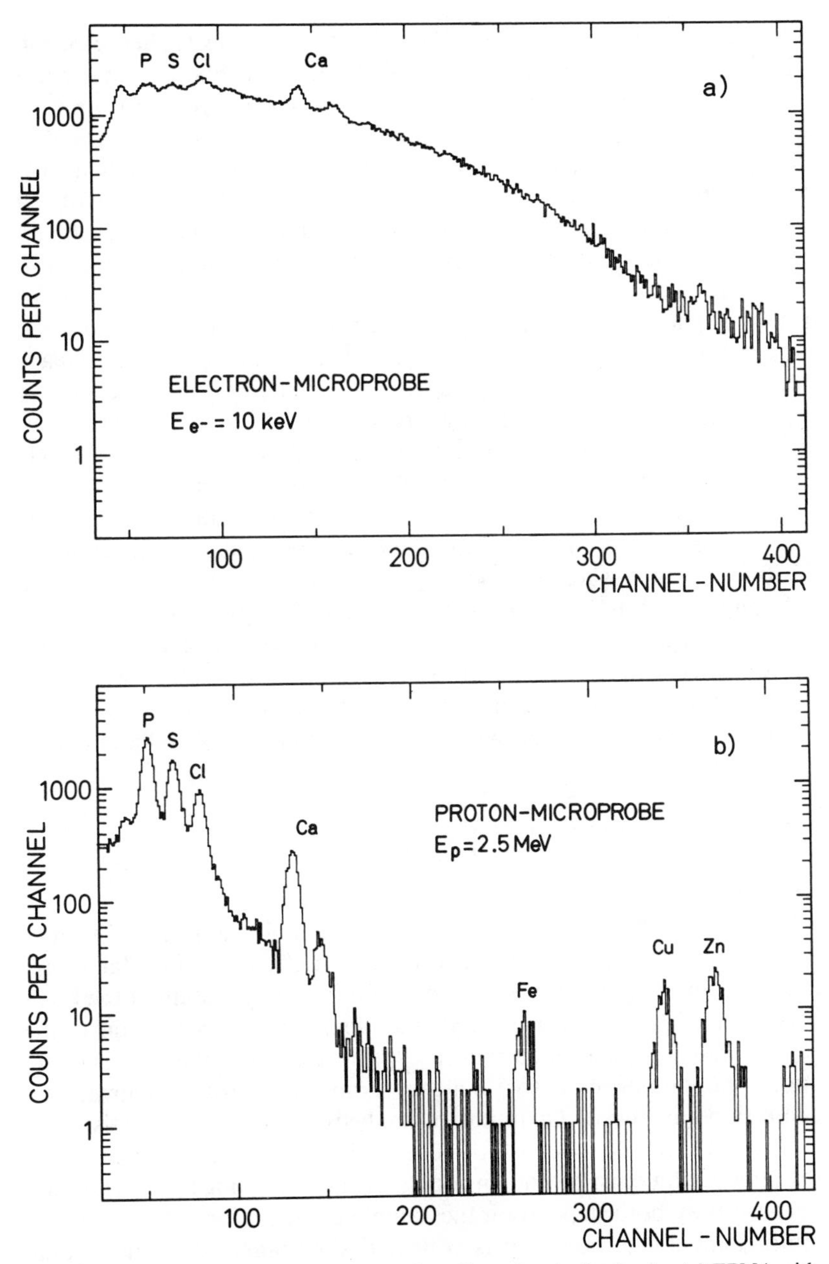

Figure 1 A comparison of x-ray spectra of a pollen tube obtained using (*a*) EPMA with 10-keV electrons, and (*b*) PIXE with 2.5-MeV protons.

For a K-shell vacancy a Kα x ray is emitted if an L-shell electron fills the vacancy; or if an M- or N-shell electron fills the vacancy, a more energetic Kβ x ray is emitted. Similarly Lα, Lβ, and Lγ x rays are caused by an L-shell vacancy filled by an electron from a higher shell. X rays are usually measured with a lithium drifted silicon (SiLi) detector in PIXE analysis, which has a typical energy resolution of 150 eV. This means that all elements heavier than sodium are resolved from each other, but this energy resolution is insufficient to measure any microstructure of the α, β, and γ lines (19). The SiLi detector is usually separated from the target chamber vacuum by an 8-μm thick Be window, and because x rays emitted from the sample must pass through this window, the SiLi detection efficiency for low energy x rays is poor as they suffer strong attenuation. This light element x-ray attenuation and energy resolution of the SiLi limits the use of PIXE to the detection and measurement of elements heavier than sodium.

The fraction of K-shell vacancies that give rise to an x ray, and not to an auger electron, is called the fluorescence yield ω_K, and similarly the fraction of L-shell vacancies giving rise to an x ray is ω_L. The fraction of these K or L x rays that give the α, β, or γ line is given by the branching ratio b. Numerical values for x-ray energies, fluorescence yields, and branching ratios are tabulated in References 12, 20, and 21. When a thin uniform homogeneous target of areal density D, containing a fraction p_z by weight of an element (A, Z) is bombarded with N-MeV ions, the x-ray yield from each element is

$$Y(Z) = \left[\frac{N_{av}\sigma_z(E_0)b_z\epsilon_z\omega_z}{A_z}\right] Np_zD, \qquad 2.$$

where $\sigma_z(E_0)$, b_z, ω_z, ϵ_z, and N_{av}, are the ionization cross section, branching ratio, fluorescent yield, detection efficiency for that x ray, and Avogadro's number respectively. If the x-ray yield under the bombardment of N ions is measured, the elemental concentration in the sample can be determined with a typical accuracy of 10%. This accuracy can be improved to 5% if thin sample standards of known concentrations are used. Quantitative analysis of thick samples is more difficult (13, 22–24) because the energy lost in the sample and the x-ray attenuation must be known accurately and knowing them is difficult because they both depend on the sample composition. In conclusion, multi-elemental information is collected simultaneously with an analytical sensitivity of less than 0.1 ppm at best for elements heavier than sodium.

1.2 *Rutherford Backscattering Spectrometry (RBS)*

MeV ions lose most of their energy through interaction with outer valence electrons of the sample atoms, but occasionally an ion approaches closely enough to interact either elastically or inelastically with a sample nucleus. In RBS the energy of an elastically backscattered incident ion, typically 2-MeV α particles, is measured (7), and this is a powerful analytical technique for determining the stoichiometry and depth distribution of the sample.

The elastic scattering is caused by Coulomb repulsion between the ion and the sample nucleus, and the scattered incident ion retains all its energy except that lost making the sample nucleus recoil. The energy of an ion scattered through θ degrees is given in nonrelativistic form by

$$E_1 = E_0 \frac{m_1{}^2}{(m_1 + m_2)^2} \left[\cos\theta + \left(\frac{m_2{}^2}{m_1{}^2} - \sin^2\theta \right)^{1/2} \right]^2, \qquad 3.$$

where m_1 and m_2 are the masses of the incident and sample nuclei and E_0 is the energy of the incident ion scattered at a given depth in the sample. The cross section of scattering through θ degrees (4) is given by

$$\frac{d\sigma_R}{d\Omega} = 1.296 \left(\frac{zZ}{E_0} \right)^2 \left[\sin^{-4}\frac{\theta}{2} - 2\left(\frac{m_1{}^2}{m_2} + \cdots \right) \right] \text{mb sr}^{-1}, \qquad 4.$$

where z and Z are the charges of the incident ion and the sample nucleus. From Equation 4 the backscattering cross section increases as the incident ion energy decreases and also as the charge on the incident ion or sample nucleus increases. Because the scattering cross section depends on $\sin^{-4}(\theta/2)$, there is progressively less beam scattered through larger backward angles. Since the measured ion energy depends on the sample atomic number and the rate of energy lost by the recoiling ion along its inward and outward paths, the composition and depth distribution of the sample can be determined. Figure 2 shows a typical RBS energy spectrum measured with a silicon surface barrier detector using a monoenergetic ion beam scattered through 170° from a progressively thicker sample containing three different elements. There is a peak from each element at the energy calculated from Equation 3, and the heavier element causes ions to backscatter with higher energy. The areas of the peaks will be in the ratios determined by the relative concentrations in the sample and the cross sections given by

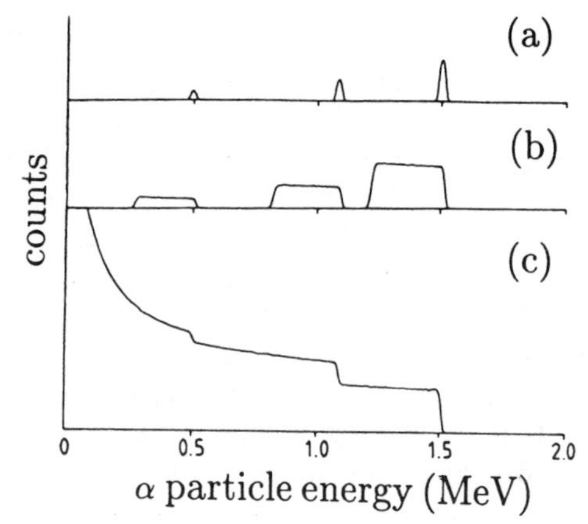

Figure 2 RBS spectra obtained when 2-MeV α particles are backscattered from a progressively thicker sample of Co, Al, and C in equal amounts by weight.

Equation 4. If the beam fluence and surface barrier detector solid are known, then the sample composition may be determined from the measured RBS energy spectrum. The width of the signal from each element in the energy spectrum increases as the sample thickness increases because of the additional energy lost by ions recoiling from deeper in the sample; thus the measured RBS energy spectrum can be used to create depth profiles for elements.

The depth resolution attainable with RBS depends on the accuracy to which the energy of an ion recoiling from a given depth can be measured, and this depth resolution can be degraded in several ways. The energy straggle of the ion beam moving through the sample is one limitation on determining the depth at which the ion was backscattered. From Equation 3 there is also a kinematic energy spread due to the range of scattering angles accepted by the surface barrier detector. This is minimized by having the detector at steep backward angle close to 180°, and by accepting only a narrow angular range of recoiling ions. The energy resolution of the surface barrier detector and its electronics is typically 15 keV for a backscattered 2-MeV α particle. Using 2-MeV α particles at a steep backward angle with a narrow angular acceptance, a depth resolution of 20 nm for about the top 0.5 μm of the sample is attainable.

1.3 *Nuclear Reaction Analysis (NRA)*

Several types of MeV ions can overcome the Coulomb barrier height and interact with the sample nucleus (1, 4, 25). This inelastic collision causes structural changes to the sample nucleus, and reaction products such as a particles, protons, neutrons (27), or γ rays (28, 29) are emitted. The cross section for an incident MeV light ion interacting with the sample nucleus depends on the ion energy and on the internal properties of the nuclei involved in a complex manner, but a simple rule is that the ion will react with the nucleus only if

$$E > \frac{zZ}{(a^{1/3} + A^{1/3})} \text{ MeV}, \qquad 5.$$

where E is the ion energy, and a and A are the atomic weights of the incident ion and sample nuclei. The probability of interaction thus drops off with increasing sample atomic weight, because of the increased Coulomb barrier height. In practice, the interaction of incident ion and sample nucleus is very sensitive to the incident beam energy, leading to sharp resonances in the reaction cross section as the beam energy is varied. Loosely bound ions such as the deuteron and ^{3}He nucleus, however, can interact with the light sample elements at a lower energy than given by Equation 5. A typical NRA measurement uses a 1–10-MeV incident light ion beam to generate MeV protons or α particles as charged particle reaction products from light elements present in the heavier sample matrix. The energy of these charged reaction products is measured using a surface barrier detector, and since the only ions entering the detector are from the particular nuclear reaction being excited, there is very little background in the measured spectrum, so NRA can achieve sensitivities of 0.01 ppm. Light elements can also be depth profiled using NRA by measuring the energy loss of the charged particle reaction products during their transit out of the sample, in a manner similar to RBS. NRA has cross sections typically 1000 times smaller than either PIXE or RBS, so it is not practical in most cases to generate an NRA image of the sample because it takes too long. Most Nuclear Microprobe applications of NRA thus usually involve measuring light element concentrations at specific points on the sample surface.

1.4 *Scanning Transmission Ion Microscopy (STIM)*

STIM was developed specifically as a Nuclear Microprobe technique for imaging the density distribution of thin samples, and not for elemental analysis. An MeV light ion beam is sufficiently energetic to pass through samples less than about 50 μm thick, and the energy lost by the ion beam at each position in the scanned area depends on the elemental composition and thickness. In STIM the transmitted energy spectrum from the beam scanning over the sample surface is measured with a surface barrier detector and divided into energy windows. Areas of different energy loss by the beam are imaged by measuring the distribution of counts in each window.

STIM has until recently only been used as an imaging technique to locate sample features for subsequent analysis with PIXE or RBS (30–32), but it shows great promise as an analytical technique in its own right because of its rapid imaging capability. In STIM, the energy of nearly all the incident ions is measured, so a very small beam current (<1 fA) is used for analysis. Since this beam current is about 10^{-5} of that needed for PIXE, RBS, and NRA analyses, STIM can achieve a higher spatial resolution, and a probe size of 50 nm diameter has been reported (33). There is no reason why even higher resolution STIM probes cannot be attained when limitations due to ions scattered from the Microprobe apertures and stray magnetic fields are eliminated.

If a STIM image of the thin sample is recorded at many different sample orientations, then a three-dimensional image of the specimen density can be reconstructed. This technique is called Ion Microtomography (IMT) (34–36) and is developing into a useful tool for nondestructive materials characterization.

1.5 *Ion Channeling*

If the incident MeV light ion beam is collimated such that it has an angular convergence of ≤0.1°, and if it is aligned with a crystal axis of the sample, then the lattice planes steer the ion beam through the crystal. This channeling of the ion beam results in the sample nuclei and inner shell electrons being shielded from the beam, which reduces the production cross sections for the various analytical signals. The channeled ions suffer less energy loss because of this, until at some depth in the crystal the thermal vibration of the lattice causes the beam to dechannel. The change in measured yield between channeled and nonchanneled alignment gives information on crystal quality, epitaxy, and the lattice position of interstitial elements (37, 38). Most uses of ion channeling measure the change in yield of backscattered ions to give depth-resolved channeling information, but the measured yield using PIXE (39, 40) and NRA (41) also decreases in channeling alignment, and both of these techniques have also been used to identify interstitial impurity elements using ion channeling.

The Nuclear Microprobe has a unique channeling capability in that it can image variations of the channeled yield over the sample using the measured RBS or PIXE yield, to give information on variations in crystal quality across the surface. This technique is called Channeling

Contrast Microscopy (CCM) (42). Because of the reduced energy lost in a channeled compared to a nonchanneled orientation, the energy of channeled beam transmitted through thin crystals also gives information on the crystalline quality through the full sample thickness. This technique of measuring the energy lost by an MeV ion beam transmitted through thin crystals is called channeling STIM (CSTIM) (43). Variations in crystal quality can be imaged with a beam current of less than 1 fA using CSTIM, and this is developing into a very useful technique for imaging dislocations in bulk specimens.

1.6 *Ion-Beam-Induced Charge (IBIC)*

An MeV ion beam generates electron-hole (eh) pairs in semiconducting material as it passes through. If there is an electric field in the semiconductor, such as a pn junction or a Schottky barrier, then charge carriers are detected and the variation in the measured intensity of charge carriers in a region scanned by the beam shows the position of the active junction regions. This technique is similar to EBIC (Electron-Beam-Induced Current), where a keV electron beam is used to generate eh pairs (44), and it is widely used to measure the semiconductor material diffusion length and to image electrically active dislocations. However, a keV electron beam cannot penetrate the thick passivation layers and metallizations of microelectronic devices without severe loss of spatial resolution, whereas an MeV ion beam can because of its high penetrating power. Hence, this is the only technique capable of imaging the active region of complete microelectronic devices without the need to strip away the metallization and passivation layers, and it has great potential for device analysis (44a).

1.7 *Elastic Recoil Detection Analysis (ERDA)*

A heavy MeV incident ion can eject lighter ions from the sample. In ERDA the energy of a forward recoiling sample ion is measured using a glancing angle geometry to give quantitative elemental analysis and depth profiles with about 5-nm resolution for light elements present (45). However, there are few examples of Nuclear Microprobe ERDA (46) because of the very low cross sections for forward scattering sample ions, which necessitate using a high beam current, so good spatial resolution is difficult to achieve.

1.8 *Secondary Electron Imaging*

An incident MeV light ion beam ionizes a large number of sample atoms. The electrons released have a maximum energy of about 100 eV, which means that only those produced very close to the surface can escape from the sample. The detection of these secondary electrons generated by the ion beam scanning over the sample using either a channel electron multiplier (channeltron) or a photomultiplier tube allows the surface topography to be imaged (47–49). No elemental analysis is possible, but this technique is valuable for sample positioning. The spatial resolution of about 0.5 μm in a secondary electron image using MeV ions is much worse than the 5 nm attainable using keV electrons, because the secondary electrons generated beneath the surface from the highly scattered incident electrons do not escape from the sample and degrade spatial resolution.

2. COMPARISON WITH OTHER TECHNIQUES

The analytical techniques available with the Nuclear Microprobe have been described, but there are many other methods of surface and thin film analysis (50), each with its own strengths and drawbacks, and the relative merits of the Nuclear Microprobe can only be appreciated when compared with these other techniques. More detailed reviews and comparisons of these techniques may be found in References 51–54, and also in Section 12 of Reference 55.

Table 1 shows a comparison of the most relevant analytical requirements of the MeV ion beam techniques and secondary ion mass spec-

Table 1 Comparison of analytical techniques

Method	Measured signal	Spatial resolution (μm)	Depth resolution (μm)	Detectable elements (Z)	Detection sensitivity (wppm)[a]	Elemental quantity (%)
PIXE	X rays	0.3	5	>11	0.1	5
RBS	Backscattered incident ions	0.5	0.02	>2	1	3
NRA	Reaction products	1	0.005	all low Z	0.01	3
ERDA	Forward scattered incident ions	>1	0.005	<15	500	3
SIMS	Secondary ions	0.05	0.005	all	0.001–10	50
AES	Auger electrons	0.1	0.001	>2	1000	50
XPS	Photoelectrons	1000	0.002	>2	1000	50
ESCA	Photoelectrons	500	0.003	>2	1000	50
LEIS	Backscattered ions	100	0.0003	>2	1000	
EPMA	X rays	0.5	1	>6	500	1
LAMMA	Ions	0.5	0.1	all	0.07–20	50
XRF	X rays	1000	5	>11	1	5
STEM	X rays	0.0005	—	>11	500	5

[a] wppm = weight parts per million.

trometry (SIMS) (56), auger electron spectroscopy (AES) (10), x-ray photoelectron spectroscopy (XPS) (57), electron spectroscopy for chemical analysis (ESCA) (57), low energy ion scattering (LEIS) (58), electron probe microanalysis (EPMA) (59), laser microprobe mass analysis (LAMMA) (60), x-ray fluorescence (XRF) (61), and scanning transmission electron microscopy (STEM) (62). The values given can only be considered as approximate, since the capabilities of each technique vary widely depending on operating conditions and the analytical requirements of the particular sample. The depth and spatial resolutions are given in microns, elemental quantitivity in percentage, and the elemental detection sensitivity in wppm. Here elemental quantitivity is defined as the uncertainty in the measured elemental concentration. The MeV techniques STIM, IBIC, and secondary electron imaging are not considered here as they are not capable of elemental analysis.

Unlike SIMS, XPS, LAMMA, and AES, MeV ion techniques give information independent of the chemical bonding in the sample. This is a weakness in that no chemical information about the sample may be obtained, but it is also the greatest strength of MeV ion techniques because it is this insensitivity to chemical effects that allows quantitative elemental analysis. EPMA is more quantitative than PIXE for elemental concentrations greater than 500 ppm, but it is much less sensitive. The spatial resolution attainable with PIXE and RBS is comparable with EPMA, but it is worse than the 0.05 μm that is attainable with high resolution SIMS. The main advantages of the Nuclear Microprobe over EPMA are thus its quantitivity for light elements using NRA, its multi-elemental high sensitivity using PIXE, and its depth information obtained using RBS. The advantages of the Nuclear Microprobe over SIMS are its quantitivity and its depth-profiling capability using RBS and NRA, which do not require erosion of the sample needed for SIMS, erosion that can introduce problems of diffusion and pitting as well as uncertainties in the erosion rate.

3. PROBE FORMATION AND OPERATION

A brief outline of the production and limitations of focused MeV light ion beams using a Nuclear Microprobe is given here; References 53, 55, 64, and 65 contain a wealth of more detailed information. A comprehensive account of quadrupole optics for Nuclear Microprobe uses is given in Reference 20, and in Reference 53 many other aspects of MeV ion probe formation are considered in detail.

3.1 *Production of an MeV Ion Beam*

A low energy ion beam (typically 20 keV) is first extracted from either a radiofrequency (rf) or a duoplasmatron ion source (67). It is then injected into a megavolt electrostatic Van de Graaff or tandem accelerator and emerges as an MeV ion beam. In a Van de Graaff accelerator, a positive ion beam of charge q is created at the positive accelerator terminal at a voltage $+V$, and is accelerated towards ground potential, where it has an energy qV. In a tandem accelerator, a low energy negative ion beam is injected and accelerates towards the positive terminal voltage $+V$. Here it is stripped of its electrons and turned into a positive ion beam $+q$, which then accelerates away from the positive terminal and emerges with energy $(1+q)V$. A tandem accelerator thus needs a smaller terminal voltage to produce a given beam energy than does a Van de Graaff, which makes it cheaper to purchase and operate. With a tandem accelerator, the ion source is outside the accelerator pressure vessel, so it is easier to maintain; in a Van de Graaff accelerator, the ion source is inside the pressure vessel. With a tandem, a positive ion beam from the source is changed into a negative beam for injection into the accelerator, and then back to a positive ion beam at the accelerator terminal. There is an increase in the emittance (67) and energy spread of the beam because of this, which degrades its quality. However, the Oxford Nuclear Microprobe uses a 1.7-MV tandem accelerator and achieves excellent spatial resolution, but the fundamental limitations of tandem accelerators for forming higher resolution probes should be borne in mind. The advantages of each type of accelerator for Microprobe use are dealt with in greater depth in Reference 68.

3.2 *Probe Formation*

The MeV ion beam from the accelerator is bent through 90° by an analyzer magnet to separate out the single ion species required. The MeV ion beam hitting a set of high and low energy slits beyond the analyzer magnet controls the voltage stability of the accelerator. If the terminal voltage increases slightly, so does the beam energy and more beam hits the high energy control slit because the beam is not bent so much by the 90° magnet field. The feedback control then causes the terminal voltage to be reduced slightly to equalize the current hitting the high and low energy control slits. The beam passing through these energy stabilization slits is focused onto the Microprobe object slits. The divergence of the beam that passes through the object slits is collimated just before the beam enters the quadrupole probe-forming lenses; this is done in order to reduce the effect of lens aberrations, as discussed in Section 3.3.

Because the energy and mass of MeV ions are much higher compared with those of keV electrons, quadrupole lenses are used to focus the beam (20, 69, 70). Electrostatic quadrupoles are difficult to construct to withstand the high voltages necessary to focus the beam to a short working distance in the Nuclear Microprobe, so magnetic quadrupoles lenses are more commonly used. A single quadrupole lens can only form a line rather than a point focus, so more than one quadrupole lens of alternating polarity and different strengths is required to produce a point focus on the sample surface. Most Nuclear Microprobes use a combination of two, three, or four magnetic quadrupole lenses, and the merits of different lens configurations are discussed in Reference 20.

3.3 *Limitations on Spatial Resolution*

Unlike other instruments that use ion beams, such as SIMS machines, ion implanters, and Accelerator Mass Spectrometry machines, Nuclear Microprobes suffer virtually no space-charge problems. This is because of the low current (about 100 pA) and high beam energy used. Both these considerations diminish the degrading effects of space charge on the final probe size.

The amount of current on the sample depends on the normalized brightness of the beam at the object slits. This is defined for a current of I pA as $b_n = I/(x_0 y_0 \theta_0 \phi_0 V)$, where x_0, y_0 is the size of the aperture in the object slits in μm, θ_0, ϕ_0 is the beam divergence into the lenses in mrad, and V is the beam energy in MeV. A typical value of normalized brightness in these units is 5 pA/(μm^2 mrad^2 MeV), and this depends on the source emittance and beam transmission through the accelerator. The focused probe thus contains more current if the object slits and divergence slits are opened, but this is at the expense of increasing the probe size.

There are two factors that determine the focused probe size on the sample surface. The first is the geometric size, which depends on the size of the object aperture and the demagnifications of the lenses. In general the lens demagnifications should be as large as possible to reduce the geometric probe size. It is difficult to generate a strong magnetic quadrupole field needed for a short working distance without saturating the pole piece iron, which can introduce parasitic fields. The problems associated with high field magnetic quadrupole lenses led to an attempt to develop a superconducting round lens (71) and a plasma lens (72) to focus the MeV ion beam, but these Microprobes have not yet attained good spatial resolution.

The second contribution to the image size on the sample surface comes from beam-divergence-dependent *aberrations* through the quadrupole lenses (for a detailed discussion of the effect of each type of aberration, see 20, 53, 73). Most tandem and Van de Graaff accelerators have an energy spread of about 0.05%, with the beam stabilized in the control slits. Since the focusing action of a magnetic quadrupole depends on the ion velocity, this gives rise to *chromatic* aberration. The aberration can be reduced either by improving the accelerator voltage stability or by using a combined electrostatic and magnetic quadrupole lens (74), in which the focusing action is insensitive to variations in ion velocity. These achromatic quadrupoles suffer, however, from large parasitic fields that limit their use for probe formation. *Spherical* aberration is intrinsic to the focusing action of the quadrupole (69, 73, 75), and it arises from the boundary conditions when the Laplace equation is solved for an ideal lens. It results in slightly different forces on ions traveling at different angles to the lens axis, and it increases with divergence angle. There is also a finite ripple on the power supplies either supplying the current to a magnetic quadrupole or the voltage to an electrostatic quadrupole. This gives rise to an *excitation* aberration and results in the ion beam being continually over- or underfocused in the plane of the sample surface. The current supplies to the magnetic quadrupoles of the Oxford Nuclear Microprobe have a stability of 10 ppm, which is sufficient for submicron spatial resolution. Since more than one quadrupole lens is required to focus the ion beam, there can be *translational* or *rotational* misalignments between each lens. Rotational misalignment is much more serious; it is a result of departure from round lens symmetry, where there is no comparable aberration. To obtain submicron resolution, the lenses must be rotationally aligned to within about 0.01 mrad for the Oxford Nuclear Microprobe. Quadrupole lenses are difficult to manufacture accurately, and any deviations from field symmetry resulting from a radial or transverse displacement of a pole piece generates *parasitic fields*, which perturb the quadrupole field. The most serious type resulting from any deviations of field symmetry is the sextupole field. These parasitic fields can be corrected by altering the current flowing through one or two pole piece coils of a magnetic lens (76, 77), or the voltage on one or two pole pieces of an electrostatic lens. However, recent advances in quadrupole lens construction technology have resulted in lenses with very low parasitic sextupole field strengths (78, 79). This has enabled spatial resolution attainable with the Oxford Nuclear Microprobe at a beam current of 100 pA to be about $0.3 \times 0.3\ \mu m^2$, which is the current state of the art.

4. APPLICATIONS OF THE NUCLEAR MICROPROBE

Applications of the Nuclear Microprobe in microelectronics, semiconductor analysis, metallurgy, medicine, biology, and geology are discussed in this section, and the advantages of Nuclear Microprobe analysis over other techniques available are given in each case. Only a few examples of the myriad of different applications are given here. Much more comprehensive accounts of Nuclear Microprobe applications in each of these fields can be found elsewhere (53, 55, 64, 65).

4.1 *Biology and Medicine*

Trace elements play a fundamental role in the biological processes (80). Many clinical and pathological disorders arise from an imbalance of these trace elements, and any technique giving information on trace element distributions in tissue provides an insight into the workings of biological systems. The most important Nuclear Microprobe technique in biology and medicine is PIXE because of its high elemental sensitivity, allowing the distribution of trace elements to be quantified and imaged. Reviews of Nuclear Microprobe and PIXE analyses in biology and medicine are given in References 81–83.

One biological application was the study of the elemental variations in a germinating fungus, which showed different distributions for Ca and Zn, and very similar distributions for P, S, Cl, K, Mn, Fe, and Cu (84). A study of the hippocampus in a rat brain showed the accumulation of Ca and the loss of K in specific cell layers (85). Legge et al have imaged a single human red blood cell (86); the elements P, S, Cl, K, and Fe were detected using PIXE, and the C distribution was measured using RBS. Human macrophage tumor cells (87), which showed characteristic heavy element aggregation within the cell, have also been studied.

4.1.1 CALCIUM AND BARIUM UPTAKE IN SPIROGYRA

Spirogyra is a common green filamentous alga that lives in small, stagnant bodies of water. The filaments are composed of cylindrical cells with firm cellulose walls and an outer mucous coat. The chloroplasts are in the form of helical bands lying against the internal wall of the cell. Spirogyra accumulates Ba in the form of $BaSO_4$, and recently it was found that its cells may contain crystals of $CaCO_3$. Filaments of spirogyra collected from water

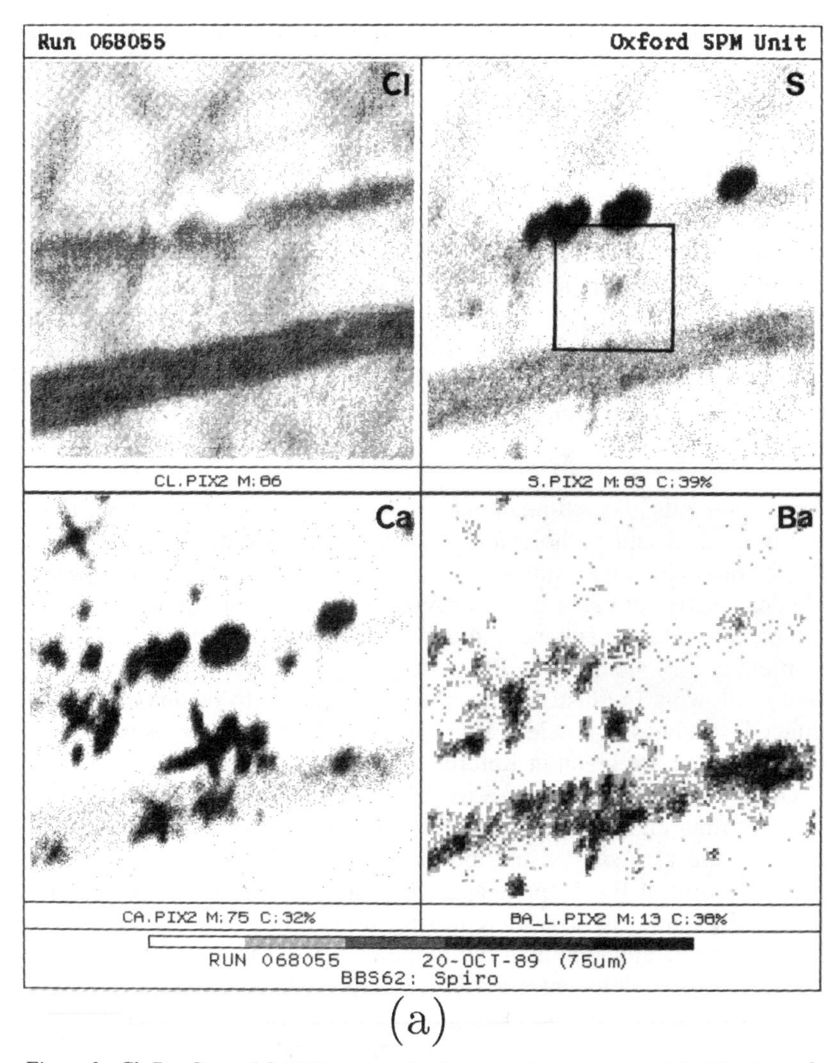

Figure 3 Cl, Ba, Ca, and S PIXE maps of a filament of spirogyra, with (*a*) 75×75 μm². (*b*) 25×25-μm² scan size from within the box shown in (*a*).

containing high Ca and Mg levels (89) were air dried onto a 1-μm polypropylene film as the only sample preparation. PIXE maps for the elements Ba, S, Ca, and Cl with scan sizes of 75×75 and 25×25 μm² are shown in Figure 3 (90). The 75×75-μm² Cl map shows the intercrossing spiral chloroplasts, which can also be seen less clearly in the S map. Ca is widely distributed throughout the scanned area, and oc-

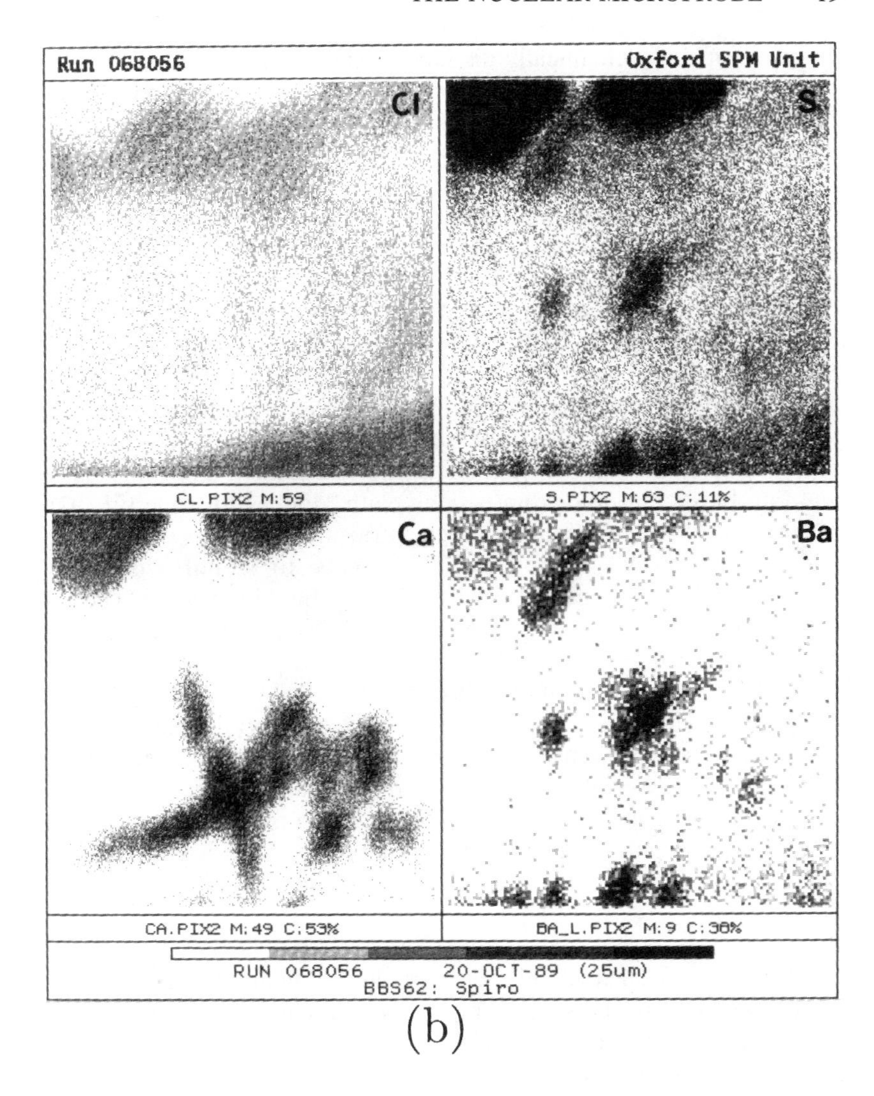

(b)

curs in regions also containing S, which suggests the presence of $CaSO_4$ crystals, a finding previously unrecorded in spirogyra. Similarly, Ba also occurs in the cells as $BaSO_4$, although the crystals appear much smaller than those of $CaSO_4$. Evidence for $CaCO_3$ crystals can be seen in the 25×25-μm^2 Ca map, where they show a characteristic star-shaped structure and no correlation with the S distribution. The occurrence of three chemically distinct crystals within one cell has not been previously reported, and it was the large penetrating power of a 3-MeV proton beam that made it possible to identify these crystals and discover their distribution without sectioning the tissue.

4.1.2 ANALYSIS OF MENKES' DISEASE TISSUE Menkes' disease is an X-linked genetic disorder associated with defective Cu metabolism. Atomic absorption spectroscopy has been used to study the average Cu content of both normal and Menkes' fibroblasts, but it does not give any information on the elemental distribution of a single cell. EPMA can give single-cell information but the poor trace element sensitivity limits its use. The Nuclear Microprobe has both the necessary spatial resolution and elemental sensitivity, and it has been used to study trace elemental distributions in normal human skin fibroblasts and fibroblasts cultured from patients with Menkes' disease (91, 92). Menkes' cells recorded an average intracellular Cu level six times higher than normal fibroblasts. This ability to identify individual Menkes' cells by means of their Cu content should allow the Nuclear Microprobe to play an important rôle in the detection of carriers of Menkes' disease.

The high spatial resolution attainable with STIM was also used to image these fibroblasts in stained tissue, in order to examine density variations due to internal structure within the cell not detectable in PIXE maps. Figure 4 (91) shows a STIM image of a Menkes' cell; there is sufficient energy loss contrast and spatial resolution to highlight clearly the cell outline, nuclear membrane, and three possible alveoli about 3 μm in diameter.

4.1.3 IDENTIFICATION OF SENILE PLAQUES It is usually necessary to stain tissue in order to identify the pathological features present before elemental analysis. This inevitably leads to a reduction in the integrity of the sample, introducing problems both of contamination and the possible loss of elements of interest from the tissue. The use of STIM to identify pathological features in the brains of patients with Alzheimer's disease (93), for subsequent analysis with PIXE and RBS, is described here. A neuritic or senile plaque is one of the primary pathological features of Alzheimer's disease, and its elemental composition has been a subject of much debate (94). Senile plaques are largely composed of amyloid protein, which is more dense than the surrounding tissue, so it is possible to image this variation in density distribution of the unstained tissue using STIM. Once identified, the elemental composition of the unstained plaque can be determined using PIXE and

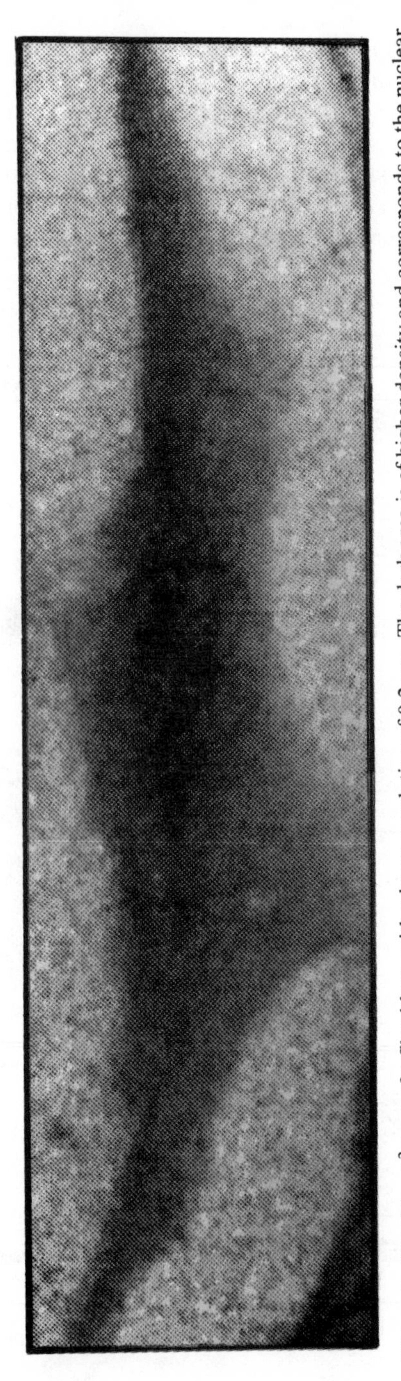

Figure 4 90×24-μm^2 scan of a fibroblast, with a beam resolution of 0.2 μm. The dark area is of higher density and corresponds to the nuclear membrane.

RBS. Figure 5 is a STIM image of an unstained plaque. The dark region is the plaque core, and the characteristic diffuse halo associated with plaques can also be seen. This same region was then analyzed using PIXE and RBS and showed the elemental distribution characteristic of senile plaques in stained tissue (93). This work illustrates how the combined techniques of STIM, PIXE, and RBS can image and analyze the structure and elemental composition of unstained tissue, which is a significant advance in biological microanalysis.

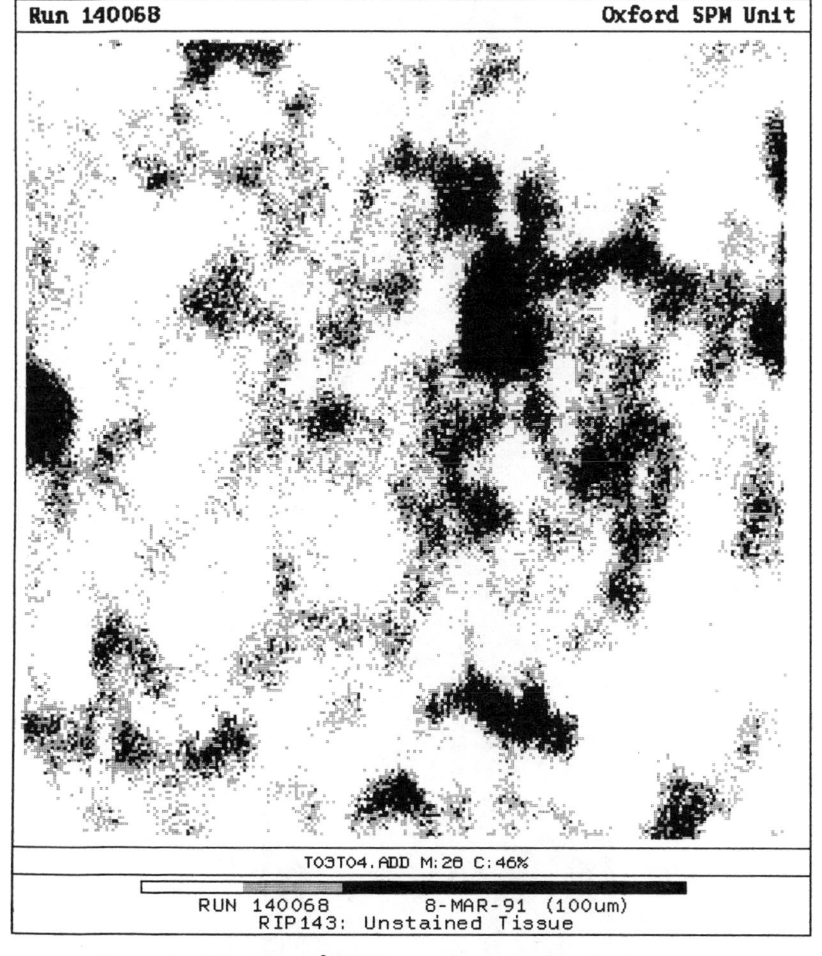

Figure 5 100×100-μm^2 STIM map of unstained senile plaque core.

4.2 *Geology, Art, and Archaeology*

Geological materials are often complex assemblies of minerals containing many elements in a wide variety of chemical states. Many contain inclusions and individual mineral grains down to micron dimensions. The concentration of major and minor trace elements in a rock is determined by factors such as the chemical environment, initial temperature, and rate of cooling, and it is subsequently modified by processes such as diffusion, leaching, and dissolution. By studying the compositon and distribution of elements within a sample, clues to the genesis and physicochemical history of the rock can be obtained and related to the overall geological history of the region.

To enable microstructure to be investigated and to avoid errors in the measurement of trace element concentrations caused by analyzing the many pits and inclusions present, a high spatial resolution technique is required. EPMA has sufficient spatial resolution but its typical sensitivity is 500 ppm, and only 100 ppm at best is insufficient for analysis to image trace element distributions. SIMS has higher spatial resolution and sensitivity than PIXE, and has great potential for analytical geochemistry. However, it is difficult to quantify elemental concentrations measured with SIMS, which is why PIXE analysis has developed into an important technique in analytical geochemistry (96, 97). Applications of the Nuclear Microprobe in this field include the measurement of trace element concentrations in coal (98, 99). Quantitative concentrations of the major and minor elements Na, Mg, Al, Si, S, Cl, Ca, and Ti were measured, as were the trace elements V, Cr, Mn, Ni, Cu, Zn, Ga, Ge, As, Se, Br, and Sr with concentrations of less than 100 ppm. Use of a Nuclear Microprobe allowed the heterogeneity to be studied, and allowed the detection of many subsurface inclusions. Another Nuclear Microprobe application is the mapping of deuterium, carbon, and nitrogen in oilfield rocks using NRA (100), which showed that they are all preferentially absorbed at grain boundaries.

The Nuclear Microprobe has been used in art and archaeology to study manuscripts, paintings, bones, glass, pottery, bronze, coins, and gold jewelry (for discussion, see Chapter 10 of Reference 53). Such applications rely heavily on the elemental sensitivity of PIXE. Swann (101) has tabulated the detection sensitivity of PIXE for a variety of elements in matrices commonly encountered in archaeology, such as glasses and bronze, and it is typically 100 times better than EPMA. The nondestructive nature of Nuclear Microprobe analysis is also essential for archaeological applications because samples such as ancient manuscripts are both extremely delicate and irreplaceable. Much of the work uses an external beam facility (102) whereby the MeV ion beam is transmitted out of vacuum through a thin foil to atmosphere. This is vital, again because of the delicate nature of some samples such as manuscripts, which cannot be placed in vacuum, and also because other samples such as paintings are too large to fit in the vacuum chamber.

Applications in archaeology include a study of the Gutenburg Bible (103). The measurement of the relative concentrations of Ca, Mn, and Fe present allowed researchers to distinguish the paper made by different craftsmen, whereas a study of the different inks relied on the measurement of the Cu and Pb ratio using PIXE. PIXE is also valuable in studies of gold and silver artifacts because with PIXE there is no corrosion; surface analysis therefore gives information on the bulk material properties. Uses include the study of fabrication techniques (104) and the detection of fakes and repairs.

4.2.1 TRACE ELEMENT DISTRIBUTIONS ACROSS GRAIN BOUNDARIES The PIXE maps in Figure 6 show the distribution of Sr, Ni, Cr, and Mn in a piece of garnet (105). These elemental maps show an angulated grain of orthopyroxene surrounded by Ni-rich olivine. Whereas the Cr, Ni, and Mn are distributed homogeneously in the mineral phases, the map of the Sr distribution shows that it is localized along narrow zones on grain boundaries and along cracks. This pattern of Sr concentration implies that the Sr was introduced after the texture of the rock had formed, and the existence of serpentinized olivine along the cracks strongly suggests that the Sr was introduced as an aqueous fluid. This example demonstrates why the high elemental sensitivity of PIXE is important in determining geochemical properties.

4.3 *Semiconductor Analysis*

Much of the early work on Nuclear Microprobe semiconductor analysis was carried out at the University of Melbourne (for review, see 53, Chapter 9; see also 106, 107). Semiconductors are commonly doped with either group 3 or group 5 elements to modify the crystal properties, such as the energy band gap and conductivity. Knowing the location of these dopants in the semiconductor lattice is vital since the location affects these properties, and so ion channeling has become an important investigative tool in semiconductor research. Thornton et al used ion

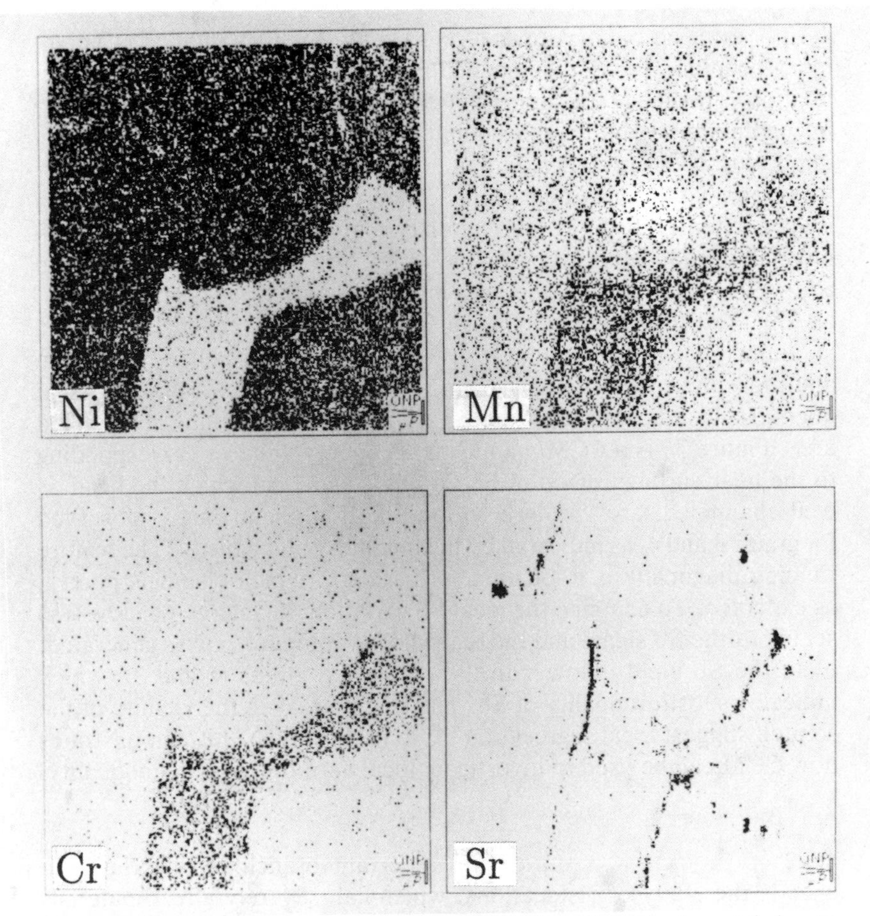

Figure 6 4×4-mm² PIXE maps showing the distribution of Ni, Mn, Sr, and Cr in a 30-μm thick section of garnet.

channeling in conjunction with a Nuclear Microprobe to show poor crystalline growth of $NiSi_2$ on a diode structure (108). A silicon-on-insulator device fabricated by zone melt recrystallization of a surface Si layer by selective epitaxial regrowth from seed windows in a SiO_2 layer has also been studied (109). Ion channeling in 5×5-μm² seed windows aligned with the ⟨100⟩ axis of the Si indicated alignment of the overlying layer but suggested the presence of defects, as indicated by an increase in dechanneling from these areas. Other semiconductor applications of the Nuclear Microprobe include studies by Kinomura et al (110).

4.3.1 CHANNELING IN POLYSILICON LAYERS Ion implantation is a common technique for doping semiconductors for integrated circuit fabrication. The need to examine the damage distribution introduced during implantation, and its subsequent removal by heat treatment, is very important. CCM provides a unique way of examining the damage distributions within individual grains of ion-implanted polysilicon, which is used extensively in the fabrication of semiconductor devices. A commercially produced polycrystalline wafer with grain sizes ranging from a few microns to about 1 mm was implanted with Sb ions and annealed at 650° for 30 minutes. CCM was carried out by channeling with a stationary beam in the grain of interest, and then scanning the beam and measuring the RBS spectrum (107, 111). Figure 7*a* is an optical micrograph indicating the typical grain structure within the scanned area. Figure 7*b* is a CCM map using an energy window corresponding to the near surface region of Si. Grains 1 and 2, in which the beam is well channeled, are the dark regions. Partial channeling is observed for grains 3 and 4, as indicated by the medium point density. The feature running diagonally is a location scribe mark, which channels poorly, as expected. To examine the local Sb variation, an energy window was set about the Sb signal and the map shown in Figure 7*c* was generated. Note the Sb yield is lower in the channeled grains (1 and 2), which indicates substitutionality of Sb. The bright areas in the vicinity of the scratch suggest local segregation of Sb in this area. This demonstrates how CCM can be used to investigate local variations in crystal quality.

4.3.2 DISLOCATION IMAGING An important branch of materials science is the study of dislocations, which can severely affect materials performance. Transmission electron microscopy (TEM) is the major technique used for investigating and imaging dislocations in thin samples because of its high spatial resolution. However, sample preparation for TEM analysis is very time-consuming. It is difficult to analyze precise areas only microns across, and sample preparation is even more difficult when the sample has features on the surface. Because of the higher penetrating power of MeV ions, CSTIM can image dislocations in specimens 20 μm thick, and the presence of surface features does not affect dislocation contrast.

The sample studied (112) was 20-μm thick (001) Si substrate with 2.5-μm high mesas etched in it, with their edges aligned with the [110] and [$1\bar{1}0$] planes. A 0.75-μm thick layer of $Si_{0.85}Ge_{0.15}$ has been grown over this, and the lattice mismatch between the two layers gives rise to misfit dislocations.

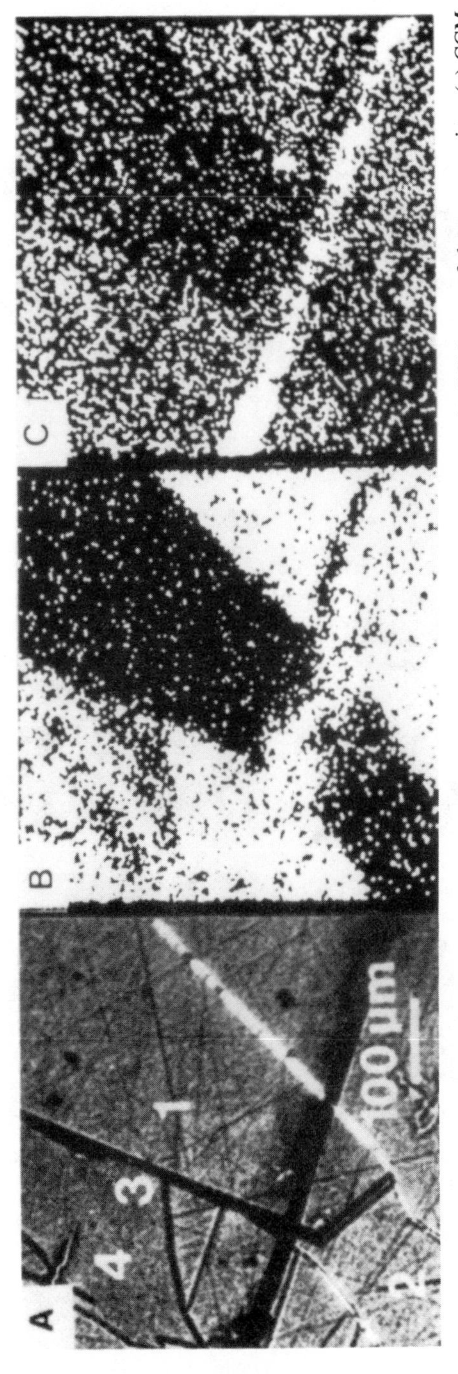

Figure 7 (*a*) Optical image of a $500 \times 500\text{-}\mu m^2$ region of polysilicon, showing individual grains. (*b*) CCM map of the same region. (*c*) CCM map of the Sr in this region.

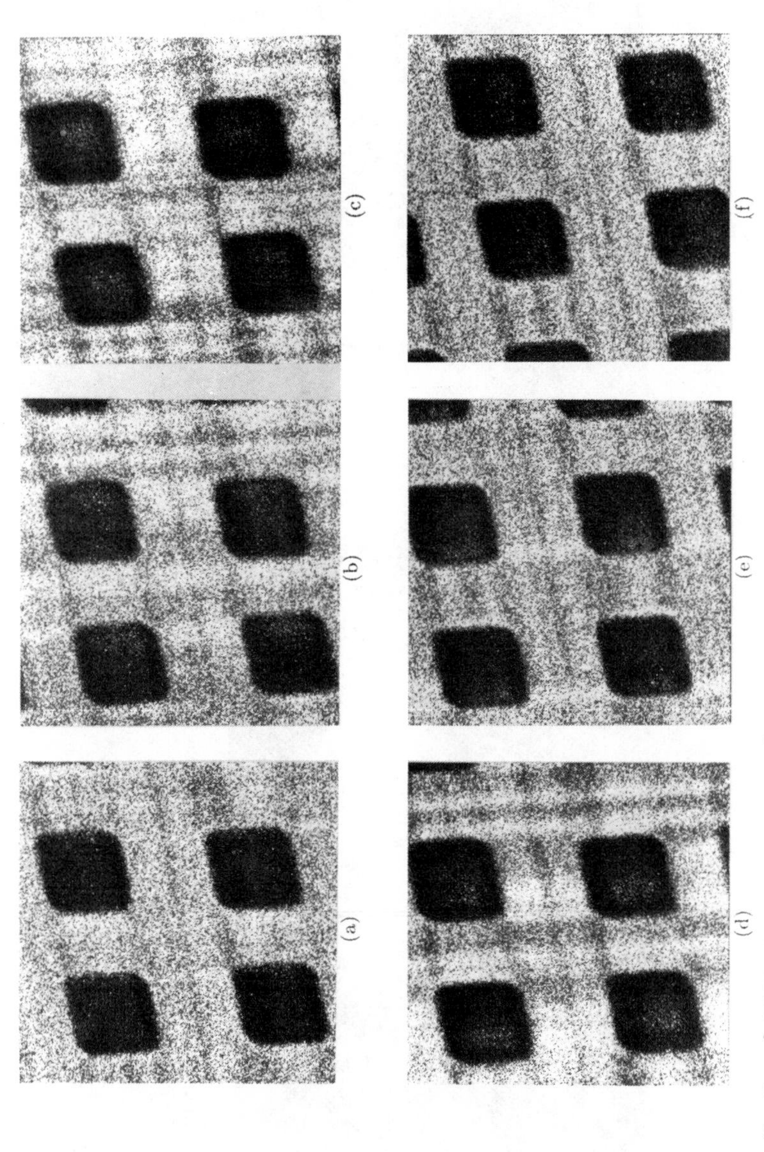

Figure 8 40×40-μm² CSTIM images of four 10-μm² mesas. The orientation about the x axis is fixed in the $[1\bar{1}0]$ planes and the sample is tilted about the y axis in 0.2° steps through the [110] planes. (*a*) −0.4°, (*b*) −0.2°, (*c*) channeled, (*d*) +0.2°, (*e*) +0.4°, (*f*) +0.6°.

Figure 8 shows CSTIM images of the central group of four 10-μm^2 mesas (112). As with STIM images, the thick mesa areas appear darker because more energy is lost here. The dislocations appear as dark bands because the ion beam dechannels from them, and so the transmitted beam loses more energy in these areas. The sample is aligned about the x axis (channeling in the $[1\bar{1}0]$ planes), and these six images show the variation in contrast as the sample is tilted about the y axis (through the [110] planes), through channeling alignment in the (001) axis, in steps of 0.2°. There is very little vertical dislocation contrast when the sample is tilted 0.4° off the [110] planes, as can be seen in Figure 8*a*, because the beam is not well enough channeled in any region of the scanned area. As the sample is tilted about the y axis towards the (001) axis, the vertical dislocation contrast becomes stronger, because the beam is now able to channel well in [110] planes aligned with the beam. Areas that are well channeled in one image are poorly channeled in the other. The weak channeling contrast in the mesas themselves also changes position with different sample orientation. Figure 8 demonstrates how CSTIM can image dislocation bunches typically 2 μm wide.

There is no reason why a spatial resolution of about 10 nm should not be achieved within the next few years using CSTIM, and this would make it a valuable technqiue for imaging individual dislocations throughout the full thickness of bulk specimens.

4.4 *Microelectronics*

The continually shrinking feature size of microelectronic devices means that many techniques cannot adequately resolve the smallest details. However, both PIXE and RBS can achieve 0.5-μm spatial resolution, and a 50-nm STIM resolution has been reported (33), so all three techniques are capable of analyzing very small device features. Furthermore, the composition, thickness, and distribution of buried layers in microcircuit structures are difficult to determine with optical and electron beam techniques without etching the surface layers of the sample, because of the limited penetration depth. PIXE and RBS both have the penetrating power to analyze through the device layer structure. If the device substrate is mechanically polished away from the rear until the remaining layer thickness is about 40 μm or less, a 3-MeV proton beam has sufficient energy to travel through the various layers. STIM can thus be used as a rapid method of determining the uniformity and distribution of metallization layers, even if they are buried under other metal layers (113, 114).

The Albany Microprobe group has studied the metallization thickness and incomplete etching of devices (115, 116), and Doyle has studied the W metallization of porous silicon layers of integrated circuits (117). Another MeV ion beam technique with great promise for microelectronics analysis is Ion Microtomography (IMT) (118, 119), in which a three-dimensional image can be reconstructed by recording STIM images at many different sample orientations. The data set can then be manipulated to view the internal sample structure from any orientation, and to slice off the surface layers to image detail below. It has been shown that there is adequate energy loss contrast to produce different STIM images from a thinned microelectronic device (114), so IMT would make it possible to image the three-dimensional internal density structure of devices.

4.4.1 DETECTION OF METALLIZATION DEFECT The test device studied here has regions of a 0.7-μm thick first metallization layer of W/Ti overlaid with a 0.9-μm thick second metallization layer of Al in some areas (113, 114). There is a 1.1-μm thick SiO_2 passivation layer over all the surface apart from the contact pads, which makes the circuit difficult to analyze with electron beam techniques without etching away layers. STIM images of this device with scan sizes 200 and 60 μm^2 are shown in Figure 9*a* and 9*b*. The tracks of W produce the greatest energy loss and thus the strongest contrast in these STIM images. The individual contact holes visible on the 60-μm^2 image have higher energy loss around their edges because the thickness of tungsten traversed by the beam increases over this region. On the 60-μm^2 image there is indication of a fault in the metallization. The upper of the two faint lines indicated by the arrows is slightly displaced from the center of the gap in the first metallization layer. This defect has been detected with STIM, but PIXE is used to determine precisely the nature of the fault giving rise to the contrast variation. Figure 9*c* shows 20-μm^2 W L and Figure 9*d* shows the Ti K PIXE image of this same area. The W distribution is as expected for the first metallization, but the Ti distribution is indeed displaced in the upper 1-μm wide track, possibly because of a fault in the mask used to lay down this layer. This example demonstrates how PIXE and STIM can image the device layer distribution. There is also great potential for submicron RBS analysis in precisely located regions of the device to show localized variations in layer stoichiometry using the Nuclear Microprobe (113).

4.4.2 DEVICE IMAGING USING IBIC Electron-Beam-Induced Current (EBIC) is widely used to image device depletion regions (44). However, a keV electron beam cannot penetrate thick passivation and metallization layers without severe degradation of spatial resolution. An MeV

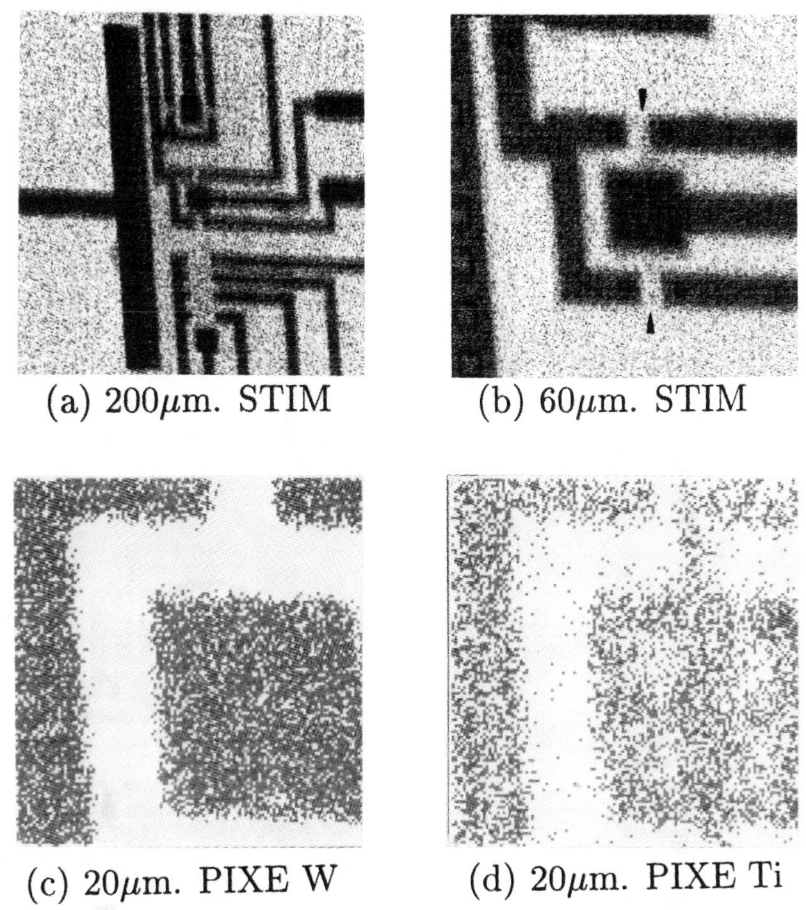

Figure 9 (*a*) 200 × 200-μm² STIM image, (*b*) 60 × 60-μm² STIM image of the test device. (*c*) 20 × 20-μm² PIXE W image, (*d*) 20 × 20-μm² PIXE Ti image showing the displaced track.

ion beam has a much higher penetrating power, and suffers much less degradation of resolution through scattering. IBIC can image depletion regions in complex devices without the need to strip off surface passivation and metallization layers (44a, 120). Figure 10*a* shows how the source and drains of two enhancement field effect transistors (FETs) are connected together, and Figure 10*b* shows how their drains and sources are laid out in the analyzed area. There is a 1-μm thick Al metallization layer over the drain and source regions, and the device is covered with a 1.5-μm thick passivation layer. Figure 10*c* and *d*

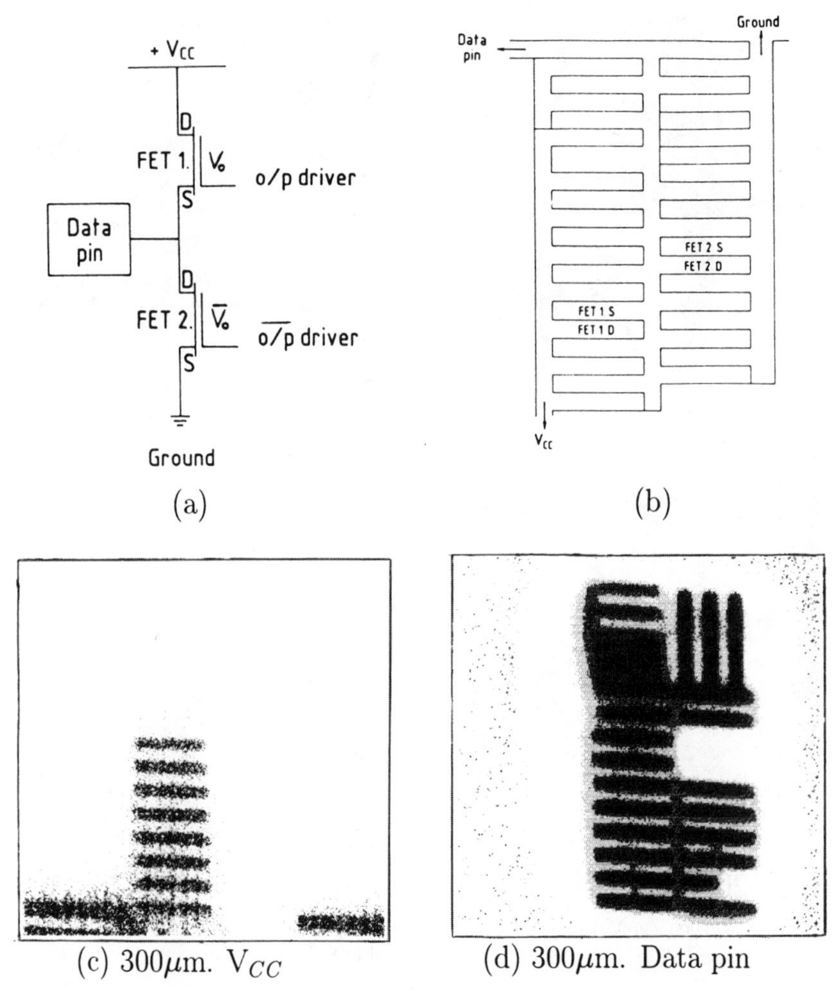

Figure 10 (*a, b*) Schematic device layout for the two FETs imaged with IBIC. (*c, d*) $300 \times 300\text{-}\mu m^2$ IBIC images. (*c*) Measuring between the supply voltage pin, with V_{CC} = +5 V and ground. (*d*) Measuring between the data pin and ground.

shows two IBIC images of this area (44a), with the carrier current generated by the ion beam in different parts of the circuit being displayed. Dark indicates high measured current in the IBIC image. Figure 10*c* shows the IBIC image measured across the supply voltage pin V_{CC} and ground, and Figure 10*d* shows the IBIC image measured between the data pin and ground. These images were measured simultaneously using two different sets of detection electronics. The source of FET 1 and the drain of FET 2 are connnected to the detection electronics measuring at the data pin, so the currents generated in these areas appear on the IBIC image of Figure 10*d* and not in Figure 10*c*, where they have to travel through the FET 1 channel to be measured. Figure 10*c* thus shows only the FET 1 drain region. This demonstrates how different parts of the device can be imaged using IBIC without the need to strip off surface layers.

4.5 *Metallurgy*

NRA is widely used for Nuclear Microprobe applications in metallurgy because of its ability to pick out light elements, which have a major influence on the properties of metals (1, 121, 122). The major constituents of most metals are fairly heavy elements so they give little background in the NRA spectrum and can usually withstand the high beam fluence needed for NRA. By comparison, EPMA cannot give quantitative measurements of light elements in the heavier matrices because of the strong attenuation of soft x rays. Chapter 8 of Reference 53 lists almost 40 applications of Microprobe NRA for measuring the main isotopes of all elements up to $Z = 13$, apart from ^{6}Li and ^{20}Ne, using incident proton, deuteron, ^{3}He, ^{4}He, and ^{11}B beams, and detecting charged particle reaction products. Much of the work is devoted to measuring light element distributions in Fe- and Ni-based alloys, and also quantitative C and N measurements using (d,p) reactions.

Other applications include the measurement of γ rays produced by a 1.8-MeV deuteron beam from the Nuclear Microprobe to study the corrosion of concrete grout used around geothermal extraction pipes. The problem begins when Ca is removed by CO_2 (123). Heck (124) has studied the behavior of Li in stainless steel, and the choice of $^{7}Li(p,\alpha)$ reactions using a 3-MeV proton beam allowed the simultaneous measurement of C and O using RBS and of heavier elements using PIXE. Maps for C and O using RBS and Cr, Fe, Ni, and Mo from a scanned area of corrosion were measured, together with maps for the Li for different depths. The results indicated that Cr was concentrated at grain boundaries together with Li and O, possibly in the form of a lithium chromate compound. Another application is the study of welding problems of stainless steel (125). C, N, and O in the weld were measured when various gas mixtures were used, with the conclusion that CO_2 should not be used in the welding of ferritic stainless steel.

4.5.1 CARBON PROFILING IN METALS A very common use of NRA in metallurgy is the measurement of carbon profiles using the reaction $^{12}C(d,p)$. An MeV deuteron beam is used and high energy protons emitted from the sample are measured. Typical energy spectra obtained from this (d,p) reaction are shown in Figure 11 (126). Spectrum *b* is from a sample with a uniform C content; the spectrum has a peak about 0.5 MeV wide where the shape depends on the varying cross section for deuterons reacting with ^{12}C at different depths in the sample. Spectrum *a* is from quartz with a thin C layer on top, which gives a sharp peak from the ^{12}C, and lower energy peaks from various oxygen and silicon nuclear reactions. The unfolding of the NRA energy spectrum to determine the depth distribution to accuracies of about 0.1 μm over depths of a few microns is a well-established technique (127).

One of the few Nuclear Microprobe NRA applications in which the sample was imaged used the $^{12}C(d,p)$ reaction, and is shown in Figure 12 (128). Figure 12*a* is a photomicrograph showing the surface of the

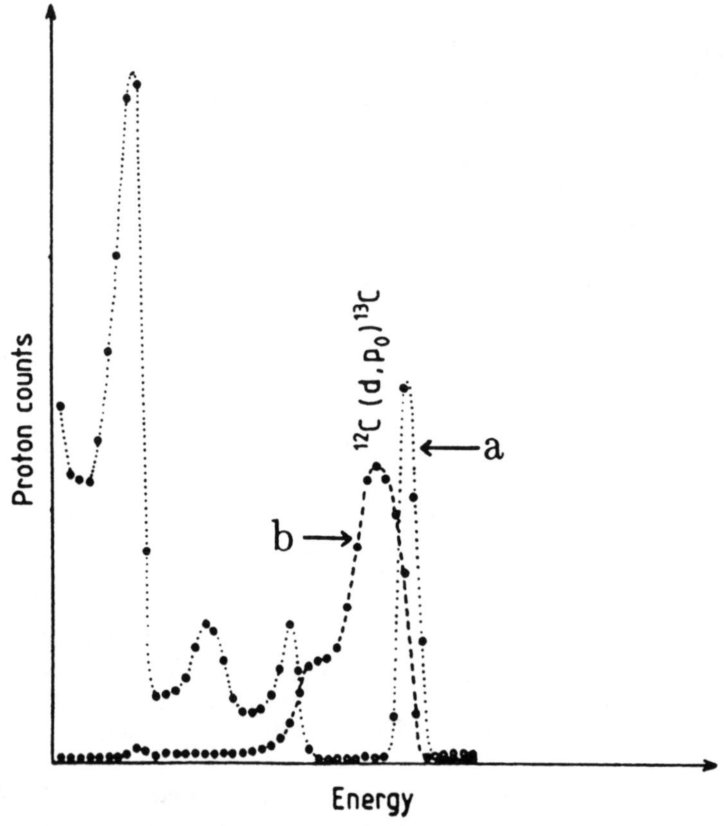

Figure 11 Measured proton spectra for NRA analysis on two different samples bombarded with 1.3-MeV deuterons: (*a*) quartz with a thin surface C film, (*b*) standard steel.

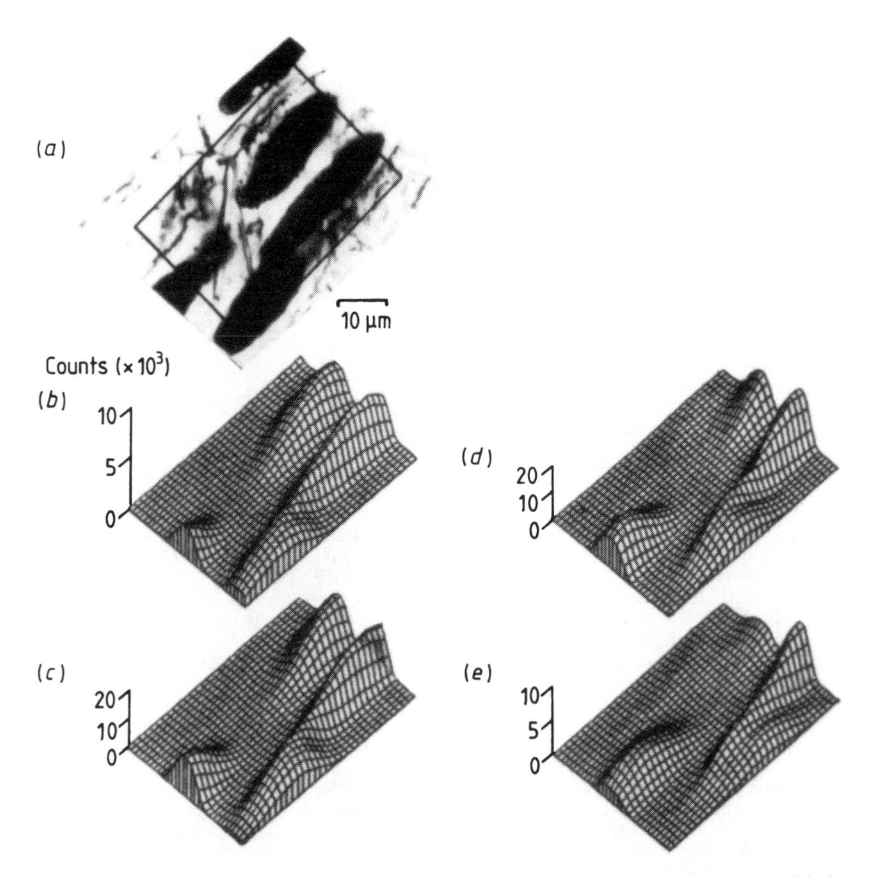

Figure 12 (*a*) Optical image showing the surface of the sample with C fibers embedded in a metal matrix. The rest of the figure shows the strength of the carbon signal from measured protons corresponding to C at depths of (*b*) 0–0.8 μm, (*c*) 0.8–2.1 μm, (*d*) 2.1–3.2 μm, and (*e*) 3.2–4.4 μm.

sample with carbon fibers embedded in a metal matrix. The rest of Figure 12 shows the strength of the carbon signal from measured protons corresponding to C at different depths, and demonstrates the ability of the Nuclear Microprobe to generate depth-resolved images using NRA.

CONCLUSION

The Nuclear Microprobe has developed into an important analytical technique in a wide variety of fields because of the high sensitivity and quantitivity of PIXE, RBS, and NRA. Newly developed techniques such as STIM, CCM, CSTIM, and IBIC will play an important role because of their large depth of analysis compared to other analytical methods. Other analytical techniques have better spatial resolution, but have reached their fundamental limits, whereas there is no reason why the current best spatial resolution of 0.5 μm for PIXE and RBS should not decrease to 100 nm or less, and a STIM resolution of 10 nm should be possible within a few years. This higher resolution capability would give the Nuclear Microprobe an increasingly important analytical role in the next decade.

Literature Cited

1. Mayer, J. W., Rimini, E., eds. *Ion Beam Handbook for Materials Analysis*. New York: Academic (1977)
1a. Bird, J. R., Williams, J. S., eds., *Ion Beams for Materials Analysis*. New York: Academic (1989)
2. Deconninck, G., *Introduction to Radioanalytical Physics*. Amsterdam: Elsevier Scientific (1978)
3. Bird, J. R., Duerden, P., Wilson, D. J., *Nucl. Sci. Appl.* B1:357 (1983)
4. Marion, J. B., Young, F. C., *Nuclear Reaction Analysis Graphs and Tables*. Amsterdam: North-Holland (1968)
5. Northcliffe, L. C., Schilling, R. F., *Nucl. Dat. Tables* A7:223 (1970)
6. Ziegler, J. F., ed., *The Stopping and Ranges of Ions in Matter*. Oxford: Pergamon (1985)
7. Chu, W.-K., Mayer, J. W., Nicolet, M. A., *Backscattering Spectrometry*. New York: Academic (1978)
8. Johansson, S. A. E, Johansson, T. B., *Nucl. Instrum. Methods* 137:473–516 (1976)
9. Garcia, J. D., *Phys. Rev* A1:280 (1970)
10. van Oostrom, A., *Surf. Sci.* 89:615 (1979)
11. Khan, M. R., Crumpton, D., *CRC Crit. Rev. Anal. Chem.* 11:103, 61 (1981)
12. Mitchell, I. V. Barfoot, K. M., *Nucl. Sci. Appl.* 1:99 (1981)
13. Johansson, S. A. E, Campbell, J. L., *PIXE: A Novel Technique for Elemental Analysis*. New York: Wiley (1988)
14. Cookson, J. A., *Nucl. Instrum. Methods* 181:115–24 (1981)
15. Legge, G. J. F., *Nucl. Instrum. Methods* B3:561–71 (1984)
16. Folkmann, F., *Ion Beam Surface Layer Analysis*, Vol. 2, ed. O. Meyer, et al. New York: Plenum (1976)
17. Bosch, F., et al, Max Planck Inst. Kernphys. Intern. Rep. MPIH-1979-V29 (1979)
18. Legge, G. J. F., McKenzie, C. D., Mazzolini, A. D., *J. Microsc.* 117:185 (1979)
19. Raman, S., Vane, C. R., *Nucl. Instrum. Methods* B3:71–77 (1984)
20. Grime, G. W., Watt, F., *Beam Optics of Quadrupole Probe-Forming Systems*. Bristol: Adam Hilger (1984)
21. Bearden, J. A., *Rev. Mod. Phys.* 39: 78–142 (1967)
22. Grime, G. W., et al, *Nucl. Instrum. Methods* B54:353–62 (1991)
23. Campbell, J. L., Cookson, J. A., *Nucl. Instrum. Methods* B3:185–97 (1984)
24. Cookson, J. A., *Nucl. Instrum. Methods* 165:477–508 (1979)
25. Jarjis, R. A., Nuclear cross-section data for surface analysis. Univ. Manchester Rep. (1979).
26. Deleted in proof
27. McMillan, J. W., et al, *Nucl. Instrum. Methods* 149:83–91 (1978)
28. Dieumegaard, D., Maurel, B., Amsel, G., *Nucl. Instrum. Methods* 168:93–103 (1980)
29. Engelmann, C., *At. Energy Rev. Suppl.* 2:107 (1981)
30. Overley, J. C., et al, *Nucl. Instrum. Methods* 218:39–42, 43–46 (1983)
31. Lefevre, H. W., et al, *Nucl. Instrum. Methods* B54:363–70 (1991)
32. Sealock, R. M., Mazzolini, A. P., Legge, G. J. F., *Nucl. Instrum. Methods* 218:217–20 (1983)
33. Bench, G. S., Legge, G. J. F., *Nucl. Instrum. Methods* B40/41:655–58 (1989)
34. Pontau, A. E., et al, *Nucl. Instrum. Methods* B40/41:646–50 (1989)
35. Ito, A., Koyama-Ito, H., *Nucl. Instrum. Methods* B3:584–88 (1984)

36. Bench, G., et al, *Nucl. Instrum. Methods* B54:390–96 (1991)
37. Morgan, C. V., ed., *Channelling Theory, Observations and Applications*. New York: Wiley (1973)
38. Feldman, L. C., Mayer, J. W., Picraux, S. T., *Materials Analysis by Ion Channelling*. New York: Academic (1982)
39. Ecker, K. H., *Nucl. Instrum. Methods* B3:283–87 (1984)
40. Romano, L. T., et al, *Phys. Rev.* B44(13):6927–31 (1991)
41. Bugeat, J. P., Chami, A. C., Ligeon, E., *Phys. Lett.* 58A:127–30 (1976)
42. McCallum, J. C., et al, *Appl. Phys. Lett.* 42:827–29 (1983)
43. Cholewa, M., et al, *Nucl. Instrum. Methods* B54:397–400 (1991)
44. Leamy, H. J., *J. Appl. Phys.* 53:R51–R80 (1982)
44a. Breese, M. B. H., Grime, G. W., Watt, K., *Oxford Nuclear Physics Rep.* OUNP-91-33
45. L'Ećuyer, J., et al, *J. Appl. Phys.* 47:381–82 (1976)
46. Sofield, C. J., Bridwell, L. B., Wright, C. J., *Nucl. Instrum. Methods* 191:379 (1981)
47. Younger, P. A., Cookson, J. A., *Nucl. Instrum. Methods* 158:193–98 (1979)
48. Traxel, K., Mandel, A., *Nucl. Instrum. Methods* B3:594–97 (1984)
49. Legge, G. J. F., *Nucl. Instrum. Methods* B15:669–74 (1986)
50. Feldman, L. C., Mayer, J. W., *Fundamentals of Surface and Thin Film Analysis*. Amsterdam: North-Holland (1986)
51. Cookson, J. A., *Nucl. Instrum. Methods* 197:255–58 (1982)
52. Reuter, W., *Nucl. Instrum. Methods* 218:391–99 (1983)
53. Watt, F., Grime, G. W., eds., *Principles and Applications of High Energy Ion Microbeams*. Bristol: Adam Hilger (1987)
54. Shaffner, T. J., *Scanning Electron Microsc.* 1:11–24 (1986)
55. Second International Conference on Nuclear Microprobe Technology and Applications. *Nucl. Instrum. Methods* B54:1–440 (1991)
56. Wittmaack, K., *Nucl. Instrum. Methods* 168:343–56 (1980)
57. Briggs, D., ed., *Handbook of X-ray and Ultraviolet Photo-electron Spectroscopy*. London: Heyden (1977)
58. Haeussler, E. N., *Surf. Interface Anal.* 2:134 (1980)
59. Heinrich, K. F. J., *Electron Beam X-ray Microanalysis*. New York: Van Nostrand (1981)
60. Wechsung, R., et al, *Scanning Electron Microsc.* 2:279–90 (1979)
61. Goulding, F. S., Jaklevic, J. M., *Nucl. Instrum. Methods* 142:323–32 (1977)
62. Hren, J. J., Goldstein, J. I., Joy, D. C., eds., *Analytical Electron Microscopy*. Plenum: New York (1979)
63. Deleted in proof
64. Second Int. Conf. on Chemical Analysis. *Nucl. Instrum. Methods,* Vol. 197 (1982)
65. First Int. Conf. on Nuclear Microprobe Technology and Applications. *Nucl. Instrum. Methods,* Vol. B30 (1988)
66. Deleted in proof
67. Banford, A. P., *The Transport of Charged Particle Beams*. London: E and F. N. Spon (1966)
68. Grime, G. W., Watt, F., Jamieson, D. N., *Nucl. Instrum. Methods* B45:508–12 (1990)
69. Hawkes, P. W., *Quadrupole Optics. Springer Tracts in Modern Physics 42*. Berlin: Springer-Verlag (1966)
70. Regenstreif, E., *Focusing of Charged Particles*, ed. A. Septier. New York: Academic (1967)
71. Koyama-Ito, H., Grodzins, L., *Nucl. Instrum. Methods* 174:331–40 (1980)
72. Booth, R., Lefevre, H. W., *Nucl. Instrum. Methods* 151:143–47 (1978)
73. Jamieson, D. N., PhD thesis, Univ. Melbourne (1985), unpublished
74. Martin, F. W., Goloskie, R., *Nucl. Instrum. Methods* 197:111–16 (1982)
75. Wollnik, H., *Nucl. Instrum. Methods* 103:479–84 (1972)
76. Breese, M. B. H., Jamieson, D. N., Cookson, J. A., *Nucl. Instrum. Methods* B47:443–52 (1990)
77. Breese, M. B. H., Cookson, J. A., *Nucl. Instrum. Methods* B61:343–47 (1991)
78. Jamieson, D. N., Grime, G. W., Watt, F., *Nucl. Instrum. Methods* B40/41:669–74 (1989)
79. Grime, G. W., et al, *Nucl. Instrum. Methods* B54:52–63 (1991)
80. Williams, R. J. P. *Pure Appl. Chem.* 55:1089 (1983)
81. Vis, R. D., *The Proton Micropobe: Applications in the Biomedical Field*. New York: CRC (1985)
82. Malmqvist, K. G., *Scanning Electron Microsc.* 3:821–45 (1986)
83. Lindh, U., *Nucl. Instrum. Methods* B54:160–71 (1991)
84. Mazzolini, A. P., et al, *Nucl. Instrum. Methods* B54:151–55 (1991)

85. Themner, K., et al, *Nucl. Instrum. Methods* B30:424–29 (1988)
86. Legge, G. J. F., *Nucl. Instrum. Methods* B3:561–71 (1984)
87. O'Brien, P. M., et al, *Aust. Phys. Eng. Sci. Med.* 5:30 (1982)
88. Deleted in proof
89. Hudson, J. W., Crompton, K. J., Whitton, B. A., *Vasculum* 56:38 (1971)
90. Watt, F., et al, *Nucl. Instrum. Methods* B54:123–43 (1991)
91. Allan, G. L., PhD thesis, Univ. Melbourne (1989), unpublished
92. Allan, G. L., Camakaris, J., Legge, G. J. F., *Nucl. Instrum. Methods* B54:175–79 (1991)
93. Watt, F., Landsberg, J. P., Grime, G. W., McDonald, B., In *Proc. X-ray Microanalysis in Biology, Manchester*. Cambridge Univ. Press (1991), to be published
94. Candy, J. M., et al, *Senile Dementia of the Alzheimer Type*. Berlin: Springer-Verlag, p. 197 (1985)
95. Deleted in proof
96. Maggiore, C. J., et al, *IEEE Trans. Nucl. Sci.* NS-30:1224 (1983)
97. Sie, S. H., et al, *Nucl. Instrum. Methods* B54:284–91 (1991)
98. Chen, J. R., et al, *Nucl. Instrum. Methods* 181:151–57 (1981)
99. Minkin, J. A., et al, *Scanning Electron Microsc.* 1:175–84 (1982)
100. Toulhoat, N., et al, *Nucl. Instrum. Methods* B54:312–16 (1991)
101. Swann, C. P., *IEEE Trans. Nucl. Sci.* NS-30:1298–1301 (1983)
102. Cookson, J. A., Piling, F. D., *Phys. Med. Biol.* 21:965 (1976)
103. Kusko, B. H., et al, *Nucl. Instrum. Methods* B3:689–94 (1984)
104. Demortier, G., Decroupet, D., Mathot, S., *Nucl. Instrum. Methods* B54:334–45 (1991)
105. Fraser, D. G., et al, *Nature* 312:352 (1984)
106. Williams, J. S., McCallum, J. C., Brown, R. A., *Nucl. Instrum. Methods* B30:480–85 (1988)
107. McCallum, J. C., PhD thesis, Univ. Melbourne (1989), unpublished
108. Thornton, J., Harper, R. E., Albury, D. M., *Nucl. Instrum. Methods* B29: 515–20 (1987)
109. Jamieson, D. N., et al, *Inst. Phys. Conf. Ser.* 100:87–92 (1989)
110. Kinomura, A., Takai, M., Namba, S., *Nucl. Instrum. Methods* B45:523–26 (1990)
111. McCallum, J. C., McKenzie, C. D., Williams, J. S., *IEEE Trans. Nucl. Sci.* NS-30:1228–31 (1983)
112. Breese, M. B. H., et al, *J. Appl. Phys.* (1992) to be published
113. Breese, M. B. H., et al, *Inst. Phys. Conf. Ser.* 117:101–4 (1991)
114. Breese, M. B. H., et al, *Nucl. Instrum. Methods* B64:505–10 (1992)
115. Morris, W. C., Bakhru, H., Haberl, A. W., *Nucl. Instrum. Methods* B15: 661 (1986)
116. Morris, W. G., Fesseha, S., Bakhru, H., *Nucl. Instrum. Methods* B24/25: 635–37 (1987)
117. Doyle, B. L., Sandia Natl. Lab. Rep. SAND-87–1138C (1987)
118. Pontau, A. E., Antolak, A. J., Morse, D. H., *Nucl. Instrum. Methods* B45: 503–7 (1990)
119. Pontau, A. E., Antolak, A. J., Morse, D. H., *Nucl. Instrum. Methods* B54: 383–89 (1991)
120. Angell, D., et al, *Nucl. Instrum. Methods* B44:172–78 (1989)
121. McMillan, J. W., *Nucl. Instrum. Methods* B30:474–79 (1988)
122. Cookson, J. A., Breese, M. B. H., *Nucl. Instrum. Methods* B50:208–16 (1990)
123. Coote, G. E., Vickridge, I. C., Aldridge, L. P., Inst. Nucl. Sci. INS-R-385. Lower Hutt, New Zealand: DSIR (1988), p. 6
124. Heck, D., *Nucl. Instrum. Methods* B30:486–90 (1988)
125. Vickridge, I. C., Sanders, M. C., see Ref. 123, p. 6
126. Pierce, T. B., et al, *Nucl. Instrum. Methods* 118:115–24 (1974)
127. Bird, J. R., Campbell, B. L., Price, P. B., *At. Energy Rev.* 12:275
128. Heck, D., *Atomkernenerg.-Kerntech.* 46:187 (1986)

Annu. Rev. Nucl. Part. Sci. 1992. 42:39–76

EXPLOSIVE HYDROGEN BURNING

A. E. Champagne

Department of Physics and Astronomy, University of North Carolina at Chapel Hill, Chapel Hill, North Carolina 27599; and Triangle Universities Nuclear Laboratory, Duke University, Durham, North Carolina 27706

M. Wiescher

Department of Physics, University of Notre Dame, Notre Dame, Indiana 46556

KEY WORDS: nucleosynthesis, hot CNO cycles, rp-process

CONTENTS

0163-8998/92/1201-0039$02.00

1. INTRODUCTION

Under astrophysical conditions of extreme temperature and density, nuclear reactions can occur with rates that are comparable to a dynamical free-fall time, i.e. on the order of seconds. Explosive hydrogen burning will take place in a number of sites, most notably in cataclysmic binary systems (including novae and some x-ray bursts), and during Type II supernovae. The former objects are thought to be thermonuclear outbursts triggered by mass accretion onto the surface of either a white dwarf or a neutron star. A knowledge of the relevant nuclear reaction cross sections is therefore critical to our understanding of the outburst phenomenon itself. In particular, this information determines the amount of energy generated to power the outburst and the inventory of elements produced. These results may be compared with direct observations to provide a tight constraint on models of these systems. Much useful information can be obtained from the nucleosynthesis itself. For example, measurements of absolute and relative elemental abundances, combined with nucleosynthesis calculations, could yield information regarding the physical conditions during the explosion. Also, if a detectable amount of a gamma-ray emitter such as ^{22}Na or ^{26}Al is produced, then observations of the gamma-ray flux over time may shed some light on how the ejecta are mixed into the interstellar medium. The nuclei of active galaxies are thought to be powered by a similar mechanism, in this case mass accretion by massive black holes. As a result, much of what we learn by studying cataclysmic binaries might also be applied to models of active galaxies. In the case of supernovae, explosive hydrogen burning is incidental to the underlying explosion mechanism. However, nucleosynthetic yields are again useful in probing the explosion. In addition, since supernovae are major sources of heavy elements, explosive nucleosynthesis is responsible for much of galactic chemical evolution.

The goal of this article is to review the status of the nuclear reaction data base used for calculations of explosive hydrogen burning. We begin with a brief discussion of the astrophysical sites followed by a general description of the relevant reaction networks and thermonuclear reaction rates. Following this is a detailed summary of the data for the two primary reaction networks, which are known as the Hot

CNO cycles and the r(apid)p(roton)-process. Finally, we review the results of recent network calculations.

1.1 *Novae*

A nova event is generally understood to result from mass transfer within a close-binary system. In this picture, envelope material from a star that has filled its Roche lobe is captured by a white-dwarf companion (1–3). The white dwarf is typically composed of C-O (with masses on the order of 1 $M_{\odot}$) or, in the case of more evolved progenitors, O-Ne-Mg [with masses in the range 1.2–1.4 $M_{\odot}$ (4)]. In a simple picture, the infalling material carries a great deal of angular momentum and builds up into an accretion disk before spiraling to the surface of the white dwarf. However, formation of an accretion disk does not always accompany mass transfer. If the orbital radius is sufficiently small, the two stars will share a common envelope. This can occur either on the way to or following the thermonuclear runaway (5, 6). The response of the white dwarf to mass transfer depends upon its mass (typically on the order of 1 $M_{\odot}$), its temperature (or degree of degeneracy), and the rate of mass transfer, $\dot{M}$ (7). For modest mass-transfer rates ($\dot{M} \approx 10^{-9}$ $M_{\odot}$/yr), the infalling material has an opportunity to diffuse into the white dwarf, cool and achieve some measure of degeneracy. Eventually, the density of accreted material will exceed 10^3 g cm^{-3}, which triggers nuclear reactions on a rapid time scale (8, 9). The course of the resulting thermonuclear runaway is determined by the degree of degeneracy (8). In a fully degenerate gas, the pressure is determined by density rather than by temperature. Under these reactions, nuclear burning is unstable: Although the energy released by nuclear reactions leads to a sharp rise in temperature, there is little or no corresponding rise in outward pressure, which would tend to cool the gas. Partially degenerate material will burn violently, but will not be ejected from the white dwarf. Complete mass ejection, which characterizes a classical nova outburst, occurs if the material is degenerate enough to delay ejection while the thermonuclear runaway strengthens. Under these conditions, peak temperatures range between $T = 10^8$ and 2×10^8 K ($T_9 = 0.1$–0.2) for a C-O white dwarf and $T_9 = 0.4$–0.5 for an O-Ne-Mg white dwarf. The thermonuclear runaway occurs over the course of 100–200 s. During this time, the two stars will spiral apart in order to conserve angular momentum. However, following the explosion, mass will be transferred at an accelerated rate as radiation heating causes the envelope of the companion star to expand (10). Finally, the system will settle down into a period of low $\dot{M}$ until friction and/or gravitational radiation brings the stars back into close contact. An outburst may reoccur after an interval of 10^3–10^5 yr.

1.2 *X-Ray Bursts*

In general, the observed spectral features of x-ray bursts suggest some interaction involving neutron stars (11 and references therein). Some sources have been observed in binary systems and in these cases, it is believed that there is a superficial resemblance between the mechanism responsible for an x-ray burst and that producing novae. In particular, it is assumed that mass from a companion star is transferred to the surface of a neutron star where it is ignited under degenerate conditions. Because this material falls through a much stronger gravitational potential than was the case for novae, nuclear burning starts at a comparatively greater density (10^6–10^8 g cm^{-3}) and produces more energy. A typical event may last only 5–10 s, but peak temperatures may be in the range $T_9 = 0.7$–1.5. However, the strong gravitational potential prevents much material from being ejected, unlike the case for novae. Thus, x-ray bursts are not important contributors to galactic nucleosynthesis.

1.3 *Supernovae*

Following core collapse in a massive star, a shock front is generated and expands through the stellar envelope and ultimately ejects it into the interstellar medium. During passage of the shock, the envelope will experience a dramatic increase in pressure and density that will trigger a burst of nuclear burning. Although the bulk of the nucleosynthetic yield comes from the inner silicon- and carbon-rich zones, calculations performed for the collapse of a 25-$M_\odot$ star (12) indicate that the outer hydrogen- and helium-rich layers will also undergo some shock processing. The duration of this compression and heating is 5–10 s, during which time the material will experience peak temperatures of $T_9 =$ 0.2–1.0 and densities of $\rho = 10^2$-10^4 g cm^{-3}.

2. NUCLEAR REACTIONS DURING AN OUTBURST

2.1 *Reaction Networks*

The production sites considered above are characterized by temperatures in the range $T_9 = 0.1$–1.5 and densities of $\rho = 10^3$–10^8 g cm^{-3}. For seed material consisting of C and O, the nuclear reaction network will resemble the well-known CNO cycles of hydrostatic hydrogen burning. However, an important difference between hydrostatic and explosive burning is that for the latter case, the time scales for nuclear reactions are comparable to or shorter than typical beta-decay lifetimes. Hence, following the $^{12}C(p,\gamma)^{13}N$ reaction, the beta decay of ^{13}N ($T_{1/2}$ = 9.97 m) will be bypassed by the $^{13}N(p,\gamma)^{14}O$ reaction, which leads to the Hot CNO (HCNO) cycles (13–17). These cycles are diagrammed in Figure 1. From ^{16}O, the route into the HCNO cycles is by the sequence $^{16}O(p,\gamma)^{17}F(\beta^+)^{17}O(p,\alpha)^{14}N(p,\gamma)^{15}O$ at the lower temperatures and by $^{16}O(p,\gamma)^{17}F(p,\gamma)^{18}Ne$, etc as the temperature increases. While the HCNO cycles are in operation, higher-mass material, e.g. Ne or Mg will be burned via (p,γ) reactions operating near the line of beta stability.

For temperatures up to $T_9 \approx 0.2$, the cycle

$$^{12}C(p,\gamma)^{13}N(p,\gamma)^{14}O(\beta^+)^{14}N(p,\gamma)^{15}O(\beta^+)^{15}N(p,\alpha)^{12}C \qquad 1.$$

is thought to dominate the reaction flow and would therefore be the energy source for classical novae. Within this main cycle, the rate of energy production is limited by the β^+ decays at $^{14,15}O$ ($T_{1/2}$ = 71 s and 122 s, respectively). Therefore, in equilibrium most of the catalytic

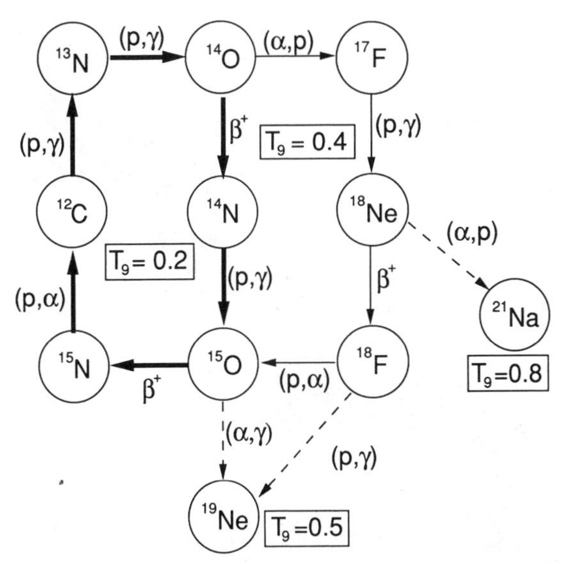

Figure 1 The Hot CNO cycles. The dominant reaction path is denoted by the heavy arrows. At higher temperatures, the sequence initiated by the $^{14}O(\alpha,p)^{17}F$ reaction becomes important (*light arrows*). At still higher temperatures, it may be possible to leave the Hot CNO cycle entirely (*dashed arrows*).

material will exist as $^{14,15}O$ and ultimately as $^{14,15}N$ in novae ejecta. In addition, the abundance of ^{15}N will be enhanced relative to ^{14}N owing to the longer half life of ^{15}O versus ^{14}O. As the temperature approaches $T_9 = 0.4$, alpha-induced reactions may begin to come into play. For example, the waiting point at ^{14}O may be bypassed by the sequence $^{14}O(\alpha,p)^{17}F(p,\gamma)^{18}Ne(\beta^+)^{18}F(p,\alpha)^{15}O$, which will increase the rate of energy generation. At still higher temperatures, it may be possible to break out of the HCNO cycles via the $^{15}O(\alpha,\gamma)^{19}Ne$, $^{18}F(p,\gamma)^{19}Ne$, or $^{18}Ne(\alpha,p)^{21}Na$ reactions. At this point, ^{19}Ne and ^{21}Na can become the catalysts for a further series of rapid proton- and alpha-induced reactions (Figure 2) known as the rp-process (18, 19). However, note that an rp-process may be triggered by a sufficient pre-explosion abundance of Ne, Na, or Mg without direct feeding from the HCNO cycles. These reactions may be the source of the enhancements of Ne, Na, Mg, and Al observed in the ejecta of novae Cyg 1975 (20), Aquilae 1982 (21), CrA81 (22), and Vulpeculae (23).

The path of the rp-process lies between the line of stability and the

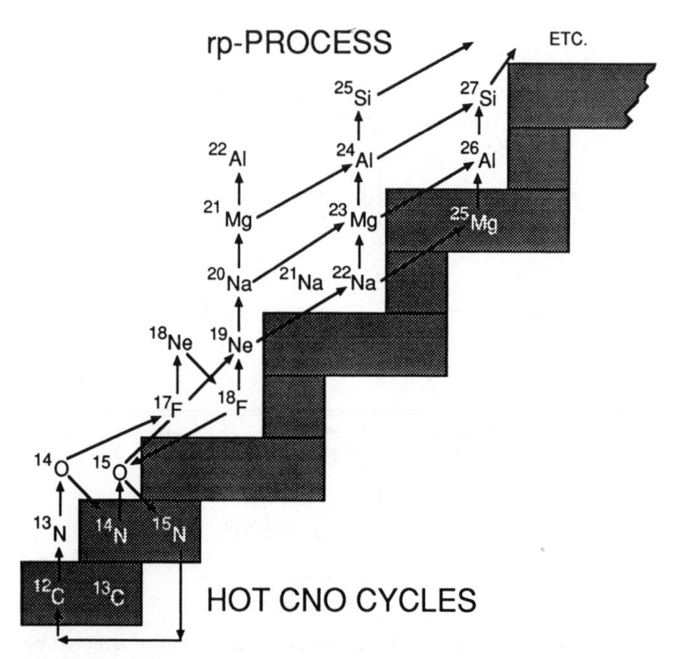

Figure 2 The transition from the Hot CNO cycles to the rp-process at high temperatures. The shaded region comprises the stable nuclei. More detailed reaction paths may be found in Section 5.

proton drip line and is determined by a competition between charged-particle reactions and beta decay. The reaction rates have an exponential temperature dependence whereas the beta-decay rates are, to first approximation, temperature independent. At low temperatures, these rates are comparable and so the rp-process runs close to or along the line of stability. At higher temperatures, where reactions occur more rapidly than beta decay, the reaction path will run close to the proton drip line. The flow of material toward heavier masses is governed by the rates of the slowest reactions along the rp-path as well as by any other effects that may impede the reaction flow. These include beta-decay waiting points, photodisintegration reactions, and reaction cycles. All of these effects must be taken into consideration when calculating nucleosynthesis within the rp-process. However, crude estimates of the reaction rates in the rp-process indicate that nucleosynthesis can occur up to or beyond the iron-peak elements and that energy is produced at a rate of about 100 times that of the main HCNO cycle (18). This rate may be sufficient to power an x-ray burst.

Measurements of the reactions of the HCNO cycles and in the rp-process are necessary to fully understand explosive hydrogen burning. However, since the majority of the reactions of interest involve short-lived nuclei, these measurements have proven to be challenging. Only the $^{13}N(p,\gamma)^{14}O$ reaction rate has been measured with reasonable precision (in this case, to $\pm 13\%$ accuracy) and the rate of the $^{15}O(\alpha,\gamma)^{19}Ne$ reaction has been determined within an uncertainty of a factor of 2. The rates of many of the reactions of interest were estimated in surveys of explosive hydrogen burning by Wagoner et al (24) and Wagoner (25). An enormous amount of nuclear spectroscopy has been completed since then and the majority of these early estimates have been superseded. However, the latest estimates, based on available experimental and theoretical nuclear structure information, are still very uncertain (by factors of 5–1000). Because of these large uncertainties, models of cataclysmic binaries rely on rather suspect nuclear reaction input.

2.2 *Thermonuclear Reaction Rates*

The thermonuclear reaction rate (26–29) is the product of the reaction cross section σ and the center-of-mass velocity v, averaged over the distribution of velocities (assumed to be Maxwellian):

$$\langle \sigma v \rangle = \left(\frac{8}{\pi\mu}\right)^{1/2} (kT)^{-3/2} \int \sigma(E) E \exp(-E/kT)\, dE, \qquad 2.$$

where μ is the reduced mass and k is Boltzmann's constant. In general,

$\sigma(E)$ will contain contributions from nonresonant direct reactions (or direct capture in the case of radiative reactions) and resonant capture. For charged-particle reactions, the energy dependence of the nonresonant cross section is dominated by the penetrability factor for the combined Coulomb and centrifugal barriers. At low energies, this factor is approximately proportional to $\exp[-(E_G/E)^{1/2}]$, where E_G is a constant defined by

$$E_G = (2\pi e^2 Z_1 Z_2/\hbar)^2 \mu/2 \qquad 3.$$

and Z refers to the nuclear charge. With this energy dependence for the cross section, the integral in Equation 2 is nonnegligible over a restricted energy region (known as the Gamow peak). The location and width of the Gamow peak roughly define the region of interest for cross-section measurements. For (p,γ) reactions in the HCNO cycles, this region spans 25 keV to 1.1 MeV in the center of mass and 150 keV to 2.5 MeV near the upper end of the rp-process. It is sometimes useful to express the nonresonant cross section in the following simple parameterization

$$\sigma(E) = \frac{S(E)}{E} \exp[-(E_G/E)^{1/2}]. \qquad 4.$$

The factor of $1/E$ arises from the $\lambda\!\!\!^{-2}$ term in the nuclear cross section and the remaining term is known as the astrophysical S-factor. The S-factor contains the nonkinematic portions of the cross section and is usually a slowly varying function of energy. This permits the reaction rate to be determined by a straightforward numerical integration of Equation 2 or by simple analytic approximations (26–29).

In most cases, considerations of the level densities of the relevant compound nuclei suggest that the reaction rates for both the HCNO cycles and the rp-process are determined primarily by resonant capture. If the cross section is dominated by a resonance at an energy E_r, then $\sigma(E)$ takes on the familiar Breit-Wigner form

$$\sigma(E) = \pi \lambda\!\!\!^{-2} \omega \frac{\Gamma_a \Gamma_b}{(E - E_r)^2 + \Gamma^2/4}. \qquad 5.$$

Here $\lambda\!\!\!^{-}$ is the reduced Compton wavelength; ω is the spin-statistical factor; and Γ_a, Γ_b, and Γ are the partial widths for the incoming and outgoing channels and the total width, respectively. For a resonance that is isolated and narrow (i.e. if $\Gamma \ll E_r$), Equation 2 may be integrated to yield

$$\langle \sigma v \rangle = \left(\frac{2\pi}{\mu k T}\right)^{3/2} \hbar^2 \omega\gamma \exp(-E_r/kT). \qquad 6.$$

The factor $\omega\gamma$ ($\omega\gamma = \omega\Gamma_a\Gamma_b/\Gamma$) is the resonance strength. The contributions of individual resonances are simply summed. If, however, a resonance is broad, or if several resonances overlap, then the shape of the cross section may be appreciably different from the Lorentzian shape assumed in obtaining Equation 6. In this case, Equation 2 must be evaluated numerically in order to obtain the reaction rate. In situations where many, possibly overlapping, resonances contribute to the reaction, the cross section will take on a statistical character. Here, the properties of individual resonances are less important than level densities, masses and energetics. For stable nuclei, Hauser-Feshbach calculations accurately reproduce these cross sections (30, 31).

For most of the reactions of interest, a direct measurement of $\sigma(E)$ would require the production of either a radioactive ion beam or a radioactive target. Of these two approaches, the radioactive-beam method is preferable: In general, the radioactive-target technique suffers from severe background problems for targets with half-lives less than several hours, i.e. for most of the targets of interest. A complimentary approach, relying upon traditional nuclear spectroscopy involves producing the relevant compound nucleus using stable beams and stable targets. Measurements of excitation energies, spins and parities, widths, and decay branches determine all of the resonance information contained in Equation 6. Both of these approaches have been brought to bear on the study of the HCNO cycles and of the rp-process.

3. THE HOT CNO CYCLES

3.1 *Reactions Involving Stable Targets:* $^{12}C(p,\gamma)^{13}N$, $^{14}N(p,\gamma)^{15}O$, $^{15}N(p,\alpha)^{12}C$, *and* $^{16}O(p,\gamma)^{17}F$

Measurements of the ^{12}C(p,γ)^{13}N, ^{14}N(p,γ)^{15}O, ^{15}N(p,α)^{12}C, and ^{16}O(p,γ)^{17}F reactions have been carried out at energies throughout the range of interest for the HCNO cycles. The relevant nuclear data may be found in recent tabulations (32, 33), and thermonuclear reaction rates derived from these data were listed by Caughlin & Fowler (34). Overall, the data and reaction rates are well understood. However, there are a few ambiguities that merit some discussion.

The ^{12}C(p,γ)^{13}N reaction has been measured over the energy range 150 keV to 2.5 MeV by Rolfs & Azuma (35) and the recommended reaction rate (34) is based upon these results. More recent work (36) indicates that a resonance located at a laboratory energy E_p = 1.699 MeV in the earlier study is actually at E_p = 1.689 MeV. Another measurement (37) agrees with the earlier value. However, since this resonance is too weak to be astrophysically significant, there is no meaningful uncertainty in the reaction rate for the temperatures of interest.

The rate of the $^{15}N(p,\alpha)^{12}C$ reaction (34) is based upon an average of two data sets (38, 39) and is assumed to be accurate at low temperatures. In contrast, the rate for $T_9 \geqslant 1$ includes uncertain contributions from two presumed resonances at E_p = 710 and 1050 keV. The former state possesses a spin-parity $J^\pi = 0^-$. Consequently, alpha-particle emission to either ground state or first-excited state of ^{12}C is parity-forbidden and this state should not be considered as a possible (p,α) resonance. The latter state appears to be too weak to play a role in the HCNO cycles. Therefore, the lower limit compiled for the $^{15}N(p,\alpha)^{12}C$ reaction (34) should be used in network calculations of explosive hydrogen burning.

3.2 *The $^{13}N(p,\gamma)^{14}O$ Reaction*

The $^{13}N(p,\gamma)^{14}O$ reaction is dominated by s-wave capture to a single $J^\pi = 1^-$ resonance at a center-of-mass energy $E_{cm} = 541 \pm 2$ keV (corresponding to an excitation energy E_x = 5.17 MeV) (32, 40). The total width of this state, $\Gamma = 38.1 \pm 1.8$ keV, was measured (40, 41) via the $^{14}N(^3He,t)^{14}O$ reaction. Because the radiative width, Γ_γ, is on the order of a few eV and is therefore much smaller than the proton width, Γ_p, the strength of this resonance is determined primarily by Γ_γ.

3.2.1 CALCULATIONS OF Γ_γ Mathews & Dietrich (42) considered $^{13}N+p$ resonant capture as a single-particle transition on top of an excited ^{13}N core. This calculation predicts Γ_p = 34.7 keV and Γ_γ = 2.44 eV for the state of interest in ^{14}O. Although the core-polarization coupling was adjusted to reproduce Γ_γ for $^{13}N(2.365)$, the calculation overpredicts Γ_γ for ^{13}C. A microscopic-potential model was used by Langanke et al (43) to predict Γ_p = 40.1 keV and Γ_γ = 1.5 eV. This work was extended (44) to include the effects of core excitation [specifically $^{13}N(3.51)$] in the capture process, which did not lead to much difference in the final results: Γ_p = 53 keV, Γ_γ = 1.6 eV. It was noted by Barker (45) that the mechanism for populating the resonance in ^{14}O should be similar to that for s-wave capture in the $^{12}C(p,\gamma)^{13}N$ reaction where an R-matrix approach is known to work well. Such a calculation for ^{14}O yields Γ_p = 35.9 keV and a best value of Γ_γ = 1.2 eV (different channel radii lead to values of Γ_γ in the range 1.2–2.4 eV). Both this result and those of Langanke and colleagues (43, 44) depend sensitively upon the choice of potential or radius parameters. With this in mind, Descouvement & Baye (46) employed a generator-coordinate method that was thought to produce more realistic excited-state configurations. With this technique, they obtained Γ_p = 66 keV and Γ_γ = 4.1 eV and were able to do a reasonable job of reproducing the known $^{13}C + n$ scattering length. However, the method underpredicts the (n,γ) thermal cross section.

In summary, a number of very different types of calculations indicate $1.2 \leq \Gamma_\gamma \leq 4.1$ eV. No single method is obviously superior to the others and this simply underscores the difficulty involved with detailed predictions of the properties of individual excited states (particularly as the state in question decays via an E1 transition).

3.2.2 MEASUREMENTS OF Γ_γ In principle, the most direct way to obtain the $^{13}N(p,\gamma)^{14}O$ cross section is through a measurement of the $^{1}H(^{13}N^{14}O)\gamma$ reaction. Since this approach requires the use of an accelerated radioactive beam and since all of the relevant resonance parameters except for Γ_γ had already been measured, initial studies focused upon a measurement of Γ_γ/Γ using indirect nuclear spectroscopy. The calculations summarized above predict $3.1 \times 10^{-5} \leq \Gamma_\gamma/\Gamma \leq 1.1 \times 10^{-4}$. At this level, branching ratio measurements are quite difficult. Wang (40) obtained an upper limit $\Gamma_\gamma/\Gamma \leq 4.5 \times 10^{-4}$ from a $^{14}N(^{3}He,t)^{14}O^{*}(^{14}O)\gamma$ recoil-coincidence measurement. Using $\Gamma = 38.1$ keV, this branching ratio corresponds to $\Gamma_\gamma \leq 17$ eV. Two subsequent measurements of the $^{13}C(^{3}He,n,\gamma)^{14}O$ reaction have produced definite results: The ratio of coincident gamma rays to neutrons emitted from population of the 5.71-MeV state in each case implies $\Gamma_\gamma = 2.7 \pm 1.3$ eV (47) and 7.6 ± 3.8 eV (48). A fourth indirect measurement (49) involved the $^{1}H(^{14}N,^{14}O)n$ reaction. In this experiment, ^{14}O recoils were detected in singles. Therefore, in order to extract Γ_γ from this information, a measurement of the $^{14}N(p,n)^{14}O_{0,1}$ cross section was also required. A synthesis of these results is consistent with $\Gamma_\gamma = 12 \pm 7$ eV, which is somewhat higher than the previous two results.

Recently, two very different types of measurements were completed using radioactive beams. At Louvain-la-Neuve, an accelerated ^{13}N beam was used to produce the $^{1}H(^{13}N,^{14}O)\gamma$ reaction (50). The gamma-ray branching ratio was extracted from the yield of 5.17-MeV gamma rays measured in coincidence with ^{13}N recoils, in anticoincidence with cosmic rays, and in phase with the cyclotron RF. In this manner, the effects of prompt and delayed beam-induced background were minimized. The resulting gamma-ray yield is consistent with $\Gamma_\gamma = 3.8 \pm 1.2$ eV. An analysis of $^{13}N + p$ scattering (51) was also used to check the excitation energy and width. A width of 37.0 ± 1.1 keV is obtained, which is in agreement with the earlier $(^{3}He,t)$ result (41). [Delbar et al point out that the calculated value of Descouvemont & Baye (46) would also be consistent with these results if the level-shift parameter had been included in the calculation.] However, the resonance energy, $E_{cm} = 526 \pm 1$ keV is in serious disagreement with the accepted value of 541 ± 2 (40). The second technique involved breakup of an ^{14}O projectile in the Coulomb field of ^{208}Pb (52). In principle, a measurement of the breakup of ^{14}O into $^{13}N+p$ in kinematic coincidence could be used to extract the $^{13}N+p$ cross section directly (53). However, since the fragments have different Z/A ratios, they will undergo a differential Coulomb acceleration upon breakup and thus the measured kinematics will not represent the true breakup kinematics. Instead, the fragment yield was measured as a function of opening angle and compared to a simple Coulomb-plus-nuclear excitation calculation. The overall normalization between theory and experiment implies $\Gamma_\gamma = 3.1 \pm 0.6$ eV. The precision of this result reflects the fact that nuclear excitation was found to be negligible compared to the easily calculable effects of Coulomb excitation. A second measurement (54) of ^{14}O breakup yields similar results, in this case $\Gamma_\gamma = 2.4 \pm 0.9$ eV.

3.2.3 RECOMMENDED REACTION RATE A weighted average of all of these measurements is $\Gamma_\gamma = 3.1 \pm 0.4$ eV. In calculating the reaction rate, it is necessary to account for the energy dependence of the partial widths over the width of the resonance. This procedure is outlined by Fowler et al (27). We have also adopted the calculation of the direct-capture cross section from Fernandez et al (47), which is based upon the measured analog reaction $^{13}C(p,\gamma)^{14}N(2.313)$. At present, the uncertainty in the reaction rate is dominated by the uncertainty in Γ_γ and not by the disagreement in the resonance energy.

3.3 *The Structure of ^{18}Ne and the $^{14}O(\alpha,p)^{17}F$ and $^{17}F(p,\gamma)^{18}Ne$ Reactions*

Unlike the case of the $^{13}N(p,\gamma)^{14}O$ reaction, little experimental information exists regarding either the $^{14}O(\alpha,p)^{17}F$ or the $^{17}F(p,\gamma)^{18}Ne$ reactions. The compound nucleus in each reaction, ^{18}Ne, is the analog to ^{18}O, which by comparison is well studied. Thus, the properties of states in ^{18}Ne may be inferred from those of analogous states in ^{18}O. The level structure of these nuclei, based upon information compiled by Ajzenberg-Selov (55), is shown in Figure 3. As is clear from a comparison of these level schemes, several excited states remain to be discovered in ^{18}Ne and these may have some bearing on the cross sections under consideration. In addition, the resonance properties of the states that are known have not been determined because a definite analog correspondence remains to be established.

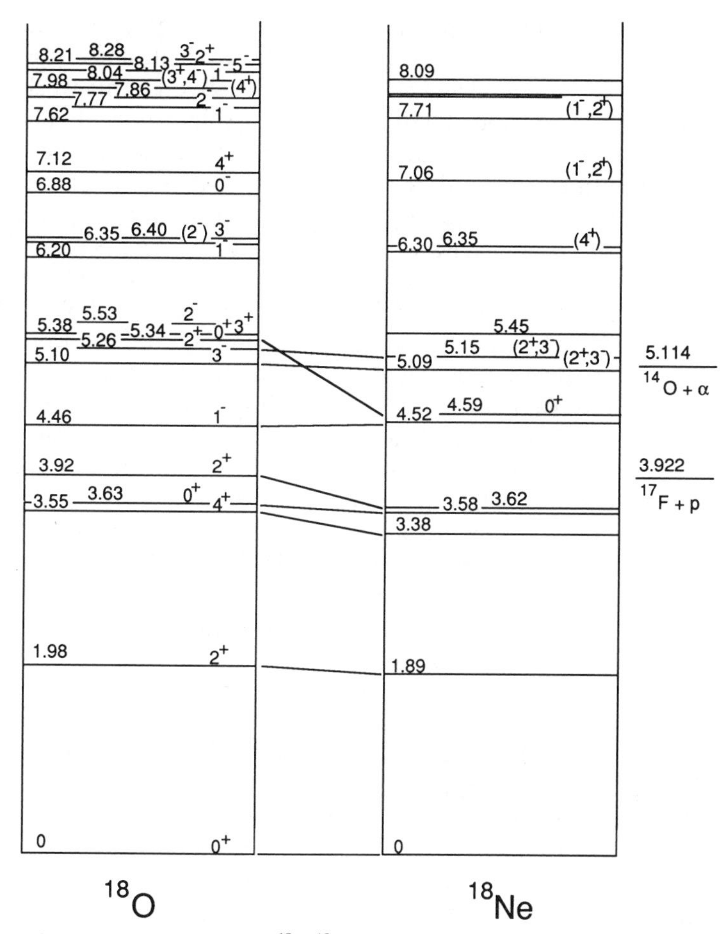

Figure 3 Level structure of the ^{18}O-^{18}Ne analog pair. Known analog connections are indicated.

3.3.1 THE $^{14}O(\alpha,p)^{17}F$ REACTION Both projectile and target in $^{14}O + \alpha$ reaction are 0^+ particles and therefore only natural-parity states will play a role in the $^{14}O(\alpha,p)^{17}F$ reaction. No such states are known to exist within about 1 MeV of the reaction threshold. It is possible that some of the undiscovered states could thus be important and efforts are under way to locate them via the $^{12}C(^{12}C,^{6}He)^{18}Ne$ and $^{20}Ne(p,t)^{18}Ne$ reactions (K. I. Hahn and A. E. Champagne, private communication). At present, the reaction rate is based upon theoretical treatments.

In their estimate of the (α,p) rate, Wallace & Woosley (18) considered two resonances at E_{cm} = 1.19 and 1.24 MeV, which correspond to known states at E_x = 6.30 and 6.35 MeV, respectively. This work laid the basis for further studies (56) that used a calculation of A = 18 Thomas-Ehrman shifts to deduce the analog correspondence between ^{18}O and ^{18}Ne. In this procedure, it was noted that several of the tentative J^π assignments (55) are inconsistent with the analog assignments. With this caveat, a new state was placed at E_x = 6.13 MeV in ^{18}Ne, which is the analog to the 6.198-MeV (J^π = 1^-) state in ^{18}O. Proton and alpha partial widths were obtained from measured $^{17}O(d,p)^{18}O$ (55) and $^{14}C(^6Li,d)^{18}O$ (57) spectroscopic factors and when these were not available, from shell-model calculations (58). Radiative widths were estimated from those known in ^{18}O. From this information, resonance strengths were calculated and the nonresonant, direct-reaction component was treated under the assumption of direct capture to α-bound, but p-unbound states. This mechanism appears to be a relatively unimportant part of the entire cross section as the reaction rate was found to be determined almost entirely by two J^π = 1^- resonances at E_{cm} = 1.01 MeV (E_x = 6.13 MeV) and E_{cm} = 1.94 MeV (E_x = 7.06 MeV). Furthermore, the competing (α,γ) reaction was found to play an insignificant role in the destruction of ^{14}O.

Subsequent to this work, it was pointed out that the direct-reaction mechanism is more complicated than simple direct capture to proton-unbound levels (59). Therefore, a microscopic-potential model was employed to calculate the (α,p) cross section. Included in this calculation were the effects of interference between resonant and nonresonant amplitudes. Because the spins of states above the alpha-capture threshold are uncertain at best, no spin information was used to constrain the calculation. Consequently, the lowest J^π = 1^- resonance is associated with the known state at E_x = 6.30 MeV rather than a new level at 6.13 MeV. Where this calculation produced results in accord with the previous work (56), the same resonance parameters were chosen. Otherwise, calculated properties of ^{18}O (60) were used. The final reaction rate differs considerably from that of Wiescher et al (56): For $T_9 \leqslant 0.35$, the new rate is more than 100 times larger while for $T_9 \geqslant 0.6$, it is about a factor of 3 smaller. Much of the difference at low energy stems from the influence of a J^π = 2^+ resonance at E_{cm} = 34 keV (identified with a known state at E_x = 5.15 MeV). The contribution from this state was treated using a narrow-resonance formalism by Wiescher et al (56) and found to be insignificant. In contrast, the latter calculation (59) accounts for the measured width of this state explicitly and finds a large effect. In further studies with an enlarged model space

(61), this enhancement is reduced by a factor of 3.6, but the high-temperature reaction rate is essentially unchanged from that of Funck & Langanke (59).

3.3.2 RECOMMENDED REACTION RATE It is difficult to recommend a reaction rate on the basis of such conflicting results and sparse experimental input. As was pointed out by Funck & Langanke (59), the previous calculation (56) may have used an oversimplified picture of the reaction mechanism. The later rate relies more heavily on calculated properties of ^{18}Ne because of ambiguities in the nuclear data, but also ignores some experimental input. For example, a calculated reduced alpha width was used for the lowest $J^{\pi} = 1^{-}$ resonance, but an experimental value exists for the analog state. The theoretical value is about a factor of 3 larger than that derived from experiment, but in this instance, there is little effect on the final rate. Fortunately, the largest disagreement among these calculations occurs for $T_9 \leqslant 0.35$ where the (α,p) reaction is probably too slow to compete with the beta decay of ^{14}O. At temperatures where the (α,p) reaction predominates, the calculations agree to within a factor of 2. At high temperatures, the rate calculated by Wiescher et al (56) includes a statistical-model estimate of the effects of high-energy resonances, a contribution to the cross section that was not examined by Funck & Langanke (59, 61). Because the high-temperature reaction rate from Wiescher et al (56) may be the most reliable, it is used in the network calculations presented in Section 5.

3.3.3 THE ^{17}F(p,γ)^{18}Ne REACTION In marked contrast to the situation regarding the ^{14}O(α,p)^{17}F reaction, the ^{17}F(p,γ)^{18}Ne reaction is reasonably well understood. A comparison of $T = 1$ states in the $A = 18$ system shows that analogous $J^{\pi} = 3^{+}$ states exist at $E_x = 5.38$ MeV in ^{18}O and at 6.16 MeV in ^{18}F (55; see Figure 3). However, until recently the corresponding state in ^{18}Ne has not been observed. Because this state could be populated by s-wave proton capture, it could be an important (p,γ) resonance. However, the possible effect of this state was not included in initial calculations (62). Later work predicted $E_x = 4.33$ MeV on the basis of a simple Thomas-Ehrman shift calculation (63). This energy would place the state in question at an astrophysically interesting location, 406 keV above the ^{17}F+p threshold. Because $\Gamma_p \approx \Gamma$ for this state, the resonance strength is $\omega\gamma \approx \omega\Gamma_{\gamma}$. Its radiative width was calculated using the decay of the mirror state in ^{18}O as a starting point, with the result that this state was indeed the most important contributor to the reaction rate over the entire temperature range of interest. However, a different calculation (64), using a more realistic nuclear potential, placed this state approximately 200 keV higher in energy where it would have a lesser influence on the reaction rate.

Experimental verification of this prediction is difficult because population of unnatural-parity states via direct two-proton or two-neutron transfer is forbidden. Consequently, this state can only be observed in reactions such as $^{16}O(^{3}He,n)^{18}Ne$ and $^{20}Ne(p,t)^{18}Ne$ if the reaction mechanism is more complicated than direct transfer. Using the (^{3}He,n) reaction, García et al in fact observed a weak state in the region near $E_x = 4.5$ MeV (64). By a process of elimination, this state has been associated with the missing 3^+ state. A line-shape analysis yields $E_x = 4.561 \pm 6 \pm 7$ MeV, where the first error is statistical and the second accounts for systematic effects. Because the resulting resonance energy is actually $E_{cm} = 639$ keV rather than 406 keV, its effect on the $^{17}F(p,\gamma)^{18}Ne$ reaction rate is much less pronounced than was previously believed.

3.3.4 RECOMMENDED REACTION RATE On the basis of an experimentally determined location for the $J^{\pi} = 3^+$ resonance, the reaction rate reported by García et al (64) is clearly superior to the earlier estimate (63) and is about a factor of 10 lower at the temperatures of interest.

3.4 *The $^{18}Ne(\alpha,p)^{21}Na$ Reaction*

Operation of the $^{18}Ne(\alpha,p)^{21}Na$ reaction may provide a link between the HCNO cycles and the rp-process. Unfortunately, little is known about the structure of ^{22}Mg above the $^{18}Ne + \alpha$ threshold [$E_{th} = 8.140$ MeV (65)]. However, in ^{22}Ne there are 22 known states in the region of excitation between 8.14 and 10.14 MeV (65). As a rule, level densities $\geqslant 10$ per MeV imply that a statistical-model treatment of the cross section is warranted (66, 67). Hence, the $^{18}Ne(\alpha,p)^{21}Na$ cross section has been obtained by a Hauser-Feshbach calculation (F.-K. Thielemann, private communication) without detailed knowledge of the level structure of ^{22}Mg. However, without this information, it is also difficult to make an a priori estimate for the accuracy of this result.

3.5 *The Structure of ^{19}Ne and the $^{15}O(\alpha,\gamma)^{19}Ne$, $^{18}F(p,\alpha)^{15}O$, and $^{18}F(p,\gamma)^{19}Ne$ Reactions*

A second link between the HCNO cycles and the rp-process is through the $^{15}O(\alpha,\gamma)^{19}Ne$ reaction. Its rate was estimated by assuming dimensionless reduced alpha widths $\theta_{\alpha}^2 = 0.02$ for all states above the $^{15}O + \alpha$ threshold (18). Radiative widths were adopted from presumed analog states in ^{19}F. This basic approach was also used by Langanke et al (68). Since the alpha widths determine the resonance strengths for energies below $E_{cm} \approx 1$ MeV, these widths were estimated with somewhat more care by using experimental or theoretical values for θ_α^2 that were not available earlier. However, the rate remained essentially unchanged. One important discovery of this work was that the direct-capture process makes an insignificant contribution to the reaction rate. This assertion was confirmed qualitatively in other calculations (60).

Because the $^{15}O+\alpha$ cross section appears to be dominated by the contributions of isolated resonances, it is possible to determine the cross section experimentally by procedures similar to those used for the $^{13}N(p,\gamma)^{14}O$ reaction. The experimental situation is further simplified by the fact that the analog states in ^{19}F have been identified (55). Measurements of the $^{15}N(\alpha,\gamma)^{19}F$ reaction (55, 69, 70) have been used to determine θ_α^2 for some of these analog pairs. After correcting for the difference in Coulomb barriers between ^{15}N and ^{15}O, these measurements determine Γ_α for states $E_x \geqslant 4.549$ MeV in ^{19}Ne. Four states lie below this level and could not be treated in this manner. Of these, the $E_x = 4.033$-MeV state ($E_{cm} = 504$ keV) is potentially the most important owing to its low spin (and hence low-angular-momentum transfer in alpha capture). The alpha width of this state was estimated from $^{15}N(^6Li,d)^{19}F$ data (71). Finally, the alpha decays of many of these states were measured via the $^{19}F(^3He,t\alpha)^{15}O$ reaction (71). Given Γ_α, this measurement was used to deduce Γ_γ. Also, with values for Γ_γ inferred from ^{19}F, the alpha-branching ratios provided an independent check on Γ_α. An intercomparison of these results provides reasonably accurate resonance strengths for resonances at $E_{cm} = 1020$, 1071, and 1183 keV as well as an estimate for the 504-keV resonance.

A similar approach is planned (M. Wiescher, private communication) in order to examine the competition between the $^{18}F(p,\alpha)^{15}O$ and $^{18}F(p,\gamma)^{19}Ne$ reactions. At present, these rates are only estimated (62) based on the locations of known states in ^{19}Ne and possible analogs in ^{19}F. Although the states involved occur approximately 3 MeV higher than those taking part in the $^{15}O+\alpha$ reaction, the level density is still low enough so that isolated resonances are expected to be important. However, resonance parameters have only been guessed at because little spectroscopic information exists.

3.5.1 RECOMMENDED REACTION RATES It is interesting to note that the experimentally derived (71) $^{15}O(\alpha,\gamma)^{19}Ne$ reaction rate differs from the theoretical estimate (68) by no more than $\pm 20\%$ at any temperature of interest. However, the calculations do not accurately represent any of the individual resonance strengths. For this reason, the experimental rate should be used as the basis for future calculations. Unfortunately, for $T_9 \leq 0.5$, (where the 504-keV resonance is important) this rate is uncertain by approximately a factor of 2. A measurement of Γ_α/Γ for this state would be a challenge because it is expected to be quite small ($\approx 10^{-4}$). The effect of this uncertainty on nucleosynthesis calculations depends upon what temperature and density are required for the (α,γ) reaction to compete with the β^+ decay of ^{15}O. This in turn is governed by the details of the astrophysical model. An improvement to our understanding of this rate may have to await the development of a radioactive beam. The resonance parameters obtained for the ^{18}F(p,α)^{15}O and ^{18}F(p,γ)^{19}Ne reactions were based on reasonable assumptions about the structure of ^{19}Ne (62). Unfortunately, individual resonances may deviate greatly from systematic behavior and thus the rates of these reactions are highly uncertain. Until new experimental information becomes available, these rates will have to be relied upon.

3.6 *Differences Between Laboratory Rates and Stellar Rates*

3.6.1 CONTRIBUTIONS FROM EXCITED STATES The above discussion implicitly assumed that the reactions of interest involved nuclei in their ground states. However, in an astrophysical environment, these nuclei will be immersed in a hot photon bath and may instead be found in a thermal population of excited states. Because energy thresholds and centrifugal barriers will be altered for reactions on excited states versus those for a target in its ground state, the latter reaction may bear little resemblance to the stellar reaction. Also, if excited states can beta decay, then beta half-lives will be modified by thermal excitation. Unless the excited state in question is isomeric (and therefore might be made into a beam), a direct measurement of the reaction cross section is virtually impossible. Reliable estimates can be made only when an analog reaction can be studied.

In equilibrium, the ratio of nuclei in an excited state to those in the ground state is given by the Boltzmann formula (28)

$$\frac{N_x}{N_0} = \frac{2J_x + 1}{2J_0 + 1} \exp(-E_x/kT), \qquad 7.$$

where the subscript "x" refers to the excited state and 0 refers to the ground state. For the HCNO cycles, all of the targets of interest will be primarily in their ground states throughout the relevant temperature range. Only $^{17,18}F$ will have any noticeable excited-state population (0.7% and 0.2% in the first-excited state at $T_9 = 1.5$, respectively). Of course, the possibility exists that a reaction occurring on an excited target might be much stronger than the ground-state reaction and this could offset the difference in population to some degree. However, based upon the available spectroscopic information, there is no reason to expect this situation for $^{17,18}F+p$.

Excited states are a serious consideration for rates in the rp-process (where level densities are generally greater). For example, ^{19}Ne will be in its first-excited state 1–47% of the time for $T_9 = 0.5–1.5$. As improved experimental information becomes available, these effects will need to be examined more closely.

3.6.2 STELLAR BETA-DECAY RATES Decay via positron emission is largely unaffected by the astrophysical environments considered here. In contrast, continuum electron capture will be enhanced with increasing density and temperature. For allowed and first-forbidden decays, the ratio of stellar to terrestrial half lives is approximately proportional to $T_{1/2}/\rho$ (72). Of the nuclei in the HCNO cycles, only ^{18}F has a branching ratio for electron capture [3.1% (55)] large enough for its stellar half life to depart significantly from its measured value. This difference will become important as densities approach 10^5 g cm^{-3} (for $T_9 = 1$). The actual magnitude of this effect will also depend upon the mean molecular weight of the gas and must be examined within the framework of a model calculation. Again, this effect will become more important with increasing mass (i.e. in the rp-process) as electron capture becomes progressively more favored relative to positron decay.

4. THE rp-PROCESS

The nuclear reactions in the CNO cycles have been determined experimentally over a wide energy range and the HCNO reactions have, at least, been constrained by experiment. In contrast, many of the proton-capture rates along the suggested rp-process path (including those involving stable nuclei) are quite uncertain. Also, unlike the HCNO cycles, the path of the rp-process is a very complicated function of the physical conditions governing the explosion. By necessity, network calculations of the rp-process have employed reaction rates that may be highly suspect. In this section, we review our present understanding of the few reactions that have been studied in any detail.

4.1 *Reaction Rates in General*

Proton capture reactions on the $A > 20$ nuclides that lie near the line of stability typically have fairly high Q-values, (on the order of 5 MeV or more). Consequently, level densities at these excitation energies in the compound nuclei are probably high enough to justify reaction rates based on Hauser-Feshbach calculations. For proton capture reactions on stable nuclei, excellent agreement has been obtained between the calculated and the experimentally derived rates (31). However, this approach is not necessarily valid for capture reactions on unstable nuclei (18). For reactions occurring near the proton drip line or near closed shells, compound nuclei are formed at very low excitation energies and therefore at low level densities. In these instances, the reaction rates are determined by the contributions of individual resonances as well as by nonresonant direct capture. As was the case for reactions in the HCNO cycles, these rates need to be derived from a direct measurement of the reaction or calculated from the spectroscopic properties of the compound nucleus. Direct measurements of nuclear reaction cross sections on unstable nuclei require the use of either a radioactive target or a radioactive beam. While a few attempts have been made to measure capture reactions on long-lived nuclei by the radioactive-target technique, the radioactive beam technique is, for the most part, still being developed (74–77). Although the first radioactive-beam measurements on light nuclei have succeeded, the cross sections for capture reactions on the heavy nuclei along the rp-process path are significantly smaller, owing to their larger Coulomb barriers. Therefore, much of what we know about the rp-process is the result of indirect nuclear spectroscopy.

4.2 *The $^{19}Ne(p,\gamma)^{20}Na$ Reaction*

Because the ^{19}Ne(p,γ)^{20}Na reaction is a trigger for the rp-process, a great deal of effort has been spent on determining its rate. Until recently, little was known about the level structure of ^{20}Na above the ^{19}Ne+p threshold and this made reliable estimates of the (p,γ) cross section difficult to obtain. Wallace & Woosley (18) used the known spectrum of states in ^{20}F (without any level shifts) to construct an analogous level scheme for ^{20}Na. An improved level scheme was proposed (68) from calculations of Thomas-Ehrman shifts. Resonance properties were obtained from the respective analog states. This calculation assumed single-particle wavefunctions appropriate for a square well, but constrained the magnitudes of the shifts on the basis of single-particle reduced widths derived from a diffuse potential. Thus there is a systematic uncertainty in their final resonance energies and proton widths.

Initial measurements of the excitation energies of proton-unbound states in ^{20}Na were carried out by Lamm et al (78) using the ^{20}Ne(^{3}He,t)^{20}Na reaction. Higher-resolution studies of this reaction and of the ^{20}Ne(p,n)^{20}Na reaction followed (79), and the results were generally in agreement with the earlier work. However, some of the states seen in the first measurement were not observed in the second and have since been attributed to contaminant reactions (80). In general, there was little correspondence between the measured excitation energies and the predictions (18, 68). With this new information, a revised reaction rate was presented (79) that was approximately 100 times that of the theoretical estimate (68) at $T_9 = 0.5$. This dramatic difference was a result of the different resonance energies used in each case. The major contributor to this new rate was a resonance at 445 keV ($E_x = 2.644$ MeV). Its spin and parity, $J^\pi = 1^+$ was deduced from (^{3}He,t) and (p,n) angular distributions. But resonance parameters in (79) were taken directly from the work of Langanke et al (68), without correcting for the difference in resonance energies between the two studies. Consequently, the proton widths, which are extremely energy dependent, are in error in this later study.

The results of these measurements have been confirmed in three other studies (80–82). An implanted ^{20}Ne target was used by Smith et al (82) to obtain better resolution and yields the best energies for the various possible resonances: $E_{cm} = 447 \pm 5$, 658 ± 5, 787 ± 5, and 857 ± 5 keV. Again, the lowest-energy resonance was found to be the most important contributor to the reaction rate. It has been assumed (80, 82) that this state is the $J^\pi = 1^+$ analog to the 3.173-MeV particle-hole state in ^{20}F and its resonance parameters were obtained from the (d,p) spectroscopic factor and calculated E1 strength of this latter state. For this and the other resonances, the resonance strength was determined primarily by the radiative width. However, it was pointed out in that the systematics of particle-hole states in this mass region do not permit simple estimates of E1 decay strengths, even for cases where the analog strength is known (82). As a result, the reaction rate may be uncertain by several orders of magnitude.

More recently, the $J^\pi = 1^+$ assignment for the 447-keV resonance has been called into question. Using a three-cluster microscopic model, Descouvemont & Baye (83) predict $J^\pi = 1^-$ for the 3.173-MeV state in ^{20}F and its analog in ^{20}Na (still taken to be the 447-keV resonance). B. A. Brown et al (private communication) argue that, on the basis of available spectroscopic information and shell-model calculations, J^π

$= 0^-$ is favored. If this assertion is correct, then the reaction rate will increase by at least a factor of 10 over what has been accepted (80, 82) because of an enhanced M1 decay strength versus a hindered E1 strength for $J^\pi = 1^+$.

It is clear that the rate of the ^{19}Ne(p,γ)^{20}Na reaction cannot be estimated with any degree of confidence until J^π for the 447-keV resonance has been established. This state is not observed to be populated in the beta decay of ^{20}Mg (84) nor is its presumed analog populated in the decay of ^{20}O (85). While these results do not allow a definitive J^π assignment, they would naturally follow from $J^\pi = 0^-$ or 1^-. With the recent development of an accelerated ^{19}Ne beam (86), a direct measurement of the (p,γ) cross section may be possible in the near future. Resonances have been observed at E_{cm} = 839 and 904 keV in the ^{1}H(^{19}Ne,p)^{19}Ne reaction. The lowest resonance might correspond to that inferred from (^{3}He,t) studies at E_{cm} = 857 ± 5 keV (82). However, the upper state is only seen in one of the indirect measurements (80). More work will be required before these direct measurements can be pushed down to lower energies. It should also be remembered that the correct astrophysical reaction rate must also include contributions from excited states of ^{19}Ne as well as that from the ground state.

4.3 *The Early rp-Process: ^{20}Na(p,γ)^{21}Mg, ^{22}Na(p,γ)^{23}Mg, and ^{26}Al(p,γ)^{27}Si*

4.3.1 THE ^{20}Na(P,γ)^{21}Mg REACTION Following the sequence ^{15}O(α,γ)^{19}Ne(p,γ)^{20}Na, which links the HCNO cycles with the rp-process, the ^{20}Na(p,γ)^{21}Mg reaction is expected to continue the reaction flow to higher masses. The level structure of ^{21}Mg has been investigated (87) via the ^{24}Mg(^{3}He,^{6}He)^{21}Mg reaction. The level shifts in the ^{21}F-^{21}Mg analog pair are larger than what had been assumed in earlier calculations of the rp-process (18, 19). Because the states that could play a role in the (p,γ) reaction are now shifted to lower energies than before, the reaction rate may be two orders of magnitude smaller than had been anticipated.

4.3.2 THE ^{22}Na(p,γ)^{23}Mg AND ^{26}Al(p,γ)^{27}Si REACTIONS The rates of the ^{22}Na(p,γ)^{23}Mg and ^{26}Al(p,γ)^{27}Si reactions are of particular interest because they regulate the amount of ^{22}Na and ^{26}Al produced by an explosion. Both of these isotopes are long-lived ($T_{1/2}$ = 2.61 yr and 7.2 x 10^5 yr, respectively) and thus may be observable as gamma-ray sources in the interstellar medium by satellite-based gamma-ray detectors such as the Compton Observatory or the Nuclear Astrophysics Explorer. Several observations of ^{26}Al have already been reported (88–

91) and further measurements would provide tight constraints on calculations of explosive nucleosynthesis, provided that the net production rates are calculable. These reactions are two cases in which the radioactive-target technique has been used successfully.

Measurements of the $^{24}Mg(^{3}He,\alpha)^{23}Mg$ (92) and $^{24}Mg(p,d)^{23}Mg$ (93) reactions have yielded accurate excitation energies and J^{π} values (94) for states in the vicinity of the $^{22}Na + p$ threshold[Q = 7.554 MeV (65)]. The high density of levels observed implied that the reaction rate could be calculated by using a statistical model. An initial attempt to measure the $^{22}Na(p,\gamma)^{23}Mg$ directly (95) used an enriched, evaporated ^{22}Na target. However, no resonances were observed because of substantial target impurities and high background radiation from target activity. A second attempt (96) was more successful: With an implanted ^{22}Na target, the $^{22}Na(p,\gamma)^{23}Mg$ reaction was investigated in the energy range 0.17 to 1.29 MeV, using high efficiency NaI and D_2O detectors (97). Several resonances were observed at energies above 0.29 MeV and the resulting resonance energies are in excellent agreement with values derived from the transfer studies. From these results, the reaction rate could be calculated accurately for the temperature range $0.3 \leq T_9 \leq 1.5$. As expected from the level density, this rate is well reproduced by a Hauser-Feshbach calculation (31).

The $^{26}Al(p,\gamma)^{27}Si$ reaction was first measured directly by Buchmann et al (98). In this work, seven isolated resonances were discovered at energies $276 \leq E_{cm} \leq 893$ keV. Seven additional resonances were discovered over a somewhat larger energy range in work by Vogelaar (99). From these measurements, the (p,γ) reaction rate is well understood for temperatures $T_9 \geq 0.15$. A number of additional states have been identified in spectroscopic studies (100–102) and these may have a large effect on the reaction rate at temperatures below $T_9 = 0.15$. A calculation of the low-temperature rate, based on this information and on shell-model calculations is in progress (A. E. Champagne, private communication). However, it should be noted that all of this work is relevant only to reactions on ^{26}Al in its ground state. For temperatures $T_9 \geq 0.4$, ^{26}Al will be found in a thermal equilibrium between its ground state and its first-excited state (103) and the effects of these thermal excitations must be estimated in the final reaction rate.

4.4 *The rp-Process for* A = *30–60*

For temperatures $T_9 \leq 0.5$, the reaction flow in the A = 30–50 mass range is expected to proceed via (p,γ) reactions on the T = 0 even-even nuclei ^{28}Si, ^{32}S, ^{36}Ar, and ^{40}Ca. Because these reactions are characterized by low Q-values and low level densities in the compound nucleus, they are expected to be dominated by the contributions of isolated resonances. These reactions have all been measured by bombarding a solid target with a high-intensity proton beam (65, 104, 105) and the reaction rate has been determined accurately from the measured excitation function.

In the iron region, the reaction flow is expected to pass through ^{56}Ni. This nucleus is short-lived terrestrially ($T_{1/2}$ = 6.1 days), but is essentially stable during the short duration of an explosion. Hence, it may be an important link to masses beyond Fe-Ni. Although a direct measurement of the ^{56}Ni(p,γ)^{57}Cu reaction is not currently feasible, some relevant information has been obtained from nuclear spectroscopy: The reaction Q-value (Q = 740 keV) has been obtained from recent mass measurements of ^{57}Cu (106–108). From measurements of the ^{58}Ni(^{7}Li,^{8}He)^{57}Cu (107) and ^{58}Ni(^{14}N,^{15}C)^{57}Cu (108) reactions, two proton-unbound levels are observed at E_x = 1.04 and 2.52 MeV. Calculations of Thomas-Ehrman shifts for well-known states in the analog nucleus ^{57}Ni indicate that each of these ^{57}Cu states is a doublet (108). All of these states are expected to be strong single-particle states. Proton widths were calculated from estimates of the single-particle spectroscopic factors for these states and these are found to be the major contributors to the reaction rate. However, this result and others derived from nuclear spectroscopy are uncertain by at least an order of magnitude. This is because the nuclear structure is not known well enough to permit resonance parameters to be calculated with complete confidence. To verify these rates, a direct measurement, using radioactive-beam techniques, is necessary.

4.5 *Impedance Effects*

On the basis of Coulomb barriers, Woosley (73) estimated how far rp-nucleosynthesis would proceed in charge and mass as a function of time. However, the reaction flow may be substantially delayed by impedance effects in which material is stored over a period of time within a certain mass range before it can be processed further.

In principle, four impedance effects may operate within the rp-process: waiting points, photodisintegration, bottlenecks, and reaction cycles. In the mass range A = 20–40, the nuclear structure systematics underlying these impedance effects; are fairly well understood. However, in the mass range A = 40–70, the lack of experimental information about the nuclei near the proton drip line makes general predictions difficult.

4.5.1 WAITING POINTS Once the reaction flow gets to within one proton of the proton drip line, further proton captures are not possible because the compound nuclei are particle unbound. Therefore these isotopes can only be depleted by β^+ or electron-capture decay. Nucleosynthesis toward heavier masses then has to wait for these decays. These waiting point nuclei can be identified by mapping the proton drip line and by measuring lifetimes. This information is essential for a determination of the endpoint of rp-process nucleosynthesis at high temperatures. For the A = 20–40 isotopes, the proton drip line has been established at ^{24}Si, ^{29}S, ^{33}Ar, ^{37}Ca, ^{38}Ca, [^{41}Ti], etc (65). At higher masses and particularly at ^{43}V, ^{47}Mn, ^{50,51}Co, ^{60}Ga, ^{55}As, ^{59}Br, and ^{73}Rb, various atomic-mass predictions (109) disagree as to whether some nuclei are proton bound or unbound. A recent systematic study of proton-rich nuclei in the A = 50–100 mass region indicates that ^{61}Ga, ^{65}As, and ^{69}Br have lifetimes of at least 150 ns (110). Whether or not these nuclei are reached in the rp-process is governed by their binding energies. However, ^{73}Rb was not observed and this implies a waiting point at ^{72}Kr. Because ^{72}Kr has a half-life of 17.2 s, an rp-process of less than 17 s duration will stop at this point.

4.5.2 PHOTODISINTEGRATION Proton capture reactions on even-even, T = 1 nuclei, (^{22}Mg, ^{26}Si, ^{30}S, ^{34}Ar, ^{42}Ti, ^{46}Cr, etc) typically have very small Q-values ($Q \leq 0.5$ MeV) and are strongly hindered at high temperatures by the inverse photodisintegration of the T = 3/2 compound nucleus. If the photodisintegration rate is larger than the proton-capture rate, then further reactions are halted pending the β decay of the T = 1 nucleus. Typically, the T = 1 and T = 3/2 nuclei will maintain an equilibrium abundance and therefore a second proton capture may bridge the T = 3/2 nucleus. The latter situation occurs during conditions of high temperature and density. Because of the low proton thresholds in the T = 3/2 compound nuclei, (p,γ) reactions will usually involve isolated resonances, corresponding to the first and second excited states, and by the direct capture to the ground state. However, current experimental knowledge about the levels in T = 3/2 nuclei, including the excitation energies of the low-lying excited states, is very limited.

As an example, the first excited level in ^{23}Al, which is relevant to the ^{22}Mg(p,γ)^{23}Al reaction, was observed via the ^{24}Mg(^{7}Li,^{8}He)^{23}Al reaction (111) at an excitation energy of 460 ± 60 keV (corresponding to E_{cm} = 360 ± 60 keV). The spin and parity for this level were adopted from the mirror state in ^{23}Ne, and the single-particle spectroscopic factor and B(E2)-strength for γ decay to the ground state were calculated in terms of the Brown-Wildenthal sd-model (112). However, experimental verification of these level parameters would seem to be warranted.

4.5.3 BOTTLENECK REACTIONS The reaction Q-values for proton capture on even-odd $T = 1/2$ nuclei (^{23}Mg, ^{27}Si, ^{31}S, ^{35}Ar, ^{39}Ca, ^{43}Ti, ^{55}Ni) are also usually low ($Q \leq 2.0$ MeV) and again the reaction rates are determined by the influence of isolated resonances and by nonresonant direct capture. Clearly, the Hauser-Feshbach approach is not justified here. Because they represent the only possible reaction link towards heavier elements for temperatures below $T_9 = 1$, these reactions are referred to as bottlenecks. Their reaction rates therefore limit the overall reaction flow, and small rates may lead to a substantial delay for further reactions.

The $^{31}S(p,\gamma)^{32}Cl$ reaction ($Q = 1.574$ MeV) is an important branch out of the SiP-cycle to heavier nuclei. Three states have been located in the energy region between threshold and $E_{cm} = 1$ MeV via $^{32}S(^{3}He,t)^{32}Cl$ studies (113, 114). The excitation energies from these two measurements disagree by about 20 keV. However, the latter results are in excellent agreement with the energies obtained from a measurement of the beta-delayed γ decay of ^{32}Ar. Resonance parameters were obtained from the properties of the well-known analog states in ^{32}P (65). The reaction rate for temperatures characteristic of explosive hydrogen burning appears to be entirely determined by the contribution of the d-wave resonance at 568 keV ($E_x = 2.142$ MeV, $J^{\pi} = 3^+$). At these temperatures and for densities of $\rho \leq 10^6$ g cm^{-3}, the β decay of ^{31}S is significantly faster than the proton capture reaction. Therefore, this reaction is a bottleneck in the route to higher masses. Again, it must be stressed that a measurement of the level parameters or a direct study of the resonance strength is highly desirable.

4.5.4 REACTION CYCLES For temperatures $T_9 \leq 0.4$, unstable even-odd $T = 1/2$ nuclei such as ^{23}Mg, ^{27}Si, ^{31}S, ^{35}Ar, ^{39}Ca, and ^{43}Ti will most likely β decay to the odd-even $T = 1/2$ nuclei ^{23}Na, ^{27}Al, ^{31}P, ^{35}Cl, ^{39}K, and ^{43}Sc, respectively, rather than undergo further reactions. For proton capture on these decay products, both (p,γ) and (p,α) reaction channels are open. The (p,γ) branch will continue processing toward higher masses, whereas the competing (p,α) reaction will complete a cyclic reaction sequence, e.g. NeNa, MgAl, SiP, SCl, ArK, and CaSc cycles. Depending on the reaction branching, $b = \langle p,\alpha\rangle/\langle p,\gamma\rangle$, the reaction flow is hindered by the storage of material in the cycle over b cycle-times τ and the cycle acts as a flow impedance with a temperature-dependent time constant $b \cdot \tau$. To study the time scales in explosive stellar burning processes it is therefore important to determine the cycle times (which at low temperatures are determined by the proton capture rates and at high temperatures by the β^+ decay rates) and to study the various (p,α)-(p,γ) reaction branching b.

In all cases, the proton capture populates a $T = 0$ compound nucleus at fairly high excitation energies ($E_x \approx 8$–12 MeV). Because of the high level density in this region of excitation, and because all populated resonances can decay into the Γ_γ channel, a Hauser-Feshbach approach should be applicable for the (p,γ) reaction rate. In contrast, the competing (p,α) reaction channel is open only for natural parity, $T = 0$ levels. The density of these levels may not be high enough to allow a reliable calculation of the reaction rate by the Hauser-Feshbach approximation. This restriction requires that we investigate the $T = 0$ nuclei in order to determine the level density of natural parity states, and the energies and resonance strengths in the (p,α) reaction channel.

Measurements of reaction branching for the NeNa, MgAl, and SiP cycles have recently been reported (115–119): Competition between the ^{23}Na(p,γ)^{24}Mg and ^{23}Na(p,α)^{20}Ne reactions was examined by populating ^{23}Na + p resonances directly and measuring both reaction channels simultaneously (115, 116). The resulting reaction-rate ratio indicates that for $T_9 \geq 0.2$, a large fraction of the NeNa material is recycled via the ^{23}Na(p,α)^{20}Ne reaction.

A similar measurement has been carried out for the MgAl cycle where the ^{27}Al(p,γ)^{28}Si reaction competes with the ^{27}Al(p,α)^{24}Mg reaction (117). A careful search for low-energy resonances in the (p,α) channel produced restrictive upper limits on the strengths of several resonances that are known to be strong (p,γ) resonances. These results were confirmed by a fundamentally different technique in which the ^{27}Al(^{3}He,d)^{28}Si reaction was used to populate the states of interest and decay protons and alphas were detected in coincidence with deuterons (118). In contrast to the NeNa cycle, these results indicate that the MgAl cycle actually cycles only 0.1–1% of the time for $T_9 \approx 0.2$ and about 10% of the time at higher temperatures. Thus, most of the material in the MgAl cycle is processed to higher masses.

The closure of the SiP cycle was investigated by measuring the ^{31}P(p,γ)/(p,α) branching ratio (119) in a manner similar to the previous measurements (115–117). While several resonances have been observed in the ^{31}P(p,γ)^{32}S reaction at energies above 250 keV, only a single strong resonance was identified (at 383 keV) in the ^{31}P(p,α)^{28}Si reaction. The reaction rates for both reaction channels can be calculated reliably for $T_9 \geq 0.4$. At lower temperatures, some uncertainty results from the unknown spin of the E_x = 9.060-MeV ($J^\pi = 0^- - 2^-$) state observed in transfer reaction studies. For natural parity (1^-), the ^{31}P(p,α) channel is expected to dominate; for unnatural parity (0^-, 2^-) the α channel is closed and only in ^{31}P(p,γ) channel is a strong resonance expected.

5. NETWORK CALCULATIONS

Nucleosynthesis in the rp-process can be described via a large-scale network calculation (18, 19, 31), in which the net reaction flow $F_{i,j}$ between two nuclei, i and j, is defined by

$$F_{ij} = \int [\dot{Y}(i \rightarrow j) - \dot{Y}(j \rightarrow i)]\, dt, \qquad 8.$$

where $\dot{Y}(i \leftrightarrow j)$ is the change in the isotopic abundance Y_i ($Y = X/A$, i.e. mass fraction divided by atomic number) with time, induced by all reactions converting nucleus i to j. The total time evolution of the isotopic abundances is calculated from all depleting and producing reaction rates λi as a function of temperature and density

$$\dot{Y}_i = \sum_j \lambda_j Y_j + \sum_{j,k} \lambda_{j,k} Y_j Y_k. \qquad 9.$$

The first term in the equation includes β^+ decays and photodisintegation of all nuclei j to i, while the second term describes two-particle reaction processes between nuclei j and k leading to i.

Recently, reaction network calculations have been performed to investigate the reaction flow in the rp-process up to the mass range A = 80 for various conditions of temperature and density (114, 120). In these calculations, the reaction network used by Wallace & Woosley (18) for reaction flow up to the mass range A = 40 has been updated and extended. The new network contains 216 stable and unstable nuclei and 946 nuclear interactions—including β^+ electron capture weak interaction processes, as well as (p,γ), (p,α), (α,γ), and (α,p) reactions and the various inverse reactions.

The reaction flow of the rp-process will be discussed in the following for three temperature, density and time-scale conditions: T_9 = 0.3, ρ = 10^3 g cm^{-3}, calculated over a period of t = 10 s; T_9 = 0.4, ρ = 10^4 g cm^{-3}, calculated over a period of t = 1000 s; and T_9=1.5, ρ = 10^6 g cm^{-3}, calculated over a period of t = 10 s. These parameters represent the peak conditions of explosive hydrogen burning in the supernova shock wave, in the thermal runaway in novae and in the thermal runaway in x-ray bursts, respectively. All calculations have been performed for constant temperature and density conditions. For the initial elemental-abundance distribution, solar isotopic abundances have been used (121).

5.1 *Low Temperatures and Densities*

Figure 4 shows the reaction flow pattern in the mass range $A = 20$–60, calculated over a period of $t = 10$ s for temperature and density conditions of $T_9 = 0.3$ and $\rho = 10^3$ g cm^{-3}. The connecting lines indicate the time integrated reaction flux $F_{i,j}$. Clearly indicated is the reaction flux in the HCNO cycles that leads to an enhancement in the abundances of the waiting-point nuclei ^{14,15}O and ^{18}Ne by two proton captures on the initial isotopes ^{12,13}C and ^{16}O. However, there is no connection between the HCNO cycles and the NeNa region. This indicates that an rp-process will occur only if there is preexisting $A \geqslant 20$ material. In the NeNa mass region, the flow pattern indicates the hot NeNa cycle

$$^{20}\text{Ne}(p,\gamma)^{21}\text{Na}(p,\gamma)^{22}\text{Mg}(\beta^+\nu)^{22}\text{Na}(p,\gamma)^{23}\text{Mg}(\beta^+\nu)^{23}\text{Na}(p,\alpha)^{20}\text{Ne}. \quad 10.$$

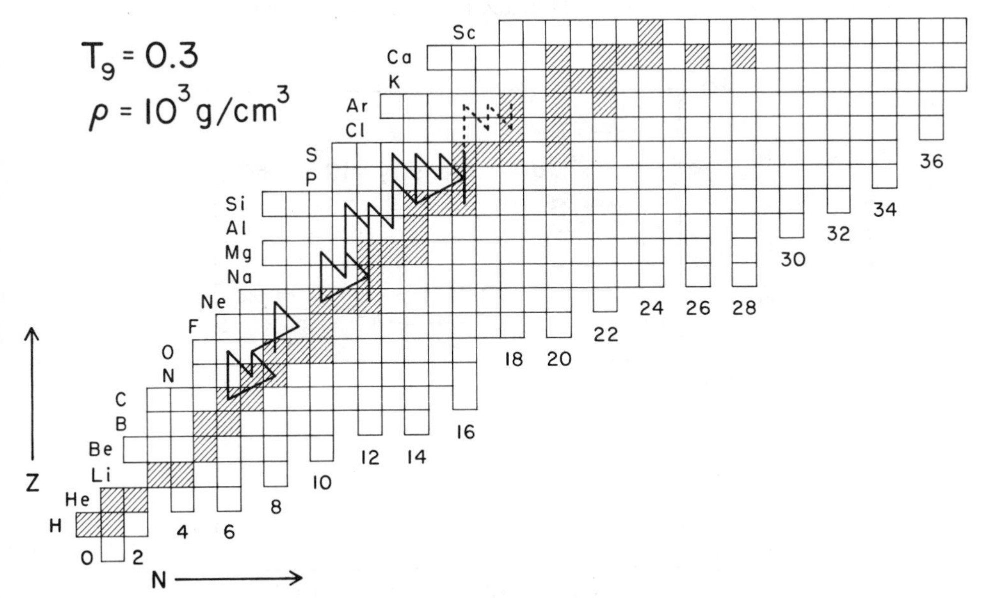

Figure 4 The rp-process reaction path, calculated for $T_9 = 0.3$ and $\rho = 10^3$ g cm^{-3}. Stable nuclei are shown as shaded squares. The solid line indicates the main reaction flow, integrated over a period of 10 s. The dashed line indicates additional processing after a period of 100 s.

The flow towards heavier masses passes through the two bottleneck reactions $^{23}Mg(p,\gamma)^{24}Al$ and $^{24}Mg(p,\gamma)^{25}Al$. Because of the large time constant for breakout from the NeNa cycle, the bulk of the material will stay confined in the NeNa region over the time period of the calculation. A continuous reaction sequence

$$^{24}Al(p,\gamma)^{25}Si(\beta^+\nu)^{25}Al(p,\gamma)^{26}Si(\beta^+\nu)^{26}Al(p,\gamma)^{27}Si(p,\gamma)^{28}P(\beta^+\nu)^{28}Si \qquad 11.$$

leads into the Si,P,S region. Here the flow pattern is characterized by the hot SiP cycle

$$^{28}Si(p,\gamma)^{29}P(p,\gamma)^{30}S(\beta^+\nu)^{30}P(p,\gamma)^{31}S(\beta^+\nu)^{31}P(p,\alpha)^{28}Si. \qquad 12.$$

A significant percentage of this material is processed toward higher masses via the reaction sequence $^{31}P(p,\gamma)^{32}S(p,\gamma)^{33}Cl$. Only a weak flux characterizes further reactions up to the endpoint, ^{40}Ca.

Figure 5 shows the abundances of various isotopes along the process path as a function of time. The initial ^{20}Ne is processed mainly into ^{22}Mg, where it remains confined because of its long half-life ($T_{1/2}$ = 3.86 s). Preexisting ^{24}Mg and ^{28}Si are processed into ^{26}Si ($T_{1/2}$ = 2.23 s) and ^{30}S ($T_{1/2}$ = 1.18 s). The decay of these isotopes leads to an

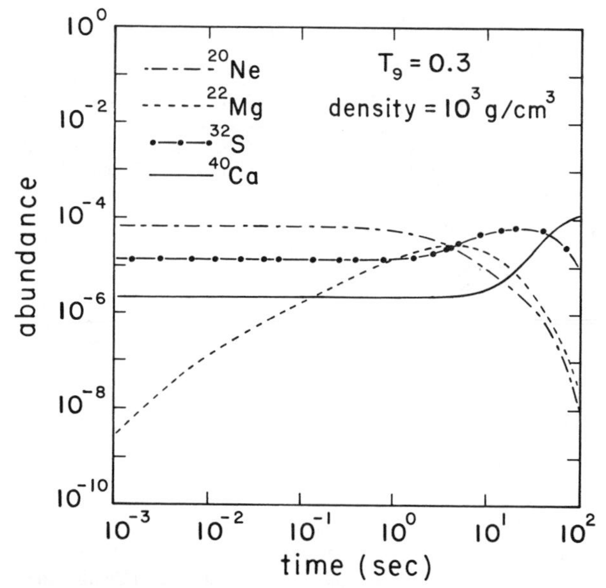

Figure 5 The time evolution of ^{20}Ne, ^{22}Mg, ^{32}S, and ^{40}Ca for T_9 = 0.3 and ρ = 10^3 g cm^{-3}. Solar abundances (121) have been used for initial values.

enrichment of ^{31}S ($T_{1/2}$ = 2.57 s). As the reaction-flow calculations show, ^{31}S primarily decays to ^{31}P, which is recycled by a (p,α) reaction to ^{28}Si where it essentially remains distributed in the SiP region (^{31}S). The time constant for the reaction cycle (which impedes further processing) is determined by the lifetime of ^{31}S and the $^{31}P(p,\gamma)/(p,\alpha)$ reaction rate ratio, $\tau_{SiP} = \tau(^{31}S)\langle p,\gamma\rangle/\langle p,\alpha\rangle$. Because the production rate for ^{32}S via the $^{31}P(p,\gamma)$ reaction is approximately three orders of magnitude larger at these temperatures than the ^{32}S depleting reaction $^{32}S(p,\gamma)^{33}Cl$, processing with each cycle by the $^{31}P(p,\gamma)^{32}S$ branch leads to a final enhancement of ^{32}S. The nucleus ^{32}S, therefore is the endpoint of nucleosynthesis for initial material in the mass range A = 20–32 after a burning time of 10 s. For a longer burning period, an endpoint is reached at ^{40}Ca. It should be pointed out that this result differs from previous network calculations based on the old $^{32}S(p,\gamma)^{33}Cl$ rate by Wallace & Woosley (18), which was predicted to be two orders of magnitude larger than the rate recently measured by Iliadis et al (104).

5.2 *Intermediate Temperatures and Densities*

For typical novae, peak temperatures will be near T_9 = 0.2 and hydrogen burning will be dominated by the primary HCNO cycle. However, for energetic novae ($T_9 \approx 0.4$), rp-process nucleosynthesis will also be important. Network flow calculations, appropriate for extreme nova peak conditions, $T_9 = 0.4$, $\rho = 1 \times 10^4$ g cm^{-3}, are shown in Figure 6. These calculations have been performed for a burning period of Δt = 100 s. The flow pattern indicates a considerable flux of material from the CNO region towards the mass region $A \geqslant 20$ via the reaction sequence

$$^{15}O(\alpha,\gamma)^{19}Ne(p,\gamma)^{20}Na(p,\gamma)^{21}Mg(\beta^+,\nu)^{21}Na(p,\gamma)^{22}Mg. \qquad 13.$$

At the "waiting point" ^{22}Mg, the flow branches into two sequences bypassing the NeNa and the MgAl cycles. The first is initiated by the $^{22}Mg(p,\gamma)^{23}Al$ reaction,

$$^{22}Mg(p,\gamma)^{23}Al(p,\gamma)^{24}Si(\beta^+,\nu)^{24}Al(p,\gamma)^{25}Si(\beta^+,\nu)^{25}Al(p,\gamma)\text{-}$$
$$^{26}Si(\beta^+,\nu)^{26}Al(p,\gamma)^{27}Si, \qquad 14.$$

and the second follows the β decay of ^{22}Mg,

$$^{22}Mg(\beta^+,\nu)^{22}Na(p,\gamma)^{23}Mg(p,\gamma)^{24}Al(p,\gamma)^{25}Si, \text{ etc.} \qquad 15.$$

A further series of reactions runs close to the proton drip line, feeding the SiP mass range by $^{27}Si(p,\gamma)^{28}P(p,\gamma)^{29}S(\beta^+\nu)^{29}P(p,\gamma)^{30}S$. The mass region $A \geqslant 40$ is characterized by a sequence of three reaction cycles,

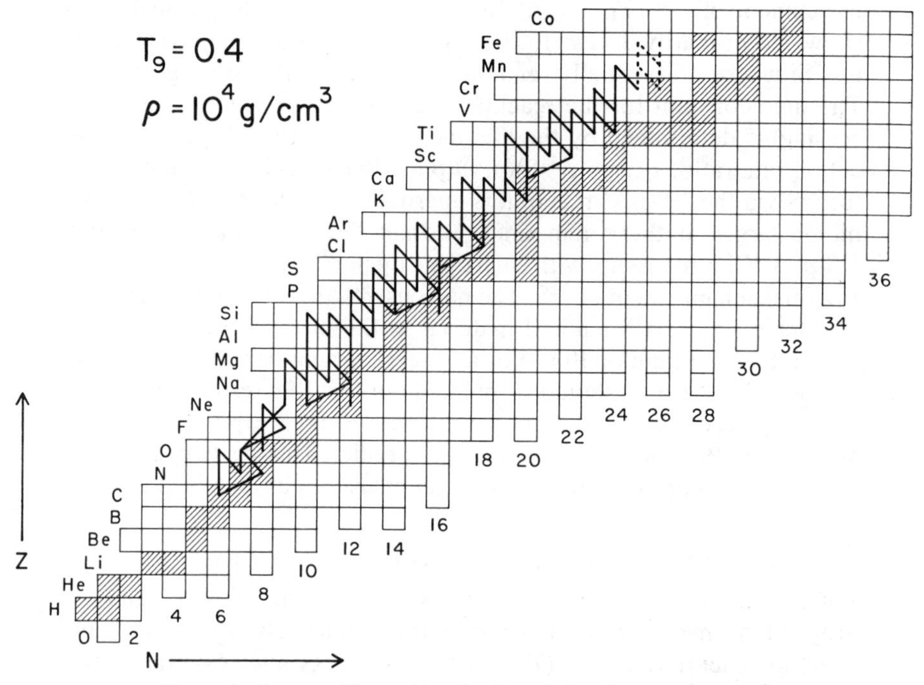

Figure 6 Same as Figure 4 but for $T_9 = 0.4$ and $\rho = 10^4$ g cm^{-3}.

the SiP, SCl, and the ArK cycles, which branch into each other via the bottleneck reactions $^{31}S(p,\gamma)^{32}Cl$, $^{35}Ar(p,\gamma)^{36}K$. Progress towards $A > 40$ is channeled via two weak proton-capture reactions, $^{39}Ca(p,\gamma)^{40}Sc$ and $^{40}Ca(p,\gamma)^{41}Sc$ (122), into the CaSc cycle. This stage of nucleosynthesis is reached after approximately 20 s of burning. In the $A = 50$ region, the reaction flow leaks out of the CaSc cycle via $^{43}Ti(p,\gamma)^{44}V(\beta^+\nu)^{44}Ti(p,\gamma)^{45}V$ and $^{43}Sc(p,\gamma)^{44}Ti(p,\gamma)^{45}V$. Further reactions lead to an endpoint at ^{52}Fe.

Figure 7 shows the corresponding time evolution of isotopic abundances along the process path. At this temperature, the original ^{20}Ne is depleted within the first second of the process and converted into ^{22}Mg(3.86 s). Because ^{22}Mg can be destroyed by a (p,γ) reaction, its effective half-life is shorter, $T_{1/2\text{eff}} = 2.4$ s, and within that time period, material is converted into the waiting-point nuclei ^{34}Ar and ^{40}Ca. The abundances of these nuclei reach an equilibrium value between production and decay after 10 s. However, since the production flow will cease, the net abundances drop. Also during this period, ^{40}Ca will be significantly enhanced owing to its very weak depletion reaction

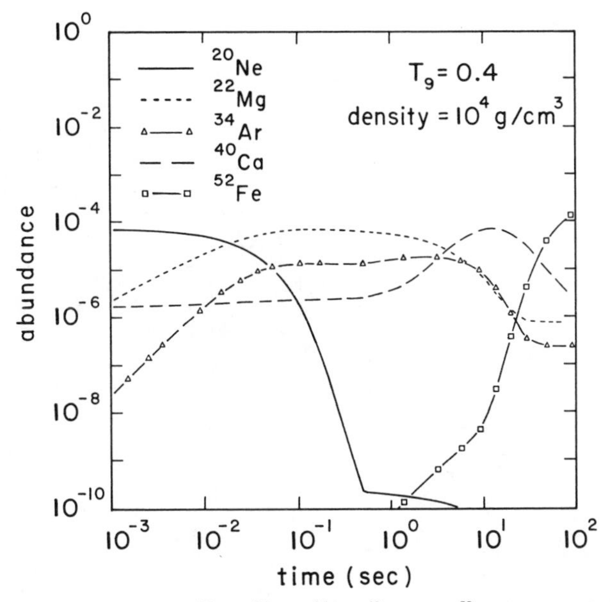

Figure 7 The time evolution of ^{20}Ne, ^{22}Mg, ^{34}Ar, ^{40}Ca, and ^{52}Fe for $T_9 = 0.4$ and $\rho = 10^4$ g cm^{-3}.

^{40}Ca(p,γ)^{41}Sc. Further processing will completely convert ^{40}Ca into ^{52}Fe. Because of its long β-decay half-life ($T_{1/2} = 8.28$ h), and its weak depletion rate via the ^{52}Fe(p,γ)^{53}Co reaction, ^{52}Fe is the endpoint of the rp-process under these conditions.

5.3 *High Temperatures and Densities*

The third calculation deals with nucleosynthesis during an x-ray burst where the temperature and density are chosen as $T_9 = 1.5$ and $\rho = 10^6$ g cm^{-3}. These conditions are expected to last for 10 s or less (11). The reaction flux (Figure 8) shows a continuous flow from ^{4}He to ^{73}Kr, which is the endpoint of the reaction network. At the start of the outburst, ^{12}C is produced by the triple reaction and then quickly converted to ^{14}O. The light waiting-point nuclei, ^{14}O, ^{18}Ne, ^{22}Mg, ^{28}Si, ^{30}S, and ^{34}Ar are bridged by a sequence of (α,p) reactions [known as the αp-process (73)]. Beyond $A = 38$, the rp-process path is characterized by proton-capture reactions and p decays along the proton drip line, up to the Ni region. Because of an increasing Coulomb barrier, further (α,p) reactions will be too slow to compete with β decay. The nucleus ^{54}Ni is predicted to be particle unbound (123), consequently the re-

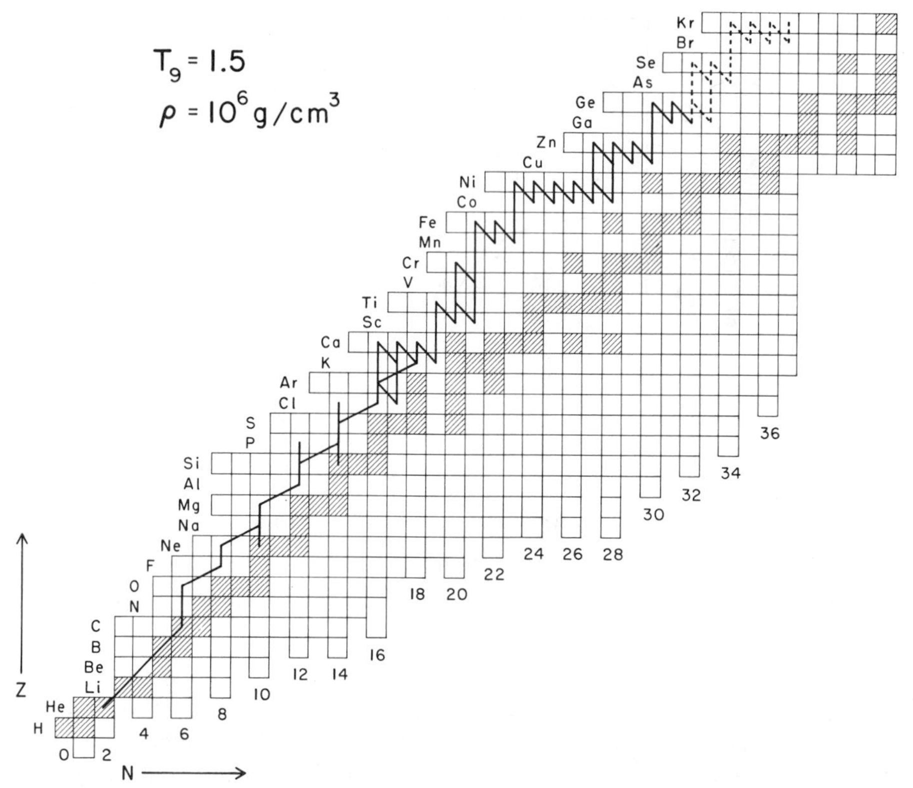

Figure 8 Same as Figure 4 but for $T_9 = 1.5$ and $\rho = 10^6$ g cm^{-3}.

action flow is channeled along the path

$$^{53}\text{Co}(\beta^+)^{53}\text{Fe}(p,\gamma)^{54}\text{Co}(p,\gamma)^{55}\text{Ni}(p,\gamma)^{56}\text{Cu}(\beta^+)^{56}\text{Ni}(p,\gamma)^{57}\text{Cu}. \quad 16.$$

The (p,γ) reactions on ^{55,56}Ni have low Q-values (Q = 0.459 and 0.767 MeV, respectively) and will be hindered by photodisintegration (120). Reactions will continue up to ^{64}Ge(63.7 s), which is essentially stable during the course of the outburst. If ^{65}As is proton unbound, then ^{64}Ge will terminate the rp-process during an x-ray burst. However, if ^{65}As is stable against proton decay, then the reaction path will continue along

$$^{65}\text{As}(p,\gamma)^{66}\text{Se}(\beta^+)^{66}\text{As}(p,\gamma)^{67}\text{Se}(\beta^+)^{67}\text{As}(p,\gamma)^{68}\text{Se} \quad 17.$$

and

$$^{68}\text{Se}(p,\gamma)^{69}\text{Br}(p,\gamma)^{70}\text{Kr}(\beta^+)^{70}\text{Br}(p,\gamma)^{71}\text{Kr}(\beta^+)^{71}\text{Br}(p,\gamma)^{72}\text{Kr}, \quad 18.$$

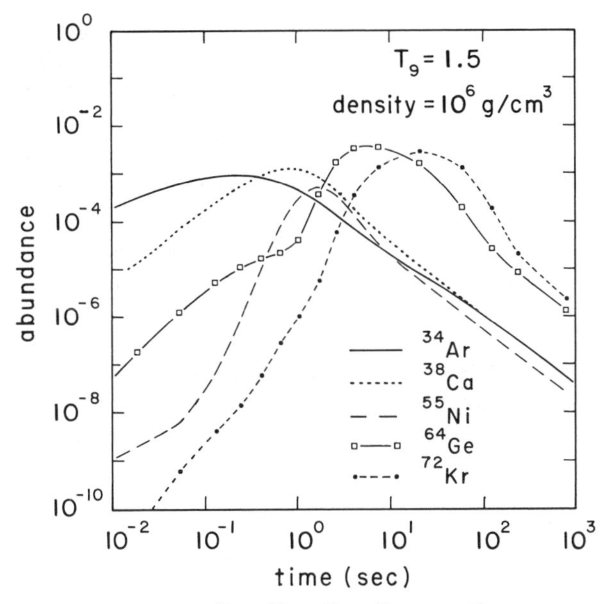

Figure 9 The time evolution of ^{34}Ar, ^{38}Ca, ^{55}Ni, ^{64}Ge, and ^{72}Kr for $T_9 = 1.5$ and $\rho = 10^6$ g cm^{-3}.

provided that ^{69}Br is also particle stable. Since recent measurements indicate that ^{73}Rb is particle unbound (110), the rp-process will terminate at ^{72}Kr(17.2 s).

Figure 9 shows the resulting isotopic abundances along the process path. After 0.5 s, preexisting CNO and NeNa material is converted into ^{34}Ar. After 1 s, the (αp-process will produce ^{38}Ca. The abundances of ^{55,56}Ni will reach a maximum at 1.5 s, and after 5 s, the bulk of the original material will reach ^{64}Ge. The abundance of ^{64}Ge will decrease slowly because both destruction processes, ^{64}Ge(p,γ)^{65}As and ^{64}Ge(β^+)^{64}Ga are weak. Ultimately, ^{64}Ge will be converted into ^{72}Kr, which will reach a maximum abundance after 20 s.

6. CONCLUDING REMARKS

Although an impressive amount of effort has been devoted to understanding nucleosynthesis during explosive hydrogen burning, a great deal of work remains to be done. The reactions described in this review are only a few of the reactions of interest. All are uncertain to some degree. Reactions that have not been mentioned may be equally important but have rates that are essentially unknown. Impressive network calculations have been performed, but based on our knowledge of reaction rates, it is fair to ask if we have not built a house of cards. There is no substitute for measured rates and this effort will require the continued development of radioactive beams as well as old-fashioned nuclear spectroscopy. Nuclear spectroscopy will also be important as a source of other necessary nuclear input such as masses, lifetimes, etc. However, in order to push these measurements to the proton drip line, radioactive beams will be required. We look forward to continued progress in this field for some time to come.

ACKNOWLEDGMENTS

We would like to thank P. D. Parker, J. Görres, and F.-K. Thielmann for some useful conversations. The authors are supported in part by the United States Department of Energy and in part by the National Science Foundation.

Literature Cited

1. Gallagher, J. S., Starrfield, S., *Annu. Rev. Astron. Astrophys.* 16:171–214 (1978)
2. Truran, J. W., in *Essays in Nuclear Astrophysics,* ed. C. A. Barnes, D. D. Clayton, D. N. Schramm. Cambridge Univ. Press. 562 pp. (1982), pp. 467–93
3. Starrfield, S., Sparks, W. M., Truran, J. W., *Astrophys. J.* 291:136–46 (1985)
4. Nomoto, K., *Astrophys. J.* 277:791–805 (1984)
5. Paczyński, B., *Science* 225:275–80 (1984)
6. Livio, M., et al, *Astrophys. J.* 356:250–54 (1990)
7. Kovetz, A., Prialnik, D., *Astrophys. J.* 291:812–21 (1985)
8. Starrfield, S., in *Advances in Nuclear Astrophysics,* ed. E. Vangioni-Flam, et al. Singapore: Kim Hup Lee (1986), pp. 221–41
9. Prialnik, D., *Astrophys. J.* 310:222–37 (1986)
10. Shara, M. M., et al, *Astrophys. J.* 311:163–71 (1986)
11. Tamm, R. E., *Annu. Rev. Nucl. Part. Sci.* 35:1–23 (1985)
12. Arnould, M., et al, *Astrophys. J.* 237:931–50 (1980)
13. Hoyle, F., Fowler, W. A., in *Quasi-Stellar Sources and Gravitational Collapse,* ed. I. Robinson, A. Schild, E. L. Schucking. Univ. Chicago Press. 475 pp. (1965), pp. 12–27
14. Audouze, J., Truran, J. W., Zimmerman, B. A., *Astrophys. J.* 184:493–516 (1973)
15. Audouze, J., Lazareff, B., in *Novae and Related Stars,* ed. M. Friedgung. Dordrecht: Reidel. 228 pp. (1977), pp. 205–15
16. Caughlan, G. R., in *CNO Isotopes in Astrophysics,* ed. J. Audouze. Dordrecht: Reidel (1977), pp. 121–31
17. Lazareff, B., et al, *Astrophys. J.* 228:875–80 (1979)
18. Wallace, R. K., Woosley, S. E., *Astrophys. J. Suppl.* 45:389–420 (1981)
19. Wiescher, M., et al, *Astron. Astrophys.* 160:56–72 (1986)
20. Ferland, G. J., Shields, G. A., *Astrophys. J.* 226:172–85 (1978)
21. Snijders, M. A. J., et al, *Mon. Not. R. Astron. Soc.* 211:7P–13P (1984)
22. Williams, R. E., et al, *Mon. Not. R. Astron. Soc.* 212:753–66 (1985)
23. Gehrz, R. D., et al, *Astrophys. J.* 308:L63–L66 (1986)
24. Wagoner, R. V., Fowler, W. A., Hoyle, F., *Astrophys. J.* 148:3–49 (1967)
25. Wagoner, R. V., *Astrophys. J.* 179:343–60 (1973)
26. Fowler, W. A., Caughlan, G. R., Zimmerman, B. A., *Annu. Rev. Astron. Astrophys.* 5:525–70 (1967)
27. Fowler, W. A., Caughlan, G. R., Zimmerman, B. A., *Annu. Rev. Astron. Astrophys.* 13:69–112 (1975)
28. Clayton, D. D., *Principles of Stellar Evolution and Nucleosynthesis.* Univ. Chicago Press. 612 pp. (1983), pp. 283–361
29. Rolfs, C. E., Rodney, W. S., *Caldrons in the Cosmos.* Univ. Chicago Press. 560 pp. (1988), pp. 150–89

30. Woosley, S. E., et al, *At. Data Nucl. Data Tables* 22:371–442 (1978)
31. Cowan, J. J., Thielemann, F.-K., Truran, J. W., *Phys. Rep.* 208:267–394 (1991)
32. Ajzenberg-Selove, F., *Nucl. Phys.* A460:1–148 (1986)
33. Ajzenberg-Selove, F., *Nucl. Phys.* A523:1–196 (1991)
34. Caughlan, G. R., Fowler, W. A., *At. Data Nucl. Data Tables* 40:284–334 (1988)
35. Rolfs, C., Azuma, R. E., *Nucl. Phys.* A227:291–308 (1974)
36. Kiss, Á. Z., Koltay, E., Somorjai, E., *Acta Phys. Hung.* 65:277–86 (1989)
37. Poyarkov, V. A., Sizov, I. V., *Sov. J. Nucl. Phys.* 45:940–43 (1987)
38. Zyskind, J. L., Parker, P. D., *Nucl. Phys.* A320:404–12 (1979)
39. Redder, A., et al, *Z. Phys.* A305:325–33 (1982)
40. Wang, T. F., PhD Thesis, Yale Univ., unpublished (1986)
41. Chupp, T. E., et al, *Phys. Rev.* C31:1023–25 (1985)
42. Mathews, G. J., Dietrich, F. S., *Astrophys. J.* 287:969–76 (1984)
43. Langanke, K., Van Roosmalen, O. S., Fowler, W. A., *Nucl. Phys.* A435:657–68 (1985)
44. Funck, C., Langanke, K., *Nucl. Phys.* A464:90–102 (1987)
45. Barker, F., *Aust. J. Phys.* 38:757 (1985)
46. Descouvemont, P., Baye, D., *Nucl. Phys.* A500:155–67 (1989)
47. Fernandez, P. B., Adelberger, E. G., García, A., *Phys. Rev.* C40:1887–1900 (1989)
48. Aguer, P., et al, in *Proc. Int. Symp. Heavy-Ion Phys. and Nucl. Astrophys. Problems,* ed. S. Kubono, M. Ishihara, T. Nomura. Singapore: World Scientific (1989), p. 107
49. Smith, M. S., PhD Thesis, Yale Univ., unpublished (1990)
50. Decrock, P., et al, *Phys. Rev. Lett.* 67:808–11 (1991)
51. Decrock, P., et al, in *Radioactive Nuclear Beams, 1991, Proc. 2nd Int Conf. on Radioactive Nuclear Beams,* ed. Th. Delbar. Bristol/Philadlephia/New York: Adam Hilger (1992), pp. 281–86
52. Motobayashi, T., et al, *Phys. Lett.* B264:259–63 (1991)
53. Bauer, G., Bertulani, C. A., Rebel, H., *Nucl. Phys.* A458:188–204 (1986)
54. Kiener, J., et al, see Ref. 51, pp. 311–16
55. Ajzenberg-Selove, F., *Nucl. Phys.* A475:1–198 (1987)
56. Wiescher, M., et al, *Astrophys. J.* 316:162–71 (1987)
57. Cunsolo, A., et al, *Phys. Rev.* C24:476–87 (1981)
58. Ellis, P. J., Engeland, T., *Nucl. Phys.* A144:161–90 (1972)
59. Funck, C., Langanke, K., *Nucl. Phys.* A480:188–204 (1988)
60. Descouvemont, P., Baye, D., *Nucl. Phys.* A463:629–43 (1987)
61. Funck, C., Grund, B., Langanke, K., *Z. Phys.* A332:109–10 (1989)
62. Wiescher, M., Kettner, K.-U., *Astrophys. J.* 263:891–901 (1982)
63. Wiescher, M., Görres, J., Thielemann, F.-K., *Astrophys. J.* 326:384–91 (1988)
64. García, A., et al, *Phys. Rev.* C43:2012–19 (1991)
65. Endt, P. M., *Nucl. Phys.* A521:1–830 (1990)
66. Holmes, J. A., et al, *At. Data Nucl. Data Tables* 18:305–412 (1976)
67. Mahaux, C., Weidenmüller, H. A., *Annu. Rev. Nucl. Part. Sci.* 29:1–31 (1979)
68. Langanke, K., et al, *Astrophys. J.* 301:629–33 (1986)
69. Rogers, D. W. O., Aitken, J. H., Litherland, A. E., *Can. J. Phys.* 50:268–77 (1972)
70. Magnus, P. V., et al, *Nucl. Phys.* A470:206–12 (1987)
71. Magnus, P. V., et al, *Nucl. Phys.* A506:332–45 (1990)
72. Bahcall, J. N., *Astrophys. J.* 139:318–38 (1964)
73. Woosley, S. E., see Ref. 75, pp. 4–27
74. Boyd, R. N., ed., *Proc. Workshop on Radioactive Ion Beams and Small Cross Sections* (1981)
75. Buchmann, L., d'Auria, J., eds., *Proc. Accelerated Radioactive Beams Workshop,* TRI-85-1, 403 pp. (1985)
76. Myers, W. D., Nitschke, J. M., Norman, E. B., eds., *Radioactive Nuclear Beams* (1990)
77. Duggan, J. L., Morgan, I. L., eds., *Proc. 11th Int. Conf. Application of Accelerators in Research and Industry, 1990, Nucl. Instrum. Methods* B56:1–671 (1991)
78. Lamm, L. O., et al, *Z. Phys.* A327:239–40 (1987)
79. Kubono, S., et al, *Astrophys. J.* 344:460–63 (1990)
80. Lamm, L. O., et al, *Nucl. Phys.* A510:503–17 (1990)
81. Clarke, N. M., et al, *J. Phys.* G16:1547–52 (1990)
82. Smith, M. S., et al, *Nucl. Phys.* A536: 333–48 (1992)

83. Descouvemont, P., Baye, D., *Nucl. Phys.* A517:143–58 (1990)
84. Kubono, S. et al, see Ref. 51, pp. 317–22
85. Alburger, D. E., Wang, G., Warburton, E. K., *Phys. Rev.* C35:1479–84 (1987)
86. Galster, W., et al, see Ref. 51, pp. 375–90
87. Kubono, S., et al, *Z. Phys.* A334: 511–12 (1989)
88. Mahoney, W. A., et al, *Astrophys. J.* 286:578 (1984)
89. Share, G. H., et al, *Astrophys. J.* 292: L61–L65 (1985)
90. MacCallum, C. J., et al, *Astrophys. J.* 317:877–80 (1987)
91. Von Ballmoos, P., Diehl, R., Schönfelder, V., *Astrophys. J.* 318:654–63 (1987)
92. Schmalbrock, P., et al, in *Capture Gamma Ray Spectroscopy,* AIP Proc. 125, ed. S. Raman. New York: Am. Inst. Phys. (1985), pp. 785–88
93. Miller D. W., et al, *Phys. Rev.* C20: 2008–24 (1979)
94. Wiescher, M., Langanke, K., *Z. Phys.* A325:309–15 (1986)
95. Görres, J., et al, *Phys. Rev.* C39:8–13 (1989)
96. Seuthe, S., et al, *Nucl. Phys.* A514: 471–502 (1990)
97. Seuthe, S., et al, *Nucl. Instrum. Methods* A272:814–24 (1988)
98. Buchmann, L., et al, *Nucl. Phys.* A415:93–113 (1984)
99. Vogelaar, R. B., PhD Thesis, Calif. Inst. Technol., unpublished (1989)
100. Schmalbrock, P., et al, *Nucl. Phys.* A457:182–88 (1986)
101. Lickert, M., et al, *Z. Phys.* A331: 409–32 (1988)
102. Wang, T. F., et al, *Nucl. Phys.* A499: 546–64 (1989)
103. Ward, R. A., Fowler, W. A., *Astrophys. J.* 238:266–86 (1980)
104. Iliadis, C., et al, *Nucl. Phys.* A539: 97–111 (1992)
105. Iliadis, C., et al, *Phys. Rev. C,* (1992), in press
106. Shinozuka, T., et al, *Phys. Rev.* C30: 2111–14 (1984)
107. Sherrill, B., et al, *Phys. Rev.* C31: 875–79 (1985)
108. Stiliaris, E., et al, *Z. Phys.* A326:139–46 (1987)
109. Haustein, P. E., *At. Data Nucl. Data Tables* 39:185–393 (1988)
110. Mohar, M. F., et al, *Phys. Rev. Lett.* 66:1571–74 (1991)
111. Wiescher, M., et al, *Nucl. Phys.* A484:90–97 (1988)
112. Brown, B. A., Wildenthal, B. H., *At. Data Nucl. Data Tables* 33:347–404 (1985)
113. Jeanperrin, C., et al, *Nucl. Phys.* A503:77–89 (1989)
114. Van Wormer, L., PhD Thesis, Univ. Notre Dame, unpublished (1991)
115. Zyskind, J., Rios, M., Rolfs, C., *Astrophys. J.* 243:L53–L56; 245:L57 (1981)
116. Görres, J., Wiescher, M., Rolfs, C., *Astrophys. J.* 343:365–68 (1989)
117. Timmermann, R., et al, *Nucl. Phys.* A477:105–19 (1988)
118. Champagne, A. E., et al, *Nucl. Phys.* A487:433–41 (1988)
119. Iliadis, C., et al, *Nucl. Phys.* A533: 153–69 (1991)
120. Wiescher, M., et al, see Ref. 51, pp. 353–58
121. Anders, E., Ebihara, M., *Geochim. Cosmochim. Acta* 46:2363 (1982)
122. Wiescher, M., Görres, J., *Astrophys. J.* 346:1041–44 (1989)
123. Masson, P. J., Jänecke, J., *At. Data Nucl. Data Tables* 39:273–80 (1988)

Annu. Rev. Nucl. Part. Sci. 1992. 42:77–100

HADRONIC INTERFEROMETRY IN HEAVY-ION COLLISIONS

Wolfgang Bauer and Claus-Konrad Gelbke

National Superconducting Cyclotron Laboratory and Department of Physics and Astronomy, Michigan State University, East Lansing, Michigan 48824-1321

Scott Pratt

Department of Physics, University of Wisconsin, Madison, Wisconsin 53706

KEY WORDS: intensity interferometry, Hanbury Brown and Twiss effect, two-particle correlation functions, transport theory

---

CONTENTS

0163-8998/92/1201-0077$02.00

1. INTRODUCTION AND SUMMARY

1.1 *Interferometry*

In 1801, Young provided a proof for the wave nature of light by using amplitude interference in his famous double-slit experiment. In this experiment, coherent light from a distant source was passed through two slits separated by a distance d, and interference maxima and minima were observed on a distant screen. From elementary geometrical considerations, one can show that constructive interference occurs (at normal incidence) when

$$d \sin \theta_n = n\lambda \qquad \text{for } n = 1, 2, \ldots, \tag{1}$$

where θ_n is the angle of the nth order maximum. Originally this equation was used to measure the wavelength λ of monochromatic light. In principle, however, it can also be used to determine an unknown slit separation ("source size"), d, by measuring θ_n and using light of a known wavelength.

For applications in astronomy, amplitude interference measurements are complicated by the fact that the relative phase of the two paths is not just given by the length difference, $d \sin \theta$, but contains also uncontrollable contributions from atmospheric distortions. Nevertheless, in 1920 Michelson was able to determine the angular diameter of seven stars by using a stellar interferometer to determine the degree of coherence of light from distant stars illuminating two slits (1).

Hanbury Brown and Twiss overcame the limitations of stellar amplitude interferometry and developed two-photon intensity interferometry as a technique for astronomical distance measurements (2–4). In this technique, which is now commonly referred to as the Hanbury Brown/Twiss (HBT) effect, one records the two-photon correlation function for incoming coincident photons as a function of their relative momentum. This correlation function can be written as

$$R(\mathbf{k}_1, \mathbf{k}_2) = \frac{\langle n_{12} \rangle}{\langle n_1 \rangle \langle n_2 \rangle} - 1. \tag{2}$$

Here $\langle n_{12} \rangle$ is the probability of detecting two coincident photons of wave number $\mathbf{k}_1$ and $\mathbf{k}_2$ in detectors 1 and 2, and $\langle n_i \rangle$ is the probability of detecting a photon of momentum $\mathbf{k}_i$ in detector i ($i = 1, 2$). Equation 2 contains only count rates, which are proportional to the absolute squares of the amplitudes. As a consequence, HBT interferometry is insensitive to phase shifts introduced by atmospheric disturbances. It can be used with very large base lines and delivers superior resolution. This was first shown by measuring the angular diameter of Sirius (4).

The physical basis of the HBT effect is that two photons have a nonzero correlation function because of the symmetrization of their wave functions, a consequence of the quantum statistics for identical particles. This can be understood by considering the simplified case of simultaneous photon emission from two distant point sources located at $\mathbf{r}_a$ and $\mathbf{r}_b$. Assuming propagation in vacuum, the coincidence rate is then proportional to the symmetrized two-photon wave function, normalized by the incident fluxes:

$$\begin{aligned} n(\mathbf{k}_1, \mathbf{k}_2) &= \tfrac{1}{2} \left| \exp(i\mathbf{k}_1\mathbf{r}_a + i\mathbf{k}_2\mathbf{r}_b) + \exp(i\mathbf{k}_1\mathbf{r}_b + i\mathbf{k}_2\mathbf{r}_a) \right|^2 \\ &= \tfrac{1}{2} \left| \exp[i\tfrac{1}{2}(\mathbf{k}_1 - \mathbf{k}_2)(\mathbf{r}_a - \mathbf{r}_b)] + \exp[i\tfrac{1}{2}(\mathbf{k}_1 - \mathbf{k}_2) \right. \\ &\quad \left. \times (\mathbf{r}_b - \mathbf{r}_a)] \right|^2 \\ &= 2\cos^2[\tfrac{1}{2}(\mathbf{k}_1 - \mathbf{k}_2)(\mathbf{r}_a - \mathbf{r}_b)]. \end{aligned} \qquad 3.$$

The correlation function depends on the relative momentum of the two photons, $\mathbf{q} = \frac{1}{2}(\mathbf{k}_1 - \mathbf{k}_2)$, and on the spatial separation of the two sources, $\Delta\mathbf{r} = (\mathbf{r}_a - \mathbf{r}_b)$. The correlation function corresponding to simultaneous emission from an extended and incoherently emitting source is obtained by integrating this equation over the spatial extent of the source. In this simplified case, the correlation function is given by the Fourier transform of the source function.

1.2 *Interferometry in Subatomic Physics*

The ideas of Hanbury Brown and Twiss can also be applied to other particles. The generalization to other pairs of identical particles is straightforward, both for bosons (such as photons, pions, and other mesons) and fermions (such as nucleons).

It was soon realized that intensity interferometry could also be used for size determinations in subatomic physics. Goldhaber et al (5, 6) studied angular distributions of π mesons in proton-antiproton annihilation processes and found that the emission probability of coincident identical pions was strongly affected by their Bose-Einstein statistics, which (in analogy to Equation 3) causes an enhancement of the correlation function at zero relative momentum, $q = 0$. The width of the maximum at $q = 0$ depends on the radius of the interaction volume (6). Later, Shuryak (7) pointed out that pion correlations are not only sensitive to the spatial dimensions of the source, but also to the time dependence of the emission process. In that sense, intensity interferometry is sensitive to the space-time characteristics of the emitting system. Other work (8–15) has developed pion intensity interferometry into a quantitative tool for the investigation of subatomic pion sources.

Pion interferometry has attracted renewed interest in the field of relativistic heavy-ion collisions of beam energies per nucleon around 1 GeV. Experiments have collected information on the apparent pion source size (16–22), and their results have been compared to the results of intranuclear cascade calculations (23).

Intensity interferometry is now strongly pursued at ultrarelativistic energies as well. Topics of particular interest are the interplay between source dynamics and final-state interaction (24–27), pion correlations from an exploding source (28–30), pion correlations calculated via microscopic space-time models (31, 32), and pion interferometry as a probe for the possible formation of a quark-gluon plasma (33–36). Several sets of data have been taken (37–41), and their analysis should greatly clarify the situation.

Other authors have worked on the relationship between short-range correlations and intermittency (42, 43), three and more pion correlations (44, 45), pion interferometry and antideuteron production (46), and coherent production of pions (47–50). Intensity interferometry using kaons has also been investigated (51–53), because kaons probe different parts of the space-time geometry of heavy-ion reactions as a result of their longer mean free path in hadronic matter and their lower contamination from long-lived resonances. First results on $2K^+$ correlations from central $^{28}Si + Au$ collisions at 14.6 *A* GeV have now been reported by the E802 collaboration (54, 55).

The future for this subfield at the Relativistic Heavy-Ion Collider (RHIC) looks very bright. Several thousand pions are expected to be produced in a single central collision. This large number will decrease the statistical uncertainty associated with the measurement, allowing all six dimensions of information in the two-pion correlation function to be probed. In fact, a reasonable correlation function could be constructed from a single event.

Even though the emission probability of high-energy photons in heavy-ion collisions is small, the two-photon HBT effect was also studied theoretically for this case (56). The interpretation of two-photon correlation functions in nuclear collisions is complicated by the fact that neutral pions decay into two energetic photons and thus provide a very strong background signal. Nevertheless, first experimental results have already been obtained by using the Two-Arm-Photon-Spectrometer (TAPS) (57).

1.3 *Interferometry for Fermions*

One can also use identical particles obeying the Fermi-Dirac statistics for intensity interferometry studies. Koonin proposed to use two-proton intensity interferometry to obtain pictures of high-energy nuclear collisions (58).

There are several advantages of using protons for intensity interferometry in heavy-ion collisions. First, they are already present in nuclei and can therefore be readily emitted without requiring large energy expenditures for the creation of their rest mass as is the case for pions and other mesons. Hence, two-proton emission in intermediate-energy collisions imposes less severe phase-space constraints than two-meson emission. Therefore, protons can be used as a probe at much lower energies. Second, the two-proton relative wave function contains the prominent ^{2}He-"resonance," which leads to enhanced sensitivity of the correlation function to the source size. Lastly, protons are easy to detect with the required resolution.

For heavy-ion collisions, two-proton interferometry has been studied intensively during the last few years. Investigations have focused on the spatio-temporal extension of the source (59–66), on the dependence of the extracted source radii on the "violence" of the collision (67–69), on lifetime effects in evaporation from compound nuclei (66, 70–75), and on longitudinal and transverse correlation functions (65, 66, 76). Correlation functions have also been studied as a function of projectile and target mass (62, 77, 78) and kinetic energy of the proton pairs (60–62, 64–66, 69, 76, 78–80).

Recent progress has been centered around the theoretical computation of two-proton correlation functions from nuclear transport theory (64–66, 78). In this framework, it is now possible to understand the dependence of the correlation functions on the parameters discussed above. Comparisons of this theory to experimental data (64, 66, 78) have now established two-proton intensity interferometry as a quantitative tool to study heavy-ion reaction dynamics. We discuss these new results in the main part of this review.

There have also been theoretical (81–84) and experimental (85–95) investigations of large-angle proton-correlations in heavy-ion collisions. At these large relative angles, one finds that the two-proton correlation function is dominated by direct proton knockout (85, 86) and by the effects of the conservation laws of total energy (82), particle number (96), momentum (81, 83, 84, 87, 93), and angular momentum (97).

Recently, several groups have begun to study two-neutron correlations (97–100). This probe has the advantage that there is no Coulomb interaction between the two neutrons or between the neutrons and the emitting source. However, bigger experimental obstacles arise from the relatively small neutron detection efficiencies of good-resolution

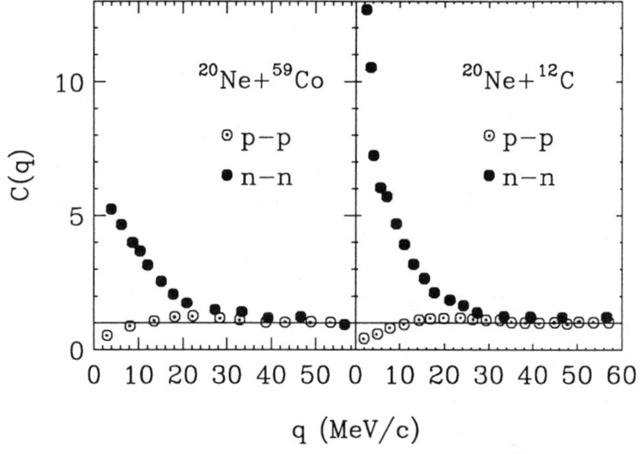

Figure 1 Correlation functions for pp and nn correlations in ^{20}Ne+C and ^{20}Ne+Co reactions at E/A = 30 MeV. The data are taken from (100).

detectors and from problems associated with "cross-talk" between neighboring detectors. The difference between the two-proton and two-neutron correlation functions is illustrated (100) in Figure 1. For small source radii, the two-neutron correlation function exhibits a strong maximum at zero relative momentum, caused by the same resonance that brings about the maximum in the p-p correlation function. However, in the n-n case, there is no Coulomb repulsion, which causes the p-p correlation function to go to zero at $q = 0$.

Finally, there is also a history of studies of fragment-fragment correlation functions (61, 62, 68, 69, 80, 93, 101–108). Here, however, the HBT effect due to quantum statistics is irrelevant, and only Coulomb repulsion (102, 106–108) between the fragments and decays of excited prefragments or particle-unstable states (72, 109–113) are important.

In the remainder of this review, we present the status of two-proton intensity interferometry in heavy-ion collisions, where much progress was made during the last two years. For a current review of the use of other probes the reader is referred to a paper by Boal et al (114). In general, we did not quote unrefereed work such as conference proceedings or preprints. However, there are some recent conference proceedings published in book form that contain additional papers and references on the subjects of interferometry and transport theory (115–117).

2. THEORETICAL BASIS

2.1 *Interferometry with Strongly Interacting Probes*

The original HBT effect is solely caused by the quantum statistics of the two identical particles. Final-state interactions of the two outgoing particles were considered negligible. For photons, this assumption is exactly justified. For charged pions, one must also consider their mutual Coulomb interaction, while it is still a reasonable approximation to neglect their strong interaction, because low-energy $\pi\pi$ scattering is dominated by a slowly varying $I = 2$ phase shift (114). There are, however, some recent investigations of the role of the strong interaction in pion interferometry (118, 119). An unwanted contribution of the strong interaction to charged-pion interferometry is given by resonance decays such as $\eta' \rightarrow \eta\pi^+\pi^-$ and $\eta \rightarrow \pi^0\pi^+\pi^-$, in which two like-charged pions are dynamically correlated (120).

For proton emission, the effects of the strong interaction are very important and were studied in early works (58, 121, 122). The shape of the correlation function at small relative angle reflects then the interplay between the short-range nuclear interaction [including the ^{2}He-"resonance" (123)], the long-range Coulomb interaction, and the Pauli exclusion principle (58, 124).

2.2 *Formalism*

The theoretical expression for the two-proton correlation function, $C(\mathbf{P},\mathbf{q})$, can be written as (28, 58, 65)

$$C(\mathbf{P},\mathbf{q}) = R(\mathbf{P},\mathbf{q}) + 1 = \frac{\Pi_{12}(\mathbf{p}_1,\mathbf{p}_2)}{\Pi_1(\mathbf{p}_1)\Pi_1(\mathbf{p}_2)}$$

$$= \frac{\int d^4x_1\, d^4x_2\, g(\tfrac{1}{2}\mathbf{P}, x_1) g(\tfrac{1}{2}\mathbf{P}, x_2) \left| \phi\left[\mathbf{q}, \mathbf{r}_1 - \mathbf{r}_2 + \frac{\mathbf{P}(t_2 - t_1)}{2m}\right]\right|^2}{\int d^4x_1\, g(\tfrac{1}{2}\mathbf{P}, x_1) \int d^4x_2\, g(\tfrac{1}{2}\mathbf{P}, x_2)}, \qquad 4.$$

where $\mathbf{P}$ and $\mathbf{q}$ are the total and relative momenta, $\mathbf{P} = \mathbf{p}_1 + \mathbf{p}_2$ and $\mathbf{q} = \frac{1}{2}(\mathbf{p}_1 - \mathbf{p}_2)$, respectively, and ϕ is the relative wave function. The terms x_1 and x_2 are the space-time points of the emission of protons 1 and 2. Π_1 is the single-particle and Π_2 is the two-particle emission probability; $g(\mathbf{p},x) \equiv g(\mathbf{p},\mathbf{x},t)$ is the single-proton phase-space emission function.

Gong et al (65) derived Equation 4 based upon the assumptions that the final-state interaction between the two detected protons dominates, that final-state interactions with all remaining particles can be neglected, that the correlation functions are determined by the two-body density of states as corrected by the interactions between the two particles, and that the emission function $g(\mathbf{p},x)$ varies slowly as a function of momentum $\mathbf{p}$, i.e. $g(\mathbf{p},x) \approx g(\mathbf{p} \pm \mathbf{q}, x)$.

Within this approximation, the correlation function depends only on the final relative positions of all particles with momentum $\frac{1}{2}\mathbf{P}$. This can be illustrated more clearly by rewriting Equation 4 as

$$C(\mathbf{P}, \mathbf{q}) = \int \mathrm{d}^3 r F_{\mathbf{P}}(\mathbf{r}) \mid \phi(\mathbf{q}, \mathbf{r}) \mid^2. \qquad 5.$$

Here $\mathbf{r} = \mathbf{r}_1 - \mathbf{r}_2$ is the relative coordinate of the two emitted particles and the function $F_{\mathbf{P}}(\mathbf{r})$ is defined as

$$F_{\mathbf{P}}(\mathbf{r}) = \frac{\int \mathrm{d}^3 R f(\frac{1}{2}\mathbf{P}, \mathbf{R} + \frac{1}{2}\mathbf{r}, t_>) f(\frac{1}{2}\mathbf{P}, \mathbf{R} - \frac{1}{2}\mathbf{r}, t_>)}{\left(\int \mathrm{d}^3 r f(\frac{1}{2}\mathbf{P}, \mathbf{r}, t_>)\right)^2}, \qquad 6.$$

where $\mathbf{R} = \frac{1}{2}(\mathbf{r}_1 + \mathbf{r}_2)$ is the center-of-mass coordinate of the two particles, and the Wigner function $f(\mathbf{p},\mathbf{r},t_>)$ is the phase-space distribution of particles with momentum $\mathbf{p}$ and position $\mathbf{r}$ at some time $t_>$ after both particles have been emitted:

$$f(\mathbf{p}, \mathbf{r}, t_>) = \int_{-\infty}^{t_>} \mathrm{d}t \, g(\mathbf{p}, \mathbf{r} - \mathbf{p}(t_> - t)/m, t). \qquad 7.$$

For a given momentum $\mathbf{P}$, the correlation function has three degrees of freedom, which are a function of $F_{\mathbf{P}}(\mathbf{r})$. Therefore correlation function measurements should allow the extraction of $F_{\mathbf{P}}(\mathbf{r})$, the normalized probability of two protons with the same momentum $\mathbf{P}/2$ being separated by $\mathbf{r}$.

Alternatively, one may use correlation function measurements to test various theoretical models capable of predicting $g(\mathbf{p},\mathbf{r},t)$ and thus making specific predictions about the correlation functions. This approach is more realistic in its goals since a full six-dimensional determination of $C(\mathbf{P},\mathbf{q})$ is very difficult in practice.

2.3 *Sensitivity of the Two-Proton Correlation Function*

In this section, we follow the discussion of Gong et al (65). The two-proton relative wave function appearing in Equation 5, $\phi(\mathbf{q},\mathbf{r})$, is influenced by identical particle interference, short-range hadronic interaction, and Coulomb repulsion. For noninteracting identical particles, the squared wave function has the form

$$|\phi(\mathbf{q},\mathbf{r})|^2 \propto 1 \pm \cos(2\mathbf{q}\cdot\mathbf{r}), \qquad 8.$$

where the plus sign stands for bosons and the minus sign is for fermions of identical spin projection. (For spin-half particles with random spin orientations, the correlation function is reduced to one half at $|q| = 0$ and returns to unity with a width $\Gamma_q \approx 1/R$.)

The Coulomb interaction reduces the correlation function to 0 as $|\mathbf{q}| \rightarrow 0$. Since typical nuclear sources are small compared to the two-proton Bohr radius, $R_b = 2\hbar^2/(e^2M_p) = 57.6$ fm, the shape of this Coulomb dip depends only on the Gamow penetration factor (15)

$$G(|\,\mathbf{q}\,|) = \frac{2\pi\alpha M_p}{|\,\mathbf{q}\,|\,[\exp(2\pi\alpha M_p/|\,\mathbf{q}\,|) - 1]}. \qquad 9.$$

The strong interaction is an excellent size gauge for nuclear systems in the case of two-proton interferometry. This is because of the appearance of the ^{2}He-"resonance," a bump in the two-proton spectrum at $|\mathbf{q}| \approx 20$ MeV/c, the height of which is roughly proportional to R^{-3} (65, 110, 114). With no Coulomb suppression at $q = 0$, two-neutron correlation functions exhibit even greater sensitivities.

Gong et al (65) calculated the relative wave function by numerically solving the Schrödinger equation with the Reid soft core potential (125) for the $l = 0$ and $l = 1$ partial waves. They used the full solution Coulomb waves, $\phi_c(\mathbf{q},\mathbf{r})$, and added the modification $\delta\phi_c(\mathbf{q},\mathbf{r})$ due to the strong interaction in the two lowest l channels. In principle, $|\phi|^2$ is dependent on the six variables $\mathbf{q}$ and $\mathbf{r}$. Because of rotational symmetries, however, $|\phi|^2$ can be completely determined by three independent variables, for which we choose $r = |\mathbf{r}|$, $q = |\mathbf{q}|$, and $\theta = \cos^{-1}(\mathbf{r}\cdot\mathbf{q}/rq)$. In Figure 2, we display $|\phi|^2$ as a function of q and r for $\theta = 45°$. Because of the dominance of the s-wave interaction, pictures for other values of θ look similar to Figure 2 in the coordinate and momentum space intervals considered here. At larger relative momenta and coordinates, one can see differences at different angles θ due to the p-wave interaction.

It is instructive to compute the two-proton correlation function in a simple source geometry to illustrate its sensitivity to the three effects discussed above. Gong et al (65) chose a zero-lifetime Gaussian source parameterization

$$g_0(\mathbf{p},\mathbf{r},t) = \rho_0 \exp(-r^2/r_0^2)\delta(t - t_0), \qquad 10.$$

where r_0 is the radius of the source. Figure 3 illustrates the effects of the different contributions to the two-proton final-state interaction for

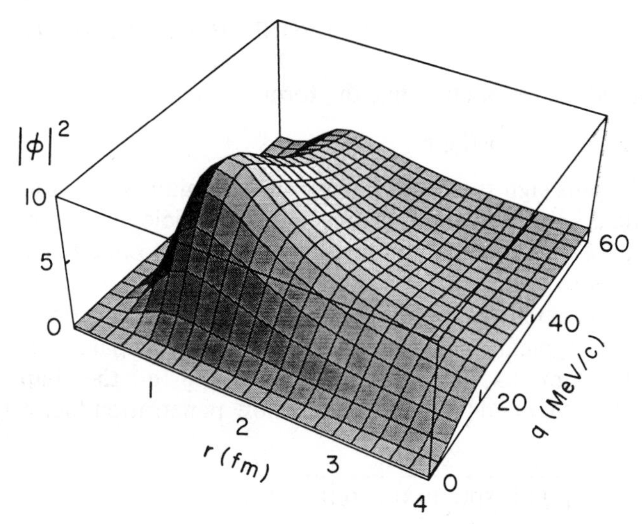

Figure 2 Two-proton relative wave function, $|\phi^2|$, as a function of the two-proton distance r and relative momentum q for $\theta = 45°$.

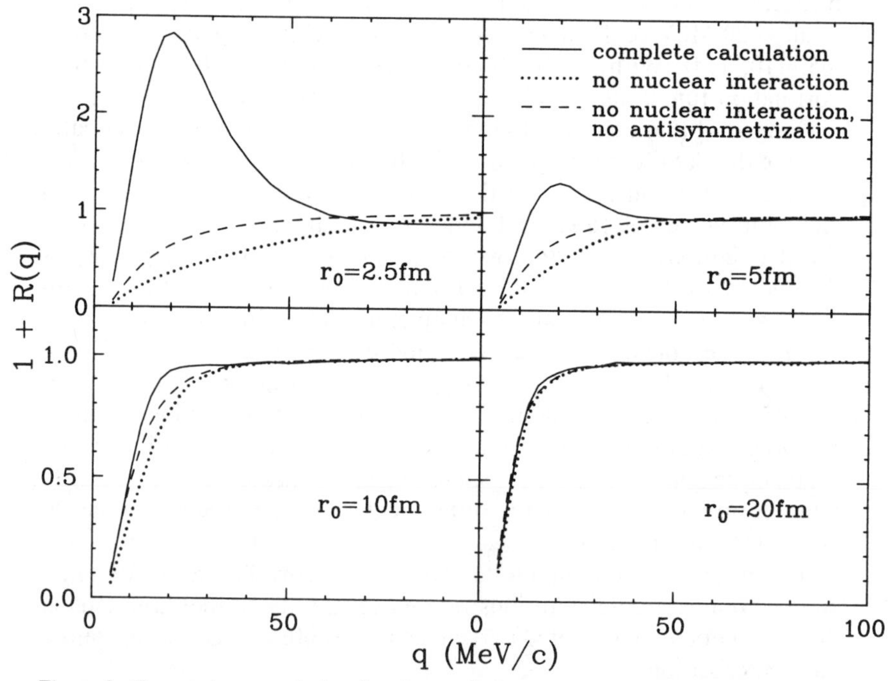

Figure 3 Two-proton correlation functions calculated with a source parameterization according to Equation 10. The solid lines represent the complete calculations including the effects of quantum statistics and of the Coulomb and strong interaction. The dashed lines represent Coulomb interaction only, and the dotted line is for Coulomb interaction plus the effect of the Fermi-Dirac statistics for the two protons. [After (65).]

sources of different radii. The Coulomb interaction completely dominates the shape of the correlation function for very large source radii. For $r_0 < 20$ fm, the correlation function becomes increasingly sensitive to the effects of antisymmetrization and the strong interaction, which clearly dominates at $r_0 = 2.5$ fm because of the prominent ^{2}He resonance.

One may illustrate the sensitivity of the shape of the correlation function on the lifetime and/or the temperature of the emitting source by choosing a simple parameterization,

$$g_t(\mathbf{p}, \mathbf{r}, t) = g_0 \theta(R_s - r)\theta(t - t_0) p \exp\left(-\frac{p^2}{2M_p T} - \frac{t - t_0}{\tau}\right). \qquad 11.$$

Here, T and τ denote the (constant) source temperature and lifetime, respectively, R_s is the fixed source radius, and t_0 is the (fixed, but arbitrary) time at which the source starts to radiate. Calculations based on this parameterization show that the two-proton correlation functions are very sensitive to the lifetime τ in the interval [0,100] fm/c, but do not change noticeably for lifetimes greater than 300 fm/c. For this study, the fixed parameters were chosen to have values $T = 6$ MeV and $R_s = 5$ fm.

One may also investigate the sensitivity of the correlation functions to the relative angle between the sum momentum $\mathbf{P}$ and the relative momentum $\mathbf{q}$ of the pair,

$$\psi = \cos^{-1}\left(\frac{\mathbf{P}\cdot\mathbf{q}}{Pq}\right). \qquad 12.$$

Pratt predicted that the transverse or sideward, $C(\psi = 90°)$, and longitudinal or outward, $C(\psi = 0°)$, correlation functions should exhibit a characteristic difference for long-lived sources (24, 28). [Such a long-lived source could possibly signal the hadronization of a quark-gluon plasma created in ultrarelativistic heavy-ion collisions (28, 34).] Bertsch et al showed in ultrarelativistic cascade calculations that this effect yields measurably different longitudinal and transverse correlation functions (33, 34). The JANUS group has experimentally observed a directional dependence of the two-pion correlation function for light projectile and target systems at the Bevalac (22). But the beam energy in this case ($E/A < 2$ GeV) was definitely too low to expect this observation to have any connection with the physics of the quark-gluon plasma. For intermediate-energy heavy-ion collisions and realistic cuts in ψ, however, Gong et al (65) find that the differences between transverse and longitudinal two-proton correlation functions are typically only of the order of about 10% for lifetimes between 30 and 100 fm/c, and even lower outside this interval.

3. EXPERIMENTAL DETERMINATION OF CORRELATION FUNCTIONS

3.1 *Normalization*

For collisions at fixed impact parameter, the correlation function $1+R(\mathbf{P},\mathbf{q})$ is related to the single- and two-particle yields, $Y(\mathbf{p})$ and $Y(\mathbf{P},\mathbf{q})$:

$$Y(\mathbf{P},\mathbf{q}) \equiv Y(\mathbf{p}_1,\mathbf{p}_2) = \lambda[1+R(\mathbf{P},\mathbf{q})]Y(\mathbf{p}_1)Y(\mathbf{p}_2), \qquad 13.$$

where $\mathbf{p}_1$ and $\mathbf{p}_2$ denote the momenta of the two detected particles, $\mathbf{P} = \mathbf{p}_1+\mathbf{p}_2$ is the total momentum of the particle pair, and $\mathbf{q}$ is the momentum of relative motion (defined in the center-of-momentum frame of the particle pair, where $\mathbf{P} = 0$; nonrelativistically, $\mathbf{q} = \mu\mathbf{v}_{rel}$). The constant λ can be determined from the condition that $R(\mathbf{P},\mathbf{q}) = 0$ for sufficiently large relative momenta for which modifications of the two-particle phase-space density arising from quantum statistics or final-state interactions become negligible.

In general, experimental determinations of two-particle correlation functions involve averages over impact parameter as well as implicit integrations over some of the six variables $\mathbf{P}$, $\mathbf{q}$ (15, 126). Most experimental correlation functions are determined according to the relation (114, 126)

$$\sum Y(\mathbf{p}_1, \mathbf{p}_2) = \lambda'[1 + R(\zeta)] \sum Y'(\mathbf{p}_1, \mathbf{p}_2). \qquad 14.$$

In Equation 14, $Y'(\mathbf{p}_1,\mathbf{p}_2)$ is the "background" yield, λ' is a normalization constant that ensures proper normalization at large relative momenta, and ζ denotes the variables for which the explicit dependence of the correlation function is evaluated (the most common choice is $\zeta = |\mathbf{q}|$). For each experimental gating condition (representing implicit integrations over a number of variables), the sums on both sides of Equation 14 are extended over all energy and detector combinations corresponding to the given bins of ζ. The experimental correlation function is defined in terms of the ratio of these two sums. Comparisons with theoretical results must take this definition into account (66, 126).

Two different approaches are commonly used for the construction of the background yield. In the "singles technique," the background yield is taken as proportional to the product of the single-particle yields, measured with the same external trigger conditions as the true two-particle coincidence yield (59–61, 64, 66, 69, 71, 103, 127, 128):

$$Y'(\mathbf{p}_1,\mathbf{p}_2) \propto Y(\mathbf{p}_1)Y(\mathbf{p}_2). \qquad 15.$$

In the "event-mixing technique," the background yield is generated by mixing particle yields from different coincidence events (12, 20, 67, 77, 95, 129–132):

$$Y'(\mathbf{p}_1, \mathbf{p}_2) = \sum_{n \neq m} \delta^3(\mathbf{p}_1 - \mathbf{p}_{1,n})\delta^3(\mathbf{p}_2 - \mathbf{p}_{2,m}). \qquad 16.$$

Here, the indices n and m label the nth and mth recorded two-particle coincidence events, and $\mathbf{p}_{1,n}$ and $\mathbf{p}_{2,m}$ denote the momenta of particles 1 and 2 recorded in events n and m, respectively. In most analyses, the index n runs over all recorded coincidence events and the index m is varied according to $m = n+k$, with typically $0 < k < 1000$. For large data sets, there is no need to eliminate the diagonal elements in the summation in Equation 16, and one may sum over all combinations of n and m (126).

If single- and two-particle data represent very different averages of impact parameter, the use of the singles technique may lead to serious distortions of the correlation function (133). Furthermore, less interesting correlations, resulting for example from phase-space constraints due to conservation laws (60, 83, 84, 87, 89, 93, 96), may be suppressed by using the event-mixing technique. These advantages of the event-mixing technique have to be weighed against its disadvantage that it attenuates the very correlations one wishes to measure (20, 126). The degree of this attenuation depends on the experimental apparatus and the magnitude of the correlations. Hence, quantitative analyses require careful Monte Carlo simulations. For the extraction of undistorted correlation functions, iterative procedures have been developed (20).

For statistical emission processes in which the emission of a single particle has negligible effect on further emissions, single- and two-particle yields should originate from similar regions of impact parameters. In such instances, the singles technique appears to be the preferable choice (126).

3.2 *Source Size*

It is customary to compare different correlation functions obtained from experiments or theories by fitting them to correlation functions generated with the aid of Equation 4 from simple zero-lifetime Gaussian sources parameterized according to Equation 10. The "source size" r_0 is a free parameter in these fits. One should, however, keep in mind that the assumption of zero lifetime is unrealistic. Therefore one extracts generally different source sizes for the same emitting system, when fitting to correlation functions generated for different bins in **P**, the sum momentum of the two emitted protons.

Nevertheless, it is instructive to perform such fits because they simplify the comparison of correlation functions observed for different systems. A first systematic compilation of extracted source radii from two-pion correlation measurements was performed by Bartke (21), who showed that the extracted radius parameter scales with the radius of the projectile.

Another systematic study of this kind is shown in Figure 4. The data points are the values of r_0 extracted from parameterizations with Equation 10. They are displayed as a function of v_p/v_{beam}, where v_{beam} is the beam velocity, and $v_p = \frac{1}{2}(p_1+p_2)/M_p$ is the average velocity of the coincident proton pair. In this comparison the "source sizes" for ^{14}N and ^{16}O projectiles track each other almost completely. They are interpolated by the solid line. The dashed and dot-dashed lines are obtained by scaling the solid line by factors of $(3/14)^{1/3}$ and $(40/14)^{1/3}$, respectively. For the emission of very energetic protons (v_p/v_{beam} >

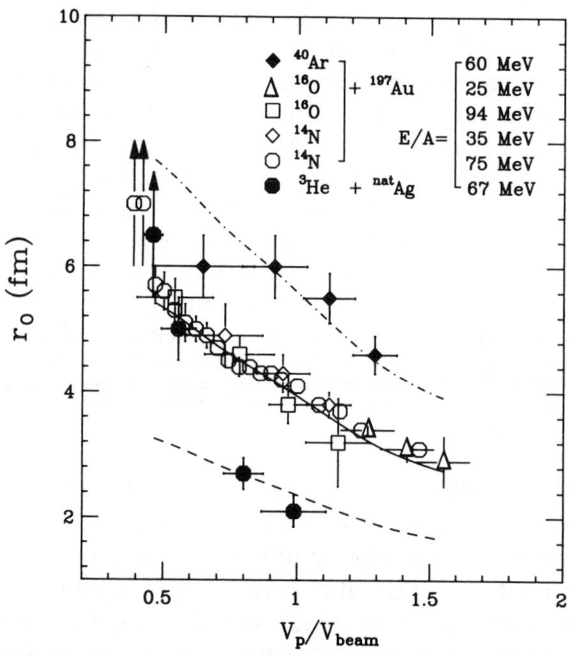

Figure 4 Systematics of Gaussian source radii extracted for a variety of reactions. [After (78).]

0.5), the radius parameter scales approximately with projectile radius. For the emission of lower energy protons, however, lifetime effects dominate the correlation functions, and a simple geometrical interpretation in terms of the radii of target and projectile becomes meaningless.

One effect that can be responsible for the observed pair-momentum dependence of the extracted source radius is cooling of the source. As a consequence of the falling temperature, the characteristic lifetime for high-energy particles is much shorter than the one for low-energy ones (65). This effect was first studied in a numerical simulation of two-proton correlation functions by Boal & DeGuise (79). Meaningful interpretations of the two-proton correlation function require, in general, calculations capable of predicting the full space-time dependence of the emission function. This type of calculation is introduced in the following.

4. CALCULATION OF CORRELATION FUNCTIONS VIA NUCLEAR TRANSPORT THEORY

4.1 *Heavy-Ion Transport Theory*

Equation 4 only requires knowledge of the proton one-body distribution function. Therefore it is possible in principle to generate small-angle correlation functions with any theory that calculates the time evolution of the one-body distribution function. Such calculations have recently been performed by Gong et al (64–66, 78). The microscopic theory used there is based on the Boltzmann-Uehling-Uhlenbeck (BUU) equation. This equation was first postulated by Nordheim (134, 135) and later worked on by Uehling & Uhlenbeck (136). It was first applied to the heavy-ion transport problem by Bertsch et al (137–139), by Stöcker et al (140–143), and by Grégoire et al (144–147). Several review articles on nuclear transport theories have been published (148–153).

The BUU equation describes the time evolution of the nuclear one-body density Wigner distribution $f(\mathbf{r},\mathbf{p},t)$ under the influence of the nuclear mean field (left side) and individual nucleon-nucleon collisions (right side):

$$\frac{\partial}{\partial t} f(\mathbf{r}, \mathbf{p}, t) + \frac{\mathbf{p}}{m} \nabla_{\mathrm{r}} f(\mathbf{r}, \mathbf{p}, t) - \nabla_{\mathrm{r}} U \nabla_{\mathrm{p}} f(\mathbf{r}, \mathbf{p}, t)$$

$$= \frac{g}{2\pi^3 m^2} \int \mathrm{d}^3 q_{1'}\, \mathrm{d}^3 q_2\, \mathrm{d}^3 q_{2'}$$

$$\times\ \delta\left[\frac{1}{2m}(p^2 + q_2^2 - q_{1'}^2 - q_{2'}^2)\right]\cdot \delta^3(\mathbf{p} + \mathbf{q}_2 - \mathbf{q}_{1'} - \mathbf{q}_{2'})\cdot\frac{d\sigma}{d\Omega}$$
$$\cdot\{f(\mathbf{r},\mathbf{q}_{1'},t)f(\mathbf{r},\mathbf{q}_{2'},t)[1 - f(\mathbf{r},\mathbf{p},t)][1 - f(\mathbf{r},\mathbf{q}_2,t)]$$
$$- f(\mathbf{r},\mathbf{p},t)f(\mathbf{r},\mathbf{q}_2,t)[1 - f(\mathbf{r},\mathbf{q}_{1'},t)][1 - f(\mathbf{r},\mathbf{q}_{2'},t)]\}. \qquad 17.$$

In all the current numerical implementations of the solution of Equation 17, the test particle method (154) is used to describe the propagation of the one-body distribution under the influence of the mean field potential $\mathbf{U}$. It is supplemented by an intranuclear cascade to solve the collisions' integral (155–160). The test particle collisions respect the Pauli exclusion principle because the factors $1 - f$ are present, factors that are numerically implemented via a Monte Carlo rejection method.

4.2 *Large-Angle Correlations*

The solution of Equation 17 represents the time evolution of the single-particle distribution function f. In the limit of $t \to \infty$ it is therefore possible to predict all single-particle observables such as proton spectra (138) in this theory. It is also possible to predict the production cross sections of secondary particles (137, 139, 140).

Predictions for two-particle correlations can only be made if these correlations are simply consequences of the conservation laws for momentum, energy, angular momentum, and particle number. Such a calculation was first performed by Bauer (83). It was shown that two-proton correlation functions measured at large angles (89–91, 93) can be successfully reproduced by the BUU theory, provided that total momentum conservation is correctly taken into account. In a second study, Ardouin et al found that the variation of the large-angle correlation function with polar angle θ is largely due to angular momentum effects (161, 162).

4.3 *Small-Angle Correlations*

Using the convolution techniques of Equation 4, one can use the output of Equation 17 and calculate two-particle correlation functions for small relative momenta. This was first done by Gong et al (64–66, 78).

As an example, Figure 5 shows correlation functions generated from $^{14}N + ^{27}Al$ collisions at $E/A = 75$ MeV. The measured two-proton correlation function is indicated by the plot symbols and is displayed for three different gates on the total momentum, $\mathbf{P}$, of the proton pair. One can see the expected features: A maximum of $C(q)$ at $q \approx 20$ MeV/c, reduction of $C(q)$ as $q \to 0$, and $C(q) \to 1$ as $q \to \infty$.

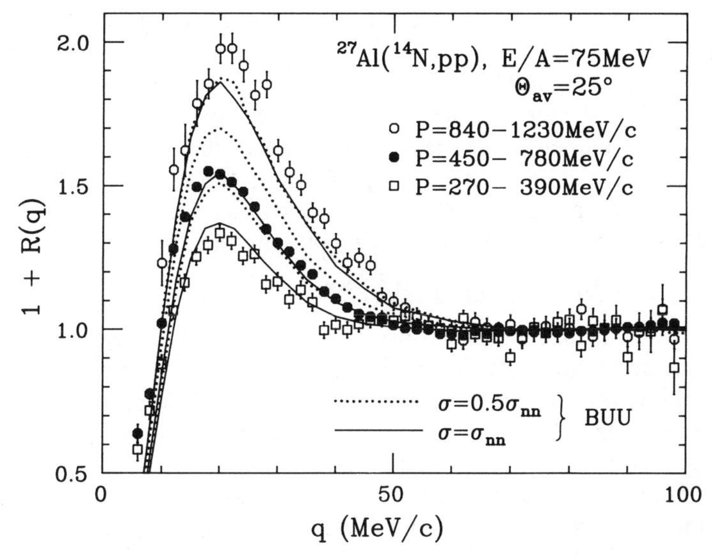

Figure 5 Two-proton correlation functions as a function of relative momentum **q**, measured for the reaction ^{14}Ne + ^{27}Al at E/A = 75 MeV for three different gates on total momentum of the proton pair **P** (*plot symbols*) are compared to the calculation based on the BUU theory (*lines*) for two different values of the in-medium nucleon-nucleon cross section. [After (66).]

The solid and dotted lines in Figure 5 represent calculations based on the BUU theory. For the dotted lines, it was assumed that the in-medium nucleon-nucleon cross section is one half of the free value; and for the solid lines the full free cross section was used. (In both cases, the corrections due to the Pauli exclusion principle were also taken into account). Medium modifications of the nucleon-nucleon cross sections were recently suggested by Cugnon et al (163), based on a formalism developed by Botermans & Malfliet (164) and Lejeune et al (165). However, from Figure 5 one can see, in particular for the two lower momentum bins, that the free space value is preferred. Gong et al (66) find that at E/A = 75 MeV the two-proton correlation function in the intermediate momentum bin (pair velocity ≈ one half of beam velocity) are very sensitive to the value of the in-medium nucleon-nucleon cross section, and their extracted value is

$$\sigma^{NN}_{\mathrm{medium}} \approx \sigma^{NN}_{\mathrm{free}}\,. \qquad 18.$$

This finding is in agreement with theoretical and experimental studies on the disappearance of nuclear collective flow, where Ogilvie et al report similar numbers (166, 167).

The calculations shown in Figure 5 are able to reproduce the experimental fact that the correlation function has a different height for different P-bins, which gives rise to the apparent pair-velocity dependence of the extracted Gaussian source size discussed in Section 3.2. From this agreement we have to conclude that the effect is due to the reaction dynamics, and that a simple zero-lifetime Gaussian fit to extract the apparent source size must not be associated with the spatial extent of a stationary source of negligible lifetime.

4.4 *Initial-State Correlations*

Up to now, our discussion has assumed that the calculated correlation functions are completely determined by the final-state interactions between the two outgoing protons, and that initial-state correlations are completely negligible. Thus, a chaotic source was assumed. However, correlations in the initial state may not always be negligible. The two most important initial-state correlations discussed in the literature are coherent particle production and initial-state spin correlations.

The idea of coherent production of particles in nuclear collisions via a bremsstrahlung-type process was introduced by Heisenberg (168). The consequences of coherent particle production on two-pion interferometry have been studied in several theoretical approaches (15, 44, 47, 48, 50). This leads to the introduction of the degree of coherence

$$D(\mathbf{P}) = \frac{n_0(\mathbf{P})}{n_0(\mathbf{P}) + n_{ch}(\mathbf{P})}, \qquad 19.$$

where n_0 is the number of coherently produced pions, and n_{ch} is the number of chaotically produced pions (15). The two-pion correlation function at $q = 0$ is then given by $C(P,q = 0) = 2 - [D(\frac{1}{2}P)]^2$. (Another commonly used variable is the chaoticity parameter λ, where $\lambda = 1 - D^2$.) Today, however, we can say that there is no experimental evidence for a collective production of pions or high-energy photons in intermediate-energy ($E/A \approx 100$ MeV to 1 GeV) heavy-ion collisions. Instead, all production cross sections of secondary particles seem to be explained in terms of incoherent nucleon-nucleon collisions (151, 169, 170). In other energy regimes the picture is not quite as clear, and the interpretation of the chaoticity parameter in ultrarelativistic heavy-ion collisions is still a topic of current research.

A second source of initial-state correlations is due to the spin, which has been studied for two-proton correlation functions from the reaction ^{3}He + Ag (78). The two-proton wave function can in general be written as

$$|\phi(\mathbf{q},\mathbf{r})|^2 = \alpha|\phi_s(\mathbf{q},\mathbf{r})|^2 + (1 - \alpha)|\phi_t(\mathbf{q},\mathbf{r})|^2, \qquad 20.$$

where ϕ_s and ϕ_t are the singlet and triplet spatial wave functions. In large systems with many protons, one has to good approximation α = 1/4, the statistical weight. Zhu et al (78) find, however, that they have to set α = 0.45 to reproduce their data in the ^{3}He + Ag reaction by calculations based on the BUU theory. This was attributed to contributions from breakup in which some fraction of the two-proton coincidences in this system arise from proton pairs in ^{3}He with their initial (singlet) spin correlation undisturbed.

4.5 *Impact Parameter Selection*

Until recently, only very few experimental attempts (67–69) were made to study the impact parameter dependence of two-particle correlation functions, and no quantitative comparisons with dynamical calculations have yet been performed.

Details of theoretical Wigner functions strongly depend upon the impact parameter of the collision. As a consequence, much information is lost in inclusive experiments. Gong et al (65) calculated this loss of information for the ^{14}N + ^{27}Al reaction at E/A = 75 MeV. Their results are summarized in Figure 6. The circular and square-shaped points in the figure show the heights of the maxima of two-proton correlation functions as predicted by the BUU theory for various cuts on the total

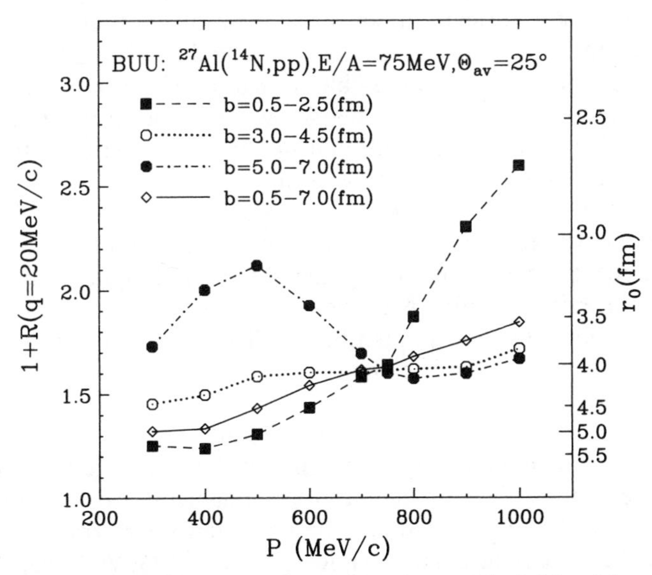

Figure 6 Momentum and impact parameter dependence of the height of the maximum of the two-proton correlation function for the reaction ^{14}Ne + ^{27}Al at E/A = 75 MeV, as predicted by calculations based on the BUU theory. [After (65).]

momentum, **P**, of the proton pairs and on impact parameter, b. The diamond-shaped points show predictions for impact-parameter-averaged measurements. (For orientation, the right-hand scale gives the corresponding Gaussian source parameters.) For different ranges of impact parameter, the predicted correlation functions exhibit qualitatively different dependences on the total momentum of the emitted proton pairs. Particularly striking is the strong momentum dependence predicted for central collisions. The calculations indicate that one should be able to extract a wealth of information about the space-time evolution of the reaction zone by detailed investigations of the momentum and impact parameter dependence of two-proton correlation functions. Such experiments would require high statistics and efficient impact parameter selections.

5. SUMMARY AND OUTLOOK

The field of intensity interferometry for heavy-ion collisions has moved from qualitative descriptions to quantitative results obtained from experiment and theory.

At the ultrarelativistic energies ($E/A \geqslant 10$ GeV), present interest is focused on two-pion correlations. Here the primary goal of intensity interferometry is to learn about the lifetime of the fireball created in the reaction. From this information, it should be possible to reach further conclusions on the possible formation of quark-gluon plasma. Future experiments at RHIC energies will be able to utilize the very large number of pions created, and one may be able to extract correlation functions on an event-by-event basis.

At intermediate energies ($E/A \approx 100$ MeV), the two-proton intensity interferometry technique is preferred and is the main topic of our review. Recent experiments have been able to yield quantitative information on the space-time extension of nuclear reactions, i.e. source radius, lifetime, and energy distribution. This is accomplished by comparing experimental data to the results of nuclear transport calculations.

We have only begun to realize the overall potential of intensity interferometry techniques in heavy-ion collisions. Limited information on, for example, the in-medium nucleon-nucleon cross section was obtained. In the future, it should be possible to perform calculations and measurements at beam energies of $E/A \approx 1$ GeV. Here, there is potential to obtain further information on the response of nuclear matter to compression, which in turn should lead to additional information on the high-density, high-temperature part of the nuclear equation of state.

To reach this goal two challenges must be met. One is of an experimental nature: The next generation of experiments should provide information on impact parameter dependence of the correlation functions. The other is theoretical: Additional work needs to be done to arrive at reliable calculations simulating the reaction dynamics in a consistent fashion for both two-particle interferometry and the emission of clusters of nucleons.

ACKNOWLEDGMENTS

The authors express their thanks to W. G. Gong for numerous collaborations and discussions on the subject of this review. Discussions with G. F. Bertsch, P. Danielewicz, W. G. Lynch, and G. D. Westfall are also acknowledged. This work was partially supported by grants PHY-8913815 and PHY-9017077 from the National Science Foundation.

Literature Cited

1. Born, M., Wolf, E., *Principles of Optics*. Oxford: Pergamon (1970)
2. Hanbury Brown, R., Twiss, R. Q., *Philos. Mag.* 45:663 (1954)
3. Hanbury Brown, R., Twiss, R. Q., *Nature* 177:27 (1956)
4. Hanbury Brown, R., Twiss, R. Q., *Nature* 178:1046 (1956)
5. Goldhaber, G., et al, *Phys. Rev. Lett.* 3:181 (1959)
6. Goldhaber, G., et al, *Phys. Rev.* 120:300 (1960)
7. Shuryak, E. V., *Phys. Lett.* B44:387 (1973)
8. Kopylov, G., Podgoretski, M., *Sov. J. Nucl. Phys.* 15:219 (1972)
9. Cocconi, G., *Phys. Lett.* B49:459 (1974)
10. Kopylov, G., Podgoretski, M., *Sov. J. Nucl. Phys.* 18:336 (1974)
11. Kopylov, G., Podgoretski, M., *Sov. J. Nucl. Phys.* 19:215 (1974)
12. Kopylov, G. I., *Phys. Lett.* B50:472 (1974)
13. Yano, F. B., Koonin, S. E., *Phys. Lett.* B78:556 (1978)
14. Gyulassy, M., *Phys. Rev. Lett.* 48:454 (1982)
15. Gyulassy, M., Kauffmann, S. K., Wilson, L. W., *Phys. Rev.* C20:2267 (1979)
16. Fung, S. Y., et al, *Phys. Rev. Lett.* 41:1592 (1978)
17. Lu, J. J., et al, *Phys. Rev. Lett.* 46:898 (1981)
18. Beavis, D., et al, *Phys. Rev.* C28:2561 (1983)
19. Beavis, D., et al, *Phys. Rev.* C27:910 (1983)
20. Zajc, W. A., et al, *Phys. Rev.* C29:2173 (1984)
21. Bartke, J., *Phys. Lett.* B174:32 (1986)
22. Chacon, A. D., et al, *Phys. Rev.* C43:2670 (1991)
23. Humanic, T. J., *Phys. Rev.* C34:191 (1986)
24. Pratt, S., *Phys. Rev. Lett.* 5:3:1219 (1984)
25. Kohlemainen, K., Gyulassy, M., *Phys. Lett.* B180:203 (1986)
26. Padula, S., Gyulassy, M., *Nucl. Phys.* B339:378 (1990)
27. Kulka, K., Lörstad, B., *Z. Phys.* C45:581 (1990)
28. Pratt, S., *Phys. Rev.* D33:1:314 (1986)
29. Pratt, S., Csörgő, T., Zimányi, J., *Phys. Rev.* C42:2646 (1990)
30. Stock, R., *Ann. Phys. (Leipzig)* 48: 195 (1991)
31. Padula, S., Gyulassy, M., Gavin, S., *Nucl. Phys.* B329:357 (1990)
32. Csörgő, T., et al, *Phys. Lett.* B241: 301 (1990)
33. Bertsch, G., Gong, M., Tohyama, M., *Phys. Rev.* C37:1896 (1988)
34. Bertsch, G. F., *Nucl. Phys.* A489:173c (1989)
35. Sinyukov, Yu. M., *Nucl. Phys.* A498:151c (1989)
36. Sinyukov, Yu. M., Averchenkov, V.

A., Lörstad, B., *Z. Phys.* C49:417 (1991)
37. Bamberger, A., et al (NA35 collaboration), *Phys. Lett.* B203:320 (1988)
38. Humanic, T. J., et al (NA35 collaboration), *Z. Phys.* C38:79 (1988)
39. Harris, J., et al (NA35 collaboration), *Nucl. Phys.* A498:133c (1989)
40. Peitzmann, T., et al (WA80 collaboration), *Nucl. Phys.* A498:397c (1989)
41. Takahashi, Y., et al (EMU05 collaboration), *Nucl. Phys.* A498:529c (1989)
42. Carruthers, P., Sarcevic, I., *Phys. Rev. Lett.* 63:1562 (1989)
43. Carruthers, P., et al, *Phys. Lett.* B222:487 (1989)
44. Weiner, R. M., *Phys. Lett.* B232:278 (1989)
45. Lörstad, B., *Z. Phys.* A334:355 (1989)
46. Mrówczyński, S., *Phys. Lett.* B248:459 (1990)
47. Fowler, G. N., Weiner, R. M., *Phys. Lett.* B70:201 (1977)
48. Fowler, G. N., Stelte, N., Weiner, R. M., *Nucl. Phys.* A319:349 (1979)
49. Fowler, G. N., Weiner, R. M., *Phys. Rev. Lett.* 55:1373 (1985)
50. Vourdas, A., Weiner, R. M., *Phys. Rev.* D38:2209 (1988)
51. Padula, S., Gyulassy, M., *Nucl. Phys.* A525:339c (1989)
52. Gyulassy, M., Padula, S., *Phys. Rev.* C41: R21 (1990)
53. Padula, S., Gyulassy, M., *Nucl. Phys.* A525:339c (1991)
54. Zajc, W. A., et al (E802 collaboration), Proc. Quark-Matter '91. *Nucl. Phys.* (1992), to be published
55. Steadman, S. G., et al (E802 collaboration), see Ref. 117
56. Neuhauser, D., *Phys. Lett.* B182:289 (1986)
57. Ostendorf, R., Schutz, Y., see Ref. 115, p. 108
58. Koonin, S. E., *Phys. Lett.* B70:43 (1977)
59. Zarbakhsh, F., et al, *Phys. Rev. Lett.* 46:1268 (1981)
60. Lynch, W., et al, *Phys. Rev. Lett.* 51:1850 (1983)
61. Pochodzalla, J., et al, *Phys. Rev.* C35:1695 (1987)
62. Chen, Z., et al, *Phys. Rev.* C36:2297 (1987)
63. Beauvais, G. E., Boal, D. H., Wong, J. C. K., *Phys. Rev.* C35:545 (1987)
64. Gong, W. G., et al, *Phys. Rev. Lett.* 65:2114 (1990)
65. Gong, W. G., et al, *Phys. Rev.* C43:781 (1991)
66. Gong, W. G., et al, *Phys. Rev.* C43:1804 (1991)
67. Gustafsson, H., et al, *Phys. Rev. Lett.* 53:544 (1984)
68. Kyanowski, A., et al, *Phys. Lett.* B181:43 (1986)
69. Chen, Z., et al, *Phys. Lett.* B186:280 (1987)
70. Ardouin, D., et al, *Nucl. Phys.* A495:57c (1989)
71. Gong, W. G., et al, *Phys. Lett.* B246:21 (1990)
72. Dabrowski, H., et al, *Phys. Lett.* B247:223 (1990)
73. Goujdami, D., et al, *Z. Phys.* A339:293 (1991)
74. Korolija, M., et al, *Phys. Rev. Lett.* 67:572 (1991)
75. Erazmus, B., Carjan, N., Ardouin, D., *Phys. Rev.* C44:2663 (1991)
76. Awes, T. C., et al, *Phys. Rev. Lett.* 61:2665 (1988)
77. Cebra, D. A., et al, *Phys. Lett.* B227:336 (1989)
78. Zhu, F., et al, *Phys. Rev.* C44:R582 (1991)
79. Boal, D. H., DeGuise, H., *Phys. Rev. Lett.* 57:2901 (1986)
80. Pochodzalla, J., et al, *Phys. Lett.* B174:36 (1986)
81. Knoll, J., Randrup, J., *Phys. Lett.* B103:264 (1981)
82. Awes, T. C., Gelbke, C. K., *Phys. Rev.* C27:137 (1983)
83. Bauer, W., *Nucl. Phys.* A471:604 (1987)
84. Bauer, W., et al, *Phys. Rev.* C37:664 (1988)
85. Tanihata, I., et al, *Phys. Lett.* B97:363 (1980)
86. Tanihata, I., et al, *Phys. Lett.* B100:121 (1981)
87. Lynch, W. G., et al, *Phys. Lett.* B108:274 (1982)
88. Tsang, M. B., et al, *Phys. Rev. Lett.* 52:1967 (1984)
89. Tsang, M. B., et al, *Phys. Lett.* B148:265 (1984)
90. Kristianson, P., et al, *Phys. Lett.* B155:31 (1985)
91. Fox, D., et al, *Phys. Rev.* C33:1540 (1986)
92. Carlén, G., et al, *Phys. Scripta* 34:475 (1986)
93. Chitwood, C. B., et al, *Phys. Rev.* C34:858 (1986)
94. Fox, D., et al, *Phys. Rev.* C36:640 (1987)
95. Fox, D., et al, *Phys. Rev.* C38:146 (1988)
96. Knoll, J., *Nucl. Phys.* A34:3:511 (1980)

97. Koonin, S. E., Bauer, W., Schäfer, A., *Phys. Rev. Lett.* 62:1247 (1989)
98. Bayukov, Yu. D., et al, *Phys. Lett.* B189:291 (1987)
99. Dünnweber, W., et al, *Phys. Rev. Lett.* 65:297 (1990)
100. Jakobsson, B., et al, *Phys. Rev.* C44:R1238 (1991)
101. Pochodzalla, J., et al, *Phys. Lett.* B161:256 (1985)
102. Pochodzalla, J., et al, *Phys. Lett.* B174:275 (1986)
103. Chitwood, C. B., et al, *Phys. Lett.* B172:27 (1986)
104. Gelbke, C. K., Boal, D. H., *Prog. Part. Nucl. Phys.* 19:33 (1987)
105. Uckert, J., et al, *Phys. Lett.* B206:190 (1988)
106. Barz, H. W., et al, *Phys. Lett.* B244:161 (1990)
107. Kim, Y. D., et al, *Phys. Rev.* C45:338 (1992)
108. Kim, Y. D., et al, *Phys. Rev.* C45:387 (1992)
109. Bernstein, M. A., Friedman, W. A., Lynch, W. G., *Phys. Rev.* C29:132 (1984); C30:412 (1984) (erratum)
110. Jennings, B. K., Boal, D. H., Shillcock, J. C., *Phys. Rev.* C33:1303 (1986)
111. Boal, D. H., Shillcock, J. C., *Phys. Rev.* C33:549 (1986)
112. Chen, Z., Gelbke, C. K., *Phys. Rev.* C38:2630 (1988)
113. Nayak, T. K., et al, *Phys. Rev. Lett.* 62:1021 (1989)
114. Boal, D. H., Gelbke, C. K., Jennings, B. K., *Rev. Mod. Phys.* 62:553 (1990)
115. Ardouin, D., ed., *Proc. Corinne 90, Int. Workshop on Particle Correlations and Interferometry in Nuclear Collisions*. Singapore: World Sci. (1990)
116. Bauer, W., Kapusta, J., eds., *Advances in Nuclear Dynamics, Proc. 7th Winter Workshop on Nuclear Dynamics*. Singapore: World Sci. (1991)
117. Bauer, W., Back, B., eds. *Advances in Nuclear Dynamics, Proc. 8th Winter Workshop on Nuclear Dynamics*. Singapore: World Sci. (1992)
118. Suzuki, M., *Phys. Rev.* D:35:3359 (1987)
119. Bowler, M. G., *Z. Phys.* C39:81 (1988)
120. Kulka, K., Lörstad, B., *Nucl. Instrum. Methods* A295:443 (1990)
121. Nakai, T., Yokomi, H., *Prog. Theor. Phys.* 66:1328 (1981)
122. Ernst, D. J., Strayer, M. R., Umar, A. S., *Phys. Rev. Lett.* 55:584 (1985)
123. Bernstein, M. A., et al, *Phys. Rev. Lett.* 54:402 (1985)
124. Pratt, S., Tsang, M. B., *Phys. Rev.* C36:2390 (1987)
125. Reid, R. V. Jr., *Ann. Phys. NY* 50:411 (1968)
126. Lisa, M. A., et al, *Phys. Rev.* C44:2865 (1991)
127. Chitwood, C. B., et al, *Phys. Rev. Lett.* 54:302 (1985)
128. Kim, Y. D., et al, *Phys. Rev. Lett.* 67:14 (1991)
129. Dupieux, P., et al, *Phys. Lett.* B200:17 (1988)
130. DeYoung, P. A., et al, *Phys. Rev.* C39:128 (1989)
131. DeYoung, P. A., et al, *Phys. Rev.* C41: R1885 (1990)
132. Elmaani, A., et al, *Phys. Rev.* C43: R2474 (1991)
133. Rebreyend, D., et al, *Proc. Roy. Soc.* A119:689 (1928)
135. Kikuchi, S., Nordheim, L., *Z. Phys.* 60:652 (1930)
136. Uehling, E. A., Uhlenbeck, G. E., *Phys. Rev.* 43:552 (1933)
137. Bertsch, G. F., Kruse, H., Das Gupta, S., *Phys. Rev.* C29:673 (1984)
138. Aichelin, J., Bertsch, G., *Phys. Rev.* C31:1730 (1985)
139. Bauer, W., et al, *Phys. Rev.* C34:2127 (1986)
140. Kruse, H., Jacak, B. V., Stöcker, H., *Phys. Rev. Lett.* 54:289 (1985)
141. Kruse, H., et al, *Phys. Rev.* C31:1770 (1985)
142. Molitoris, J. J., Stöcker, H., *Phys. Rev.* C32:346 (1986)
143. Molitoris, J. J., Stöcker, H., *Phys. Lett.* B162:47 (1986)
144. Grégoire, C., et al, *Nucl. Phys.* A436:365 (1985)
145. Remaud, B., et al, *Nucl. Phys.* A447:555c (1985)
146. Remaud, B., et al, *Phys. Lett.* B180:198 (1986)
147. Grégoire, C., et al, *Nucl. Phys.* A465:317 (1987)
148. Stöcker, H., Greiner, W., *Phys. Rep.* 137:277 (1986)
149. Bertsch, G. F., Das Gupta, S., *Phys. Rep.* 160:189 (1988)
150. Schuck, P., et al, *Prog. Part. Nucl. Phys.* 22:181 (1989)
151. Cassing, W., Mosel, U., Prog. in Part. *Nucl. Phys.* 25:235 (1990)
152. Wang, S. J., et al, *Ann. Phys. NY* 209:251 (1991)
153. Li, B.-A., Bauer, W., *Phys. Rev.* C44:450 (1991)
154. Wong, C. Y., *Phys. Rev.* C25:1460 (1982)

155. Cugnon, J., Mizutani, T., Vandermeulen, J., *Nucl. Phys.* A352:505 (1981)
156. Bertsch, G., Cugnon, J., *Phys. Rev.* C24:2514 (1981)
157. Cugnon, J., Kinet, D., Vandermeulen, J., *Nucl. Phys.* A379:553 (1982)
158. Kitazoe, Y., et al, *Phys. Lett.* B166:35 (1986)
159. Kitazoe, Y., et al, *Phys. Rev. Lett.* 58:1508 (1987)
160. Cugnon, J., Lemaire, M.-C., *Nucl. Phys.* A489:781 (1988)
161. Ardouin, D., et al, *Z. Phys.* A329:505 (1988)
162. Ardouin, D., et al, *Nucl. Phys.* A514:564 (1990)
163. Cugnon, J., Lejeune, A., Grangé, P., *Phys. Rev.* C35:861 (1987)
164. Botermans, W., Malfliet, R., *Phys. Lett.* B171:22 (1986)
165. Lejeune, A., et al, *Nucl. Phys.* A453:189 (1986)
166. Ogilvie, C. A., et al, *Phys. Rev.* C42: R10 (1990)
167. Krofcheck, D., et al, *Phys. Rev. Lett.* 63:2028 (1989)
168. Heisenberg, W., *Z. Phys.* 126:569 (1949)
169. Bauer, W., *Phys. Rev.* C40:715 (1989)
170. Nifenecker, H., Pinston, J. A., *Annu. Rev. Nucl. Part. Sci.* 40:113 (1990)

Annu. Rev. Nucl. Part. Sci. 1992. 42:101–145

LIGHT-QUARK, HEAVY-QUARK SYSTEMS

Benjamín Grinstein[1]

Superconducting Super Collider Laboratory, 2550 Beckleymeade Avenue, Dallas, Texas 75237

KEY WORDS: form factors, decay constants, effective field theory

---

CONTENTS

[1] On leave of absence from Harvard University.

0163-8998/92/1201-0101$02.00

1. INTRODUCTION

1.1 *Motivation*

There are at least three good reasons for studying in great detail the physics of B and D mesons. First, the Standard Model predicts small but observable signals of *CP* violation in decays of B mesons. Measurement of these asymmetries would give a check on the Kobayashi-Maskawa (KM) accounting of *CP* violation in the Standard Model (for recent reviews, see 1, 2). It is worth noting that, inasmuch as only one *CP*-violating parameter is known, namely, ϵ, almost any model of *CP* violation can account for it by fixing its free parameters.

Second, the rates for rare decays of heavy mesons are sensitive to departures from the Standard Model. These rare processes are good probes for new physics since they start at one-loop order in the Standard Model. For example, the partial widths for $B \to K^*\gamma$ and $B \to K^*\ell^+\ell^-$ in models with two Higgs doublets can easily differ from the Standard Model's by an order of magnitude. Less (or not at all) rare, but no less interesting, is B^0-$\overline{B}^0$ mixing. This depends strongly on the top-quark mass. The observation of B^0-$\overline{B}^0$ mixing with a large parameter ($r \sim 0.2$) in 1987 was the first evidence that the top quark was really very heavy (3).

Third, precise determination of the elements of the KM matrix is naturally done through study of decays of heavy mesons or baryons (2). This is important on two counts. The more precisely known these KM elements are, the more strongly constrained models that address the family problem will be. Also, the Standard Model predictions for *CP* asymmetries and rare decay rates, as described above, depend on the KM matrix[2].

For these reasons, it is necessary and important to make precise Standard Model predictions, in terms of Standard Model parameters, of rates for semileptonic and rare decays, and of *CP* asymmetries. Discouragingly, these calculations encounter the usual difficulties as-

[2] Although, it must be said in all fairness, in rare processes, the ratios $\Gamma(\overline{B} \to K^*\gamma)/\Gamma(\overline{B} \to De\overline{\nu})$ and $\Gamma(\overline{B} \to K^*e^+e^-)/\Gamma(\overline{B} \to De\overline{\nu})$ are fairly independent of KM angles.

sociated with hadronic matrix elements[3]: strong interactions render perturbation theory useless, and we know of no alternative calculational tool[4]. None, that is, until recently, when Isgur & Wise discovered new symmetries of QCD (5).

This review describes these recent developments. We introduce the Heavy-Quark Effective Theory (HQET) and then derive the new symmetries of QCD. We then put these symmetries to use in a variety of ways. In particular, we find that the form factors for semileptonic $\overline{\mathrm{B}}$ decays to D or D* mesons can be determined at the point of maximum momentum transfer (that is, when the resulting D or D* meson is at rest in the $\overline{\mathrm{B}}$ rest frame). We also discuss systematic corrections to these results.

In preparing this chapter I have, for the sake of clarity, departed badly from the chronological order in which these developments took place. While the present burst of interest in the field stems from the seminal work of Isgur and Wise, the main ingredients of their work were already available. Even before the discovery of open charm, i.e. of D mesons, De Rújula, Georgi & Glashow considered a model that incorporated the observation that the spin of a heavy quark decouples in a heavy-light system (6); see below. They estimated the mass difference of D* and D mesons to be much less than that of the corresponding light quark ρ and π mesons, and of the order of the pion mass. Eichten & Feinberg were led to consider a large mass $1/m$ expansion in their investigations of spin-dependent forces in heavy $q\bar{q}$ bound states (7). Almost a decade later Eichten (8) and Lepage & Thacker (9) suggested applying the same techniques to heavy-light systems, thus opening the way towards an effective theory with explicit spin symmetries. Fueled by experimental progress in the measurement of semileptonic decays of B mesons, several theoretical calculations of the rates for these decays appeared in the mid 1980s (10–13). One of them, by Isgur, Wise and me, pointed out that the matrix element for $\overline{\mathrm{B}} \rightarrow \mathrm{D}e\nu$ was most reliably calculated at the kinematic point where the D

[3] *CP* asymmetries in decays of B^0 ($\overline{\mathrm{B}}^0$) mesons to *CP* eigenstates are, in some cases, independent of nonperturbative matrix elements. Nevertheless, if the asymmetry is to be predicted, the KM angle must be previously extracted from, say, semileptonic decays, for which understanding of nonperturbative form factors is needed (see Section 5.1, below).

[4] Save for numerical simulations of lattice QCD. These, however, convey little physical insight, and are at present technologically limited in their ability to produce precise results; for example, few results are known with dynamical fermions. The methods that are the subject of this paper are valuable on the lattice (for a recent review, see 4).

meson does not recoil in the rest frame of the B meson (12). Nussinov & Wetzel noted this independently (11). Moreover, they gave a physical explanation that was close in spirit to the modern argument in terms of a HQET, namely, that the state of light degrees of freedom in heavy-light systems is independent of the mass of the heavy quark. They had discovered a new flavor symmetry of these heavy-light systems. Unfortunately, their work was, until recently, largely ignored. Influential to Isgur and Wise was the work of Voloshin & Shifman, who proposed a theoretical limit ($m_c \to m_b$), now called the Shifman-Voloshin limit, at which the matrix element for $\overline{B} \to De\nu$ is exactly calculable (14). The last element used by Isgur and Wise comes from the work of Voloshin & Shifman (15) and of Politzer & Wise (16), who extracted QCD calculable violations to the flavor symmetry predictions in the form of logarithms of the ratio of heavy quark masses. Moreover, Politzer & Wise then attempted to reproduce these calculations working directly within the context of an effective theory (17).

This review evolved from Reference 18, but it differs from it in two important ways. It is more condensed, so many detailed calculations have been omitted; and it is more complete, for it touches on more recent developments and has more extensive references.

1.2 *Physical Intuition*

The central idea of the HQET is so simple it can be described without reference to a single equation. It should prove useful to refer back to the simple intuitive notion, presented below, wherever the formalism and corresponding equations become abstruse.

The HQET is useful when dealing with hadrons composed of one heavy quark and any number of light quarks. More precisely, the quantum numbers of the hadrons are unrestricted as far as isospin and strangeness, but are ± 1 for either B or C number. In what follows we refer imprecisely to these as heavy hadrons.

The successes of the constituent quark model reveal that, inside hadrons, strongly bound quarks exchange momentum of magnitude a few hundred MeV. We can think of the typical amount Λ by which the quarks are off shell in the nucleon as $\Lambda \approx m_p/3 \approx 330$ MeV. In a heavy hadron the same intuition can be imported, and again the light quark(s) is (are) very far off shell, by an amount of order Λ. But if the mass M_Q of the heavy quark Q is large, $M_Q \gg \Lambda$, then in fact this quark is almost on shell. Moreover, interactions with the light quark(s) typically change the momentum of Q by Λ, but change the velocity of Q by a negligible amount, of the order of $\Lambda/M_Q \ll 1$. It therefore makes sense to think of Q as moving with constant velocity, and this velocity is, of course, the velocity of the heavy hadron.

In the rest frame of the heavy hadron, the heavy quark is practically at rest. The heavy quark effectively acts as a static source of gluons. It is characterized by its flavor and color SU(3) quantum numbers, but not by its mass. In fact, since spin-flip interactions with Q are of the type of magnetic moment transitions, and these involve an explicit factor of g_s/M_Q, where g_s is the strong interaction coupling constant, the spin quantum number itself decouples in the large M_Q case. Therefore, the properties of heavy hadrons are independent of the spin and mass of the heavy source of color.

The HQET is nothing more than a method for giving these observations a formal basis. It is useful because it gives a procedure for making explicit calculations. More importantly, it turns the statement "M_Q is large" into a systematic perturbative expansion in powers of Λ/M_Q. Each order in this expansion involves QCD to all orders in the strong coupling, g_s. Also, the statement of mass and spin independence of properties of heavy hadrons appears in the HQET as approximate internal symmetries of the Lagrangian.

Before closing this section, we point out that these statements apply just as well to a very familiar and quite different system: the atom. The rôle of the heavy quark is played by the nucleus, and that of the light degrees of freedom by the electrons (and the electromagnetic field)[5]. That different isotopes have the same chemical properties simply reflects the nuclear mass independence of the atomic wavefunction. Atoms with nuclear spin s are $2s + 1$ degenerate; this degeneracy is broken when the finite nucleon mass is accounted for, and the resulting hyperfine splitting is small because the nucleon mass is so much larger than the binding energy (playing the rôle of Λ). It is not surprising that, using M_Q independence, the properties of B and D mesons are related, and using spin independence, those of B and B* mesons are related, too.

[5] An obvious distinction between the atomic and hadronic systems is that in the latter the configuration of the light degrees of freedom cannot be computed because of the difficulties afforded by the nonperturbative nature of strong interactions. The methods we are describing circumvent the need for a detailed knowledge of the configuration of light degrees of freedom. The price paid is that the range of predictions is restricted. To emphasize the noncomputable aspect of the configuration of light degrees of freedom, Nathan Isgur informally referred to it as "brown muck," and the term has been partially accepted into the literature (sometimes in modified form: J. D. Bjorken has used the term "brown gunk").

2. THE HEAVY-QUARK EFFECTIVE THEORY

2.1 *The Effective Lagrangian and Its Feynman Rules*

We focus our attention here on the calculation of Green functions in QCD, with a heavy-quark line, its external momentum almost on shell. The external momentum of gluons or light quarks can be far off shell, but not much larger than the hadronic scale Λ. This region of momentum space is relevant for physical quantities e.g. decay S-matrix elements. As stated in the introduction, we expect to see approximate symmetries of Green functions in that region that are not symmetries away from it. That is, these are approximate symmetries of the S matrix, but not of the Lagrangian.

The effective Lagrangian $\mathcal{L}_{\text{eff}}$ is constructed so that it will reproduce these Green functions, to leading order in Λ/M_Q. It is given, for a heavy quark of velocity v_μ ($v^2 = 1$), by (19)

$$\mathcal{L}_{\text{eff}}^{(v)} = \overline{Q}_v iv\cdot DQ_v, \qquad 1.$$

where the covariant derivative is

$$D_\mu = \partial_\mu + ig_s A_\mu^a T^a,$$

and the heavy-quark field Q_v is a Dirac spinor that satisfies the constraint

$$\left(\frac{1+\not{v}}{2}\right) Q_v = Q_v. \qquad 2.$$

In addition, it is understood that the usual Lagrangian $\mathcal{L}_{\text{light}}$ for gluons and light quarks is added to $\mathcal{L}_{\text{eff}}^{(v)}$.

We can see how this arises at tree level, as follows (20). Consider first the tree-level 2-point function for the heavy quark

$$G^{(2)}(p) = \frac{i}{\not{p} - M_Q}. \qquad 3.$$

We are interested in momentum representing a quark of velocity v_μ slightly off shell:

$$p_\mu = M_Q v_\mu + k_\mu.$$

Here, "slightly off shell" means k_μ is of order Λ and is independent of M_Q. Substituting in Equation 3, and expanding in powers of Λ/M_Q, we obtain, to leading order,

$$G^{(2)}(p) = i\left(\frac{1+\not{v}}{2}\right)\frac{1}{v\cdot k} + O\left(\frac{\Lambda}{M_Q}\right).$$

We recognize the projection operator of Equation 2 and the propagator of the Lagrangian in Equation 1.

Similarly, the 3-point function (a heavy quark and a gluon) is given by

$$G_{\mu}^{(2,1)a}(p,q) = \frac{i}{\not p - M_Q}(-ig_s T^a \gamma^\nu)\frac{i}{\not p + \not q - M_Q}\Delta_{\nu\mu}(q),$$

where $\Delta_{\nu\mu}(q)$ is the gluon propagator. Expanding as above, we have

$$G_{\mu}^{a(2,1)}(p,q) = \left(\frac{1+\not v}{2}\right)\frac{i}{v\cdot k}(-ig_s T^a v^\nu) \times \frac{i}{v\cdot(k+q)}\Delta_{\mu\nu}(q) + O\left(\frac{\Lambda}{M_Q}\right),$$

where we have used

$$\left(\frac{1+\not v}{2}\right)\gamma_\nu\left(\frac{1+\not v}{2}\right) = \left(\frac{1+\not v}{2}\right)v_\nu.$$

Again, this corresponds to the vertex obtained from the effective Lagrangian in Equation 1. It is straightforward to extend these results to arbitrary tree-level Green functions, provided only one heavy quark is considered and all other (light) particles carry momentum of order Λ.

The effective Lagrangian describes light quarks and gluons in the presence of a source of color of fixed velocity v_μ. However, for fixed v_μ it is not Lorentz covariant. This is not a surprise since we have expanded the Green functions about one particular velocity: in boosted frames, the expansion in powers of Λ/M_Q becomes invalid because the boosted momentum k_μ can become arbitrarily large. Lorentz covariance is recovered, however, if we boost the velocity

$$v_\mu \rightarrow \Lambda_{\mu\nu}v_\nu$$

along with everything else. It will prove useful to keep this simple observation in mind[6].

2.2 *What is an Effective Theory?*

In the previous section an effective Lagrangian $\mathcal{L}_{\text{eff}}^{(v)}$ was introduced such that Green functions $\tilde{G}_v(k;q)$ calculated from it agreed, at tree

[6] In an alternative method, championed by Georgi (21), the effective Lagrangian $\mathcal{L}_{\text{eff}}$ consists of a sum over the different velocity Lagrangians, $\mathcal{L}_{\text{eff}}^{(v)}$, of Equation 1. Lorentz invariance is recovered at the price of "integrating in" the heavy degrees of freedom. This does not lead to overcounting of states because the sectors of different velocity do not couple to each other, a fact that Georgi refers to as a "velocity superselection rule" (see also 22).

level, with corresponding Green functions $G(p; q)$, in the original field theory (that is, QCD) to leading order in the large mass

$$G(p; q) = \tilde{G}_v(k; q) + O(\Lambda/M_Q) \qquad \text{(tree level)}. \qquad 4.$$

Here, Λ stands for any component of k_μ or of the q's, or for a light-quark mass, and $p = M_Q v + k$.

The remarkable thing about Equation 4 is that while the left-hand side depends on M_Q, and generally in a complicated way, the first term of the right side is independent of M_Q and is a good approximation to the left side if $M_Q \gg \Lambda$. Albeit remarkable, this fact is useless unless extended beyond tree level.

Does Equation 4 hold beyond tree level? The answer is a resounding "NO" but the correct version is still close in form to Equation 4, and more importantly, it is useful:

$$G(p; q; \mu) = \tilde{C}(M_Q/\mu, g_s)\tilde{G}_v(k; q; \mu) + O(\Lambda/M_Q) \qquad \text{(beyond tree level)}. \qquad 5.$$

This equation is proved in the following section. The Green functions G and $\tilde{G}_v$ are renormalized, so they depend on a renormalization point μ. The function $\tilde{C}$ is independent of momenta or light-quark masses: it is independent of the dynamics of the light degrees of freedom. It is there because the left-hand side has some terms that grow logarithmically with the heavy mass, $\ln(M_Q/\mu)$. The beauty of Equation 5 is that all of the logarithmic dependence on the heavy mass factors out. Better yet, since $\tilde{C}$ is dimensionless it is a function of the ratio M_Q/μ only, and not of M_Q and μ separately[7]. To find the dependence on M_Q it suffices to find the dependence on μ. This in turn is dictated by the renormalization group equation. This is discussed in more detail below.

Equation 5 is useful only to the extent the $\tilde{G}$ is really independent of M_Q. One should be careful to use M_Q-independent renormalization conditions in the effective theory. This might seem like a trivial point, but in proving Equation 5 we use an intermediate renormalization that is M_Q dependent. Also, in general the renormalization scheme and point, μ, need not be the same on both sides of Equation 5. The additional generality translates into practical complications and it is best avoided. One is therefore led to choose a mass-independent subtraction scheme on both sides of Equation 5. In practice, it is convenient to use dimensional regularization with an MS scheme.

[7] Actually, additional μ dependence is implicit in the definition of the renormalized coupling constant g_s. This reflects itself in the explicit form of $\tilde{C}$; see Section 2.4.

It is instructive to note the similarities of the HQET and the more usual kind of effective theory—call it "normal"—in which a heavy particle is "integrated out." Take, for example, the case of weak interactions at low energies, that is, when all the momenta involved are much smaller than the mass of the W boson. We can account for the effects of the W boson by adding to the Lagrangian terms of the form

$$\Delta \mathcal{L}_{\text{eff}} = \frac{1}{M_{\text{W}}^2} \kappa \mathcal{O},$$

where $\mathcal{O}$ is a 4-fermion operator and κ contains mixing angles and factors of the weak coupling constant. This is simply the statement that a Green function G of the original theory (the Standard Model including QCD) can be approximated by a Green function $\tilde{G}_{\mathcal{O}}$ of the effective theory (a gauge theory of QCD and electromagnetism) with an insertion of the effective Lagrangian:

$$G = \frac{1}{M_{\text{W}}^2} \kappa \tilde{G}_{\mathcal{O}} + \ldots .$$

The ellipses stand for terms suppressed by additional powers of $(M_{\text{W}})^{-2}$. This equation is very similar to Equation 4. It replaces the task of computing the more complicated left side, which depends on M_{W}, by the computation in the effective theory, which is independent of M_{W} and indeed completely free of the W-boson dynamical degrees of freedom. On the right-hand side, the factor of $1/M_{\text{W}}^2$ gives the dependence on the W-boson mass simply and explicitly—and incorrectly! Just as above, the full theory has logarithmic dependence on M_{W} that has not been made explicit. The correct version is (23)

$$G = \frac{1}{M_{\text{W}}^2} \kappa C(M_{\text{W}}/\mu, g_{\text{s}}) \tilde{G}_{\mathcal{O}} + \ldots .$$

The function C is, in this case, also known as the "short distance QCD effect" first calculated by Altarelli & Maiani (24) and Gaillard & Lee (25).

In summary, an effective theory (of either the "normal" or the heavy-quark type) is a method for extracting explicitly the leading large mass dependence of amplitudes. Moreover, the rules of computation of the effective theory are completely independent of the large mass.

2.3 *The Effective Theory Beyond Tree Level*

In Section 2.1 we established the validity of the HQET at tree level, and in Section 2.2 we saw that beyond tree level things must get complicated. Here we describe how the HQET works and establish the equivalence between the full and effective theories, as given by Equation 5, to 1-loop. The generalization to all orders in the loop expansion is straightforward and not particularly enlightening (see 20).

Consider a Green function, both in the full and effective theories, for a heavy quark and $n \geq 2$ gluons. It suffices to prove the equivalence for one-particle irreducible (1PI) functions. In Figure 1 the left side is calculated in the full theory and the right side in the HQET. The double line stands for the heavy propagator in the HQET.

We can prove the validity of the equation represented in Figure 1, diagram by diagram (there are several diagrams that contribute to each side of the equation). Consider, for definiteness, the diagrammatic equation in Figure 2.

The equation would trivially hold if we could make the propagator replacement

$$\frac{i}{\not{p} + \not{l} - M_Q} \rightarrow \left(\frac{1 + \not{v}}{2}\right) \frac{i}{v(k + l)}$$

even inside the loop integral. Here $p = M_Q v + k$, and l is the loop momentum. In other words, in the right-hand side of Figure 2 we take the limit $M_Q \rightarrow \infty$ and then integrate while on the left side we first integrate and then take the limit. If both integrals converge, then they agree. And that is the case for Figure 2, and indeed, it is also the case for any one-loop integral with a heavy quark and $n \geq 2$ external gluons. We have established Figure 1 for $n \geq 2$.

We are left with the 2-point ($n = 0$) and 3-point ($n = 1$) functions. These are different from the $n \geq 2$ functions in two ways. First, they receive contributions at tree level. Second, they are divergent at 1-loop. Choose some method of regularization. Dimensional regularization is particularly useful as it preserves gauge invariance (or more precisely, BRST invariance). The comparison between full and effective theories is simplest if the same gauge and regularization choices are made. For concreteness, consider Figure 3.

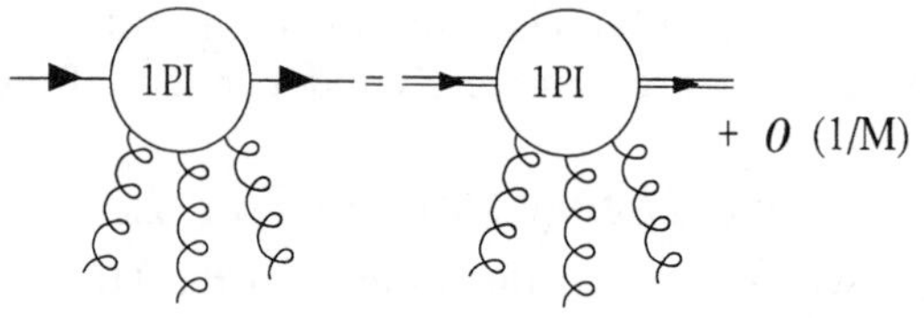

Figure 1

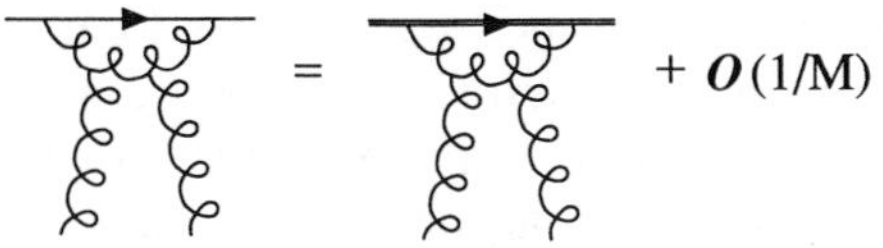

+ $O(1/M)$

Figure 2

Since both sides are regulated, and therefore finite, we can argue as before. But we run into trouble when we try to remove the regulator. One must renormalize the Green functions by adding counter terms, but there is no guarantee that the counter terms satisfy the same relation as the regulated Green functions of Figure 3. To elucidate the relation between counter terms, take a derivative on both sides of Figure 3 with respect to either the residual momentum, k_μ, or the gluon external momentum, q_μ. This makes the diagrams finite and the regulator can be removed. Thus, at one-loop, the relations

$$\frac{\partial}{\partial k_\mu} G^{(2,1)} = \frac{\partial}{\partial k_\mu} \tilde{G}_v^{(2,1)} + O(\Lambda/M_Q) \tag{6.}$$

and

$$\frac{\partial}{\partial q_\mu} G^{(2,1)} = \frac{\partial}{\partial q_\mu} \tilde{G}_v^{(2,1)} + O(\Lambda/M_Q) \tag{7.}$$

hold. The counter terms, or at least the difference between them, are k_μ and q_μ independent. It is a simple algebraic exercise to show that the difference between counter terms is of the form

$$aG^{(2,1)0} + b\tilde{G}_v^{(2,1)0} \tag{8.}$$

where the superscript 0 stands for tree level, and a and b are infinite constants, i.e. independent of k_μ and q_μ. Thus, one can subtract the one-loop Green functions by standard counter terms and establish the equality of Figure 3.

A similar argument can be constructed for the 2-point function. One must take two derivatives with respect to k_μ, but that is as it should be since the counter terms are linear in momentum.

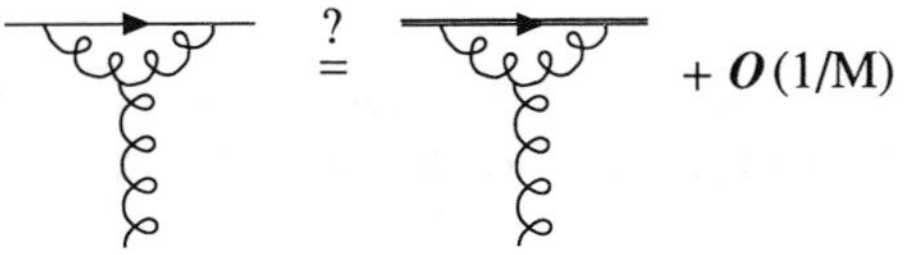

$\stackrel{?}{=}$ + $O(1/M)$

Figure 3

We have therefore established that, to one-loop, the renormalized Green functions in the full and effective theories agree. The alert reader must be puzzled as to the fate of the function $\tilde{C}(M_Q/\mu, g_s)$ of Equation 5. What has happened is that the constants a and b in the counter terms in Equation 8 are, in general, M_Q dependent. Indeed, if we take derivatives with respect to M_Q, as in Equations 6 or 7, the degree of divergence is not changed, and one cannot argue that a or b are M_Q independent. The relation between renormalized Green functions that we have derived contains hidden M_Q dependence in the renormalization prescription for the Green functions in the HQET.

Given two different renormalization schemes, the corresponding renormalized Green functions $\tilde{G}$ and $\tilde{G}'$ are related by a finite renormalization

$$\tilde{G} = z(\mu, g_s)\tilde{G}'.$$

Choosing $\tilde{G}$ to be the mass-independent subtracted Green function, and $\tilde{G}'$ the one in our peculiar subtraction scheme, we find that the relation between full and effective theories becomes

$$G^{(2,1)}(p, q; \mu) = \tilde{C}(M_Q/\mu, g_s)\tilde{G}_v^{(2,1)}(k, q; \mu) + O(\Lambda/M_Q),$$

as declared in Section 2.2. Here, $\tilde{C}$ is nothing but this finite renormalization $z(\mu, g_s)$. That we can use the same function $\tilde{C}$ for all Green functions can be established by using the same wavefunction renormalization for gluons in the full and effective theories. Otherwise, an additional factor of $z_A^{n/2}$ would have to be included in the relation between $G^{(2,n)}$ and $\tilde{G}^{(2,n)}$. This completes the argument.

It is worth mentioning that the discussion above assumes the renormalizability, preserving BRST invariance, of the effective theory. Although to my knowledge this has not been established, there is no obvious reason to doubt that the standard techniques apply in this case.

2.4 *External Currents*

We are often interested in computing Green functions with an insertion of a current. Consider, the current

$$J_\Gamma = \bar{q}\Gamma Q \qquad 9.$$

in the full theory, where Γ is some Dirac matrix, and q a light quark. In the effective theory, this is replaced according to

$$J_\Gamma(x) \rightarrow \exp(-iM_Q v\cdot x)\tilde{J}_\Gamma(x), \qquad 10.$$

where

$$\tilde{J}_\Gamma = \bar{q}\Gamma Q_v, \tag{11.}$$

and it is understood that in $\tilde{J}_\Gamma$ the heavy quark is that of the HQET, satisfying, in particular, $\not{v}Q_v = Q_v$. The exponential factor in Equation 10 reminds us to take the large momentum out through the current, allowing us to keep the external momentum of light quarks and gluons small. The relation between full and effective theories takes the form of an approximate equation between Green functions—and eventually amplitudes—of insertions of these currents:

$$G_{J_\Gamma}(p, p'; q; \mu) = \tilde{C}^{1/2}(M_Q/\mu, g_s)\tilde{C}_\Gamma(M_Q/\mu, g_s)\tilde{G}_{v,\tilde{J}_\Gamma}(k, k'; q; \mu) + O(\Lambda/M_Q), \tag{12.}$$

where p and p' are the momenta of the heavy quark and the external current, k and k' the corresponding residual momenta, $p = M_Q v + k$, $p' = M_Q v + k'$, and q stands for the momenta of the light degrees of freedom. The factor $\tilde{C}^{1/2}\tilde{C}_\Gamma$ accounts for the logarithmic mass dependence, as explained above. We see that an additional factor, namely $\tilde{C}_\Gamma$, is needed in this case to account for the different scaling behavior of the currents in the full and effective theories. It is convenient to think of the replacement of currents, not as given by Equation 10, but rather by

$$J_\Gamma(x) \rightarrow \exp(-iM_Q v\cdot x)\tilde{C}_\Gamma(M_Q/\mu, g_s)\tilde{J}_\Gamma(x). \tag{13.}$$

In fact, Equation 12, and therefore the replacement in Equation 13, are not quite correct. To reproduce the matrix elements of the current J_Γ of Equation 9, it is necessary to sum over matrix elements of several different "currents" in the effective theory. The operator $\tilde{J}_\Gamma$ of Equation 11 is just one of them. In addition, one may have to introduce such operators as $\bar{q}\not{v}\Gamma Q_v$. The correct replacement is therefore

$$J_\Gamma(x) \rightarrow \exp(-iM_Q v\cdot x) \sum_i \tilde{C}_\Gamma^{(i)}(M_Q/\mu, g_s)\tilde{\mathbb{O}}^{(i)}(x). \tag{14.}$$

Here $\tilde{\mathbb{O}}^{(i)}(x)$ is the collection of the operators of dimension 3 with appropriate quantum numbers. The first operator in the sum, call it $\tilde{\mathbb{O}}^{(0)}$, is there even at tree level and corresponds to the operator $\tilde{J}_\Gamma$ of Equation 11.

Another case of interest is that of the insertion of a current of two heavy quarks

$$J_\Gamma = \bar{Q}'\Gamma Q.$$

The replacement now is

$$J_\Gamma(x) \to \exp(-iM_Q v\cdot x + iM_{Q'}v'\cdot x) \times \sum_i \hat{C}_\Gamma^{(i)}\left(\frac{M_Q}{\mu}, \frac{M_{Q'}}{M_Q}, v\cdot v', g_s\right) \hat{\mathcal{O}}_\Gamma^{(i)}(x). \quad 15.$$

Again, $\hat{\mathcal{O}}^{(i)}(x)$ stands for the complete list of operators of dimension 3 in the effective theory with the right quantum numbers. Also, the operator $\hat{\mathcal{O}}^{(0)} = \overline{Q}'_{v'}\Gamma Q_v$ appears in the sum at tree level.

Equation 15 deserves some explanation. The Green functions now include two heavy quarks, and to properly establish the validity of Equation 15 we should begin by considering Green functions with two heavy quarks without insertion of a current. The function $\hat{C}$ connecting these full and effective Green functions will now, in general, depend on both M_Q and $M_{Q'}$. Moreover, we cannot argue that $\hat{C}$ is independent of the velocities v and v'. In fact, this was true of the simpler case considered in Section 2.3; but there $\tilde{C}$ could only depend on v_μ through $v^2 = 1$. In the case at hand there is an additional invariant on which $\hat{C}$ can depend, namely $v\cdot v'$. These observations apply just as well to the correction factors $\hat{C}_\Gamma^{(i)}$ in Equation 15.

The explicit functional dependence on M_Q in the functions $\tilde{C}_\Gamma$ and $\hat{C}_\Gamma$ can be obtained from a study of their dependence on the renormalization point μ. For clarity of presentation we neglect operator mixing for now. When necessary, this can be incorporated without much difficulty. Taking a derivative $d/d\mu$ on both sides of Equations 5 and 13, we find

$$\mu \frac{d}{d\mu} \tilde{C}_\Gamma = (\gamma_\Gamma - \tilde{\gamma}_\Gamma)\tilde{C}_\Gamma,$$

where γ_Γ and $\tilde{\gamma}_\Gamma$ are the anomalous dimensions of the currents J_Γ and $\tilde{J}_\Gamma$ in the full and effective theories, respectively. Of particular interest are the cases $\Gamma = \gamma^\mu$ and $\Gamma = \gamma^\mu\gamma_5$. These correspond in the full theory to conserved and partially conserved currents, and therefore the corresponding anomalous dimensions vanish, giving

$$\mu \frac{d\tilde{C}_\Gamma}{d\mu} = -\tilde{\gamma}_\Gamma \tilde{C}_\Gamma \qquad (\Gamma = \gamma^\mu, \gamma^\mu\gamma_5). \quad 16.$$

Before we solve this equation, we recall that

$$\mu \frac{d}{d\mu} = \mu \frac{\partial}{\partial\mu} + \beta(g_s) \frac{\partial}{\partial g_s}.$$

Here β is the QCD β function, with perturbative expansion

$$\frac{\beta(g)}{g} = -b_0 \frac{g^2}{16\pi^2} + b_1 \left(\frac{g^2}{16\pi^2}\right)^2 + \ldots,$$

and

$$b_0 = 11 - \frac{2}{3} n_f,$$

where n_f is the number of quarks in the theory. For our purposes, n_f should not include the heavy quark. This is explained in the famous paper by Appelquist & Carrazone (26); it simply reflects the fact that the logarithmic scaling of g_s is not affected by heavy-quark loops since these are suppressed by powers of M_Q. Now, the solution to Equation 16 is standard:

$$\tilde{C}_\Gamma(\mu, g_s) = \exp\left[-\int_{\bar{g}_s(\mu_0)}^{g_s} dg' \frac{\tilde{\gamma}_\Gamma(g')}{\beta(g')}\right] \tilde{C}_\Gamma[\mu_0, \bar{g}_s(\mu_0)],$$

where $\bar{g}_s$ is the running coupling constant defined by

$$\mu' \frac{d\bar{g}_s(\mu')}{d\mu'} = \beta[\bar{g}_s(\mu')], \qquad \bar{g}_s(\mu) = g_s. \qquad 17.$$

Choosing $\mu_0 = M_Q$, and restoring the dependence on M_Q, we have then

$$\tilde{C}_\Gamma(M_Q/\mu, g_s) = \exp\left[-\int_{\bar{g}_s(M_Q)}^{\bar{g}_s(\mu)} dg' \frac{\tilde{\gamma}_\Gamma(g')}{\beta(g')}\right] \tilde{C}_\Gamma[1, \bar{g}_s(M_Q)].$$

The problem of determining $\tilde{C}_\Gamma(M_Q/\mu, g_s)$ therefore breaks down into two parts. One is the determination of the anomalous dimension $\tilde{\gamma}_\Gamma$. The other is the calculation of $\tilde{C}_\Gamma[1, \bar{g}_s(M_Q)]$. Both can be done perturbatively, and $\tilde{C}_\Gamma(M_Q/\mu, g_s)$ can thus be computed, provided μ and M_Q are large enough so that $\bar{g}_s(\mu)$ and $\bar{g}_s(M_Q)$ are small. One finds, for example, that in leading order $\tilde{C}_\Gamma$ is Γ independent and there is no mixing:

$$\tilde{C}_\Gamma(M_Q/\mu, g_s) = \left[\frac{\bar{\alpha}_s(M_Q)}{\bar{\alpha}_s(\mu)}\right]^{a_I},$$

where $\bar{\alpha}_s \equiv \bar{g}_s^2/4\pi$, and $a_I \equiv -c_1/2b_0 = -6/(33 - 2n_f)$ (15, 16).

We now turn to the computation of the coefficient $\hat{C}_\Gamma$ for the current of two heavy quarks in Equation 15. A new difficulty arises. Because $\hat{C}_\Gamma$ depends on three dimensionful quantities, namely the masses M_Q and $M_{Q'}$ and the renormalization point μ, its functional dependence is not determined from the renormalization group equation (even if we neglect the implicit dependence of g_s on μ). Two different approximations have been developed to deal with this problem:

1. Treat the ratio $M_{Q'}/M_Q$ as a dimensionless parameter, and study the dependence of $\hat{C}_\Gamma$ on $M_{Q'}/\mu$ through the renormalization group (27). This is just like what was done for the heavy-light case, so we can transcribe the results:

$$\hat{C}_\Gamma\left(\frac{M_{Q'}}{\mu}, \frac{M_{Q'}}{M_Q}, v\cdot v', g_s\right)$$
$$\approx \exp\left[-\int_{\bar{g}_s(M_{Q'})}^{\bar{g}_s(\mu)} dg' \frac{\hat{\gamma}_\Gamma(g')}{\beta(g')}\right]\hat{C}_\Gamma\left[1, \frac{M_{Q'}}{M_Q}, v\cdot v', \bar{g}_s(M_{Q'})\right].$$

Again

$$\hat{C}\left[1, \frac{M_{Q'}}{M_Q}, v\cdot v', \bar{g}_s(M_{Q'})\right] = 1 + O[\bar{\alpha}_s(M_{Q'})].$$

But now, the correction of order $\bar{\alpha}_s(M_{Q'})$ is a function of $M_{Q'}/M_Q$. This method has the advantage that the complete functional dependence on $M_{Q'}/M_Q$ is retained, order by order in $\bar{\alpha}_s(M_{Q'})$. Nevertheless, it fails to re-sum the leading logarithms between the scales $M_{Q'}$ and M_Q, i.e. it does not include the effects of running of the QCD coupling constant between M_Q and $M_{Q'}$. Therefore, this method is useful when $M_{Q'}/M_Q \sim 1$, or eqivalently, when $[\bar{\alpha}_s(M_{Q'}) - \bar{\alpha}_s(M_Q)]/\bar{\alpha}_s(M_Q) \ll 1$.

2. Treat the ratio $M_{Q'}/M_Q$ as small. Expand first in a HQET, treating Q as heavy and Q' as light. The corrections are not just of order Λ/M_Q but also $M_{Q'}/M_Q$, but this is assumed to be small (even if much larger than Λ/M_Q). Then expand from this HQET, in powers of $\Lambda/M_{Q'}$, by constructing a new HQET where both Q and Q' are heavy (28). The calculation of $\hat{C}_\Gamma$ then proceeds in two steps. The first gives a factor just like that of the heavy-light current in Equation 17:

$$\exp\left[-\int_{\bar{g}_s(M_Q)}^{\bar{g}_s(\mu)} dg' \frac{\tilde{\gamma}_\Gamma(g')}{\beta(g')}\right] \tilde{C}_\Gamma[1, \bar{g}_s(M_Q)].$$

The second factor is as in Method 1 above, but neglecting $M_{Q'}/M_Q$. Moreover, the current $\tilde{J}_\Gamma$ is not conserved, so the anomalous dimension to be used is not $-\hat{\gamma}_\Gamma$ but $\tilde{\gamma}_\Gamma - \hat{\gamma}_\Gamma$. Finally, we must make explicit the fact that in the first and second steps the appropriate β functions differ in the number of active quarks. We therefore label the one in the second step β' and the corresponding running coupling constant $\bar{g}'_s$. The second factor is

$$\exp\left[\int_{\bar{g}_s(M_{Q'})}^{\bar{g}_s(\mu)} dg' \frac{\tilde{\gamma}_\Gamma(g')}{\beta(g')} - \int_{\bar{g}'_s(M_{Q'})}^{\bar{g}'_s(\mu)} dg'' \frac{\hat{\gamma}_\Gamma(g'')}{\beta'(g'')}\right] \hat{C}_\Gamma[1, 0, v\cdot v', \bar{g}_s(M_{Q'})].$$

Combining factors gives

$$\hat{C}_\Gamma\left(\frac{M_{Q'}}{\mu}, \frac{M_{Q'}}{M_Q}, v\cdot v', g_s\right) \approx \exp\left[-\int_{\bar{g}_s(M_Q)}^{\bar{g}_s(M_{Q'})} dg' \frac{\tilde{\gamma}_\Gamma(g')}{\beta(g')} - \int_{\bar{g}'_s(M_{Q'})}^{\bar{g}'_s(\mu)} dg'' \frac{\hat{\gamma}_\Gamma(g'')}{\beta'(g'')}\right] \times \tilde{C}_\Gamma[1, \bar{g}_s(M_Q)]\hat{C}_\Gamma[1, 0, v\cdot v', \bar{g}_s(M_{Q'})].$$

The advantage of Method 2 over Method 1 is that it includes the effects of running between M_Q and $M_{Q'}$. The disadvantge is that it neglects powers of $M_{Q'}/M_Q$. (Actually, the result can be improved by reincorporating the $M_{Q'}/M_Q$ dependence, as a power series expansion in this ratio.)

For example, in Method 2 Equation 15 for the weak vector current for $b \rightarrow cW^-$ becomes, in leading order (28),

$$\bar{c}\gamma^\mu b \rightarrow \left[\frac{\bar{\alpha}_s(m_b)}{\bar{\alpha}_s(m_c)}\right]^{a_I} \left[\frac{\bar{\alpha}'_s(m_c)}{\bar{\alpha}'_s(\mu)}\right]^{a_L} \bar{c}_{v'}\{(1 + \kappa)\gamma^\mu + [\lambda_b - \lambda_c(v.\cdot v')]\not{v}\gamma^\mu\}b_v \quad 18.$$

and for the corresponding axial current,

$$\bar{c}\gamma^\mu\gamma_5 b \rightarrow \left[\frac{\bar{\alpha}_s(m_b)}{\bar{\alpha}_s(m_c)}\right]^{a_I} \left[\frac{\bar{\alpha}'_s(m_c)}{\bar{\alpha}'_s(\mu)}\right]^{a_L} \bar{c}_{v'}\{(1 + \kappa)\gamma^\mu\gamma_5 - [\lambda_b + \lambda_c(v\cdot v')]\not{v}\gamma^\mu\gamma_5\}b_v, \quad 19.$$

where

$$\lambda_b = \frac{\alpha_s(m_b)}{3\pi}, \qquad \lambda_c(v\cdot v') = \frac{2\alpha_s(m_c)}{3\pi} r(v\cdot v'),$$

$$a_L(v\cdot v') = \frac{8}{33 - 2n_f}[v\cdot v' r(v\cdot v') - 1],$$

$$r(x) \equiv \frac{1}{\sqrt{x^2 - 1}} \ln(x + \sqrt{x^2 - 1}),$$

and κ is of order α_s but is a subleading logarithm.

3. SYMMETRIES

3.1 *Flavor SU*(N)

The Lagrangian for N species of heavy quarks, all with velocity v, is

$$\mathcal{L}_{\text{eff}}^{(v)} = \sum_{j=1}^{N} \overline{Q}_v^{(j)} iv\cdot DQ_v^{(j)}.$$

This Lagrangian has a $U(N)$ symmetry (5, 11, 14). The subgroup $U(1)^N$ corresponds to flavor conservation of the strong interactions and was a good symmetry in the original theory. The novelty in the HQET is then the nonabelian nature of the symmetry group. This leads to relations between properties of heavy hadrons with different quantum numbers. Please note that these will be relations between hadrons of a given velocity, even if of different momentum (since typically $M_{Q_i} \neq M_{Q_j}$ for $i \neq j$). Including the b and c quarks in the HQET, so that $N = 2$, we see that the B and D mesons form a doublet under flavor SU(2). This flavor SU(2) is an approximate symmetry of QCD. It is a good symmetry to the extent that

$$m_c \gg \Lambda \qquad \text{and} \qquad m_b \gg \Lambda.$$

These conditions can be met even if $m_b - m_c \gg \Lambda$. This is in contrast to isospin symmetry, which holds because $m_d - m_u \ll \Lambda$.

In the atomic physics analogy described in the introduction, this symmetry implies the equalty of chemical properties of different isotopes of an element.

3.2 *Spin SU(2)*

The HQET Lagrangian involves only two components of the spinor Q_v. Recall that

$$\left(\frac{1 - \not{v}}{2}\right) Q_v = 0.$$

The two surviving components enter the Lagrangian diagonally, i.e. there are no Dirac matrices in

$$\mathcal{L}_{\text{eff}}^{(v)} = \overline{Q}_v iv\cdot DQ_v.$$

Therefore, there is an SU(2) symmetry of this Lagrangian that rotates the two components of Q_v among themselves (5, 7–9).

Please note that this "spin" symmetry is actually an internal symmetry. That is, for the symmetry to hold no transformation on the coordinates is needed when a rotation among components of Q_v is made. On the other hand, to recover Lorentz covariance, one does the usual transformation on the light sector, including Lorentz transformations of both the coordinates and the velocity v_μ. A spin SU(2) transformation can be added to this procedure to mimic the original action of Lorentz transformations.

To make it plain that this symmetry has nothing to do with "spin" in the usual sense, consider the large mass limit for a vector particle (29). Using, again, $p = mv + k$, and expanding the propagator

$$-i\,\frac{g_{\mu\nu} - p_\mu p_\nu/m^2}{p^2 - m^2} = -i\left(\frac{1}{2m}\right)\frac{g_{\mu\nu} - v_\mu v_\nu}{v\cdot k} + O\left(\frac{1}{m^2}\right),$$

we see that the Lagrangian for the HVET (Heavy-Vector Effective Theory) is

$$\mathcal{L}_{\rm eff}^{(v)} = A_{v\mu}^\dagger iv\cdot D A_{v\mu},$$

with the constraint

$$(v_\mu v_\nu - g_{\mu\nu})A_{v\nu} = A_{v\mu}.$$

We have rescaled the vector field by $\sqrt{2m}$, so the field has mass dimensions 3/2. The effective Lagrangian is invariant under an SU(3) group of transformations, rotating the three components of the vector field among themselves. Note that the "spin" symmetry is not associated with SU(2) in this case.

The symmetry of the theory is larger than the product of the flavor and spin symmetries. If there are $N_{\rm S}$, $N_{\rm F}$, and $N_{\rm V}$ species of heavy scalars, fermions, and vectors, respectively, all transforming the same way under color SU(3), then the symmetry of the effective theory is $SU(N_{\rm S} + 2N_{\rm F} + 3N_{\rm V})$.

3.3 *Spectrum*

The internal symmetries of the effective Lagrangian are explicitly realized as degeneracies in the spectrum and as relations between transition amplitudes. In this section we consider the spectrum of the theory (30).

Keep in mind that momenta, and therefore energies and masses, are measured in the HQET relative to $M_{\rm Q}v_\mu$. Therefore, when we state that in the HQET the B and D mesons are degenerate, the implication is that the physical mesons differ in their masses by $m_{\rm b} - m_{\rm c}$.

For now let us specialize to the rest frame $v = (1, \mathbf{0})$. The total angular momentum operator **J**, i.e. the generator of rotations, can be written as

$$\mathbf{J} = \mathbf{L} + \mathbf{S},$$

where **L** is the angular momentum operator of the light degrees of freedom, and **S**, the angular momentum operator for the heavy quark, agrees with the generator of spin SU(2). Since **J** and **S** are separately conserved, **L** is also separately conserved. Therefore, the states of the theory can be labeled by their **L** and **S** quantum numbers $(l, m_l; s, m_s)$. Of course, $s = 1/2$, so m_s is 1/2 or $-1/2$ only.

The simplest state has $l = 0$ and, therefore, $J = 1/2$. We refer to it as the Λ_Q, by analogy with the nonrelativistic potential constituent quark model of the Λ baryon, where the strange quark combines with an $l = 0$ combination of the two light quarks.

Next is the state with $l = 1/2$. It leads to $J = 0$ and $J = 1$. We deduce that there is a meson and a vector meson that are degenerate. For the b quark, the B and B* fit the bill. They are the lowest lying $B = -1$ states. The lowest lying $C = 1$ states are the D and D* mesons. These again can very well be assigned to our $J = 0$ and $J = 1$ multiplet. The difference $M_{D^*} - M_D = 145$ MeV is reasonably smaller than the splitting between the D* and the next state, the D_1, with $M_{D_1} - M_{D^*} = 410$ MeV.

The splittings of B and B* and of D and D* result from symmetry-breaking effects. These must be corrections of order Λ/M_Q to the HQET predictions. Therefore, one must have $M_{B^*} - M_B = \Lambda^2/m_b$ and analogously for the D-D* pair. Therefore

$$\frac{M_{B^*} - M_B}{M_{D^*} - M_D} = \frac{m_c}{m_b}.$$

Approximating m_c and m_b by M_D and M_B, respectively, we get ~1/3 on the right side, in remarkable agreement with the left side. Although these results also follow from potential models of constituent quarks, it is important that they can be derived in this generality, and this simply.

The states with $l = 3/2$ have $J = 1$ and 2. The D_1 and D_2^*, with $M_{D_2^*} - M_{D_1} = 40$ MeV, are remarkably closely spaced (and of course, appear to have the appropriate quantum numbers to form a spin multiplet).

While in the infinite mass limit, states $|\, l, m_l; s, m_s\rangle$ have sharp L^2, L_z, S^2, and S_z, these are not good quantum numbers for physical states. Regardless of how small spin-symmetry-breaking effects may be, they force states into linear combinations of sharp J^2, J_z, L^2, and S^2,

$|J, m_J; l, s\rangle$. SU(2)-spin transformations connect states of $J = l + 1/2$ with those of $J = L - 1/2$. Now

$$|J, m_J, l, s\rangle = \sum |l, m_l; s, m_s\rangle C^{Jm_J}_{lm_l,sm_s},$$

where $C^{Jm_J}_{lm_l,sm_s} = C(lm_l; sm_s \mid Jm_J)$ are Clebsch-Gordan coefficients. The decomposition is useful because we know how the states on the right transform under spin SU(2). The inverse expression

$$|l, m_l; s, m_s\rangle = \sum |J, m_J; l, s\rangle (C^{Jm_J}_{lm_l,sm_s})^*,$$

gives the linear combinations of physical states with definite spin SU(2) numbers.

For example, for the B and B* multiplet, the $m_l = 1/2$ and $m_l = -1/2$ states that form spin doublets are, respectively

$$\psi_{1/2} = \begin{pmatrix} B^*(+) \\ \dfrac{B^*(0) + B}{\sqrt{2}} \end{pmatrix} \quad \text{and} \quad \psi_{-1/2} = \begin{pmatrix} \dfrac{B^*(0) - B}{\sqrt{2}} \\ B^*(-) \end{pmatrix}.$$

Rotations mix components among these doublets. We can combine them into a matrix $\Psi_{\alpha a} \equiv (\psi_a)_\alpha$. If $\mathcal{D}^{(l)}(R)$ stands for a $2l + 1$ dimensional representation of the rotation R, then the action of spin SU(2) alone is $\Psi \to \mathcal{D}^{(1/2)}(R)\Psi$, while a rotation is $\Psi \to \mathcal{D}^{(1/2)}(R)\Psi\mathcal{D}^{(1/2)}(R)^\dagger$.

This is easily generalized. For arbitrary l there are $2l + 1$ doublets of spin SU(2), ϕ_a, $a = -l, \ldots, l$. They can be assembled into a $2 \times (2l + 1)$ matrix $\Phi_{\alpha a} = (\phi_a)_\alpha$, which transforms as $\Phi \to \mathcal{D}^{(1/2)}(R)\Phi\mathcal{D}^{(l)}(R)^\dagger$ under rotations. The linear combination of physical states in $\Phi_{\alpha a}$ can be written as a sum of at most two terms:

$$\Phi_{\alpha a} = \chi^{(+)A}_{\alpha a} + \chi^{(-)A}_{\alpha a},$$

where $\chi^{(+)A}_{\alpha a}$ $(\chi^{(-)A}_{\alpha a})$ is the state with $J = l + 1/2$ $(J = l - 1/2)$, $A = m_J = \alpha + a$, weighted by the corresponding Clebsch-Gordan coefficient, $C^{l\pm 1/2 A}_{la,1/2\alpha}$.

3.4 *Strong Transitions*

As an example of the use of the symmetries of the HQET to dynamical processes (30), consider the amplitudes for the strong decays of any member of the $J = l \pm \frac{1}{2}$ multiplet to the $J = l' \pm \frac{1}{2}$ multiplet, and a light hadron h with orbital angular momentum $\mathbf{K}_h$ about the (static) heavy quark, and with total angular momentum $\mathbf{J}_h$, i.e. if the spin of h is $\mathbf{S}_h$ then $\mathbf{J}_h = \mathbf{K}_h + \mathbf{S}_h$.

Using the results of the previous section, we can represent the members of the $J = l \pm \frac{1}{2}$ multiplet by $\chi^{(\pm)A}_{\alpha a}$, and those of the $J = l' \pm \frac{1}{2}$ multiplet by $\chi'^{(\pm)B}_{\beta b}$. The spin SU(2) symmetry implies that the amplitude must be proportional to

$$\sum_{\alpha} [\chi'^{(\pm)B}_{\alpha b}]^* \chi^{(\pm)A}_{\alpha a}.$$

In the transition of the "brown muck" with angular momentum l to that with l' and h we must combine the angular momentum of the products to give that of the originating state. To combine l' and h into l we multiply by $C^{la}_{J_h m_h, l' b}$, set $m_h + b = a$ and sum over m_h. Furthermore, to combine the light hadron h and the heavy final-state hadron χ' into a state of angular momentum $J = l \pm \frac{1}{2}$, we multiply by $C^{l \pm 1/2 A}_{J_h m_h, l' \pm 1/2 B}$, set $m_h + B = A$ and sum over B. Thus we have

$$\mathcal{A}[\chi \to (\chi' h)_{J_h K_h}] = \mathcal{A}(l', l, K_h, J_h) \times \sum (C^{la}_{J_h m_h, l' b})^* C^{l \pm 1/2 A}_{J_h m_h, l' \pm 1/2 B} [\chi'^{(\pm)B}_{\alpha b}]^* \chi^{(\pm)A}_{\alpha a}, \quad 20.$$

where $\mathcal{A}(l', l, K_h, J_h)$ is the reduced matrix element. The sum is over a and b, with $\alpha = A - a$, $m_h = a - b$, and $B = A - a + b$.

As an application, consider the decays of the states in the $l = 3/2$ multiplet—say the D_1 and D_2^*—to those in the $l = 1/2$ multiplet—the D and D*—and one pion. From Equation 20 it is easy to check that D_2^* has decay amplitudes in the proportions $\sqrt{2/5}:\sqrt{3/5}$ to the $K_h = 2$ states $D\pi$ and $D^*\pi$, while its multiplet partner decays at the same total rate exclusively to $D^*\pi$. After including kinematic and phase-space factors, these predictions work remarkably well (30).

3.5 *Covariant Representation of States*

In the sections that follow we are interested in extracting the consequences of the spin and flavor symmetries of the HQET for a variety of processes. It is convenient to develop a formalism that automatically extracts this information for us (28). I follow the simple presentation of Falk (31).

A prototypical example of an application is the computation of relations between form factors in semileptonic $\overline{B}$ to D and D* decays. There one needs to study the matrix elements

$$\langle D(v) \mid \bar{c}_v \Gamma b_{v'} \mid B(v') \rangle \quad \text{and} \quad \langle D^*(v), \epsilon \mid \bar{c}_v \Gamma b_{v'} \mid B(v') \rangle. \quad 21.$$

We would like to represent these $l = 1/2$ mesons as the product

$$u_Q \bar{v}_q, \quad 22.$$

where u_Q is a spinor representing the heavy quark, $\not{v} u_Q = u_Q$, and v_q is an anti-spinor representing the light stuff with $l = 1/2$, satisfying $\bar{v}_q \not{v} = \bar{v}_q$. The product $u_Q \bar{v}_q$ is a superposition of states with $J = 0$ and 1. To identify the pseudoscalar meson P and the vector meson $V(\epsilon)$

with polarization ϵ, $\epsilon\cdot v = 0$, we must form appropriate linear combinations of the spin-up and spin-down spinors. This is most easily done in the rest frame $v = (1, \mathbf{0})$; the result will be generalized to arbitrary v by boosting. In the Dirac representation the spin operator is $\mathbf{S} = \gamma^5\gamma^0\boldsymbol{\gamma}/2$ so that the spinor basis $u^{(1)}_\alpha = \delta_{1\alpha}$ and $u^{(2)}_\alpha = \delta_{2\alpha}$ corresponds to spin up and spin down, and the anti-spinor basis $v^{(1)}_\alpha = -\delta_{3\alpha}$ and $v^{(2)}_\alpha = -\delta_{4\alpha}$ corresponds to spin down and spin up. With $\mathbf{S}(u\bar{v}) = (\mathbf{S}u)\bar{v} + u(\mathbf{S}\bar{v})$ it is easy to check that the combination

$$u^{(1)}_{\mathrm{Q}}\bar{v}^{(1)}_{\mathrm{q}} + u^{(2)}_{\mathrm{Q}}\bar{v}^{(2)}_{\mathrm{q}} = \begin{pmatrix} 0 & I \\ 0 & 0 \end{pmatrix} = \left(\frac{1+\gamma^0}{2}\right)\gamma^5$$

has zero spin, while

$$u^{(1)}_{\mathrm{Q}}\bar{v}^{(2)}_{\mathrm{q}} = \frac{1}{\sqrt{2}}\begin{pmatrix} 0 & \sigma_1 + i\sigma_2 \\ 0 & 0 \end{pmatrix} = \left(\frac{1+\gamma^0}{2}\right)\not{\epsilon}^{(+)}$$

$$u^{(1)}_{\mathrm{Q}}\bar{v}^{(1)}_{\mathrm{q}} - u^{(2)}_{\mathrm{Q}}\bar{v}^{(2)}_{\mathrm{q}} = \begin{pmatrix} 0 & \sigma_3 \\ 0 & 0 \end{pmatrix} = \left(\frac{1+\gamma^0}{2}\right)\not{\epsilon}^{(0)}$$

$$u^{(2)}_{\mathrm{Q}}\bar{v}^{(1)}_{\mathrm{q}} = \frac{1}{\sqrt{2}}\begin{pmatrix} 0 & \sigma_1 - i\sigma_2 \\ 0 & 0 \end{pmatrix} = \left(\frac{1+\gamma^0}{2}\right)\not{\epsilon}^{(-)}$$

with $\epsilon^{(\pm)} = (0, 1, \pm i, 0)$ and $\epsilon^{(0)} = (0, 0, 0, 1)$, have total spin 1, with third component 1, 0, and -1, respectively. Thus, for arbitrary velocity v one obtains the representation for pseudoscalar and vector mesons:

$$\tilde{M}(v) = \left(\frac{1+\not{v}}{2}\right)\gamma^5 \qquad \tilde{M}^*(v,\epsilon) = \left(\frac{1+\not{v}}{2}\right)\not{\epsilon}.$$

By construction, the spin symmetry acts on this representation only on the first index of the matrices $\tilde{M}(v)$ and $\tilde{M}^*(v, \epsilon)$.

The power of this machinery can now be displayed. Consider the matrix elements in Equation 21. Using the above representation of states and noting that the result should transform under the spin symmetry just as the matrix Γ, we have

$$\langle D(v) \mid \bar{c}_v\Gamma b_{v'} \mid \bar{B}(v')\rangle = -\xi(v\cdot v') \operatorname{Tr} \bar{\tilde{D}}(v)\Gamma\tilde{B}(v') \qquad \text{23a.}$$

$$\langle D^*(v)\epsilon \mid \bar{c}_v\Gamma b_{v'} \mid \bar{B}(v')\rangle = -\xi(v\cdot v') \operatorname{Tr} \bar{\tilde{D}}^*(v, \epsilon)\Gamma\tilde{B}(v'), \qquad \text{23b.}$$

where $\bar{X} = \gamma^0 X^\dagger\gamma^0$. The common factor $-\xi(v\cdot v')$ plays the role of the reduced matrix element in the Wigner-Eckart theorem. We explore the consequences of Equations 23a and 23b in depth in Section 5.2.

An even simpler case is that of the $l = 0$ multiplet. In this case the states must transform as a spinor and are obviously represented by

$u^{(s)}$, a Dirac spinor satisfying $\not{v}u = u$. This formalism can be extended to deal with multiplets of arbitrary l.

4. MESON DECAY CONSTANTS

4.1 *Preliminaries*

The pseudoscalar decay constant is one of the first physical quantities studied in the context of HQETs. For a heavy-light pseudoscalar meson X of mass M_X, the decay constant f_X scales like $1/\sqrt{M_X}$ (see below). This was known before the formal development of HQETs, although the arguments relied on models of strong interactions. The HQET will give us a systematic way of obtaining this result. Moreover, it will give us the means of studying corrections to this prediction.

The decay constant f_X is defined through

$$\langle 0 \mid A_\mu^{(0)} \mid X(p)\rangle = f_X p_\mu,$$

where $A_\mu = \bar{q}\gamma_\mu\gamma_5 Q$ is the heavy-light axial current, and the meson has the standard relativistic normalization

$$\langle X(p') \mid X(p)\rangle = 2E\delta^{(3)}(\mathbf{p} - \mathbf{p}'). \qquad 24.$$

Thus, the states have mass dimension -1. Analogous definitions can be made for other mesons. For example, for the vector meson X^* (the $l = 1/2$ partner of X), has

$$\langle 0 \mid V_\mu(0) \mid X^*(p, \epsilon)\rangle = f_{X^*}\epsilon_\mu. \qquad 25.$$

Note that the mass dimensions of f_X and f_{X^*} are 1 and 2, respectively.

4.2 *Pseudoscalar Decay Constant in the HQET*

Consider the decay constant of the meson state in the HQET. The effective pseudoscalar decay constant $\tilde{f}_X$ is defined by

$$\langle 0 \mid \tilde{A}_\mu(0) \mid \tilde{X}(v)\rangle = \tilde{f}_X v_\mu. \qquad 26.$$

The state in the HQET, $\mid \tilde{X}\rangle$, is normalized to $2E/M_X$ rather than to $2E$:

$$\langle \tilde{X}(v') \mid \tilde{X}(v)\rangle = 2v^0\delta^{(3)}(\mathbf{v} - \mathbf{v}'). \qquad 27.$$

Actually, defining states in the HQET requires some care; for a detailed analysis, I refer the interested reader to the literature (22). Obviously, since the normalization of states and the dynamics are M_Q independent, so is $\tilde{f}_X$. To relate $\tilde{f}_X$ to the physical f_X, simply multiply Equation 26 by $\sqrt{M_X}$ to restore the normalization of states of Equation 24 and write $v_\mu = p_\mu/M_X$. Thus we arrive at

$$f_X = \tilde{f}_X/\sqrt{M_X}\left[\frac{\overline{\alpha}_s(M_Q)}{\overline{\alpha}_s(\mu)}\right]^{a_I},$$

where the last factor comes from the relation between currents in the full and effective theories. The "constant" $\tilde{f}_X$ is in fact a function of the renormalization point μ; the combination $\tilde{f}_X\overline{\alpha}_s(\mu)^{-a_I}$ is independent of μ to leading-logarithmic order.

We have obtained $f_X \sim 1/\sqrt{M_X}$ plus a logarithmic correction. A useful way of quoting the result is, for the physical case of B and D mesons,

$$\frac{f_B}{f_D} = \sqrt{\frac{M_D}{M_B}}\left[\frac{\overline{\alpha}_s(M_B)}{\overline{\alpha}_s(M_D)}\right]^{a_I} \qquad 28.$$

4.3 *Vector Meson Decay Constant*

As a simple application of the spin symmetry, consider the pseudoscalar decay constant f_{X^*}. Using the 4×4 notation of Section 3.5, the matrix element in Equation 26 that defines the pseudoscalar constant is proportional to

$$\mathrm{Tr}[\gamma^\mu\gamma_5\tilde{M}(v)] = \mathrm{Tr}\left[\gamma^\mu\gamma_5\left(\frac{1+\not{v}}{2}\right)\gamma_5\right] = -2v^\mu.$$

The matrix element

$$\langle 0 \mid \tilde{V}^\mu(0) \mid \tilde{X}^*(v)\epsilon\rangle = \tilde{f}_{X^*}\epsilon^\mu$$

is proportional to

$$\mathrm{Tr}[\gamma^\mu\tilde{M}^*(v, \epsilon)] = \mathrm{Tr}\left[\gamma^\mu\left(\frac{1+\not{v}}{2}\right)\not{\epsilon}\right] = 2\epsilon^\mu$$

with the same constant of proportionality. Therefore

$$\tilde{f}_{X^*} = -\tilde{f}_X.$$

The sign is unimportant since it can be absorbed into a phase redefinition of either state. It is the magnitude that matters. Multiplying by $\sqrt{M_{X^*}} \approx \sqrt{M_X}$ to restore to the standard normalization, we have

$$f_{X^*} = -f_X M_X. \qquad 29.$$

The predictions in Equations 28 and 29 have not been tested experimentally. The difficulty is the small expected branching fraction for the decays $X \rightarrow \mu\nu$ or $X^* \rightarrow \mu\nu$, for X = B and D. Alternatively, the decay constants f_X and f_{X^*} can be measured in Monte Carlo simulations of lattice QCD. There are indications from such simulations that the $1/M_Q$ corrections to Equation 28 are large (32).

5. FORM FACTORS IN $\overline{B} \rightarrow De\nu$, $\overline{B} \rightarrow D^*e\nu$, AND $\Lambda_b \rightarrow \Lambda_c e\nu$

5.1 *Preliminaries*

The semileptonic decays of a $\overline{B}$ meson to D or D* mesons offer the most direct means of extracting the mixing angle $|V_{cb}|$. In order to extract this angle from experiment, theory must provide the form factors for the $\overline{B} \rightarrow D$ and $\overline{B} \rightarrow D^*$ transitions. Several means of estimating these form factors can be found in the literature. A popular method consists of estimating the form factor at one value of the momentum transfer $q^2 = q_0^2$, and then introducing the functional dependence on q^2 in some arbitrary, and presumably reasonable, way. The estimate of the form factor at q_0^2 is obtained from some model of strong interactions, such as the nonrelativistic constituent quark model.

The HQET gives the form factor at the maximum momentum transfer, $q^2 = q_{\rm max}^2 = (M_B - M_D)^2$—the point at which the resulting D or D* does not recoil in the rest frame of the decaying B meson. While the functional dependence on q^2 is a nonperturbative problem, it is already progress to have a prediction of the form factor at one point. Moreover, the HQET gives relations between the form factors. One may study these relations experimentally to test the accuracy of the HQET predictions.

The standard definition of form factors in semileptonic $\overline{B}$ meson decays is

$$\langle D(p') \mid V_\mu \mid \overline{B}(p)\rangle = f_+(q^2)(p + p')_\mu + f_-(q^2)(p - p')_\mu \qquad \text{30a.}$$

$$\langle D^*(p')\epsilon \mid A_\mu \mid \overline{B}(p)\rangle = f(q^2)\epsilon^*_\mu + a_+(q^2)\epsilon^*\cdot p(p + p')_\mu \qquad \text{30b.}$$

$$+ a_-(q^2)\epsilon^*\cdot p(p - p')_\mu$$

$$\langle D^*(p')\epsilon \mid V_\mu \mid \overline{B}(p)\rangle = ig(q^2)\epsilon_{\mu\nu\lambda\sigma}\epsilon^{*\nu}(p + p')^\lambda(p - p')^\sigma. \qquad \text{30c.}$$

Here, the states have the standard normalization (Equation 24) and $q^2 \equiv (p - p')^2$. The contributions to the decay rates from the form factors f_- and a_- are suppressed by m_ℓ^2/M_B^2, where m_ℓ is the mass of the charged lepton, and therefore they are usually neglected.

5.2 *Form Factors in the HQET*

In the effective theory, we would like to compute the matrix elements of the effective currents $\tilde{V}_\mu$ and $\tilde{A}_\mu$ between states of the $l = \frac{1}{2}$ multiplet. We can take advantage of the flavor and spin symmetries to write these matrix elements in terms of generalized Clebsch-Gordan coefficients and reduced matrix elements, i.e. we use the Wigner-Eckart theorem. We have already introduced the relevant machinery in Section 3.5. The matrix elements of the operator $\tilde{G} = \bar{c}_{v'}\Gamma b_v$ between B and D or D* states are given by (c.f. Equations 23)

$$\langle D(v') \mid \tilde{G} \mid \overline{B}(v)\rangle = -\xi(v\cdot v') \operatorname{Tr} \bar{\tilde{D}}(v')\Gamma\tilde{B}(v) \tag{31a.}$$

$$\langle D^*(v')\epsilon \mid \tilde{G} \mid \overline{B}(v)\rangle = -\xi(v\cdot v') \operatorname{Tr} \bar{\tilde{D}}^*(v', \epsilon)\Gamma\tilde{B}(v). \tag{31b.}$$

Before expanding Equations 31, we note that the flavor symmetry implies that the B-current form factor between $\overline{\mathrm{B}}$ meson states is given by the same reduced matrix element:

$$\langle \overline{B}(v') \mid \bar{b}_{v'}\Gamma b_v \mid \overline{B}(v)\rangle = -\xi(v\cdot v') \operatorname{Tr} \bar{\tilde{B}}(v')\Gamma\tilde{B}(v). \tag{32.}$$

Using $\Gamma = \gamma^0$ and recalling that B number is conserved, one finds that ξ is fixed at $v' = v$. With the normalization of states appropriate to the effective theory (Equation 27) and expanding Equation 32 at $v = v'$, one has

$$\xi(1) = 1. \tag{33.}$$

The reduced matrix element ξ is the universal function that describes all of the matrix elements of operators $\tilde{G}$ between $l = \frac{1}{2}$ states. It is known as the Isgur-Wise function after the discoverers of the relations in Equations 31 and 32. It is quite remarkable that the Isgur-Wise function describes both timelike form factors (as in $\overline{\mathrm{B}} \to \mathrm{D}e\nu$) and spacelike form factors (as in $\overline{\mathrm{B}} \to \overline{\mathrm{B}}$). The point, of course, is that in both cases it describes transitions between infinitely heavy sources at fixed "velocity-transfer" $(v - v')^2$.

Expanding Equations 31 for $\Gamma = \gamma^\mu$ or $\gamma^\mu\gamma_5$, we have

$$\langle D(v') \mid \tilde{V}_\mu \mid \overline{B}(v)\rangle = \xi(v\cdot v')(v_\mu + v'_\mu) \tag{34a.}$$

$$\langle D^*(v')\epsilon \mid \tilde{A}_\mu \mid \overline{B}(v)\rangle = -\xi(v\cdot v')[\epsilon^*_\mu(1 + v\cdot v') - v'_\mu\epsilon^*\cdot v] \tag{34b.}$$

$$\langle D^*(v')\epsilon \mid \tilde{V}_\mu \mid \overline{B}(v)\rangle = -\xi(v\cdot v')[-i\epsilon_{\mu\nu\lambda\sigma}\epsilon^{*\nu}v^\lambda v^\sigma]. \tag{34c.}$$

It remains to express the physical form factors in terms of the Isgur-Wise function. We must introduce the coefficient functions $\hat{C}_\Gamma$ of Equation 15, which in the leading-logarithm approximation are given in Equations 18 and 19. Also, we must multiply by $\sqrt{M_\mathrm{D}M_\mathrm{B}}$ to restore to the standard normalization of states, and express Equations 34 in terms of momenta using $v = p/M_\mathrm{B}$ and $v' = p'/M_\mathrm{D}$. For example, one has,

$$\langle D(p') \mid V_\nu \mid B(p)\rangle$$

$$= \left[\frac{\bar{\alpha}_s(m_b)}{\bar{\alpha}_s(m_c)}\right]^{a_\mathrm{I}} \left[\frac{\bar{\alpha}'_s(m_c)}{\bar{\alpha}'_s(\mu)}\right]^{a_\mathrm{L}} \xi(v\cdot v')\sqrt{M_\mathrm{B}M_\mathrm{D}}\left(\frac{p_\nu}{M_\mathrm{B}} + \frac{p'_\nu}{M_\mathrm{D}}\right).$$

It follows that

$$f_{\pm}(q^2) = \left[\frac{\overline{\alpha}_s(m_b)}{\overline{\alpha}_s(m_c)}\right]^{a_I} \left[\frac{\overline{\alpha}_s(m_c)}{\overline{\alpha}_s(\mu)}\right]^{a_L} \xi(v\cdot v') \left(\frac{M_D \pm M_B}{2\sqrt{M_B M_D}}\right). \qquad 35.$$

Similarly, f, $a_{\pm}$, and g can all be written in terms of $\xi(v\cdot v')$. Moreover, at $v\cdot v' = 1$, one has $q^2 = (M_B v - M_D v)^2 = (M_B - M_D)^2 = q^2_{max}$ and $\xi = 1$; Equation 35 yields

$$f_{\pm}(q^2_{max}) = \left[\frac{\overline{\alpha}_s(m_b)}{\overline{\alpha}_s(m_c)}\right]^{a_I} \left(\frac{M_D \pm M_B}{2\sqrt{M_B M_D}}\right). \qquad 36.$$

We have used $a_L(v\cdot v') = 0$ at $v\cdot v' = 1$. This is as it should be, for the physical quantity $f_{\pm}$ is μ independent. It should be emphasized that there is no μ dependence of $f_{\pm}$ in Equation 35: the explicit dependence through $[\overline{\alpha}_s(\mu)]^{a_L}$ is cancelled by the implicit dependence on μ of the Isgur-Wise function, $\xi(v\cdot v') = \xi(v\cdot v', \mu)$.

5.3 *Heavy-Baryon Semileptonic Decays*

The same methods can be used to obtain relations among, and normalizations of, the form factors relevant to semileptonic decays of heavy baryons. As an example, I consider here the simplest case, that of transitions between the $l = 0$ states. The case of transitions involving higher l states can be found elsewhere (31, 33).

There are three form factors, F_i, for the matrix element of the vector current between Λ_b and Λ_c states, and three more, G_i, for the matrix element of the axial current:

$$\langle\Lambda_c(v', s')\,|\,\bar{c}\gamma_\mu b\,|\,\Lambda_b(v,s)\rangle = \overline{u}^{(s')}(v')[\gamma_\mu F_1 + v_\mu F_2 + v'_\mu F_3]u^{(s)}(v),$$

$$\langle\Lambda_c(v', s')\,|\,\bar{c}\gamma_\mu\gamma_5 b\,|\,\Lambda_b(v, s)\rangle = \overline{u}^{(s')}(v')[\gamma_\mu G_1 + v_\mu G_2 + v'_\mu G_3]\gamma_5 u^{(s)}(v).$$

It is remarkable that all six are given in terms of one universal "Isgur-Wise" function (33). This can be derived from arguments similar to those of Section 5.2. From the discussion at the end of Section 3.5 one finds that, in the effective theory, the matrix element of the current is given by

$$\langle\Lambda_c(v', s')\,|\,\bar{c}_{v'}\Gamma b_v\,|\,\Lambda_b(v, s)\rangle = \zeta(v\cdot v')\overline{u}^{(s')}(v')\Gamma u^{(s)}(v), \qquad 37.$$

and that ζ is fixed at one point: $\zeta(1) = 1$. Expanding Equation 37, one gets

$$F_1 = G_1 \quad \text{and} \quad F_2 = F_3 = G_2 = G_3 = 0.$$

Moreover, $G_1(1) = [\overline{\alpha}_s(m_b)/\overline{\alpha}_s(m_c)]^{a_I}$.

5.4 *Form Factors in Order* α_s

The predicted relations between form factors, and normalizations at q^2_{max}, are only approximate. Indeed, several approximations were made in obtaining those results. Corrections that arise from subleading order in the $1/M$ expansion are considered in Section 6. Here we discuss corrections of order α_s.

As observed in Section 2.4, the vector and axial-vector currents of the full theory, $\bar{c}\Gamma b$, match onto a linear combination of "currents," i.e. dimension-3 operators, in the effective theory. At one loop, the correspondence between vector and axial currents in the full and effective theories is given by Equations 18 and 19. The constant λ_b and the function λ_c arise only from one-loop matching, and are scheme independent. The constant κ receives contributions both from matching at one-loop and from two-loops anomalous dimensions. Omitting the latter would give a meaningless, scheme-dependent result. Although κ has been computed, it is interesting to note that predictions can be made solely from the one-loop matching computation.

Indeed, comparing Equation 34b with Equation 30b, we see that at zeroth order in $\bar{\alpha}_s(m_b)$ or $\bar{\alpha}_s(m_c)$ we have

$$a_+ + a_- = 0.$$

Substituting Equation 19 into Equation 31b, we see that to order $\bar{\alpha}_s(m_c)$ and $\bar{\alpha}_s(m_b)$ there is a computable correction to this combination of form factors, namely (28)

$$\frac{a_+ + a_-}{a_+} = -4\,\frac{m_c}{m_b}\left[\frac{\bar{\alpha}_s(m_b)}{3\pi} + \frac{2\bar{\alpha}_s(m_c)}{3\pi}\,r(v\cdot v')\right].$$

The constant κ although difficult to compute, does not change the relations between form factors since it simply rescales the leading-order predictions in Equations 34 by the common factor of $(1 + \kappa)$. It does, however, affect the predicted normalization of form factors at q^2_{max}. Since at $v' = v$ the effective vector current is again $\bar{c}_{v'}\gamma_\mu b_v$, but rescaled by $[1 + \kappa + \lambda_b - \lambda_c(1)]$, the correction to Equation 36 is

$$f_\pm(q^2_{max}) = [1 + \kappa + \lambda_b - \lambda_c(1)]\left[\frac{\alpha_s(m_b)}{\alpha_s(m_c)}\right]^{a_I}\left(\frac{M_D \pm M_B}{2\sqrt{M_B M_D}}\right).$$

For a calculation of κ see the work of Ji & Musolf (34).

6. $1/M_Q$

6.1 *The Correcting Lagrangian*

One of the main virtues of the HQET is that, in contrast to models of the strongly bound hadrons, it lets us study systematically the corrections arising from the approximations we have made. To be sure, we have made several approximations already, even within the zeroth-order expansion in Λ/M_Q. For example, we computed the logarithmic dependence on M_Q, i.e. the functions $\tilde{C}_\Gamma^{(i)}$ and $\hat{C}_\Gamma^{(i)}$ of Equations 14 and 15 using perturbation theory. In this section we turn to the corrections of order Λ/M_Q.

The HQET Lagrangian was derived (Section 2.1) by putting the heavy quark almost on shell and expanding in powers of the residual momentum, k_μ, or light-quark or gluon momentum, q_μ, over M_Q, which we generally wrote as Λ/M_Q. Let us again derive the effective Lagrangian, keeping track this time of the terms of order Λ/M_Q.

We will rederive $\mathcal{L}_{\text{eff}}^{(v)}$, including $1/M_Q$ corrections, working directly in configuration space (35). The heavy-quark equation of motion is

$$(i\not{D} - M_Q)Q = 0.$$

We can put the quark almost on shell by introducing the redefinition

$$Q = \exp(-iM_Q v\cdot x)\tilde{Q}_v \qquad 38.$$

In terms of $\tilde{Q}_v$, the equation of motion is

$$[i\not{D} + M_Q(\not{v} - 1)]\tilde{Q}_v = 0. \qquad 39.$$

If we separate the $(1 + \not{v})$ and $(1 - \not{v})$ components of $\tilde{Q}_v$, we see that, as expected, the latter is very heavy and decouples in the infinite mass limit. To project out the components,

$$\tilde{Q}_v = \tilde{Q}_v^{(+)} + \tilde{Q}_v^{(-)},$$

where

$$\tilde{Q}_v^{(\pm)} = \left(\frac{1 \pm \not{v}}{2}\right)\tilde{Q}_v,$$

we multiply Equation 39 by $(1 \pm \not{v})/2$. Thus we have the equations

$$iv\cdot D\tilde{Q}_v^{(+)} = -\left(\frac{1 + \not{v}}{2}\right) i\not{D}\tilde{Q}_v^{(-)} \qquad 40.$$

and

$$iv\cdot D\tilde{Q}_v^{(-)} + 2M_Q\tilde{Q}_v^{(-)} = \left(\frac{1 - \not{v}}{2}\right) i\not{D}\tilde{Q}_v^{(+)}. \qquad 41.$$

These equations can be solved self-consistently by assuming that $\tilde{Q}_v^{(+)}$ is order $(M_Q)^0$ while $\tilde{Q}_v^{(-)}$ is order M_Q^{-1}. A recursive solution follows. From Equation 41

$$\tilde{Q}_v^{(-)} = \frac{1}{2M_Q}\left(\frac{1-\not{v}}{2}\right) i\not{D}\tilde{Q}_v^{(+)} - i\,\frac{v\cdot D}{2M_Q}\,\tilde{Q}_v^{(-)}.$$

Substituting into Equation 40 and dropping terms of order $1/M_Q^2$, we have

$$iv\cdot D\tilde{Q}_v^{(+)} = -\left(\frac{1+\not{v}}{2}\right) i\not{D}\,\frac{1}{2M_Q}\left(\frac{1-\not{v}}{2}\right) i\not{D}\tilde{Q}_v^{(+)}.$$

The right side involves

$$\left(\frac{1-\not{v}}{2}\right)\not{D}\left(\frac{1-\not{v}}{2}\right)\not{D}\left(\frac{1+\not{v}}{2}\right)$$

$$= \left(\frac{1+\not{v}}{2}\right)[D^2 - (v\cdot D)^2 + \tfrac{1}{2}g_s\sigma^{\mu\nu}G_{\mu\nu}]\left(\frac{1+\not{v}}{2}\right),$$

where $\sigma^{\mu\nu} = (i/2)[\gamma^\mu, \gamma^\nu]$ and $G_{\mu\nu} = (1/ig_s)[D_\mu,D_\nu]$ is the QCD field strength tensor. This equation of motion is obtained from the Lagrangian

$$\mathcal{L}_{\text{eff}}^{(v)} = \overline{Q}_v iv\cdot DQ_v + \frac{1}{2M_Q}\,\overline{Q}_v\left[D^2 - (v\cdot D)^2 + \frac{g_s}{2}\,\sigma^{\mu\nu}G_{\mu\nu}\right]Q_v. \quad 42.$$

Here I have reverted to the notation Q_v for $\tilde{Q}_v^{(+)}$. How to include higher-order terms in the $1/M_Q$ expansion into $\mathcal{L}_{\text{eff}}^{(v)}$ should be clear.

The $1/M_Q$ term in $\mathcal{L}_{\text{eff}}^{(v)}$ is treated as small. If it is not, it does not make sense to talk about a HQET in the first place. It is therefore appropriate to use perturbation theory to compute its effects. In this perturbative expansion, the corrections of order $1/M_Q$ to Green functions, and therefore to physical observables, are computed by making a single insertion of the perturbation

$$\Delta\mathcal{L} = \frac{1}{2M_Q}\,\overline{Q}_v\left[D^2 - (v\cdot D)^2 + \frac{g_s}{2}\,\sigma^{\mu\nu}G_{\mu\nu}\right]Q_v.$$

The symmetries of the HQET, discussed at length in Sections 1 and 3, are broken by $\Delta\mathcal{L}$. Under the SU(N_f) flavor symmetry, $\Delta\mathcal{L}$ transforms as a combination of the adjoint and singlet representations, while only the chromomagnetic moment operator

$$\overline{Q}_v\sigma^{\mu\nu}G_{\mu\nu}Q_v$$

breaks the spin SU(2) symmetry: it transforms as a triplet of spin SU(2).

A single insertion of $\Delta\mathcal{L}$ does include all orders in QCD, and it will often prove difficult to make precise calculations of $1/M_Q$ effects. Since

$\Delta\mathcal{L}$ is treated as a simple insertion in Green functions, its treatment in the HQET is entirely analogous to that of current operators of Section 2.3. There are coefficient functions that connect the HQET results with the full theory. It is convenient to include them directly into the effective Lagrangian (35–37) as

$$\Delta\mathcal{L} = \frac{1}{2M_Q}\bar{Q}_v[c_1 D^2 + c_2(v\cdot D)^2 + \tfrac{1}{2}c_3 g_s \sigma^{\mu\nu}G_{\mu\nu}]Q_v. \qquad 43.$$

Here

$$c_i = c_i(M_Q/\mu,\, g_s)$$

can be determined through the methods discussed extensively in Section 2. In leading logarithm, one finds

$$c_1 = -1$$

$$c_2 = 3\left[\frac{\bar{\alpha}_s(\mu)}{\bar{\alpha}(M_Q)}\right]^{-8/(33-2n_f)} - 2$$

$$c_3 = -\left[\frac{\bar{\alpha}_s(\mu)}{\bar{\alpha}_s(M_Q)}\right]^{-9/(33-2n_f)}.$$

6.2 *The Corrected Currents*

Just as the Lagrangian is corrected in order $1/M_Q$, any other operator is, too. In particular, the current operators studied in Sections 2–5 are modified in this order. At tree level, these corrections are given by the change of variables of the previous section:

$$J_\Gamma = \bar{q}\Gamma Q \rightarrow \bar{q}\Gamma \exp(-iM_Q v\cdot x)\left[Q_v + \frac{1}{2M_Q}\left(\frac{1-\not{v}}{2}\right)i\not{D}Q_v\right].$$

Beyond tree level, this sum of two terms has to be replaced by a more general sum over operators of the right dimensions and quantum numbers. The replacement is

$$J_\Gamma \rightarrow \exp(-iM_Q v\cdot x)\left[\sum_i \tilde{C}_\Gamma^{(i)}\bar{q}\Gamma_i Q_v + \frac{1}{2M_Q}\sum_j \tilde{D}_\Gamma^{(j)}\mathbb{O}_j\right], \qquad 44.$$

where $\mathbb{O}_j$ are operators of dimension 4 that include, for example, the operators

$$\bar{q}\Gamma i\not{D}Q_v, \qquad \bar{q}\Gamma i(v\cdot D)Q_v, \qquad \bar{q}\not{v}\Gamma i\not{D}Q_v.$$

A complete set of operators, and the corresponding coefficients, $\tilde{D}_\Gamma^{(i)}$, for the cases $\Gamma = \gamma^\mu$ and $\Gamma = \gamma^\mu\gamma_5$ in the leading-logarithmic approximation can be found elsewhere (36, 38).

The case of two heavy currents is similar. A straightforward calculation gives

$$J_\Gamma = \overline{Q}'\Gamma Q \to \exp(-iM_Q v\cdot x + iM_{Q'}v'\cdot x)\Bigg[\overline{Q}'_{v'}\Gamma Q_v + \frac{1}{2M_Q}\overline{Q}'_{v'}\Gamma\left(\frac{1-\not{v}}{2}\right) i\not{D} Q_v + \frac{1}{M_{Q'}}\overline{Q}'_{v'} i\overleftarrow{\not{D}}\left(\frac{1-\not{v}'}{2}\right)\Gamma Q_v\Bigg]. \quad 45.$$

Again, beyond tree level we must replace this expression by a more general sum over operators of dimension four,

$$J_\Gamma \to \exp(-iM_Q v\cdot x + iM_{Q'}v'\cdot x)\Bigg[\sum_i \hat{C}_\Gamma^{(i)}\overline{Q}'_{v'}\Gamma_i Q_v + \frac{1}{2M_Q}\sum_j \hat{D}_\Gamma^{(j)}\mathbb{O}_j + \frac{1}{2M_{Q'}}\sum_j \hat{D}'_\Gamma(j)\mathbb{O}_j\Bigg]. \quad 46.$$

It is worth pointing out that, in the computation of the coefficient functions $\tilde{D}_\Gamma^{(j)}$, $\hat{D}_\Gamma^{(j)}$ and $\hat{D}_\Gamma'^{(j)}$, there is a contribution from the term of order $(1/M_Q)^0$. In computing the coefficient functions to order $1/M_Q$ one must not forget graphs with one insertion of the zeroth-order term in the current and one insertion of the first-order term in the HQET Lagrangian.

6.3 *Corrections of Order* m_c/m_b

In the case of semileptonic decays of a beauty hadron to charmed hadron, we introduced earlier an approximation method (Method 2 in Section 2.4) in which m_c/m_b was treated as a small parameter. Now, $m_c/m_b \sim 1/3$ and you may justifiably worry that this is not a good expansion parameter. This section shows that the corrections are actually of the order of $(\alpha_s/\pi)(m_c/m_b)$ and are therefore small. Moreover, they are explicitly calculable.

The strategy is to look at those corrections of order $1/m_b$ that may be accompanied by a factor of m_c (36). In the first step of the approximation scheme, we construct a HQET for the b quark, treating the c quark as light. We must, of course, keep terms of order $1/m_b$ in this first step. The second step is to go over to a HQET in which the m_c quark is also heavy. For now, we care only about terms in this HQET that have positive powers of m_c.

In the first step, the hadronic current $\bar{c}\Gamma b$, with $\Gamma = \gamma^\mu$ or $\Gamma = \gamma^\mu\gamma_5$, is replaced according to Equation 44. The question is, which terms in Equation 44 can give factors of m_c when we replace the c quark by a HQET quark, $c_{v'}$. Recall that, once we complete the second step, all of the m_c dependence is explicit. The answer is that any operators in

Equation 44 that have a derivative acting on the c quark will give a factor of m_c. From Equation 38 we see that a derivative $i\partial_\mu$ acting on the charm quark becomes, in the effective theory, the operation $m_c v'_\mu + i\partial_\mu$. So the prescription is simple: take J_Γ in Equation 44 and replace

$$i\partial_\mu \rightarrow m_c v'_\mu$$

in those terms where $i\partial_\mu$ is acting on the charm quark.

For example, if the operator

$$\frac{1}{m_b}\,\bar{c}i\overleftarrow{\not{D}}\Gamma b_v$$

is generated at some order in the loop expansion, it gives an operator

$$-\frac{m_c}{m_b}\,\bar{c}_{v'}\not{v}'\Gamma b_v = -\frac{m_c}{m_b}\,\bar{c}_{v'}\Gamma b_v \qquad 47.$$

after step two is completed.

It is interesting to note that the resulting correction does not introduce any new unknown form factors. For example, the matrix element of the operator in Equation 47 between a $\overline{B}$ and a D is given by Equation 31a only with an additional factor of $-m_c/m_b$ in front.

The calculation described here has been performed in the leading-logarithmic approximation by Falk & Grinstein (36). The correction to the vector current is

$$\Delta V_\mu = \frac{m_c}{m_b}\,\bar{c}_{v'}(a_1\gamma_\mu + a_2 v_\mu + a_3 v'_\mu)b_v,$$

where the coefficients $a_i = a_i(\mu)$, written in terms of

$$z = \frac{\bar{\alpha}_s(m_c)}{\bar{\alpha}_s(m_b)},$$

are, for $n_f = 4$,

$$a_1 = \frac{5}{9}(v\cdot v' - 1) - \frac{1}{18}z^{-6/25} + \frac{2v\cdot v' + 12}{27}z^{-3/25}$$

$$- \frac{34v\cdot v' - 9}{54}z^{6/25} - \frac{8}{25}v\cdot v' z^{6/25}\ln z$$

$$a_2 = \frac{5}{9}(1 - 2v\cdot v') - \frac{13}{9}z^{-6/25} - \frac{44v\cdot v' - 6}{27}z^{-3/25}$$

$$- \frac{14v\cdot v' - 18}{27}z^{6/25}$$

$$a_3 = \frac{15}{9} - \frac{2}{3} z^{-3/25} - z^{6/25}.$$

In particular, this gives a contribution to the form factor, at $v = v'$, of

$$\frac{m_c}{m_b}(a_1 + a_2 + a_3)\big|_{v \cdot v' = 1} \simeq 0.07$$

This is not negligible! It is reassuring that this type of correction can be extracted explicitly. On the other hand, it should be remembered that both corrections of order $(m_c/m_b)^2$ and of subleading-logarithmic order can still be considerable and should be, but have not been, computed.

6.4 *Corrections of Order* $\overline{\Lambda}/m_c$ *and* $\overline{\Lambda}/m_b$

Corrections to the form factors for semileptonic decays of B's and Λ_b's that arise from the terms of order $1/m_c$ in the effective Lagrangian (Equation 42) and the currents (Equations 44 and 46) are, in principle, as large or larger than those considered in the previous section. It is a welcome surprise that the corrections to the combination of form factors that contribute to the semileptonic decay vanish at the endpoint $v \cdot v' = 1$. Thus, the predicted normalization of form factors persists, but not so the relations between form factors (see below).

The decay $\Lambda_b \rightarrow \Lambda_c e\nu$ (29) is simpler to analyze than the decay $\overline{B} \rightarrow De\nu$ and $\overline{B} \rightarrow D^*e\nu$ (40). Moreover, for the baryonic decay some relations between form factors survive at this order. For these reasons, we present here the baryonic case. We briefly return to the decay of the meson at the end of this section, where we describe the result.

There are two types of corrections to consider, coming from either the modified Lagrangian or from the modified current (39). We start by considering the former. The c_1 and c_2 terms in the effective Lagrangian (Equation 43) transform trivially under the spin symmetry, contributing to the form factors in the same proportion as the leading term in Equation 37. This effectively renormalizes the function ζ but does not affect relations between form factors.

Moreover, the normalization at the symmetry point $v \cdot v' = 1$ is not affected. This is a straightforward application of the Ademollo-Gatto theorem. If j_μ is a symmetry-generating current of a Hamiltonian H_0, then corrections to the matrix element of the current, at zero momentum, from a symmetry-breaking perturbation to the Hamiltonian ϵH_1 are of order ϵ^2. In the case at hand, the Ademollo-Gatto theorem implies that corrections to the normalization of ζ at the symmetry point are of order $(1/m_c)^2$.

The chromomagnetic moment operator in Equation 43 does not contribute at all. The spin symmetries imply

$$\langle\Lambda_c(v', s') \mid T \int d^4x(\bar{c}_{v'}\sigma^{\mu\nu}G_{\mu\nu}c_{v'})(x)(\bar{c}_{v'}\Gamma b_v)(0) \mid \Lambda_b(v, s)\rangle$$

$$= \zeta_{\mu\nu}(v, v')\bar{u}^{(s')}(v')\sigma^{\mu\nu}\left(\frac{1 + \not{v}}{2}\right)\Gamma u^{(s)}(v)$$

The function $\zeta_{\mu\nu}$ must be an antisymmetric tensor and must therefore be proportional to $v'_\mu v_\nu - v'_\nu v_\mu$. But

$$\left(\frac{1 + \not{v}'}{2}\right)\sigma^{\mu\nu}\left(\frac{1 + \not{v}'}{2}\right)v'_\mu = 0.$$

This, we see, is an enormous simplification. There is no analogous simple reason for the matrix element of the chromomagnetic moment operator to vanish in a meson transition. The chromomagnetic matrix element in that case gives uncalculable corrections to the relations between form factors.

We turn next to the contribution from the modification to the current. We need the matrix element of the local operators of order $1/m_c$ in Equation 46. Since the coefficients $\hat{D}_\Gamma^{(j)}$ in Equation 46 are known only to leading-logarithmic order, let us concentrate on the operators that arise from tree-level matching, Equation 45. Consider the matrix element

$$\langle\Lambda_c(v', s') \mid \bar{c}_{v'}i\overleftarrow{D}_\mu\Gamma b_v \mid \Lambda_b(v, s)\rangle = \bar{u}^{(s')}(v')\Gamma u^{(s)}(v)[Av_\mu + Bv'_\mu],$$

where the form of the right-hand side follows again from the spin symmetries. The form factors A and B are not independent. Rather, they are given in terms of ζ. To see this, note that, contracting with v'_μ, and using the equations of motion,

$$B = -v\cdot v'A. \qquad 48.$$

Also, if the mass of the $l = 1/2$ state in the effective theory is $\bar{\Lambda}$, then

$$\langle\Lambda_c(v', s') \mid i\partial_\mu(\bar{c}_{v'}\Gamma b_v) \mid \Lambda_b(v,s)\rangle$$

$$= \bar{\Lambda}(v_\mu - v'_\mu)\zeta(v\cdot v')\bar{u}^{(s')}(v')\Gamma u^{(s)}(v).$$

Contracting with v_μ, using the equations of motion and Equation 48, we have

$$A[1 - (v\cdot v')^2] = \bar{\Lambda}(1 - v\cdot v')\zeta.$$

Therefore, the matrix element of interest is

$$\langle\Lambda_c(v', s') \mid \bar{c}_{v'}i\overleftarrow{\not{D}}\Gamma b_v \mid \Lambda_b(v, s)\rangle$$

$$= \bar{\Lambda}\zeta(v\cdot v')\,\frac{v_\nu - (v\cdot v')v'_\nu}{1 + v\cdot v'}\,\bar{u}^{(s')}(v')\gamma^\nu\Gamma u^{(s)}(v),$$

where $\Gamma = \gamma^\mu$ or $\gamma^\mu\gamma_5$. Putting it all together, one finds

$$F_1 = G_1\left[1 + \frac{\bar{\Lambda}}{m_c}\left(\frac{1}{1 + v\cdot v'}\right)\right]$$

$$F_2 = G_2 = -G_1\,\frac{\bar{\Lambda}}{m_c}\left(\frac{1}{1 + v\cdot v'}\right)$$

$$F_3 = G_3 = 0.$$

Moreover,

$$G_1(1) = \left[\frac{\bar{\alpha}_s(m_b)}{\bar{\alpha}_s(m_c)}\right]^{a_1}$$

as before. Up to an unknown constant, $\bar{\Lambda}$, there are still five relations among six form factors. We can estimate $\bar{\Lambda}$ by writing $\bar{\Lambda} = M_{\Lambda_c} - m_c = (M_{\Lambda_c} - M_D) + (M_D - m_c)$. If the "constituent" quark mass in the D meson is $\simeq 300$ MeV, then $\bar{\Lambda} \simeq 700$ MeV. With this, we can estimate the next-order corrections to be of the order of $(\bar{\Lambda}/2m_c)^2 \sim 5\%$. There are, of course, additional computable corrections, of order $\bar{\Lambda}/2m_b$ and $\alpha_s(m_c)/\pi(\bar{\Lambda}/2m_c)$.

The result of $1/m_c$ corrections to the mesonic transitions is quite different. There, both the matrix elements of the correction to the current and of the time-ordered product with the chromomagnetic moment operator lead to new form factors. The result is that there are incalculable corrections, of order $\bar{\Lambda}/2m_c$, to all the leading-order relations between form factors. Even if $\bar{\Lambda}$ is smaller in this case, presumably $\bar{\Lambda} \sim 300$ MeV, these corrections may be large, say 10–20%. Remarkably, at the symmetry point, $v'\cdot v = 1$, there are no corrections of order $\bar{\Lambda}/2m_c$ to the leading-order predictions. Thus, one may still extract the mixing angle $|V_{cb}|$ with high precision from measurements at the end of the spectrum of the semileptonic decay rates for $B \rightarrow De\nu$ and $B \rightarrow D^*e\nu$.

7. FURTHER DEVELOPMENTS

In this section we introduce the reader to some other applications of the HQET. Space limitations and the fast pace of development of the field prevent a complete and thorough compilation of new developments; we can only choose a few from a large selection of topics.

Moreover, these are not treated in any detail. The interested reader is urged to consult the references for further study.

7.1 *Inclusive Semileptonic Decay Rates*

It has long been held that the inclusive semileptonic decay rate of a $\overline{\mathrm{B}}$ into charmed hadronic states is well approximated by the underlying quark decay width

$$\sum_{X_c} \Gamma(\overline{\mathrm{B}} \rightarrow X_c \mathrm{e}\overline{\nu}) \approx \Gamma(\mathrm{b} \rightarrow \mathrm{ce}\overline{\nu}). \tag{49.}$$

The HQET provides a derivation of this statement (41). The result is even finer than the doubly integrated result in Equation 49. It can be shown that

$$\left\langle \sum_{X_c} \frac{\mathrm{d}^2\Gamma}{\mathrm{d}x\,\mathrm{d}y} (\overline{\mathrm{B}} \rightarrow X_c \mathrm{e}\overline{\nu}) \right\rangle_f \approx \left\langle \frac{\mathrm{d}^2\Gamma}{\mathrm{d}x\,\mathrm{d}y} (\mathrm{b} \rightarrow \mathrm{ce}\overline{\nu}) \right\rangle_f \tag{50.}$$

where $x = q^2/M_{\mathrm{B}}^2 = (p_{\mathrm{e}} + p_{\bar{\nu}})^2/M_{\mathrm{B}}^2$ and $y = p_{\mathrm{B}} \cdot p_{\mathrm{e}}/M_{\mathrm{B}}^2$, and the averaging is defined by

$$\langle F(x, y) \rangle_f \equiv \int_{y_{\min}}^{y_{\max}} \mathrm{d}y\, f(y) F(x, y).$$

Here f must be a smooth function, and, in particular, $f = 1$ is a possible choice, leading to Equation 49. Because this is derived systematically, corrections of order Λ/m_{b} and of order $\overline{\alpha}_{\mathrm{s}}(m_{\mathrm{b}})/\pi$ can be systematically studied.

It is very important to understand the rôle played by smoothing over y in Equation 50. Although frequently used, the corresponding identity without averaging, i.e. Equation 50 with $f(y) = \delta(y - y_0)$, is not valid. This is most easily seen by considering the region of y close to $y_{\max}$, which is dominated by resonances, e.g. the D and D*. This means, in particular, that the corresponding formula for $\mathrm{b} \rightarrow \mathrm{ue}\overline{\nu}$, should not be trusted close to the end of the electron energy spectrum. It is not a good idea to extract $|V_{\mathrm{ub}}|$ from a study of this kinematic region that uses the free-quark model.

7.2 $\overline{B} \rightarrow \pi e \overline{\nu}$ *and* $\overline{B} \rightarrow \omega e \overline{\nu}$

The decays $\overline{\mathrm{B}} \rightarrow \mathrm{De}\overline{\nu}$ and $\overline{\mathrm{B}} \rightarrow \mathrm{D^*e}\overline{\nu}$ can be used, as we have seen, to extract the mixing angle $|V_{\mathrm{cb}}|$ with high precision. The determination of $|V_{\mathrm{ub}}|$ is far more complicated. Experimentally, the fact that $|V_{\mathrm{ub}}|/|V_{\mathrm{cb}}| \ll 1$ and that charm decays fast to light hadrons makes the process $\mathrm{b} \rightarrow \mathrm{ue}\overline{\nu}$ difficult to observe. While the experimental effort has so far concentrated mainly on establishing the occurrence of the inclusive process $B \rightarrow Xe\nu$ in the kinematic regime inaccessible to the underlying $b \rightarrow c$ transition, it is expected that it will shift toward the determination of exclusive modes, such as $\overline{B} \rightarrow \omega e\overline{\nu}$. These will give clean, unquestionable evidence of the observation of the underlying $b \rightarrow u$ transition. Theoretically, the calculation of either inclusive or exclusive rates is very untrustworthy, the existing results (mainly from phenomenological models) varying widely and depending sensitively on the choice of parameters.

The HQET suggests a method that may afford higher accuracy in the extraction of $|V_{ub}|$ from exclusive decays (42). At the very least, since the method follows from the HQET, the corrections to the lowest-order predictions can be studied. This should give us some idea of the uncertainty in the determination of the CKM angle (something that can hardly be said about the existing alternatives).

The idea is simple. As opposed to what was done in the $\overline{B} \rightarrow D$ case, the symmetries of the HQET cannot be used to relate the initial and final states anymore: the "brown muck" of the $\overline{B}$ is not at all the same as that of the ρ (or the π, or any other light-quark resonance, for that matter). Nevertheless, the HQET flavor symmetry can be used to relate, for example, the $\overline{B} \rightarrow \rho$ and $D \rightarrow \rho$ matrix elements. Measure the form factors for the latter and use them in the former. One could even use the light-quark flavor SU(3) symmetry to relate the form factors for $D \rightarrow \rho$ and the Cabibbo-allowed $D \rightarrow K^*$.

There are altogether six form factors in the $\overline{B} \rightarrow \pi$ and $\overline{B} \rightarrow \rho$ matrix elements, defined in a way entirely analogous to Equations 30, and which we label with a B superscript, e.g. $f^{B}_{\pm}$. The corresponding D form factors are labeled with the superscript D. The HQET gives

$$\langle \pi(p') \,|\, j_\mu \,|\, \overline{B}(v)\rangle = \langle \pi(p') \,|\, j_\mu \,|\, D(v)\rangle$$

$$\langle \rho(p')\epsilon \,|\, j_\mu \,|\, \overline{B}(v)\rangle = \langle \rho(p')\epsilon \,|\, j_\mu \,|\, D(v)\rangle,$$

Here, j_μ stands for either V_μ or A_μ. It is straightforward to find the relations between form factors. For example,

$$f^{B}_{\pm}(p\cdot p') = \frac{1}{2}\left(\sqrt{\frac{M_D}{M_B}} \pm \sqrt{\frac{M_B}{M_D}}\right) f^{D}_{+}\left(\frac{M_D}{M_B}\, p\cdot p'\right) + \frac{1}{2}\left(\sqrt{\frac{M_D}{M_B}} \mp \sqrt{\frac{M_B}{M_D}}\right) f^{D}_{-}\left(\frac{M_D}{M_B}\, p\cdot p'\right),$$

where we have written $f^{X}_{\pm}$ as a function of $p\cdot p'$ rather than $q^2 = -2p\cdot p' + M_X^2 + M_\pi^2$. Moreover, $f^{X}_{\pm}$ scale with the heavy mass as $f^{X}_{+} \pm f^{X}_{-} \sim M_X^{\mp 1/2}$, and therefore $f^{X}_{-} \approx -f^{X}_{+}$.

There is a practical difficulty with this proposal. The end of the spectrum for the B decay, $(p \cdot p')_{max} = (M_B^2 + M_\pi^2)/2$, goes well beyond the end of the spectrum for the D decay. Neglecting the pion mass for simplicity, we see that $f_\pm^B(p \cdot p')$ cannot be inferred from $f_\pm^D$ for $M_D M_B/2 \leq p \cdot p' \leq M_B^2/2$.

To our knowledge, the $1/m_c$ corrections to this process have not been studied.

7.3 *Rare $\overline{B}$ Decays*

As mentioned in the introduction, rare $\overline{B}$ decays are believed to be a good probe of new physics. For "calibration" it is important to obtain precise predictions from the Standard Model. Unfortunately, this often involves hadronic matrix elements that, needless to say, we cannot compute. The HQET gives us a handle on this problem (42). Again, the trick is to relate the matrix element of interest, say $\overline{B} \rightarrow \overline{K}e^+e^-$, to a more easily measured process, like $D \rightarrow \overline{K}e\nu$. In fact, it is easy to analyze this example in some detail, for we only need the matrix element of the vector current between the heavy and light pseudoscalar meson states. But this is precisely the same problem as studied in the previous section. Now, though, the semileptonic D decay is not Cabibbo suppressed, so one can do much better!

The processes $\overline{B} \rightarrow \overline{K}^*\gamma$ and $\overline{B} \rightarrow \overline{K}^*e^+e^-$ receive contributions from a transition magnetic moment operator, i.e. one needs to compute

$$\langle \overline{K}^*(p')\epsilon \mid \bar{s}_L \sigma^{\mu\nu} b_R \mid \overline{B}(p) \rangle.$$

It is remarkable that the spin SU(2) symmetry of the HQET allows us to relate this to the matrix element of a current (which is itself related to semileptonic D decay). To see how this goes, consider for example the $\mu = 0$, $\nu = i$ terms, i.e. the matrices σ^{0i}. These are proportional to $\gamma^0\gamma^i - \gamma^i\gamma^0 = -2\gamma^i\gamma^0$. Now, in the rest frame of the B meson, $v = (1, \mathbf{0})$, and the projection operator in the HQET is $(1 + \not{v})/2 = (1 + \gamma^0)/2$. So in the HQET, one can substitute

$$\bar{s}_L \sigma^{0i} b_R \rightarrow -i\bar{s}_L \gamma^i b_L.$$

Unfortunately, the two-body decay $\overline{B} \rightarrow \overline{K}^*\gamma$ has a fixed $\overline{K}^*$ momentum outside the kinematic range of the appropriately rescaled momentum in the corresponding semileptonic D decay.

7.4 *$e^+e^- \rightarrow B\overline{B}$*

In e^+e^- annihilation into a pair of heavy quarks, the HQET can be used to relate cross sections for different exclusive processes (27, 43).

For example, the flavor symmetry can be used to relate $\sigma(e^+e^- \to B\bar{B})$, at a center-of-mass energy of $\sqrt{s} = m_b\sqrt{(v + v')^2}$, to $\sigma(e^+e^- \to D\bar{D})$, at $\sqrt{s} = m_D\sqrt{(v + v')^2}$. Also, the spin symmetry can be used to relate $\sigma(e^+e^- \to B\bar{B})$, $\sigma(e^+e^- \to B^*\bar{B})$, and $\sigma(e^+e^- \to B^*\bar{B}^*)$. They are in the ratios

$$1 + h : s/2m_b^2 : 3(1 + s/3m_b^2 + h),$$

where

$$h = -\frac{2\alpha_s}{3\pi}\sqrt{1 - 4m_b^2/s}\,\log\left(s/2m_b^2 - 1 + s/2m_b^2\sqrt{1 - 4m_b^2/s}\right).$$

7.5 $\Lambda_b \to \Lambda_c D_s$ *vs* $\Lambda_b \to \Lambda_c D_s^*$

Here is an example of an application to a purely hadronic weak decay (44). To exploit the symmetries of the HQET, one constructs an effective Hamiltonian for the underlying process $b \to c\bar{c}s$. It is a sum of four quark operators, roughly of the form

$$[\bar{s}\gamma^\mu(1 - \gamma_5)c_{\boldsymbol{v}}][\bar{c}_{v'}\gamma_\mu(1 - \gamma_5)b_v],$$

where $c_{\boldsymbol{v}}$ is the charm antiquark field with velocity $\boldsymbol{v}$ in the HQET. From this, one can study the implications of the spin and flavor symmetries on the amplitudes for, for example, $\Lambda_b \to \Lambda_c D_s$.

There are two independent terms in the amplitude for $\Lambda_b \to \Lambda_c D_s$, and two for $\Lambda_b \to \Lambda_c D_s^*$. The symmetries give the latter in terms of the former. The amplitude $\mathcal{A}[\Lambda_b(v) \to \Lambda_c(v')D_s(\bar{v})]$ is given by

$$\bar{u}(v', s')[S + P\gamma_5]u(v, s),$$

where S and P are respectively amplitudes for the D_s to be in an S-wave and P-wave orbital angular momentum state. The amplitude $\mathcal{A}[\Lambda_b(v) \to \Lambda_c(v')D_s^*(\bar{v},\epsilon)]$ is then given by

$$\tfrac{1}{2}\bar{u}(v', s')(1 + \gamma_5)[(A + 2Bv\cdot v')\not{\epsilon}^* - 2B(\epsilon^*\cdot v)\not{\bar{v}} + B\not{\bar{v}}\not{\epsilon}^*]u(v, s),$$

where

$$A = S - P \qquad \text{and} \qquad B = -\left(\frac{m_b}{m_c}\right)(S - P) - (S + P).$$

7.6 *Factorization*

Factorization in two-body decays of heavy pseudoscalar mesons was resurrected a few years ago as a means of estimating their rate, using existing calculations of semileptonic decay form factors. For the factorization assumption no justification was given. Here by factorization

I mean, for example,

$$\langle D\pi \mid j^{\mu} J_{\mu} \mid \overline{B} \rangle \approx \langle D \mid J_{\mu} \mid \overline{B} \rangle \langle \pi \mid j^{\mu} \mid 0 \rangle, \qquad 51.$$

where J_{μ} and j_{μ} are the V-A currents for $b \rightarrow c$ and $u \rightarrow d$, respectively. Comparing the rate for $\overline{B} \rightarrow D\pi$ to the semileptonic rate, using this identity naïvely, one finds

$$\frac{\Gamma(\overline{B} \rightarrow D\pi)}{\left.\frac{d\Gamma(\overline{B} \rightarrow De\overline{\nu})}{dm^2_{e\nu}}\right|_{m^2_{e\nu}=m^2_{\pi}}} = 6\pi^2 f^2_{\pi}. \qquad 52.$$

The problem with Equation 51 is that it makes no sense. The left and right sides each have a different dependence on the renormalization point. Such a relation cannot have physical content. The HQET, together with mild extensions of the method, furnish a way of correcting Equation 51 (45). Of course, it is more than that, because the corrected version can actually be shown to hold. In fact, as far as we know, this and the large-N_{color} version are the only proofs of factorization in some limit of the underlying theory. The result is a slight modification of Equation 52:

$$\frac{\Gamma(\overline{B} \rightarrow D\pi)}{\left.\frac{d\Gamma(\overline{B} \rightarrow De\overline{\nu})}{dm^2_{e\nu}}\right|_{m^2_{e\nu}=m^2_{\pi}}} = A^2 6\pi^2 f^2_{\pi},$$

where A is a calculable correction factor. In the leading-logarithmic approximation $A \approx 1.05$, in remarkable agreement with experiment (46). The violations to factorization are expected to be of the order of $\Lambda/(M_B/2) \sim \Lambda/2M_D \sim 10\%$, and of order $\overline{\alpha}_s(m_b)/\pi \sim 10\%$ (47).

Just as interesting is the fact that the same method cannot be used to prove factorization in some other processes, such as $B \rightarrow \pi\pi$ and $B \rightarrow D\overline{D}$. In fact, it suggests that factorization does not hold for them. A striking confirmation (or refutation) of these ideas could be provided by studying $B \rightarrow D\pi\pi$. The methods of Dugan & Grinstein (45) suggest factorization holds only in the limit of collinear pions, and it would be interesting to plot the decay rate as a function of the angle between the pions, in units of the rate computed by assuming factorization.

7.7 *Concluding Remarks*

As this manuscript was being completed, a workshop and a topical conference on "Heavy Quark Symmetry: Theory and Applications" were held at the Institute for Theoretical Physics in Santa Barbara, California. A considerable number of new results were presented, and many interesting issues were debated. A sampler follows.

1. An effective Lagrangian has been written for heavy mesons and light pseudoscalars that incorporates both the spin and flavor symmetries of heavy quarks and the light-quark chiral symmetries (48). This allows a study of semileptonic decays $\overline{B} \rightarrow DXe\overline{\nu}$ and $\overline{B} \rightarrow D^*Xe\overline{\nu}$ where X is a low momentum state of one or more pseudo-Goldstone bosons. The leading contributions, in chiral perturbation theory, to $f_{D_s}/f_D - 1$ and $B_{B_s}/B_B - 1$ have been computed (49).

2. Shifman has proposed a new theoretical limit for factorization (50). His argument relies on quark-hadron duality. Turning it around, he criticizes the work of Dugan & Grinstein (45) as being inconsistent with quark-hadron duality. Discussions during the ITP workshop resulted in the suggestion that duality is saturated in channels for which the assumptions of Dugan & Grinstein are not met. Specifically, they assume the final state created by the light-quark current has a wavefunction that is not dominated by quarks or gluons with small momentum fraction (45). Duality is presumably saturated by jet-like final states, which have fragmentation functions that peak at zero momentum. More work is needed to verify this. On the experimental side, there is now strong experimental evidence that purely hadronic two-body decays of D and D_s mesons do not factorize (51), as suggested by Dugan & Grinstein (45).

3. Isgur (51a) has given a simple model of hadronization that highlights the nonperturbative nature of the high energy end of the spectrum for inclusive semileptonic B decays. There are many interesting questions about the justification for quark-hadron duality described in Section 7.1. How are these nonperturbative effects accounted for? In computing the inclusive rate, is the meson mass, M_B, rather than the quark mass, m_b, the one that fixes the limits of phase space?

4. Can nonperturbative effects ruin the perturbative construction of the HQET in subleading orders, much as nonperturbative effects may ruin the perturbative calculation of Wilson coefficients in the OPE for R (52)? Even if this is not the case, nonperturbative renormalization of the operators that appear in subleading orders in the $1/M$ expansion may be necessary (53).

5. So far there is little that the HQET predicts (or "postdicts") that had not been predicted (or "postdicted") by hadronic models, such as the constituent nonrelativistic quark model. The normalization of form factors for $\overline{B} \rightarrow De\overline{\nu}$ and $\overline{B} \rightarrow D^*e\overline{\nu}$ of Section 5 has been seen to hold approximately in such models. It is becoming clear that these models are far too predictive and can be incorrect. Experimentally there is evidence that the branching fractions for semileptonic $\overline{B}$ decays to excited charmed-meson states (e.g. the $l = 3/2$ states D_1 and D_2^*) are grossly underestimated by such models (54). This represents no prob-

lem for the HQET. Quark-hadron duality, as in Equation 50, together with the normalization of form factors in $\overline{B} \rightarrow De\overline{\nu}$ and $\overline{B} \rightarrow D^*e\overline{\nu}$ and an assumption about the smoothness of the form factors, leads to an upper bound on the branching fractions into other states, but this is consistent with experiment (see also 55). It has been suggested that HQET ideas may be successfully integrated into superior phenomenological models (56). Burdman & Donoghue have argued that the model of Bauer, Stech & Wirbel (10) is inconsistent with scaling results of the HQET (56).

6. The most burning question of them all: what scale defines the onset of the large mass limit? The answer depends on which quantity one looks at. Lattice computations will be crucial in resolving this issue.

The reader can see that this is the subject of much current research and is undergoing rapid growth. This review is therefore necessarily incomplete, but I hope it is timely and will help the reader join in the development of the subject at this early stage.

ACKNOWLEDGMENTS

I am indebted to N. Isgur and M. Wise for suggestions and comments, and to them and A. Falk, M. Dugan, M. Golden, and H. Georgi for many useful conversations. I would like to thank the Alfred P. Sloan Foundation for partial support. This work is supported in part by the Department of Energy under contract DE-AC35-89ER40486.

Literature Cited

1. Nir, Y., Quinn, H. R., *Annu. Rev. Nucl. Part. Sci.* 42:211–50 (1992)
2. Nir, Y., in *Proc. Theor. Adv. Study Inst. in Elementary Part. Phys.*, Boulder, Colo., June 2–8, 1991. Singapore: World Scientific (1992), to be published; SLAC-PUB-5676, WIS-91/73/Oct-PH, Oct. (1991)
3. Albrecht, H., et al (ARGUS Collaboration), *Phys. Lett.* B192:245 (1987)
4. Eichten, E., *Nucl. Phys.* B(Proc. Suppl.)20:475 (1990)
5. Isgur, N., Wise, M. B., *Phys. Lett.* B232:113 (1989); *Phys. Lett.* B237:527 (1990)
6. De Rújula, A., Georgi, H., Glashow, S. L., *Phys. Rev.* D12:147 (1975)
7. Eichten, E., Feinberg, F. L., *Phys. Rev. Lett.* 43:1205 (1979); *Phys. Rev.* D23:2724 (1981)
8. Eichten, E., *Nucl. Phys.* B(Proc. Suppl.)4:170 (1988)
9. Lepage, G., Thacker, B. A., *Nucl. Phys.* B(Proc. Suppl.)4:199 (1988)
10. Altarelli, G., et al, *Nucl. Phys.* B208: 365 (1982); Suzuki, M., *Nucl. Phys.* B258:533 (1985); Bauer, M., Stech, B., Wirbel, M., *Z. Phys.* C29:637 (1985); Korner, J. G., Schuler, G. A., *Z. Phys.* C38:511 (1988), erratum *Z. Phys.* C41: 690 (1989)
11. Nussinov, S., Wetzel, W., *Phys. Rev.* D36:130 (1987)
12. Grinstein, B., et al, *Phys. Rev. Lett.* 56:298 (1986)
13. Altomari, T., Wolfenstein, L., *Phys. Rev. Lett.* 58:1583 (1987); Isgur, N., et al, *Phys. Rev.* D39:799 (1989)
14. Voloshin, M. B., Shifman, M. A., *Sov. J. Nucl. Phys.* 47:511 (1988)
15. Voloshin, M. B., Shifman, M. A., *Sov. J. Nucl. Phys.* 45:292 (1987)
16. Politzer, H. D., Wise, M. B., *Phys. Lett.* B206:681 (1988)
17. Politzer, H. D., Wise, M. B., *Phys. Lett.* B208:504 (1988)
18. Grinstein B., Lectures on Heavy Quark Effective Theory. To be pub-

lished in *High Energy Phenomenology*, Proc. Workshop, Mexico City 1–12 July 1991, ed. R. Huerta, M. A. Pérez. Singapore: World Scientific; SSCL-Preprint-17
19. Eichten, E., Hill, B., *Phys. Lett.* B234: 511 (1990)
20. Grinstein, B., *Nucl. Phys.* B339:253 (1990)
21. Georgi, H., *Phys. Lett.* B240:447 (1990)
22. Dugan, M., Golden, M., Grinstein, B., HUTP-91/A045, BUHEP-91-18, SSCL-Preprint-12, Nov. (1991)
23. Witten, E., *Nucl. Phys.* B122:109 (1977)
24. Altarelli, G., Maiani, L., *Phys. Lett.* 52B:351 (1974)
25. Gaillard, M. K., Lee, B. W., *Phys. Rev. Lett.* 33:108 (1974)
26. Appelquist, T., Carrazone, J., *Phys. Rev.* D11:2856 (1975)
27. Falk, A. F., Grinstein, B., *Phys. Lett.* B249:314 (1990)
28. Falk, A., et al, *Nucl. Phys.* B343:1 (1990)
29. Carone, C., *Phys. Lett.* 253B:408 (1991)
30. Isgur, N., Wise, M. B., *Phys. Rev. Lett.* 66:1132 (1991)
31. Falk, A. F., SLAC-PUB-5689, Nov. (1991)
32. Boucaud, P., et al, *Phys. Lett.* B220: 219 (1989); Allton, C. R., et al, *Nucl. Phys.* B349:598 (1991)
33. Isgur, N., Wise, M. B., *Nucl. Phys.* B348:276 (1991); Georgi, H., *Nucl. Phys.* B348:293 (1991); Mannel, T., Roberts, W., Ryzak, Z., *Nucl. Phys.* B355:38 (1991), *Phys. Lett.* B255:593 (1991)
34. Ji, X., Musolf, M. J., *Phys. Lett.* B257: 409 (1991)
35. Falk, A., Grinstein, B., Luke, M., *Nucl. Phys.* B357:185 (1991)
36. Falk, A., Grinstein, B., *Phys. Lett.* B247:406 (1990)
37. Eichten, E., Hill, B., *Phys. Lett.* B243: 427 (1990)
38. Golden, M., Hill, B., *Phys. Lett.* B254: 225 (1991)
39. Georgi, H., Grinstein, B., Wise, M. B., *Phys. Lett.* 252B:456 (1990)
40. Luke, M. E., *Phys. Lett.* B252:447 (1990)
41. Chay, J., Georgi, H., Grinstein, B., *Phys. Lett.* B247:399 (1990); Bjorken, J. D., talk presented at Les Recontré de Physique de la Vallee d'Acoste, La Thuile, Italy, March 18-24, 1990, SLAC-PUB-5278 (1990), unpublished
42. Isgur, N., Wise, M. B., *Phys. Rev.* D42:2388 (1990)
43. Mannel, T., Ryzak, Z., *Phys. Lett.* B247:412 (1990)
44. Grinstein, B., et al, *Nucl. Phys.* B363: 19 (1991)
45. Dugan, M. J., Grinstein, B., *Phys. Lett.* B255:583 (1991)
46. Bortoletto, D., Stone, S., *Phys. Rev. Lett.* 65:2951 (1990)
47. Politzer, H. D., Wise, M. B., *Phys. Lett.* B257:399 (1991)
48. Burdman, G., Donoghue, J. F., UMHEP-365; Wise, M. B., CALT-68-1765
49. Grinstein, B., et al, CALT-68-1768, SSCL-Preprint-25, UCSD/PTH 92-05 (1992)
50. Shifman, M. A., TPI-MINN-91-46-T (1991)
51. Daoudi, M., et al, (CLEO Collaboration), CLNS-91-1108, Sep. (1991)
51a. Isgur, N., CEBAF-TH-92-02 (1992)
52. Andrei, N., Gross, D. J., *Phys. Rev.* D18:468 (1978)
53. Maiani, L., Martinelli, G., Sachrajda, C. T., *Nucl. Phys.* B368:281 (1992)
54. Henderson, S., et al, (CLEO Collaboration), *Phys. Rev.* D45:2212 (1992)
55. Isgur, N., Wise, M. B., *Phys. Rev.* D43:819 (1991)
56. Burdman, G., Donoghue, J. F., UMHEP-358, Jan. (1992)

Annu. Rev. Nucl. Part. Sci. 1992. 42:147–75

DELAYED FISSION[1]

Howard L. Hall

Nuclear Chemistry Division L-233, Lawrence Livermore National Laboratory, Livermore, California 94550

Darleane C. Hoffman

Department of Chemistry, University of California, Berkeley, California 94720; and G. T. Seaborg Institute for Transactinium Science, Lawrence Livermore National Laboratory, Livermore, California 94550

KEY WORDS: fission barriers, total kinetic energy (TKE) distributions, mass-yield distributions, fission fragment kinetic energy, half-lives, low-energy fission, shape isomers, fission isomers, superdeformation, charged-particle reactions, nuclear explosions, nucleosynthesis, radiochemical separations, detection systems

---

CONTENTS

INTRODUCTION

Delayed fission is a nuclear decay process that couples β decay and fission. In the delayed fission (DF) process, a parent nucleus undergoes β decay or electron capture (EC) and thereby populates excited states in the daughter. If these states are of energies comparable to or greater than the fission barrier of the daughter, then fission may compete with other decay modes (primarily γ decay or neutron emission) of the excited states in the daughter. This process is schematically illustrated in Figure 1.

Beta decay or electron capture from a parent with a Q-value smaller than the height of the daughter's fission barrier should initially populate

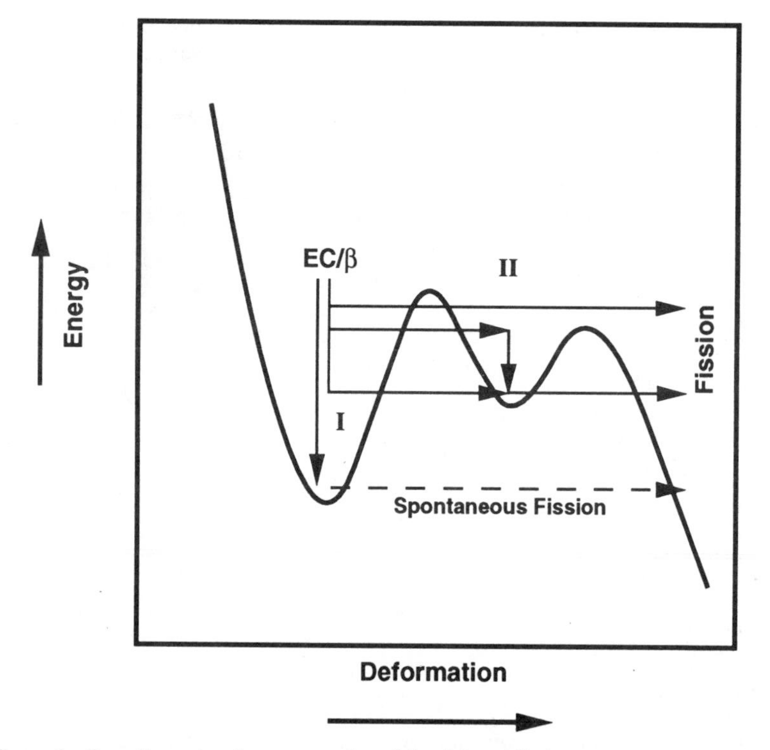

Figure 1 Two-dimensional representation of the delayed fission process. The potential energy of the daughter nucleus as a function of deformation is shown, displaying the double-humped fission barrier prevalent in the actinides. Region I is the inner, or ground-state, well (first minimum); region II is the outer, or shape isomer, well (second minimum).

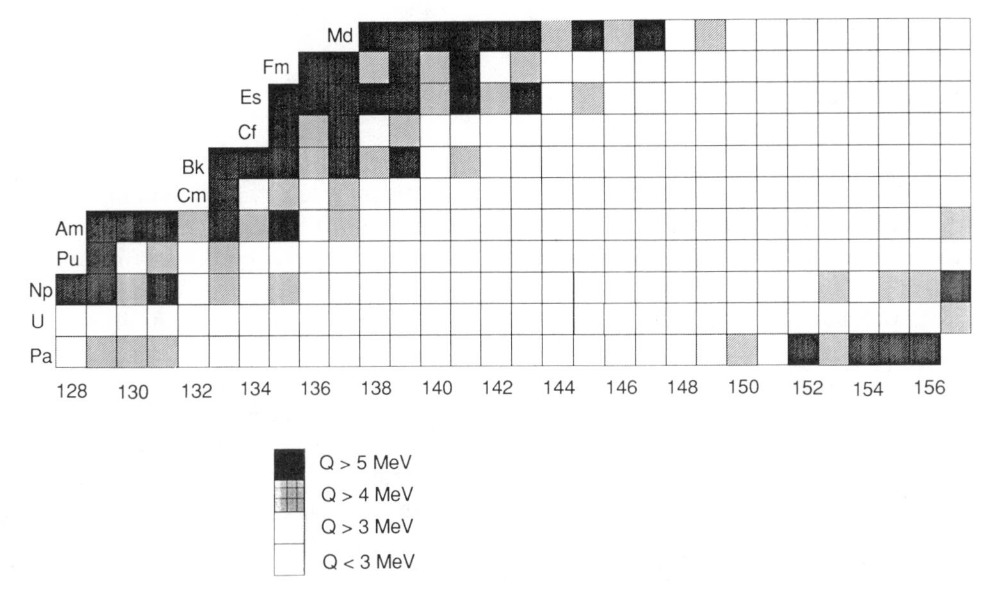

Figure 2 Regions in which delayed fission may be a significant decay mode.

states in the inner, or ground-state, well of the daughter, since direct feeding into the second well would require a collective rearrangement of many nucleons in the daughter simultaneous with the β or EC decay. Once the high-lying states in the inner well are populated, tunneling through the inner barrier must compete against γ decay to populate states in the second well. Once in the second well, tunneling through the outer barrier (fission) must again compete with γ decay and tunneling back through the inner barrier.

To determine what regions of the chart of the nuclides might have significant delayed fission branches, it is necessary to consider the basic criterion for delayed fission to occur; that is, the Q-value for the transition must be comparable to or greater than the fission barrier. This leads to the conclusion that the likely sites for delayed fission will be in the heavy elements (where fission barriers are small) and fairly far from the line of of β stability (where Q-values become large), as shown in Figure 2. This, of course, means that nuclei decaying by delayed fission are challenging to observe experimentally because of the small cross sections for producing actinides far from stability and the short half-lives concomitant with high Q-values. In fact, experiments using accelerator-produced nuclei have been limited to cases in which the delayed fission probability (P_{DF}) is less than 0.01. P_{DF} is defined as the ratio of the number of β decays followed by fission to the total number of β decays,

$$P_{DF} = \frac{N_{\beta f}}{N_{\beta}}. \qquad 1.$$

Delayed fission is believed to play a significant role in the production of heavy or superheavy elements in nature (1–5). All the naturally occurring actinides are produced in r-process nucleosynthesis (6), which consists of rapid neutron capture to make extremely neutron-rich nuclei in the heart of a supernova. These nuclei then sequentially β decay toward stability to produce heavier elements. In this manner, the region of short-lived nuclei between bismuth and the natural actinides can be bypassed. Delayed fission is thought to affect this process in two ways: first, as the neutron-rich nuclei β decay, delayed fission depopulates the mass chains, changing the final distribution of elements and isotopes from that expected from pure neutron capture; furthermore, if there were no delayed fission, one would expect the same sort of bypass mechanism to avoid the short-lived transplutonium nuclei and produce superheavy elements in nature. The distribution of the actinides and the utter lack of experimental evidence for cosmogenic superheavy elements indicates that delayed fission may have an effect on nucleosynthesis. However, terrestrial experiments with nuclear explosives indicate that some theoretical predictions may have overestimated this effect and are therefore difficult to reconcile with data.

EXPERIMENTAL METHODS

To review experimental methods used for the study of delayed fission, it is convenient to address the relevant publications in the sequence of the general stages that occur in delayed fission experiments. These are the production of the nuclei of interest, separations used to isolate them, and the measurements performed on them.

Production

There are three basic methods of production that have been used in delayed fission studies: charged-particle reactions at accelerators, fast neutron capture, and terrestrial nucleosynthesis in nuclear explosions. Charged-particle reactions form the greatest volume of work in this field, primarily because of the relative ease in producing nuclei far from β stability. However, charged-particle reactions are predominantly limited to the production of neutron-deficient nuclei—the region of electron-capture-delayed fission (ECDF). The more physically interesting phenomenon is β-delayed fission (βDF), which occurs in the neutron-rich nuclei. Neutron capture, whether of reactor- or accelerator-produced neutrons, is usually limited to nuclei only one or two neutrons removed from the target material. This means that nuclei with high P_{DF}'s, the extremely neutron-rich ones, are not accessible by "usual" neutron capture. They can be reached in nuclear explosions, where the highly intense neutron flux in the supercritical region of the device leads to the capture of many neutrons. However, the local environment of nuclear explosions does not lend itself readily to the collection and isolation of individual nuclides. One must instead seek to glean information from the longer-lived products found in and isolated from the debris left after the explosion.

CHARGED-PARTICLE REACTIONS The first production of nuclei undergoing delayed fission was reported by Kuznetsov et al (7, 8) in 1966–1967 using the reactions $^{209}Bi + ^{22}Ne$ and $^{230}Th + ^{10}B$. The fission activity was produced by the irradiation of obliquely mounted targets that were subsequently moved sequentially past track detectors, which recorded the fission decays. The use of oblique targets, or targets mounted so the incident beam strikes them at a low angle, greatly increases the amount of target material irradiated but allows fission fragments to escape from a relatively thin layer of material. In the case of $^{230}Th + ^{10}B$, the target consisted of 250 $\mu g\ cm^{-2}$ thorium mounted at 12 degrees to the beam axis. This creates an effective target thickness of 1.2 $mg\ cm^{-2}$ for the irradiation while the fission fragments only have to pass through 250 $\mu g\ cm^{-2}$ of material to exit the target. In addition, the target system was mounted on the inside of the accelerator. This maximizes the amount of beam available for irradiation, as none of the beam is lost during extraction. Internal target arrangements generally have at least as much or more beam current than external (beam line) targets. However, internal targets are also more difficult to use—they are in an area that becomes highly radioactive by activation, and to change or repair the targets often requires opening the main cyclotron vacuum tank.

Initially, the fission activities discovered by Kuznetsov et al were mistakenly attributed to a new form of fission isomerism. In 1969, Berlovich & Novikov (9) demonstrated that these fission activities met the theoretical conditions required for delayed fission, although the activities were not specifically attributed to a delayed fission process until 1972 (10). Definitive proof that delayed fission was responsible for these activities was not obtained until our work in 1989 (11).

Gangrskii et al reported βDF in $^{236,238}Pa$ produced in an accelerator experiment (12). In this case, the accelerator was used to produce neutrons by the reaction of deuterons on beryllium. These neutrons were used to irradiate uranium foils, and thus made Pa isotopes by fast neutron capture. This experiment was repeated by Baas-May et al (13), and no fission was observed in radiochemically isolated Pa isotopes. Gangrskii has also used a number of charged-particle reactions to produce delayed fissioning isotopes of Bk, Es, and Md, in some cases using inclined targets, and in others thin targets with catcher foils (14).

Lazarev et al used an internal rotating drum target in the exciting discovery of delayed fission in preactinide nuclei (15). This study involved the irradiation of internal targets with intense beams of ^{40}Ar and ^{40}Ca. Lazarev's target materials were coated on the outside of a water-cooled Cu drum, which rotated at a controlled rate past a set of mica track detectors. This system allowed the use of approximately 1 pμA of beam, at least an order of magnitude higher than the amount of beam available for external targets. With this higher beam current, Lazarev was able to make this discovery. A schematic of such a target system is shown in Figure 3.

Habs et al produced ^{232}Am by the $^{237}Np(\alpha,9n)$ reaction on a single target of ^{237}Np (16). Light-ion reactions such as these are attractive, since they generally have higher cross sections than heavy-ion reactions and most accelerators can produce a higher light-ion flux than for heavier ions. The disadvantage of light-ion bombardments is that, being light ions, the beam particles impart very little recoil momentum to reaction products. As a result, only a very thin layer of target material is effective (such that the recoils can escape). We and others have also used light-ion irradiations employing a novel target system that allows irradiation of up to 23 targets at one time, with the activities being collected by KCl aerosols (17). This system, shown in Figure 4, exploits the low dE/dx of light ions to compensate for the small effective target thickness. This target system also has the advantage that it suppresses the collection efficiency for fission products, since the targets

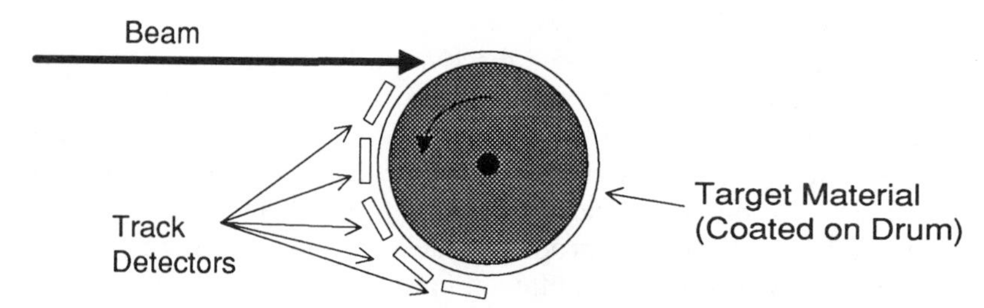

Figure 3 Schematic illustration of the operation of a rotating internal target.

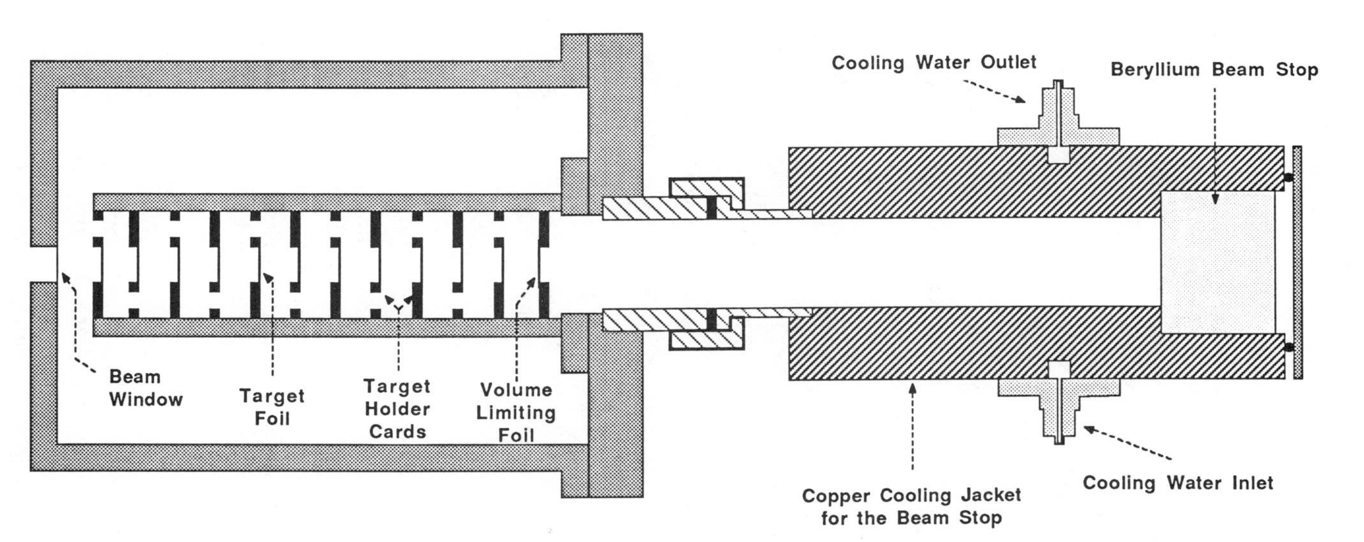

Figure 4 Cross-sectional view of the multiple target system developed by Hall et al for light-ion irradiations (17). Reprinted with permission.

are so closely spaced that the fission fragments will stop in the next target rather than attach to aerosols.

We have also used the ^{254}Es(t,p) stripping reaction to study βDF in ^{256m}Es (18). While this is also a light-ion reaction, it predated the development of the multiple target system (17). The multiple target system, even if available then, would not have improved this experiment because the supply of ^{254}Es was extremely limited and the bombarding energy in this irradiation was low (16 MeV).

Mezilev et al used spallation reactions to search for delayed fission in volatile heavy elements (19). In this experiment, a 1-GeV proton beam was used to irradiate a molten UC_X target, forming neutron-rich francium and actinium isotopes by spallation. The volatile activities were mass-separated and measured with solid-state photon and charged-particle detectors.

NUCLEAR EXPLOSIONS In a nuclear explosion, neutron fluxes many orders of magnitude higher than obtainable from reactors or accelerators can be sustained for a brief instant. During that time, the heavy elements in the device can undergo rapid multiple neutron capture to form extremely neutron-rich nuclides that must β decay toward stability. As an example, the United States' "Mike" explosion in 1952 showed evidence of 17-neutron capture residues (this test also led to the discovery of Es and Fm) (20). This "terrestrial r-process" is the only way to reach the extremely neutron-rich nuclides. βDF depopulates the mass chains during the β decay toward stability, so the heavy-element yields from a nuclear explosion contain data on the effect of βDF. Hoff has examined such data to compare theoretical predictions to the observed yield distributions (21, 22).

Measurement and Detection Systems

MEASUREMENTS WITHOUT SEPARATIONS The earliest measurement systems were simple track detectors, as used by Kuznetsov et al (7, 8) and Gangrskii et al (14). In these systems, either the irradiated target or the catcher foils were measured with track detectors at a preset interval. Since the time per exposure is known, the distribution of fission fragments in the track detector set yields the half-life of the fissioning species. The advantages of track detectors are that they are insensitive to mildly ionizing radiation (so can be used with internal accelerator targets), simple to use, and form a permanent record once developed. Their disadvantages are that they are passive (requiring developing and scanning off-line) and, as used, they provide only an integrated number of fissions with no data on fragment energy. The most recent work with track detectors is that of Lazarev et al (15).

More advanced systems rely on silicon semiconductor detectors (either surface-barrier or passivated ion-implanted). These detectors are very sensitive to high fluxes of ionizing radiation, so the samples to be counted have to be removed from the vicinity of the accelerator beam. Habs et al (16) used a system in which the recoiling ^{232}Am was embedded in a very thin metal foil. After a suitable irradiation period, this foil was shuttled to a position between two surface barrier detectors, which detected the coincident fission fragments. Kreek et al (23) and the authors (24, 25) have used a system in which the recoiling reaction products are captured by KCl aerosols in a flowing stream of He gas. The activity-laden aerosols are transported out of the irradiation vault and collected on thin foils that are mounted on the periphery of a wheel in a vacuum chamber. The wheel is moved at preset intervals, passing the sample between six pairs of silicon detectors. The advantages of silicon detectors are good energy resolution, the ability to detect both fission fragments and α particles, good timing resolution (<1 ns), and the ability to perform coincident measurements. Disadvantages are their inability to provide good data in highly radioactive environments, ease of physical damage, and the assorted instabilities associated with electronics.

We have also combined Si detectors with intrinsic Ge γ-ray detectors (11, 24, 25). This system allows simultaneous measurement of the K-capture x rays from electron capture with the subsequent delayed fission. This system also uses the He/KCl transport system, but samples are collected on a low-Z support that is then placed between the two detectors. Such a system has to be located very far from the irradiation area, since the Ge detector will respond to the plethora of high-energy photons produced near a light-ion irradiation. These experiments involved He/KCl transport over a distance of about 80 m to the collection site, with transport efficiencies of 50% or better.

MEASUREMENTS WITH SEPARATION Hingmann et al (26) report the observation of delayed fission in ^{242}Es using the velocity filter SHIP to separate the products from the ^{205}Tl + ^{40}Ar reaction. In SHIP, the reaction products are separated from the higher velocity beam particles by their greater deflection in an electric field. The separated reaction products were implanted into position-sensitive Si detectors. A total of three fission events was observed, along with α particles from the new isotopes ^{241}Es and ^{242}Es.

A number of radiochemical experiments have also been performed for delayed fission studies. Baas-May et al (13) developed an automated system to examine Gangrskii's claim (12) of βDF in ^{236,238}Pa. The radiochemical separation involved the removal of I and Br (delayed neutron precursors), the extraction of the Pa(V) (the pentavalent state of protactinium) into diisobutylcarbinol from 9-M HCl, and the back-extraction of the Pa(V) into 12-M HCl/0.25-M HF. The Pa was coprecipitated with a small amount of iron hydroxide that was then counted for fissions with polycarbonate track detectors. The total time for preparing a sample for counting was 2.3 minutes. A schematic of the system developed for this experiment is shown in Figure 5. Using this apparatus, Baas-May et al observed no fission tracks and were able to set an upper limit for delayed fission in ^{238}Pa an order of magnitude lower than that reported by Gangrskii. Baas-May et al postulated that the fissions observed by Gangrskii et al arose from delayed-neutron-induced fission or photofission of nuclei in the uranium foil, which was counted without separations.

We have also used extensive radiochemical separations in our study of ^{256m}Es (18). The ^{256m}Es was collected with a gold catcher foil during a series of approximately seven-hour tritium irradiations of the ^{254}Es target. After irradiation, the catcher foil was dissolved and a radiochemically pure Es fraction was generated after either three or four ion exchange chromatographic separations. The chemical separations required over three hours to complete, but produced samples with only trace amounts of impurities (as assayed by γ spectroscopy).

We have used rapid radiochemical separations in our studies of light americium isotopes (11, 24, 25, 27). Two procedures were developed, one for Z identification and one for γ spectroscopy. The procedure for Z identification was quite simple since the sample did not have to be separated from fission products, only from other actinides. Since Am is the only trivalent actinide that can be produced in the ^{237}Np(α,xn) reaction, a single small anion exchange column was sufficient to isolate americium from the lighter actinides. The total time for this procedure was about 90 seconds. The second procedure involved both anion and cation exchange, followed by an optional coprecipitation with CeF_3. The total time for this separation was 90 seconds without the coprecipitation, 4 minutes with it.

Kreek et al (23) developed a rapid separation procedure for berkelium as part of their study of ^{238}Bk. In this procedure, berkelium produced by the α irradiation of multiple ^{241}Am targets is oxidized to the +4 oxidation state, extracted into a hexane solution of bis(2-ethylhexyl)orthophosphoric acid, then reduced and back-extracted into HCl. Cerium is the only major contaminant in this procedure, and it can be removed with a small cation-exchange column.

In ongoing work, we are using anion-exchange separations as part of a study of the fission properties of ^{228}Np (28). In this separation the

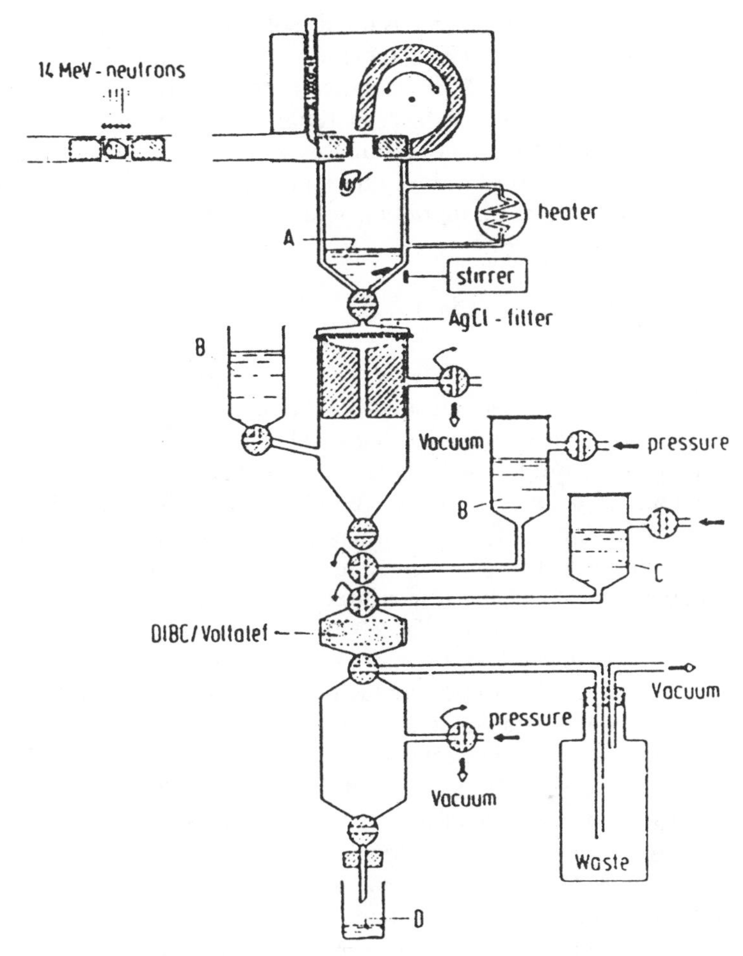

Figure 5 Schematic illustration of the automated chemical apparatus used by Baas-May et al (13) in their search for delayed fission from ^{238}Pa. Solution *A* is 0.5-M $HNO_3 + SO_2$, *B* is 12-M HCl/O.1-M oxalic acid, *C* is 12-M HCl/0.25-M HF, and *D* is boric acid + H_2O_2. Reprinted with permission.

Np is absorbed on an anion exchange column while most fission products are washed through. After washing, the Np is eluted and the liquid sample counted by γ spectrometry. This procedure requires about two minutes for completion.

Of course, the most elaborate radiochemical studies of delayed fission are those used to obtain the data reviewed by Hoff (21, 22) from underground nuclear explosions. These separations require obtaining debris from the rubble of the explosion, dissolving the material (mostly rock), and performing a series of separations to yield fractions of sufficient purity to measure all the isotopes present in the sample. The time scale is days to weeks for the preparation of a sample for counting, so short-lived radioactivities are lost in this process by decay. Selected radiochemical procedures used for this application have been reported by Los Alamos National Laboratory, and can also be found in many radiochemistry texts (29).

EXPERIMENTAL RESULTS

ECDF Activities

ECDF activities form the greatest part of the reported work on delayed fission, primarily because nuclei that display ECDF are much more amenable to accelerator-based studies than βDF activities. Most early measurements of delayed fission have only reported fission activities with an associated half-life. Z identification was not unequivocal, and the P_{DF}, if reported, was generally estimated from systematics. More recent studies report half-lives, measured P_{DF}, and post-scission properties of the fission fragments. These studies often include radiochemical identification of the Z of the parent.

The first reports of fissions [eventually attributed to delayed fission (10)] were those of Kuznetsov et al (7, 8). Fission activities in the light americium and neptunium regions were observed with half-lives on the order of minutes. At the time, Kuznetsov et al postulated that these activities may be a new form of shape (fission) isomers with remarkably long half-lives. This is quite understandable in light of the discovery of fission isomers and the excitement it generated a few years earlier (see 30 for a discussion of the timeline of fission developments). However, it became apparent that these activities could not be fission isomers as more was discovered about the underlying physics of shape isomerism. Berlovich & Novikov (9) published a theoretical analysis of these fission activities in 1969, and noted that the nuclei in question met all the necessary criteria for delayed fission. In 1972, Skobelev (10) reexamined the data of Kuznetsov et al and assigned the observed fissions to delayed fission activities in $^{232,234}Am$ and ^{228}Np.

Somerville et al (31) confirmed the fission activity of ^{234}Am with a half-life of 2.6 minutes in 1977. Habs et al (16) confirmed ^{232}Am in 1978, and used the P_{DF} value they derived to estimate the height of the daughter's fission barrier. The experiment of Habs et al marks the first application of silicon charged-particle detectors to delayed fission studies.

A number of cases of ECDF were discovered by Gangrskii et al (14) in 1980. This paper marks the expansion of delayed fission to the heavier neutron-deficient actinides and the establishment of delayed fission as a decay mode common to potentially all actinide nuclei with Q-values of 4 MeV or more. As with the earlier papers of Habs et al (16) and Gangrskii et al (12) (discussed in the βDF section), this work used rudimentary theoretical calculations to attempt to elucidate the height of the fission barriers of the daughter nuclei.

In 1987 an extremely exciting discovery was made by Lazarev et al: the first observation of delayed fission in preactinide nuclei (15). This experiment, although made very difficult by the low production cross sections and the short half-life of the activity produced (0.7 sec), demonstrated the possibility of extending delayed fission studies into a completely new area of the chart of the nuclides. It may also be possible to study the low-energy fission properties of these nuclei and greatly expand the data available for understanding fission.

Since 1989 we have been studying the light americium and neptunium regions in depth (11, 17, 24, 25, 27, 28). Through the innovation of the multiple target system for light-ion bombardments, we have been able to perform experiments wherein thousands of fission events have been observed rather than the tens to hundreds of earlier experiments. Our studies have led to better measurements of the half-lives of the nuclei studied, experimental measurements of the P_{DF}, and other advances discussed in the following sections. Kreek is expanding these in-depth studies to the light berkelium region (23). A summary of reports and selected measurements is given in Table 1.

βDF Activities

Compared to the number of cases of ECDF, βDF has a paucity of individual cases studied but a plethora of "composite" results. Only three nuclides have measured βDF branches, and two of those have been shown to be the likely result of a systematic experimental error. In composite results, the studies of Hoff (21, 22) have tested theoretical predictions against experimental results from nuclear weapons tests. Indeed, one can consider the distribution of the heavy elements in nature to be a test of delayed fission theories. As is discussed in the theory section, delayed fission tempers the production of the actinides, and hence affects actinide chronometers and the estimates of the age of the universe derived from them.

The first reports of experimentally observed βDF were those of Gangrskii et al (12), in the isotopes $^{236,238}Pa$. These were produced by n,p reactions on uranium foils, which were exposed without any separa-

Table 1 Reports on delayed fission

Nuclide[a]	$t_{1/2}$	P_{DF}	Measurements	Reference
^{250}Md	52 sec	2×10^{-4}	$t_{1/2}$[b]	14
^{256m}Es	7.6 hr	2×10^{-5}	$t_{1/2}$, P_{DF}, levels	18
^{248}Es	28 min	3×10^{-7}	$t_{1/2}$	14
^{246}Es	8 min	3×10^{-5}	$t_{1/2}$	14
^{244}Es	37 sec	10^{-4}	$t_{1/2}$	14
^{242}Es?	5 - 25 sec	$(1.4\pm0.8)\times10^{-2}$	$t_{1/2}$, P_{DF}	26
^{240}Bk	4 min	10^{-5}	$t_{1/2}$	14
^{238}Bk	144 ± 6 sec	NR	$t_{1/2}$, XF[d]	23
^{234}Am	2.6 ± 0.2 min	6.95×10^{-5}	$t_{1/2}$	10 & 32
	2.6 ± 0.2 min	NR[c]	$t_{1/2}$	31
	2.32 ± 0.08 min	$(6.6\pm1.8)\times10^{-5}$	$t_{1/2}$, P_{DF}, KE, MY, TKE, XF, RC	24
^{232}Am	1.4 ± 0.25 min	6.96×10^{-2}	$t_{1/2}$	10 & 32
	0.92 ± 0.12 min	$1.3^{+4}_{-0.8}\times10^{-2}$	$t_{1/2}$	16
	1.31 ± 0.04 min	$(6.9\pm1.0)\times10^{-4}$	$t_{1/2}$, P_{DF}, KE, MY, TKE, XF, RC	25
^{228}Np	60 ± 5 sec	NR	$t_{1/2}$	10
	64.2 ± 1.8 sec	$\approx10^{-4}$	$t_{1/2}$, P_{DF}, KE, MY, TKE, RC	28
^{180}Tl	$0.70^{+0.12}_{-0.09}$ sec	$\approx10^{-6}$	$t_{1/2}$	15

[a]The nuclide undergoing β or EC decay is given.

[b]P_{DF} for nuclei not marked with "P_{DF}" in the measurements column were obtained from systematics or evaporation codes.

[c]Not reported

[d]KE: fission fragment (ff) kinetic energy distribution; MY: ff mass yield distribution; TKE: ff total kinetic energy distribution; XF: x-ray -- fission coincidence; RC: radiochemical confirmation of Z.

tions to track detectors. Fission activities were observed and the half-lives were consistent with those of the two Pa isotopes, although the number of events was too small to measure the half-life accurately. In 1985, Baas-May et al (13) examined the ^{238}Pa claim in an experiment utilizing rapid chemical separations to isolate the protactinium. Baas-May et al observed no fissions and were able to set an upper limit on delayed fission from ^{238}Pa approximately an order of magnitude lower than that reported by Gangrskii et al. Baas-May et al postulated that the fission activity attributed to ^{238}Pa was actually delayed-neutron-induced fission or photofission of nuclei in the uranium targets. Gangrskii et al had considered these sources of background and taken steps to minimize their effects, but the work of Baas-May et al indicates that these steps were insufficient. On the basis of the work of Baas-May et al, ^{236}Pa and ^{238}Pa have been excluded from Table 1.

We have studied βDF in ^{256m}Es and observed a weak delayed fission branch (18). In this experiment, the amount of ^{256m}Es produced was limited by the amount of target material, ^{254}Es. Only 0.1 μg of ^{254}Es was available for this study. Two fissions attributed to delayed fission were observed in coincidence with the β particle from the decay of ^{256m}Es to ^{256}Fm. A unique difficulty in this experiment was that the daughter, ^{256}Fm, decays partly by spontaneous fission so a simple count of fission events would not be useful. To eliminate this problem, the fissions were required to be in coincidence with the β particle. Adding to the difficulty is the fact that all fissions will register in coincidence with conversion electrons from the fission fragments. Therefore, it was necessary to measure the level structure of ^{256}Fm to see if the fissioning level could be identified. It was determined that a 7^- level at 1425 keV was predominately populated and had a half-life of 70 ns. Since this was the only high-lying level with a significant delay, we searched for fissions with timing consistent with decay from this level. Two events were observed. Statistical analysis of the experiment suggested that there is less than a 1% chance that these events were random or background. This experiment is the only report in which the discrete level leading to fission has been identified. With the work of Baas-May disproving the Pa isotopes, it is the only case of βDF experimentally measured. Unfortunately, the extremely small amount of ^{254}Es available limited the sensitivity of the experiment.

Hoff's comparison of the effects of delayed fission on the yields of heavy elements in nuclear tests is probably one of the most stringent tests of delayed fission theory performed to date (21, 22). While accelerator-based work cannot reach the very heavy nuclei for which βDF is significant, the reactions in an exploding nuclear device can. Analysis of the debris gives data that can be compared to theoretical estimates in much the same way as isotopic abundances can be compared with theoretical calculations for actinide cosmochronometers; however, a nuclear explosion has many fewer variables than the formation and evolution of the universe. Hoff's study of these data indicates that the theories available then heavily overestimated the effect of delayed fission (see Figure 6).

The most recent search for nuclei decaying by βDF was performed by Mezilev et al (19). Although no delayed fission was observed in the

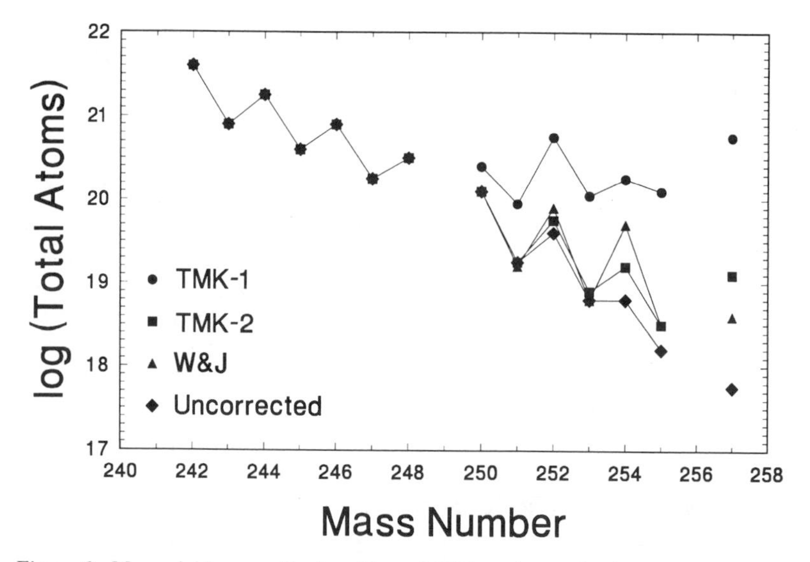

Figure 6 Mass-yield curves for the effect of βDF on the production of heavy elements in the nuclear explosion "Hutch" (21). TMK-1 refers to the TMK (5) model, TMK-2 is the TMK model using the fission barriers and masses used by Meyer et al (4), W&J is from Reference 1, and the last entry is the yield uncorrected for βDF. The starred points represent the superpositions of the various symbols at the same *x*, *y* point. Reprinted with permission.

nuclei studied ($^{228,230,232}Fr$ and ^{232}Ac), upper limits of P_{DF} for these nuclei were established in the range of 10^{-4} to 10^{-7}.

P_{DF} *and Other Delayed Fission Properties*

Central to any attempt to extract nuclear structure data from delayed fission is the delayed fission probability, P_{DF}. However, P_{DF} is not trivial to measure. The number of fissions can be readily measured with track detectors or other charged-particle detectors, but the total number of β decays cannot be easily measured in most of the studies reported to date. As a result, most reported P_{DF} values are based at least partially on systematics or models (such as cross-section codes).

In the works of Gangrskii et al (14), Kuznetsov (32), and Lazarev et al (15), P_{DF} values were obtained from the observed fission cross section and a calculated production cross section with an electron capture branch estimated from systematics. With both a cross section and a branching ratio calculated, one might expect considerable uncertainty in the resulting value for P_{DF}. The values obtained are given for each nuclide in Table 1.

In the report by Habs et al (16) for ^{232}Am, somewhat less uncertainty is involved in the calculation of P_{DF}. Habs et al measured both the fission from ECDF of ^{232}Am and its α decay branch. For their calculation of ECDF, the α branching ratio was taken from systematics and the remainder of the decay was assumed to be either EC or ECDF. It should be noted, however, that systematic calculations of α branching ratios are very sensitive to the α-particle energy and the overall half-life of the nuclide.

Hall et al have determined P_{DF} in two ways. In the case of ^{256m}Es, it was possible to determine the ratio of delayed fissions to the total number of β decays, since the β decay always leads to ^{256}Fm (which has a spontaneous fission branch) (18). This yields P_{DF}. In studies of the light americium nuclei (11, 24, 25, 27, 28), a different method was required for direct measurement of P_{DF}. In these nuclei, the EC branch leads to long-lived nuclei also decaying by EC, so no convenient charged particle was available for EC branch measurements. To determine the EC branch directly, it was necessary to chemically separate the Am from other species so the K-capture x rays could be measured. These separations were performed repeatedly (with ^{241}Am as a tracer) on an alternating basis with gross fission measurements. In this way, all the experimental variables that could affect P_{DF} would cancel, provided they varied with a period that was long compared to the rate of the measurements (for a discussion of this method and the affiliated equations, see 27). P_{DF} values obtained with this method are listed, along with the others, in Table 1. Figure 7 shows the behavior of P_{DF} as the EC Q-value of the parent varies for ECDF.

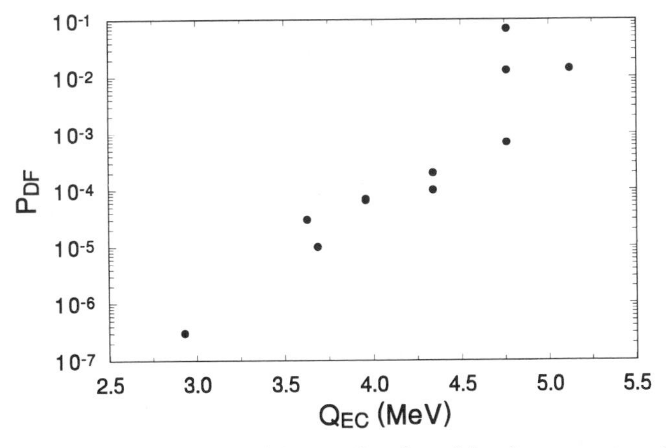

Figure 7 Delayed fission probabilities as a function of the electron capture Q-value.

We have also measured the time correlation between EC and the subsequent delayed fission for ^{232}Am and ^{234}Am (11, 24, 25, 27). These experiments, which provided the first direct proof of the ECDF process, were performed by measuring the coincidence between the Pu K x rays (emitted when an Am nucleus captures a 1s orbital electron) and the fission fragments. This time correlation unequivocally ties the fission activity to the EC, verifying that the observed fissions are due to ECDF.

The increased number of fission events observable per unit time using the multiple target system has also allowed the exploitation of delayed fission for the study of the low-energy fission properties of the light americium (24, 25, 27) and neptunium (28) isotopes. With 1000–2000 observed fissions, it is possible to make fairly accurate measurements of the fragment kinetic energy distributions, the mass yield, and the total kinetic energy distribution. An example of such measurements is shown in Figure 8. These measurements, along with the work of Gangrskii et al (14) showing that ECDF is a common decay mode in

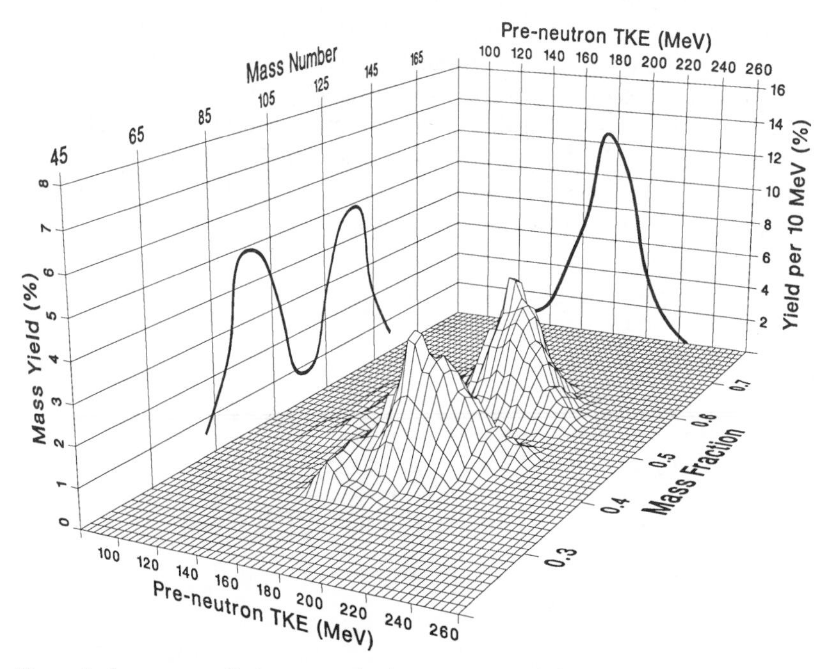

Figure 8 Low-energy fission properties (pre-neutron emission total kinetic energy, and mass-yield distributions) of ^{234}Pu as determined from the ECDF of ^{234}Am. Reprinted with permission.

the neutron-deficient actinides, demonstrate that ECDF has the potential to expand greatly the number of nuclei whose low-energy fission properties can be measured.

THEORY

Background

There has been considerable theoretical interest in β-delayed fission processes because they may significantly affect the final abundances of the heavy elements produced in the astrophysical r-process and in other multiple neutron capture processes taking place in very high neutron fluxes, such as those existing in thermonuclear explosions. The originally produced neutron-rich heavy-element nuclides can decay toward stability via β^- emission to still higher Z elements. However, βDF or neutron emission, as well as prompt fission and spontaneous fission, can deplete or terminate the decay of a given mass chain, and thus significantly affect the final abundance distribution of the heavy elements. The actinide production ratios calculated for the r-process are of special interest because they can be compared with measured abundance ratios to give information on the age of the galaxy and chemical evolution. As early as 1957, Burbidge et al (6) proposed that βDF could deplete the yield of heavy elements produced in supernovas. Much later, Wene (1, 2) suggested that βDF might be part of the reason why superheavy elements are not found in nature.

Early calculations neglected the effect of βDF and neutron emission on actinide yields, but with the availability of improved calculations for beta strength functions and fission barriers, a number of calculations of r-process abundances have been made (1, 5, 36). In 1983, Thielemann, Metzinger & Klapdor (TMK) (5) computed new values for the actinide production ratios that led to a galactic age of nearly 21 Gyr, much greater than the previous estimates (33) of 13 to 15 Gyr. Recently, Meyer et al (4, 34) used different calculations of the β-strength function and obtained much lower βDF probabilities than those of TMK. Actual experimental data on delayed fission probabilities would be particularly valuable in assessing the validity of the various theoretical models.

Calculations of Delayed Fission Probabilities

The probability for delayed fission decay, P_{DF}, can be expressed in terms of experimentally measurable quantities as

$$P_{\mathrm{DF}} = \frac{N_{\mathrm{if}}}{N_{\mathrm{i}}} = \frac{\sigma_{\mathrm{if}}}{\sigma_{\mathrm{i}}}, \qquad 2.$$

where N_i is the number of the type of decays of interest (e.g. β or EC) and N_{if} is the number of those decays leading to delayed fission. Similarly, σ_{if} and σ_i are the corresponding cross sections. P_{DF} can also be derived from theoretical considerations as

$$P_{DF} = \frac{\int_0^{Q_i} W_i(Q_i - E) \frac{\Gamma_f}{\Gamma_{tot}} (E)\, dE}{\int_0^{Q_i} W_i(Q_i - E)\, dE}, \quad 3.$$

where $W_i(Q_i - E)$ is the transition probability function for the decay of interest, $(\Gamma_f/\Gamma_{tot})(E)$ is the ratio of the fission width of excited levels within the daughter nucleus to the total depopulation width of these states, E is the excitation energy of the daughter nucleus, and Q_i is the Q-value for the decay of interest.

It is often assumed that no decay channels are open to the excited nucleus except fission and γ decay. As a result, the term $(\Gamma_f/\Gamma_{tot})(E)$ is taken as being equal to $[\Gamma_f/(\Gamma_f+\Gamma_\gamma)](E)$. To be strictly correct, $\Gamma_{tot}(E)$ should include terms for particle (neutron) emission as well as fission and γ decay, but they are neglected in this approximation.

The transition probability function, $W_i(E)$, can be approximated as the product of the Fermi function, f, and the beta strength function, S_β, giving

$$W_i(Q_i - E) \approx f(Q_i - E,Z)S_\beta(E). \quad 4.$$

The Fermi function may be approximated as

$$f \approx \begin{cases} (Q_\epsilon - E)^2 & \text{for EC decay} \\ (Q_\beta - E)^5 & \text{for } \beta \text{ decay} \end{cases}, \quad 5.$$

for the calculation of P_{DF} in Equation 3.

S_β can be treated in several different ways. It can be taken as being proportional to the nuclear level density (1, 35), it can be generated from the gross theory of β decay (36), or it can be taken as a constant above a certain energy (37, 38). Klapdor et al (39) have pointed out that all three of these common techniques ignore low-lying structure in the beta strength function. Klapdor found that inclusion of low-lying structure in the calculation of P_{DF} has a significant effect on the value obtained. For a qualitative understanding of P_{DF}, S_β can treated as constant above a specified cut-off energy.

It is important to remember that the structural effects ignored by this approach can significantly influence P_{DF}. For example, we have observed βDF to occur in ^{256m}Es with a P_{DF} of 2×10^{-5} from a single level 1425 keV above the ground state (18). The assumption of a constant S_β would predict no delayed fission branch, but in this case γ decay from the 1425-keV level is highly hindered (level half-life = 70 ns) and fission can compete. Likewise, a nucleus with a high Q-value would be expected to have a large P_{DF}, but if decay to the daughter's ground state is superallowed ($\Delta I^{\Delta\pi} = 0^{no}$), no high-lying states may be populated, hence P_{DF} in this case may be nearly zero.

The large dependence of P_{DF} on the energy available for the decay and the structure of the fission barrier arises primarily from the fission-width term, $[\Gamma_f/(\Gamma_f+\Gamma_\gamma)](E)$. $\Gamma_\gamma(E)$, the width for γ decay, can be estimated (14) from the probability for γ transitions, P_γ, as

$$\Gamma_\gamma = \frac{P_\gamma}{2\pi\rho} = \frac{C_\gamma \Theta^4 \exp(E/\Theta)}{2\pi\rho}, \qquad 6.$$

where ρ is the nuclear level density, C_γ is a constant with the value 9.7×10^{-7} MeV^{-4}, and Θ is the nuclear temperature (0.5–0.6 MeV). The fission width, $\Gamma_f(E)$, derived from the penetrability of the fission barrier in a similar fashion, yields

$$\Gamma_f(E) = \frac{P_f(E)}{2\pi\rho}, \qquad 7.$$

where P_f is the penetrability of the fission barrier.

Since the fission barrier in the region of the actinide nuclei is rather complex (14, 16), the penetrability through the entire two-humped barrier is often simplified and approximated by

$$P_f \approx P_A(E) R_B, \qquad 8.$$

where $P_A(E)$ is the penetrability for tunneling through the inner barrier and R_B is the transmission coefficient for fission from the lowest state in the second well. This, in effect, requires the nuclear motion in the second well to be strongly damped, i.e. fission from the second well is not allowed to occur before γ decay to the lowest-lying state. Hence, the calculation of $P_f(E)$ becomes much simpler. Transmission through the inner barrier B_A is then approximated as a simple parabolic barrier problem using the Hill-Wheeler (40) formalism,

$$P_A = \left\{1 + \exp\left[\frac{2\pi(B_f - E)}{h\omega_f}\right]\right\}^{-1} \qquad 9.$$

where B_f is the height of the barrier and $h\omega_f$ is the energy associated with the barrier curvature. This allows Γ_f to be expressed as

$$\Gamma_f \approx \frac{R_B}{2\pi\rho}\left\{1 + \exp\left[\frac{2\pi(B_f - E)}{h\omega_f}\right]\right\}^{-1} \qquad 10.$$

and

$$\frac{\Gamma_f}{\Gamma_f+\Gamma_\gamma}(E) \approx \frac{R_B\left\{1+\exp\left[\frac{2\pi(B_f-E)}{h\omega_f}\right]\right\}^{-1}}{C_\gamma\Theta^4\exp(E/\Theta)+R_B\left\{1+\exp\left[\frac{2\pi(B_f-E)}{h\omega_f}\right]\right\}^{-1}}, \quad 11.$$

which illustrates the strong dependence of this term on the energy available for decay and the structure of the fission barrier.

With these approximations, Equation 3 for electron capture becomes simply

$$P_{\epsilon DF} \approx \frac{\int_C^{Q_\epsilon} (Q_\epsilon - E)^2 \frac{\Gamma_f}{\Gamma_\gamma + \Gamma_f}(E)\,dE}{\int_C^{Q_\epsilon} (Q_\epsilon - E)^2\,dE}, \quad 12.$$

where C is the cut-off energy below which S_β is presumed to be zero. This value has been given (37) as $C = 26A^{-1/2}$ MeV. The integral in the denominator is trivial and may be readily evaluated directly to give a normalization function $N_\epsilon(A)$,

$$[N_\epsilon(A)]^{-1} \equiv \int_C^{Q_\epsilon} (Q_\epsilon - E)^2\,dE = \frac{(Q_\epsilon - 26A^{-1/2})^3}{3}. \quad 13.$$

For βDF, $N_\beta(A)$ can be likewise evaluated to yield

$$[N_\beta(A)]^{-1} \equiv \int_C^{Q_\beta} (Q_\beta - E)^5\,dE = \frac{(Q_\beta - 26A^{-1/2})^6}{6}. \quad 14.$$

$P_{\epsilon DF}$ then becomes

$$P_{\epsilon DF} \approx N_\epsilon(A)\int_C^{Q_\epsilon} (Q_\epsilon - E)^2 \frac{\Gamma_f}{\Gamma_\gamma + \Gamma_f}(E)\,dE, \quad 15.$$

which is exponentially dependent on the difference between the fission barrier and the Q-value for electron-capture decay.

Similarly, the probability for βDF can be derived as

$$P_{\beta DF} \approx N_\beta(A)\int_C^{Q_\beta} (Q_\beta - E)^5 \frac{\Gamma_f}{\Gamma_\gamma + \Gamma_f}(E)\,dE. \quad 16.$$

Hence, for delayed fission to become a prominent decay mode in the actinide region (where fission barriers are on the order of 4–6 MeV), it is necessary to choose nuclei in which Q is comparable to the fission barrier. This requires study of nuclei far from the valley of β-stability (see Figure 2), which introduces a number of experimental difficulties in the production and characterization of these nuclei.

It should be noted that the term in the integral arising from the Fermi function, $(Q_{\beta} - E)^5$, shows that ECDF is more likely to occur than βDF, all other things being equal. This is because the Fermi function for β decay approaches zero faster than that for electron capture at high energies. Since the high-energy states have higher penetrabilities, the overall P_{DF} would be higher for ECDF than for βDF. As a result, ECDF in general should be easier to study than βDF, even if β-delayed fissile species were not so difficult to produce experimentally.

It should also be emphasized that the form of the delayed fission probability developed in Equations 15 and 16 gives only an estimate of the probabilities for delayed fission. More quantitative calculations of P_{DF} require a rigorous treatment of the β strength function S_{β} as it appears in Equation 4, e.g. including the low-lying energy levels in deformed nuclei (39). A quantitative model must also include a more realistic treatment of transmission through the fission barriers and avoid the assumption of complete damping required for the approximation in Equation 8.

The probability of β-delayed fission (and β-delayed neutron emission) can be calculated more rigorously if the appropriate daughter-nucleus fission-barrier heights, B_f, neutron-separation energies, S_n, and β strength functions are known. Meyer et al (4) calculated βDF and neutron-emission probabilities for 19 neutron-rich nuclei in the region of interest for the production of the actinide cosmochronometers. They used the Gamow-Teller β strength functions derived from microscopic calculations in which nuclear deformation was treated explicitly. They pointed out the importance of nuclear deformation and the uncertainties resulting from inaccuracies in the calculated fission barriers and in the β strength distributions. Using standard statistical models, they calculated the probabilities for fission, neutron emission, and γ deexcitation following β decay. They found that the greatest uncertainty in their calculation of these branching ratios arose from uncertainties in the model for the β strength function. The two major sources of uncertainty in calculating the strength functions were their dependence on the accuracy of the level spectrum of the underlying single-particle model and the neglect of first-forbidden β decay. Inclusion of the latter factor would make the probability for βDF even smaller. In order to compare their results with those of TMK (5), they compared the effects of using the complete damping approximation (used by TMK) and the more realistic Wentzel-Kramers-Brillouin (WKB) barrier penetration

Table 2 Comparison of the results from Meyer et al (4) to the results of TMK (5)

				Complete Damping		WKB Barrier Penetration		
Parent	Q_β	B_f	S_n	TMK	Meyer		(Meyer)	
Nucleus		(MeV)		P_{DF}	P_{DF}	P_{DF}	P_n	P_γ
^{234}Fr	5.07	8.16	5.70	3%	0%	0%	0%	100%
^{244}Fr	7.90	6.73	4.64	93%	0%	0%	43%	%57
^{252}Fr	9.51	5.13	3.96	82%	68%	0%	83%	17%
^{246}Ac	6.79	6.03	4.96	84%	0%	0%	13%	87%
^{248}Ac	7.23	5.53	4.78	92%	3%	0%	37%	63%
^{252}Ac	8.20	4.59	4.33	96%	58%	9%	63%	28%
^{264}Ac	10.30	4.84	3.54	10%	83%	2%	90%	8%
^{250}Pa	6.24	4.80	5.10	89%	8%	10%	1%	89%
^{252}Pa	6.65	4.29	4.90	83%	36%	34%	2%	64%
^{254}Pa	7.28	3.83	4.66	95%	54%	40%	11%	49%
^{260}Pa	8.34	3.82	4.49	100%	92%	97%	0%	3%
^{270}Pa	10.50	5.15	3.51	2%	81%	14%	77%	9%
^{252}Np	5.09	4.16	5.42	32%	2%	5%	0%	95%
^{254}Np	5.56	4.00	5.23	50%	19%	13%	0%	87%
^{276}Np	10.41	4.46	3.31	25%	83%	9%	84%	7%
^{251}Am	1.90	5.03	4.75	0%	0%	0%	0%	100%
^{258}Am	5.13	4.09	5.29	35%	12%	18%	0%	82%
^{264}Am	6.19	3.28	5.11	100%	85%	100%	0%	0%
^{277}Am	8.04	4.14	2.70	47%	47%	19%	75%	6%

calculation (see Table 2). They found somewhat lower probabilities for βDF with the WKB calculation than for the complete damping approximation and much lower probabilities than calculated by TMK, especially in the region around mass number 250, which is of particular interest. Meyer et al also studied the effects of changing the fission barrier heights by ±1 MeV. As might be expected, large decreases in βDF probability occurred with increased barriers, again emphasizing the sensitivity of delayed fission to the barrier heights. They concluded that when nuclear deformation is treated consistently in the models used to compute β strength functions, fission-barrier heights, and nuclear masses, the resulting probability for βDF is usually less than previously calculated.

Comparison with Experimental Data

Estimates of the probabilities for β-delayed fission can be obtained by careful analysis of the actinide yields from thermonuclear explosions and comparison with the yields calculated using different βDF probabilities. Such analyses led Hoff (21, 22) to conclude that the larger βDF rates calculated by TMK seriously overestimate the effect of βDF in the uranium decay chains with $A = 252$ to 257. This also indicates a large uncertainty in the very long 21-Gyr galactic age that TMK derived using these large βDF probabilities.

Calculations using the recent lower βDF probabilities of Meyer et al (4) for the actinide cosmochronometers gave ratios that were lower by factors of 2 to 3 from previous results and would thus lead to a younger galactic age. Calculations by Staudt et al (41) also indicate lower βDF probabilities. Fowler (42) recalculated the actinide production ratios with no inclusion of βDF and obtained 11.0 Gyr. It would obviously be of great help in estimating the galactic age if experimental βDF values could be obtained. Unfortunately, the neutron-rich actinide nuclides that are predicted to have large βDF probabilities are 10 or more neutrons beyond the known isotopes and appear to be currently unattainable. Therefore, experimental values for the upper limits to βDF and information derived from ECDF will have to be used to develop and test new models of delayed fission.

FUTURE PROSPECTS

New Regions of Delayed Fission

To date, delayed fission has been observed in the actinides and one case has been seen in ^{180}Tl. It is likely that more cases in the actinides will be studied intensely if the proper projectile-target combination can be determined and used. There are certainly enough cases already known in the neutron-deficient actinides to occupy experimenters for a few years, and new ones may still be found. The study of delayed fission in the pre-Pb region is just in its infancy. Although this region is very difficult to study, there are undoubtedly more examples of ECDF waiting to be discovered and characterized.

Another potential site for ECDF is in the nuclei having $Z > 82$ and $N < 126$, or in other words, the nuclides beyond lead but below the

$N = 126$ closed neutron shell. No cases of delayed fission have been reported here. It may be that the short α half-lives of the highly neutron-deficient nuclides in this region overwhelm EC to the extent that ECDF cannot be observed. Without EC, of course, there can be no ECDF. It remains to be seen if experiments can be devised to successfully study delayed fission in this area.

The prospects of future study of βDF are encouraging. Although the very neutron-rich nuclei are quite difficult to produce in any experiment short of a nuclear explosion, we may be able to explore the fringes of this region by employing radioactive nuclear beams (43). With this technique, an accelerator is used to make radioactive neutron-rich nuclei (for example, ^{20}O), which are separated and accelerated in a secondary beam for use in production reactions. This technique should extend the region of nuclides accessible by at least 2–4 nuclides beyond the heaviest now known. While this will not reach nuclei that have P_{DF}'s of 50%, it may reach P_{DF}'s comparable to those studied by ECDF and allow comparable studies. This would be critically important to theoretical predictions, since the extrapolation from ECDF to βDF may involve errors of which we are as yet unaware.

Fission Properties Far From Stability

As we pointed out earlier in this review, investigations of delayed fission make it possible to expand greatly the number of nuclides whose low-energy fission properties may be measured. Typically, the excited daughter products, which are the species undergoing fission, have such a small probability for undergoing spontaneous fission decay that their low-energy fission properties cannot otherwise be studied in detail. The half-lives in the case of ECDF may be as long as minutes, and chemical separations can be performed to reduce interference from unwanted activities and to make positive atomic number assignments for new species. Potentially, studies of delayed fission will permit detailed measurements of kinetic energy and mass-yield distributions, γ deexcitation of nuclear levels in the second well, structure and height of fission barriers, and probabilities of fission, γ deexcitation, and neutron emission for a host of otherwise unreachable nuclides far from stability.

Fission Isomers, Superdeformation, and Nuclear Structure

An anomalously short (14-ms half-life) spontaneously fissioning species was first reported in 1962 by Polikanov et al (44) and Pereligyn et al (45) to be formed in bombardments of uranium and plutonium targets with ^{16}O and ^{22}Ne ions. Later it was established that it was an isomer of ^{242}Am, and many other spontaneously fissioning isomers with half-lives ranging from 5 ps to 14 ms have since been found in the region from U through Bk. [See the recent review of Poenaru, Ivascu & Mazilu (46).] The main features of these isomers could be explained by Strutinsky's theory (47) of shell corrections to the liquid drop model. They were associated with the second minimum in the now famous plots of deformation energy showing the double-humped fission barrier (see Figure 1). Subsequent experimental evidence confirmed that these fission isomers are indeed "shape" isomers and thus constitute the first known examples of superdeformed nuclei. They are characterized by very large prolate deformations with a major to minor axis ratio of about 2.

Superdeformation in lighter nuclei was first reported in 1986 by Twin et al (48) for ^{152}Dy, in which they observed gamma deexcitation of levels with spins as high as 60. Such superdeformed shapes in the lighter nuclei are found only at very high spins, where the superdeformed shape is stabilized by the centrifugal energy. Although fission is much less probable in these lighter nuclei than in the heavy elements, it will still limit the amount of deformation (angular momentum) that can be accommodated in the nucleus, and will probably restrict the study of the gamma decay of superdeformed states to spins of around 60 as fission quickly becomes the dominant process.

During the delayed fission process, excited states in the fission shape isomer may be populated. It should then be possible to study these fission shape isomers, whose half-lives are only milliseconds or less, in an "out-of beam" environment, where the background is significantly less and chemical separations can be performed to reduce it further. This prospect for studying the level structure of fission isomers is particularly intriguing. For species in which γ decay occurs after penetration of the first barrier and before scission, γ decay to lower-lying nuclear levels can be investigated and information about the level structure (and deformation) of these shape isomers can be obtained by measuring the γ rays in coincidence with the K x rays from the EC decay of the precursor. Gamma rays in coincidence with delayed fission can also be measured. The same high-efficiency, multiple detector arrays, such as HERA (49) and Gammasphere (50), that have been (or will be) developed for the studies of superdeformation in lighter elements can be used, especially in order to detect γ rays in the presence of the large background from the prompt γ emission associated with fission of the isomeric state. It should also be possible to measure β strength functions by summing the energies of all gamma rays and x rays occurring when a K x ray from electron capture of the parent is detected. Decay of the shape isomer and branching ratios for decay back to the first well and for fission can be investigated and will give information about the relative penetrability and heights of the inner and outer barriers. The possible population of K-isomers, which are at higher energies within the second well and decay through K-forbidden γ transitions to the ground state, can also be explored. Use of specially designed multiple detector arrays can be envisioned for either "off-line" or "on-line" studies, depending on the half-lives of the species involved.

Development of Comprehensive Theory and Predictive Models

The challenge of creating a theoretical model for calculating delayed fission probabilities is comparable to that of formulating a comprehensive, dynamic theory of spontaneous fission and the fission process itself. In order to formulate realistic β strength functions and to calculate the probabilities for fission, γ deexcitation, and particle (neutron) emission, it is essential to consider the effects of nuclear deformation. The effects of deformation and excitation on the fission barrier heights must also be considered in a consistent manner as the fission probability is exquisitely sensitive to these effects as well as to the nuclear masses. However, the substantial progress made in developing new models for calculating masses, ground-state and excited-state deformations, and fission barriers gives promise that more accurate calculations of delayed fission probabilities will be available in the not too distant future. These, together with additional experimental results for both ECDF and βDF, should enable us to develop a much better understanding not only of delayed fission probabilities, but of the mechanism of the fission process itself—for which a detailed dynamic theoretical model is not yet available, even more than fifty years after its discovery!

ACKNOWLEDGMENT

The authors are indebted to Drs. Frank S. Stephens and Kenneth E. Gregorich for many helpful conversations. This work was performed in part under the auspices of the US Department of Energy at the Lawrence Livermore National Laboratory under Contract No. W-7405-ENG-48.

Literature Cited

1. Wene, C.-O., Johansson, S. A. E., *Phys. Scripta* 10A:156–62 (1974)
2. Wene, C.-O., *Astron. Astrophys.* 44: 233 (1975)
3. Klapdor, H. V., et al, *Z. Phys.* A299: 213–29 (1981)
4. Meyer, B. S., et al, *Phys. Rev.* C39: 1876–82 (1989)

5. Thielemann, F.-K., Metzinger, J., Klapdor, H. V., *Z. Phys.* A309:301–17 (1983)
6. Burbidge, E. M., et al, *Rev. Mod. Phys.* 29:547 (1957)
7. Kuznetsov, V. I., Skobelev, N. K., Flerov, G. N., *Yad. Fiz.* 4:279–81 (1966) [*Sov. J. Nucl. Phys.* 4:202–3 (1967)]
8. Kuznetsov, V. I., Skobelev, N. K., Flerov, G. N., *Yad. Fiz.* 5:271–73 (1967) [*Sov. J. Nucl. Phys.* 5:191–92 (1967)]
9. Berlovich, E. E., Novikov, Yu. P., *Dok. Akad. Nauk SSSR* 185:1025–27 (1969) [*Sov. Phys. Dokl.* 14:349–50 (1969)]
10. Skobelev, N. K., *Yad. Fiz.* 15:444–47 (1972)[*Sov. J. Nucl. Phys.* 15:249–50 (1972)]
11. Hall, H. L., et al, *Phys. Rev. Lett.* 63: 2548–50 (1989)
12. Gangrskii, Yu. P., et al, *Yad. Fiz.* 27: 894–99 (1978) [*Sov. J. Nucl. Phys.* 27: 475–78 (1978)]
13. Baas-May, A., Kratz, J. V., Trautmann, N., *Z. Phys.* A322:457–62 (1985)
14. Gangrskii, Yu. P., et al, *Yad. Fiz.* 31: 306–17 (1980) [*Sov. J. Nucl. Phys.* 31: 162–68 (1980)]
15. Lazarev, Yu. A., et al, *Europhys. Lett.* 4:893–98 (1987)
16. Habs, D., et al, *Z. Phys.* A285:53–57 (1978)
17. Hall, H. L., Nurmia, M. J., Hoffman, D. C., *Nucl. Instrum. Methods* A276: 649–51 (1989)
18. Hall, H. L., et al, *Phys. Rev.* C39: 1866–75 (1989)
19. Mezilev, K. A., et al, *Z. Phys.* A337: 109 (1990)
20. Combined Radiochemistry Group, *Phys. Rev.* 148:1192 (1966)
21. Hoff, R. W., in *Weak and Electromagnetic Interactions in Nuclei,* ed. H. V. Klapdor. Heidelberg: Springer-Verlag (1986), pp. 207–12
22. Hoff, R. W., *Inst. Phys. Conf. Ser. No. 88; J. Phys. G: Nucl. Phys.* 14(Suppl.):S343–56 (1988)
23. Kreek, S. A., et al, *Lawrence Berkeley Lab. Nucl. Sci. Division 1989–1990 Annual Rep.,* LBL-30798 (1992), p. 59
24. Hall, H. L., et al, *Phys. Rev.* C41:618–30 (1990)
25. Hall, H. L., et al, *Phys. Rev.* C42: 1480–88 (1990)
26. Hingmann, R., et al, Gesellschaft für Schwerionenforschung, Darmstadt, Rep. GSI 85-1 (1985), p. 88
27. Hall, H. L., *Delayed fission properties of neutron-deficient americium nuclei.* PhD thesis, Univ. Calif. (1989)
28. Hall, H. L., et al, see Ref. 23, p. 58
29. Kleinberg, J., ed., *Collected Radiochemical and Geochemical Procedures.* Los Alamos Natl. Lab. Rep. LA-1721, 5th ed. (1990)
30. Hall, H. L., Hoffman, D. C., *J. Radioanal. Nucl. Chem.* 142:53–78 (1990)
31. Somerville, L. P., et al, *Lawrence Berkeley Lab. Nucl. Sci. Division Annual Rep., 1976–1977,* LBL-6575 (1977), p. 39
32. Kuznetsov, V. I., *Yad. Fiz.* 30:321–29 (1979)[*Sov. J. Nucl. Phys.*30:166–71 (1979)]
33. Symbalisty, E. M. D., Schramm, D. N., *Rep. Prog. Phys.* 44:293–328 (1981)
34. Meyer, B. S., et al, *Nuclei Off the Line of Stability,* ACS Symp. Ser. 324. Washington, DC: ACS (1986), pp. 149–52
35. Shalev, S., Rudstam, G., *Nucl. Phys.* A275:76–92 (1977)
36. Kodama, T., Takahashi, K., *Nucl. Phys.* A239:489–510 (1975)
37. Kratz, K. L., Herrmann, G., *Z. Phys.* 263:435–42 (1973)
38. Hornshoj, P., et al, *Nucl. Phys.* A239: 15–28 (1975)
39. Klapdor, H. V., et al, *Z. Phys.* A 292: 249–55 (1979)
40. Hill, D. L., Wheeler, J. A., *Phys. Rev.* 89:1102–45 (1953)
41. Staudt, A., et al, *Z. Phys.* A334:47–57 (1989)
42. Fowler, W. A., *Q. J. R. Astron. Soc.* 28:87 (1987)
43. Casten, R. F., et al, eds., *The IsoSpin Laboratory: Research Opportunities with Radioactive Nuclear Beams,* Brookhaven Natl. Lab. Rep. LALP 91-51 (1992)
44. Polikanov, S. M., et al, *Zh. Eksp. Teor. Fiz.* 42:1464–71 (1962) [*Sov. Phys. JETP* 15:1016–21 (1962)]
45. Pereligyn, V. P., et al, *Zh. Eksp. Teor. Fiz.* 42:1472–74 (1962) [*Sov. Phys. JETP* 15:1022–23 (1962)]
46. Poenaru, D. N., Ivascu, M. S., Mazilu, D., in *Particle Emission from Nuclei,* ed. D. N. Poenaru, M. S. Ivascu. Boca Raton, Fla: CRC (1989), 3:41–61
47. Strutinsky, V. M., *Nucl. Phys.* A95: 420–42 (1967)
48. Twin, P. J., et al, *Phys. Rev. Lett.* 57: 811–14 (1986); Nolan, P. J., Twin, P. J., *Annu. Rev. Nucl. Part. Sci.* 38: 533–62 (1988)
49. Diamond, R. M., see Ref. 34
50. Delaplanque, M.-A., Diamond, R. M., eds., *GAMMASPHERE—A Proposal for a National Gamma Ray Facility,* Lawrence Berkeley Lab. PUB-5202 (1988)

Annu. Rev. Nucl. Part. Sci. 1992. 42:177–210

CLASSICAL SOLUTIONS IN QUANTUM FIELD THEORIES[1]

Erick J. Weinberg

Physics Department, Columbia University, New York, New York 10027; and School of Natural Sciences, Institute for Advanced Study, Princeton, New Jersey 08540

KEY WORDS: soliton, instanton, magnetic monopole, tunneling

CONTENTS

1. INTRODUCTION

Realistic quantum field theories are difficult to solve because they are governed by nonlinear operator equations. In the usual perturbative treatment, one begins with the solution to the linearized (free-field) version of the theory and then incorporates the effects of the interactions as a power series expansion in some small coupling. A complementary approach, in which the nonlinearity of the system is retained at all stages in the calculation, is based on an expansion about solutions of classical field equations; the higher-order corrections in

[1] This work was supported in part by the US Department of Energy, by the Monell Foundation, and by NASA under grant NAGW-2381.

0163–8998/92/1201–0177$02.00

this approach are obtained as a power series in the same small coupling. This approach has led to new insight into the properties of quantum field theories. In this article I give an overview of these methods and describe some of the most important results.

In many cases one finds that the classical field equations have solutions that suggest a particle interpretation. They are localized, with their energy density concentrated within a fairly well-defined region of space. Outside this region, the fields rapidly approach their vacuum values. These solutions are stable and maintain their form as time goes on. Finally, they can be boosted to give linearly moving solutions. These carry linear momentum and display the proper relationships among mass, momentum, and energy.

The existence of these objects, known as solitons,[2] depends crucially on the nonlinear nature of the field equations. This is reflected in their nonanalytic behavior as the coupling constants of the theory approach zero. In particular, the soliton mass typically diverges in this limit, behaving as an inverse of a coupling constant.

The quantum theory contains particle states corresponding to these classical objects. These states are not accessible to ordinary perturbation theory. Nevertheless, they are most easily studied in the weak coupling limit, where the Compton wavelength of the massive soliton is much smaller than the spatial extent of the classical solution. This makes it possible to localize the soliton and allows one to use the classical field configuration as the basis for a description of the internal structure of the particle.

How do we know that a particular theory has soliton solutions? One way, of course, is to obtain analytic solutions to the field equations. However, this is feasible only in a very few cases, most of them idealized models rather than phenomenologically relevant theories. In any case, simply displaying the solution does not explain the physical basis for its existence.

In many theories with spontaneous symmetry breaking, topological arguments can be used to establish the existence of solutions. In these topological solitons, the fields approach different degenerate vacua as one approaches spatial infinity in different directions. These vacua are chosen in such a fashion that they cannot be continuously deformed to a single vacuum. This choice guarantees the stability of the soliton and gives rise to a new type of conserved quantum number, known as the topological charge.

[2] This usage differs from that in other fields, where the term soliton is applied only to localized solutions that maintain their form even after scattering. Except in the sine-Gordon theory, none of the classical solutions encountered in high energy physics are solitons in this more restricted sense.

A second class of solutions, nontopological solitons, are also stabilized by a conserved charge carried by the soliton. However, in this case the charge is of the same kind as that carried by the elementary particles of the theory. For stable solitons to exist, their mass-to-charge ratio must be small enough to prevent decay by emission of these elementary charged particles.

For the solutions discussed above, the quantum interpretation is a straightforward extension of the classical meaning. Such is not the case for instantons, which are solutions of the Euclidean field equations, i.e., the equations obtained by continuing to imaginary time. These are associated with quantum mechanical tunneling. This connection can be motivated by recalling the WKB treatment of one-dimensional barrier penetration, where the exponent in the tunneling amplitude is of the form of an action $\int p\, dq$, but with p given by $\sqrt{2\, m[V(q) - E]}$. The nonstandard signs are just what one would obtain by doing mechanics with imaginary time. The one-dimensional WKB approximation can be extended to systems with many degrees of freedom, where the barrier is in a multidimensional configuration space. The tunneling amplitude is then obtained by considering all possible paths through the barrier and finding the one for which the one-dimensional tunneling probability is the greatest. In a field theory, each point along this path corresponds to a specification of a field configuration over all of three-dimensional space. The path, being a sequence of such configurations, can itself be viewed as a configuration in a four-dimensional space. This configuration turns out to be given by a solution of the Euclidean field equations, namely the instanton. Thus, the instanton is not an object existing in real space, but rather a device for calculating a quantum mechanical amplitude.

Such solutions can be applied to the decay of a classically stable, but quantum mechanically metastable, state. This situation arises in some cosmological scenarios, since the early universe could have been for a time in a "false vacuum" state, corresponding to a local, but not global, minimum of the field potential $V(\phi)$. Such a state would have decayed by the nucleation of bubbles of the true vacuum, with the nucleation occurring (at low temperatures) through quantum mechanical tunneling. The Euclidean solution associated with this process is often referred to as a bounce solution.

In nonrelativistic quantum mechanics, tunneling plays a role in the treatment of systems that have two or more degenerate classical minima separated by potential energy barriers. The standard example is that of a particle in a double well potential where, because of tunneling, the ground state is a linear combination of the ground states of the two wells, and the splitting of the ground state from the first excited state is obtained from the WKB tunneling amplitude. Similar phenomena can also occur in field theory. In particular, non-Abelian gauge theories can be viewed as having multiple vacua separated by finite energy barriers. Tunneling between these vacua, described by the Yang-Mills instanton solution, has a number of important consequences for both QCD and the standard electroweak theory.

Of course, there is more than one way to get to the other side of a potential energy barrier. At high temperature, a system may have enough kinetic energy to go over the barrier without the need for quantum tunneling. In a multidimensional configuration space this process will occur most readily across the point where the barrier is lowest. The saddle point, i.e. the high point on the lowest path over the barrier, is a stationary point of the potential energy and thus a static, although unstable, solution of the field equations. Such solutions have come to be known as sphalerons.

To illustrate some of the features of solitons and the issues involved in going from the classical solution to the quantum theory, I begin in Section 2 by discussing the kink, a soliton in one space dimension. Although not of direct physical interest (except as a model for some cosmological domain walls), this example has the advantage of being simple enough that much of the analysis can be done explicitly. I then go on in Section 3 to discuss more complex solitons, both topological and nontopological. As examples of the former I consider two-dimensional vortices (which find applications in three dimensions as models for magnetic flux tubes and cosmic strings) and three-dimensional magnetic monopoles. In Section 4 I discuss the application of Euclidean solutions to the treatment of tunneling phenomena, and I briefly describe sphalerons.

In a review of this size many aspects of the field must be omitted. Reviews containing a fuller discussion, including further references, of many of the topics covered here include several on magnetic monopoles (1–5), nontopological solitons (6–8), and instantons (9, 10), as well as some covering a broader range of topics (11–16). A recently discovered family of solitons in Chern-Simons theories is reviewed by Jackiw & Pi (17).

2. THE KINK—A ONE-DIMENSIONAL EXAMPLE

A useful illustrative example is provided by a theory containing a single scalar field in one space and one time dimension. While containing many of the features encountered in more physically interesting cases, this model has the advantage that much of the analysis can be done explicitly. The theory is governed by the Lagrangian

$$\mathcal{L} = \frac{1}{2}(\partial_\mu \phi)^2 - \frac{\lambda}{4}\left(\phi^2 - \frac{m^2}{\lambda}\right)^2. \tag{1.}$$

There is a symmetry $\phi \to -\phi$, but this is spontaneously broken by the existence of two degenerate vacua at $\phi = \pm m/\sqrt{\lambda}$. In either of these vacua the elementary particles of the quantum theory have mass $\sqrt{2}m$.

The classical field equation following from this Lagrangian is

$$\ddot{\phi} - \phi'' = \lambda\phi\left(\phi^2 - \frac{m^2}{\lambda}\right), \tag{2.}$$

where dots and primes refer to time and space derivatives, respectively. Of interest to us are solutions that remain localized in space. In principle, these could be static or they could have a periodic or quasiperiodic time dependence. In practice, however, it turns out to be rather difficult to find solutions with any but the simplest time dependence. Restricting ourselves therefore to static solutions, we can drop the $\ddot{\phi}$ term and multiply both sides of Equation 2 by ϕ'. The resulting equation can be rewritten as

$$\frac{\mathrm{d}}{\mathrm{d}x}\left[\frac{1}{2}\phi'^2 - \frac{\lambda}{4}\left(\phi^2 - \frac{m^2}{\lambda}\right)^2\right] = 0, \tag{3.}$$

which implies that the quantity in brackets must be an x-independent constant. Since any finite energy solution has $\phi = \pm m/\sqrt{\lambda}$ at spatial infinity, this constant must vanish. It follows that

$$\phi' = \pm\sqrt{\frac{\lambda}{2}}\left(\phi^2 - \frac{m^2}{\lambda}\right). \tag{4.}$$

This can be integrated to give

$$\phi(x) = \pm\frac{m}{\sqrt{\lambda}}\tanh\left[\frac{m}{\sqrt{2}}(x - a)\right], \tag{5.}$$

where a is a constant of integration.

Taking the upper choice of sign gives the so-called kink solution (18–20). (The solution with the opposite sign may be termed an antikink.) The kink is rather localized in that it deviates from a vacuum solution only within a region of width $\sim m^{-1}$ centered about the point $x = a$; the fact that it approaches different vacua to the left and to the right is irrelevant to any local observations, since the two vacua are indistinguishable. We may think of it as a kind of classical particle—a soliton—whose mass is given by the energy of the classical solution. Integration of the energy density

$$\mathscr{E}(x) = \frac{1}{2}(\phi')^2 + \frac{\lambda}{4}\left(\phi^2 - \frac{m^2}{\lambda}\right)^2 = \frac{m^4}{2\lambda}\,\mathrm{sech}^4\left[\frac{m}{\sqrt{2}}(x-a)\right] \qquad 6.$$

obtained from the Lagrangian in Equation 1 gives the mass

$$M_{\mathrm{cl}} = \frac{2\sqrt{2}}{3}\frac{m^3}{\lambda}\,. \qquad 7.$$

The dependence on the coupling should be noted. The kink mass is of order m^2/λ times greater than that of the elementary excitation of the theory; as the coupling approaches zero, the kink mass becomes infinite. Similarly, the magnitude of the field in the kink solution (measured relative to either one of the two vacua) grows inversely with the square root of the strength of the coupling. Thus, the kink is an essentially nonperturbative phenomenon—it cannot be seen by studying small fluctuations about the vacuum in the weak coupling limit.

The discussion up to this point has been purely classical. The transition to the quantum theory can be investigated by a variety of techniques (18, 20–26). The essential result that one obtains, at least for weak coupling, is that these classical objects survive the quantization process relatively unscathed. Thus, not only does the quantum theory possess particle states corresponding to these solitons, but the classical solution gives a good first approximation to the properties of these particles. The underlying reason is that the Compton wavelength $1/M$ of the kink is much smaller than its classical size $1/m$. Thus, without having to give it a very high energy, the kink can be sufficiently localized that the quantum fluctuations in its position do not smear out the classical solution.

Furthermore, the quantum fluctuations of $\phi(x)$ are small enough that they do not obscure the classical field profile. This statement needs some explanation. In any quantum field theory, the fluctuations in the field at a given point are infinite. However, it is only averages of the field over finite volumes that are actually measurable. It turns out that the quantities $\phi_L(x)$ obtained by averaging the field over a region of width L centered about the point x have fluctuations of the order of $(\ln mL)^{-1/2}$. [In $d > 2$ space-time dimensions, with $L \leq m^{-1}$, the fluctuations would be of order $L^{-(d-2)/2}$.] It is therefore possible to choose

L to be much smaller than the width of the kink and yet still have the magnitude of the fluctuations much less than the overall variation $2m/\sqrt{\lambda}$ of the classical solution.

In the weak coupling limit the quantum corrections to the kink mass can be calculated perturbatively. To a first approximation this can be done by treating the kink as a fixed stationary object, which provides a classical background for the quantum field theory. This is achieved by writing the operator field $\phi(x)$ as

$$\phi(x) = \phi_{\text{kink}}(x) + \psi(x), \tag{8.}$$

where $\phi_{\text{kink}}(x)$ is a c-number field given by Equation 5 (with some fixed value of a) and $\psi(x)$ is an operator field. Substitution of this decomposition into the Hamiltonian gives

$$\begin{aligned} H = {} & \frac{2\sqrt{2}}{3}\frac{m^3}{\lambda} + \int \mathrm{d}x \left(\frac{1}{2}\dot{\psi}^2 + \frac{1}{2}\psi'^2 \right. \\ & \left. + \frac{m^2}{2}\left\{3\tanh^2\left[\frac{m}{\sqrt{2}}(x-a)\right] - 1\right\}\psi^2 \right) \\ & + \int \mathrm{d}x \left\{ m\sqrt{\lambda}\tanh\left[\frac{m}{\sqrt{2}}(x-a)\right]\psi^3 + \frac{\lambda}{4}\psi^4 \right\}. \end{aligned} \tag{9.}$$

The first term, independent of ψ, is just the classical kink mass. The terms cubic and quartic in ψ are suppressed by factors of $\sqrt{\lambda}$ and λ, respectively, and may be treated as perturbations. This leaves a quadratic Hamiltonian that can be rewritten as a sum of harmonic oscillator Hamiltonians by expanding $\psi(x)$ in terms of normal modes, i.e. the solutions of

$$\left(-\frac{\mathrm{d}^2}{\mathrm{d}x^2} + m^2\left\{3\tanh^2\left[\frac{m}{\sqrt{2}}(x-a)\right] - 1\right\}\right)\psi_j = \omega_j^2\,\psi_j(x). \tag{10.}$$

These modes can all be found explicitly. There are two normalizable modes, with frequencies 0 and $\sqrt{3/2}\,m$, and a continuum of nonnormalizable modes beginning at $\omega = \sqrt{2}\,m$.

The ground state of this Hamiltonian, in which each of the normal modes is in its lowest state, corresponds to the kink. In addition to the classical kink energy, there is a contribution of the form $\frac{1}{2}\sum \omega_j$ from the zero-point energies of the oscillators. This sum (or more properly, integral, because of the continuous part of the spectrum) is divergent. However, two additional effects need to be included. First, since the kink mass should be measured relative to the vacuum, we must subtract a similar sum containing the frequencies ω_j^{vac} of the normal modes about the classical vacuum. Second, there is a contribution from the counter terms needed to cancel the divergences of the quantum theory. The mass counter term δm^2 arises at order λ in ordinary perturbation theory. However, because the kink field is itself of order $1/\sqrt{\lambda}$, this term also contributes to the lowest-order quantum correction. Adding (with appropriate regularization) these three divergent terms gives

$$\begin{aligned} M_{\text{kink}} &= M_{\text{cl}} + \frac{1}{2}\sum_j \omega_j - \frac{1}{2}\sum_j \omega_j^{\text{vac}} + \int \mathrm{d}x\, \delta m^2 \phi_{\text{kink}}^2(x) + O(\lambda) \\ &= \frac{2\sqrt{2}}{3}\frac{m^3}{\lambda} + c_0 + O(\lambda), \end{aligned} \qquad 11.$$

where c_0 is finite, of order unity, and calculable (18). The term of order λ has also been calculated (27) and the calculation can, at least in principle (28, 29), be continued to arbitrary order in λ.

The excited states of the Hamiltonian are obtained by giving nonzero occupation numbers to some of the normal modes. Physically, they correspond to a kink plus a number of elementary bosons, with the states of the latter determined by which of the normal modes are occupied. The continuum modes approach plane waves far from the kink, and they correspond to unbound bosons scattering off the kink. The discrete mode with $\omega = \sqrt{3/2}\, m$ corresponds to a state with an elementary boson bound to the kink.

The zero-frequency mode, however, is quite different. This mode arises because the kink breaks the translational symmetry of the Lagrangian. It is given explicitly by

$$\begin{aligned} \psi_0(x) &= N \frac{m^2}{\sqrt{2\lambda}} \operatorname{sech}^2\left[\frac{m}{\sqrt{2}}(x - a)\right] \\ &= N \frac{\mathrm{d}\phi_{\text{kink}}(x)}{\mathrm{d}x}, \end{aligned} \qquad 12.$$

where N is a normalization constant. An infinitesimal deformation of the kink of the form $\phi_{\text{kink}} \rightarrow \phi_{\text{kink}} + \epsilon\psi_0$ is equivalent to a displacement of the kink to the right by an amount ϵN. The existence of the zero mode reflects the fact that such a displacement leaves the energy unchanged.

This mode must be treated differently from the modes with nonzero frequencies since, after all, a zero-frequency harmonic oscillator is not really an oscillator. To treat this mode, one introduces a "collective coordinate" associated with translation of the kink (23, 28, 30). Essentially, this amounts to promoting the integration constant a to a time-dependent dynamical variable. The corresponding conjugate momentum P is determined by the way in which $\dot{a}$ enters the Lagrangian. Equation 1 gives

$$L = \frac{1}{2}\dot{a}^2 \int \mathrm{d}x \left(\frac{\mathrm{d}\phi_{\mathrm{kink}}}{\mathrm{d}x}\right)^2 + \cdots$$
$$= \frac{1}{2} M_{\mathrm{cl}}\dot{a}^2 + \cdots . \qquad 13.$$

Here the ellipses represent terms that do not contain $\dot{a}$, as well as higher-order effects arising from the possible deviation of $\phi(x)$ from $\phi_{\mathrm{kink}}(x)$ due to the excitation of the nonzero-frequency modes. The integral on the first line of Equation 13 is precisely equal to the classical kink mass (an explanation for this apparent coincidence is given in the next section), so to lowest order $P = M_{\mathrm{cl}}\dot{a}$, and the corresponding contribution to the Hamiltonian is $P^2/(2M_{\mathrm{cl}})$. At higher orders in perturbation theory, the M_{cl} in the denominator is replaced by the expansion (Equation 11) for the exact mass M_{kink} and the relativistic corrections to the kinetic energy begin to appear.

3. SOLITONS IN MORE THAN ONE SPATIAL DIMENSION

Solitons in more spatial dimensions present no issues of interpretation beyond those encountered in the study of the kink. The difficulty is in actually finding solutions. Even for static solutions the field equations are partial differential equations, and hence much harder to solve. Further, there is a result known as Derrick's theorem that forbids static solitons in scalar field theories with Lagrangians of the standard form (31). This result follows from the fact that a static solution of the field equations must be a stationary point of the energy function in n spatial dimensions,

$$E[\phi(x)] = \int \mathrm{d}^n x[(\partial_i\phi)^2 + V(\phi)]$$
$$\equiv T[\phi(x)] + U[\phi(x)]. \qquad 14.$$

Here $V(\phi)$ is understood to vanish at its minimum. In particular, let us assume that $\phi(x)$ is such a solution, and consider the family of field configurations $\tilde{\phi}_\beta(x) = \phi(\beta x)$ obtained by rescaling its length scale. If the original configuration is indeed a solution, then the energy functional

$$E[\tilde{\phi}_\beta(x)] = \beta^{n-2}T[\phi(x)] + \beta^n U[\phi(x)] \qquad 15.$$

must be stationary at $\beta = 1$. This implies that

$$T[\phi(x)] = \frac{n}{2 - n} U[\phi(x)]. \tag{16.}$$

For $n = 1$ we have $T = U = E/2$, thus explaining the apparently fortuitous equality, found in Equation 13, between $P/\dot{a}$ and the kink mass. However, if $n \geq 2$ this relation cannot be satisfied with T and U both finite and positive, and hence the assumed solution cannot exist. Derrick's theorem is easily extended to the case of many scalar fields.

To find solutions, then, we must either introduce time dependence or else go to theories with a more complicated structure. In either case the task of solving the field equations becomes more difficult. Rather than a brute force approach, one needs an understanding of the mechanisms that can give rise to particle-like solutions. Two broad classes of solutions, topological and nontopological solitons, have been found. In both cases the existence and stability of the soliton can be traced to a conserved charge that it carries.

3.1 *Topological Solitons*

One class of solitons is based on the possibility of topologically nontrivial field configurations. An example of this is the kink solution studied in the previous section. The theory described by Equation 1 has two degenerate vacua, $\phi = \pm m/\sqrt{\lambda} \equiv v$. In any configuration of finite energy, $\phi(x)$ must approach one or the other of these vacua as $x \to \pm\infty$. One can therefore divide all such configurations into four classes, with $(x(-\infty), x(\infty))$ equal to (v, v), $(-v, -v)$, $(-v, v)$, and $(v, -v)$. While any two configurations within the same class can be smoothly deformed one into the other, it is not possible to go continuously from one class to another without passing through configurations of infinite energy. Now recall that any local minimum of the energy functional gives a solution of the static field equations. It should therefore be possible to obtain static solutions by minimizing the energy within each of the four classes of configurations. The minima within the first two classes are just the two vacua. For the last two classes the minima clearly cannot be vacuum solutions, and so must be solitons; they are the kink and the antikink, respectively. Note the power of this argument: even if we had not been able to obtain an analytic expression for the kink solution, we would still be assured that a soliton did in fact exist.

In this theory there can be multisoliton configurations containing a number of kinks and antikinks, although these will not be static solutions. By annihilation of kink-antikink pairs, such a configuration can evolve into one containing either a single kink, a single antikink, or no solitons at all, according to whether the number of kinks less the number of antikinks is 1, -1, or 0. (Because kinks and antikinks necessarily alternate in position, these are the only possible values.) This can be stated more formally by defining a topological charge $Q = [\phi(\infty) - \phi(-\infty)]/(2v)$, so that kinks and antikinks have $Q = 1$ and $Q = -1$, respectively. From the above remarks it is evident that Q is conserved, and that the topological charge of a multisoliton configuration is the sum of the charges of the individual solitons.

These ideas are readily extended to more spatial dimensions. As an illustration, I discuss below a two-dimensional example, the vortex solution, and then describe the three-dimensional magnetic monopole solutions. A somewhat different type of topological soliton, the skyrmion, is discussed elsewhere in this volume (32).

3.1.1 THE VORTEX—A SOLITON IN TWO DIMENSIONS A two-dimensional topological soliton (33) occurs in the Abelian Higgs model, in which a U(1) gauge symmetry is broken by a complex Higgs field ϕ. Instead of the two degenerate minima of the theory of Equation 1, the Higgs potential now has a continuous family of minima, given by $\phi = ve^{i\alpha}$. This is just what we need, since in two dimensions spatial infinity is not just the two points $x = \pm\infty$, but rather an infinite set of points that may be viewed as a circle at $r = \infty$. A configuration can be classified topologically by the behavior of ϕ around this circle. Thus, let $\phi(r = \infty, \theta) = ve^{i\alpha(\theta)}$. In order that the field be continuous, the net change in the phase α as θ varies from 0 to 2π must be equal to 2π times some integer n, which may be defined to be the topological charge of the configuration. This charge acquires further significance from the observation that for the energy to be finite the covariant derivative $D_i\phi = (\partial_i - ieA_i)\phi$ must fall sufficiently rapidly at large distance. Because the θ dependence of ϕ implies that $\partial_i\phi$ falls only as $1/r$, there must be a nonzero vector potential $A_i = -(i/e)\partial_i(\ln \phi)$ at large r. Although this potential, being a pure gradient, gives a vanishing magnetic field F_{12} at large distance, the application of Stokes' theorem shows that the total magnetic flux $\Phi = \int d^2x\, F_{12}$ must be equal to $2\pi n/e$.

To obtain a soliton, we look for the configuration of minimum energy among those with unit topological charge; by appropriate choice of gauge, this configuration can be taken to have $\phi(r = \infty, \theta) = ve^{i\theta}$. In contrast to the kink, this solution cannot be obtained analytically. Instead, the field equations must be solved numerically. One finds that the solution is centered about a point where $\phi(x)$ vanishes. (The existence of such a point is a consequence of the boundary conditions at spatial infinity.) In a region of radius $\sim 1/(ev)$ about this point, A_i is not simply a gradient, and gives the required nonzero magnetic flux. This solution is known as the vortex.

By adding one more spatial dimension, z, and taking the fields to be z independent, one can turn this two-dimensional particle-like solution into a three-dimensional string-like solution with magnetic flux flowing along the string. Recall that the Abelian Higgs model in three space dimensions is essentially the same as the Landau-Ginzburg model for superconductivity, with the symmetric and asymmetric vacua corresponding to the normal and superconducting phases, respectively. Then we can see that the vortex solution gives a model for magnetic flux tubes in a superconductor, with the quantization of the topological charge being equivalent to the quantization of the magnetic flux.

An antivortex solution, with topological charge -1, can be obtained from the vortex by the substitutions $\phi(x) \rightarrow \phi^*(x)$ and $A_i(x) \rightarrow -A_i(x)$. One can also seek solutions with $n > 1$. Topological arguments alone cannot determine whether these exist, since the configuration that minimizes the energy could well be n widely separated unit vortices. It turns out that rotationally invariant solutions with topological charge $n > 1$ do exist, but their stability depends on the details of the Higgs potential. In terms of the superconductivity analogy, stable higher-charged vortices exist if the superconductor is Type I, but not if it is Type II.

In order to extend these arguments to other two-dimensional theories, and to motivate the generalization to the three-dimensional case, it is helpful to formulate them in more general terms.[3] Let us define two loops in a manifold M to be equivalent if one loop can be continuously deformed into the other. On a simply connected manifold, where any loop can be continuously shrunk to a point, all loops are equivalent. If the manifold is not simply connected, there are a number of equivalence classes, which are the elements of the first homotopy group, $\Pi_1(M)$. Multiplication in this group corresponds to tracing out one loop after the other, while the identity element corresponds to the trivial loop containing a single point.

In a theory in which a symmetry group G is spontaneously broken to a subgroup H, the values of the scalar field that minimize the potential form a manifold M, which may be identified with the quotient group G/H. As one goes around a loop at spatial infinity, the fields $\phi(r = \infty, \theta)$ trace out a loop in M. The topological charge of the con-

[3] For a fuller treatment of the topological methods used in this section, see Refs. (1, 4, 13, 16).

figuration can be identified with the corresponding element of $\Pi_1(M)$. Topologically stable solitons exist whenever $\Pi_1(M)$ is nontrivial.

Thus, in the Abelian Higgs model the manifold M is topologically equivalent to a circle, S^1. The fundamental group is $\Pi_1(M) = \Pi_1(S^1) = Z$, the additive group of the integers. This reflects both the quantization of the topological charge and the fact that the topological charge of a multivortex configuration is the sum of the charges of its component vortices. Other possibilities arise in other gauge groups. For example, one can find theories in which $\Pi_1(M) = Z_2$, the group formed by addition modulo two. In these theories, topological charge is added modulo two, so that a two-vortex configuration has the same topological charge as the vacuum. Thus, two unit vortices can annihilate one another.

3.1.2 THE SU(2) MAGNETIC MONOPOLE The ideas discussed above are readily generalized to three dimensions, where spatial infinity can be viewed as a two-sphere, S^2. Instead of tracing out a loop in M, the scalar fields at infinity map out a closed two-dimensional surface in M. Instead of $\Pi_1(M)$, the relevant quantity is now the second homotopy group, $\Pi_2(M)$, which is the group of equivalence classes of maps from S^2 to M. Topological solitons exist in theories in which $\Pi_2(M)$ is nontrivial.

The simplest example (19, 34) occurs in an SU(2) gauge theory with the symmetry spontaneously broken to U(1) by a triplet Higgs field; I will describe this U(1) with the language of electromagnetism. The Lagrangian is

$$\mathcal{L} = \frac{1}{2}(D_\mu\phi)^2 - \frac{1}{4}F_{\mu\nu}^2 - \frac{\lambda}{4}(\phi^2 - v^2)^2. \qquad 17.$$

The elementary particles of the theory include a massless photon, two vectors with mass $m_V = ev$ and charges $\pm e$, and a neutral scalar with mass $m_S = \sqrt{2\lambda}v$.

For this theory the manifold M consists of the set of all isovectors with length v and is topologically equivalent to a two-sphere. Because $\Pi_2(S^2) = Z$, the theory should have solitons carrying an additive topological charge, which (with a suitable normalization) can take on any integer value.

The solution with unit topological charge can be obtained by requiring that at large distances the Higgs field approach the "hedgehog" configuration $\phi^a(r) = vr^a/r \equiv v\hat{r}^a$ in which the direction of ϕ in the internal symmetry space is correlated with the angle in physical space. [Superscripts on fields are SU(2) indices, while subscripts denote spatial components.] Because of the changing direction of the Higgs field, $\partial_i\phi$ falls only as $1/r$ at large distances. To have finite energy, the covariant derivative $D_i\phi$ must fall more rapidly. As with the vortex, this is achieved by having a suitable long-range vector potential. Here, the appropriate choice is

$$A_i^a \sim \epsilon_{aij} \frac{\hat{r}^j}{er}, \qquad 18.$$

which gives a field strength

$$F_{ij}^a \sim \epsilon_{ijk}\, \hat{r}^a \hat{r}^k \frac{1}{er^2}. \qquad 19.$$

This field is parallel to ϕ in internal space and thus should be interpreted as purely electromagnetic. It is in fact the Coulomb magnetic field corresponding to a magnetic monopole with magnetic charge $Q_M = 1/e$. Similarly, one can show that any configuration with topological charge n carries magnetic charge $Q_M = n/e$. [There, however, are no static solutions with multiple magnetic charge, except in the mathematically interesting, but unphysical, limit of vanishing scalar meson mass (35, 36).]

The full monopole solution can be obtained by multiplying the asymptotic forms of the fields by functions of r:

$$\phi^a = \hat{r}^a v h(r) \qquad 20.$$

$$A_i^a = \epsilon_{aij}\, \hat{r}^j \frac{1 - u(r)}{er}. \qquad 21.$$

Substituting this ansatz into the field equations, one obtains two coupled differential equations. The boundary conditions are that $h(\infty) = 1$ and $u(\infty) = 0$ (to agree with the presumed asymptotic behavior) and that $h(0) = 0$ and $u(0) = 1$ (so that the fields are nonsingular at the origin).

Although these equations can only be solved numerically, a rough qualitative picture yields some of the essential features of the solution. In this picture the monopole is viewed as having a core of radius R_{core} in which $h \neq 1$ and $u \neq 0$, with the fields having their asymptotic form for $r > R_{\text{core}}$. The mass of the monopole can then be divided into a contribution from the core and one from the Coulomb magnetic field outside the core. If the energy density inside the core is approximated by a constant ρ_0, we obtain

$$M_{\text{mon}} \approx \frac{4\pi}{3} \rho_0 R_{\text{core}}^3 + \frac{2\pi}{e^2 R_{\text{core}}}. \qquad 22.$$

Minimizing with respect to R_{core} leads to $M_{\text{mon}} \approx 8\pi/(3e^2R_{\text{core}})$. The only distance scales in the problem are the Compton wavelengths of the massive elementary particles. If these are roughly equal, we would expect $R_{\text{core}} \sim 1/m_{\text{V}} = 1/(ev)$, and hence $M_{\text{mon}} \approx 8\pi v/(3e)$. By comparison, numerical solution of the differential equations gives $M_{\text{mon}} = 4\pi vC/e$, where C ranges from 1 to 1.787 as λ/e^2 varies from 0 to ∞ (37).

Further understanding of the structure of the monopole can be gained by applying a singular gauge transformation to make the orientation of ϕ uniform. The fields can then be written in the form

$$\phi^a = \delta^{a3}vh(r) \qquad 23.$$

$$W_j \equiv A_j^1 + iA_j^2 = f_j(\theta, \phi)\,\frac{u(r)}{er} \qquad 24.$$

$$A_j^3 = \epsilon_{ij3}r_i\,\frac{1 - \cos\theta}{er^2\sin^2\theta}\,. \qquad 25.$$

In this gauge the unbroken U(1) defined by ϕ corresponds to the 3-direction in internal space. The fields A_j^1 and A_j^2 correspond to the massive charged vector bosons of the theory; to emphasize this, they have been combined into a single complex vector field. The electromagnetic vector potential A_j^3 is singular along the negative z axis; this singularity is precisely the Dirac string singularity of a U(1) magnetic monopole (38). Within the U(1) theory this string is unavoidable, but it is unobservable as long as $2qQ_{\text{M}}$ is an integer for any possible electric charge q; the remarkable effect of embedding the U(1) in a larger gauge group is that the string can be eliminated completely.[4]

In this gauge it is easy to show that a new solution can be obtained by a phase rotation of the form $W_j(x) \rightarrow e^{i\alpha}W_j(x)$. This leads to a zero-frequency mode when the quantum corrections to the monopole are calculated. As with the translational zero modes, this zero-frequency mode must be treated by introducing a collective coordinate, which in this case is a periodic variable specifying the overall phase of the solution (39–41). The momentum conjugate to this variable is proportional to the electric charge Q. Because the overall phase is a periodic variable, its conjugate momentum is quantized; a detailed calculation shows the unit of charge to be e. Just as the linear momentum gives a contribution $P^2/(2M)$ to the energy, the charge gives an additional energy of the form $Q^2/(2I)$. Here I, which can be thought of as a moment

[4] Since all the charged elementary particles of the theory have unit electric charge, one might have expected a monopole with $Q_{\text{M}} = 1/(2e)$. However, we could have included in the model an isospinor field, whose particles would have charges $\pm e/2$; this would require that the minimum magnetic charge be $1/e$, which is indeed what we find.

of inertia in internal space, is obtained from a spatial integral of $| W_j |^2$; it is of the order of $1/(ev)$. An energy of a similar magnitude is associated with the Coulomb electric field of the dyon. Thus, built upon the monopole is a series of dyons carrying both electric and magnetic charge. Their masses, like that of the monopole, are of order v/e, while the mass splitting between successive dyons is of order e^3v.

Although the dyons appear here as time-dependent solutions, they can be made time independent by an appropriate gauge transformation; it is in this form that they were first found (42).

3.1.3 MONOPOLES IN LARGER GAUGE GROUPS Monopole solutions can also occur in gauge theories with larger gauge groups and with a variety of scalar field representations. The only requirement is that the full gauge group G and the unbroken subgroup H be such that $\Pi_2(M) = \Pi_2(G/H)$ is nontrivial. In particular, it can be shown that if G is simple or semisimple then $\Pi_2(G/H) = \Pi_1(H)$, and hence that monopoles exist if H is not simply connected.[5] This applies in particular to any grand unified theory. By its very definition, such a theory has a simple gauge group G. To agree with experiment, the unbroken gauge group must be the SU(3) $\times$ U(1) of the strong and electromagnetic interactions; because of the U(1) factor, the gauge group is not simply connected, and $\Pi_2(G/H) = \Pi_1[\mathrm{SU}(3) \times \mathrm{U}(1)] = Z$. Thus, any grand unified theory must contain magnetic monopoles.

The mass of these monopoles is determined by the symmetry-breaking scale at which a nontrivial Π_2 first appears. In the simple SU(5) model an adjoint Higgs field ϕ with $\langle\phi\rangle = v_{\mathrm{GUT}} \sim 10^{15}$–$10^{16}$ GeV breaks the symmetry to SU(3) $\times$ SU(2) $\times$ U(1). It is further broken to SU(3) $\times$ U(1) by a second scalar field χ, in the fundamental representation, with $\langle x\rangle = v_{\mathrm{EW}} \sim 250$ GeV. The first level of symmetry breaking gives rise to monopoles (43) with unit magnetic charge $1/e$ and mass $\sim v_{\mathrm{GUT}}/e$. Spherically symmetric solutions with two and three times the unit charge also exist, but are unstable against dissociation into unit monopoles (44); stable but less symmetric solutions with up to six units of magnetic charge occur for certain ranges of parameters (45).

Other possibilities arise in more complicated models. For example, there are SO(10) models in which the symmetry is first broken to SO(6) $\times$ SO(4) by a scalar field ϕ_1 at a scale v_{GUT} (46, 47). An explicit U(1) factor appears only at a subsequent stage of symmetry breaking, when a second field ϕ_2 acquires a vacuum expectation value $v_2 \ll v_{\mathrm{GUT}}$. The first symmetry breaking gives rise to monopoles of mass $\sim v_{\mathrm{GUT}}/e$.

[5] A technical point: this result assumes that G is simply connected. This requirement can always be satisfied by taking G to be the covering group of the Lie algebra.

Because $\Pi_2[SO(10)/SO(6) \times SO(4)] = Z_2$, these monopoles would have a Z_2 topological charge, with monopoles and antimonopoles equivalent, if the SO(6) × SO(4) symmetry remained unbroken. When the U(1) factor appears at the scale v_2, the homotopy group Π_2 is enlarged to Z, and the previous Z_2 monopoles acquire an ordinary magnetic charge of magnitude $1/e$; for these monopoles both ϕ_1 and ϕ_2 twist in a topologically nontrivial manner. However, a second type of monopole, in which only ϕ_2 twists, also occurs. These monopoles carry magnetic charge $2/e$ and have a much smaller mass, $\sim v_2/e$.

Since magnetic monopoles are one of the definite predictions of grand unification, it is of considerable interest to know if any actually exist. Because of their great mass, they cannot be produced in any conceivable accelerator. However, the energies required were available in the very early universe. Indeed, fairly straightforward arguments based on standard cosmology suggest that not only would monopoles have been produced at early times, but that enough would have survived to the present to far exceed the rather stringent upper bounds on the present-day monopole abundance (48, 49). One of the motivations of the inflationary universe scenario (50) was to provide a solution to this primordial monopole problem. For a discussion of other approaches to the problem, and of the astrophysical and observational bounds on the monopole abundance, see the reviews by Preskill (4) and Giacomelli (5).

3.1.4 ADDING FERMIONS—THE CALLAN-RUBAKOV EFFECT The coupling of fermions to a magnetic monopole leads to a number of unusual phenomena, including, for example, the existence of objects with fractional fermion number (51). Perhaps the most important of these effects is the Callan-Rubakov effect (52–56) by which baryon number conservation is violated in the scattering of fermions by certain types of magnetic monopoles.

Angular momentum considerations give the first hint that scattering by a monopole might have unusual properties. A system containing both a particle with electric charge e and one with magnetic charge g has, in addition to any contributions from orbital motion or spin, an angular momentum of magnitude eg directed along the line from the electrically charged particle to the magnetically charged one. A classical electric charge moving directly toward the center of a magnetic monopole could not pass through to the other side, because to do so would require a sudden reversal of this angular momentum. The quantum mechanical analogue of such a radial trajectory is s-wave scattering by a monopole, and indeed, examination of the s-wave scattering states reveals a mismatch between incoming and outgoing modes (57, 58). This is seen, for example, in the solutions of the Dirac equation for a massless isodoublet fermion in the background of the SU(2). The incoming solutions with vanishing total angular momentum and fermion number 1 are either left-handed with positive charge or right-handed with negative charge, while the outgoing states are either left-handed with negative charge or right-handed with positive charge; the charges are reversed for the modes with fermion number -1. It is evident, then, that one of the conserved quantum numbers of the fermion must change. A first guess might be that the fermion would change its electric charge, with the monopole compensating for the change by becoming a dyon. However, the mass splitting between the dyon and monopole is too great to allow such charge transfer in low energy scattering. Instead, the issue is resolved by the effects (59, 60) of the triangle anomaly (61, 62). Recall that the chiral current, whose conservation would appear to be guaranteed by the symmetry of the Lagrangian, acquires a nonzero divergence through one-loop quantum effects. For the case at hand this divergence is proportional to $\mathbf{E} \cdot \mathbf{B}$, the scalar product of the electric and magnetic fields. In the presence of the classical magnetic field of a monopole, the quantum fluctuations in the electric field can generate such a term and lead to a change in the total chiral charge. The net effect is that in the scattering of a massless fermion by the monopole there is a large amplitude for chirality nonconservation.

In more complicated theories, analogous processes lead to nonconservation of other anomalous charges. In particular, violation of baryon number conservation can occur in grand unified theories. Of course, these theories have baryon number violation mediated by superheavy gauge bosons, even in the absence of monopoles. However, in ordinary low energy scattering of nucleons this violation is suppressed by a factor of $(E/M_{\mathrm{GUT}})^4$ and is thus essentially unobservable. No such factor enters in the scattering of a nucleon by a monopole, and so the amplitude for baryon number violation in such scattering can be large (52–56). Although a precise calculation has not been performed, it is estimated that the cross section for baryon-number-changing nucleon-monopole scattering is essentially geometrical in size, with $\sigma_{\Delta \mathrm{B}} \sim 1/E^2$, where E is the energy of the nucleon (for a review, see 63).

3.2 *Nontopological Solitons*

A second type of soliton arises in theories in which an unbroken symmetry gives rise to a conserved charge Q. These nontopological solitons are localized solutions with nonzero charge. Because this charge is of the same type as that carried by the elementary excitations of the theory, there is the possibility that it might be lost through emission of charged elementary particles. The stability of the soliton depends on whether or not such emission is energetically allowed and is therefore sensitive to the values of the parameters of the theory. In contrast with topological solitons, which in most cases occur only for a few low values of the topological charge, nontopological solitons (in three or more space dimensions) typically exist only if Q is greater than some minimum charge and often have no upper limit on their charge or mass. [The last property raises the possibility of solitons of truly astronomical size—soliton stars (64–68).]

A wide variety of nontopological solitons have been found (for some early examples, see 69–73). Perhaps the simplest example (69, 74) occurs in a theory involving a single complex scalar field ϕ in three space dimensions. The Lagrangian is of the form

$$\mathcal{L} = \frac{1}{2} | \partial_\mu \phi |^2 - V(| \phi |), \qquad 26.$$

where V reaches its minimum at $\phi = 0$, so that the symmetry is unbroken; it is convenient to set $V(0) = 0$. This Lagrangian is invariant under the transformation $\phi(x) \rightarrow e^{i\alpha}\phi(x)$, with the corresponding conserved charge

$$Q = \int d^3x \, \mathrm{Im}(\phi^* \dot{\phi}), \qquad 27.$$

where the dot signifies a time derivative. At the classical level, this charge can take on any value. However, in the quantum theory Q is quantized (see the discussion of dyons above) and only takes on integer values.

If the solitons are to possess this charge, they clearly cannot be static. However, it is not hard to show that for fixed charge Q the solution that minimizes the energy has the quasistatic form

$$\phi(\mathbf{x}, t) = f(\mathbf{x}) e^{i\omega t} \qquad 28.$$

with $f = | \phi |$ real. For such solutions, we find

$$Q = \omega \int d^3x \, | \phi |^2 \qquad 29.$$

and so the energy can be written as

$$E = \int d^3x \left[\frac{1}{2} (| \nabla\phi |)^2 + V(| \phi |) \right] + \frac{Q^2}{2 \int d^3x \, | \phi |^2} . \qquad 30.$$

For these quasistatic configurations the field equation reduces to

$$\frac{d^2\phi}{dr^2} + \frac{2}{r}\frac{d\phi}{dr} = \frac{d}{d\phi}\left(V - \frac{1}{2}\omega^2 \mid \phi \mid^2\right), \qquad 31.$$

where the last term arises from $\ddot{\phi} = -\omega^2\phi$. This is just the equation for a static soliton in a theory with scalar field potential

$$\hat{V} \equiv V - (1/2)\omega^2 \mid \phi \mid^2. \qquad 32.$$

At large r the soliton must approach the vacuum solution $\phi = 0$. If it is to be stable, this value must be a minimum of $\hat{V}$, which implies that $\omega^2 < m^2 = (d^2V/d\phi^2)_{\phi=0}$, where m is the mass of the elementary charged particles of the theory. Furthermore, Derrick's theorem shows that such a soliton is possible only if $\hat{V}$ is negative for some values of ϕ, implying that $\hat{V}$ must have a second, deeper, minimum at some nonzero value of ϕ. In the region where $\hat{V}$ is negative we have $(2V/\mid \phi \mid^2) < \omega^2 < m^2 = (2V/\mid \phi \mid^2)_{\phi=0}$. Thus, the function $(2V/\mid \phi \mid^2)$ must achieve its minimum value ν^2 at some nonzero value of ϕ. It turns out (74, 75) that the existence of such a minimum is essentially all that is needed for the existence of nontopological solitons in this model, and that solutions exist for all values of ω in the range $\nu < \mid \omega \mid < m$.

Matters become particularly simple in the limit of large Q. In this case the soliton is a sphere of radius R, inside of which $\mid \phi \mid$ has some constant value ϕ_0, surrounded by a surface region of thickness $\delta \sim m^{-1} \ll R$ in which ϕ goes to its vacuum value $\phi = 0$. There is a uniform charge density $\omega\phi_0^2$ in the interior of the soliton, giving a total charge

$$Q = \frac{4\pi}{3}\omega\phi_0^2 R^3, \qquad 33.$$

which I assume here to be positive. R and ϕ_0 are determined by minimizing the energy (Equation 30) with Q held fixed. For sufficiently large R (and Q) the contribution to the energy from the surface region can be neglected relative to that from the interior, and we can write

$$E = \frac{4\pi}{3}V(\phi_0)R^3 + \frac{3}{8\pi}\frac{Q^2}{\phi_0^2 R^3} + \cdots, \qquad 34.$$

where the ellipses represent terms that can be ignored in the limit of large R. Minimizing with respect to R, with ϕ_0 held fixed, gives

$$R = \left(\frac{3}{4\pi}\right)^{1/3}[2V(\phi_0)\phi_0^2]^{-1/6}Q^{1/3} \qquad 35.$$

and hence

$$E = \sqrt{\frac{2V(\phi_0)}{\phi_0^2}}\, Q + \cdots . \qquad 36.$$

For E to be a minimum, ϕ_0 must be the value of the field at which $2V/|\phi^2|$ achieves its minimum value ν^2, and hence

$$E = \nu\,|\,Q\,| + \cdots . \qquad 37.$$

The stability condition $\nu < m$ noted above can now be seen as the requirement that the mass-to-charge ratio of the soliton be less than that of the elementary particles of the theory. From Equations 33 and 35 we now find that $\omega = \sqrt{2V(\phi_0)/\phi_0^2} = \nu$. Thus, in the large Q limit the fields in the bubble interior are independent of Q; solitons of this sort have been termed Q-balls (74).

We can now go back and include the effects of the surface energy, which gives a contribution of the form σR^2 to Equation 34. To leading order, this gives a correction of order $Q^{2/3}$ to the energy, and increases ω^2 above ν^2. A more detailed analysis reveals that ω increases as Q is decreased. Since ω is bounded from above by m, this leads to the existence of a minimum charge $Q_{\rm min}$ below which the soliton solution ceases to exist. Further, the soliton is stable only if its charge is greater than a value $Q_{\rm stab} > Q_{\rm min}$.

Nontopological solitons need not be Q-balls. It is possible to construct solutions whose energy, in contrast to that of a Q-ball, grows less than linearly with charge; e.g. in the theory of a charged scalar coupled to a neutral scalar with a broken discrete symmetry (76). A similar phenomenon occurs for bosonic field configurations that are stabilized by coupling to a fermionic field carrying a conserved charge; such objects (77–79), which because of the presence of the fermionic field are not truly classical solutions, are related to the bag models (80, 81) for hadrons. Finally, there are Q-balls with a massless gauge field coupled to the soliton charge (82); in this case the Coulomb energy places an upper limit on Q.

4. EUCLIDEAN SOLUTIONS AND BARRIER PENETRATION

One can also find localized classical solutions in Euclidean space-time. These solutions do not correspond to particles, as do their Minkowskian counterparts, but are instead related to quantum mechanical tunneling. The starting point for this connection is the WKB formula for the tunneling amplitude through a one-dimensional barrier:

$$\mathcal{A}_{\mathrm{WKB}} \sim \exp\left\{-\int_{x_1}^{x_2} \mathrm{d}x \sqrt{2m[V(x) - E]}\right\}, \qquad 38.$$

where the integral ranges over the entire classically forbidden region. To generalize this result to a system with more than one degree of freedom (83, 84), one considers all possible paths through the multi-dimensional barrier, calculates a tunneling probability for each path using the one-dimensional formula, and then maximizes this amplitude to find the most probable path. The leading approximation to the tunneling amplitude is given by the one-dimensional integral along the most probable path. Thus, for a system of N particles, all with mass m, with coordinates $q_1, q_2, \cdots, q_{3N}$, one must find the path $q_j(s)$ through configuration space that minimizes the integral

$$I = \int \mathrm{d}s \left[\sum_j \left(\frac{\mathrm{d}q_j}{\mathrm{d}s}\right)^2\right]^{1/2} \sqrt{2m[V(q) - E]}. \qquad 39.$$

Reversing the signs in front of $V(q)$ and E would yield the principle of least action, which determines the trajectory of a classical mechanical system with fixed energy. But we know that this variational principle is equivalent to Hamilton's principle, which tells us to minimize the action $S = \int \mathrm{d}t(T - V)$ and which leads to the Lagrangian equations of motion. The appropriate sign changes can be obtained by working with an imaginary time $t = ix_4$. Doing so, and then retracing the steps relating the two variational principles, one finds that the path minimizing I is also a stationary point of the Euclidean action

$$S_{\mathrm{E}} = \int \mathrm{d}x_4 \left[\frac{1}{2m} \sum_j \left(\frac{\mathrm{d}q_j}{\mathrm{d}x_4}\right)^2 + V(q_j)\right] \qquad 40.$$

and is given by a solution of the Euler-Lagrange equations in imaginary time. Furthermore, for this path $I = S_{\mathrm{E}}$. The end points of this path are on the surfaces on either side of the barrier where $E = V$. Hence, at these end points the "velocities" $\mathrm{d}q_j/\mathrm{d}x_4$ vanish.

This method can be carried over to field theory (85–89). The coordinates q_j are replaced by the values of the field at each point in space, $\phi(\mathbf{x})$, and the path $q_j(x_4)$ becomes a sequence of three-dimensional field configurations, $\phi(\mathbf{x}, x_4)$, which may itself be viewed as a field configuration in a four-dimensional Euclidean space. It should be stressed that x_4 is not a time in any physical sense, but simply a variable parameterizing a path through configuration space.

4.1 *Vacuum Decay by Tunneling*

One application of this method is to the decay of an unstable vacuum (85). An example arises in a theory with a scalar field governed by a potential $V(\phi)$ with two unequal minima, one a "false vacuum" at $\phi = \phi_f$ and the other a deeper "true vacuum" at $\phi = \phi_t$. The false vacuum state is stable classically, but quantum mechanically it can decay via tunneling through the barrier in the potential energy

$$U = \int d^3x \left[\frac{1}{2} (\nabla\phi)^2 + V(\phi) \right]. \qquad 41.$$

This tunneling cannot go directly from the homogeneous false vacuum to a homogeneous state with $\phi \approx \phi_t$ because the volume integral makes the barrier between these infinite. Instead, the tunneling is to a state in which a bubble of true vacuum is embedded in a false vacuum background. A field configuration corresponding to such a bubble is shown in Figure 1. Varying R while keeping the field profile in the wall region fixed gives a one-parameter family of configurations whose energy is plotted in Figure 2. It is the sum of negative contribution $(4\pi/3)(\Delta V)R^3$, arising from the replacement of false vacuum by true in the interior, and a wall energy of the form $4\pi\sigma R^2$, arising from both the gradient terms in the energy and the barrier in $V(\phi)$, which is traversed by the field as it passes through the bubble wall. These precisely cancel when $R = 3\sigma/(\Delta V)$, which would be the end point of the tunneling path if the system were constrained to this one-parameter set of configurations. At this point one finds $\partial E/\partial R < 0$, which indicates that the bubble

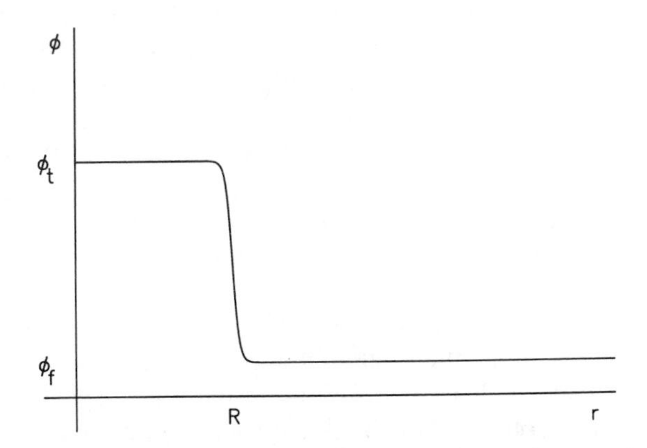

Figure 1 A field profile describing a true vacuum bubble of radius R in a false vacuum background.

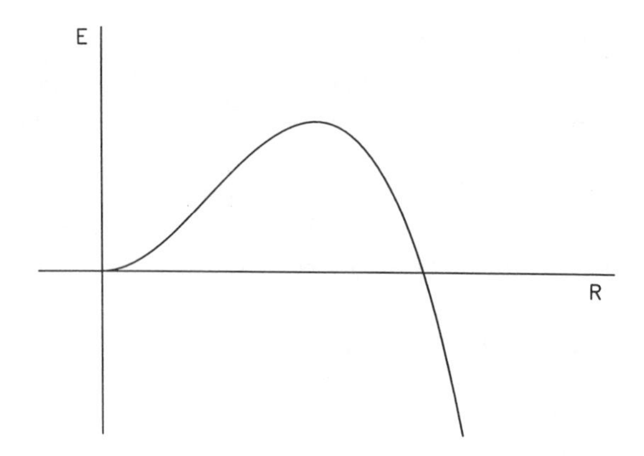

Figure 2 The energy, as a function of the bubble radius, of the bubble depicted in Figure 1.

would expand once it was nucleated. (In actuality, the bubble profile changes somewhat as one goes along the optimal path through the potential energy barrier.)

The rate for this process, as well as the optimal sequence of bubble profiles, can be obtained by solving the Euclidean field equations

$$\sum_{i=1}^{4} (\partial_i \phi)^2 = \frac{dV}{d\phi} . \qquad 42.$$

The boundary conditions are that $\phi(x) = \phi_f$ at the initial value x_4^i, while the configuration at the final value x_4^f is a bubble embedded in false vacuum. The interval between these is in fact semi-infinite; i.e. we must take $x_4^i = -\infty$, while x_4^f has a finite value that, by x_4-translation invariance, can be chosen to be 0. Because $\partial_4 \phi$ vanishes at the end points of the tunneling path, the reflection $x_4 \rightarrow -x_4$ yields another solution, running from $x_4 = 0$ to $x_4 = \infty$, with the same action. Patching these two solutions together gives what is known as the bounce solution, whose Euclidean action $S_E = B$ is twice that of the original solution. This factor of two is the same as that arising when the tunneling amplitude is squared to obtain the tunneling probability, which is thus proportional to e^{-B}.

There are actually an infinite number of Euclidean solutions and tunneling paths, since the final bubble could equally well be located at any point in space. It is therefore more natural to speak of the probability per unit volume. This bubble nucleation rate per unit volume is of the form $\lambda = Ae^{-B}$. The prefactor A may be viewed as probing the energy barrier in directions orthogonal to the optimal tunneling path.

If small derivations from this path have little effect on the one-dimensional tunneling amplitude, A is large, and conversely. An expression for A in terms of a functional determinant can be derived by using path integral methods (90). However, in realistic applications it is seldom possible to evaluate this determinant, and one simply argues on dimensional grounds that $A \sim M^4$, where M is a typical mass scale of the theory.

4.2 *Yang-Mills Instantons*

Quantum mechanical tunneling can also arise within the context of a stable vacuum. The most important example of this is the Yang-Mills instanton (91), which describes a tunneling process in which both the initial and final field configurations are classical ground states.

For a Yang-Mills theory the classical energy is clearly minimized when the field strength $F_{\mu\nu}$ vanishes. (Here, and throughout this section, the field strength $F_{\mu\nu}$ and gauge potential A_μ should be understood to be matrices that are linear combinations of the generators of the gauge group.) This does not imply that the gauge potential also vanishes, but only that it must be gauge equivalent to zero, i.e. of the form $A_\mu = U^{-1}\partial_\mu U$, where $U(x)$ is an element of the gauge group. This degeneracy is greatly reduced when a gauge condition is imposed. In some gauges (e.g. axial gauge, $A_3 = 0$) one can impose conditions such that $A_\mu = 0$ is the unique classical ground state (13). However, in non-Abelian theories this can be done only at the (aesthetic) cost of allowing finite energy configurations for which the potentials do not vanish at spatial infinity. In other gauges (e.g. temporal gauge, $A_0 = 0$) these configurations are avoided, but an infinite set of degenerate classical ground states remains (92, 93). These ground states correspond to gauge functions U_n, which can be characterized by the integer winding number

$$n = \frac{1}{24\pi^2} \int d^3x \, \epsilon_{ijk} \, \mathrm{Tr} \, U_n^{-1} \, \partial_i U_n U_n^{-1} \, \partial_j U_n U_n^{-1} \, \partial_k U_n. \qquad 43.$$

It is impossible classically to go from one of these degenerate ground states to the next (i.e. to change the winding number by one unit). However, quantum tunneling between the states is possible. Such tunneling mixes the vacuum states of definite winding number, so that the true vacuum is a linear combination of these. The amplitude for this tunneling can be calculated by finding a solution of the Euclidean field equations such that the configurations at the initial and final values of x_4 (which turn out to be $\pm\infty$) are the two classical vacua; this solution is the instanton.

Tunneling also occurs in gauges with unique classical vacua (94). In these gauges the instanton describes a process in which the field starts at the vacuum and then tunnels through a potential energy barrier simply to get back to where it started. This is somewhat analogous to the case of a particle constrained to move on a vertically oriented ring, with the energy of the particle being too small to overcome the gravitational potential energy barrier at the top of the ring. Classically the particle will stay at the bottom of the ring, but quantum mechanically it can go around the ring by tunneling. Although its description is rather different in these two classes of gauges, the observable consequences of this instanton-induced tunneling are, of course, the same in all gauges (95, 96).

The instanton is in fact a four-dimensional topological soliton. As noted above, the configurations at $x_4 \to \pm\infty$ must be the desired initial and final configurations, which are of the form $U^{-1}\,\partial_\mu U$. Since the tunneling proceeds via finite energy configurations, the fields must also be of this form at spatial infinity. The instanton thus assigns an element $U(x)$ of the gauge group G to every point on the three-sphere at Euclidean infinity, $\mathbf{x}^2 + x_4^2 = \infty$. Any such assignment gives an element of the group $\Pi_3(G)$, which for any simple non-Abelian gauge group turns out to be the group of the integers.

Thus any such Euclidean configuration can be assigned an integer, called the Pontryagin index, given by

$$k = \int d^4x \left[\frac{1}{16\pi^2}\,\mathrm{Tr}\,F^{\mu\nu}\,\tilde{F}_{\mu\nu}\right], \qquad 44.$$

where the dual field strength $\tilde{F}_{\mu\nu} = (1/2)\epsilon_{\mu\nu\alpha\beta}F^{\alpha\beta}$. Although written here as a volume integral, the expression on the right-hand side depends only on the fields at infinity, because the integrand can be written as the divergence of a current j_μ. Moreover, for vacuum solutions the spatial integral of j_0 reduces to Equation 43 for the winding number. Rewriting Equation 44 as a surface integral and working in temporal gauge, it is easy to show that the contributions from the surfaces at spatial infinity vanish, so that k is equal to the difference of the winding numbers of the initial and final configurations. Tunneling between adjacent classical vacua is therefore described by a solution with unit Pontryagin number. By analogy with the arguments ensuring the existence of the vortex and monopole solutions, we know that minimizing the Euclidean action among the set of configurations with unit Pontryagin number will give such a solution.

In fact, an analytic expression for this solution can be found. The

SU(2) instanton (which is easily extended to larger gauge groups) may be written as

$$A_\mu(x) = \frac{-i}{g}\frac{(x-a)^2}{(x-a)^2+\lambda^2}U^{-1}(x-a)\partial_\mu U(x-a), \qquad 45.$$

where $U(y) = (y_0 - i\mathbf{y}\cdot\boldsymbol{\sigma}))/\sqrt{y^2}$. Its Euclidean action is $S_{\text{instanton}} = 8\pi^2/g^2$, where g is the gauge coupling. This solution depends on five real parameters. Four of these, the components of the vector a_μ, are a consequence of the translation invariance of the theory. The fifth parameter λ determines the size of the instanton. The freedom to choose λ arbitrarily reflects the scale invariance of the classical Yang-Mills theory and means that there are an infinite number of physically inequivalent (i.e. unrelated by either gauge transformation or simple spatial translation) tunneling paths with the same tunneling action. An instanton of size λ specifies a path through configuration space involving field configurations whose spatial extent is of the order of λ. Along this path the maximum potential energy (i.e. the maximum height of the barrier) is of the order of $1/(\lambda g^2)$. While this barrier height decreases as one goes to larger instantons, the length of the path through the barrier grows with λ in just such a manner that the tunneling action is unchanged.

When calculating the effects of tunneling, one must take into account the possibility of several successive tunnelings, or of several roughly simultaneous tunneling processes taking place at widely separated points in space. This can be done by considering Euclidean configurations containing a "gas" of separated instantons and anti-instantons (97).[6] The higher action of these configurations can be outweighed by the increase in "entropy" arising from the freedom to choose the positions of the instantons and anti-instantons independently. Stated differently, the amplitude for tunneling via the path described by a configuration with two instantons and one anti-instanton is far smaller than that for tunneling by the path corresponding to a single instanton (a factor of $e^{-3S_{\text{instanton}}}$ compared to $e^{-S_{\text{instanton}}}$). However, because there are far more paths of the former type than of the latter (in a space of finite volume Ω, one is proportional to Ω^3 and the other to Ω), the former can dominate. If λ were fixed, the dominant contribution would be from configurations with a density of instantons in Euclidean space-

[6] The anti-instanton, which has Pontryagin index of -1, is obtained by changing the sign of the fourth component of the vector $x - a$; i.e. by making the substitution $(x - a)_4 \rightarrow -(x - a)_4$ in Equation 45.

time of the order of $\lambda^{-4}e^{-S_{\text{instanton}}}$. For weak coupling $S_{\text{instanton}}$ is large and we have a dilute gas in which the separation between instantons is large compared to their size. The problem is that λ is not fixed, and we must therefore integrate over the sizes of the individual instantons. This leads to an integral that naively appears to diverge at both large and small λ. When one-loop quantum corrections (98–101) are taken into account, the g in the action becomes the running gauge coupling evaluated at a momentum of order $1/\lambda$. This ensures the convergence of the integral as $\lambda \rightarrow 0$, but only makes the problem of calculating the effects of large instantons worse. This greatly complicates the task of obtaining reliable quantitative calculations of instanton effects in QCD.

On the other hand, the addition of a Higgs field, as in the electroweak theory, breaks the scale invariance at the classical level (98). This ensures the convergence of the integral over λ, which is now dominated by the contributions from instantons of a single size.

4.3 *Physical Consequences of Vacuum Tunneling*

4.3.1 THETA PARAMETER AND *CP* VIOLATION The interference between instanton and noninstanton paths in the path integral can lead to *CP*-violating effects. This is most easily seen by working in a gauge with a unique classical vacuum, and by recalling the analogy with the particle on a ring. In this analogous problem, suppose that a term of the form $(\alpha/2\pi)\dot{\theta}$ is added to the Lagrangian. Being a total time derivative, this term will not affect the classical equation of motion. However, in the quantum theory it introduces a relative phase factor of $e^{i\alpha}$ between the amplitudes for paths that tunnel through the gravitational potential energy barrier at the top of the ring and those that do not. Unless α is 0 or π, this gives parity-violating effects even though the classical equations of motion are parity invariant. In a similar fashion, the addition of the total divergence

$$\Delta\mathscr{L} = \frac{\theta}{16\pi^2}\,\mathrm{Tr}\, F_{\mu\nu}\tilde{F}_{\mu\nu} \qquad 46.$$

to the Yang-Mills Lagrangian density has no effect classically, but it gives an extra phase factor $e^{i\theta}$ to trajectories that proceed by instanton-induced tunneling.[7]

If θ is neither 0 nor π, the interference effects from this additional term are both parity violating and *CP* violating. This might suggest that

[7] In gauges with multiple vacua, where the true vacuum is a linear combination of states with definite winding number, the effects of $\Delta\mathscr{L}$ can be mimicked by assigning appropriate phases to the expansion coefficients. The resulting states are called θ-vacua.

the inclusion of such a term in the QCD Lagrangian could provide an alternative explanation of the observed *CP* violation. However, it turns out that a θ large enough to account for the parity violation in the K meson system implies a neutron electric dipole moment well above the experimental upper limits. The essence of the difficulty is that the latter is now a purely strong interaction effect, while the former still involves weak interactions. By contrast, when *CP* violation is attributed to phases in the Kobayashi-Maskawa matrix, both effects suffer the same weak interaction suppression.

To give an acceptably small neutron dipole moment, θ must be less than about 10^{-9}. It would seem that the simplest way to meet this constraint would be to set $\theta = 0$ and omit the entire term from the Lagrangian. However, when fermions are included in the theory one finds that a chiral rotation of the fermion fields is equivalent to a shift in θ (92, 93). As a consequence, the effective value of θ receives a contribution from the phases in the fermion mass matrix. These phases in turn depend on the phase of the scalar vacuum expectation value responsible for the mass generation. The result is that zero is not a particularly natural value for the final θ_{eff}. If there were at least one exactly massless quark, θ_{eff} could always be set to zero by a chiral rotation. Since this appears not to be true, one must seek an explanation for the otherwise fortuitous fact that θ_{eff} is so small. One possible solution (102, 103) to this "strong *CP* problem" is to add fields to the Lagrangian in such a way that a vanishing θ_{eff} is chosen dynamically. There is then an approximate new symmetry, whose breaking by instanton effects gives rise to the hypothetical axion (104, 105).

4.3.2 VIOLATION OF ANOMALOUS CONSERVATION LAWS Additional effects come into play if massless (or very light) fermions are present in the theory. In a fixed gauge field background, the fermions can be described by finding the eigenfunctions of the Dirac equation and specifying which modes were occupied. If the background is varied, these eigenfunctions and their eigenvalues will change. If we regard the instanton-mediated tunneling process as the passage through a sequence of such gauge field configurations and follow the changes in the fermion eigenmodes over the course of the process, we find that the net effect is to shift the spectrum in such a manner that one left-handed mode is shifted from positive energy to negative energy, while one right-handed mode goes from negative to positive energy (97). If this process were sufficiently slow that the occupation numbers of the various eigenmodes did not change, the result would be the creation and annihilation of various particles. For example, if initially all the negative energy modes were filled and all the positive energy modes empty, then the final state would contain one unfilled left-handed negative energy state and one filled right-handed positive energy state. This would correspond to the creation of a right-handed antifermion and a right-handed fermion, thus changing the chiral charge by two units (92, 93). The same change in chirality is found for other initial states; furthermore, this result can be shown to be exact (106) and not depend on the adiabatic approximation. (The underlying reason for this is that the integrand in Equation 44 for the Pontryagin index is essentially the same as the anomalous divergence of the chiral current.)

Applying this result to QCD resolves the so-called U(1) problem. It is well known that the eight light pseudoscalar mesons can be interpreted as the Goldstone bosons of an approximate chiral SU(3) symmetry that is spontaneously broken. However, the Lagrangian describing the coupling of three light quarks to the color gauge field has an approximate chiral U(3) symmetry. This larger symmetry would lead one to expect a ninth light pseudoscalar meson; the absence of such a particle is the U(1) problem. The issue is resolved once it is recognized that the instanton effects just described violate the extra symmetry, but not the chiral SU(3).

A second application is to the electroweak theory. Because the baryon number current has an anomalous divergence, the shifts in the fermion energy levels caused by SU(2) $\times$ U(1) instantons will lead to nonconservation of baryon number. However, the amplitude for tunneling is proportional to $e^{-S_{\text{instanton}}} \sim e^{-2\pi/\alpha_{\text{weak}}}$. Since the exponent is of the order of 200, the probability of this ever actually happening would appear to be negligible. (There is no enhancement from the integration over the instanton size, because this is fixed by the Higgs field.)

Recently, Ringwald (107) and Espinosa (108) argued that in high energy scattering it may be possible to overcome this exponential suppression. The essential idea is that the tunneling induces an effective Lagrangian for baryon number violation that has interaction terms involving the product of the light fermion fields and arbitrary powers of the Higgs and gauge boson fields. These interactions are point-like and thus lead to amplitudes for processes with many bosons in the final state that grow as powers of the energy. Moreover, if one calculates an inclusive quark-quark scattering cross section, the sum over the number of final-state bosons gives an exponential dependence on energy (109). The calculation suggests that baryon number violation becomes large at energies of the order of m_W/α_{weak}, and thus could be observable at the SSC. However, corrections to this result must be significant, because otherwise extrapolation to still higher energy would lead to violation of unitarity. At present, the magnitude of the corrections remains unclear (for a recent review of the situation, see 110).

4.4 *Thermal Fluctuations and Sphalerons*

At finite temperature, quantum tunneling through a potential energy barrier must compete with "classical" barrier crossing by means of thermal fluctuations. When the temperature is small compared to the height of the barrier, the latter process proceeds primarily via paths that traverse the barrier near its lowest point. Associated with these paths is a Boltzmann factor $e^{-E_{s.p.}/T}$, where $E_{s.p.}$ is the energy of the saddle-point configuration lying at the high point on the lowest path across the barrier. Although other paths become important as the temperature approaches and then exceeds $E_{s.p.}$, knowledge of the saddle point is still the first step in the analysis of the problem.

For the decay of a metastable false vacuum, the saddle-point configuration is one with a true vacuum bubble of critical size embedded within a false vacuum background. The radius of the critical bubble is such that the outward pressure from the true vacuum interior is just balanced by the inward push of the surface tension. It corresponds to the maximum of the curve of $E(R)$ in Figure 2. (Actually, at high temperature one should use the free energy, rather than the energy, for determining the critical radius; qualitatively the picture is unchanged.)

Thermal fluctuations can also cross the barriers separating the degenerate classical vacua of Yang-Mills theory. For an unbroken gauge symmetry the analysis of the problem is complicated by the fact that the scale invariance of the classical theory implies that there is no saddle point. Instead, the barrier height decreases monotonically as one goes along a direction in configuration space corresponding to field configurations of increasing spatial extent. (Roughly speaking, these correspond to cross sections through instantons of increasing scale size.) While this would tend to favor fluctuations with larger spatial extent, the temperature provides an infrared cutoff and thus suppresses the largest configurations.

Matters are simpler for a spontaneously broken symmetry, where the Higgs field breaks the scale invariance. A saddle point now exists, and is known as the sphaleron. It is a solution of the static field equations but, because it is a saddle point, it is unstable. In an SU(2) theory with the symmetry broken by an isodoublet Higgs field ϕ with vacuum expectation value v, the sphaleron solution has the form (111, 112)

$$A_i^a = \epsilon_{aij}\,\hat{r}^j\,\frac{f(r)}{r} \qquad 47.$$

$$\phi = \frac{iv}{\sqrt{2}}\,h(r)\hat{r}^j\sigma^j\psi_0, \qquad 48.$$

where the functions f and h vary from 0 to 1 as r ranges from 0 to ∞ and ψ_0 is a constant isospinor. [This solution was studied previously in a different context (113–115).] The extension to the full SU(2) × U(1) electroweak theory can be obtained by expanding about $\sin^2\theta_W = 0$ (116). The sphaleron energy is then found to be a few times m_W/α_{weak}; i.e. of the order of 10 TeV. [To actually use the sphaleron to calculate high temperature barrier crossing, one must take into account the existence of a symmetry-restoring phase transition and the thermal variation of the gauge boson mass. These and other effects are discussed by Arnold & McLerran (117).]

Now recall from the previous section that in the electroweak model the vacuum tunneling described by the instanton leads to baryon number nonconservation. A similar violation of baryon number conservation occurs when thermal fluctuations carry the system over the potential energy barrier (118) and may have important consequences for the generation of the baryon asymmetry in the early universe (for a further discussion, see 119).

ACKNOWLEDGMENTS

I thank Piet Hut, Kimyeong Lee, and Alfred Mueller for helpful comments on the manuscript. I would also like to acknowledge the hospitality of the Theoretical Physics and Theoretical Astrophysics Groups at Fermilab, where part of this review was written.

Literature Cited

1. Goddard, P., Olive, D. I., *Rep. Prog. Phys.* 41:1357 (1978)
2. Rossi, P., *Phys. Rep.* 86:317 (1983)
3. Coleman, S., in *The Unity of the Fundamental Interactions*, ed. A. Zichichi. New York: Plenum (1983), p. 21
4. Preskill, J., *Annu. Rev. Nucl. Part. Sci.* 34:461 (1984)
5. Giacomelli, G., *Riv. Nuovo Cimento*, Vol. 7, No. 12 (1984)
6. Lee, T. D., Pang, Y., Columbia preprint CU-TP-506 (1991), *Phys. Rep.* (1992), in press
7. Wilets, L., *Nontopological Solitons*. Singapore: World Sci. (1989), 154 pp.
8. Birse, M. C., *Prog. Part. Nucl. Phys.* 25:1 (1990)
9. Coleman, S., in *The Whys of Subnuclear Physics*, ed. A. Zichichi. New York: Plenum (1979), pp. 805–916; also in Coleman, S., *Aspects of Symmetry*. Cambridge Univ. Press (1983), pp. 265–330
10. Olive, D., Sciuto, S., Crewther, R. J., *Riv. Nuovo Cimento* 12: No. 8 (1979)
11. Rajaraman, R., *Phys. Rep.* 21:227

(1975)
12. Jackiw, R., *Rev. Mod. Phys.* 49:681 (1977)
13. Coleman, S., in *New Phenomena in Subnuclear Physics,* ed. A. Zichichi. New York: Plenum (1977), pp. 297–407; also in Coleman, S., *Aspects of Symmetry*. Cambridge Univ. Press (1983), pp. 185–264
14. Rajaraman, R., *Solitons and Instantons*. Amsterdam: North-Holland (1982), 409 pages.
15. Actor, A., *Rev. Mod. Phys.* 51:461 (1979)
16. Goddard, P., Mansfield, P., *Rep. Prog. Phys.* 49:725 (1986)
17. Jackiw, R., Pi, S. Y., *Prog. Theor. Phys.* (1992), in press
18. Dashen, R. F., Hasslacher, B., Neveu, A., *Phys. Rev.* D10:4130 (1974)
19. Polyakov, A. M., *JETP. Lett.* 20:194 (1974)
20. Goldstone, J., Jackiw, R., *Phys. Rev.* D11:1486 (1975)
21. Dashen, R. F., Hasslacher, B., Neveu, A., *Phys. Rev.* D10:4114 (1974)
22. Cahill, K., *Phys. Lett.* 53B:174 (1974)
23. Christ, N. H., Lee, T. D., *Phys. Rev.* D12:1606 (1975)
24. Tomboulis, E., *Phys. Rev.* D12:1678 (1975)
25. Klein, A., Krejs, F. R., *Phys. Rev.* D12:3112 (1975)
26. Creutz, M., *Phys. Rev.* D12:3126 (1975)
27. de Vega, H. J., *Nucl. Phys.* B115:411 (1976)
28. Callan, C. G., Gross, D. J., *Nucl. Phys.* B93:29 (1975)
29. Gervais, J. L., Jevicki, A., Sakita, B., *Phys. Rev.* D12:1038 (1975)
30. Gervais, J. L., Sakita, B., *Phys. Rev.* D11:2943 (1975)
31. Derrick, G. H., *J. Math. Phys.* 5:1252 (1964)
32. Oka, M., Hosaka, A., *Annu. Rev. Nucl. Part. Sci.* 42:333–65 (1992)
33. Nielsen, H. B., Olesen, P., *Nucl. Phys.* B61:45 (1973)
34. t'Hooft, G.C., *Nucl. Phys.* B79:276 (1974)
35. Prasad, M. K., Sommerfield, C. H., *Phys. Rev.* Lett. 35:760 (1975)
36. Bogomol'nyi, E., *Sov. J. Nucl. Phys.* 24:449 (1975)
37. Kirkman, T., Zachos, C. K., *Phys. Rev.* D24:999 (1981)
38. Dirac, PA. M., *Proc. R. Soc. (London)* A133:60 (1931)
39. Rajaraman, R., Weinberg, E. J., *Phys. Rev.* D11:2950 (1975)
40. Tomboulis, E., Woo, G., *Nucl. Phys.* B107:221 (1976)
41. Christ, N. H., Guth, A. H., Weinberg, E. J., *Nucl. Phys.* B114:61 (1976)
42. Julia, B., Zee, A., *Phys. Rev.* D11:2227 (1975)
43. Dokos, C., Tomoras, T., *Phys. Rev.* D16:1221 (1977)
44. Schellekens, A. N., Zachos, C. K., *Phys. Rev. Lett.* 50:1242 (1983)
45. Gardner, C. L., Harvey, J. A., *Phys. Rev. Lett.* 52:879 (1984)
46. Lazarides, G., Shafi, Q., *Phys. Lett.* 94B:149 (1980)
47. Lazarides, G., Magg, M., Shafi, Q., *Phys. Lett.* 97B:87 (1980)
48. Zel'dovich, Ya. A., Khlopov, M. Y., *Phys. Lett.* 79B:239 (1978)
49. Preskill, J., *Phys. Rev. Lett.* 43:1365 (1978)
50. Guth, A. H., *Phys. Rev.* D23:347 (1981))
51. Jackiw, R., Rebbi, C., *Phys. Rev.* D13:3398 (1976)
52. Rubakov, V. A., *JETP. Lett.* 33:644 (1981)
53. Rubakov, V. A., *Nucl. Phys.* B212:391 (1982)
54. Callan, C. G., *Phys. Rev.* D25:2141 (1982)
55. Callan, C. G., *Phys. Rev.* D26:2058 (1982)
56. Callan, C. G., *Nucl. Phys.* B212:391 (1982)
57. Dereli, T., Swank, J. H., Swank, L. J., *Phys. Rev.* D11:3541 (1975)
58. Kazama, Y., Yang, C. N., Goldhaber, A. S., *Phys. Rev.* D15:2287 (1977)
59. Blaer, A. S., Christ, N. H., Tang, J. F., *Phys. Rev. Lett.* 47:1364 (1981)
60. Blaer, A. S., Christ, N. H., Tang, J. F., *Phys. Rev.* D25:2128 (1982)
61. Bell, J. S., Jackiw, R., *Nuovo Cimento* 60A:47 (1969)
62. Adler, S. L., *Phys. Rev.* 177:2426 (1969)
63. Rubakov, V. A., *Rep. Prog. Phys.* 51:189 (1988)
64. Lee, T. D., *Phys. Rev.* D35:3637 (1987)
65. Lee, T. D., *Comments Nucl. Part. Phys.* XVII:225 (1987)
66. Friedberg, R., Lee, T. D., Pang, Y., *Phys. Rev.* D35:3640 (1987)
67. Friedberg, R., Lee, T. D., Pang, Y., *Phys. Rev.* D35:3658 (1987)
68. Lee, T. D., Pang, Y., *Phys. Rev.* D35:3678 (1987)
69. Rosen, C., *J. Math. Phys.* 9:996, 999 (1968)
70. Kaup, D. J., *Phys. Rev.* 172:1331

(1968)
71. Ruffini, R., Bonazzola, S., *Phys. Rev.* 187:1767 (1969)
72. Vinciarelli, P., *Nuovo Cimento Lett.* 4:905 (1972)
73. Lee, T. D., Wick, G. C., *Phys. Rev.* D9:2291 (1974)
74. Coleman, S., *Nucl. Phys.* B262:263 (1985)
75. Lee, T. D., in *Multiparticle Dynamics,* ed. A. Giovanni, W. Kittel. Singapore: World Sci. (1990), p. 743
76. Friedberg, R., Lee, T. D., Sirlin, A., *Phys. Rev.* D13:2739 (1976)
77. Friedberg, R., Lee, T. D., *Phys. Rev.* D15:1694 (1977)
78. Friedberg, R., Lee, T. D., *Phys. Rev.* D16:1096 (1977)
79. Friedberg, R., Lee, T. D., *Phys. Rev.* D18:2623 (1978)
80. Chodos, A., et al, *Phys. Rev.* D9:3471 (1974)
81. Bardeen, W. A., et al, *Phys. Rev.* D11:1094 (1975)
82. Lee, K., et al, *Phys. Rev.* D39:1665 (1989)
83. Banks, T., Bender, C. M., Wu, T. T., *Phys. Rev.* D8:3346 (1973)
84. Banks, T., Bender, C. M., *Phys. Rev.* D8:3366 (1973)
85. Coleman, S., *Phys. Rev.* D15:2929 (1977)
86. Gervais, J. L., Sakita, B., *Phys. Rev.* D16:3507 (1977)
87. Bitar, K., Chang, S. J., *Phys. Rev.* D17:486 (1978)
88. Bitar, K., Chang, S. J., *Phys. Rev.* D18:435 (1978)
89. de Vega, H. J., Gervais, J. L., Sakita, B., *Nucl. Phys.* B143:125 (1978)
90. Callan, C. G., Coleman, S., *Phys. Rev.* D16:1762 (1977)
91. Belavin, A. A., et al, *Phys. Lett.* 59B: 85 (1975)
92. Callan, C. G., Dashen, R. F., Gross, D. J., *Phys. Lett.* 63B:334 (1976)
93. Jackiw, R., Rebbi, C., *Phys. Rev. Lett.* 37:172 (1976)
94. Bernard, C. W., Weinberg, E. J., *Phys. Rev.* D15:3656 (1977)
95. Wadia, S., Yoneya, T., *Phys. Lett.* 66B:341 (1977)
96. Rothe, K. D., Swieca, J. A., *Nucl. Phys.* B138:26 (1978)
97. Callan, C. G., Dashen, R. F., Gross, D. J., *Phys. Rev.* D17:2717 (1978)
98. t Hooft, G., *Phys. Rev.* D14:3432 (1976)
99. Belavin, A. A., Polyakov, A. M., *Nucl. Phys.* B123:429 (1977)
100. Chadha, S., et al, *Phys. Lett.* 72B:103 (1977)
101. Ore, F., *Phys. Rev.* D16:2577 (1977)
102. Peccei, R. D., Quinn, H. R., *Phys. Rev. Lett.* 38:1440 (1977)
103. Peccei, R. D., Quinn, H. R., *Phys. Rev.* D16:1791 (1977)
104. Weinberg, S., *Phys. Rev. Lett.* 40:223 (1978)
105. Wilczek, F., *Phys. Rev. Lett.* 40:279 (1978)
106. Christ, N. H., *Phys. Rev.* D21:1591 (1980)
107. Ringwald, A., *Nucl. Phys.* B330:1 (1990)
108. Espinosa, O., *Nucl. Phys.* B343:310 (1990)
109. McLerran, L., Vainshtein, A., Voloshin, M., *Phys. Rev.* D42:171 (1990)
110. Mattis, M. P., Los Alamos preprint LA-UR-91-2926 (1991), *Phys. Rep.* (1992), in press
111. Manton, N. S., *Phys. Rev.* D28:2019 (1983)
112. Forgacs, P., Horvath, Z., *Phys. Lett.* 138B:397 (1984)
113. Dashen, R. F., Hasslacher, B., Neveu, A., *Phys. Rev.* D10:4138 (1974)
114. Boguta, J., *Phys. Rev. Lett.* 50:148 (1983)
115. Burzlaff, J., *Nucl. Phys.* B233:262 (1984)
116. Klinkhamer, F. R., Manton, N. S., *Phys. Rev.* D30:2212 (1984)
117. Arnold, P., McLerran, L., *Phys. Rev.* D36:581 (1987)
118. Kuzmin, V. A., Rubakov, V. A., Shaposhnikov, M. E., *Phys. Lett.* 155B:36 (1985)
119. Dolgov, A. D., Kyoto preprint YITP/K-940 (1991), *Phys. Rep.* (1992), in press

Annu. Rev. Nucl. Part. Sci. 1992. 42:211–250

CP VIOLATION IN B PHYSICS[1]

Yosef Nir[2]

Physics Department, Weizmann Institute of Science, Rehovat 76100, Israel

Helen R. Quinn

Stanford Linear Accelerator Center, Stanford University, Stanford, California 94309, USA

KEY WORDS: particle physics, Standard Model, B meson decays, CKM matrix, penguin diagrams

---

CONTENTS

[1] Work supported by the Department of Energy, contract DE-AC03-76SF00515. The US Government has the right to retain a nonexclusive royalty-free license in and to any copyright covering this paper.
[2] Incumbent of the Ruth E. Recu Career Development Chair. Supported in part by the Israel Commission for Basic Research, by the United States–Israel Binational Science Foundation, and by the Minerva Foundation.

1. INTRODUCTION

1.1 CP *is Not a Symmetry of Nature*

It was long thought that *CP* symmetry was exact in nature and that only theories that had this property were viable descriptions of the observed world. The observation of the decay $K_L \rightarrow \pi\pi$ by Christenson, Cronin, Fitch & Turlay in 1964 (1) changed this view dramatically. *CP* violation was unambiguously demonstrated by this decay. Since that time much effort has gone into studying the nature of *CP* violation. We have understood that *CP* violation is a crucial feature of any theory that attempts to explain the observed asymmetry between matter and antimatter in the universe starting from initially symmetric conditions (2). We have also found that it is a natural feature of the three-generation Standard Model (3). So far the kaon system has been the only laboratory for the observation of *CP* violation, and we have been unable to perform calculations and measurements of sufficient precision to thoroughly test the Standard Model predictions of *CP* violation.

The observation of baryon asymmetry in the universe can be explained if some *CP*-conjugate pairs of processes have different rates. If the initial conditions of the universe were baryon symmetric, then the asymmetry should be generated by dynamical baryogenesis, which requires three ingredients (2): (*a*) there must exist baryon-number-violating processes, (*b*) these processes must go out of equilibrium sometime during the history of the universe, and (*c*) *C* and *CP* must be violated. While *CPT* requires that the total decay rates for a particle and its antiparticle be equal, *CP* symmetry requires that the partial rates of *CP*-conjugate processes be equal. If this were always the case, then for any process that violates baryon number there would be a *CP*-conjugate process of equal rate and no asymmetry could be generated. Thus, *CP* violation seems to be a necessary ingredient of any theory of elementary particles. Moreover, detailed analyses of baryogenesis imply that sources of *CP* violation beyond the Standard Model are required (for a recent review, see 4).

The possibility remains that processes not included in the Standard Model may play an important role in *CP* violation. This makes the investigation of *CP* violation in the B-meson system extremely interesting. At the very least, it will allow us to measure some of the remaining parameters of the Standard Model, parameters as fundamental as the quark masses themselves. If we are lucky it could do a lot more: if the results are inconsistent with Standard Model predictions then they may provide some clues about physics beyond the Standard Model. We have precious few ways to seek these clues, so a source of B mesons with luminosity high enough to study *CP* violation would be a truly exciting physics facility.

1.2 CP *Violation in the Neutral Kaon System*

To date, three *CP*-violating processes have been measured (5). The results are parameterized as follows:

$$|\eta_{+-}| = \left[\frac{\Gamma(K_L \to \pi^+\pi^-)}{\Gamma(K_S \to \pi^+\pi^-)}\right]^{1/2} = (2.268 \pm 0.023) \times 10^{-3},$$

$$|\eta_{00}| = \left[\frac{\Gamma(K_L \to \pi^0\pi^0)}{\Gamma(K_S \to \pi^0\pi^0)}\right]^{1/2} = (2.253 \pm 0.024) \times 10^{-3}, \quad 1.$$

$$\delta = \frac{\Gamma(K_L \to \pi^-\ell^+\nu) - \Gamma(K_L \to \pi^+\ell^-\nu)}{\Gamma(K_L \to \pi^-\ell^+\nu) + \Gamma(K_L \to \pi^+\ell^-\nu)}$$

$$= (3.27) \pm 0.12) \times 10^{-3}. \quad 2.$$

These three parameters measure *CP* violation due to mixing of $\overline{K}^0$ and K^0, that is, *CP* is violated in $\Delta S = 2$ processes. The neutral kaon mass eigenstates, K_L and K_S, are not *CP* eigenstates but instead

$$|K_L\rangle = \frac{1+\epsilon}{\sqrt{2(1+|\epsilon|^2)}}|K^0\rangle + \frac{1-\epsilon}{\sqrt{2(1+|\epsilon|^2)}}|\overline{K}^0\rangle,$$

$$|K_S\rangle = \frac{1+\epsilon}{\sqrt{2(1+|\epsilon|^2)}}|K^0\rangle - \frac{1-\epsilon}{\sqrt{2(1+|\epsilon|^2)}}|\overline{K}^0\rangle. \quad 3.$$

The term ϵ parameterizes the deviation from the *CP* eigenstates. Were K_L a pure *CP*-odd state, it could not decay into two pions, which are here in a *CP*-even state ($J = 0$), and it would decay into leptons of opposite charges at equal rates. If $|\epsilon| = 0$ the observables in Equations 1 and 2 would vanish. Instead, they are all compatible with the single value

$$|\epsilon| \approx 2.26 \times 10^{-3}. \quad 4.$$

There is yet another *CP*-violating parameter in the neutral kaon system. The values of $|\eta_{00}|$ and $|\eta_{+-}|$ may differ slightly from each other, and this difference is parametrized by ϵ':

$$\eta_{+-} \approx \epsilon + \epsilon', \qquad \eta_{00} \approx \epsilon - 2\epsilon', \quad 5.$$

where

$$\epsilon' \approx \frac{i}{\sqrt{2}}\,\mathrm{Im}(a_2/a_0)e^{i(\delta_2-\delta_0)}. \quad 6.$$

Here a_I is the amplitude for K^0 to decay into a final two-pion state of isospin I, with the strong phase $e^{i\delta_I}$ factored out. (We explain the term strong phases in the next paragraph.) If *CP* violation could be attributed exclusively to the $\Delta S = 2$ mixing, then it would not depend on the final state and ϵ' would vanish. A nonzero value of ϵ' would signify direct *CP* violation, that is *CP* violation in $\Delta S = 1$ (decay) processes. The most recent measurements (6) give

$$\epsilon'/\epsilon = \begin{cases} (2.3 \pm 0.7) \times 10^{-3} & \text{NA31} \\ (0.6 \pm 0.7) \times 10^{-3} & \text{E731.} \end{cases} \qquad 7.$$

Thus, there is as yet no compelling evidence for direct *CP* violation: the measurements are consistent with Standard Model estimates, but the weighted average for ϵ' is only two standard deviations from zero.

Before proceeding, let us briefly discuss the mechanism for *CP* violation: how does a complex phase in the Lagrangian translate into a prediction of a *CP*-violating observable? Physical amplitudes for any decay or scattering process are generally complex, even when the Lagrangian itself is real and *CP* conserving. Phases arise from the fact that there are coupled channels in most physical processes. Amplitudes acquire phases from the absorptive parts associated with these coupled channels. These phases are here referred to as strong phases, since the rescatterings are dominated by strong interaction processes. The strong phases of a pair of *CP*-conjugate processes are always of the same magnitude and sign. The *CP*-violating Lagrangian phases are generally referred to as weak phases because they appear in the weak interaction parts of the Lagrangian. The weak phases of a pair of *CP*-conjugate amplitudes are always of the same magnitude but of opposite sign.

If there were only a single term in the amplitude for a process, then the *CP*-conjugate process would proceed at an identical rate, despite the fact that the amplitude has a different phase. To see a *CP*-violating effect, one needs two different contributions to the amplitude. Then interference terms are sensitive to the difference between the phases and hence can give *CP*-violating effects, that is rates that are different for a process and for its *CP*-conjugate.

In the neutral K (or B) system, the particle can decay directly to a given final state or it can mix to its *CP*-conjugate and then decay to the same final state. Thus mixing provides the necessary second contribution to the amplitude. In addition, the Standard Model suggests that there are two different mechanisms at work in the direct decay, generically referred to as tree and penguin decays (see Section 4). These mechanisms can contribute to the amplitude with different weak and strong phases, thus resulting in direct *CP* violation. Note, however, that ϵ' depends not only on the weak phase difference, $\arg(a_2/a_0)$, but also on the strong phase difference, $\delta_2 - \delta_0$, and on the magnitude of the amplitudes, $|a_2/a_0|$. This dependence on hadronic processes is a common feature of direct *CP* violation effects, which explains why clean theoretical predictions for these effects are not available.

One beauty of the B system is the great variety of decays that can be studied. A second advantage is that, because the b quark is more massive than the s quark, we can more reliably calculate certain strong corrections by perturbative techniques, since they are more dominated by short-distance physics. Thus in B decay the relation between the measured asymmetries and Standard Model parameters has fewer uncertainties.

1.3 CP *Violation in the Standard Model*

Under what conditions is a Lagrangian *CP* conserving? The general answer is this: whenever all the coupling and mass terms in the Lagrangian can be made real by an appropriate set of field redefinitions. Within the Standard Model, the most general theory with only two quark generations and a single Higgs multiplet is of this type. However, when we add a third quark generation then the most general quark mass matrix allows for *CP* violation. Similarly, when we extend the fermion sector in various other ways or extend the scalar sector beyond the single Higgs doublet of the minimal Standard Model, additional parameters appear that cannot all be made simultaneously real by field redefinitions and hence allow further *CP*-violating effects.

The three-generation Standard Model with a single Higgs multiplet has only a single nonzero phase. It appears in the matrix that relates weak eigenstates to mass eigenstates, commonly known as the CKM (Cabibbo-Kobayashi-Maskawa) matrix (3, 7). This matrix must be unitary, a constraint that provides relationships between its elements. With relatively few further assumptions this translates into very specific predictions for the relationships between the parameters measured in different B-decay processes. This makes the B decays an ideal tool to probe for physics beyond the Standard Model; theories with other types of *CP*-violating parameters generally predict different relationships (see Section 3).

Here we do not consider *CP* violations that arise from the terms induced in the Lagrangian by instanton effects. For the weak SU(2) gauge theory, such a term can always be removed by appropriate rephasings of lepton fields. For the strong SU(3) gauge theory, such a term gives strong *CP* violation even in the two-generation case. Experimentally, the bound on strong *CP* violation from the absence of an electric dipole moment of the neutron [$d_n \leq 12 \times 10^{-26}$ e-cm (5)] is $\theta \lesssim 10^{-9}$. Any effect of such a term on the processes discussed here is completely negligible.

Let us now discuss *CP* violation from quark mixing in more detail. In the Standard Model, the charged current interactions are given by

$$-\mathcal{L}_W = \frac{g}{\sqrt{2}} (\overline{u_L^I}\ \overline{c_L^I}\ \overline{t_L^I})\, \gamma^\mu \begin{pmatrix} d_L^I \\ s_L^I \\ b_L^I \end{pmatrix} W_\mu^+ + \text{h.c.} \qquad 8.$$

The superscript "I" denotes interaction eigenstates. The mass matrices M_D and M_U are not simultaneously diagonal in this basis. However, any 3×3 Hermitian matrix can be diagonalized via a bi-unitary transformation. Thus

$$V_{dL} M_D V_{dR}^\dagger = M_D^{diag}; \qquad V_{uL} M_U V_{uR}^\dagger = M_U^{diag}, \qquad 9.$$

where the M_Q^{diag} matrices are real and diagonal. The matrices V_{qL} and V_{qR} define the transformation from the interaction eigenstates to the mass eigenstates. Thus the interactions of Equation 8 can be rewritten in the mass eigenbasis:

$$-\mathcal{L}_W = \frac{g}{\sqrt{2}} (\overline{u_L}\ \overline{c_L}\ \overline{t_L})\, \gamma^\mu\, (V_{uL} V_{dL}^\dagger) \begin{pmatrix} d_L \\ s_L \\ b_L \end{pmatrix} W_\mu^+ + \text{h.c.} \qquad 10.$$

The matrix $(V_{uL} V_{dL}^\dagger)$ is the mixing matrix for three quark generations. It is a 3×3 unitary matrix. It contains nine parameters, of which three can be chosen as real, namely the angles, θ_{12}, θ_{23}, and θ_{13}, and six are phases. We can reduce the number of phases by redefining the phases of the quark mass eigenstates. Then

$$(V_{uL} V_{dL}^\dagger) \rightarrow V = P_u (V_{uL} V_{dL}^\dagger) P_d^*, \qquad 11.$$

with P_u and P_d unitary diagonal matrices. For three generations, there are five independent phase differences between the elements of P_u and those of P_d, while there are six phases in $(V_{uL} V_{dL}^\dagger)$. Consequently, the mixing matrix V contains one physically meaningful phase, which we denote by δ (3). The three-generation Standard Model predicts *CP* violation unless $\delta = 0$.

The standard parametrization of V is (5, 8)

$$V = \begin{pmatrix} c_{12}c_{13} & s_{12}c_{13} & s_{13}e^{-i\delta} \\ -s_{12}c_{23} - c_{12}s_{23}s_{13}e^{i\delta} & c_{12}c_{23} - s_{12}s_{23}s_{13}e^{i\delta} & s_{23}c_{13} \\ s_{12}s_{23} - c_{12}c_{23}s_{13}e^{i\delta} & -c_{12}s_{23} - s_{12}c_{23}s_{13}e^{i\delta} & c_{23}c_{13} \end{pmatrix}, \qquad 12.$$

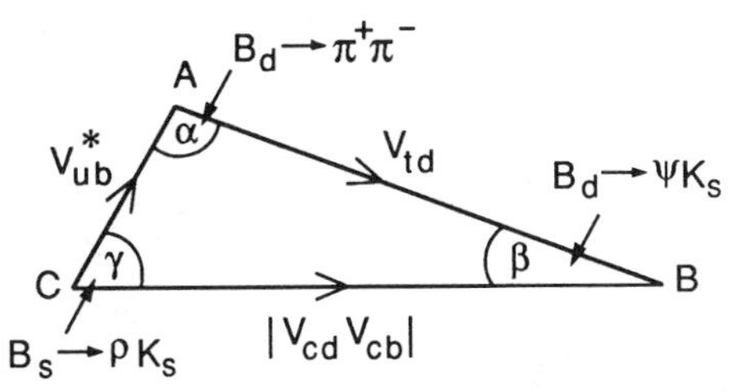

Figure 1 The unitarity triangle is a representation in the complex plane of the triangle formed by the CKM matrix elements $V_{ud}V_{ub}^*$, $V_{cd}V_{cb}^*$, and $V_{td}V_{tb}^*$.

where $c_{ij} \equiv \cos\theta_{ij}$ and $s_{ij} \equiv \sin\theta_{ij}$. The matrix V for the three-generation mixing is called the Cabibbo-Kobayashi-Maskawa (CKM) matrix.

The unitarity of the CKM matrix leads to relations such as

$$V_{ub}^* V_{ud} + V_{cb}^* V_{cd} + V_{tb}^* V_{td} = 0. \qquad 13.$$

The unitarity triangle is a geometrical representation of this relation in the complex plane: the three complex quantities $V_{ub}^* V_{ud}$, $V_{cb}^* V_{cd}$, and $V_{tb}^* V_{td}$ should form a triangle, as shown in Figure 1.

Rescaling the sides of the triangle by $1/|V_{cb}^* V_{cd}|$ and choosing a phase convention where $V_{cb}^* V_{cd}$ is real (this holds to a very good approximation for the parametrization in Equation 12), we obtain the coordinates of the three vertices A, B, and C as follows:

$$A\left[\frac{\mathrm{Re}\, V_{ub}}{s_{12}\,|V_{cb}|}, -\frac{\mathrm{Im}\, V_{ub}}{s_{12}\,|V_{cb}|}\right], \qquad B(1, 0), \qquad C(0, 0). \qquad 14.$$

Another commonly used parametrization is that of Wolfenstein (9), in which the coordinates of the vertex A are denoted by (ρ, η). The unitarity triangle gives a relation between the two most poorly determined entries of the CKM matrix, V_{ub} and V_{tb}. Thus it is convenient to present constraints on the values of the CKM parameters as bounds on the coordinates of the vertex A. Furthermore, the Standard Model predictions for the *CP* asymmetries in neutral B decays into certain *CP* eigenstates are fully determined by the values of the three angles of the unitarity triangle, α, β, and γ. Their measurement will test these Standard Model predictions and consequently provide a probe for physics beyond the Standard Model.

2. *CP* VIOLATION IN NEUTRAL B DECAYS

2.1 *General Formalism*

We consider a neutral meson B^0 and its antiparticle $\overline{B}^0$. The two mass eigenstates are B_H and B_L (H and L stand for Heavy and Light, respectively):

$$|B_L\rangle = p\,|B_0\rangle + q\,|\overline{B}^0\rangle, \qquad |B_H\rangle = p\,|B_0\rangle - q\,|\overline{B}^0\rangle. \tag{15.}$$

The eigenvalue equation is

$$\left(\mathbf{M} - \frac{i}{2}\mathbf{\Gamma}\right)\begin{pmatrix} p \\ \pm q \end{pmatrix} = \left(M_{L,H} - \frac{i}{2}\Gamma_{L,H}\right)\begin{pmatrix} p \\ \pm q \end{pmatrix}. \tag{16.}$$

Here $\mathbf{M}$ (the mass matrix) and $\mathbf{\Gamma}$ (which describes the exponential decay of the system) are 2×2 Hermitian matrices. Since $\Delta\Gamma \equiv \Gamma_H - \Gamma_L$ is produced by channels with branching ratios of $O(10^{-3})$ that contribute with alternating signs (10), we have $\Delta\Gamma \ll \Gamma$ and therefore may safely set $\Gamma_H = \Gamma_L \equiv \Gamma$. We define $M \equiv (M_H + M_L)/2$ and $\Delta M \equiv M_H - M_L$. Furthermore, $\Gamma_{12} \ll M_{12}$ (see Section 3) gives $|q/p| = 1$. The amplitudes for the states B_H or B_L at time t can be written as

$$A_H(t) = A_H(0)\exp[-(\Gamma/2 + iM_H)t] \qquad A_L(t) = A_L(0)\exp[-(\Gamma/2 + iM_L)t]. \tag{17.}$$

The proper time evolution of states that at time $t = 0$ were either pure B^0 $[A_L(0) = A_H(0) = 1/(2p)]$ or pure $\overline{B}^0$ $[A_L(0) = -A_H(0) = 1/(2q)]$ states is given by

$$|B^0_{phys}(t)\rangle = g_+(t)\,|B_0\rangle + (q/p)g_-(t)\,|\overline{B}^0\rangle \qquad |\overline{B}^0_{phys}(t)\rangle = (p/q)g_-(t)\,|B_0\rangle + g_+(t)\,|\overline{B}^0\rangle, \tag{18.}$$

where

$$g_+(t) = \exp(-\Gamma t/2)\exp(-iMt)\cos(\Delta M t/2) \qquad g_-(t) = \exp(-\Gamma t/2)\exp(-iMt)i\sin(\Delta M t/2). \tag{19.}$$

We are interested in the decays of neutral B's into a *CP* eigenstate (11, 12), which we denote by f_{CP}. We define the amplitudes for these processes as

$$A \equiv \langle f_{CP}|\mathcal{H}|B^0\rangle, \qquad \overline{A} \equiv \langle f_{CP}|\mathcal{H}|\overline{B}^0\rangle. \tag{20.}$$

We further define

$$\lambda \equiv \frac{q}{p}\frac{\bar{A}}{A}. \qquad 21.$$

Then

$$\langle f_{CP} \mid \mathcal{H} \mid B^0_{\text{phys}}(t)\rangle = A[g_+(t) + \lambda g_-(t)]$$
$$\langle f_{CP} \mid \mathcal{H} \mid \bar{B}^0_{\text{phys}}(t)\rangle = A(p/q)[g_-(t) + \lambda g_+(t)]. \qquad 22.$$

The time-dependent rates for initially pure B^0 or $\bar{B}^0$ states to decay into a final *CP* eigenstate at time t is given by:

$$\Gamma[B^0_{\text{phys}}(t)\to f_{CP}] = |A|^2 e^{-\Gamma t}\left[\frac{1+|\lambda|^2}{2} + \frac{1-|\lambda|^2}{2}\cos(\Delta M t) - \text{Im}\lambda\sin(\Delta M t)\right]$$
$$\Gamma[\bar{B}^0_{\text{phys}}(t)\to f_{CP}] = |A|^2 e^{-\Gamma t}\left[\frac{1+|\lambda|^2}{2} - \frac{1-|\lambda|^2}{2}\cos(\Delta M t) + \text{Im}\lambda\sin(\Delta M t)\right]. \qquad 23.$$

We define the time-dependent *CP* asymmetry as

$$a_{f_{CP}}(t) \equiv \frac{\Gamma[B^0_{\text{phys}}(t) \to f_{CP}] - \Gamma[\bar{B}^0_{\text{phys}}(t) \to f_{CP}]}{\Gamma[B^0_{\text{phys}}(t) \to f_{CP}] + \Gamma[\bar{B}^0_{\text{phys}}(t) \to f_{CP}]}. \qquad 24.$$

Then

$$a_{f_{CP}}(t) = \frac{(1 - |\lambda|^2)\cos(\Delta M t) - 2\text{Im}\lambda\sin(\Delta M t)}{1 + |\lambda|^2}. \qquad 25.$$

The quantity Imλ that can be extracted from $a_{f_{CP}}(t)$ can be directly related to CKM matrix elements in the Standard Model.

2.2 *Modes that Measure the Angles of the Unitarity Triangle*

The measurement of the *CP* asymmetry (Equation 24) will determine Imλ through Equation 25. If $|A/\bar{A}| = 1$ (as well as $|q/p| = 1$), then Equation 25 simplifies considerably:

$$a_{f_{CP}}(t) = -\text{Im}\lambda \sin(\Delta M t). \qquad 26.$$

Moreover, in this case Imλ depends on electroweak parameters only, without hadronic uncertainties. The condition that guarantees $|A/\overline{A}| = 1$ is easy to find (13). In general we can write

$$A = \sum_i A_i \exp(i\delta_i) \exp(i\phi_i), \quad \overline{A} = \sum_i A_i \exp(i\delta_i) \exp(-i\phi_i), \qquad 27.$$

where A_i are real, ϕ_i are CKM phases, and δ_i are strong phases. Thus, $|A| = |\overline{A}|$ if all amplitudes that contribute to the direct decay have the same CKM phase, which we denote by ϕ_D: $\overline{A}/A = \exp(-2i\phi_D)$. For $\Gamma_{12} \ll M_{12}$, we have $q/p = \sqrt{M_{12}^*/M_{12}} = \exp(-2i\phi_M)$, where ϕ_M is the CKM phase in the B-$\overline{\text{B}}$ mixing. Thus

$$\lambda = \exp[-2i(\phi_M + \phi_D)] \rightarrow \text{Im}\lambda = -\sin 2(\phi_M + \phi_D). \qquad 28.$$

Note that Imλ is independent of phase convention and does not depend on any hadronic parameters. In what follows, we concentrate on those processes that, within the Standard Model, are dominated by amplitudes that have a single CKM phase. For some cases in which there are two contributions with different weak phases, one can still extract the CKM parameters through isospin analysis, if sufficient data are available on the full set of isospin-related channels. This is discussed in Section 4.3.

For mixing in the B_d [B_s] system, we have $M_{12} \propto (V_{tb}V_{td}^*)^2$ [$(V_{tb}V_{ts}^*)^2$] and consequently,

$$\left(\frac{q}{p}\right)_{B_d} = \frac{V_{tb}^* V_{td}}{V_{tb} V_{td}^*}; \qquad \left(\frac{q}{p}\right)_{B_s} = \frac{V_{tb}^* V_{ts}}{V_{tb} V_{ts}^*}. \qquad 29.$$

For decays via quark subprocesses $b \rightarrow \bar{u}_i u_i d_j$, which are dominated by tree diagrams,

$$\frac{\overline{A}}{A} = \frac{V_{ib} V_{ij}^*}{V_{ib}^* V_{ij}}. \qquad 30.$$

Thus, for B_{d_j} decaying through $\bar{b} \rightarrow \bar{u}_i u_i \bar{d}_j$,

$$\text{Im}\lambda = \sin\left[2 \arg\left(\frac{V_{ib} V_{ij}^*}{V_{tb} V_{tj}^*}\right)\right]. \qquad 31.$$

For decays with a single K_S (or K_L) in the final state, K-$\overline{\text{K}}$ mixing plays an essential role since $B^0 \rightarrow K^0$ and $\overline{B}^0 \rightarrow \overline{K}^0$. Interference is possible only because of K-$\overline{\text{K}}^0$ mixing. for these modes

$$\lambda = \left(\frac{q}{p}\right)\left(\frac{\overline{A}}{A}\right)\left(\frac{q}{p}\right)_K, \qquad \left(\frac{q}{p}\right)_K = \frac{V_{cs} V_{cd}^*}{V_{cs}^* V_{cd}}. \qquad 32.$$

Decay processes $b \to \bar{s}sd_j$ are dominated by penguin diagrams. For these

$$\frac{\bar{A}}{A} = \frac{V_{tb}V_{tj}^*}{V_{tb}^*V_{tj}}. \qquad 33.$$

Note that the sign of (Imλ) depends on the *CP* transformation properties of the final state. The analysis above corresponds to *CP*-even final states. For *CP*-odd states, Imλ has the opposite sign. In what follows, we specify Imλ of *CP*-even states, regardless of the *CP* assignments of specific hadronic modes discussed.

CP asymmetries in decays to *CP* eigenstates, $B^0 \to f_{CP}$, provide a way to measure the three angles of the unitarity triangle (see Figure 1) defined by

$$\alpha \equiv \arg\left(-\frac{V_{td}V_{tb}^*}{V_{ud}V_{ub}^*}\right), \qquad \beta \equiv \arg\left(-\frac{V_{cd}V_{cb}^*}{V_{td}V_{tb}^*}\right),$$

$$\gamma \equiv \arg\left(-\frac{V_{ud}V_{ub}^*}{V_{cd}V_{cb}^*}\right). \qquad 34.$$

The aim is to perform enough independent measurements of the sides and angles that this triangle is overdetermined and thus one can test the validity of the Standard Model. We now give three explicit examples for asymmetries that measure the three angles α, β, and γ.

2.2.1 MEASURING SIN(2β) IN $B \to \psi K_S$ The mixing phase in the B_d system is given in Equation 29. With a single kaon in the final state, one has to take into account the mixing phase in the K^0 system given in Equation 32. The decay phase (Equation 30) in the quark subprocess $b \to c\bar{c}s$ is $\bar{A}/A = (V_{cb}V_{cs}^*)/(V_{cb}^*V_{cs})$. Thus

$$\lambda(B \to \psi K_S) = \left(\frac{V_{tb}^*V_{td}}{V_{td}V_{td}^*}\right)\left(\frac{V_{cs}^*V_{cb}}{V_{cs}V_{cb}^*}\right)\left(\frac{V_{cd}^*V_{cs}}{V_{cd}V_{cs}^*}\right)$$

$$\to \text{Im}\,\lambda = -\sin(2\beta). \qquad 35.$$

(As ψK_S is a $CP = -1$ state, there is an extra minus sign in the asymmetry that we suppress here.) In addition, there is a small penguin contribution to $b \to c\bar{c}s$. However, it depends on the CKM combination $V_{tb}V_{ts}^*$, which has, to a very good approximation, the same phase (modulo π) as the tree diagram, which depends on $V_{cb}V_{cs}^*$. Hence, only a single weak phase contributes to the decay. Other examples of final hadronic states in B_d decays that measure sin(2β) are χK_S, ϕK_S, $\eta_c K_S$, ωK_S, D^+D^-, $D^0\bar{D}^0$, and similar modes with K_L instead of K_S. In addition, vector-vector modes such as ΨK^* and $D^{*+}D^{*-}$ can be used with angular analysis (see Section 2.4).

2.2.2 MEASURING SIN(2α) IN $B \rightarrow \pi^+\pi^-$ Using Equation 29 and $(\bar{A}/A) = (V_{ub}V_{ud}^*)/(V_{ub}^*V_{ud})$ (from Equation 30), we get

$$\lambda(B \rightarrow \pi^+\pi^-) = \left(\frac{V_{tb}^*V_{td}}{V_{tb}V_{td}^*}\right)\left(\frac{V_{ud}^*V_{ub}}{V_{ud}V_{ub}^*}\right) \rightarrow \text{Im}\lambda = \sin(2\alpha). \qquad 36.$$

In this case, the penguin contribution is also expected to be small, but it depends on the CKM combination $V_{td}^*V_{tb}$, which has a phase different from that of the tree diagram. Uncertainties due to the penguin contribution can be eliminated using isospin analysis (14) (see Section 4.3). Other examples of final hadronic states in B_d decays that measure $\sin(2\alpha)$ are $\rho\pi^0$, $\pi^0\pi^0$, $\omega\pi^0$, and, with angular analysis, $p\bar{p}$ and $\rho\rho$.

2.2.3 MEASURING SIN(2γ) IN $B_s \rightarrow \rho K_S$ The mixing phase in the B_s system is given in Equation 29. Because of the final-state K_S, the mixing phase for the K^0 system has to be taken into account. The quark subprocess is the same as in $B_d \rightarrow \pi\pi$, namely $b \rightarrow u\bar{u}d$. Thus we get

$$\lambda(B_s \rightarrow \rho K_S) = \left(\frac{V_{tb}^*V_{ts}}{V_{tb}V_{ts}^*}\right)\left(\frac{V_{ud}^*V_{ub}}{V_{ud}V_{ub}^*}\right)\left(\frac{V_{cs}^*V_{cd}}{V_{cs}V_{cd}^*}\right)$$
$$\rightarrow \text{Im}\,\lambda = -\sin(2\gamma). \qquad 37.$$

Other examples for final hadronic states in B_s decays that measure $\sin(2\gamma)$ are ωK_S and similar modes with K_L instead of K_S.

These three examples demonstrate that the three angles of the unitarity triangle can, in principle, be measured independently of each other. In Tables 1 and 2 we list *CP* asymmetries for various channels in B_d and B_s decays, respectively. The channels are classified by the quark subprocess, denoted by i ($i = 1, \ldots, 6$) for different decaying

Table 1 *CP* asymmetries in B_d decays

Class (iq)	Quark subprocess	Final state (example)	SM prediction
1d	$\bar{b} \rightarrow \bar{c}c\bar{s}$	ψK_S	$-\sin(2\beta)$
2d	$\bar{b} \rightarrow \bar{c}c\bar{d}$	D^+D^-	$-\sin(2\beta)$
3d	$\bar{b} \rightarrow \bar{u}u\bar{d}$	$\pi^+\pi^-$	$\sin(2\alpha)$
4d	$\bar{b} \rightarrow \bar{s}s\bar{s}$	ϕK_S	$-\sin(2\beta)$
5d	$\bar{b} \rightarrow \bar{s}s\bar{d}$	K_SK_S	0
6d	$\bar{b} \rightarrow \bar{c}u\bar{s}$, $\bar{u}c\bar{s}$	$D^0_{CP}K^{*0}$	$\sin(\gamma)$

Table 2 *CP* asymmetries in B_s decays

Class (*i*q)	Quark subprocess	Final state (example)	SM prediction
1s	$\bar{b} \to \bar{c}c\bar{s}$	$\psi\phi$	0
2s	$\bar{b} \to \bar{c}c\bar{d}$	ψK_S	0
3s	$\bar{b} \to \bar{u}u\bar{d}$	ρK_S	$-\sin(2\gamma)$
4s	$\bar{b} \to \bar{s}s\bar{s}$	$\eta'\eta'$	0
5s	$\bar{b} \to \bar{s}s\bar{d}$	ϕK_S	$\sin(2\beta)$

mesons B_q (q = d, s). One possible hadronic final state for each class is listed as an example. Other states may be more favorable experimentally. We always quote the *CP* asymmetry for *CP*-even states, regardless of the specific hadronic state listed.

Perhaps the most difficult angle to measure will be γ, since at e^+e^- machines it is difficult to achieve a high production rate of B_s, and in hadron experiments a mode such as ρK_S is plagued by large backgrounds. An alternative way to measure γ (15, 16), using B decays into non-*CP* eigenstates, is denoted as 6d in Table 1. One has to measure the rates for B_d decays into $D^0_{CP}K^{*0}$, D^0K^{*0}, and $\overline{D}^0K^{*0}$, and the three *CP*-conjugate $\overline{B}_d$ decays (16). Here D^0_{CP} denotes a decay of a D^0 or $\overline{D}^0$ to a *CP* eigenstate such as $\pi^+\pi^-$. The flavor of the initial B can be identified from a flavor-tagging decay of the K^{*0}. The six rates can be used to extract the value of $|\sin\gamma|$ up to some discrete ambiguity. The feasibility of this method depends on branching ratios and techniques to separate the modes of interest from backgrounds. A similar method, using charged B decays, was suggested in by Gronau & Wyler (15). Studies of *CP* asymmetries in $B \to DK$ are presented elsewhere (17, 18).

Finally, we mention that the sign of the various asymmetries is predicted within the Standard Model (and not just the relative signs between various asymmetries). Measuring the signs of several asymmetries will test whether the CKM phase δ is indeed the source of *CP* violation (19).

2.3 *Current Constraints on Standard Model Parameters*

This section presents the current status of our knowledge of the experimental constraints on the parameters of the Standard Model (20, 21), updating and extending previous work (22–25). We use the unitarity triangle of the CKM matrix to show the constraints on the angles α, β, and γ as a function of top-quark mass. The resulting range of asymmetries allowed for a given type of B decay is evaluated. The luminosity of various colliders needed in order to guarantee a statistically significant measurement of *CP* violation in one or more types of B decay is then presented in Section 5.

We impose three constraints on the form of the unitarity triangle. [All the values of parameters quoted below are taken from a report by Nir (21), where the relevant references can be found.] First,

$$0.06 \leq |V_{ub}/V_{cb}| \leq 0.16, \qquad 38.$$

is directly measured in semileptonic B decay and is thus independent of m_t. The other two constraints depend on loop processes: *CP* violation in K-$\overline{K}^0$ mixing parametrized by ϵ, and B_d-$\overline{B}_d$ mixing parameterized by x_d. The resulting constraints depend strongly on the yet-unknown mass of the top quark, m_t. The analytical expressions are (26)

$$x_d \equiv \frac{\Delta M(B^0)}{\Gamma(B^0)} = \tau_b \frac{G_F^2}{6\pi^2} \eta_B M_B (B_B f_B^2) M_t^2 f_2(y_t) \, |V_{td}^* V_{tb}|^2, \qquad 39.$$

$$|\epsilon| = \frac{G_F^2}{12\pi^2} \frac{M_K}{\sqrt{2}\Delta M_K} (B_K f_K^2) M_W^2 \times \{\eta_1 y_c \mathrm{Im}\,[(V_{cd}^* V_{cs})^2]$$

$$+ \eta_2 y_t f_2(y_t) \mathrm{Im}\,[(V_{td}^* V_{ts})^2] + 2\eta_3 f_3(y_t) \mathrm{Im}\,(V_{cd}^* V_{cs} V_{td}^* V_{ts})\}, \qquad 40.$$

where $y_i \equiv m_i^2/M_W^2$ and

$$f_2(y_t) = 1 - \frac{3}{4}\frac{y_t(1+y_t)}{(1-y_t)^2}\left[1 + \frac{2y_t}{1-y_t^2}\ln(y_t)\right]$$

$$f_3(y_t) = \ln\left(\frac{y_t}{y_c}\right) - \frac{3}{4}\frac{y_t}{1-y_t}\left[1 + \frac{y_t}{1-y_t}\ln(y_t)\right]. \qquad 41.$$

The values of well-known quantities used here are

$$f_K = 0.165 \text{ GeV}; \quad m_c = 1.4 \text{ GeV}; \quad M_B = 5.28 \text{ GeV};$$

$$M_W = 80 \text{ GeV}; \quad G_F = 1.166 \times 10^{-5} \text{ GeV}^{-2}; \qquad 42.$$

$$|V_{us}| = \sin\theta_C = 0.22; \quad |\epsilon| = 2.26 \times 10^{-3}.$$

The QCD correction factors for ϵ (27) and x_d (28) are

$$\eta_B = 0.85; \quad \eta_1 = 0.7, \quad \eta_2 = 0.6, \quad \eta_3 = 0.4. \qquad 43.$$

We consider the range $0.038 \leq |V_{cb}| \leq 0.052$, and $90 \leq m_t \leq 200$ GeV. As examples we choose $m_t = 90$, 120, 160, and 200 GeV. The constraint on $|V_{ub}/V_{cb}|$ (see Equation 38) forces the vertex A to lie between two circles (dotted in Figure 2) centered at the vertex $C(0, 0)$.

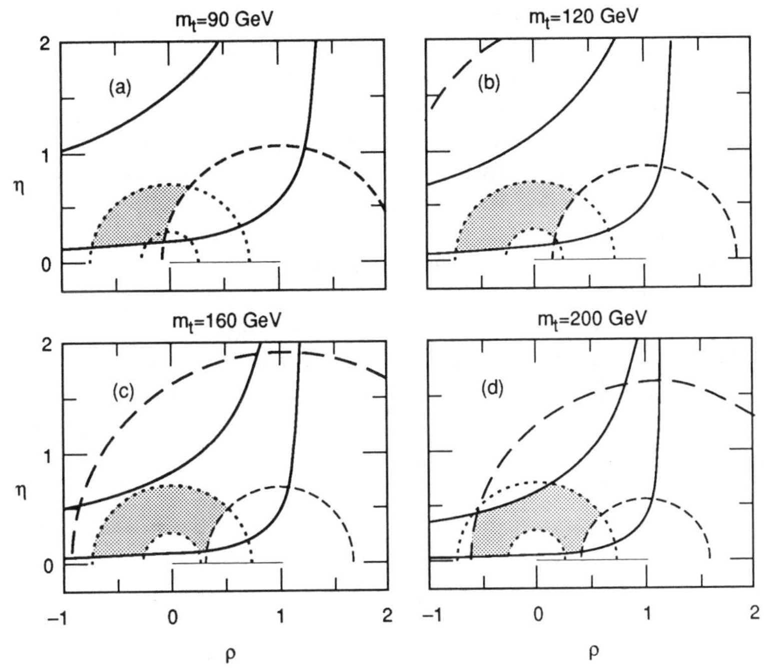

Figure 2 Constraints (21) from $|V_{ub}/V_{cb}|$ (*dotted lines*), x_d (*dashed curves*), and ϵ (*solid curves*) on the rescaled unitarity triangle for $m_t = 90$, 120, 160, and 200 GeV. The shaded region is that allowed for the vertex $A(\rho, \eta)$.

The constraint on x_d (see Equation 39) requires the vertex A to lie between two circles (dashed in Figure 2) centered at $B(1, 0)$. The width of this band arises mainly from theoretical uncertainties in $B_B f_B^2$ and, to a lesser extent, from errors in lifetime and mixing measurements:

$$(0.1\ \text{GeV})^2 \leq B_B f_B^2 \leq (0.2\ \text{GeV})^2$$
$$2.9 \times 10^9 \leq \tau_b\,|V_{cb}|^2 \leq 4.1 \times 10^9\ \text{GeV}^{-1} \qquad 44.$$
$$0.55 \leq x_d \leq 0.77.$$

The constraint on ϵ (see Eq. 40) demands that the vertex A lie between the two hyperbolas (solid curves in Fig. 2). The width of this band arises from the theoretical uncertainty in $|V_{cb}|$ and in the B_K parameter,

$$1/3 \leq B_K \leq 1. \qquad 45.$$

The resulting allowed domain for the vertex A is indicated by the shaded region in Figure 2.

The allowed value for the angles α, β, and γ can be deduced from Figure 2. Note that values of 45° or 135° correspond to a maximum *CP* asymmetry, while 90° for an angle implies that there will be no CP asymmetry in the corresponding class of B decays. However, if one angle is 90°, then *CP* violation will necessarily exhibit itself in the decays that measure the other two angles. Examining Figure 2, we see that either α or γ may be 90° for all top masses. Consequently, zero asymmetries may occur for class (3d), e.g. $B_d \to \pi^+\pi^-$, or class (3s), e.g. $B_s \to \rho K_S$ decays, respectively. In fact, for $\sin(2\alpha)$ and $\sin(2\gamma)$, all values are allowed. The possibilities range from maximum ($|\mathrm{Im}\lambda| = 1$) to vanishing ($|\mathrm{Im}\lambda| = 0$) *CP* asymmetry.

The fact that a particular interference term might vanish is disconcerting; however, failure to observe *CP* violation in just a class (3d) or just a class (3s) process would not be evidence against *CP* violation originating in the CKM matrix. Fortunately, a nonvanishing asymmetry is guaranteed in decays of classes (1d), (2d), and (4d) in the Standard Model (see Figure 3*a*), since the angle β satisfies

$$2° < \beta \leq \arcsin|V_{ub}/(V_{cd}V_{cb})| \lesssim 47°, \quad 46.$$

giving (see Figure 3*b*)

$$0.08 \leq \sin(2\beta) \leq 1. \quad 47.$$

Moreover, this is just the angle that can be most readily measured. As we discuss below, an e^+e^- B factory with a luminosity of 3×10^{33} $cm^{-2}s^{-1}$ is expected to achieve within one year of running an accuracy of ± 0.05 in the measurement of $\sin(2\beta)$, while a hadron collider such as the upgraded Tevatron at Fermilab could achieve an accuracy of about ± 0.15.

The three angles of the unitarity triangle are, of course, correlated. Thus, an experiment that measures asymmetries proportional to both $\sin(2\alpha)$ and $\sin(2\beta)$ is assured that $|\mathrm{Im}\,\lambda| \geq 0.1$–$0.2$ (the exact value depends on the top mass) for at least one of the two processes (24). Similarly, an experiment searching simultaneously for *CP* asymmetries in processes sensitive to each of the three different angles is guaranteed to find that $|\mathrm{Im}\,\lambda| \geq 0.2$–$0.3$ for at least one of the three classes of *CP*-violating asymmetries.

The allowed range for the unitarity triangle is rather sensitive to the value of f_B. Recent lattice calculations give, instead of the range in Equation 44, $f_B \sim 0.3$–0.5 GeV (29–32). It is interesting to note that a recent QCD sum rule calculation, consistent with heavy-quark symmetry constraints, yields $f_B = 0.19 \pm 0.05$ GeV (33). We believe that it is premature to include these lattice calculation results in our range for f_B since they are still being tested and refined, and it appears that finite-mass corrections will lower the values (e.g. 34, 35). However,

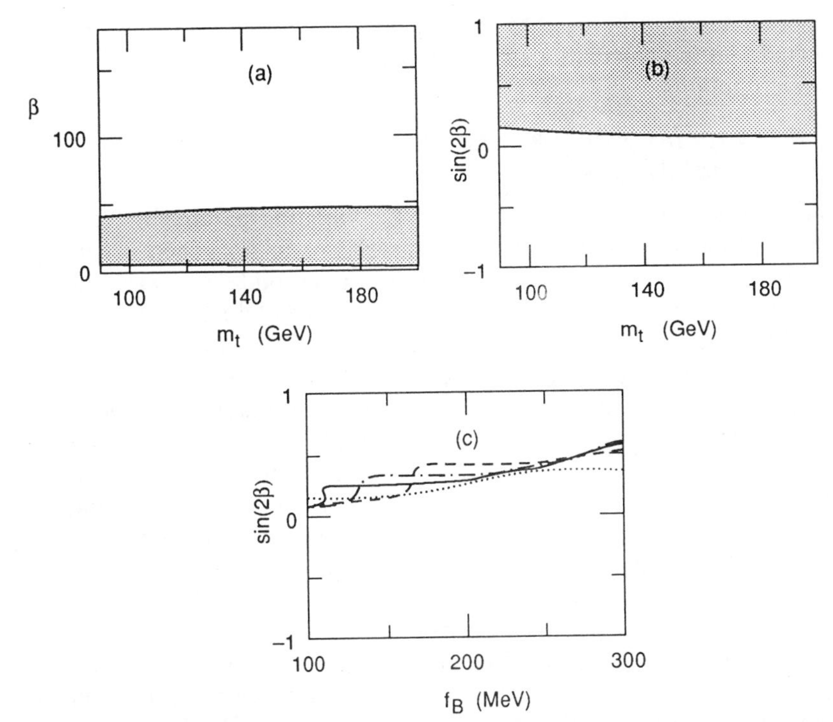

Figure 3 (*a*) The allowed range for the angle β of the unitarity triangle as a function of the top mass. (*b*) The allowed range for sin(2β) as a function of the top mass. (*c*) The lower bound on sin(2β) as a function of f_B for top masses of 90 GeV (*dotted*), 120 GeV (*dashed*), 160 GeV (*dot-dashed*), and 200 GeV (*solid*).

we note that large values of f_B would be very favorable for the measurements of *CP* asymmetries. To demonstrate that, we give in Figure 3*c* the lower bound on sin(2β) (i.e. $a_{\psi K_S}$) as a function of f_B for various values of m_t. Note that if $m_t \gtrsim 120$ GeV and $f_B \gtrsim 170$ MeV, then $\sin(2\beta) \gtrsim 0.26$.

A further comment on the program of testing the unitarity triangle is in order. Suppose we cannot measure all three angles. [As mentioned above, sin(2γ) may be difficult to study.] The triangle can be overdetermined by measuring two angles and one additional side—for example, by determining $|V_{ub}|$. Measurements of the decay $B \rightarrow \rho e \nu$ and comparison to $D \rightarrow \rho e \nu$ can be done quite accurately at a B factory. The question here is how well can theorists determine the model-dependent $1/m_c$ corrections to the heavy-quark symmetry that relates the B- and D-decay form factors (see 21 for a review and references on this topic). A priori these corrections can be large, and a variety of approaches may be used to try to achieve accurate estimates. These include lattice calculations as well as the more traditional models for wavefunctions.

2.4 *Angular Analysis and Other Ways to Treat Additional Modes*

The simplest modes to analyze for *CP*-violating parameters are the decays to a pure *CP* eigenstate such as ψK_S or $\pi^+\pi^-$. There are many other decay modes that contain both even and odd *CP* contributions, but that can still be used to extract the fundamental parameters of the CKM matrix by carefully selecting a definite *CP* contribution.

The simplest case is any collection of *CP* self-conjugate particles, where the mixture of *CP* states comes from the possibility of more than one orbital angular momentum state. Since $P \propto (-1)^L$, even- and odd-L states contribute with opposite *CP*. This apparently nonrelativistic argument can be restated in a fully relativistic form using the helicity formalism. The problem of isolating a definite *CP* contribution can then be studied in terms of helicities from an angular analysis of the decays of unstable particles with spin such as ρ or K^*. The use of such methods in the decays of a scalar to two vector particles was discussed in general by Nelson (36, 37) and for B decays by Valencia (38). A special case of this analysis in the context of a modern B factory has been pointed out, namely that selecting zero-helicity vector particles is one way to select a definite *CP* state (39). Subsequently, a study of various approaches to angular analyses for few-body B decays was carried out (40). We here summarize a few of the results.

When sufficient data are available on a set of isospin-related decay channels, for example ψK^{*0} and ψK^{*+}, the best method is to determine the helicity and isospin amplitudes by a maximum likelihood fit to all the data. The *CP* asymmetry is one of the parameters that can be extracted from such a fit. In such an analysis, small *CP*-violating contributions in charged B-decay channels that can arise from penguin processes are neglected.

When the information from related channels is not available to determine the parameters, one can still perform an angular analysis to isolate definite *CP* contributions. A particularly attractive way to do this for final vector-vector states is first to define a plane in the rest frame of the B that contains the decay products of one of the vector particles and the second vector particle and then to analyze the spin projection of the second vector particle on the direction perpendicular to this plane. This quantity, called transversity, is invariant under boosts in the plane and is directly related to the *CP* eigenvalue. One advantage of this method is that it does not require a true two-body process; the three-body state is all that is needed to define the plane and hence nonresonant production of the pair of scalars or of leptons can also be included in the statistical sample.

Dunietz et al list additional modes that could possibly be studied using an angular analysis (40). The usefulness will depend on branching ratios, which are not yet known for many of these modes. In a few cases more definite statements can be made. For example, BR(B $\rightarrow$ ψK^*) is larger than BR(B $\rightarrow$ ψK_S) by a factor of 3–6. Angular analysis applied to the decay of the K^* can yield a measure of the angle β of comparable accuracy to that obtainable from the ψK_S channel, where no such analysis is needed. If both *CP* states contribute equally to the decay, it will require about 5000 reconstructed events, or about four times as many as for the pure *CP* channel ψK_S, to achieve $\delta(\sin 2\beta) = \pm 0.05$ (W. Toki, private communication, 1990). Results from Argus indicate that this decay may be dominated by a single *CP* eigenmode, such that ψK^* measurements may prove superior to ψK_S in accuracy for a given luminosity (42). For the comparison of $\pi\pi$ with $\rho\rho$, the branching ratios are not yet known, but it is likely that the latter channel has a significantly larger branching ratio and hence again may be as important as the simpler case in extracting an accurate value for the angle α.

Clearly one would like to use as many modes as possible for extracting the underlying *CP*-violating parameters of the Standard Model. Another potentially useful class of decays (43) includes two-body decays in which the particles are not *CP* eigenstates but the underlying set of four quarks is unchanged in a *CP* transformation, e.g. $\rho^+\pi^-$. Again, the data will allow extraction of the underlying CKM parameters, provided a sufficient set of related channels can be measured. Yet another possibility is to extract *CP* asymmetries from Dalitz plot distributions (44–46).

The usefulness of any of these methods will depend on the ability of the detectors to give good angular resolution of the particle decays and good particle identification. However, there is by now quite a bag of tricks awaiting any data with sufficient statistics. At present it appears that none of the more complicated analyses will yield better results for the underlying CKM parameters than the simplest modes first studied, but they could be competitive with them in accuracy.

3. PHYSICS BEYOND THE STANDARD MODEL

CP asymmetries in B decays are a sensitive probe of new physics in the quark sector, because they are likely to differ from the Standard

Model predictions if there are sources of *CP* violation beyond the CKM phase of the Standard Model. This can contribute in two ways:

1. If there are significant contributions to B-$\overline{\text{B}}$ mixing (or B_s-$\overline{B}_s$ mixing) beyond the box diagram with intermediate top quarks; or
2. If the unitarity of the three-generation CKM matrix does not hold, namely if there are additional quarks.

Other ingredients of the analysis of *CP* asymmetries in neutral B decays are likely to hold in any model that satisfies current experimental constraints:

3. $\Gamma_{12} \ll M_{12}$. In order for this relation to be violated, one needs a new dominant contribution to tree decays of B mesons or strong suppression of the mixing compared to the Standard Model box diagram. Neither possibility is likely. The argument is particularly solid for the B_d system, as it is supported by experimental evidence: $\Delta M/\Gamma \sim 0.7$, while branching ratios into states that contribute to Γ_{12} are $\lesssim 10^{-3}$.
4. The relevant decay processes (in classes $i = 1,2,3$) are dominated by the Standard Model tree diagrams. Again, it is unlikely that new physics, which typically takes place at a high energy scale, would compete with weak tree decays.
5. The phase of the mixing amplitude for neutral kaons is approximately $\arg(V_{cd}^* V_{cs})$. This is practically guaranteed by the measurement of the ϵ parameter (47).

Within the Standard Model, both b-quark decays and B_q-$\overline{B}_q$ mixing are determined by combinations of CKM matrix elements. The asymmetries then measure the relative phases between these combinations. Unitarity of the CKM matrix directly relates these phases (and consequently the measured asymmetries) to angles of the unitarity triangle. In models with new physics, unitarity of the three-generation, charged-current, mixing matrix may be lost and consequently the relation between the CKM phases and the angles of the unitarity triangle violated. But this is not the main reason that the predictions for the asymmetries are modified. The reason is rather that if B_q-$\overline{B}_q$ mixing has significant contributions from new physics, the asymmetries measure different quantities: the relative phases between the CKM elements that determine b decays and the elements of mixing matrices in sectors of new physics (squarks, multi-Higgs, etc) that contribute to B_q-$\overline{B}_q$ mixing.

Thus, when studying *CP* asymmetries in models of new physics, we look not only for violation of the unitarity constraints:

$$V_{ud}^{*}V_{ub} + V_{cd}^{*}V_{cb} + V_{td}^{*}V_{tb} = 0$$
$$V_{us}^{*}V_{ub} + V_{cs}^{*}V_{cb} + V_{ts}^{*}V_{tb} = 0, \qquad 48.$$

but also for contributions to B_q-$\overline{B}_q$ mixing that are different in phase and at least comparable in magnitude to the Standard Model (SM) contribution:

$$M_{12}^{SM}(B_q) = \frac{G_F^2}{12\pi^2}\eta M_B(B_B f_B^2)M_W^2 y_t f_2(y_t)(V_{tb}^{*}V_{tq})^2. \qquad 49.$$

The results of a survey of models beyond the Standard Model (48) are summarized in Figure 4. Models beyond the Standard Model that have been studied include four quark generations (49–52), a multiscalar model with natural flavor conservation (NFC) and with explicit or spontaneous *CP* violation (SCPV) (53), *Z*-mediated flavor-changing neutral currents (FCNC) (54, 55), left-right symmetry (LRS) (56, 57), supersymmetry (SUSY) (58), and "real superweak" models (59). Many of these models allow substantial deviations from the Standard Model predictions.

Some relations among the asymmetries do not depend on certain assumptions and thus may hold beyond the Standard Model or, conversely, if they are violated can help pinpoint which ingredients must be added to the Standard Model (19, 47, 48). The predictions

$$\text{Im}\,\lambda_{1d} = \text{Im}\,\lambda_{2d}, \qquad \text{Im}\,\lambda_{1s} = \text{Im}\,\lambda_{2s} \qquad 50.$$

depend only on the mechanism for tree-level decays and the K^0-$\overline{K}^0$ mixing phase. Existing constraints already ensure that these will hold in any viable models. Violation of

$$\text{Im}\,\lambda_{1d} = \text{Im}\,\lambda_{4d}, \qquad \text{Im}\,\lambda_{1s} = \text{Im}\,\lambda_{4s}, \qquad 51.$$

would indicate that the second unitarity relation in Equation 48 is violated. Similarly, there are clean tests of the first relation in Equation 48. Violation of

$$\text{Im}\,\lambda_{1s} = 0 \qquad 52.$$

would indicate a new mechanism for B_s-$\overline{B}_s$ mixing. Violation of

$$\text{Im}\,\lambda_{2d} = -\sin(2\beta), \qquad \text{Im}\,\lambda_{3d} = \sin(2\alpha), \qquad 53.$$

would indicate a new mechanism for B_d-$\overline{B}_d$ mixing. Finally, we note that the three angles deduced from measurements of Im λ_{1d}, Im λ_{3d}, and Im λ_{3s} will sum to 180° whenever the amplitude for B_s-$\overline{B}_s$ mixing

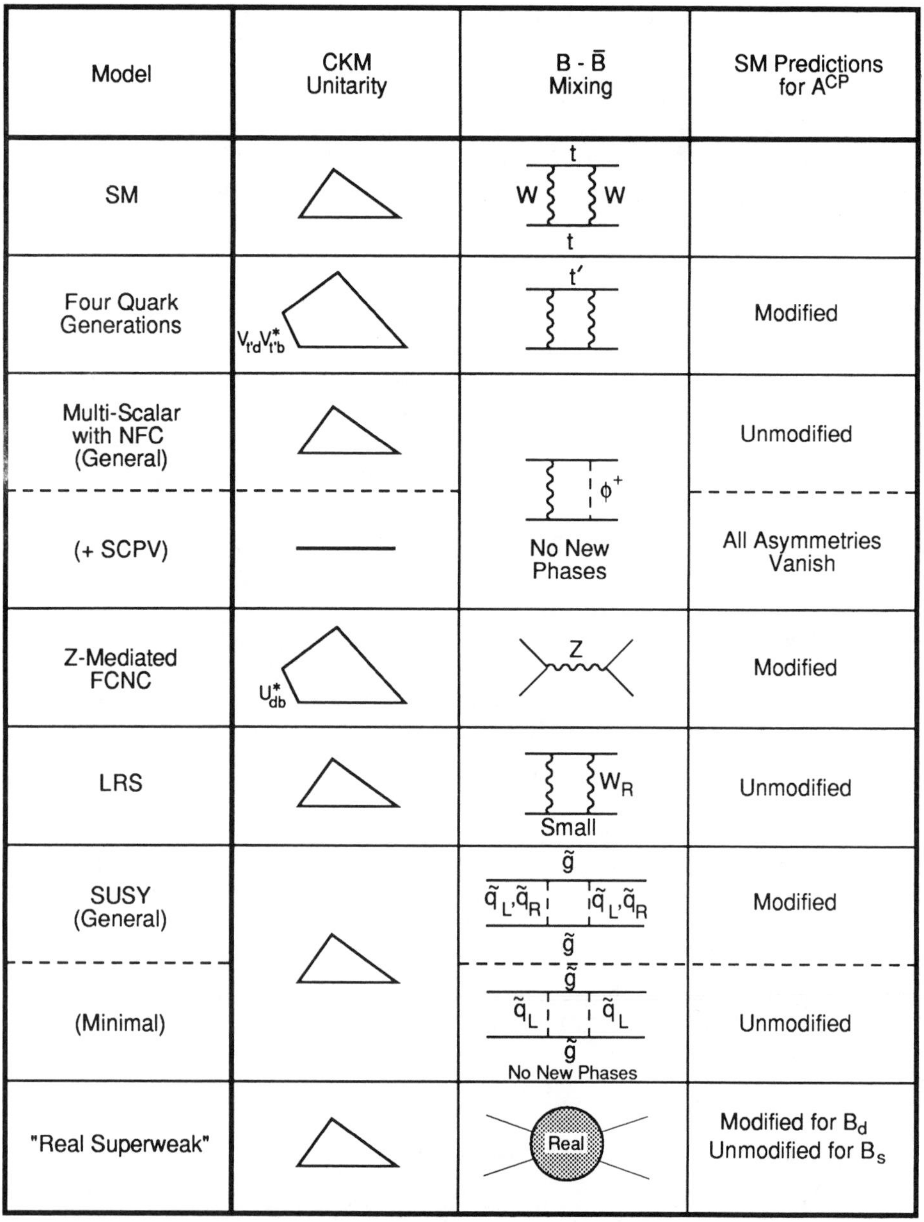

Model	CKM Unitarity	B - $\bar{B}$ Mixing	SM Predictions for A^{CP}
SM	(triangle)	W, t, t, W	
Four Quark Generations	(quadrangle) $V_{t'd}V^*_{t'b}$	t′	Modified
Multi-Scalar with NFC (General)	(triangle)	ϕ^+	Unmodified
(+ SCPV)	(line)	No New Phases	All Asymmetries Vanish
Z-Mediated FCNC	(quadrangle) U^*_{db}	Z	Modified
LRS	(triangle)	W_R Small	Unmodified
SUSY (General)	(triangle)	$\tilde{g}$, $\tilde{q}_L,\tilde{q}_R$, $\tilde{q}_L,\tilde{q}_R$, $\tilde{g}$	Modified
(Minimal)		$\tilde{g}$, $\tilde{q}_L$, $\tilde{q}_L$, $\tilde{g}$ No New Phases	Unmodified
"Real Superweak"	(triangle)	Real	Modified for B_d Unmodified for B_s

Figure 4 New physics effects on *CP* asymmetries in neutral B decays (48). The second column describes whether unitarity of the three-generation CKM matrix is maintained (*a triangle*) or violated (*a quadrangle*). The third column gives an example of a new contribution to the mixing. Unless otherwise mentioned, the contribution could be large and carry new phases.

is real (47). This is independent of whether they correspond to the angles of the unitarity triangle or not.

4. TREE AND PENGUIN PROCESSES

4.1 *General Discussion*

In the Standard Model the decays of any meson containing a heavy quark necessarily proceed via charged-current interactions. The amplitudes are generally divided into two classes, called tree or penguin type. If all the complications of strong interactions are ignored, this classification is readily explained in terms of the quark diagrams for the underlying weak transition (see Figure 5). The penguin process is one in which the W boson is reabsorbed on the same quark line from which it was emitted, and all other diagrams are tree processes; that is, they have no closed loop in the weak diagram. Tree diagrams can be further subdivided into spectator (where the light quark in the initial meson is disconnected in the weak decay diagram), exchange (where the W is exchanged between the two quarks of the initial meson), and annihilation (where the quark and antiquark of the initial meson annihilate to form the W). These subdivisions are unimportant in a discussion of *CP* violation because whenever two types of tree diagrams contribute to the same decay amplitude they do so with the same CKM matrix element and hence the same weak phase.

For the penguin contributions, we have drawn the diagram in Figure 5 without identifying the gluon which is emitted from the W-quark loop

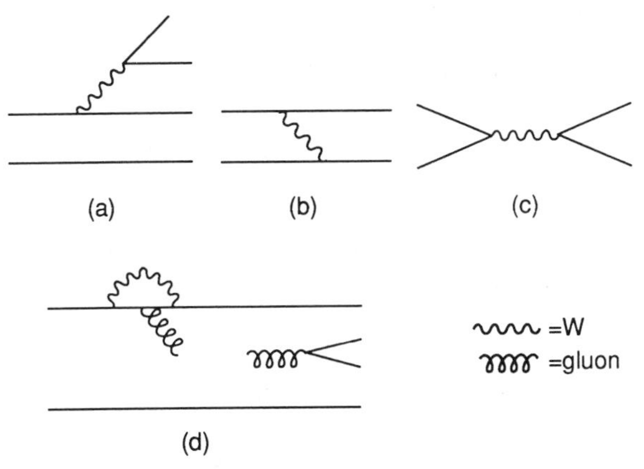

Figure 5 Tree and penguin contributions. Diagrams (*a*) spectator, (*b*) W exchange, and (*c*) annihilation are all tree contributions. Diagram (*d*) represents the penguin contribution. The gluons associated with binding are not shown.

with that which produces the additional quark-antiquark pair. We did this to stress the fact that the term penguin in principle includes all such contributions. Note that the many gluons involved in the binding are not drawn here, so the disconnected line simply means a gluon absorbed in and another produced from the general glue.

In contrast to the various tree diagrams, the penguin contributions must be treated separately in analyzing *CP* violations because, in general, they will depend on different CKM matrix elements. In fact, in the penguin transition $b \rightarrow q$ ($q = d,s$), there are three penguin terms with different quarks in the loop ($i = u,c,t$), each contributing with a different CKM combination:

$$v_{iq} = V_{ib}^{*} V_{iq}. \qquad 54.$$

(With three quark generations, unitarity requires

$$v_{uq} + v_{cq} + v_{tq} = 0, \qquad 55.$$

which allows one of these three phases to be expressed in terms of the remaining two.) Hence it becomes important to be able to calculate the relative strengths and strong phases of the tree and penguin terms.

The penguin involves a strong interaction processes, as shown in Figure 5. When the penguin contribution is evaluated perturbatively one identifies the two gluon lines in Figure 5 for the leading contribution and then adds additional gluon corrections for a higher-order calculation. The justification for this perturbative treatment is that the gluon emitted from the quark loop is quite hard because of the large mass difference between the initial b quark and the final s or d quark. We are doubtful that this approach is completely correct. For example, a contribution in which the hard gluon is absorbed by the other quark of the original meson could be comparable to the one usually calculated, particularly in inclusive rates. The standard approach is to calculate the penguin contribution perturbatively.

Such a calculation gives an estimate of the inclusive asymmetries summed over all states with a particular quark content. It is difficult to convert these numbers into reliable estimates for rates and asymmetries for particular exclusive (few-body) decays (see, for example, 60). Each configuration of final hadrons corresponds to some weighted integral over quark kinematics, but unfortunately we have no way to determine that integral reliably. Since the calculated quark-level asymmetries depend on the momentum transfer to the $q\bar{q}$ pair and since in some cases they even change sign as a function of this variable, it is very difficult to convert the quark estimates into estimates for exclusive hadron processes.

Furthermore, because of the dependence of the asymmetry on the difference of strong phases as well as that of the weak phases, calculations are sensitive to other aspects of hadronization. In the quark diagram calculation, the long-range, final-state, hadron-hadron interaction phase shifts are ignored. The only strong phases included are those that arise from the absorptive parts of penguin loop diagrams. The assumption that no additional phase shifts are caused by final-state rescattering is known as factorization. It is built into these calculations but has not yet been well tested. For a discussion of the justification of this assumption, see the report by Bjorken (45). On the other hand, it has been argued that hadronization can result in final-state phase shifts that could decrease the resulting asymmetries compared to the quark-diagram perturbative calculations (61).

In the following section we review *CP* asymmetries in charged B decays. With the exception of the special case $B \rightarrow D^0_{CP}K(+n\pi)$, the penguin processes are central to the existence of any *CP* asymmetries in these decays. Hence these asymmetries are much more difficult to predict reliably than those of the neutral B decays. If such asymmetries are observed, they will establish the existence of penguin-type processes and direct *CP* violation, but it will be difficult to extract from these measurements any of the fundamental CKM parameters.

4.2 CP *Violation in Charged B Decays*

In charged B decays, one can search for *CP*-violating differences of the form

$$a_f = \frac{\Gamma(B^+ \rightarrow f) - \Gamma(B^- \rightarrow \bar{f})}{\Gamma(B^+ \rightarrow f) + \Gamma(B^- \rightarrow \bar{f})}, \qquad 56.$$

where f is any final state and $\bar{f}$ is its *CP* conjugate. Although *CPT* symmetry requires that the total B^+ and B^- decay widths be the same, specific channels or sums of channels can contribute to asymmetries of the form in Equation 56. For this to occur, there must be interference between two separate amplitudes that contribute to the decay $B^+ \rightarrow f$ with different weak phases, $\phi_1 \neq \phi_2$, and with different strong phase shifts, $\delta_1 \neq \delta_2$. Then

$$a_f \propto \sin(\delta_1 - \delta_2)\sin(\phi_1 - \phi_2). \qquad 57.$$

It was recognized by Bander, Silverman & Soni (62) that, within the Standard Model, these conditions can readily be met in three types of B decays: CKM-suppressed, CKM-forbidden, and radiative decays.

4.2.1 CKM-SUPPRESSED B DECAYS The tree amplitudes for $b \to u\bar{u}s$ can interfere with penguin-type processes.

4.2.2 CKM-FORBIDDEN B DECAYS In the channels $b \to sd\bar{d}$, $b \to dd\bar{d}$, $b \to ss\bar{s}$, and $b \to ds\bar{s}$, which have no tree contributions, there can still be asymmetries due to the interference of penguin contributions with different charge-2/3 quarks in the loop. For example, in $b \to ds\bar{s}$

$$A_{b \to ds\bar{s}} = v_{ud}A_{u,ds\bar{s}} + v_{cd}A_{c,ds\bar{s}} + v_{td}A_{t,ds\bar{s}}, \qquad 58.$$

where v_{iq} is defined in Equation 54 and $A_{i,ds\bar{s}}$ is the corresponding penguin amplitude (including the strong phase shift). We can use the unitarity constraint (Equation 55) to eliminate v_{tq}:

$$A_{b \to ds\bar{s}} = v_{ud}\Delta_{ut}(ds\bar{s}) + v_{cd}\Delta_{ct}(ds\bar{s}). \qquad 59.$$

This then leads to a nonvanishing asymmetry,

$$a_{ds\bar{s}} \propto \mathrm{Im}\,(v_{ud}^{*}v_{cd})\ \mathrm{Im}\,(\Delta_{ut}^{*}\Delta_{ct}), \qquad 60.$$

provided the quantities Δ_{ct} and Δ_{ut} have different strong final-state phases. The phases of the penguin amplitudes can be evaluated by examining the imaginary parts of the amplitude corresponding to various possible cuts through the diagrams. If the u and c quarks were degenerate, the two contributions Δ_{ut} and Δ_{ct} would be identical and the asymmetry would vanish.

Since the penguin amplitudes are each of order α_s, the penguin-penguin interference term is of order α_s^2. Thus a consistent perturbative calculation must take into account all other order α_s^2 contributions to the rate (63).

4.2.3 RADIATIVE B DECAYS This mechanism for *CP* asymmetries is similar to that of the pure penguin cases discussed above, except that the leading contribution to the decay is an electromagnetic penguin, that is one where the gluons shown in Figure 5 are replaced by a single photon.

A special case is the channel $B^{\pm} \to D^0_{CP}K^{\pm}$, where D^0_{CP} represents the decay of a D^0 or $\bar{D}^0$ to a *CP* eigenstate such as $\pi^+\pi^-$ (15). Here interference between the D^0 and $\bar{D}^0$ tree contributions can give rise to a *CP* violation. This is an exception to the general statement that observation of *CP* violation in charged B decays requires nonvanishing penguin contributions. *CP* violation in this mode occurs because the D^0 and $\bar{D}^0$ have common decay channels, and the weak amplitudes for $B^0_d \to D^0, \bar{D}^0$ have different phases.

Detailed studies of expected asymmetries in charged B decays for

CKM-suppressed and CKM-forbidden B decays have recently been carried out by two groups (64, 65) with similar conclusions; namely, $a_{\mathrm{su\bar{u}}}$, $a_{\mathrm{sd\bar{d}}}$, and $a_{\mathrm{ss\bar{s}}}$ are a few tenths of a percent, while for the rarer processes $a_{\mathrm{dd\bar{d}}}$ and $a_{\mathrm{ds\bar{s}}}$ could be as large as a few percent. These estimates are for the inclusive sum of all channels with a given quark content, which is not readily measured. The cases $\mathrm{b} \rightarrow \mathrm{q\bar{d}d}$ would be particularly difficult to extract since many modes with a contribution from this quark content will also have $\mathrm{b} \rightarrow \mathrm{q\bar{u}u}$ contributions. (For the q = d case, such terms dominate all channels). Even to extract the $\mathrm{q\bar{d}d}$ contribution requires careful subtraction of different isospin combinations. It is unlikely that one can measure the small asymmetry in the resulting quantity. Earlier estimates based on model calculations (66) give larger asymmetries for some exclusive channels. Estimates of asymmetries in baryonic modes are given by Eilam et al (67). Asymmetries in radiative B decays have been studied by Soares (68), who found $a_{\mathrm{s}\gamma} \sim (1\text{–}10) \times 10^{-3}$ and $a_{\mathrm{d}\gamma} \sim (1\text{–}30) \times 10^{-2}$.

Some calculations suggest that the *CP* violations in charged B decays will be extremely difficult to observe, requiring $O(10^{10})$ produced B's for exclusive $\mathrm{b} \rightarrow \mathrm{s}$ modes and $O(10^9)$ B's for exclusive $\mathrm{b} \rightarrow \mathrm{d}$ modes (64, 65). This would be improved to perhaps as low as 10^7 produced B's if one could sum all two-body or quasi-two-body $\mathrm{b} \rightarrow \mathrm{ds\bar{s}}$ modes (64), but the experimental difficulties of such a semi-inclusive measurement may defeat this theoretical improvement. These estimates include neither any factors for the inefficiencies introduced by triggering and detection requirements, nor suppression due to final-state rescattering, and thus are quite optimistic. The situation is even more difficult for the radiative decays (68), which give comparable asymmetries but have lower branching ratios.

Although the uncertainties inherent in the calculations described above cannot exclude some small possibility of larger effects [see, for example, the model predictions of Chau & Cheng (66)], it appears that the calculations are sufficiently reliable that asymmetries that are an order of magnitude larger than predicted (64, 65, 68) would have to be interpreted as evidence for some *CP*-violating mechanism that arises from sources beyond the Standard Model. Various "beyond standard" models contain novel *CP*-violating decay mechanisms that could be comparable to the Standard Model penguin contributions. For example, with four quark generations there is a penguin diagram with an intermediate t' that depends on additional phases of the 4×4 mixing matrix; in models with *Z*-mediated flavor-changing neutral currents, there is a tree diagram that depends on new phases in the neutral current mixing matrix.

From the discussion in this section, we conclude that, unfortunately, with regard to the charged B decays the situation is not unlike that for ϵ' of the K system. The Standard Model predicts a small effect that will be experimentally very difficult to measure. However, any program of B experiments should certainly attempt to measure as many different asymmetries of the form in Equation 56 as possible. Any large nonzero result could provide a clue to physics beyond the Standard Model.

4.3 *Eliminating Penguin Uncertainties with Isospin Analysis*

In neutral B decays, the penguins are not an essential part of the *CP* violation mechanism because the mixing mechanism provides two paths to many final states even in the absence of any penguin contributions. The theoretical uncertainties in estimating relative strength of penguin and tree contributions can, however, lead to uncertainties in the relationship between measured asymmetries and fundamental CKM matrix parameters (69–71). In $b \to c\bar{c}s$ processes (e.g. $B \to \psi K_S$) both amplitudes carry the same CKM phase; extracting $\sin(2\beta)$ from this asymmetry is thus independent of the relative strength of tree and penguin contributions and hence free of such uncertainties. In $b \to u\bar{u}d$ processes (e.g. $B \to \pi\pi$) the two amplitudes carry different CKM phases. The perturbative estimates indicate that the contribution from the penguin amplitude is small (a few percent), but it could be larger than the naive expectation if the matrix element for the penguin operator is enhanced; thus the value of $\sin(2\alpha)$ deduced from this asymmetry suffers from hadronic uncertainties. In this section, we describe a method of isospin analysis that can test for the existence of a significant penguin contribution and, with sufficient data, would provide a clean extraction of CKM parameters even if such a contribution were large (14). For $b \to u\bar{u}s$ processes (e.g. $B \to K\pi$) the situation is even worse: not only do the tree and penguin amplitudes carry different CKM phases, but perturbative estimates suggest that they are comparable in magnitude (the tree process is strongly CKM suppressed). Extraction of CKM parameters from data on this channel would require isospin analysis. Since the expected rates are low, it is unlikely that there will be enough data accumulated, even at a high-luminosity B factory, to carry out such an analysis (72–74).

We briefly review here the analysis of the $\pi\pi$ channel (14). There are three amplitudes for B^+ and B^0 decays into final $\pi\pi$ states,

$$A^{ij} \equiv \langle \pi^i \pi^j \mid \mathcal{H} \mid B^{i+j} \rangle. \qquad 61.$$

Similarly, there are three amplitudes, $\overline{A}^{ij}$, for $\overline{\mathrm{B}}^0$ and B^- decays to two pions. We define $\overline{A}^{ij}$ to be the amplitude for the *CP*-conjugated process of A^{ij}, e.g. $\overline{A}^{+0}$ corresponds to $\mathrm{B}^- \rightarrow \pi^-\pi^0$. The $\overline{A}^{ij}$ amplitudes carry weak phases opposite to those of A^{ij}, but unchanged strong phases.

In the general case, there is one independent amplitude A_{I_t,I_f} for each possible combination of $\{I_t,I_f\}$, where I_t is the transition isospin and I_f is the final-state isospin (including the spectator quark). However, there is no $I_f = 1$ state because it is forbidden by Bose symmetry for a two-pion system with zero angular momentum. Consequently, $I_t = 3/2$ transitions lead to $I_f = 2$ states only, while $I_t = 1/2$ transitions lead to $I_f = 0$ states only. Therefore, we have only two independent amplitudes. This leads to a triangle relation within each set of decay amplitudes:

$$\sqrt{\frac{1}{2}}A^{+-} = A^{+0} - A^{00}$$
$$\sqrt{\frac{1}{2}}\overline{A}^{+-} = \overline{A}^{+0} - \overline{A}^{00}. \qquad 62.$$

For the neutral modes, the decay rates are given in Equation 23, with possibly $|\lambda| \neq 1$. Measuring the total rates for the charged and neutral B decays gives all six magnitudes, $|A^{ij}|$ and $|\overline{A}^{ij}|$, and consequently the shapes of the two triangles can be determined. In addition, from the time-dependent decay rates into $\pi^+\pi^-$, one can deduce $\mathrm{Im}\lambda^{+-} = \mathrm{Im}\,[\exp(-2i\phi_M)(\overline{A}^{+-}/A^{+-})]$. We can define $\tilde{A}^{ij} = [\exp(2i\phi_T)]\overline{A}^{ij}$, where the phase ϕ_T is the CKM phase in the tree diagram. Since the penguin diagram is purely $I_t = 1/2$, only tree diagrams contribute to A^{+0} and $\overline{A}^{+0}$, which are pure $I_t = 3/2$ transitions. Thus, the triangle formed by the $\tilde{A}$'s shares a common side with that formed by the A's, $A^{+0} = \tilde{A}^{+0}$. Any difference between the two triangles is now due to the penguin contributions. Thus we can measure the angle between A^{+-} and $\tilde{A}^{+-}$ (up to an overall ambiguity arising from the four possible orientations of the two triangles relative to their common side).

The *CP* asymmetry is

$$\mathrm{Im}\lambda^{+-} = \mathrm{Im}\left\{\exp[-2i(\phi_M + \phi_T)]\frac{\tilde{A}^{+-}}{A^{+-}}\right\}. \qquad 63.$$

Were $A_{1/2,0}$ dominated by the tree-level diagram, we would have $(\tilde{A}^{+-}/A^{+-}) = 1$, and Equation 63 would reduce to the usual $\sin 2(\phi_M + \phi_T)$ expression. However, from the triangles we know both the magnitude and the phase of $(\tilde{A}^{+-}/A^{+-})$; we need not make the assumption of a small penguin amplitude anymore. We are able to disentangle the value of $(\phi_M + \phi_T)$ without any uncertainty from the unknown penguin contribution to $A_{1/2,0}$. If we assume the Standard Model, we can in principle also use the construction described above to extract a measure of the penguin contribution to $A_{1/2,0}$ (72). If, as expected, the penguin contribution is small, then the two triangles will be the same within errors. In that case this isospin analysis will simply place an experimental bound on the errors in $\sin(2\alpha)$ due to penguin contributions.

Similar isospin analysis can in principle be applied to many other modes, e.g. ρρ, ρπ, Kπ, Kρ, and Kππ (73). It is doubtful that sufficient data will be available to allow such analyses. The number of discrete ambiguities further reduces the usefulness of this approach in all but the simplest case (74, 75).

5. EXPERIMENTAL PROSPECTS

The many interesting features of *CP* violation in B decays can only be studied with a copious source of B mesons. We review here several different experimental approaches.

5.1 *B Factories*

The term B factory is used for a high-luminosity e^+e^- collider running at the energy of the Υ(4S) resonance. The Υ(4S) decays almost exclusively into $B^0\overline{B}^0$ pairs (50%) and B^+B^- pairs (50%). The consensus of various design groups (most noticeably at SLAC and Cornell in the US and at KEK in Japan) has now settled on a design based on an "asymmetric" collider, with two rings at different energies (for detailed machine design reports, see 76). The purpose of this asymmetry in the energies of the e^+ and e^- storage rings is to produce a $B^0\overline{B}^0$ system that is moving with a significant relativistic γ factor in the laboratory. This will cause the two B mesons to decay typically far enough apart in space that the separation between their decay vertices can be measured. [In contrast, an Υ(4S) state at rest would produce B mesons almost at rest, and their decay vertices could not be resolved experimentally.] This then allows a reconstruction of the time difference between the two decays. To measure a time-dependent *CP* asymmetry, one uses events in which one of the B mesons decays to the studied mode at a time t_{CP}, and the other B decays to a mode that identifies this B as a B^0 or $\overline{B}^0$ at a time t_{tag}. The knowledge of the time difference $t = t_{CP} - t_{tag}$ between the two decays allows for a time-dependent measurement of an asymmetry. The $B^0\overline{B}^0$ pair are produced in a well-defined coherent *CP* eigenstate that does not evolve until one of the two B's decays. This decay then effectively starts the clock for the remaining particle (B or $\overline{B}$) to evolve due to mixing with its *CP* conjugate. Notice that t can be either positive or negative. Since the *CP* asymmetry is an odd function of t, $a_{f_{CP}} \propto \sin(\Delta M t)$, a time-dependent measurement is essential to any study of *CP* violation for this system. A time-integrated measurement from an e^+e^- collider will have a zero asymmetry.

The challenge is thus to design an asymmetric collider with sufficient luminosity at the Υ(4S) to allow these interesting measurements to be made. A number of studies have been performed, and the common conclusion is that a luminosity of 3×10^{33} cm^{-2} s^{-1} is sufficient for the task and can be achieved. [With this luminosity, a 3σ measurement of sin(2β) is possible within three years of running even if it is at the lower bound (Equation 47).] To study B_s decay modes, the machines could be operated at the Υ(5S) resonance, but the small cross section, $\sigma[e^+e^- \rightarrow \Upsilon(5S)] = 0.16$ μb, and the smaller branching ratio, $BR[\Upsilon(5S) \rightarrow B_s\overline{B}_s] \leq 0.1$, make it difficult to achieve a sufficient statistical sample for *CP* asymmetry measurements.

The ability to reconstruct both the tagging modes and the *CP* modes is of course as important as the luminosity in determining what can be achieved. Extensive Monte Carlo studies have been made by the proponents of these machines and preliminary detector designs exist. In Table 3 we display estimates of the accuracy with which the asymmetries in various modes can be measured (77, 78).

Table 3 δ(sin 2φ) at Υ(4S) with 30 fb^{-1}

φ	Mode	$10^4 \times$ BR[a]	δ(sin 2φ)
β	ψK_S	4.0 ± 1.4 (*E*†)	0.05
	ψK^*	37 ± 13 (*E*)	
	$\psi' K_S$	11 ± 8 (*E*†)	
	D^+D^-	1–10 (*T*)	
	$D^{*+}D^{*-}$	1–10 (*T*)	
α	ππ	0.2(*T*); ≤0.9 (*E*)	0.18
	ρπ	0.6(*T*); ≤60 (*E*)	0.12
	$a_1\pi$	0.6(*T*); ≤5.7 (*E*)	0.18

[a] *E* = direct experimental determination; *E*† = deduced from a charged channel estimate; *T* = theoretical estimate.

The estimates are made using the general formula

$$\delta(\sin 2\phi) = [(1 - 2w)d]^{-1}[2\epsilon_f\epsilon_t B_f f_0 \sigma(b\bar{b}) \int \mathcal{L}\, dt]^{-1/2} \qquad 64.$$

where ϵ_f (ϵ_t) are the detection efficiencies for the *CP* mode (the tagging mode), B_f is the branching fraction for the *CP* mode, w is the fraction of wrong tags, f_0 is the fraction of b quarks that appear as $\bar{B}^0$ mesons, and d is a dilution factor introduced by any time integration, by background contributions, by fitting procedures, and by mixing of the tagged decay. The integrated luminosity $\int \mathcal{L}\, dt$ is given in nb^{-1} for running at the Υ(4S). The estimates given here are from the SLAC detector workshop studies; similar estimates have been made by other groups. These numbers are meant as approximations only, since the precise numbers depend on details of detector design and on branching ratios that are not yet known. Note that Equation 64 can be used also to provide similar estimates for hadron machines when the appropriate values for the cross section and $\bar{B}^0$ fraction (here assumed equal to that for B^0) are used. Table 3 shows the assumed branching ratios, which are labeled as *E* for direct experimental determination and $E^\dagger$ when deduced from a charged channel (from 5), or *T* for theoretical estimates (from 79, 80). We list only the most commonly discussed modes. The estimates suggest that the *CP*-violating angles α and β can be well measured at such an e^+e^- B factory. A 3σ measurement of $\sin(2\beta) \gtrsim 0.15$ can be made within one year of running at design luminosity (see Equation 47 for the Standard Model predictions). Additional channels that require angular analysis or other techniques to isolate a particular *CP* asymmetry can give confirming measurements of comparable accuracy.

It will be more difficult to constrain the γ angle of the unitarity triangle in this way. Not only is $\sigma[e^+e^- \to \Upsilon(5S)] \ll \sigma[e^+e^- \to \Upsilon(4S)]$, but also $BR[\Upsilon(5S) \to B_s\bar{B}_s] < BR[\Upsilon(4S) \to B_d\bar{B}_d]$. In addition, B_s-$\bar{B}_s$ oscillations are very fast ($x_s \gg x_d$) and hence the dilution factor d is small for this channel. Estimates show that an integrated luminosity of 300 fb^{-1} will be needed to achieve $\delta(\sin 2\gamma) = \pm 0.05$ for $x_s = 4\pi$ (78). If x_s is larger, the measurement would become even more difficult. For current machine designs, this means many years of running and does not appear to be a feasible goal.

We note that a B factory will allow a measurement of form factors in $B \to \rho e\nu$ and $D \to \rho e\nu$. Consequently, the determination of $|V_{ub}|$ will be limited in accuracy mainly by theoretical uncertainties. As discussed in Section 2, such a measurement is useful to overdetermine the unitarity triangle even in the case that sin(2γ) cannot be measured.

A second suggestion for a study of *CP* asymmetries in B decays in an e^+e^- machine is based on the idea of a Z^0 factory, with polarized beams that would preferentially produce B^0 or $\overline{B}^0$ in certain directions (81, 82). While the idea is intriguing, it does not now seem likely that a sufficiently high luminosity will be available at either LEP (CERN) or the SLC (SLAC), nor is there any current proposal for a facility of this type to be built at any other location.

5.2 *Hadron Experiments*

The prospects for studying B physics with hadron colliders are not as clear, simply because the experimental problems are much more difficult and because the studies of what can be done with such machines are less advanced than those for the B-factory proposals. This section is based on preliminary studies and estimates for various hadron colliders, most noticeably the upgraded Tevatron, the CERN LHC, and the SSC. B production processes at hadron colliders are quite different from those at an e^+e^- B factory. Instead of production of a coherent $B^0\overline{B}^0$ state, we have the production of a pair of $b\overline{b}$ quarks, which then hadronize independently, either as B mesons (charged or neutral) or as baryons. Thus, the experimental challenges in measuring *CP* asymmetries are very different for the two types of machines.

The most significant advantage of hadron colliders in conducting B-physics research is that they will copiously produce B mesons, via either $b\overline{b}$ pair production or t-quark decays. The estimated number of $b\overline{b}$ pairs to be produced in the LHC or the SSC is $O(10^{11})$ per year, about a thousand times larger than in a lepton B factory (83, 84). A fixed-target experiment could produce $O(10^9)$ pairs. Estimates for the upgraded Fermilab Tevatron are similar (N. Roe, private communication; P. Tipton, private communication). A comparison of various machines is given in Table 4. The principal experimental problems for

Table 4 Comparison of $b\overline{b}$ production for various facilities

Quantity	Units	Tevatron	SSC	Fixed-target at SSC	Υ(4s)
Collision		$\overline{p}p$	pp	pNucleus	e^+e^-
Energy	TeV	2	40	0.2	10.63×10^{-3}
Luminosity	10^{33} cm^{-2} s^{-1}	0.01	0.1	0.1	3
$\sigma_{b\overline{b}}$	μb	20	500	2	1.2×10^{-3}
σ_{tot}	mb	50	100	10	4.8×10^{-3}
$N_{b\overline{b}}^{prod}$	(year)$^{-1}$	2×10^9	5×10^{11}	2×10^9	3.6×10^7
fraction$_{b\overline{b}}$		0.0004	0.005	0.0002	0.25

hadron colliders are the methods for triggering on B events at such high rates, and the signal-to-noise ratio caused by the additional hadrons in the underlying events. What remains to be seen is whether these problems can be overcome in a way that retains enough of the produced B's to compete in accuracy with a B factory on *CP* violation measurements.

Unlike the Υ(4S) machines, in hadron colliders there is no coherent B-$\overline{\text{B}}$ production, so time-integrated measurements of *CP* asymmetries can be made. We note, however, that time-integrated asymmetries are diluted by a factor $d_q = x_q/(1 + x_q^2)$ (q = d,s). In particular, d_s is expected to be very small. With sufficiently good position resolution on the decay vertex, time-dependent measurements will be possible.

On the other hand, the lack of coherence means that the fraction f of wrong tags here includes the mixing of a tagging B^0 or $\overline{B}^0$ to its conjugate particle. Moreover (particularly in time-integrated measurements), the observed asymmetry is affected by possible confusion of B_s decays with B_d decays since the mass resolution in these experiments may not be good enough to separate these contributions. The problem is most severe if the ratio of the contribution to a specific channel from B_s to that from B_d, denoted x in the expression below, is $O(1)$. Another source of uncertainty is that B^0 and $\overline{B}^0$ may be produced in different numbers. In a pp machine, $r_B \equiv (N_B - N_{\overline{B}})/(N_B + N_{\overline{B}}) \neq 0$ is expected because the probability for a b or $\overline{\text{b}}$ to hadronize as a baryon is affected by the population of quarks and antiquarks in its vicinity. Even in a p$\overline{\text{p}}$ machine there could be a forward-backward asymmetry. These sources of uncertainties modify the observed asymmetry:

$$a_{\text{observed}} = \frac{(1 - 2f)[d_d a(B_d) + x d_s a(B_s)] + \delta}{1 + x + \delta'}. \qquad 65.$$

Note that the dilution factors d_q can be avoided if a time-dependent measurement is made. The quantity δ vanishes if r_B vanishes, but for nonzero r_B it does not vanish even when all *CP*-violating asymmetries are zero. However, we expect r_B and hence δ to introduce only small corrections, of order a percent or so. The most severe problem will be background contributions. The quantity δ' depends on the ratio of background to B_d events and on r_B. These latter corrections depend on a number of factors such as the flavor-tagging efficiencies for baryons and for mesons and the fraction of wrong-sign tags in each case as well as on r_B. These factors are presumably not even universal but vary across phase space. Thus Equation 65 is also not universal; the corrections must be calculated for each kinematic situation. The general form of Equation 65 is given here mainly to stress that an accurate measurement of the asymmetry requires good control of backgrounds, including those from B_s decays (or B_d decays when a B_s channel is studied). Note also that the background problems eliminate many of the tagging modes that can be used in the e^+e^- environment.

Let us now examine the feasibility of measuring *CP* asymmetries in various specific modes. A mode such as $B \rightarrow \psi K_S$ that has a μ-pair signature for the ψ can probably be studied and *CP*-violating parameters extracted. For this mode, B_s contamination will be small since $B_s \rightarrow \psi K_S$ is CKM suppressed. For the upgraded Tevatron with an upgraded D0 detector, Roe (private communication) has estimated that two years running would allow an accuracy of about $\delta(\sin 2\beta) = \pm 0.15$. She included only the $\mu^+\mu^-$ decay mode of the ψ and only muon tags so this number can perhaps be improved by adding electron channels. A similar estimate for the CDF detector has been given (P. Tipton, private communication). Further upgrades to either detector to improve particle identification may make additional tagging channels viable and thus improve the data collection rate.

Purely hadronic final states such as $\pi\pi$ or ρK_S will be much more difficult to separate from backgrounds. For LHC or SSC, widely different estimates of possible efficiencies for B reconstruction and signal to noise ratio exist and further study is clearly needed (for example, compare Ref. (84) to Ref. (83)]. Typical Tevatron events are estimated to have about one K and more than one π in the underlying event; the problem becomes worse at higher energies where the multiplicities are higher. Selection of B events will require cuts in transverse momentum and other variables. The CDF group has already shown that one can pursue B physics in this hadronic environment but the requirements of *CP* violation physics are more challenging. Methods for triggers that select B events and cuts that reduce backgrounds need to be developed.

Since measuring the angle γ seems difficult for e^+e^- colliders, it is interesting to consider whether a hadron collider can do better. The production rate of B_s is comparable to that of B_d (within an order of magnitude), but as $x_s \geq 5$ the mixing is much more rapid. This means that the dilution of the asymmetry due to the time-integrated measurement, $d_s = x_s/(1 + x_s^2)$, will be a significant loss. Furthermore, for $B_s \rightarrow \rho K_S$, the hadronic backgrounds as well as background from B_d events present a severe problem. The CKM suppression of $B_d \rightarrow \rho K_S$ is compensated by a higher B_d production rate. As the two channels have different predicted asymmetries, the B_d-initiated events do provide a serious background problem. Measurement of γ by this approach does not appear feasible at this time.

Another mode of interest in B_s decays is $\psi\phi$. The Standard Model prediction is zero asymmetry, but with new physics in B_s mixing, even maximal asymmetry is possible (54). The hadronic background problems can be avoided by observing the $\psi \rightarrow \mu^+\mu^-$ decay, but the small branching ratio makes the measurement difficult (P. Tipton, private communication). Further studies of these modes and of strategies to improve signal-to-noise and/or detection efficiencies are needed.

For asymmetries in charged B decay, the problems of mixing and tagging are not relevant but the problem of hadronic backgrounds is severe. Given the small asymmetries predicted for charged B decays, it appears that they will be very difficult to observe in a hadronic environment. Studies of *CP* asymmetry measurements with baryonic modes are not yet available, but need to be pursued to evaluate the full range of physics options for a hadron collider (85).

Before concluding this section, let us mention a few more intriguing ideas concerning hadron colliders. It has been suggested that one can turn the asymmetry in production of B^0 and $\overline{B}^0$ into a useful tool, as it allows a (time-dependent) measurement of *CP* asymmetries without tagging (86). The time-dependent rate in the $|\lambda| = 1$ case, summing over both B^0 and $\overline{B}^0$ contributions, is

$$\Gamma^{(\text{no tagging})} = |A|^2 e^{-\Gamma t}[1 - r_B \operatorname{Im} \lambda \sin(\Delta M t)], \qquad 66.$$

where r_B is defined at $t = 0$. Of course, if $r_B = 0$ the untagged rate gives no asymmetry measurement. To extract useful information, r_B has to be accurately known. Present estimates vary greatly, but probably $r_B \lesssim 0.01$ even in the forward direction (86). In principle r_B can be measured by looking at flavor-tagging decays, though the effect is diluted by mixing and other sources of wrong-sign tags. An experiment with no flavor tagging requires accurate vertex reconstruction to isolate the $\sin(\Delta M t)$ contribution, which is suppressed by r_B even when $\operatorname{Im}\lambda$ is large. For example, $\operatorname{Im}\lambda \sim 0.3$ would give a fraction of a percent deviation from a purely exponential decay. Hence it does not appear to us that this method provides a feasible measurement of any *CP*-violating asymmetries.

It was recently suggested that microvertex detectors placed very close to the beam, in "roman pots" (thin-walled enclosures that are inserted into the vacuum pipe and can be retracted during machine fills and tests), could view a much larger fraction of the B mesons produced close to the forward direction and provide better vertex reconstruction (87). Unfortunately, not only the signal events but also the backgrounds are strongly forward-backward peaked and so this method does not alleviate the signal-to-noise ratio and triggering problems. The problems of data rate and of possible radiation damage to the detector in such a configuration are severe. However, the idea certainly merits further study.

Another possible way to study B physics at high-energy proton accelerators is in a fixed-target mode, using either extracted beams or a gas jet target (e.g. 88). Preliminary studies indicate that such an experiment may result in a sample of B events of comparable size and purity to that expected from colliding-beam experiments. This requires very efficient beam extraction and vertex triggers. Again this is a subject where more study is needed to reach a conclusion, but the approach could offer an alternative to the collider experiment.

To summarize, it appears to us that the only *CP*-violating asymmetry that can be readily measured in a hadron collider or a high-energy hadron fixed-target experiment is $\sin(2\beta)$ via $B_d \rightarrow \psi K_S$. For this mode, a measurement of comparable accuracy to that achievable at an e^+e^- collider is feasible at the upgraded Tevatron or the SSC. For any other mode, significant improvements in event selection methods are required before this interesting physics can be tackled.

6. SUMMARY

The physics of *CP* violation is an area in which the Standard Model makes definite predictions that have yet to be tested experimentally. B mesons provide an excellent system with which to test these predictions and thus to search for any clues to physics beyond the Standard Model. Current efforts to understand baryogenesis in the early universe suggest that there must be *CP*-violating processes beyond the Standard Model, which makes this search even more attractive. We have reviewed here the predictions of the Standard Model and have shown how the relationships between various measurements can be used to test the Cabibbo-Kobayashi-Maskawa picture of *CP* violation, to measure some remaining unknown Standard Model parameters, and to seek for clues to physics beyond the Standard Model.

Considerable effort by several groups has now been devoted to the design of asymmetric e^+e^- B factories and to studies of the physics opportunities offered by such facilities. The progress made in machine and detector design show that such a facility would be a very exciting laboratory for the program of *CP* violation studies and other aspects of B physics.

B physics at hadron colliders or in very high-energy fixed-target hadron experiments also offers very interesting possibilities. First results show that the identification of B mesons in high-energy hadron-hadron collisions is feasible. Much more work is needed on detectors particularly designed for the task, and the necessary triggering and event selection procedures must be further studied before definitive conclusions can be reached about the capability of hadron machines to carry out a competitive and complementary program to that possible at lepton machines for B decay and *CP* violation studies.

The prospects for these experiments at either type of facility are as yet quite uncertain. Although several groups around the world have been working on designs of B experiments, as yet none of these projects is funded for construction. On the hadron side, the Tevatron upgrade at Fermilab, the European LHC, and the SSC are all proceeding, but designs for detectors especially suited to study B physics are still in the preliminary study stage. Much interesting and fundamental physics waits for the experiment that can reconstruct a sufficient number of B decays in a variety of modes.

ACKNOWLEDGMENTS

We acknowledge the input of many conversations with our colleagues at SLAC, the Weizmann Institute, and elsewhere. In particular we want to thank Vera Lüth, Natalie Roe, and Art Snyder for advising and educating us about the experimental aspects of the subject. Helen Quinn wishes to acknowledge the Aspen Institute for Physics, where she had numerous discussions that helped form some of the opinions expressed here, most particularly with Isi Dunietz and Lincoln Wolfenstein. Yossi Nir wishes to acknowledge the hospitality of the SLAC theory group.

Literature Cited

1. Christenson, J. H., et al, *Phys. Rev. Lett.* 13:138 (1964)
2. Sakharov, A. D., *ZhETF Pis. Red.* 5: 32 (1967); *JETP Lett.* 5:24 (1967)
3. Kobayashi, M., Maskawa, T., *Prog. Theor. Phys.* 49:652 (1973)
4. Dine, M., et al, Stanford Linear Accelerator preprint SLAC-PUB-5741 (1992)
5. Particle Data Group, *Phys. Lett.* B239: 1 (1990)
6. Barr, G. D., "New results on *CP* violation from NA31," and Winstein, B., "Results from E731," talks presented at the 15th Int. Symp. on Lepton-Photon Interactions at High Energies, Geneva (1991); to be published in conference proceedings
7. Cabibbo, N., *Phys. Rev. Lett.* 10:531 (1963)
8. Chau,, L.-L., Keung, W.-Y., *Phys. Rev. Lett.* 53:1802 (1984)
9. Wolfenstein, L., *Phys. Rev. Lett.* 51:1945 (1983)
10. Bigi, I. I., et al, in *CP Violation,* ed. C. Jarlskog. Singapore: World Scientific (1989), p. 175
11. Carter, A. B., Sanda, A. I., *Phys. Rev. Lett.* 45:952 (1980); *Phys. Rev.* D23:1567 (1981)
12. Bigi, I. I., Sanda, A. I., *Nucl. Phys.* B193:85 (1981); B281:41 (1987)
13. Dunietz, I., Rosner, J. L., *Phys. Rev.* D34:1404 (1986)
14. Gronau, M., London, D., *Phys. Rev. Lett.* 65:3381 (1990)

15. Gronau, M., Wyler, D., *Phys. Lett.* B265:172 (1991)
16. Dunietz, I., *Phys. Lett.* B270:75 (1991)
17. Bigi, I. I., Sanda, A. I., *Phys. Lett.* B211:213 (1988)
18. Gronau, M., London, D., *Phys. Lett.* B253:483 (1991)
19. Nir, Y., Quinn, H. R., *Phys. Rev.* D42:1473 (1990)
20. Gilman, F. J., Nir, Y., *Annu. Rev. Nucl. Part. Sci.* 40:213 (1990)
21. Nir, Y., Stanford Linear Accelerator preprint SLAC-PUB-5676 (1991); in *Proc. TASI 1991 Program*. Singapore: World Scientific (1992), to be published
22. Krawczyk, P., et al, *Nucl. Phys.* B307:19 (1988)
23. Blinov, A. B., Khoze, V. A., Uraltsev, N. G., *Int. J. Mod. Phys.* A4:1933 (1989)
24. Dib, C. O., et al, *Phys. Rev.* D41:1522 (1990)
25. Kim, C. S., Rosner, J. L., Yuan, C.-P., *Phys. Rev.* D42:96 (1990)
26. Inami, T., Lim, C. S., *Prog. Theor. Phys.* 65:297 (1981); (E) 65:772 (1982)
27. Gilman, F. J., Wise, M. B., *Phys. Rev.* D27:1128 (1983)
28. Hagelin, J. S., *Nucl. Phys.* B193:123 (1981)
29. Eichten, E., *Nucl. Phys.* B(Proc. Suppl.)20:475 (1991)
30. Allton, C. R., et al, *Nucl. Phys.* B349: 598 (1991)
31. Alexandrou, C., et al, Wuppertal Univ. preprint, WU-B-90-91-2 (1990); in *Proc. Lattice '90 Conf.* (1992), to be published
32. Bernard, C., Labrenz, J., Soni, A., Santa Barbara Inst. for Theoretical Physics preprint NSF-ITP-90-216 (1990); in *Proc. Lattice '90 Conf.* (1992), to be published
33. Neubert, M., Stanford Linear Accelerator preprint SLAC-PUB-5712 (1992)
34. Maiani, L., *Helv. Phys. Acta* 64:853 (1991)
35. Hashimoto, S., Saeki, Y., *Mod. Phys. Lett.* A7:387 (1992)
36. Nelson, C. A., *Phys. Rev.* D30:1937 (1984)
37. Dell'Aquila, J. R., Nelson, C. A., *Phys. Rev.* D33:80, 101 (1986)
38. Valencia, G., *Phys. Rev.* D39:3339 (1989)
39. Kayser, B., et al, *Phys. Lett.* B237:508 (1990)
40. Dunietz, I., et al, *Phys. Rev.* D43:2193 (1991)
41. Deleted in proof
42. Danilov, M., see Ref. 6
43. Aleksan, R., et al, *Nucl. Phys.* B361:141 (1991)
44. Simonius, M., Wyler, D., *Z. Phys.* C42:471 (1989)
45. Bjorken, J. D., in *Proc. of the 18th SLAC Summer Inst. on Particle Physics,* ed. J. Hawthorne, SLAC-REPORT-378 (1990)
46. Burdman, G., Donoghue, J. F., *Phys. Rev.* D45:187 (1991)
47. Nir, Y., Silverman, D., *Nucl. Phys.* B345:301 (1990)
48. Dib, C. O., London, D., Nir, Y., *Int. J. Mod. Phys.* A6:1253 (1991)
49. Bigi, I. I., Wakaizumi, S., *Phys. Lett.* B188:501 (1987)
50. Tanimoto, M., Suetaka, Y., Senba, K., *Z. Phys.* C40:539 (1988)
51. Hasuike, T., et al, *Mod. Phys. Lett.* A4:2465 (1989); *Phys. Rev.* D41:1691 (1990)
52. London, D., *Phys. Lett.* B234:354 (1990)
53. Donoghue, J. F., Golowich, E., *Phys. Rev.* D37:2543 (1988)
54. Nir, Y., Silverman, D., *Phys. Rev.* D42:1477 (1990)
55. Silverman, D., *Phys. Rev.* D45:1800 (1992)
56. Ecker, G., Grimus, W., *Z. Phys.* C30:293 (1986)
57. London, D., Wyler, D., *Phys. Lett.* B232:503 (1989)
58. Bigi, I. I., Gabbiani, F., *Nucl. Phys.* B352:309 (1991)
59. Liu, J., Wolfenstein, L., *Phys. Lett.* B197:536 (1987)
60. Simma, H., Wyler, D., *Phys. Lett.* B273:395 (1991)
61. Wolfenstein, L., *Phys. Rev.* D43:151 (1991)
62. Bander, M., Silverman, D., Soni, A., *Phys. Rev. Lett.* 43:242 (1979)
63. Gérard, J.-M., Hou, W. S., *Phys. Rev. Lett.* 62:855 (1989)
64. Gérard, J.-M., Hou, W. S., *Phys. Lett.* B253:478 (1991); *Phys. Rev.* D43:2909 (1991)
65. Simma, H., Eilam, G., Wyler, D., *Nucl. Phys.* B352:367 (1991)
66. Chau, L.-L., Cheng, H. Y., *Phys. Rev. Lett.* 59:958 (1987)
67. Eilam, G., Gronau, M., Rosner, J. L., *Phys. Rev.* D39:819 (1989)
68. Soares, J. M., *Nucl. Phys.* B367:575 (1991)
69. London, D., Peccei, R. D., *Phys. Lett.* B223:257 (1989)
70. Gronau, M., *Phys. Rev. Lett.* 63:1451 (1989)

71. Grinstein, B., *Phys. Lett.* B229:280 (1989)
72. Nir, Y., Quinn, H. R., *Phys. Rev. Lett.* 67:541 (1991)
73. Lipkin, H. J., et al, *Phys. Rev.* D44:1454 (1991)
74. Gronau, M., *Phys. Lett.* B265:389 (1991)
75. Lavoura, L., Carnegie Mellon Univ. preprint CMU-HEP-91-22 (1991)
76. Cornell Univ. Rep. CNLS-1070 (1991); Stanford Linear Accelerator Rep. SLAC-372 (1991)
77. Lüth, V., MacFarlane, D. B., Stanford Linear Accelerator preprint SLAC-PUB-5419 (1990)
78. Burchat, P., Univ. Calif. at Santa Cruz preprint SCIPP 91/11 (1991)
79. Neubert, M., et al, Heidelberg Univ. preprint HD.THEP-91-28 (1991); in *Heavy Flavors,* ed. A. J. Buras, M. Lindner. Singapore: World Scientific (1992), to be published
80. Bauer, M., Stech, B., Wirbel, M., *Z. Phys.* C29:637 (1985)
81. Atwood, W. B., Dunietz, I., Grosse-Wiesmann, P., *Phys. Lett.* B216:227 (1989)
82. Atwood, W. B., et al, *Phys. Lett.* B232:533 (1989)
83. Fridman, A., Snyder, A., Stanford Linear Accelerator preprint SLAC-PUB-5319 (1990)
84. Nakada, T., Paul Scherrer Inst. preprint PSI-PR-91-24 (1991)
85. Dunietz, I., CERN preprint CERN 6240/91 (1992)
86. Chaichian, M., Fridman, A., CERN report, CERN-TH.6068/91 (1991)
87. Erhan, S., et al, CERN preprint CERN-PPE-91-10 (1991)
88. Rosen, J., et al, SSC Expression of Interest EOI-13 (1990); Cox, B., et al, SSC Expression of Interest EOI-14 (1990); Carboni, G., INFN PI/AE 91/04 (1991)

Annu. Rev. Nucl. Part. Sci. 1992. 42:251–289

TESTS OF PERTURBATIVE QCD AT LEP

S. Bethke

Physikalisches Institut, University of Heidelberg, W-6900 Heidelberg, Germany

J. E. Pilcher

Department of Physics and Enrico Fermi Institute, University of Chicago, Chicago, Illinois 60637 USA

KEY WORDS: strong interactions, quantum chromodynamics, hard processes, partons, quarks, gluons, LEP, experimental review

---

CONTENTS

0163–8998/92/1201–0251$02.00

1. INTRODUCTION

Recently a new level of precision has been reached in the study of the strong and electroweak interactions. The Large Electron-Positron collider (LEP), operating at energies close to the Z^0 boson, has provided large data samples of e^+e^- annihilations in a new energy regime and under excellent experimental conditions. The data probe both the electroweak interaction of quarks and leptons and the strong interaction of quarks and gluons. The electroweak studies were reviewed recently in this journal by Burkhardt & Steinberger (1). Here we summarize the approximately 40 publications from LEP dealing with the properties of the hadronic final states and their comparisons with expectations based on quantum chromodynamics (QCD).

Quantum chromodynamics is the candidate theory of the strong interaction. It describes the interaction of quarks through the exchange of an octet of massless vector gauge bosons, the gluons. The interaction is more complex than QED since the quarks and gluons carry an additional quantum number, color. For short-range, large momentum transfer interactions, the quarks and gluons are expected to interact only weakly; the strength decreases logarithmically with momentum transfer. This is the concept of asymptotic freedom (2). In this regime, perturbation theory can be used to calculate the scattering amplitudes. At larger distances the interaction strength increases and the perturbation approach fails. This is the regime of the binding of quarks and gluons into hadronic systems and of low-energy scattering. It is characterized by an energy scale of about 1 GeV. In this region, calculational difficulties have generally prevented the precise prediction of experimental observables. An important feature of LEP is that the energy scale of the interactions is far beyond the nonperturbative regime.

High-energy e^+e^- interactions are especially well suited to QCD studies. Since the interaction is an annihilation of point-like particles, the quantum numbers and center-of-mass energy of the initial state are precisely known and there are no remnant particles from the initial state to complicate the analysis. The experimental situation at LEP can be appreciated from Figure 1, where the cross sections for e^+e^- annihilation to hadrons, muon pairs, and photon pairs are shown as a function of center-of-mass energy. At low energies, the cross sections are dominated by the $1/E^2_{cm}$ behavior of the reactions $e^+e^- \rightarrow \gamma \rightarrow f\bar{f}$ (f $\equiv$ fermion) and $e^+e^- \rightarrow \gamma\gamma$. At LEP energies the fermion pair-production cross sections are dominated by the Z^0 resonance $e^+e^- \rightarrow Z^0 \rightarrow f\bar{f}$. The cross section on the resonance is approximately a thou-

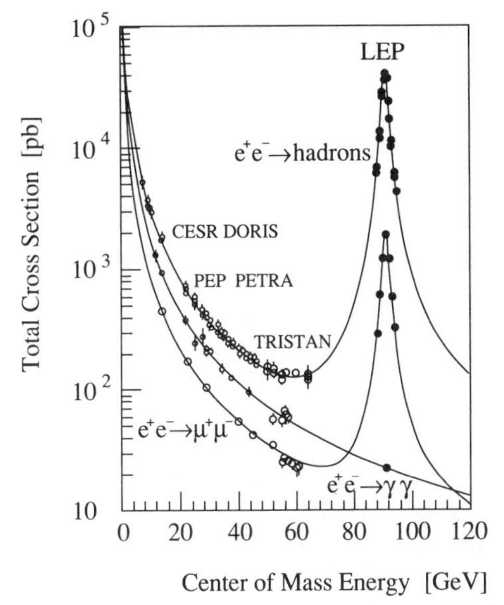

Figure 1 Total cross sections for various processes in e^+e^- annihilation and comparison with Standard Model predictions.

sand times the continuum level under the resonance and is approximately equal to that of a center-of-mass energy of 3.5 GeV. Near the resonance 88% of the visible cross section is associated with hadronic decays of the Z^0, and it is these final states that are used for the QCD studies. They correspond to the decay of the Z^0 to a quark-antiquark pair. The study of gluon radiation by the quarks is the essence of the QCD studies.

Four collaborations are active at LEP using large, general-purpose detectors. These are ALEPH (3), DELPHI (4), L3 (5), and OPAL (6). The main features of LEP (7) and of the detectors themselves were summarized by Burkhardt & Steinberger (1).

2. QCD IN e^+e^- ANNIHILATION

Asymptotic freedom is a fundamental property of QCD arising from the self-interaction of the gluons because of the color charge they carry. Unlike QED, it leads to a coupling constant α_s that decreases with increasing momentum transfer. Thus high momentum transfer interactions of quarks and gluons should be describable by perturbation theory. Since only colorless hadrons are seen, a hadronization or fragmentation process must occur to convert the colored quarks and gluons to hadrons. Unfortunately, no calculational procedure exists for this low-Q^2 regime, so a common procedure is to use phenomenological models to describe the hadronization. Clearly, any significant tests of QCD must be insensitive to the hadronization process, and the model dependence of each analysis must be carefully evaluated and accounted for as a systematic uncertainty.

For QCD plus hadronization models to give an accurate description of the reaction $e^+e^- \rightarrow \gamma, Z^0 \rightarrow$ hadrons, they must contain initial-state photon radiation, the electroweak production of a $q\bar{q}$ pair, the further radiation of gluons and quarks described by perturbative QCD, and finally, the hadronization process. This general structure is shown in Figure 2.

Two approaches have been used for the prediction of perturbative QCD. In the parton shower method, the leading and next-to-leading logarithmic terms from all orders in perturbation theory are summed and used to propagate a shower of quarks and gluons (partons) down

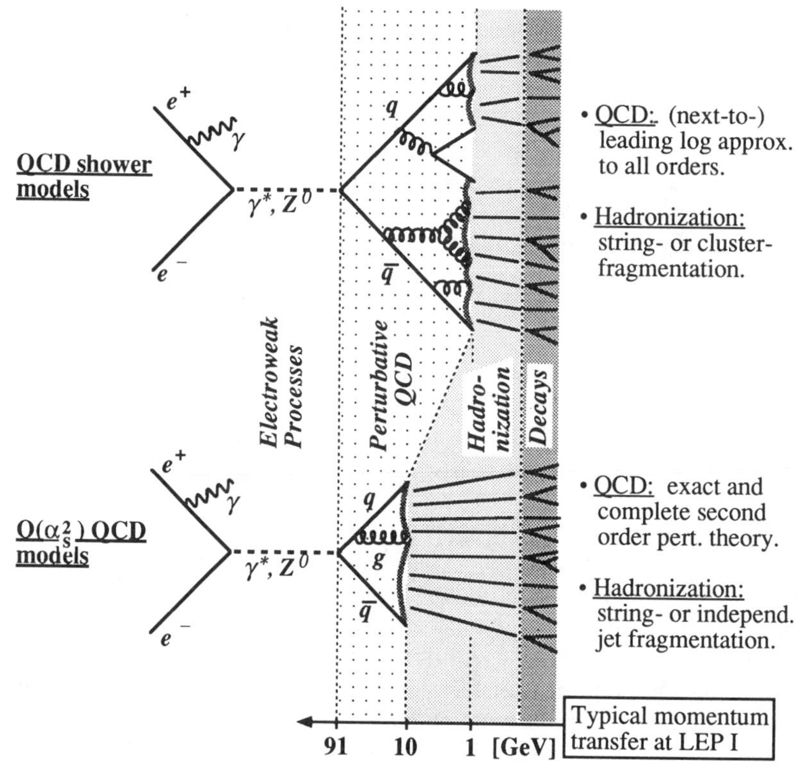

Figure 2 Basic structure of the most commonly used QCD plus hadronization models.

to some cutoff value Q_0 in virtual parton mass. At LEP energies, typically 10 final-state partons are generated for a value of $Q_0 = 1$ GeV. An alternative approach is to use the result of complete second-order [$O(\alpha_s^2)$] perturbation theory. These final states can contain no more than four partons. Thus the phase-space to be covered by the hadronization model is substantially larger in this approach.

The advantages of the $O(\alpha_s^2)$ QCD calculations are that their predictions are valid to exact second-order perturbation theory, and that the coupling strength α_s and the QCD scale parameter Λ can be well defined in a certain renormalization scheme, as for instance in the $\overline{\text{MS}}$ scheme used throughout this report (8). The energy dependence of $\alpha_s(\mu)$, in second-order perturbation theory, can be written as

$$\alpha_s(\mu) = \frac{12\pi}{(33 - 2N_f)\ln(\mu^2/\Lambda^2_{\overline{\text{MS}}})}\left\{1 - 6\,\frac{153 - 19N_f}{(33 - 2N_f)^2}\,\frac{\ln[\ln(\mu^2/\Lambda^2_{\overline{\text{MS}}})]}{\ln(\mu^2/\Lambda^2_{\overline{\text{MS}}})}\right\}, \quad 1.$$

where N_f is the number of quark flavors (five at LEP) and μ is the "typical" energy of the reaction at which the theory is renormalized. Typical observables $\langle O \rangle$ that are sensitive to gluon radiation, like the relative number of 3-jet ($q\bar{q}g$) events and event shape variables, are expected to be proportional to α_s in first-order perturbation theory, and in nth-order perturbation theory are described by

$$\langle O \rangle = \sum_{i=0}^{n} C_i(O)\alpha_s^i. \quad 2.$$

The QCD coefficients C_i depend on the detailed definition of the observable O. They have been calculated up to complete $O(\alpha_s^2)$ for many event shape parameters, jet definitions, and energy correlation measures (9). These observables are chosen such that they are insensitive to the radiation of collinear and soft gluons, where perturbation theory does not apply. Calculations in complete third-order perturbation theory [$O(\alpha_s^3)$] are available for the QCD correction to the total hadronic cross section in e^+e^- annihilation (10) and for the branching ratio of τ leptons to hadrons (11, 12).

QCD models with hadronization are typically used to study detector acceptance and resolution, to examine hadronic effects in new particle searches, and to assess the hadronization corrections in QCD analyses. Analytic QCD calculations, however, are more suited to the extraction of well-defined quantitative results such as the value of α_s. Comprehensive descriptions of these calculations and of QCD models in e^+e^- annihilation can be found in (9) and (13), respectively. Detailed reviews of the theory and of tests of perturbative QCD prior to LEP have been given by Altarelli (14, and references therein).

3. HADRONIC EVENT SHAPES AND JET PRODUCTION

Hadronic event shape variables are tools to study both the amount of gluon radiation and the details of the hadronization process, while measurements of jet production rates serve as a direct test of the underlying parton structure of hadronic final states. Since the laboratory frame in which the events are measured is essentially identical to the center-of-mass system of the annihilation, events from $q\bar{q}$ final states without hard gluon radiation result in two collimated, back-to-back jets of hadrons. The emission of one hard gluon leads to planar 3-jet events, while the emission of two or more energetic gluons can cause nonplanar multijet event structures. The LEP experiments have studied in detail global event shape distributions (15–17) and jet production rates (18–22) and compared these measurements to both QCD shower model calculations and to analytic predictions in $O(\alpha_s^2)$.

3.1 *Hadronic Event Shapes*

A typical event shape observable is thrust, T, which is defined as the normalized sum of the momentum components of all particles of a given event along a specific axis; this axis is chosen such that T is maximized. Ideal 2-jet events result in $T = 1$, while $2/3 < T < 1$ for planar 3-jet events and $T = 1/2$ for completely spherical events.

In Figure 3 the thrust distribution of hadronic Z^0 decays, measured by the DELPHI collaboration at center-of-mass energies close to 91 GeV (17), is compared to the predictions of several QCD plus hadronization models. The data are corrected for detector acceptance, resolution, and the effects of initial-state photon radiation, which are small at the Z^0 pole. The model parameters, like the QCD scale parameter Λ and several parameters specifying the hadronization process, were taken from previous adjustments of these models to provide an overall good description of hadronic Z^0 decays (16, 17, 23). In general, such studies show that most global event shape distributions can be well described by the QCD plus hadronization models studied so far. QCD shower models are somewhat superior compared to $O(\alpha_s^2)$ QCD models, presumably because the latter cannot generate events with more than four parton jets. At present, the overall best description of hadronic events is provided by the JETSET parton shower model (24).

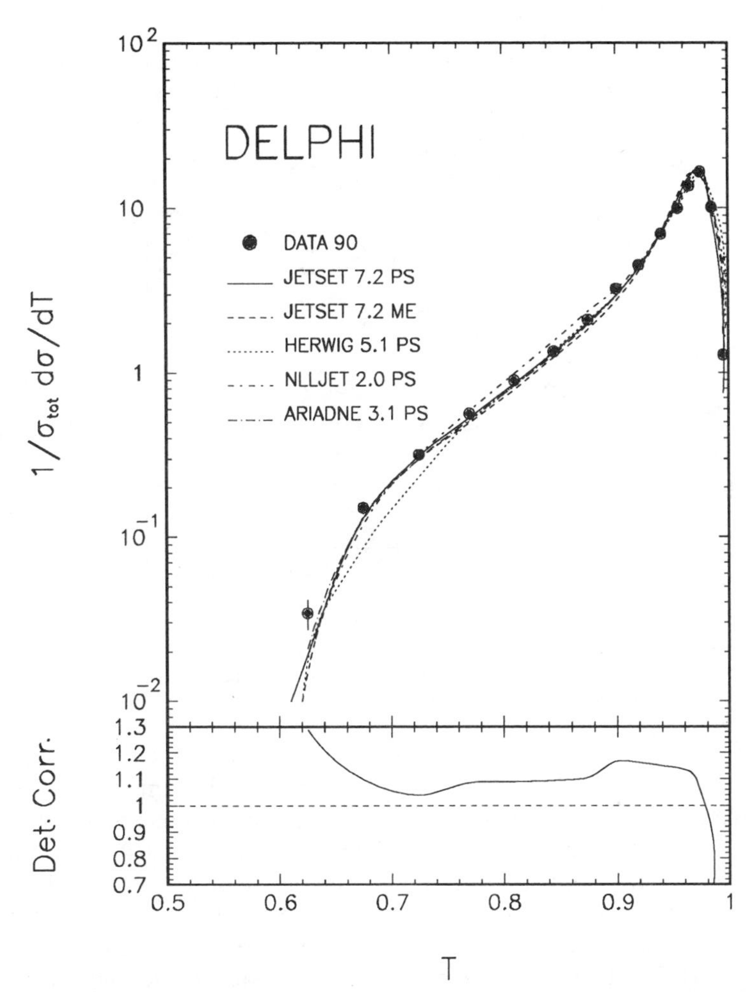

Figure 3 Thrust distribution measured at E_{cm} = 91 GeV, compared to various QCD plus hadronization models tuned to describe hadronic Z^0 decays. (Data from Reference 17.) Detector acceptance corrections applied to the data are indicated below the graph.

QCD predicts scaling violations for observables that do not depend on absolute energies or momenta, as for example the T distribution. These scaling violations are caused by the energy dependence of α_s, which determines the amount of gluon radiation. In leading order the probability of gluon radiation is proportional to α_s. It is thus expected that fewer 3-jet and multijet events should be observed at higher center-of-mass energies, and that event shape distributions evolve toward the

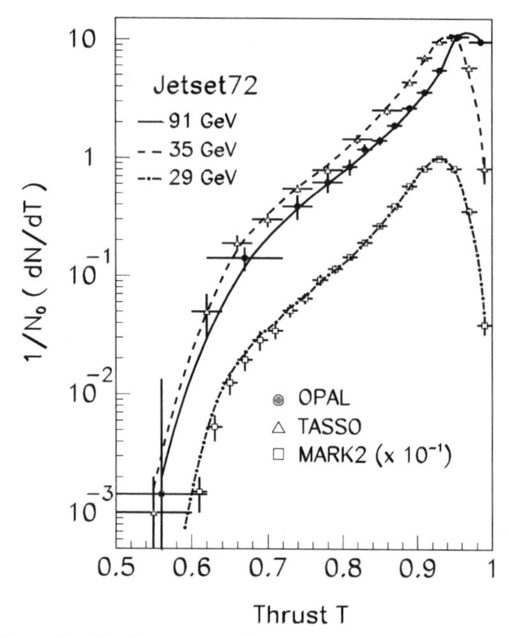

Figure 4 The thrust distribution measured at various center-of-mass energies, compared to model predictions with parameters tuned to the 91-GeV data. The figure is taken from Reference 16. Jetset 72 is described in Reference 24.

2-jet limit. The OPAL collaboration compared their measured event shape distributions (16) with similar measurements done at lower center-of-mass energies. As seen in Figure 4 for the T distribution, there are small but significant differences. To distinguish scaling violations from effects of a possible energy dependence of the hadronization process, the data at 29 and 35 GeV are also compared with the expectation of the JETSET QCD shower model. The model predictions, with no parameter changed but the center-of-mass energy, describe the data well at all energies. Within this model, the energy dependence of these distributions is thus explained by QCD scaling violations plus an energy-independent parametrization of the hadronization process. The data are consistent with a running coupling constant and asymptotic freedom.

3.2 *Jet Production Rates*

Studies of jet production rates provide the most intuitive tests of the underlying parton structure of hadronic events at high energies. While hadron jets and the corresponding jet multiplicity of individual events can often be inferred from graphical displays of hadronic events at

LEP, a quantitative study requires the exact definition of resolvable jets.

The most commonly used algorithm to define and reconstruct jets of hadrons was introduced by the JADE collaboration (25, 26). The scaled pair mass of two resolvable jets i and j, $y_{ij} = M_{ij}^2/E_{\text{vis}}^2$, is required to exceed a threshold value y_{cut}, where E_{vis} is the total visible energy of the event. In a recursive process, the pair of particles or clusters of particles n and m with the smallest value of y_{nm} is replaced by (or "recombined" into) a single jet or cluster k with four-momentum $p_k = p_n + p_m$, as long as $y_{nm} < y_{\text{cut}}$. The procedure is repeated until all pair masses y_{ij} are larger than the jet resolution parameter y_{cut}, and the remaining clusters of particles are called jets. Several methods of jet recombination and definitions of M_{ij} exist (9, 22, 27, 28); the original JADE (also called E0) scheme with $M_{ij}^2 = 2E_iE_j(1 - \cos\theta_{ij})$, where E_i and E_j are the energies of the particles and θ_{ij} is the angle between them, has the smallest hadronization corrections if jet rates at the hadron and at the parton level are compared with each other.

The agreement between parton jets and hadron jets, calculated with the JETSET parton shower model for hadronic Z^0 decays is demonstrated in Figure 5, where the relative production rates R_n for events

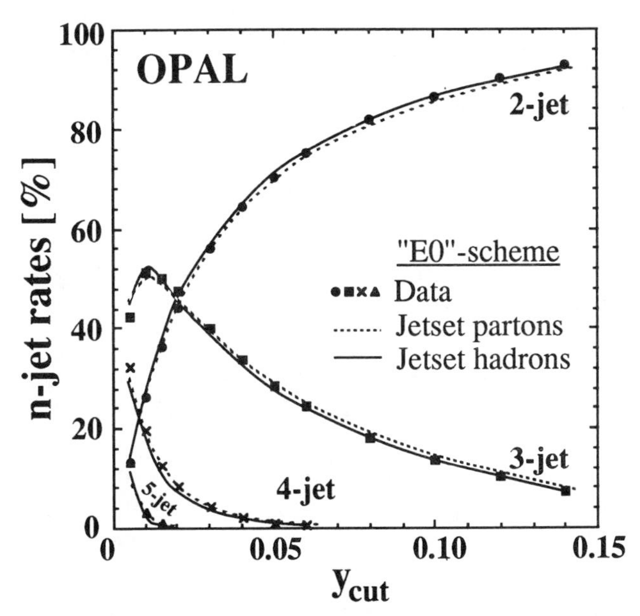

Figure 5 Measured production rates of n-jet events, compared to QCD shower model calculations before and after the hadronization process. The E0 scheme of jet definition is due to JADE (25, 26). (Data from Reference 22.)

with 2, 3, 4, and 5 or more jets are plotted as a function of y_{cut} using the JADE jet finder. Also shown are the jet rates measured by the OPAL collaboration (22), which are well described by the model calculations. The good agreement between data and model calculations was also observed at lower center-of-mass energies (27, 29, and references therein), and was also demonstrated using other jet recombination schemes (22).

Such models are therefore regarded as reliable tools to correct measured data distributions not only for the effects of limited detector acceptance and resolution, but also for the nonperturbative hadronization process. Having applied these corrections, the data can be directly compared to the predictions of analytic QCD calculations that exist for a large variety of observables like event shape parameters, jet production rates, and angular correlations between jets or particles within jets (9). Studies of this type are reviewed in the following sections.

4. DETERMINATIONS OF α_s

The determination of α_s, which is the only free parameter of the theory, has always been one of the key analyses of hadronic final states in e^+e^- annihilation, and of quantitative QCD tests in general. Comprehensive summaries of α_s measurements prior to LEP can be found in (30, 31) and more recently in (14, 29). Measurements of α_s (M_{Z^0}) at LEP were recently summarized in (32, 33) and are also updated and reviewed in the following sections.

4.1 *$\alpha_s(M_{Z^0})$ from Jet Rates, Hadronic Event Shapes, and Energy Correlations*

A large number of observables sensitive to the amount of hard gluon radiation in hadronic events have been used to determine α_s. The most common ones are listed in short (for detailed definitions, see 9):

1. Thrust (T), oblateness (O), and the C parameter, which measure the deviation from ideal two-jet event configurations.
2. The heavy jet mass $M_H^{(T)}$, which is defined to be the largest of the normalized invariant masses of the two event hemispheres defined by the plane perpendicular to the thrust direction, and $M_D^{(T)}$, which is the quadratic difference between the heavy and the light hemisphere masses.

For a given hadronic event each of these observables provides one characteristic number, and the differential distributions from a large number of events are used to extract the value of α_s. In addition to event shape observables, the relative production rates of n-jet events, R_n, are analyzed using a suitable jet definition and reconstruction algorithm, and α_s is determined in two ways:

1. From the 3-jet rate R_3 at a given value of the jet resolution parameter y_{cut}, and/or
2. From the differential 2-jet distribution $D_2(y) = [R_2(y) - R_2(y - \Delta y)]/\Delta y$, where $y \equiv y_{cut}$.

Energy and angular correlations between the final-state hadrons are also sensitive to the value of α_s. Commonly used observables of this type are as follows:

1. The energy-energy correlation distribution (EEC), which is an energy-weighted histogram of the angles χ between all particle pairs within an event, summed over many events;
2. The asymmetry of the EEC distribution, $\mathrm{AEEC}(\chi) = \mathrm{EEC}(180° - \chi) - \mathrm{EEC}(\chi)$, in which measurement errors and hadronization corrections symmetric in χ cancel out; and
3. The planar triple energy correlation distribution (PTEC), which is a two-dimensional, energy-weighted histogram of the angles between planar particle triplets within an event (34).

The $O(\alpha_s^2)$ QCD calculations for the differential distributions of these observables X are given in the form

$$\frac{1}{\sigma_0}\frac{d\sigma}{dX} = \frac{\alpha_s(\mu)}{2\pi} A(X) + \left[\frac{\alpha_s(\mu)}{2\pi}\right]^2 [A(X) 2\pi\beta_0 \ln(x_\mu^2) + B(X)], \qquad 3.$$

where σ_0 is the leading (0th) order cross section for e^+e^- annihilation into hadrons, $\beta_0 = 11 - 2N_f/3$, $x_\mu = \mu/E_{cm}$ is the renormalization scale factor, and $A(X)$, $B(X)$ are calculated functions depending on the variable in question (9, 34). In truncated order of perturbation theory, the next-to-leading and higher order coefficients of α_s depend on the choice of the renormalization scale μ, which is not unambiguously defined by the theory. This imposes a theoretical scale uncertainty. The scale dependence only vanishes in infinite order, which is why the scale uncertainty may also be regarded as an uncertainty due to the unknown higher order contributions.

The renormalization scale μ should not be confused with the momentum transfer Q in a given process, as for instance $Q^2 = E_{cm}^2$ in $e^+e^- \to q\bar{q}$. Apart from the fact that the "typical" value of Q for gluon radiation is more related to the tranverse momenta or invariant masses at the q-g vertex rather than to E_{cm}, the "optimal" choice of μ for a given observable in a given perturbative order is not strictly defined. Several theoretical procedures to optimize or fix μ can be found in the literature (35). While none of them can be proven to be the correct or optimal solution as long as the result in infinite order perturbation theory remains unknown, these procedures predict effective scales that are smaller than E_{cm} and that are different for most of the event shape, jet rate, and energy correlation observables.

To account for the theoretical scale uncertainty $\alpha_s(\mu)$, or equivalently $\Lambda_{\overline{MS}}$, is adjusted to describe the measured observable under study, for ranges of renormalization scales μ as discussed below. The resulting values of $\Lambda_{\overline{MS}}$ are then converted into values of $\alpha_s(M_{Z^0})$ using Equation 1. The spread of these results due to the variation of x_μ is taken as a systematic uncertainty.

Further systematic uncertainties arise from corrections for acceptance, resolution, and hadronization. These corrections are obtained from model calculations and must be applied before the theoretical expressions are fitted to the data. The experimental uncertainties are typically assessed by using several parametrizations of the detector simulation and/or by analyzing the data with different independent detector subsystems. The hadronization uncertainties are studied by using different QCD models, e.g. $O(\alpha_s^2)$ and QCD shower models and/or different hadronization schemes, or by varying the model parameters within one model in appropriate ranges. The detailed methods to calculate and to assign the systematic uncertainties usually differ among the four experiments, but the final results for α_s all contain rather complete investigations of these errors.

A multitude of α_s determinations is available from the four LEP experiments: ALEPH provided α_s results from measurements of the C, O, T, D_2, $M_H^{(T)}$, and $M_D^{(T)}$ distributions (20), and from the C, O, T, and EEC distributions calculated from preclustered jets instead of the final hadrons (36). The L3 collaboration presented results from studies of R_3 (21) and of the EEC and AEEC distributions (37). DELPHI recently updated and summarized measurements of $\alpha_s(M_{Z^0})$ from C, O, T, $M_H^{(T)}$, $M_D^{(T)}$, D_2, EEC, and AEEC (17). A recent update of α_s determinations from OPAL is based on thirteen event shape, jet, and energy correlation observables (38).

Analyses of α_s based on different observables and a consistent treatment of the same data offer insight into the size and structure of higher order QCD uncertainties. The OPAL group has demonstrated that the results for $\alpha_s(M_{Z^0})$—obtained from event shape, jet rates, and energy correlation measurements using a single common renormalization scale—significantly disagree with each other if only the experimental uncertainties of typically 1 to 3% are considered (38). These disagreements are likely to be due to unknown higher order contributions, which may differ for each observable. In the OPAL analysis, this assumption is supported by the good agreement of the α_s measurements when higher order uncertainties are quantified and accounted for, separately for each observable. Especially for an overall combination of all available LEP results on $\alpha_s(M_{Z^0})$, it is necessary to study and include theoretical uncertainties in a consistent manner.

The uncertainties from hadronization corrections and renormalization scale uncertainties have been intensively studied by all LEP experiments, as mentioned above. For example, the scale dependence of $\alpha_s(M_{Z^0})$ determined by DELPHI, is shown in Figure 6 (17). For all observables except the oblateness, $\alpha_s(M_{Z^0})$ decreases for scales smaller than E_{cm} and reaches a minimum at values as small as $\mu^2/M_{Z^0}^2 \sim 0.001$. Strategies to define suitable ranges of $x_\mu = \mu/M_{Z^0}$ are to vary x_μ between unity and the value giving the best fit to the data (22, 38), between unity and the b-quark mass normalized to E_{cm} (20), or within arbitrarily chosen limits like $1/4 \leq x_\mu \leq 1$ (39). For individual studies of α_s, the detailed choice of the procedure is not critical since the uncertainty is of the order of $\pm 10\%$ of $\alpha_s(M_{Z^0})$, and the estimated uncertainty typically varies by less than a factor of two among the different procedures. It should be mentioned that for most observables, the best-fit result for the scales is close to the theoretical estimates from scale optimization (18, 38, 40).

Since the four experiments have chosen different ways to determine and quote the central values of $\alpha_s(M_{Z^0})$ and the systematic uncertainties, it is mandatory to apply a consistent definition of these quantities before an overall average value of $\alpha_s(M_{Z^0})$ from LEP is quoted. For this purpose, we choose the following procedure to extract the central values of $\alpha_s(M_{Z^0})$ and the corresponding uncertainties from the original publications:

1. For each observable and from each experiment, we take the values of $\alpha_s(M_{Z^0})$ for two scale choices, for $x_\mu = 1$ and for the value of x_μ where α_s (M_{Z^0}) reaches a minimum. The location of the latter value of x_μ is determined by the QCD coefficients A and B (see Equation 3) and is largely independent from the experimental data; see the corresponding results in (17, 38).
2. We take the average of these two $\alpha_s(M_{Z^0})$ values as the central result for each particular observable, and quote half of their difference as the corresponding scale uncertainty.
3. When α_s is obtained using different model calculations to unfold

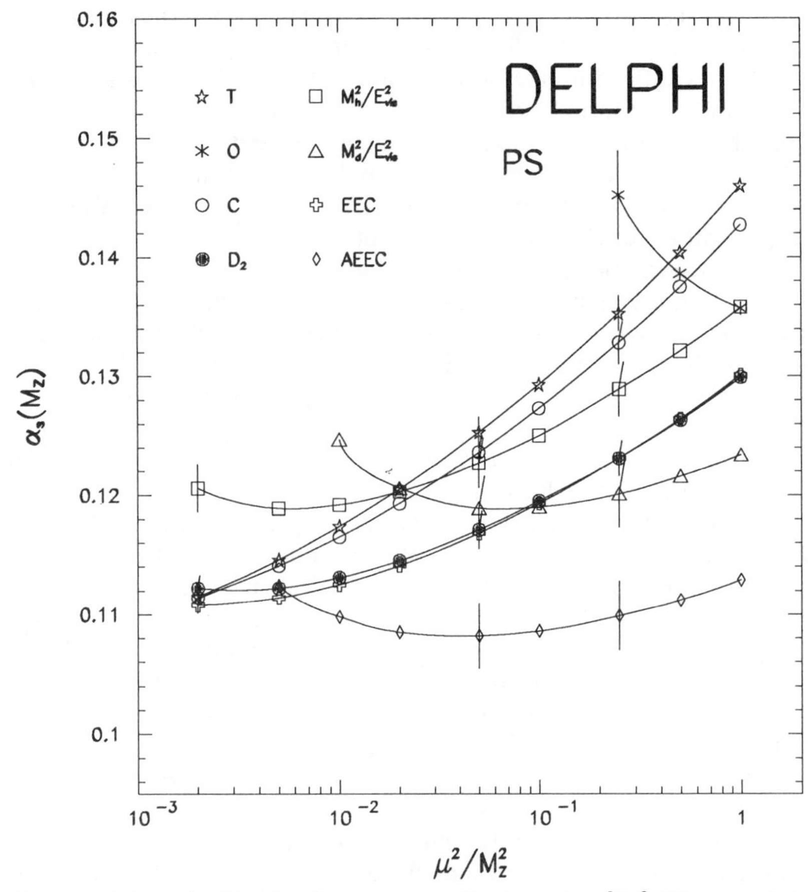

Figure 6 Values of $\alpha_s(M_{Z^0})$ for different renormalization scales μ^2/M_Z^2. The errors show typical statistical and experimental systematic errors added in quadrature. See text and Reference 9 for the definitions of the different observables used to estimate α_s (Data from Reference 17.)

hadronization effects, as for instance Abreu et al (17) did for QCD shower and $O(\alpha_s^2)$ models, we take the central value as the average from the two models and the uncertainty as half of the difference. Alternatively, when the results from different model calculations are not given in detail but are quoted as a systematic uncertainty, as for instance the "uncertainty due to the parton virtuality" (38), and the "hadronization model uncertainty" (20, 21, 38), we account for these uncertainties as independent errors on $\alpha_s(M_{Z^0})$.

4. Finally, we add in quadrature the statistical and systematic errors

quoted in the publications, together with the theoretical uncertainties mentioned above.

The results are listed in Table 1 and are shown in Figure 7. Weighted averages for $\alpha_s(M_{Z^0})$ are calculated for each observable and for each experiment, whereby the weights are defined to be the inverse squares of the uncertainty of each measurement. The individual α_s results and their uncertainties are correlated, within a single experiment as well as between experiments. For calculating the uncertainties of the averages of each observable, given in the last column of Table 1, we assume that the experimental errors between the experiments are independent, which results in combined experimental errors of typically 1 to 2% in α_s, and we add them in quadrature to the corresponding linear averages of the theoretical uncertainties. The uncertainties of the average results for each experiment (in the row labelled "Weighted

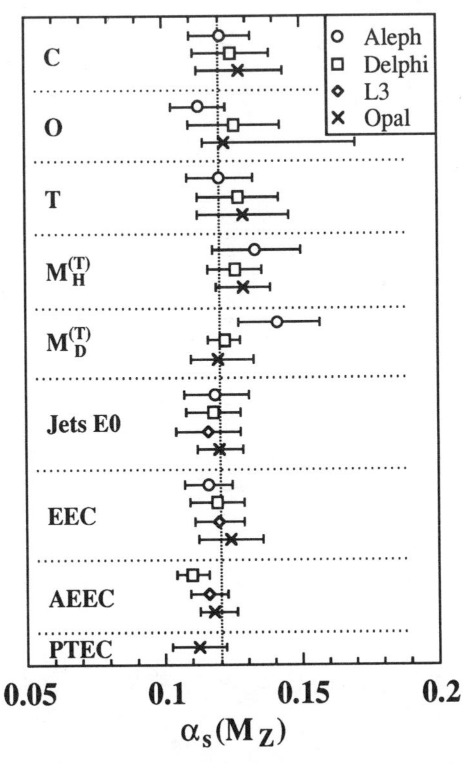

Figure 7 Compilation of measurements of α_s from event shapes, jet rates, and energy correlations, in $O(\alpha_s^2)$, from the four detectors at LEP. The errors contain the experimental and theoretical uncertainties. See text for discussion.

Table 1 Summary of measurements of $\alpha_s(M_{Z^0})$ at LEP, based on a consistent procedure to determine the central values of α_s and their theoretical uncertainties[a]

Observable	$\alpha_s(M_{Z^0})$ in $O(\alpha_s^2)$				Weighted average
	ALEPH	DELPHI	L3	OPAL	
C	0.121 ± 0.011	0.125 ± 0.014	—	0.128 ± 0.016	0.124 ± 0.013
O	0.113 ± 0.010	0.126 ± 0.017	—	$0.122 ^{+0.047}_{-0.008}$	0.117 ± 0.017
T	0.120 ± 0.012	0.127 ± 0.015	—	0.129 ± 0.017	0.129 ± 0.014
$M_H^{(T)}$	0.134 ± 0.016	0.126 ± 0.010	—	0.129 ± 0.010	0.129 ± 0.009
$M_D^{(T)}$	0.142 ± 0.015	0.122 ± 0.006	—	$0.120 ^{+0.013}_{-0.010}$	0.124 ± 0.006
Jets E0	0.119 ± 0.012	0.118 ± 0.010	0.116 ± 0.012	$0.120 ^{+0.009}_{-0.008}$	0.119 ± 0.010
EEC	0.116 ± 0.009	0.119 ± 0.010	0.120 ± 0.009	0.124 ± 0.012	0.119 ± 0.009
AEEC	—	0.110 ± 0.006	0.116 ± 0.007	$0.118 ^{+0.009}_{-0.005}$	0.114 ± 0.006
PTEC	—	—	—	0.112 ± 0.010	0.112 ± 0.010
Weighted av.	0.121 ± 0.009	0.119 ± 0.007	0.118 ± 0.009	$0.121 ^{+0.006}_{-0.005}$	0.120 ± 0.006
Official av.	$0.117 ^{+0.008}_{-0.010}$	0.113 ± 0.007	0.115 ± 0.009	$0.122 ^{+0.006}_{-0.005}$	0.117 ± 0.006

[a] The errors include the experimental statistical and systematic errors and the theoretical uncertainties added in quadrature. Weighted averages are calculated using the inverse squares of these errors; see text for details of these calculations. The final error of $\Delta\alpha_s = \pm 0.006$ was taken from determinations of the overall uncorrelated uncertainties (17, 38). The "official averages" (from 32, 38) are partly based on different averaging procedures and do not all contain the complete set of published results.

av.") are taken from investigations done by the experiments themselves; see the discussion below. The final weighted average of $\alpha_s(M_{Z^0})$, from all observables and from all experiments together, is $\alpha_s(M_{Z^0}) = 0.120 \pm 0.006$, where the overall uncertainty of ± 0.006 is taken from the most recent studies of DELPHI (17) and OPAL (38), which accounted for correlations between the various observables. Since the statistical errors are almost negligible compared to the systematic uncertainties, a further reduction of the overall error cannot be achieved by adding the results of further experiments. From Figure 7 we see that most of the individual α_s results agree with the final value within their overall uncertainties.

These uncertainties are dominated by theory in all cases, which is why such a good agreement is indeed expected, provided that the methods used to obtain the uncertainties are meaningful. Only two of the 27 individual measurements of $\alpha_s(M_{Z^0})$ deviate from the average value of 0.120 by more than their final error. The overall good agreement of the multitude of results is an important consistency check of perturbative QCD and of the general methods used to estimate theoretical, i.e. higher order, systematic uncertainties, since all data should be described by the theoretical predictions with one common value of $\alpha_s(M_{Z^0})$.

Another way to derive an overall average value of $\alpha_s(M_{Z^0})$ from the measurements at LEP was pursued in earlier works. The combined values of $\alpha_s(M_{Z^0})$ given by each of the four experiments were averaged, resulting in $\alpha_s(M_{Z^0}) = 0.115$ (32, 33). An update of the final results quoted by the experiments is given in the last row of Table 1, whereby DELPHI and OPAL determine and quote the overall uncorrelated uncertainty between all the observables they have studied as their final uncertainty (17, 38). They average to $\alpha_s(M_{Z^0}) = 0.117$. This value is smaller than our final result of $\alpha_s(M_{Z^0}) = 0.120$ presented above for two reasons. DELPHI applied a different averaging procedure to their data, giving a larger weight to those results that are obtained using small renormalization scales (17), and ALEPH and L3 elected to calculate their combined value from only a subset of their published results (T. Lohse, ALEPH Collaboration, private communication; T. Hebbeker, L3 Collaboration, private communication). Since our overall average was obtained by treating all published results in the same consistent way, we choose to quote as the final value of $\alpha_s(M_{Z^0})$, obtained in $O(\alpha_s^2)$ from event shape, jet, and energy correlation measurements at LEP:

$$\alpha_s(M_{Z^0}) = 0.120 \pm 0.006.$$

We also note that different procedures to select and average a number of single α_s results may decrease the final value of $\alpha_s(M_{Z^0})$ by as much as 0.006 (38). The freedom in choosing the averaging procedure and in calculating the uncertainties of single and combined values of α_s are mainly due to the higher order uncertainties of the $O(\alpha_s^2)$ QCD calculations.

4.2 $\alpha_s(M_{Z^0})$ *from Resummed QCD Calculations*

Recently, new calculations for the thrust (T) and heavy jet mass ($M_H^{(T)}$) variables became available. They contain, in addition to the complete $O(\alpha_s^2)$ terms, the resummation of all terms of the form $\alpha_s^n \ln^m(1 - T)$ or $\alpha_s^n \ln^m(M_H/E_{cm})$ with $m \geq n$. This resummation yields the so-called next-to-leading logarithmic approximation (NLLA) to all orders (43, 44). These $O(\alpha_s^2)$ + NLLA calculations are expected to permit a more accurate prediction of the distributions, especially at high thrust or low masses, and they should also provide a much reduced dependence on the renormalization scale if compared to the $O(\alpha_s^2)$ calculations alone.

The OPAL collaboration has performed fits of the $O(\alpha_s^2)$ + NLLA calculations to the thrust and heavy jet mass data (38). The results indeed confirm the expectations mentioned above. The data are significantly better described at high thrust and low jet masses, and the fitted scale factors x_μ are much closer to unity than when $O(\alpha_s^2)$ calculations are used alone. This corresponds to a reduced renormalization scale uncertainty in the resulting value of $\alpha_s(M_{Z^0})$ that, including all experimental and theoretical uncertainties as discussed in the previous section, is quoted by OPAL as

$$\alpha_s(M_{Z^0}) = 0.122^{+0.003}_{-0.006}.$$

The $O(\alpha_s^2)$ + NLLA fits result, however, in a worse description of the heavy jet mass distribution at large masses, where the NLLA terms apparently do not vanish sufficiently rapidly toward the kinematic limit. Various ways to suppress the NLLA terms at high masses while retaining them at low masses were explored and, while these led to an improved description of the data, the result for $\alpha_s(M_{Z^0})$ was little affected (38). Despite these imperfections, the merit of the $O(\alpha_s^2)$ + NLLA calculations in reducing the higher order uncertainties has been clearly demonstrated. Further developments such as improved merging procedures of the $O(\alpha_s^2)$ and the NLLA terms and corresponding predictions for other observables, as for instance jet rates defined in the so-called Durham jet finding scheme (28, 45, 46), will provide further important tests of the reliability of this new class of calculations.

4.3 α_s (M_{Z^0}) *from the Hadronic Partial Width of the* Z^0

An attractive way to determine $\alpha_s(M_{Z^0})$ is a precise measurement of the ratio R of the hadronic and leptonic partial widths of the Z^0,

$$R \equiv \left(\frac{\Gamma_{\text{had}}}{\Gamma_{\text{lept}}}\right)_{\text{exp}} = \left(\frac{\Gamma_{\text{had}}}{\Gamma_{\text{lept}}}\right)_0 (1 + \delta_{\text{QCD}}), \qquad 4.$$

since R is not subject to hadronization corrections and because the QCD correction δ_{QCD} has been calculated to third-order [$O(\alpha_s^3)$] perturbation theory (10). Including quark mass corrections, δ_{QCD} is of the form (47)

$$\delta_{\text{QCD}} = 1.05\left(\frac{\alpha_s}{\pi}\right) + 0.9\left(\frac{\alpha_s}{\pi}\right)^2 - 13\left(\frac{\alpha_s}{\pi}\right)^3. \qquad 5.$$

The Standard Model expectation for $(\Gamma_{\text{had}}/\Gamma_{\text{lept}})_0$, without QCD corrections, is 19.97 with only little uncertainty due to the unknown masses of the top quark and the Higgs particle (48). The measurement of R derives from counting events and is independent of nonperturbative hadronization effects. However, the precision of $\alpha_s(M_{Z^0})$ from a measurement of R is given by

$$\Delta\alpha_s \approx \frac{\Delta R}{R}\pi,$$

such that R is required to be known to about 4 per mil to reach $\Delta\alpha_s(M_{Z^0}) = \pm 0.01$.

The average value of R, summarized in Figure 8 from the measurements of the four LEP experiments, is $R = 20.91 \pm 0.12$ (49–52). This is based on a total of 571,000 hadronic and 63,000 leptonic Z^0 decays. From this result and from Equation 5 one infers

$$\alpha_s(M_{Z^0}) = 0.139 \pm 0.018 \qquad [\text{in } O(\alpha_s^3)]$$

or

$$\alpha_s(M_{Z^0}) = 0.137 \pm 0.017 \qquad [\text{in } O(\alpha_s^2)].$$

The uncertainties are dominated by the event statistics and will thus improve with time.

The significance of $\alpha_s(M_{Z^0})$ from R can be increased in a combined fit to the hadronic and leptonic Z^0 line shape and asymmetry measurements of all experiments. A combined fit of $\alpha_s(M_{Z^0})$ and of the top quark mass M_t to all available LEP data, with the additional constraint of the W boson mass from $p\bar{p}$ collider and of $\sin^2\theta_W$ from neutrino-

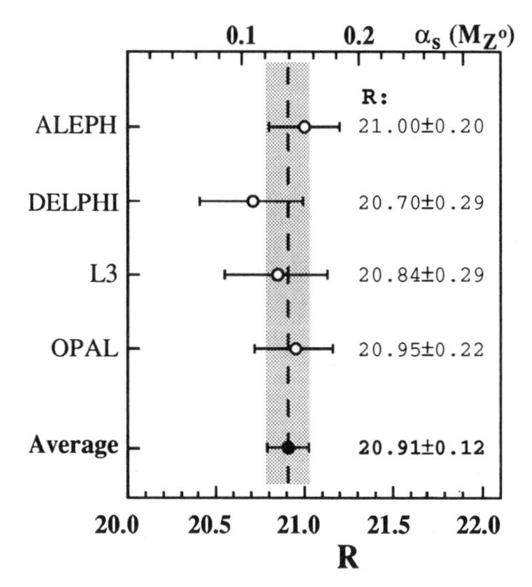

Figure 8 Compilation of measurements of $R = \Gamma_{\mathrm{had}}/\Gamma_{\mathrm{lept}}$ at LEP. The errors are statistical and systematic uncertainties, added in quadrature. The corresponding values of $\alpha_s(M_{Z^0})$, in $O(\alpha_s^3)$ and including quark mass corrections, are indicated on the upper axis. (Experimental results are from References 49–52.)

nucleon scattering experiments gives (53)

$$\alpha_s(M_{Z^0}) = 0.138 \pm 0.015 \qquad [\text{in } O(\alpha_s^3)]$$

and

$$M_t = 124^{+28}_{-31}\,{}^{+16}_{-18}\ \text{GeV},$$

where the second error of M_t is due to the unknown mass of the Higgs particle. All these results were obtained for $\mu = M_{Z^0}$. OPAL recently studied the renormalization scale dependence of α_s from Z^0 line shape fits in both $O(\alpha_s^2)$ and $O(\alpha_s^3)$ (38). For scale variations of $0.3 \leq x_\mu \leq 2.0$, where the χ^2 of the overall fit remained good and where the influence of x_μ on the fitted value of M_t was negligible, they obtained $\Delta\alpha_s = \pm 0.004$ in $O(\alpha_s^3)$. This has negligible impact if added in quadrature to the experimental errors given above.

4.4 *α_s from τ Lepton Decays*

The ratio R_τ of the hadronic and electronic branching fractions of the τ lepton,

$$R_\tau = \frac{B(\tau \to \text{hadrons} + \nu_\tau)}{B(\tau \to e\bar{\nu}_e\nu_\tau)} \equiv \frac{1 - B_e - B_\mu}{B_e}, \qquad 6.$$

which can be reliably determined by measurements of the electronic and muonic branching fractions B_e and B_μ, is theoretically expected to be given by (12)

$$R_\tau = 3.058(1.001 + \delta_{pert} + \delta_{nonpert}). \qquad 7.$$

Here, δ_{pert} and $\delta_{nonpert}$ are the perturbative and the nonperturbative QCD corrections. The perturbative QCD correction was recently calculated to $O(\alpha_s^3)$ as

$$\delta_{pert} = \frac{\alpha_s(M_\tau)}{\pi} + B\left[\frac{\alpha_s(M_\tau)}{\pi}\right]^2 + C\left[\frac{\alpha_s(M_\tau)}{\pi}\right]^3, \qquad 8.$$

with $B = 5.202$ and $C = 26.37$, assuming the production of three quark flavors ($N_f = 3$) at the energy scale of $O(M_\tau)$ (11, 12). The nonperturbative correction was estimated to be $\delta_{nonpert} = -0.007 \pm 0.004$ (12).

Measurements of the leptonic branching fractions of the τ lepton are available from ALEPH (54), L3 (55), and OPAL (56). The resulting experimental values of R_τ are summarized in Figure 9. These values have been calculated from the right-hand side of Equation 6 and assume

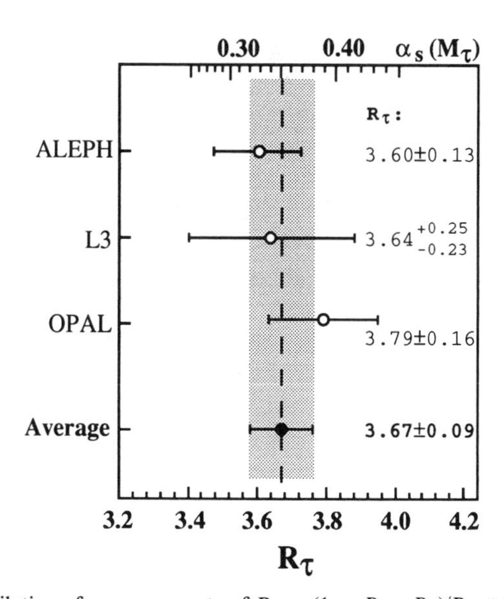

Figure 9 Compilation of measurements of $R_\tau = (1 - B_e - B_\mu)/B_e$ at LEP. The errors are statistical and systematic uncertainties, added in quadrature. The corresponding values of $\alpha_s(M_\tau)$, in $O(\alpha_s^3)$ perturbation theory, are indicated on the upper axis. (Data from References 54–56.)

lepton universality plus a phase-space factor of 0.9728 for B_μ to obtain a more precise average value of the lepton branching ratio. The statistical and experimental systematic errors have been added in quadrature. Assuming that errors are uncorrelated between the experiments, one obtains an average value of $R_\tau = 3.67 \pm 0.09$ from LEP. Applying Equations 7 and 8, this value results in $\alpha_s(M_\tau) = 0.347^{+0.029}_{-0.031}$ in $O(\alpha_s^3)$ for three quark flavors. The result, which at this level contains only the experimental uncertainties, is significantly larger than the value of $\alpha_s(M_{Z^0})$ from event shapes and jet rates measured in Z^0 decay, as might be expected for an energy-dependent α_s.

To test the running of α_s more quantitatively, the value of α_s is extrapolated from $\mu = M_\tau$ to $\mu = M_{Z^0}$ using the renormalization group equation given by Marciano (57). We follow the approximation that at a threshold, where N_f changes by one unit, α_s is a continuous function of μ (57). This results in $\alpha_s(M_{Z^0}) = 0.1185^{+0.0028}_{-0.0034}$, where the relative size of the experimental error decreases because of the logarithmic behavior of α_s with μ. A systematic uncertainty of the extrapolation arises from the fact that the exact value of μ at which the number of flavors should be changed is unclear. Following the analysis of OPAL (38) we assume that the thresholds are somewhere between M_q and $2M_q$, where q stands for either the c quark ($M_c \approx 1.5$ GeV) or the b quark ($M_b \approx 5$ GeV). If the charm quark threshold is smaller than M_τ, $N_f = 4$ is assigned at $\mu = M_\tau$, and α_s is evolved only through the bottom quark threshold. In this case, $\alpha_s(M_{Z^0})$ is increased by $+0.0029$, compared to its value when $N_f = 3$ at $\mu = M_\tau$. We take the average of these two results and quote half their difference as the systematic uncertainty of the extrapolation procedure.

Another source of theoretical uncertainty is the unknown higher order contributions to δ_{pert}, which can be parametrized by the sensitivity of the α_s result to variations of the renormalization scale. For the $O(\alpha_s^3)$ calculation, scale variations in the range $1.0 \leq \mu \leq 2.5$ GeV result in $\Delta\alpha_s(M_\tau) = {}^{+0.036}_{-0.038}$. Extrapolation to $\mu = M_{Z^0}$ gives $\Delta\alpha_s(M_{Z^0}) = {}^{+0.0033}_{-0.0042}$. We take this result as a measure of the higher order uncertainties. Adding in quadrature the experimental error, the error from the extrapolation, and the scale uncertainty gives

$$\alpha_s(M_{Z^0}) = 0.120^{+0.005}_{-0.006}.$$

We note that this result is in remarkably good agreement with the overall value from event shapes, jet rates, and energy correlations, which gave $\alpha_s(M_{Z^0}) = 0.120 \pm 0.006$ in $O(\alpha_s^2)$.

4.5 *Further Results on α_s*

To test the flavor independence of the strong interaction, the L3 collaboration measured the ratio of α_s for bottom quarks and for light quarks (58). The coupling strength was determined from the fraction of 3-jet events, measured for all hadronic events and for events with inclusive muons or electrons. The purity of primary bottom quark events was evaluated to be 22% for the first data set and 87% for the inclusive lepton sample. The ratio of α_s from these two samples was found to be $\alpha_s(b)/\alpha_s(udsc) = 1.00 \pm 0.05(stat.) \pm 0.06(syst.)$. This result is consistent with the flavor independence of α_s as predicted by QCD.

In a study of properties of multihadronic events with final-state photons, OPAL has also determined the strong coupling constant in $O(\alpha_s^2)$ by comparing the rates of two-jet events with a photon to those of three-jet events in the inclusive multihadronic sample (59). Based on the knowledge of the size of the electromagnetic coupling constant α_{em}, on the $O(\alpha_{em}\alpha_s)$ matrix elements for final-state photon production (60), and on the $O(\alpha_s^2)$ matrix elements for three-jet production in hadronic final states (9), the result is $\alpha_s(M_{Z^0}) = 0.122 \pm 0.010$. The error contains statistical and experimental systematic uncertainties, but no contributions from theoretical uncertainties.

4.6 *Summary of Measurements of α_s*

As a consistency check we summarize in Figure 10 the measurements of α_s from different high-energy particle reactions, as a function of the energy scale μ at which the determinations were performed. It should be noted that the results for α_s from deep inelastic scattering (61), from Υ decays (14), from the total hadronic cross section R in e^+e^- annihilation between center-of-mass energies of 7 GeV and 56 GeV (62, 63) [modified to reflect the corrected calculations in $O(\alpha_s^3)$ (10)], from hadronic event shapes and energy correlations in e^+e^- annihilation around 35 GeV (29), and from W production in $\bar{p}p$ collisions (63a) do not all contain estimates of the unknown higher order uncertainties as the LEP results do. We also include the value of $\alpha_s(M_\tau) = 0.304^{+0.030}_{-0.034}(exp.)\ ^{+0.036}_{-0.038}(theor.)$, which we determine from the world average (without LEP results) of leptonic branching ratios of τ leptons, $B_e = 0.177 \pm 0.004$ and $B_\mu = 0.178 \pm 0.004$ (64) using the formulae and procedures described in Section 4.4. There is good agreement between the measurements and the $O(\alpha_s^3)$ QCD predictions for the running coupling, as shown in Figure 10. Without the measurements from R_τ and in particular from LEP, the evidence was rather weak.

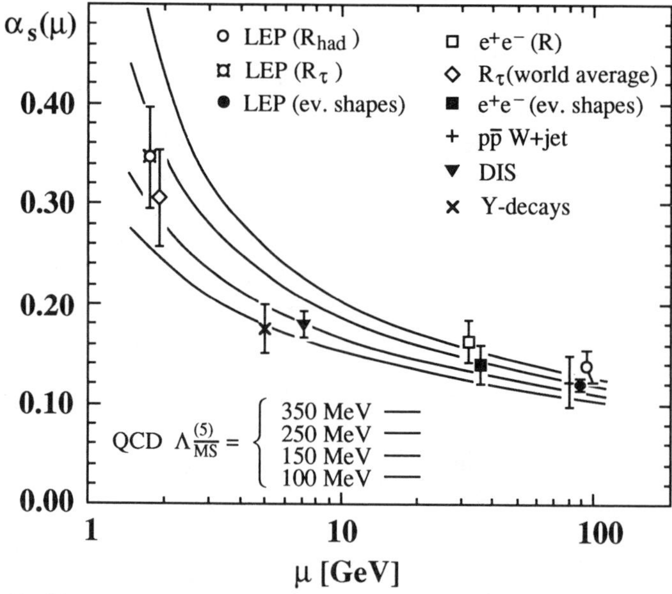

Figure 10 Measurements of α_s in the energy range from $\mu = M_\tau = 1.78$ GeV to $M_{Z^0} = 91.2$ GeV. The open squares and circles correspond to α_s determinations in next-next-to-leading order $[O(\alpha_s^3)]$ perturbation theory, the others are in next-to-leading order. The lines correspond to the QCD prediction of a running α_s for four values of $\Lambda_{\overline{MS}}$, expressed for five quark flavors above $\mu = 10$ GeV. See text for discussion and references.

5. PROPERTIES OF THE GLUON

As noted in the introduction, the gauge bosons of QCD are massless spin-one particles carrying color charge. The latter property follows from the nonabelian nature of the underlying SU(3) gauge group and leads to couplings of three and four gluons. These properties of the gluon are essential to the internal consistency of the theory and hence to many of its predictions. Although there is no viable alternative theory, it is nonetheless important to look for direct evidence of the gluon's characteristics.

5.1 *Gluon Spin from Three-Jet Events*

The LEP data are especially well suited to studying multijet events because of the good jet definition at low y_{cut}, the reduced hadronization corrections, and the high statistics. Three-jet final states are expected to correspond to $q\bar{q}g$ production with the gluon radiated from one of the quarks. From energy and momentum conservation, the parton configuration of the final state is specified by two parton energies and two angles. We concentrate here on two variables especially sensitive to the difference between vector and scalar gluons.

To first order, the three-jet cross section for vector gluon production is

$$\frac{d\sigma}{dx_1\,dx_2} \propto \frac{x_1^2 + x_2^2}{(1 - x_1)(1 - x_2)}, \qquad 9.$$

where $x_i = 2p_i/E_{\text{cm}}$, $i = 1,2,3$ are the reduced momenta of the partons and $x_1 + x_2 + x_3 = 2$ (65). The x_i are chosen to be ordered with $x_1 > x_2 > x_3$. Because of the bremsstrahlung character of gluon emission, x_3 most likely corresponds to the gluon jet and this is assumed in Equation 9. The important feature of the cross section is the singularity as x_2, and hence x_1, approach unity. The cross section for scalar (i.e. spin-zero) gluons does not contain such a singularity. The left panel of Figure 11 shows the distribution of x_2 as measured by L3 (66). The figure also shows the expectations of vector and scalar gluon models. The vector gluon hypothesis is overwhelmingly favored.

A second variable with good discrimination is the Ellis-Karliner angle λ, defined as the angle between jets 1 and 2 in the rest frame of jets 2 and 3 (67). Again, the first-order cross section for a vector gluon has a pole at $\cos\lambda = 1$, while no pole occurs for a scalar gluon. The right panel of Figure 11 shows the model comparison with the data (66). As before, the scalar gluon hypothesis is incompatible with the data.

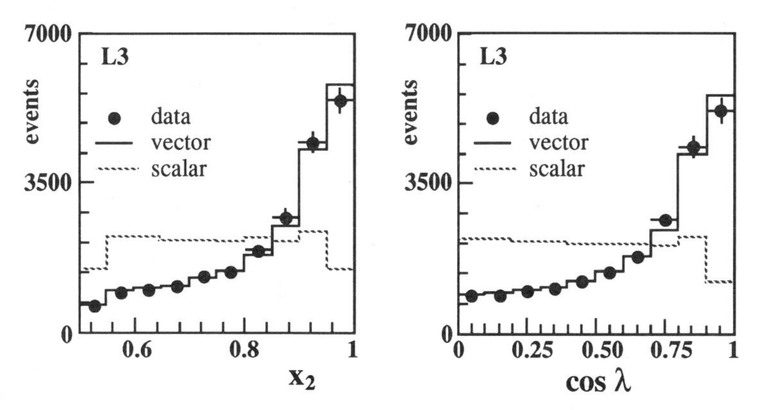

Figure 11 Distributions sensitive to gluon spin in three-jet events: (*left*) Reduced momentum of second most energetic jet, (*right*) Ellis-Karliner angle (see text for definition). (Data from Reference 66.) The data are corrected for detector effects. The solid and dashed curves show the predictions for vector and scalar gluons, respectively. They include effects of hadronization and photon radiation.

Similar results are available from OPAL (68) and DELPHI (69). Evidence for the vector nature of the gluon has also been reported from earlier studies (70), but the results from LEP are especially striking.

5.2 *Evidence of Gluon Self-Coupling*

The leading order processes producing four-jet final states are shown in Figure 12. The gluon self-coupling term of Figure 12*a* would be absent in an abelian theory. The spin of the virtual gluon g* is expected to be polarized in the $q\bar{q}g^*$ event plane. The process $g^* \rightarrow q'\bar{q}'$ (Figure 12*d*) tends to favor configurations normal to the event plane, while gluons from the process $g^* \rightarrow gg$ tend to lie in the event plane (71). In addition, for three-jet events at, for example, $y_{cut} = 0.01$, the $g^* \rightarrow q'\bar{q}'$ process should comprise only 4.7% of the four-jet final states according to QCD but 31% in an abelian theory (72). This enhancement aids in discriminating between the hypotheses.

The Bengtsson-Zerwas angle χ_{BZ} is defined between the planes of jet momenta $\mathbf{p}_1,\mathbf{p}_2$ and $\mathbf{p}_3,\mathbf{p}_4$, where the indices are assigned by energy ordering (73). On average $\mathbf{p}_1$ and $\mathbf{p}_2$ correspond to the $q\bar{q}$ system while $\mathbf{p}_3$ and $\mathbf{p}_4$ are associated with gg or $q'\bar{q}'$. Figure 13 shows a sample of 2500 four-jet events from L3 selected with $y_{cut} = 0.02$ (74). The bands show the variation from different theoretical predictions. The data are well described by QCD but are incompatible with an abelian theory. Similar analysis have been reported by other groups (75–77).

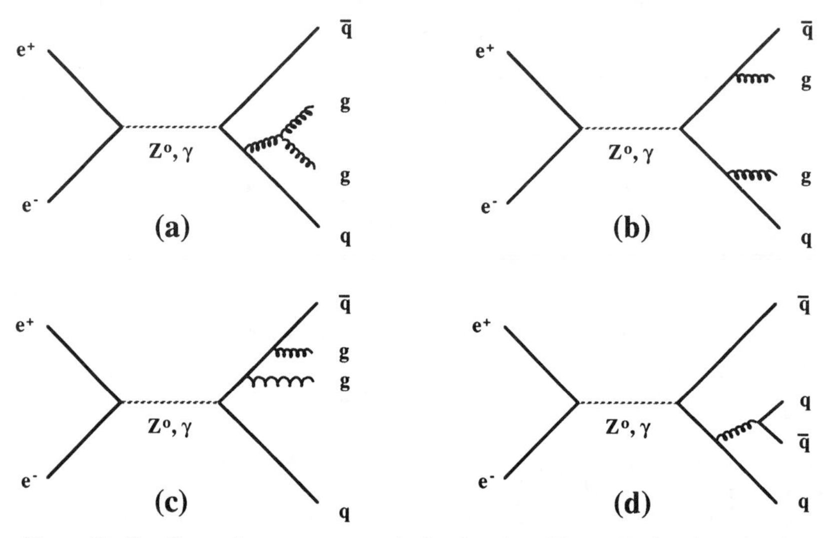

Figure 12 Leading order processes producing four jets. Figure 12*a* involves the gluon self-coupling.

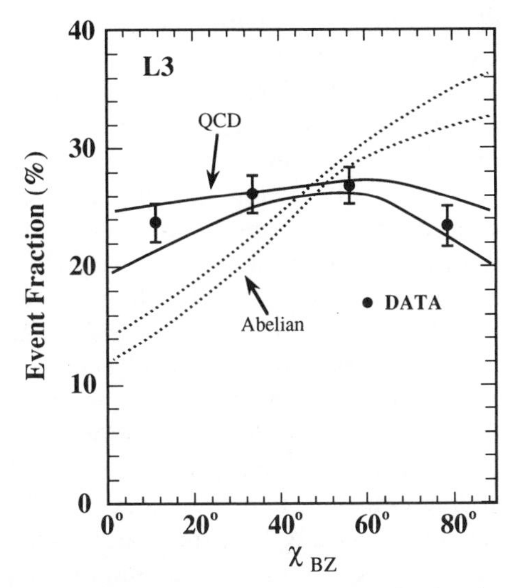

Figure 13 Distribution of Bengtsson-Zerwas angle χ_{BZ} for four-jet events. Solid lines show the range of predictions for QCD, while dashed lines show prediction for an abelian model. (Data from Reference 74.)

The DELPHI group has analyzed a two-dimensional distribution from the four-jet events in order to extract a ratio of Casimir color factors for the underlying gauge group (78). The ratio N_C/C_F corresponds to the strength of gluon splitting to two gluons relative to gluon coupling to a quark, while T_R/C_F gives the strength of gluon splitting to a quark-antiquark pair relative to gluon coupling to a quark. DELPHI obtained $N_C/C_F = 2.55 \pm 0.55 \pm 0.4 \pm 0.2$ and $T_R/C_F = 0.1 \pm 2.4$. These results are to be compared with $N_C/C_F = 2.25$ and $T_R/C_F = 1.875$ for QCD, and $N_C/C_F = 0$ and $T_R/C_F = 15$ for an abelian theory. The virtue of this analysis is its generality, with QCD and the abelian theory just two points of comparison. The result significantly restricts the possible gauge groups.

5.3 *Evidence of Asymptotic Freedom*

A fundamental consequence of the gluon self-interaction is that the effective coupling constant α_s should decrease as the energy scale of the interaction rises and the corresponding distance scale decreases. To leading order, α_s varies as $1/\ln(\mu/\Lambda_{\overline{MS}})^2$, where μ denotes the energy scale of the interaction and $\Lambda_{\overline{MS}}$ reflects a fixed renormalization point for QCD (see Equation 1). A test of this logarithmic dependence is an important goal to which the high-energy LEP data have contributed.

The fraction of hadronic events with exactly three jets, R_3, is a good monitor of the energy variation of α_s since the two are nearly proportional. This observable can be determined in a consistent, energy-independent manner at a series of center-of-mass energies and the results directly compared. This situation is in contrast to the diverse set of measurements described in Section 4.6 and Figure 10, where the sources of systematic error vary appreciably among the reactions.

Figure 14 shows the measured three-jet rate at LEP (32, 38) together with results from lower energies (75, 79). The rates are obtained with a jet definition variable $y_{cut} = 0.08$, which leads to uncertainties from hadronization effects of less than 2% for all energies above 25 GeV (27). The results of a one-parameter fit of the analytic $O(\alpha_s^2)$ QCD calculation (9, 80) using energy (renormalization) scales of $\mu = E_{cm}$ and $\mu = 0.07E_{cm}$ are shown. Also indicated is the best fit for an energy-independent coupling constant, and the expectation of an abelian alternative to QCD where, as in QED, the coupling strength increases with energy scale (27). The rise in α_s predicted by the abelian QCD is much steeper than in the QED case since the energy-dependent term is multiplied by the number of quark flavors $N_f = 5$. The results of the fits corresponding to these assumptions are given in Table 2. They

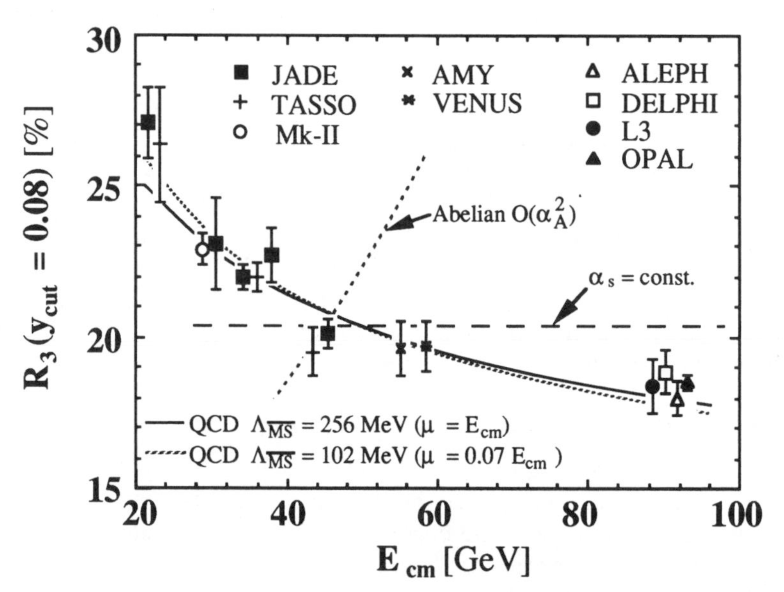

Figure 14 Energy dependence of three-jet event rates compared with fits to QCD in $O(\alpha_s^2)$; energy-independent α_s; and abelian theory in $O(\alpha_A^2)$. (Data from References 32, 38, 75, 79.)

Table 2 Results of fits to the energy dependence of three-jet rates

Theory	Fit result	χ^2/degrees of freedom
QCD, $x_\mu = 0.07$	$\Lambda_{\overline{MS}} = (102 \pm 4)$ MeV	14.2/12
QCD, $x_\mu = 1.0$	$\Lambda_{\overline{MS}} = (256^{+12}_{-11})$ MeV	8.9/12
α_s = constant	$\langle R_3 \rangle = 20.3 \pm 0.2$	76.2/12
Abelian theory	$\alpha_A(44 \text{ GeV}) = 0.26$	∞

include the value of an abelian coupling constant necessary to describe the data at $E_{cm} = 44$ GeV. The QCD prediction is in good agreement with the data, almost independent of the choice of renormalization scale, although the value of $\Lambda_{\overline{MS}}$ does depend strongly on renormalization scale. A constant or rising coupling constant is clearly excluded by the data.

An alternative form of this result is shown in Figure 15, where R_3 is plotted as a function of $1/\ln(E_{cm})$. At infinite energy $[1/\ln(E_{cm}) \rightarrow 0]$

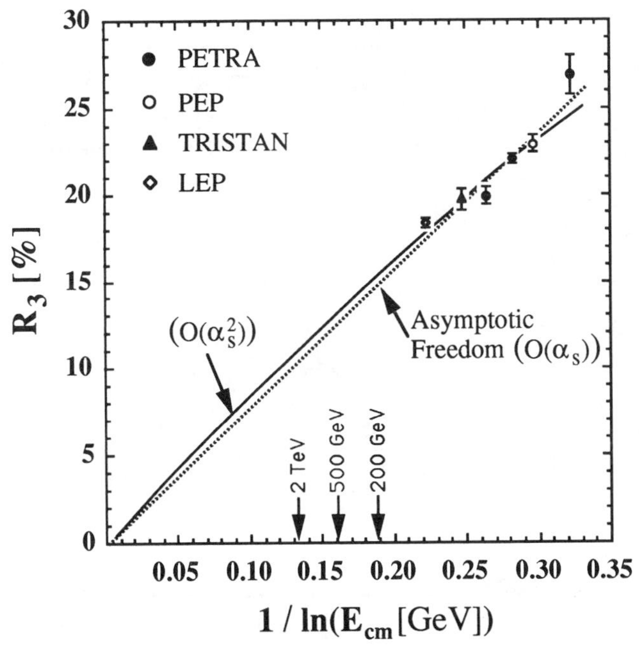

Figure 15 Three-jet production rates, combined at similar center-of-mass energies, as a function of $1/\ln(E_{cm})$, compared with the prediction of asymptotic freedom. The extrapolation can be used to infer R_3 values at future colliders. The arrows indicate three commonly discussed energy scales.

the extrapolations of R_3 and α_s are expected to vanish. The value of R_3 at future accelerator facilities can be obtained from Figure 15.

6. SOFT HADRON PHENOMENA

The boundary between perturbative QCD and the nonperturbative region associated with hadronization is not clearly defined and is an active area of study. In most approaches the parton shower is propagated down to some cutoff Q_0 and a fragmentation procedure applied to convert the partons to hadrons. The string fragmentation scheme (81), incorporated in the JETSET simulation (24), and the coherent parton shower with cluster fragmentation of HERWIG (82) are two examples of giving an excellent description of the data once parameters describing the fragmentation have been adjusted.

An alternative, purely perturbative, method has also been pursued with good success. Here a modified leading-log approach (MLLA) is used to describe the parton shower down to a cutoff of a few hundred

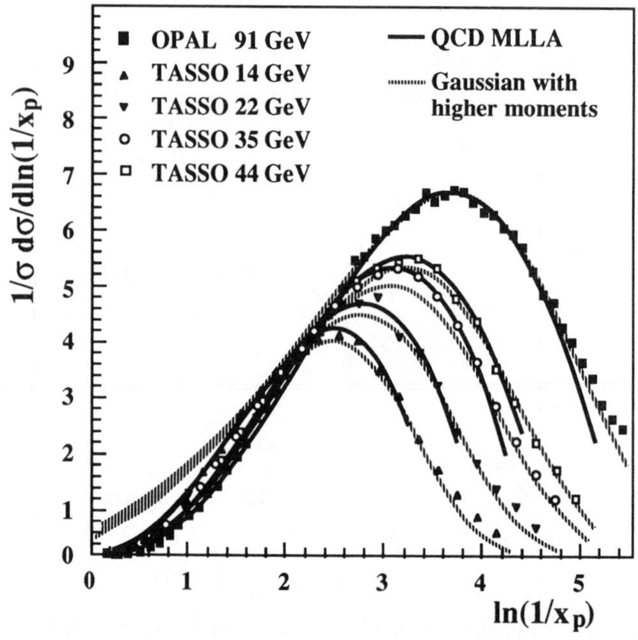

Figure 16 Inclusive hadron momentum spectrum and its energy evolution: $(1/\sigma)\, d\sigma/d\xi_p$ versus $\xi_p = \ln(1/x_p)$ at five center-of-mass energies. (Data from Reference 85. The theoretical curves are also discussed therein.)

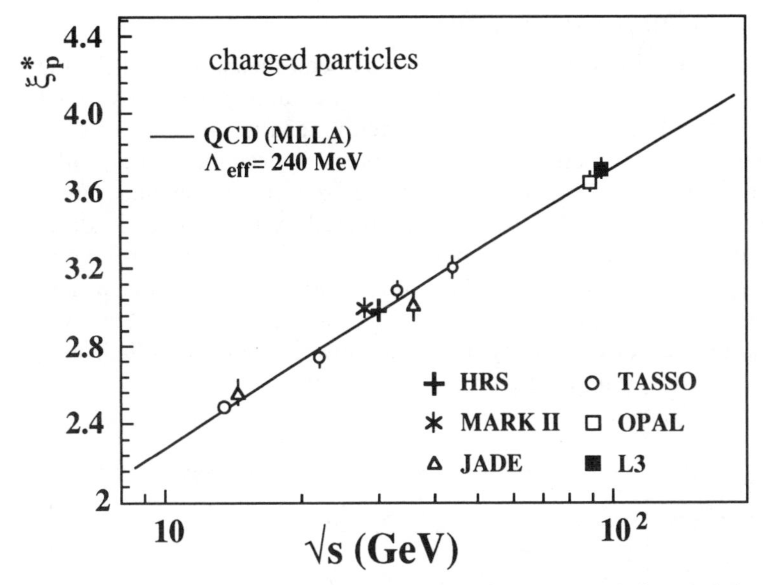

Figure 17 Inclusive hadron momentum spectrum and its energy evolution: variation of the peak position ξ_p^* of the $\ln(1/x_p)$ spectrum with E_{cm}. (The figure is from Reference 86.)

MeV (83). The partons themselves are then treated as hadrons under the hypothesis of local parton-hadron duality (84). This technique gives successful analytic descriptions of features such as multiplicity, inclusive energy spectra, and particle correlations, without using any hadronization model. The inclusion of interference effects among soft gluons is an essential element in its success.

6.1 *Particle Spectra and Soft Gluon Coherence*

The energy distribution of final-state particles is usually described in terms of $\xi_p = \ln(1/x_p)$, where x_p is the particle momentum normalized to the beam energy. Figure 16 shows the distribution of ξ_p measured at LEP and its evolution from lower energies (85). The solid curves are two-parameter fits to the analytic form of the MLLA calculation. They provide a good description of the spectra and explain the reduced production at low x_p (high ξ_p) in terms of destructive interference between soft gluons. This same calculation successfully describes the variation of the peak position with E_{cm} as shown in Figure 17 (86). The simulations of JETSET and HERWIG also give excellent descriptions of the data.

6.2 *Multiplicity Studies*

The charged-particle multiplicity has been measured at LEP (87) and compared with a number of models, including perturbative QCD. As shown in Figure 18, the dependence of mean multiplicity on center-of-mass energy is well described by several approaches. The width of the multiplicity distribution as characterized by its second moment is shown in Figure 19. The value predicted by the next-to-leading QCD calculation is found to be about 8% larger than the measured one. The large change between leading order and next-to-leading order QCD suggests that neglected higher order terms may be important. The data can be well described by a parton shower model with hadronization, as shown by the solid curve of Figure 19.

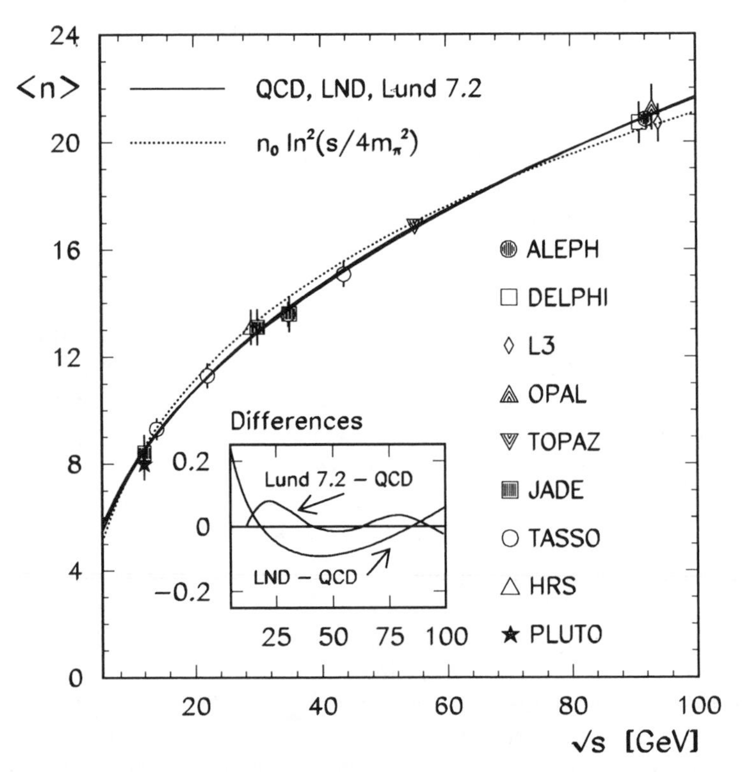

Figure 18 Mean charged particle multiplicity versus center-of-mass energy. The data are compared with a phenomenological parameterization (*dotted line*) and with expectations (*full lines*) for a log-normal distribution (LND), the next-to-leading order QCD predictions, and the Lund 7.2 parton shower model. The inset shows the differences between the QCD curve, Lund 7.2, and LND. The figure is from Decamp et al (87).

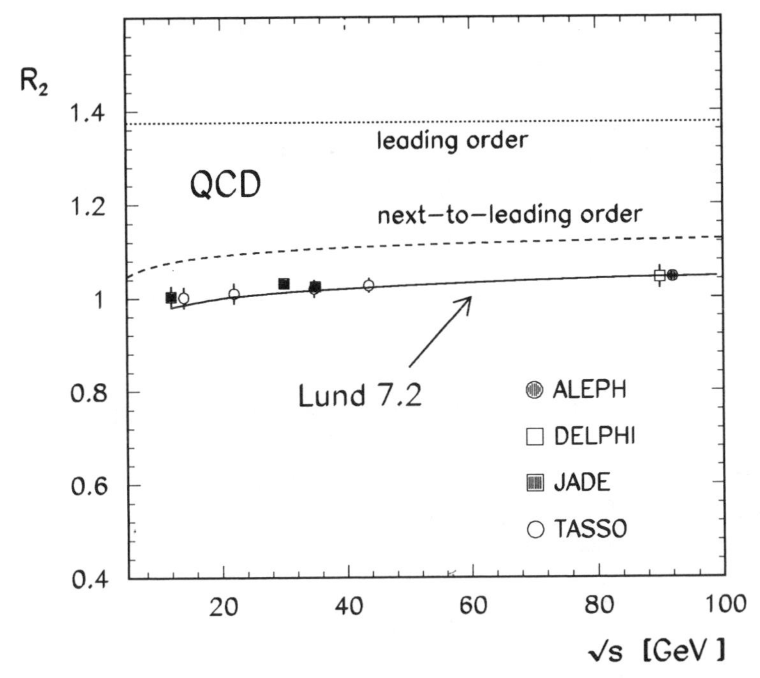

Figure 19 Properties of the charged multiplicity distribution versus $E_{\rm cm}$: second moment of fit to negative binomial distribution. Energy dependence of the second binomial moment compared with leading order (*dotted line*) and next-to-leading order (*dashed line*) QCD predictions for $\Lambda_{\rm LLA} = 0.145$ GeV. The full line is the prediction of the tuned Lund 7.2 parton shower model. The figure is from Decamp et al (87).

6.3 *Studies of the String Effect*

It has been seen at lower energies that in a three-jet event the number of particles between a quark and gluon jet is larger than between two quark jets (88). Explanations are provided by both the string fragmentation scheme (81) and by the analytic QCD calculations involving soft gluon coherence (89).

New measurements available from LEP exploit the high statistics to aid in identification of the gluon jets and to study particle flow in symmetric three-jet configurations (90). This method allows a comparison of quark-antiquark and quark-gluon systems having similar jet energies, without reference to any model calculation. Planar three-jet events are selected in which the axis of the most energetic jet bisects the angle between the other two. One of the two less energetic jets is required to contain a well-identified energetic muon or electron. This tags the

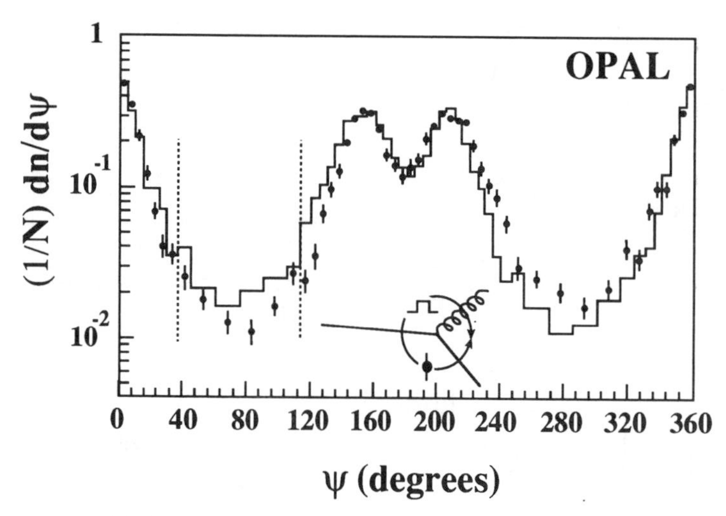

Figure 20 Particle flux in symmetric three-jet events with lepton-tagged quark jet versus angle in the event plane. The points and histogram show the same data. To compare the flux between quark and gluon jets with that between two quark jets, the histogram shows the data plotted versus ψ while the points show the data plotted versus $2\pi - \psi$ (see inset). (The data are from Reference 90.)

jet as being associated with a heavy quark. The most energetic jet is taken as the other quark jet, and the remaining jet is thus identified as the gluon jet. The expected purity of the identification is about 80%.

Figure 20 shows the measured particle flux as a function of angle in the event plane measured from the highest energy (quark) jet. The histogram corresponds to an angle increasing toward the gluon jet (q-g-q), while the points show the same data but with the angle increasing toward the quark jet (q-q-g). The comparison of the two plots shows a higher particle flux between quark and gluon jets than between two quark jets. This asymmetry implies that dynamical differences exist between quark-gluon and quark-antiquark systems, which could be explained by gluon coherence in perturbative QCD, by string fragmentation models, or by differences between quark and gluon jet fragmentation. Further studies are under way to understand the origin of the effect.

6.4 *Properties of Quark- and Gluon-Initiated Jets*

In QCD the gluons carry a larger color charge than quarks and are therefore expected to radiate more. Their jets should have a softer particle spectrum, larger angular width, and higher multiplicity than quark jets of the same energy.

The sample of tagged three-jet events discussed in Section 6.3 has been used to compare the characteristics of quark and gluon jets (91). The gluon jets are indeed seen to be broader and in their cores are found to have a softer momentum spectrum. It was also demonstrated, without relying on model calculations, that these differences apply not only to a comparison of gluon jets and heavy quark jets, but also to quark jets of average flavor content. Little difference, however, was found in the average charged multiplicities of gluon and quark jets, with $\langle n_{ch}\rangle_{gluon}/\langle n_{ch}\rangle_{quark} = 1.02 \pm 0.04^{+0.06}_{-0.00}$. A naive prediction of this ratio for virtual gluon-gluon and quark-antiquark systems at asymptotic energies is 9/4. At present, there is no clear QCD prediction that can be directly compared with this result.

7. SUMMARY AND OUTLOOK

The first three years of LEP operation have provided a large number of significant tests of perturbative QCD. Many of these studies are of such high experimental precision that theoretical uncertainties from uncomputed higher orders—from hadronization, parton masses, and virtuality—limit the final precision of the results. After careful evaluation of these uncertainties, the data are in good agreement with the predictions of perturbative QCD and no significant deviation from the Standard Model has yet been observed.

The strong coupling constant α_s has been determined from more than 30 individual analyses of approximately 15 observables. These include hadronic event shapes, jet rates, and energy correlations; all calculated in complete $O(\alpha_s^2)$ perturbation theory. If all published LEP results are treated in a consistent manner, they give a weighted average of $\alpha_s(M_{Z^0}) = 0.120 \pm 0.006$, corresponding to $\Lambda_{\overline{MS}}^{(N_f=5)} = 250^{+95}_{-70}$ MeV. The consistency of this average with the many individual determinations, each subject to different uncertainties, is an important test of our understanding.

Two determinations of α_s have been obtained using full $O(\alpha_s^3)$ calculations. The Z^0 line shape and decay asymmetries give $\alpha_s(M_{Z^0}) = 0.138 \pm 0.015$, while the hadronic branching fraction of the τ lepton gives $\alpha_s(M_{Z^0}) = 0.120^{+0.005}_{-0.006}$. These are in good agreement with the $O(\alpha_s^2)$ result. These results are also compatible with the prediction of $\alpha_s(M_{Z^0}) = 0.11 \pm 0.01$, based on a compilation of α_s measurements from lower energy studies before LEP began operation (14).

The nonabelian structure of QCD is demonstrated by several studies focusing on the energy dependence of α_s and on the gluon self-coupling process. Hadronic event shape distributions and their energy dependence, if compared with similar measurements from lower e^+e^- center-of-mass energies, can be well described, within QCD shower models, in terms of QCD scaling violations and an energy-independent parametrization of hadronization. The comparison of $\alpha_s(M_\tau) = 0.35 \pm 0.05$ from τ decays and of $\alpha_s(M_{Z^0}) = 0.120 \pm 0.006$ from Z^0 hadronic event shapes and jet rates gives a dramatic demonstration of the running of α_s between $M_\tau = 1.78$ GeV and $M_Z = 91.2$ GeV. The energy dependence of three-jet event production rates, when compared with corresponding results from lower energies, also provides convincing evidence for asymptotic freedom. Studies of four-jet final states such as angular correlations between jet axes and the experimental determination of the Casimir color factors give further evidence for the gluon self-coupling.

An active area of study at LEP is the boundary between perturbative QCD and the nonperturbative region associated with hadronization. Measurements of inclusive particle spectra are in good agreement with analytic QCD calculations based on a modified leading-log approximation and the hypothesis of local parton-hadron duality. New methods of quark and gluon jet tagging, together with the large data samples from LEP, have allowed a model-independent study of the string effect and of differences between quark and gluon jets.

While most of these results were based on data samples of fewer than 2×10^5 events per experiment, future studies using more than 10^6 events can be expected. These should include comparisons of final-state photon and gluon radiation from quarks, QCD studies of heavy quarks using vertex tagging, and measurements of identified particle spectra and correlations. For many future studies, however, new theoretical input is needed to improve the precision. This is especially true of α_s measurements, which are limited by the unknown higher order contributions to the event shapes and jet production rates. These observables provide the smallest experimental uncertainties in α_s determinations, at 3% or less. The hadronization uncertainties, inferred from models that provide realistic descriptions of the data, are in most cases smaller than the estimated uncertainties of higher order effects.

Resummation of leading and next-to-leading logarithms to all orders, as recently done for some observables, is likely to become the major improvement in the field of perturbative QCD since the advent of $O(\alpha_s^2)$ calculations more than 11 years ago. These NLLA + $O(\alpha_s^2)$ calculations indeed provide a significantly better description of the corresponding LEP data in regions that were previously classified as the nonperturbative regime. They also result in a significant reduction of the renormalization scale dependence, as theoretically anticipated. The extension of such calculations to further event-shape, jet-rate, and energy-correlation observables, as well as the availability of calculations complete to $O(\alpha_s^3)$, will be indispensable for a deeper understanding of perturbative QCD and for more precise determinations of α_s at LEP.

ACKNOWLEDGMENTS

We wish to thank the members of the four LEP experiments for their efforts and efficient collaboration, and for making their results available for this review. We are grateful to S. Komamiya for his help preparing the results of Section 4.4, and we thank T. Mori and A. Dieckmann for providing Figure 1.

Literature Cited

1. Burkhardt, H., Steinberger, J., *Annu. Rev. Nucl. Part. Sci.* 41:55–96 (1991)
2. Gross, D. J., Wilczek, F., *Phys. Rev. Lett.* 30:1343 (1973); *Phys. Rev.* D8:3633 (1973); Politzer, H. D., *Phys. Rev. Lett.* 30:1346 (1973)
3. Decamp, D., et al (ALEPH Collaboration), *Nucl. Instrum. Methods* A294:121 (1990)
4. Aarnio, P., et al (DELPHI Collaboration), *Nucl. Instrum. Methods* A303:233 (1991)
5. Adeva, B., et al (L3 Collaboration), *Nucl. Instrum. Methods* A289:35 (1990)
6. Ahmet, K., et al (OPAL Collaboration), *Nucl. Instrum. Methods* A305:275 (1991)
7. Myers, S., Picasso, E., *Contemp. Phys.* 31:387 (1990)
8. Bardeen, W. A., et al, *Phys. Rev.* D18:3998 (1978); Buras, A., *Rev. Mod. Phys.* 52:199 (1980)
9. Kunszt, Z., Nason, P., in *Z Physics at LEP 1,* ed. G. Altarelli, R. Kleiss, C. Verzegnassi, CERN Rep. 89-08 (1989)
10. Surguladze, L. R., Samuel, M. A., *Phys. Rev. Lett.* 66:560 (1991); Gorishny, S. G., Kataev, A. L., Larin, S. A., *Phys. Lett.* B259:144 (1991)
11. Braaten, E., *Phys. Rev. Lett.* 60:1606 (1988)
12. Braaten, E., Narison, S., Pich, A., CERN preprint TH.6070/91 (1991)
13. Sjöstrand, T., see Ref. 9
14. Altarelli, G., *Annu. Rev. Nucl. Part. Sci.* 39:357 (1989)
15. Decamp, D., et al (ALEPH Collaboration), *Phys. Lett.* B234:209 (1990)
16. Akrawy, M. Z., et al (OPAL Collaboration), *Z. Phys.* C47:505 (1990)
17. Abreu, P., et al (DELPHI Collaboration), CERN preprint PPE/91-181/rev. (1991), submitted to *Z. Phys.* C
18. Akrawy, M. Z., et al (OPAL Collaboration), *Phys. Lett.* B235:389 (1990)
19. Abreu, P. et al (DELPHI Collaboration), *Phys. Lett.* B247:167 (1990)
20. Decamp, D., et al (ALEPH Collaboration), *Phys. Lett.* B255:623 (1991)
21. Adeva, B., et al (L3 Collaboration), *Phys. Lett.* B248:464 (1990)
22. Akrawy, M. Z., et al (OPAL Collaboration), *Z. Phys.* C49:375 (1991)
23. de Boer, W., et al *Z. Phys.* C49:141 (1991)
24. Sjöstrand, T., *Comput. Phys. Commun.* 39:347 (1986); Sjöstrand, T., Bengtsson, M., *Comput. Phys. Commun.* 43:367 (1987)
25. Bartel, W., et al (JADE Collaboration), *Z. Phys.* C33:23 (1986)
26. Bethke, S., et al (JADE Collaboration), *Phys. Lett.* B213:235 (1988)
27. Bethke, S., in *Proc. of the Workshop on Jet Physics at LEP and HERA,* Durham, 1990; *J. Phys.* G17:1455 (1991)
28. Bethke, S., et al, *Nucl. Phys.* B370:310 (1992)
29. Bethke, S., LBL-28112 (1989)
30. Wu, S. L., *Phys. Rep.* 107:59 (1984)
31. Naroska, B., *Phys. Rep.* 148:67 (1987)
32. Hebbeker, T., Plenary talk presented at the LP-HEP91 Conf. on High Energy Physics, Geneva, 1991, Aachen preprint PITHA 91/17 (1991)
33. Bethke, S., in *Proc. of the EPS Conf. on Hadronic Structure and Electroweak Interactions,* Amsterdam, 1991; Heidelberg preprint HD-PY 91/5
34. Csikor, F., et al, *Phys. Rev.* D31:1025 (1985)
35. Stevenson, P. M, *Phys. Rev.* D23:2916 (1981); Brodsky, S. J., Lepage, G. P., Mackenzie, P. B., *Phys. Rev.* D28:228 (1983); Grunberg, G., *Phys. Lett.* B95: 70 (1980)

36. Decamp, D., et al (ALEPH Collaboration), *Phys. Lett.* B257:479 (1991)
37. Adeva, B., et al (L3 Collaboration), *Phys. Lett.* B257:469 (1991)
38. Acton, P. D. (OPAL Collaboration), CERN preprint PPE/92-18 (1992), submitted to *Z. Phys.* C
39. Magnoli, N., Nason, P., Rattazzi, R., *Phys. Lett.* B252:271 (1990)
40. Bethke, S., *Z. Phys.* C43:331 (1989)
41. Deleted in proof
42. Deleted in proof
43. Catani, S., et al, *Phys. Lett.* B263:491 (1991)
44. Catani, S., Turnock, G., Webber, B. R., *Phys. Lett.* B272:368 (1991)
45. Catani, S., et al, *Phys. Lett.* B269:432 (1991)
46. Brown, N., Stirling, W. J., preprints RAL-91-049 and DTP/91/30 (1991)
47. Hebbeker, T., Aachen Rep. PITHA 91/08 (1991) (revised version)
48. Bardin, D. Yu., et al, *Nucl. Phys.* B351:1 (1991); *Comput. Phys. Commun.* 59:303 (1990); *Z. Phys.* C44:493 (1989)
49. Decamp, D., et al (ALEPH Collaboration), *Z. Phys.* C53:1 (1992)
50. Adeva, B., et al (L3 Collaboration), *Z. Phys.* C51:179 (1991)
51. Abreu, P., et al (DELPHI Collaboration), *Nucl. Phys.* B367:511 (1991)
52. Alexander, G., et al (OPAL Collaboration), *Z. Phys.* C52:175 (1991)
53. The LEP Collaborations (ALEPH, DELPHI, L3, and OPAL), *Phys. Lett.* B276:247 (1992)
54. Decamp, D., et al (ALEPH Collaboration), CERN preprint PPE/91-186 (1991), submitted to *Z. Phys.* C
55. Adeva, B., et al (L3 Collaboration), *Phys. Lett.* B265:451 (1991)
56. Alexander, G., et al (OPAL Collaboration), *Phys. Lett.* B266:201 (1991)
57. Marciano, W. J., *Phys. Rev.* D29:580 (1984)
58. Adeva, B., et al (L3 Collaboration), *Phys. Lett.* B271:461 (1991)
59. Acton, P. D., et al (OPAL Collaboration), *Z. Phys.* C54:193 (1992)
60. Kramer, G., Lampe, B., *Phys. Lett.* B269:401 (1991)
61. Virchaux, M., Milsztajn, A., *Phys. Lett.* B274:221 (1992)
62. D'Agostini, G., de Boer, W., Grindhammer, G., *Phys. Lett.* B229:160 (1989)
63. Marshall, R., *Z. Phys.* C43:595 (1989)
63a. Alitti, J., et al (UA2 Collaboration), *Phys. Lett.* B263:513 (1991)
64. Particle Data Group, Review of Particle Properties, *Phys. Lett.* B239:1 (1990)
65. Kramer, G., *Springer Tracts Mod. Phys.* 102:39 (1984)
66. Adeva, B., et al (L3 Collaboration), *Phys. Lett.* B263:551 (1991)
67. Ellis, J., Karliner, I., *Nucl. Phys.* B148:141 (1979); Ellis, J., Karliner, I., Stirling, W. J., *Phys. Lett.* B217:363 (1989)
68. Alexander, G., et al (OPAL Collaboration), *Z. Phys.* C52:543 (1991)
69. Abreu, P., et al (DELPHI Collaboration), *Phys. Lett.* B274:498 (1992)
70. Brandelik, R., et al (TASSO Collaboration), *Phys. Lett.* B97:453 (1980); Wu, S. L., *Phys. Rep.* 107:59 (1984); Behrend, H. J., et al (CELLO Collaboration), *Phys. Lett.* B110: 329 (1982); Berger, C., et al (PLUTO Collaboration), *Phys. Lett.* B97:459 (1980); Burger, J. D., et al (Mark J Collaboration), in *Proc. 21st Int. Conf. on High Energy Physics,* Paris, 1982, *J. Phys.* 43:C3 (1982); Koller, K., Krasemann, H., *Phys. Lett.* B88:119 (1979)
71. Nachtmann, O., Reiter, A., *Z. Phys.* C16:45 (1982)
72. Bethke, S., Ricker, A., Zerwas, P. M., *Z. Phys.* C49:59 (1991)
73. Bengtsson, M., Zerwas, P. M., *Phys. Lett.* B208:306 (1988)
74. Adeva, B., et al (L3 Collaboration), *Phys. Lett.* B248:227 (1990)
75. Park, I. H., et al (AMY Collaboration), *Phys. Rev. Lett.* 62:1713 (1989)
76. Abe, K., et al (VENUS Collaboration), *Phys. Rev. Lett.* 66:280 (1991)
77. Akrawy, M. Z., et al (OPAL Collaboration), *Z. Phys.* C49:49 (1991)
78. Abreu, P., et al (DELPHI Collaboration), *Phys. Lett.* B255:466 (1991)
79. Bethke, S., et al (JADE Collaboration), *Phys. Lett.* B213:235 (1988); Braunschweig, W., et al (TASSO Collaboration), *Phys. Lett.* B214:286 (1988); Bethke, S., et al (MARK 2 Collaboration), *Z. Phys.* C43:325 (1989); Abe, K., et al (VENUS Collaboration), *Phys. Lett.* B240:232 (1990)
80. Ellis, R. K., Ross, D. A., Terrano, A. E., *Nucl. Phys.* B178:421 (1981)
81. Andersson, B., Gustafson, G., Sjöstrand, T., *Z. Phys.* C6:235 (1980)
82. Marchesini, G., Webber, B. R., *Nucl. Phys.* B238:1 (1984); Webber, B. R., *Nucl. Phys.* B238:492 (1984); Marchesini, G., Webber, B. R., *Nucl. Phys.* B310:461 (1988)
83. Dokshitzer, Y. L., Troyan, S. I., Leningrad preprint LNPI-922 (1984); Azimov, Y. I., et al, *Z. Phys.* C27:65 (1985); Khoze, V. A., Dokshitzer, Y. L., Troyan, S. I., Lund preprint LU

TP 90-12 (1990)
84. Amati, D., Veneziano, G., *Phys. Lett.* B83:87 (1979); Azimov, Y. I., et al, *Z. Phys.* C27:65 (1985)
85. Akrawy, M. Z., et al (OPAL Collaboration), *Phys. Lett.* B247:617 (1990); Braunschweig, W., et al (TASSO Collaboration), *Z. Phys.* C47:187 (1990)
86. Adeva, B., et al (L3 Collaboration), *Phys. Lett.* B259:199 (1991)
87. Decamp, D., et al (ALEPH Collaboration), *Phys. Lett.* B273:181 (1991); Abreu, P., et al (DELPHI Collaboration), *Z. Phys.* C50:185 (1991); Adeva, B., et al (L3 Collaboration), *Phys. Lett.* B259:199 (1991); Acton, P. D., et al (OPAL Collaboration), CERN preprint PPE/91-176 (1991), submitted to *Z. Phys.* C
88. Bartel, W. (JADE Collaboration), *Phys. Lett.* B101:129 (1981)
89. Azimov, Y. I., et al, *Phys. Lett.* B165: 147 (1985); *Sov. J. Nucl. Phys.* 43:95 (1986)
90. Akrawy, M. Z., et al (OPAL Collaboration), *Phys. Lett.* B261:334 (1991)
91. Alexander, G., et al (OPAL Collaboration), *Phys. Lett.* B265:462 (1991)

Annu. Rev. Nucl. Part. Sci. 1992. 42:291–332

PARTON DISTRIBUTION FUNCTIONS OF HADRONS

Joseph F. Owens[1]

Physics Department, Florida State University, Tallahassee, Florida 32306

Wu-Ki Tung[2]

Physics Department, Illinois Institute of Technology, Chicago, Illinois 60616

KEY WORDS: QCD parton model, quantum chromodynamics

CONTENTS

[1] Supported in part by the United States Department of Energy and the Texas National Laboratory Research Commission.
[2] Supported in part by the National Science Foundation and the Texas National Laboratory Research Commission.

0163-8998/92/1201-0291$02.00

1. INTRODUCTION

The standard model (SM) has undergone a great deal of development in both the electroweak interaction (1) and the strong interaction (2) sectors during the past two decades. On both frontiers, current research focuses on issues of precision and consistency far beyond those possible in the qualitative leading-order picture of the 1970s. Precision studies of the SM provide, on the one hand, ever-improving determinations of the unknown parameters of the theory and, on the other hand, a window into new physics beyond the minimal theory. These two aspects are, indeed, completely inseparable.

Quantum chromodynamics (QCD), the theory of interactions of quarks and gluons, provides the basic description of strong interactions in the standard model. The quarks also couple to leptons and vector bosons via the electroweak force. However, physical processes involve only leptons, vector bosons, and hadrons. In order to make meaningful comparisons between theory and experiment, we need a formalism relating calculable (elementary) quantities to measurable (physical) ones. For high-energy processes, perturbative QCD provides this framework through "factorization theorems": physical cross sections are factorized, to all orders of the QCD running coupling α_s, into a "hard cross section" among elementary partons and a "soft part" consisting of universal distribution functions of partons inside hadrons. In leading order (LO), or tree approximation, this formalism reduces to the simple parton model of the earlier years. We have since progressed far beyond that.

The universal (i.e. process-independent) parton distribution functions (PDFs) play a central role in SM phenomenology. Many of the most precise tests of the electroweak theory are currently limited by the lack of reliable PDFs. All quantitative calculations for signals of "new physics" at high energies, as well as standard model processes (which form significant backgrounds for the new physics), depend on our current knowledge of the PDF. Most lepton-hadron and hadron-hadron scattering processes can serve either as a source of information on the PDF, or as testing grounds of QCD predictions based on known PDFs. The systematic analysis of these processes to extract accurate parton distributions is therefore intrinsically intertwined with all aspects of high-energy physics research.

The relatively simple first generation of PDFs no longer satisfies the rigorous demands of current applications. As both theory and experiments have made significant progress in sophistication and complexity, the analyses of second- and third-generation parton distributions necessarily involve many nontrivial issues and uncertainties. Not all of these are fully taken into account in currently used PDFs. Many of the subtleties are also unfamiliar or unknown to the average user of these PDFs.

The purpose of this article is as follows: (*a*) to summarize the elements of perturbative QCD that form the basis of parton distribution analysis (Sections 2 and 3); (*b*) to review the various physical processes that contribute most significantly to this analysis (Section 4); (*c*) to describe the strategies for global analyses of parton distributions and discuss in some detail the theoretical, experimental, and phenomenological issues and uncertainties involved in the global analyses (Section 5); (*d*) to comment on the currently available distribution sets and their proper use (Section 6); and (*e*) to conclude with a list of unfinished tasks and critical challenges for further progress in this key area of high-energy physics research (Section 7).

For well-established results we only briefly summarize the most relevant elements and, wherever appropriate, refer the reader to available books and review articles for comprehensive lists of references to the original papers. Additional details on many of the topics discussed in this review can be found in the proceedings of a recent workshop (3).

2. QCD FORMALISM FOR HARD PROCESSES

2.1 *Factorization Theorems and the Parton Model*

In the original parton model, hard-scattering cross sections are written as the product of two probabilities: parton distribution functions (the probability of finding partons of a given flavor and momentum fraction inside the hadron) and the parton level cross sections (the probability of the hard-scattering subprocess to occur). With the advent of QCD, the fundamental ideas underlying the parton model received theoretical support through the systematic study of the short-distance behavior of quark and gluon scattering cross sections using perturbative techniques. In lowest order the expressions for these cross sections reproduce the simple parton model.

When higher-order terms are included, one encounters divergences that must be regularized (rendered finite) and properly subtracted (renormalized) in order to yield meaningful finite results. The systematic subtraction of ultraviolet singularities leads to the introduction of a running coupling α_s (discussed in the next section) that depends on a renormalization scale. Since $\alpha_s(\mu)$ becomes small for large μ (asymptotic freedom), we obtain a small expansion parameter by choosing μ to be of the order of a large characteristic momentum transfer in the scattering process. This justifies the use of the perturbative approach for such hard processes. In the limit of zero-mass partons, which is appropriate for leading power-law (i.e. twist-2) analyses, one also encounters collinear divergences. Physically, the subtraction of collinear divergences corresponds to removing the overlap (hence the double counting) between (*a*) that part of the next-order cross sections with almost collinear and on-mass-shell gluon lines and (*b*) the contribution of the corresponding lower-order term. The higher-order correction is ambiguous by a finite amount, depending on the fraction of the higher-order term that is reclassified as part of the lower-order term. The systematic treatment of this problem to all orders leads to the concept of parton distribution functions that depend on a factorization scale. This factorization scale serves to separate the short- and long-distance portions of the scattering process; therefore, it should also be of the same order of magnitude as the characteristic large momentum scale of the physical process. Since the factorization and renormalization scales have distinct origins, they need not be exactly the same. However, they have to be of the same order of magnitude in order to avoid logarithms of large ratios. For simplicity, in subsequent discussions we use the symbol μ for both scales and do not attempt to distinguish the two unless necessary.

The precise results of the systematic analysis of the high-energy behavior of hard cross sections are expressed in a set of factorization theorems. They provide the theoretical basis of the QCD parton formalism (for comprehensive recent reviews, see 4, 5). We quote only the most relevant results.

For a generic lepton-hadron scattering process $\ell + \mathrm{A} \rightarrow \ell' + \mathrm{C} + \mathrm{X}$, where C either represents an identified final-state particle with specific attributes (such as heavy mass or large transverse momentum) or is null in the case of total inclusive scattering, the factorization formula for the cross section (excluding the known lepton vertex) reads

$$\sigma^{\mathrm{i}}_{\mathrm{A}\rightarrow\mathrm{C}}(q, p) = \sum_{\mathrm{a}} \int_x^1 d\xi \, f^{\mathrm{a}}_{\mathrm{A}}(\xi, \mu) \hat{\sigma}^{\mathrm{i}}_{\mathrm{a}\rightarrow\mathrm{C}}(q, \xi p, \mu, \alpha_s), \qquad 1.$$

where "A" is the target hadron label, "a" is the parton label, "i" is the electroweak vector boson (weak isospin and helicity) label, (q,p)

are the momenta of the vector boson and the hadron respectively, μ is the renormalization scale, and ξ is the fractional momentum carried by the parton with respect to the hadron. This basic theorem expresses the physically measurable cross section as a convolution of a set of universal parton distribution functions f_A^a and a hard-scattering cross section $\hat{\sigma}^i_{a\to C}$.

Similarly, for a generic hard hadron-hadron collision process A + B → C + X, the corresponding factorization theorem, when applicable (4, 5), states

$$\sigma_{AB\to C}(p_A, p_B, p_C, \ldots) \qquad 2.$$
$$= \sum_{a,b} \int_{x_a}^{1} d\xi_a \int_{x_b}^{1} d\xi_b \, f_A^a(\xi_a, \mu)\hat{\sigma}_{ab\to C}(\xi_a p_A, \xi_b p_B, \mu, \alpha_s)\, f_B^b(\xi_b, \mu),$$

where ξ_a and ξ_b represent the fraction of momenta carried by the two partons with respect to the two incoming hadrons, $\hat{\sigma}_{ab\to C}$ represents the cross section of the fundamental parton process a + b → C + X, and x_a and x_b represent appropriate integration limits set by the kinematics of the process under consideration.

2.2 *Renormalization Scheme and Scale Dependence*

The factorization theorems hold to all orders in α_s; they are only subjected to power-law (higher-twist) corrections. It is important to note that the physical cross sections on the left-hand sides of the factorization theorems are independent of the renormalization scheme and the choice of the scale μ, both of which are needed to define the renormalized hard cross section $\hat{\sigma}$. Thus, the scheme and scale dependence of $\hat{\sigma}$ on the right-hand sides of these equations must be compensated by corresponding dependences of the parton distribution functions. This is the formal origin of the scheme and scale dependence of parton distributions. In particular, a change in the factorization scale μ merely amounts to a rearrangement of finite contributions between $\hat{\sigma}$ and $f_A^a(x, \mu)$ because the boundary that defines collinear and non-collinear parton lines is shifted, as mentioned above. Thus, parton distribution functions are theoretical constructs that are not "physical" (in the sense that they are unambiguously and directly measurable), as they are often perceived to be.

Truncating the perturbation series at a given order spoils the perfect compensation between the scale dependences of $\hat{\sigma}$, and $f_A^a(x, \mu)$, and hence introduces an artificial dependence on the choice of μ to the QCD prediction for the cross section. This dependence represents an intrinsic uncertainty of the perturbative approach that must be understood and brought under control in quantitative applications.

The leading-order (LO) QCD formalism consists of using the tree-level results for the hard cross section, the one-loop expression for the running coupling (Section 2.3), and parton distributions generated by one-loop evolution kernels (Section 3.1). This formalism provides a remarkably consistent description of a wide variety of large momentum transfer processes (see 6 for a review). However, at this level of approximation, the hard cross section has no μ dependence [except through a possible overall power of $\alpha_s(\mu)$, depending on the process] so that the cross section acquires a net μ dependence through $f_A^a(x, \mu)$. As the scale is not specified other than that it should be of the order of a large momentum variable, the predictions will vary considerably when the choice of this variable or the proportionality constant between μ and this variable is changed. For instance, if the relevant region of x in the parton distributions is above about 0.2, the predictions become monotonically decreasing functions of the scale. Variations in the choice of μ, e.g. $p_T/2$ or $2p_T$ in high-p_T processes, can cause the predictions to vary by a factor of two or more.

A related problem is that the scale parameter Λ in the running coupling (see Section 2.3) is, strictly speaking, not well defined in LO calculations: changing its value by any finite factor only leads to corrections of the next order, and hence can be ignored! Likewise, changes in the factorization prescription affect only subleading terms and, hence, can be made freely. All these changes, however, result in different numerical predictions. In summary, the LO formalism suffers from rather severe artificial factorization scheme and scale dependence that limits its usefulness as a quantitative model of high-energy processes.

The next-to-leading-order (NLO) formalism—involving NLO hard cross sections with two-loop α_s and two-loop-evolved parton distributions—greatly improves the situation. All three elements now acquire unambiguous meaning, and the net scale dependence of the predictions on physical cross sections is substantially reduced because the variation in one of these factors will be compensated by the others (except for corrections of even higher order). This compensation is guaranteed by theory (the renormalization group equation)—provided consistency in the choice of renormalization and factorization schemes and scales is maintained in handling all elements of the calculation. The last proviso is crucial because any mixed use of these quantities defined in different schemes (implicitly or explicitly) spoils the cancellation mechanism and, hence, gives results no better than the LO formalism.

We conclude therefore that (*a*) quantitative applications of the QCD parton formalism require at least the NLO approximation; and (*b*) at NLO, it is essential to specify explicitly the choice of renormalization and factorization scheme for both $\hat{\sigma}$ and $f_A^a(x, \mu)$, and to maintain the consistency of these. Disregarding this essential feature of the factorization theorem inevitably leads to misleading results. For instance, it was pointed out recently that the scheme dependence of the gluon and sea-quark distributions can be quite substantial (7). This can lead to important phenomenological consequences (see Section 6).

2.3 *The QCD Coupling and* Λ_{QCD}

The running coupling $\alpha_s(\mu)$ is the most basic of QCD quantities. We examine some nontrivial aspects of the QCD coupling function α_s in the presence of heavy quarks. To begin, we recall the standard formulas for α_s in LO and in NLO in the modified minimal subtraction ($\overline{\text{MS}}$) scheme for the case of all zero-mass quarks:

$$\alpha^{LO}(n_f, \mu/\Lambda) = \frac{4\pi}{\beta_0 \ln(\mu/\Lambda)^2} \tag{3.}$$

$$\alpha^{NLO}(n_f, \mu/\Lambda) = \frac{4\pi}{\beta_0 \ln(\mu/\Lambda)^2}\left[1 - \frac{\beta_1}{\beta_0^2}\frac{\ln\ln(\mu/\Lambda)^2}{\ln(\mu/\Lambda)^2}\right], \tag{4.}$$

where n_f is the number of (massless) quark flavors and it enters the right-hand side through the constants

$$\beta_0 = \frac{33 - 2n_f}{3} \qquad \beta_1 = 102 - \frac{38n_f}{3}. \tag{5.}$$

If all quarks are massless, the number n_f is fixed and the running coupling α_s is determined by a single parameter Λ—the "QCD lambda." In the presence of massive quarks, the situation is quite different. According to the decoupling theorem (8), each heavy quark "i" with mass m_i is effectively decoupled from physical cross sections at energy scales μ below a certain threshold Q_i, which is of the order m_i. Thus, the number of effective quark flavors n_f^{eff} is an increasing step function of the scale μ. Under this circumstance, the specification of the running coupling α_s and the associated Λ_{QCD} is not as simple as before. Although this point is fairly well known, there still exist confusing and ambiguous statements about these parameters in the current literature and in conference presentations. Hence, it is worthwhile to summarize the proper formulation of the problem explicitly.

The definitions of α_s and Λ_{QCD} in the presence of mass thresholds are not unique—they depend on the renormalization scheme used. A natural choice is based on the requirement that $\alpha_s(\mu)$ must be a continuous function of μ and that, between thresholds, it must reduce to the familiar $\overline{\text{MS}}$ α_s (9). This requirement leads to the condition that $Q_i = m_i$ (in contrast to $2m_i$, or even $4m_i$, as are often used). This choice has the additional desirable feature that the parton distribution functions so defined are also required to be continuous across the thresholds. If Equation 4 is to remain valid with α_s being continuous in μ, but n_f a discontinuous function of μ, it is quite obvious that the effective value of Λ must also make discontinuous jumps with μ at heavy-quark thresholds. The same remark applies if one uses the LO formula for α_s, Equation 3.

Figure 1 shows a typical α_s vs μ plot and the corresponding Λ_{QCD} as a function of μ (bottom scale) and n_f^{eff} (top scale). The left-hand panel of Figure 1 explicitly shows that the running coupling function of QCD $\alpha_s(\mu)$ can be unambiguously specified by giving its value at a (standard) scale, say M_Z. On the other hand, as shown in the right-hand panel, this same coupling function is associated with many different values of Λ_{QCD}, depending on the number of effective quark flavors and on whether the LO or NLO formula is used. Thus, if one prefers to define α_s by specifying a value of Λ_{QCD}, it is imperative that one specifies the associated n_f^{eff} and the order (LO or NLO) explicitly. In the recent literature, the second-order $\overline{\text{MS}}$ Λ_{QCD} with four flavors has increasingly become the standard choice, although that with five flavors is also used—too often without explicit notation.

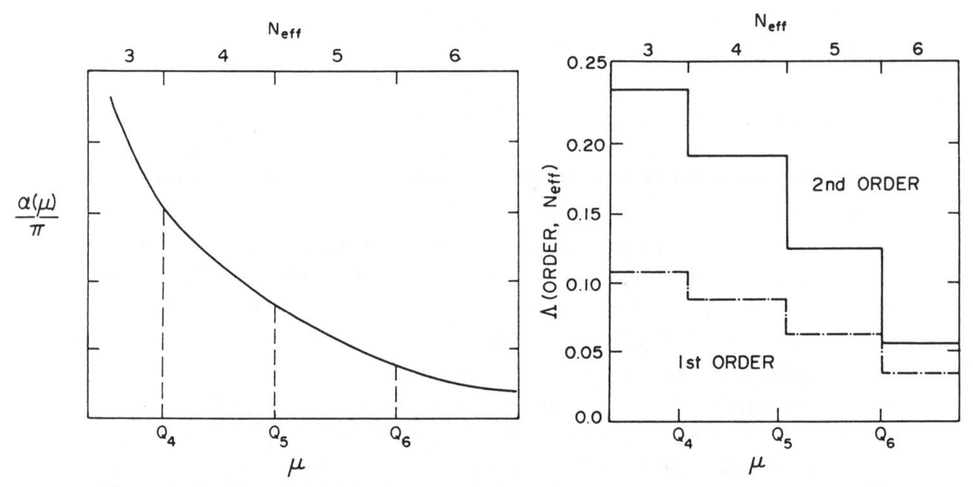

Figure 1 (*Left*) α_s and (*right*) Λ_{QCD} scale dependence of the strong coupling and n_f^{eff}.

3. PARTON DISTRIBUTION FUNCTIONS

3.1 *QCD Evolution Equations*

The scale dependence of the parton distributions in QCD is generated by the interactions of the quarks and gluons via such elementary processes as gluon emission from quarks, $q \to qg$, gluon emission by gluons, $g \to gg$, and the creation of quark-antiquark pairs by gluons, $g \to q\bar{q}$. Consider deep inelastic scattering in which one of the quarks in the target nucleon interacts with the electroweak current. This quark may have radiated gluons either prior to or subsequent to the interaction, or it may have originated from a gluon. In any case, the unobserved radiated partons must be integrated over the remaining available phase space, the scale for which is set by μ. These radiative corrections are the physical source of the logarithmic scale dependence predicted by the theory. More precisely, the scale dependence of the parton distributions is governed by a set of coupled integro-differential QCD evolution equations, valid to all orders in α_s (10, 11)

$$\frac{df^q(x,\mu)}{dt} = \frac{\alpha_s(\mu)}{2\pi}\int_x^1 \frac{dy}{y}\left[P_{qq}(y)f^q\left(\frac{x}{y},\mu\right) + P_{qg}(y)f^g\left(\frac{x}{y},\mu\right)\right] \tag{6}$$

$$\frac{df^g(x,\mu)}{dt} = \frac{\alpha_s(\mu)}{2\pi}\int_x^1 \frac{dy}{y}\left[\sum_q P_{gq}(y)f^q\left(\frac{x}{y},\mu\right) + P_{gg}(y)f^g\left(\frac{x}{y},\mu\right)\right], \tag{7}$$

where $t = \ln(\mu^2/\Lambda^2)$, and the superscript "q" is used to denote quark flavors. The kernels of these equations, $P_{ij}(z)$, correspond to splitting functions for the elementary processes mentioned before and have the physical interpretation as the probability density for obtaining a parton of type i from one of type j with a fraction z of the parent parton's momentum.

In LO (one-loop) QCD there are four such splitting functions. They can be found in all standard books and references. The NLO (two-loop) expressions for $P_{ij}(z)$ was calculated by several groups (12–15). Until recently, there had been an unresolved minor discrepancy for part of the function P_{gg} between the results obtained in a covariant gauge and those obtained using the axial gauge. This has now been clarified: a detail discussion can be found elsewhere (16).

When solving Equations 6 and 7 it is convenient to define a singlet distribution

$$\Sigma(x, \mu) = \sum_{q} [f^{q}(x, \mu) + f^{\bar{q}}(x, \mu)], \qquad 8.$$

which mixes with the gluon in two coupled evolution equations of the form given above.

The nonsinglet (or valence) distributions

$$f^{q_v}(x, \mu) = f^{q}(x, \mu) - f^{\bar{q}}(x, \mu) \qquad 9.$$

each satisfy an uncoupled equation. In LO, the kernel of this equation is simply the well-known splitting function $P_{qq}(z)$. Beyond LO, in addition to higher-order terms for $P_{ij}(z)$, there emerges another type of nonsinglet distribution:

$$f^{q_v}_{+}(x, \mu) = f^{q}(x, \mu) + f^{\bar{q}}(x, \mu) - \Sigma(x, \mu)/n_f^{\text{eff}}, \qquad 10.$$

which also satisfies an uncoupled equation with a different kernel function (12–15).

This set of equations can be solved numerically, once a set of distributions is specified at some initial value of μ, hereafter denoted by μ_0. Note that in each case the logarithmic derivative with respect to μ at a given value of x is given in terms of the parton distributions evaluated at the same value of μ with momentum fractions greater than or equal to x. This is of practical interest since experimental measurements at fixed μ and beam energy cannot reach all the way to $x = 0$.

3.2 *Systematics of Scale Dependence*

The scale dependence of parton distributions has the general feature that $f(x, \mu)$ is a decreasing function of μ at large values of x, and it becomes an increasing function at small x (17). This can be understood mathematically by considering the LO nonsinglet evolution equation as described above. Substituting in the expression for P_{qq} in the uncoupled equation yields

$$\frac{df^{q_v}(x, \mu)}{dt} = \frac{\alpha_s(\mu)}{2\pi} C_F \int_x^1 \frac{dy}{y(1 - y)} \left[(1 + y^2) f^{q_v}\left(\frac{x}{y}, \mu\right) - 2y f^{q_v}(x, \mu)\right] + \frac{\alpha_s(\mu)}{2\pi} C_F \left[\frac{3}{2} + 2\ln(1 - x)\right] f^{q_v}(x, \mu). \qquad 11.$$

The first term in Equation 11 is negative for all x [since $f^{q_v}(x, \mu)$ is a decreasing function of x and the argument of the second term is always smaller than that of the first term], while the second is negative at large x, becoming positive for $x \leq 1 - \exp(-3/4) \approx 0.53$. Therefore, at large values of x, $f^{q_v}(x, \mu)$ decreases with increasing μ. On the other hand, the integral of f^{q_v} over x from 0 to 1 must remain constant as μ varies because it counts the net number of valence quarks of a given type (the quark-number sum rule). Thus, $f^{q_v}(x, \mu)$ must increase with μ at small x to compensate for the decrease at large x. The physics behind this pattern of scaling violation is simple to understand. A parton with a given momentum fraction x can radiate another parton, thereby decreasing its momentum fraction and reducing the value of the parton density at that value of x. However, the parton density at that value of x will also receive an increase from partons with higher momentum fractions radiating partons. There is thus a type of gain-loss competition taking place. Indeed, it is possible to recast the derivation of the LO evolution equations into such a form, thereby making the gain-loss terms manifest (18).

The preceding discussion illustrates a general feature of the μ dependence of the parton distributions, namely that the values of the distributions at small x and large μ values are determined in large part by the values of the distributions at large x and small μ values. Thus, a significant fraction of the region of x and μ that is relevant for predictions for high-energy collider processes receives contributions via the evolution equations from regions well measured by current experiments.

3.3 *Extrapolations to Small* x

For a high-energy process, the most relevant range of x values is given by $x \sim \mu/\sqrt{s}$ where μ is a typical momentum scale and $\sqrt{s}$ is the center-of-mass energy. At future high-energy colliders where $\sqrt{s}$ will be very large, many interesting physics processes taking place at moderate values of μ, say 5–50 GeV, will probe parton distributions and interactions at very low values of x. At hadron colliders, processes such as the production of minijets (jets with moderate values of p_T), heavy-flavor particles, and low-mass lepton pairs all belong to this category. The behavior of parton distributions in this region is not well understood for two reasons: (*a*) theoretically, within the perturbative QCD framework, the occurrence of powers of $\log(1/x)$ (which becomes large in this region) can spoil the conventional (twist-2) formalism as described up to now; and (*b*) phenomenologically, even within the standard approach, the initial parton distributions needed in solving the evolution equation are largely unknown because existing data do not extend into this region. Unfortunately, for moderate values of μ, as we have here, the solutions to the evolution equation are sensitive to the initial distributions, unlike at large μ where their behavior is primarily driven by the leading singularity of the evolution kernel at $x = 0$.

The theoretical issues can be further differentiated into several areas: (*a*) For fixed μ, the small-x limit is analogous to the Regge limit in hadronic processes ($s \rightarrow \infty$) for which there exists certain conventional wisdom and a large amount of more recent QCD study (19) pioneered by the work of Lipatov (20). In particular, the Lipatov Equation (20) [involving resummation of large $\ln(1/x)$ factors to all orders] and its recent generalizations (21) govern the "small-x evolution" of parton distributions for fixed μ. (*b*) In the region where both $\ln(\mu/\Lambda)$ and $\ln(1/x)$ are large, a different set of resummation techniques has to be developed. The large body of work on this topic has been summarized in several recent comprehensive reviews (22). (*c*) Finally, since the parton densities increase precipitously as $x \rightarrow 0$, they will eventually saturate at some value of x when the packed partons become so dense that they interact strongly in spite of the small effective coupling (22, 23). In the saturation region, the parton picture itself breaks down; one enters an entirely new regime of QCD. Each of these three areas presents new challenges for theorists. Much progress has been made in the last ten years in developing new techniques of calculation in these regions, but they are not yet ready for routine applications. We refer interested readers to the proceedings of a recent workshop for state-of-the-art reviews on this active frontier of research (19).

All phenomenological analyses of parton distributions based on the usual QCD formalism use certain assumed parametrizations of the initial distribution functions that implicitly determine the extrapolated small-x behavior. The conventional ansatz of a power-law behavior $f(x, \mu_0) \sim x^{\gamma}$ as $x \rightarrow 0$ for the gluon and sea quarks, with γ ranging from -1 (Regge "Pomeron") to -1.5 (24), is a highly ambiguous proposition because the singular behavior near $x = 0$ is very sensitive to the unknown value of μ_0 and the effective value of γ is a rapidly varying function of μ_0 (25). Since QCD evolution due to parton radiation is mainly responsible for the rise of the parton density at small values of x, it is also natural to assume logarithmic factors of the form $\ln^{\delta}(1/x)$ at a given μ_0. The variation of the parameters γ and δ with μ_0 and the range of values of these parameters consistent with current experiments have been systematically studied by Tung (25). The parametrization of the small-x behavior of parton distributions in the context of global analysis of data is discussed in Sections 5.4 and 7.

4. REFERENCE PROCESSES

4.1 *Deep Inelastic Scattering*

Deep inelastic lepton-nucleon scattering provided the first evidence in support of the original parton model. Since then, this process has played a leading role in the determination of parton distributions because it probes the structure of the target hadron via the clean and well-understood weak or electromagnetic current. In the LO QCD parton model, the "hard cross sections" reduce to just electroweak coupling constants multiplied by an on-shell delta function that removes the convolution integral, so the physically measurable deep inelastic scattering structure functions become simple linear combinations of the parton distribution functions $f_A^a(x, \mu)$. For example, the structure function F_2 measured in eN or μN scattering has the form

$$F_2^N(x, Q) = x \sum_q e_q^2 f_N^q(x, Q) \qquad \text{(leading order)} \tag{12}$$

where e_q denotes the charge of the quark q. The factorization scale μ has been identified with the characteristic momentum scale Q (virtual mass of the exchanged vector boson) of this hard-scattering process.

The complete set of relationships for all measurable structure functions has been compiled by Tung et al (26). This kind of direct relationship has led to the widespread practice of referring to the parton distribution functions $f_A^a(x, \mu)$ also as "structure functions," which in turn has fostered the tantalizing misconception among many that parton distribution functions are "physical" (i.e. directly measurable) objects. Unfortunately, the simple connection between the two quite distinct concepts holds only in leading order. Indiscriminant mingling of the two can lead to incorrect conclusions. As emphasized throughout this review, for quantitative applications of the QCD parton formalism, it is crucial to recognize the "unphysical" aspects of the parton distributions (i.e. renormalization scheme and scale dependence) as well as their attractive physical interpretations.

In leading order, the gluon distribution affects the measurable structure functions only via the quark evolution equation (Equation 6); one must go to higher orders in order to probe the gluon distribution directly. In NLO, one has

$$F_2^N(x, Q) = x \sum_q e_q^2 f_N^q(x, Q) + \alpha_s(Q) \left[\sum_q C_2^q \otimes f_N^q + C_2^g \otimes f_N^g \right], \tag{13}$$

where the circled multiplication sign indicates a convolution. The coefficient functions $\{C_2\}_{\overline{\text{MS}}}$ were calculated early in the development of QCD using the $\overline{\text{MS}}$ subtraction scheme (27, 28). The explicit appearance of the gluon distribution on the right-hand side of this equation, however, does not necessarily give a good handle on f_N^g. First, this term is small compared to the leading quark term; more importantly, it is possible to exploit the arbitrariness of the choice of factorization scheme to define away the entire NLO term in Equation 13 (arising from the hard-scattering part in the $\overline{\text{MS}}$ scheme) by absorbing it into the LO quark term with a new definition of the quark distribution functions (the soft part of the factorization theorem) (28). In the new scheme—often referred to as the DIS scheme—the NLO formula for F_2^N is, by definition, the same as the LO formula (Equation 12), i.e. $C_{2\text{DIS}}^q \equiv C_{2\text{DIS}}^g \equiv 0$, hence there is no apparent gluon dependence. (We discuss the issues on choice of scheme below and in Sections 5.3 and 7.)

The gluon distribution does, however, contribute directly to the observable longitudinal structure function $F_L(x, Q)$ and to the rate of change of the structure function $F_2(x, Q)$ with respect to Q. Both are intrinsically order α_s effects. Specifically, we have:

$$F_L^N(x, Q) = \alpha_s(Q)\left[\sum_q C_L^q \otimes f_N^q + C_L^g \otimes f_N^g\right] \qquad 14.$$

where the coefficient functions C_L are well known (27–29). The simplicity of this equation makes this potentially the best means of measuring the gluon distribution, especially because the dominant part of the quark term can be related to the measurable $F_2^N(x, Q)$, see Equation 12 (30).

The rate of change of the structure function $F_2(x, Q)$ with respect to Q is given by

$$\frac{dF_2^N(x, Q)}{dt} = \int_x^1 \frac{dy}{y}\left[P_{qq}(y)F_2^N\left(\frac{x}{y}, Q\right) + \left(\sum_q e_q^2\right)P_{qg}(y)f_N^g\left(\frac{x}{y}, Q\right)\right] \qquad 15.$$

where $t = \ln Q^2/\Lambda^2$. Equation 15 is obtained from the first of the evolution equations, Equation 6, by multiplying with xe_q^2, and summing over q. Since $F_2(x, Q)$ is the most accurately measured structure function, this relation offers a useful method of constraining the gluon distribution, at least in the region of small-x values where the gluon term in Equation 15 makes a significant contribution.

In LO, the structure functions, F_i (i = 1, 2, 3) measured in νN or $\bar{\nu}$N scattering depend on different linear combinations of parton distributions (26). With sufficiently precise data over a wide enough kinematic region, one can separate the valence and sea-quark distributions. However, the various flavors of sea quarks cannot yet be readily differentiated because (*a*) quarks of different generations contribute to the measured quantities in the same way, and (*b*) current experimental practice is to combine the ν and $\bar{\nu}$ measurements (to increase statistics), rather than to use them separately to yield independent linear combinations of parton flavors. This issue is discussed in Section 7.

In NLO, both structure functions F_L and F_3 acquire nontrivial order α_s corrections, similar to Equation 13. This is true even in the DIS scheme, which is designed to make F_2 simple, as described above. Since F_2 has no particular significance in theory, the DIS scheme does not have any special status except an historical one. This is becoming clearer as the application of the QCD formalism expands to an ever-increasing number of physical processes beyond deep inelastic scattering.

Because at least two different schemes are commonly used in the current literature, and because this has been a continuing source of confusion (and hence misuse of the QCD parton model, see Sections 5.3 and 7), it is useful to review the definition of these schemes and specify the relation between them. The $\overline{\mathrm{MS}}$ scheme is defined by a universal prescription to calculate perturbative matrix elements independent of any physical process (27). The $\overline{\mathrm{MS}}$ scheme parton distributions are those appearing in formulas such as Equations 13 and 14 with the hard-scattering part (C_i) calculated with the $\overline{\mathrm{MS}}$ subtraction prescription. It can be shown that the quark sum rules and the momentum sum rule are automatically satisfied with this scheme. On the other hand, the DIS scheme, as described above, is specially designed to render the simple parton model deep inelastic scattering structure function formula, Equation 12, for F_2 applicable even at NLO. Comparing the NLO equations for F_2 in the two schemes, it is easy to determine the transformation formula for quark distributions defined in these two schemes:

$$f^{q}_{\mathrm{DIS}} = [1 + \alpha_s(Q)C^{q}_{2\overline{\mathrm{MS}}}] \otimes f^{q}_{\overline{\mathrm{MS}}}] + \alpha_s(Q)C^{g}_{2\overline{\mathrm{MS}}} \otimes f^{g}_{\overline{\mathrm{MS}}}/2n^{\mathrm{eff}}_{f}. \qquad 16.$$

Within the context of deep inelastic scattering, there is no obvious counterpart to Equation 16 for the gluon, since it does not couple directly to the electroweak current. Among possible choices to specify the gluon distribution in this scheme, the natural requirement that the momentum sum rule be preserved (in order to maintain the parton model interpretation) motivates the following definition:

$$f^{g}_{DIS} = [1 - \alpha_s(Q)C^{g}_{2\overline{MS}}] \otimes f^{g}_{\overline{MS}} - \alpha_s(Q) \sum_{q} C^{q}_{2\overline{MS}} \otimes f^{q}_{\overline{MS}}, \qquad 17.$$

which is adopted by most groups working with the DIS scheme. Because this latter equation is not unique, it is prudent to ascertain the precise definition of the gluon when a given set of DIS distributions is used.

As perturbative formulas, Equations 16 and 17 can easily be inverted [to $O(\alpha_s)$] in order to derive $\overline{MS}$ distributions from a given set of DIS distributions. This is safe as long as the NLO terms remain small compared to the leading term. Two cases in which this condition is not met have been noted (7) and are discussed in Section 7.

4.2 *W, Z, and Lepton Pair Production*

The production of W and Z bosons, as well as massive lepton pairs (via time-like virtual vector bosons), in hadron collisions is related to one-particle production in deep inelastic scattering by crossing one lepton and one hadron line. Here we refer to these processes generically as vector boson production and use γ^* as the representative example except when otherwise noted.

The lowest-order process in this case is the Drell-Yan (31) process $q\bar{q} \rightarrow \gamma^*$. As in the deep inelastic case, gluons contribute in the next order through the elementary processes $gq \rightarrow \gamma^*q$ and $q\bar{q} \rightarrow \gamma^*g$. Results on the Q^2 and y (or x_F) distributions can be found in the literature (32, 33). The ratio of the full $O(\alpha_s)$ result to that of the lowest order, the so-called K-factor, was found to be unexpectedly large—about two. This was largely due to two effects: (*a*) the continuation of the vector boson mass Q from space-like values in deep inelastic scattering (where the conventional parton distributions were defined) to time-like values for vector boson production introduces a factor of π^2 in the correction term; and (*b*) the phase-space boundaries for the two processes are different, which generates some additional large corrections near the kinematic boundary.

The large corrections raised questions about the reliability of perturbative calculations and inspired in-depth studies of the nature of these corrections and the means to control them. Much work has been directed toward calculating the next-higher-order terms to gain more information on the corrections. Recently, the full second-order calculation has been completed (see 34 and references therein). An alternative approach is to identify the origin of the large corrections and then to use this knowledge to resum the perturbation series, taking into account this type of correction to all orders. Steady progress has been made along this line (35). The two approaches are complementary; it has been shown that the resummed result (which exponentiates) reproduces the bulk of the next-higher-order results over most of the kinematic region—a very encouraging indication that the large corrections can be "understood" and quantitatively controlled (35). Further advances can be made in simplifying the resummed results and in expanding the kinematic regions where similar techniques can be applied.

Because the exponentiated correction factor has certain universal properties among hadron-hadron processes, Sterman suggested that it may be desirable to absorb this factor into the definition of parton distributions when analyzing hadron processes (36). The "advantage" would be that the remaining hard-scattering part would contain only intrinsic higher-order effects. This proposal would result in two classes of parton distributions—one for lepton-hadron and one for hadron-hadron processes—which would be related by known factors. Whether the proposal is adopted in practice or not, it is only a matter of convention and would not affect the underlying physics if used consistently, just like the scheme dependence discussed above.

In addition to the integrated cross section and the rapidity distribution, perturbative QCD has also been applied to study the p_T distributions of vector boson production. This requires more sophisticated resummation methods because of the presence of more than one large scale (say, Q and p_T), which give rise to additional large logarithms in regions where their ratio is large. The theory has been well established (37, 38). The first calculation of the W-vector boson p_T distribution yielded a qualitative description of measurements at the hadron colliders. A recent NLO perturbative calculation of the vector boson p_T distribution has been shown to merge well into the resummed formula obtained earlier (39). Detailed phenomenological work on the analysis of vector boson p_T distribution, especially in the context of providing information on parton distributions, has not yet been done.

4.3 *Direct Photon Production*

The production of photons with large transverse momentum in hadron-hadron collisions is known to be sensitive to the gluon distribution. The two Born processes, both of $O(\alpha\alpha_s)$, are $q\bar{q} \rightarrow \gamma g$ and $gq \rightarrow \gamma q$. In pp interactions the contribution from the $q\bar{q}$ subprocess is small, resulting in a dominant term that is directly proportional to the gluon distribution. An $O(\alpha\alpha_s)$ calculation of the invariant cross section has been performed (40), and a computer program for calculating the cross section based on this work is widely available. The theory appears to be in good agreement with the data currently available (for review, see 6, 41, 42). Recently, a higher-order calculation of the cross section for photon plus jet production has been performed (43, 44) and found to be in agreement with the available data (45).

Photons in the initial or final state—whether real or virtual—are good probes of hard-scattering processes to the extent that they participate directly in the hard-scattering process. However, in the case of photon production, this apparent advantage is diminished by some complications. Specifically, direct photons are not a priori distinguishable from radiative photons (i.e. bremsstrahlung) accompanying high-p_T partons produced in regular hadron hard scattering. Such photons will, however, tend to be produced nearly collinear with the parent parton. Indeed, one can define photon fragmentation functions that satisfy a set of evolution equations modified by the addition of an inhomogeneous term (46). The leading-logarithm contribution to the solution is independent of input boundary conditions and simple parametrizations of these are available (6). The contributions of such terms to the fully inclusive cross section are generally significant only in the small-x_T region, so that determinations of the gluon distribution in the mid- to large-x range are not affected to any large degree. However, in collider experiments it is often necessary to place rather restrictive isolation cuts on the electromagnetic triggers in order to obtain a relatively clean sample of direct photons. That is, it is necessary to limit the amount of hadronic energy in the vicinity of the photon. This cut removes a portion of the bremsstrahlung signal, but not all of it. Theoretically, this might seem to render the isolated cross section incalculable (because it is, strictly speaking, not infrared safe), but this is not necessarily the case, as is discussed by Berger & Qiu (47, 48). They showed that the isolated photon cross section can be calculated perturbatively as long as the cuts are not too restrictive.

The higher-order calculation described by Baer et al (43) can take these isolation cuts into account—the integrations are done using a combination of analytic and Monte Carlo methods in order to maximize the flexibility of the program with regard to calculating different observables. While this is very useful for generating various types of predictions to be compared with experiment, it is not adequate for incorporation into a large fitting routine because of the time required to generate accurate answers and the statistical errors associated with Monte Carlo integration. A new program for the isolated cross section is being developed that does not utilize Monte Carlo methods (E. L. Berger & J. Qui, private communication).

4.4 *Heavy-Quark Production*

The production of heavy quarks in lepton-hadron and hadron-hadron collisions provides a good test of QCD. In this case the heavy-quark mass sets the large scale needed to justify the use of perturbative techniques. Charm production in charged current lepton-hadron scattering experiments has been the main source of information on the strange-quark distribution inside the nucleon. Similarly, its production in neutral current scattering can determine the charm content of the nucleon. At the higher energies of ep colliders, bottom- and top-quark production may also become available. Heavy-flavor production in high-energy hadron collisions is sensitive to the small-x behavior of gluons and quarks.

The LO calculation of heavy-quark production in lepton-hadron scattering appears to be straightforward. The effect of the heavy-quark mass has been incorporated in all existing phenomenological applications in the form of "slow rescaling," reflecting the modified on-shell condition for the heavy quark (see 26 for a review). But none of the existing applications, both theoretical and experimental, pay full attention to mass effects in general terms applicable to both charged current and neutral current processes. In addition to the need for correcting these deficiencies, it was pointed out recently that the mixing of heavy quarks with gluons and the abundance of gluons inside the nucleon require at least the NLO formulation of this process in order to be meaningful (7, 49). NLO calculations of the hard cross section for this process have been performed recently by several groups (49–51). The implementation of the QCD formalism, even with known NLO hard cross section, is tricky in the region not far above the heavy-quark threshold where most experimental data are available. The method for calculating heavy-particle production developed by Olness & Tung (52), which is used by Aivazis et al (49), ensures physically meaningful results from the threshold to the asymptotic regime.

Heavy-flavor production in hadron collisions has not, so far, played any role in parton distribution analysis because of the experimental difficulty in measuring the relevant cross sections. Theoretical calculations on the hard cross sections have been carried out to order α_s^3 by two different groups (53, 54).

4.5 *Other High-*p_T *Processes*

Progress continues to be made on higher-order calculations for various types of hard-scattering processes. These types of calculations are important for a variety of reasons, such as performing more precise tests of the standard model and obtaining better estimates of standard model backgrounds to signals of new physics. Most notable among purely hadronic processes is the $O(\alpha_s^3)$ calculations of the single jet invariant cross section (55, 56). Phenomenological applications of these calculations can place some constraints on parton distributions or discriminate between existing distributions, especially for the gluon, which gives the dominant contribution to hadron scattering at high energies.

Another area of QCD applications where much progress has been made is multijet and W+jets production—both at tree level (57) and, recently, at the one-loop level (58). In order to gain control of these complicated calculations, many new techniques on powerful helicity amplitude analyses, color factor algorithms, and (4-dimensional) superstring technology have been developed and brought to bear (59, and detailed references cited therein). At present, these calculations can help in understanding the observed experimental features. However, the phenomenology is still quite far away from being able to quantitatively give feedback on the parton distributions.

5. GLOBAL ANALYSES

Ideally, the global analysis of parton distributions makes use of experimental data from a complete set of physical processes to extract a unique set of universal parton distribution functions. These can then be used in other applications, e.g. to make predictions for other conventional processes, to provide stringent tests of the self-consistency of the perturbative QCD framework itself or of the standard model in general, and to search for new physics. We review here the progress made toward achieving this goal. We discuss recent experimental developments and relevant experimental, theoretical, and phenomenological issues and uncertainties involved in such analyses. For this purpose, it is helpful to have a global view of the physical processes contributing to parton distribution analysis, particularly the relevant kinematic ranges covered by the various types of experiments. A "map" for such a global view was compiled at the Snowmass 90 Workshop (60).

5.1 *General Strategies and Review of Recent Developments*

Deep inelastic scattering of leptons on nucleons and nuclear targets remains the primary source of information on parton distributions. Mishra & Sciulli (61) reviewed the experimental results up to 1989.

The original high statistics, high energy data of the early 1980s, on which the first generation of parton distributions were based, have been superseded in the last few years by much more accurate data from muon-scattering [BCDMS (62)] and neutrino-scattering [CHARM (63), CDHSW (64), FMM (65a), and CCFR (65b)] experiments. The substantial changes in experimental data (up to 15–20% in some regions such as the small-x region) necessarily make the pioneering parton distribution sets obsolete for modern applications. Significant recent developments on the experimental front include the following.

1. Very accurate new and re-analyzed SLAC-MIT electron-scattering data (66) as well as the re-analyzed EMC data (67) have finally resolved the much publicized BCDMS-EMC controversy (for the controversy, see 68a; for the resolution, see 68b).
2. The new Tevatron neutrino data (65a,b), especially those of CCFR (65b), revealed impressive general agreement with QCD expectations (much like the earlier results of BCDMS for muon data).
3. New NMC results (69) yield information on differences in the neutron and proton structure functions and expand the measured kinematic range to ever-smaller values of x.

Recently a comprehensive and easily accessible database on deep inelastic scattering structure functions has been compiled by the Durham-RAL group (70). This greatly facilitates future phenomenological work on deep inelastic scattering and on global analysis in general. As clearly shown by the Durham-RAL group (70), deep inelastic scattering data are in very good shape and form the foundation for parton distribution analyses. In addition, major advances are eagerly awaited from experiments at HERA.

However, as is well known, the inclusive DIS structure functions are mostly sensitive to certain combinations of quark distributions. Hence even the most accurate DIS data do not place tight constraints on the gluon distribution at intermediate or large values of x, and they are not effective in differentiating all the quark parton flavors.

Vector boson production—including the production of lepton pairs, direct photons at large transverse momenta, and W and Z bosons—provides important complementary information on parton distributions. Lepton pair production has several unique features: (*a*) cross sections in pN collisions are directly proportional to the antiquark distributions; (*b*) cross sections in πp (and Kp) collisions offer one of the few handles on parton distribution functions in mesons; and (*c*) the measured A dependence provides complementary information on the "EMC-effect" (for review, see 71). The relatively new Fermilab E605 results (72) provide a significant improvement over earlier fixed-target experiments, both in accuracy and in kinematic coverage. Combined analysis of these data with DIS leads to much more tightly constrained parton distributions than those obtained with DIS data alone (73). Collider results for this process will offer a significant opportunity to probe the small-x region (say, $x \leq 10^{-3}$, see Section 7).

Direct photon production is particularly sensitive to the gluon distribution, and therefore can play an important role in global analyses (see Section 4.3). The measurement of the angular distribution of the photon compared to that for neutral pions in the parton-parton center-of-mass frame shows the flatter distribution expected from theory, confirming the presence of direct photons. Combining direct photon and deep inelastic data in a simultaneous analysis places much stronger constraints on the gluon distribution than using deep inelastic scattering data alone, as was shown by Aurenche et al (74). This analysis used data for the invariant cross section from the WA-70 experiment (75) at 280 GeV/c. The data were sensitive to the gluon distribution in the region of x from 0.35 to 0.55, thereby complementing the information available from deep inelastic scattering, which is sensitive to the gluon distribution at smaller values of x. The data available so far are expected to be greatly enhanced by anticipated results from the Fermilab experiment E706 (76).

Collider experiments on direct photon production probe gluons at a smaller x region. Complications due to bremsstrahlung contributions and photon isolation cuts have been brought under better control (see Section 4.3). New programs being developed will make it possible to incorporate collider data into the global analysis. Data from CDF and D0 will undoubtedly contribute to the determination of the gluon distribution.

In principle, additional sensitivity to the gluon distribution can be obtained by using data for the photon plus jet cross section. Knowledge of the four-vectors of both the jet and the photon allows for a tight constraint on the underlying parton kinematics. This reduces the amount that the parton distributions are smeared out by the convolutions in going from the parton level to the observed particles. Furthermore, the useful range in x where the data can be used to constrain the gluon distribution can be extended if sufficient rapidity coverage is available. To date, such data have been provided only by one experiment (45), and both the photon and jet were constrained to be centered on $y = 0$. However, several other experiments expect to have new data for this observable in the near future. The benefits resulting from more precise knowledge of the hard-scattering kinematics are gained at the expense of statistics since the data set will be subdivided into more bins. It remains to be seen whether sufficient statistics can be obtained to constrain the fitting process adequately.

Finally, W and Z production in colliders, especially from the increasingly accurate measurements at the Tevatron, provides additional information on the parton structure of the proton (CDF Collaboration; for references, see 77). With increasingly abundant and accurate data on the W asymmetry, W/Z ratios, etc, one will be able to determine some important aspects of parton distributions (such as the u/d ratio) from these data. Since the Q^2 value for these data is enormous compared to other processes, the relative contribution due to the charm quark is large, which makes this a potential source of information for the charm distribution (78).

5.2 *Experimental Considerations and Uncertainties*

Greatly improved experiments make it necessary for today's global analyses to take into account many details that were left out of earlier efforts. A discussion of relevant issues has been provided by the "Structure Functions and Parton Distributions" group in the 1988 Snowmass Workshop (26). We summarize the key points and refer the reader to workshop proceedings for details (3, 26).

In most modern high-statistics experiments, the experimental errors are now predominantly systematic rather than statistical. Understanding these errors is a major part of the experimental effort, and often takes years of work. Thus it is important to incorporate the systematic errors in the phenomenological analysis. Since systematic errors are usually correlated, incorporating them is much harder to do than the common practice of including only the point-to-point statistical errors. Often they cannot be properly incorporated without close cooperation between experimentalists and theorists.

In a global analysis based on data sets from a wide range of experiments, additional caution is needed. First, even among experiments of the same type, different groups often apply quite different theoretically motivated "corrections" to the data. In deep inelastic scattering structure function measurements, for example, common corrections include longitudinal structure function, "strange-sea," "isoscalar," and "slow rescaling" corrections. The discrepancies resulting from varied practices in applying these corrections can be quite significant (26) and have sometimes been responsible for troubling controversies. For instance, a substantial part of the well-known BCDMS-EMC controversy at low values of x is now attributed to the fact that EMC assumed a vanishing longitudinal structure function while

BCDMS used the QCD prediction (67). Clearly a valid phenomenological analysis must pay close attention to all potential sources of discrepancies of this type. Furthermore, certain "corrections" historically originating from the naive parton model, e.g. the slow-rescaling correction, alter the basic meaning of the measured structure functions, and hence render the latter unsuitable for precise NLO QCD analysis. For modern applications, such "corrections" must be avoided.

Global analyses usually are based on least-chi-squared or maximum likelihood methods. When combining several different types of processes, involving experiments with a wide range of statistical and systematic errors, the proper definition of overall chi-square and/or likelihood is not obvious. To obtain precise determinations of the QCD parameters and parton distributions requires careful thought and a great deal of systematic phenomenological study. Efforts along this line are only beginning.

5.3 *Theoretical Considerations and Uncertainties*

The perturbative QCD approach has certain inherent limitations and uncertainties that need attention in any systematic study. Some of these limitations can be overcome by further theoretical developments, and hence represent current frontiers in QCD research. All must be clearly recognized and are included in the assessment of uncertainties in state-of-the-art global analyses.

5.3.1 FACTORIZATION SCHEME When performing next-to-leading-order global analyses, close attention must be paid to the consistent choice of the factorization scheme used in defining the parton distributions and the hard cross sections in all processes, as emphasized in Section 2.2. Two schemes—the "DIS" and the "$\overline{\text{MS}}$" schemes—are widely used in the literature. In principle, parton distribution functions obtained in one scheme can be transformed into the other scheme. However, the transformation is not unique, as discussed in Section 4.1. Comparison of these two schemes and cautions on their proper use are discussed in Section 6.

5.3.2 RENORMALIZATION AND FACTORIZATION SCALE DEPENDENCES The truncation of the perturbation series invariably leads to renormalization and factorization scale dependence of QCD predictions, as discussed in Section 2.2. However, in almost all processes in which results beyond leading order are known, one can clearly see a decrease in the dependence of the predictions on the choice of scale as higher-order terms are included. This underlines the importance of carrying out calculations to higher orders. Unfortunately, the sensitivity to the choice of scale varies from process to process. Thus, in practice, the uncertainty associated with the choice of scale can only be assessed phenomenologically by varying the relevant scales over reasonable ranges. If significant scale dependence is found in a particular kinematic region for some process, then the usefulness of such data will be reduced until new theoretical techniques are developed to reduce that dependence. (See below.)

5.3.3 HIGHER-ORDER CORRECTIONS The uncertainty associated with higher-order corrections beyond current calculations (see Section 4) cannot be reliably estimated in general. One manifestation of this uncertainty is the scale dependence described above. That is, however, not the whole story. In special kinematic regions, relatively large corrections can be identified on physical grounds. Often the bulk of these effects can be calculated by special resummation techniques. Prominent examples are the small-x region, the threshold region (x near unity), and the small transverse momentum region (Section 4.2). Advances in these fields help to reduce the uncertainties and improve the reliability of the phenomenological analyses. Implementation of the higher-order (or "all-order") corrections, however, inevitably makes such calculations quite complicated and more removed from the simple intuitive parton model results.

5.3.4 HIGHER-TWIST EFFECTS Contributions to hard-scattering cross sections that are suppressed by powers of the relevant large momentum transfer are referred to as "higher-twist" corrections. These become insignificant at very high energies in regions far from the kinematic boundaries, but they may not be entirely negligible at moderate energy scales, where a great deal of data exist. The theory for power-law terms is known to be much more complicated than the leading-twist one (79). Nevertheless, factorization properties for the twist-4 corrections to conventional perturbative QCD have recently been established (80). The first foundation for phenomenological work in the relatively low Q^2 region appears to be laid. Recent convergence of the wide range of high-precision DIS data, spanning the entire range of Q^2 from 1 to 300 GeV^2, described above, sets the stage for detailed QCD analyses incorporating higher-twist components. The emerging new results from NMC and E665 will complement existing data, especially in the small-x region, in a comprehensive phenomenological study. Since there are many more theoretical unknowns than in the standard twist-2 formalism, realistic applications must invoke certain phenomenological ingredients in addition to the established theory. If this proves fruitful, it will help to improve the conventional twist-2 QCD parton analysis by providing a smooth transition into the low Q^2 range without an arbitrary kinematic cut (see below). An interesting example of this can be found in the analyses presented by Milsztajn and Virchaux (81).

5.4 *Phenomenological Considerations and Uncertainties*

5.4.1 KINEMATIC CUTS The outcome of a given global analysis depends on the selection of physical processes as well as on the experimental kinematic range of data. To ensure the applicability of the conventional perturbative QCD formalism, it is desirable to make reasonably high cuts in dimensional variables such as Q and W, the invariant mass of the hadronic system. On the other hand, experimental data are usually much more abundant in the low-Q region; in addition, it is much easier to get a handle on the predicted logarithmic QCD scale dependence at moderate values of Q (as it diminishes at high energies). Thus, it is important to consider the optimal choice of kinematic cuts in data selection and to determine the dependence of the results on this choice (73, 82). The inclusion of target mass corrections and phenomenological higher-twist terms can extend the useful Q range to smaller values than would otherwise be the case.

5.4.2 NUCLEAR TARGET EFFECTS Our primary concern is the parton structure of the nucleon. However, many of the key fixed-target experiments in deep inelastic scattering and lepton pair production use nuclear targets. The so-called EMC-effect (83) demonstrates that nuclear structure functions are not simple incoherent sums of the corresponding nucleon ones—the ratio shows a distinct x dependence, deviating from unity by up to 10–15%. Theoretical understanding of this phenomenon is not complete and lacks predictive power over the full x range. In practice, corrections are usually made on a phenomenological basis by using experimentally measured ratios whenever possible. Whereas this procedure should be quite reliable when applied to the same physical process, the uncertainty of using correction factors measured with one probe (muon) on another (neutrino) is not known a priori.

5.4.3 NORMALIZATIONS, SHAPES, AND RATIOS In certain situations, theoretical uncertainties in the determination of parton distributions can be reduced by focusing on specific aspects of measured cross sections such as the shape of some differential cross sections, ratios of cross sections, and asymmetries. For example, it has been shown that the shape of the direct photon p_T distribution is sensitive to the gluon distribution, whereas the normalization is more dependent on the value of the coupling constant (74). Since experimental normalization uncertainties are usually hard to pin down, it makes sense to allow an overall floating normalization factor for some or all experiments in the global analysis. Note that the overall normalization of the parton distribution functions is constrained by the quark number and momentum sum rules. Of course, within one experiment, measured cross section ratios and asymmetries always have much less associated uncertainty, and thus can be used with advantage whenever appropriate. For instance, the forward-backward asymmetry in W production and the W-to-Z production ratio at the hadron colliders, when they are more accurately determined, are expected to yield valuable information on the u/d ratio and the charm content of the nucleon.

5.4.4 EFFECTS FROM THE CHOICE OF PARAMETRIZATION OF THE INITIAL DISTRIBUTIONS There is considerable freedom in choosing the parametric form of the initial parton distributions (at scale Q_0) in making the global analysis. The parametrization must be general enough to accommodate all possible x and quark-flavor dependences; but it should not contain so many parameters that the fitting procedure becomes too underdetermined (73, 82). In practice, most groups use a functional form

$$f^a(x, Q_0) = A_0^a x^{A_1^a}(1 - x)^{A_2^a} P^a(x), \qquad 18.$$

where "a" is the flavor label (including the gluon), and $P^a(x)$ is a smooth function. The choice of $P^a(x)$ varies considerably—examples include polynomials in x of various degrees or logarithmic forms such as $\ln^{A_3^a}(1/x + 1)$ (73). To reduce the large number of parameters to a more practical number, various simplifications are usually made (such as setting the A_1^a terms to specific values and assuming flavor independence of various coefficients), based on educated guesses. The question is: how are the final results dependent on these assumptions?

It is very difficult to quantify the uncertainties resulting from the choice of parameters and the functional form for $P^a(x)$, although it is obvious that inappropriate choices can bring about misleading results. The small-x behavior of the parton distributions and the flavor dependence of the sea-quark distributions at a general scale Q are strongly dependent on the assumptions made about the relevant parameters for the initial distributions at Q_0. The only definitive way to resolve these uncertainties is to compare theory with experiment in processes sensitive to these parameters. This requires persistent work that is constantly updated following advances in both theory and experiment. Where uncertainties are present, it is very important to study systematically the limits on the relevant parameters set by existing theory and experiment. Work in this direction is only beginning (84).

One attempt to reduce the dependence upon ad hoc parametric forms is to generate as much as possible the parton distributions via the evolution equations themselves. Such distributions are generally referred to as radiatively generated parton distributions, a recent example of which is described by Glück et al (85). The basic idea is to use a minimal set of input parton distributions (say, only valence quarks with or without gluons) at a relatively small scale Q_0, typically in the range of 2Λ to 3Λ. The bulk of the sea and gluon distributions will then be arrived at by evolution, driven primarily by the valence terms. This approach is reviewed by Reya (86). It usually leads to too many partons in the small-x range.

5.5 *Parametrizations of Results*

The results on parton distributions from a global analysis can be presented in two ways. First, if the parametrization of the initial distributions and all the QCD parameters are given, any user can accurately recreate the distributions at all (x, Q) by using a QCD evolution program with these as input. This is not usually done. Instead, the common practice is to approximate the outcome of a global fit over all (x, Q) by a set of parametrized functions. The way this approximation is made varies widely between the available distribution sets, ranging from a simple interpolation formula over a large three-dimensional array (x, Q, and flavor), to Chebeschev polynomial expansions of the functions, to simple Q-dependent parametrizations of the form of Equation 18.

Although, in principle, the form of the approximation is transparent to the user of the distributions, there are circumstances under which misleading results can arise from a lack of understanding of the parametrization. For instance, in applications to very high-energy processes, the convolution integral in the QCD parton formula (Equation 2) often involves extremely small values of x—beyond the range where the parton distributions were originally derived—resulting in inadvertent extrapolations of the distributions. This can lead to quite unrealistic results if the functions behave badly beyond the originally intended range (e.g. negative, oscillatory, or discontinuous—examples of all these cases occur for some existing distribution sets). The danger is that the user will not even know that the results are unreliable, unless they are obviously absurd. This problem is eschewed with a parametrization using positive-definite functions smooth in both variables (x, Q), such as done by Morfin & Tung (73) with $P^{\rm a}(x, Q) = \ln^{A_3^{\rm a}}(Q)(1/x + 1)$ in Equation 18 (see also 25, 73 for a detailed discussion of the other advantages of this parametrization).

6. SURVEY OF RECENT PARTON DISTRIBUTIONS

The first-generation parton distribution sets, based on leading-order evolution and data of the early 1980s, have been widely used for calculation of high-energy processes (87). Experimental data have drastically improved (and substantially changed, in some cases) since then, however, and most current applications require at least next-to-leading-order (NLO) treatment. Therefore, second-generation global analyses based on NLO evolution and more current data have been made by several groups in recent years. Broadly speaking, these can be categorized into two groups: (*a*) specialized analyses focusing on some specific issue or process [such as the gluon distribution and direct photon production (74), neutrino scattering (88), and radiatively generated parton distributions (85)]; and (*b*) global analyses encompassing a wide range of processes (73, 89–91). These analyses differ considerably in many aspects on issues discussed in the previous section, e.g. the range of data used, the way experimental errors are treated, the choice of renormalization scheme, and assumptions on the initial distributions, etc. Consequently, a critical comparison of these distribution sets is quite difficult, if not impossible.

The currently available parton distribution sets, both old and new, have been compiled at CERN and were distributed recently as a program package, PDFLIB (92). We refer the reader to this package for detailed references and some comparison plots. It is important to note, however, that the very convenience resulting from having all published distributions at once also tends to encourage indiscriminant use of these distributions in the hands of inexperienced users, given the complexity of issues discussed before. In this section we highlight some of the most relevant considerations in putting these distributions in perspective.

In evaluating different sets of parton distributions or calculations of some physical quantity based on these distributions, it is essential that meaningful corresponding objects are compared. On the parton distribution level, LO, NLO-$\overline{\rm MS}$, and NLO-DIS distributions, like apples and oranges, should not be compared with one another—as is often done, for example, even in the PDFLIB document (92). To illustrate this point, we show in the left panel of Figure 2 two curves each from the MT-S2 and HMRS-B parton distribution sets corresponding to the same strange quark defined in the DIS and $\overline{\text{MS}}$ schemes, respectively. The transformation between the two functions is performed according to Equation 16. The significant difference between the two functions, especially in the small-x region, arises from the large contribution of the gluon term on the right-hand side of the equation, reflecting the quark-gluon mixing effect in changing from one scheme to the other. Since the gluon distribution is numerically one order of magnitude larger than the sea quark, the nominal order α_s term can be quite significant compared to the "leading" term. The actual size of the effect depends very much on the specific input distributions. Examples in which the effect is even larger than shown on the left in Figure 2 can easily be found (49). Likewise, on the right-hand side of Figure 2 we show the same comparison for the gluon distribution. Note the big difference, in the large-x region, between the curves for the same distribution in two different schemes, MT-S2(DIS) and MT-S2($\overline{\text{MS}}$), in contrast to the relatively small difference between the results from the two entirely independent analyses expressed in the same scheme, HMRS-B and MT-S2($\overline{\text{MS}}$). In this case, the large difference arises from mixing of the gluon with the much larger valence quarks in the large-x region (73). These results demonstrate that at moderate values of Q it is almost meaningless to talk about a leading-order sea-quark or gluon distribution; only at the NLO level do these quantities acquire unambiguous meaning, provided the choice of renormalization scheme is specified. In this regard, the situation is rather like that of the QCD coupling α_s, as discussed in Section 2.3.

On the level of physical quantities, LO, DIS, and $\overline{\text{MS}}$ distributions must be convoluted with the corresponding LO, NLO-DIS, and NLO-$\overline{\text{MS}}$ hard-scattering cross-section formulas in order to yield meaningful predictions, as emphasized in previous sections of this review. Applying parton distributions defined in different schemes to the same hard-scattering cross-section formula to make comparisons, a practice too often seen in the literature and conference reports, is simply meaningless and distracts from real physics issues.

It should be self-evident that, for use in quantitative applications of QCD, the first prerequisite parton distributions should satisfy is that they fit the well-established data. Paradoxically, this obvious point has been consistently ignored in the eagerness to present results either with a favorite distribution set or with a variety of distribution sets in order to show the supposed range of "theoretical uncertainties"—even if

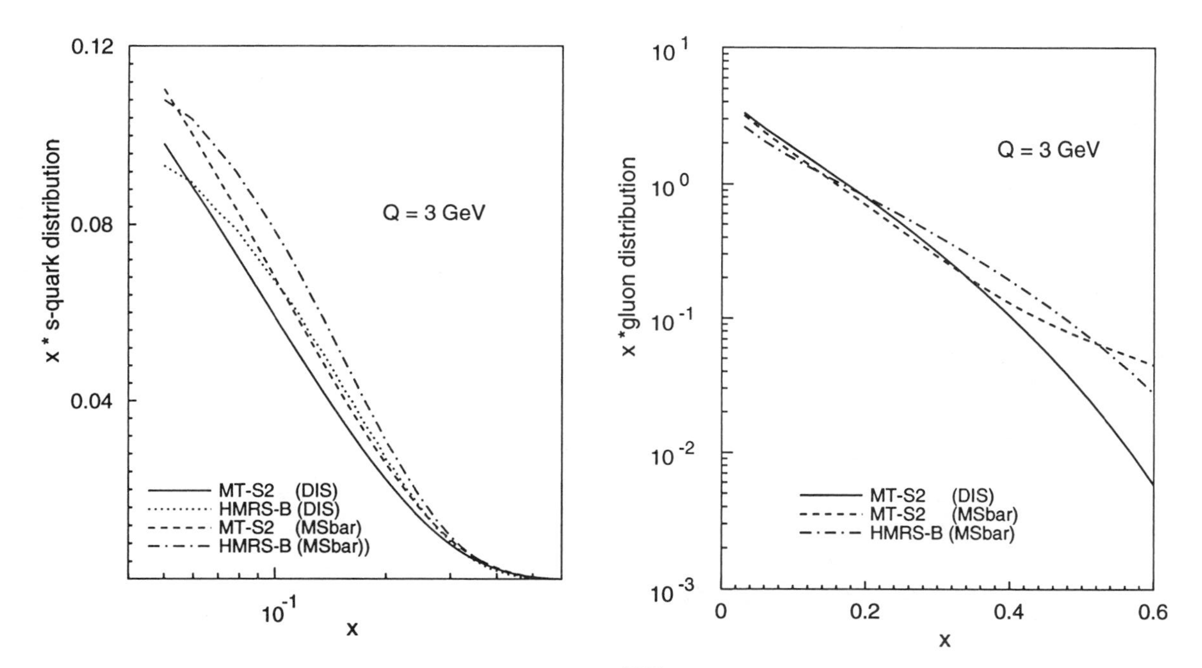

Figure 2 Comparison of parton distributions defined in the DIS and $\overline{\text{MS}}$ schemes: (*left*) the strange quark; and (*right*) the gluon. Two representative distributions from the recent analyses of Harriman et al (HMRS-B) and Morfin & Tung (MT) are shown. Conversion of one scheme to the other is done as described in the text.

these are already known to disagree with current data. It is worthwhile to point out in this regard that, contrary to general impressions, even the recent second-generation distribution sets do not necessarily fit existing high-precision data. In Figure 3 we show a representative comparison of BCDMS data with predictions from three commonly used NLO distribution sets. The significant departure of the prediction of the widely used DFLM distributions from data is due to the fact that muon-scattering data were not used in the analysis for their extraction.

There are contemporary applications of QCD, such as certain estimates for LHC and SSC physics, that do not require NLO accuracy or for which the required NLO calculations have not been performed. The LO formalism is attractive because the numerical calculation is generally at least one order of magnitude faster than the NLO one. Obviously, even in LO calculations, it is desirable to use parton distributions that agree with current data. It is for this purpose that two recent LO parton distribution sets were published (73, 93).

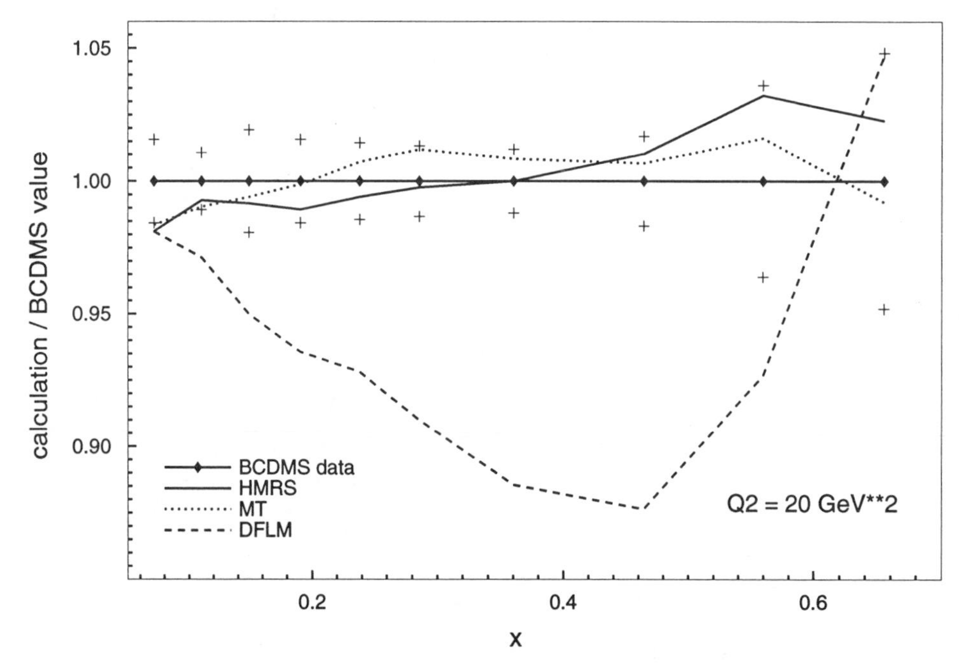

Figure 3 Comparison of representative data from the BCDMS experiment with predictions from three recent NLO parton distribution sets: DFLM, HMRS-B, and MT-S2.

7. CURRENT UNCERTAINTIES AND IMMEDIATE CHALLENGES

Given a complete set of experiments, one can in principle determine all the independent parton distributions using the QCD formalism. In practice, because of existing limitations of both theory and experiment, the program of a comprehensive global analysis of parton distributions is still evolving. In this section, we discuss how to assess uncertainties on parton distributions, identify the main areas of current uncertainty, and enumerate the challenges involved in overcoming them.

A seemingly simple way to exhibit the uncertainty on parton distributions is to plot or tabulate a large number of published distribution sets side by side and expose the range of differences. An inspection of a typical plot will show a rather wide range of variation for all the parton flavors—in apparent sharp contrast to the extreme accuracy of current experimental data on DIS structure functions from which these distributions are extracted (70). Such comparisons are, of course, totally misleading for reasons already discussed in the previous two sections. A better way, along the same line, is to make the comparison using only distribution sets defined in the same scheme, to the same order, and fitting all the relevant current data. Figures 3 and 4, discussed below, are sensible comparisons of this kind. One will see that the spread between currently viable distributions of the same type is much less than what is commonly perceived.

Even this procedure does not necessarily reveal the true range of uncertainty because different published distribution sets may invoke different assumptions and use different shortcuts in dealing with the experimental and theoretical issues in their analysis (see Section 5). In addition, in all global analyses, there is a certain arbitrariness in selecting the final distributions (for publishing or circulation) from a fairly wide range of possible fits. In order to gain a quantitative measure of current uncertainties on specific aspects of the parton distributions, only dedicated studies using clearly defined and consistent procedures make scientific sense. Some steps along this line have been taken by the various groups exploring the range of possible small-x behavior (73, 91) (see Section 7.3 below). More comprehensive investigations on the ranges for the shape parameters of the parton distributions and the QCD parameter Λ are clearly needed (84). Indeed, we anticipate the systematic exploration of this multidimensional parameter space, utilizing currently available data as well as new data, to be the natural frontier of QCD analysis and phenomenology in the near future. [A

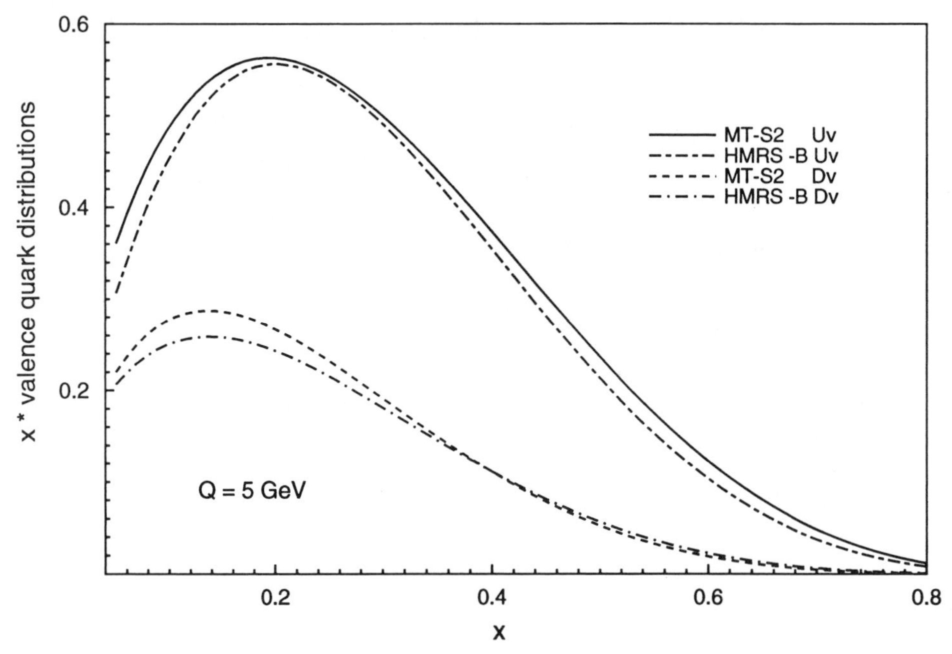

Figure 4 Comparison of the valence quark distributions from the HMRS-B and MT-S2 sets of parton distribution functions.

long-term project involving twelve theorists and experimentalists from eight institutions, Coordinated Theoretical/Experimental Project on Quantitative QCD Phenomenology and Tests of the Standard Model (CTEQ), has recently been formed to address the full range of theoretical and phenomenological problems in a systematic and consistent fashion.]

Generally speaking, the valence quark distributions and the sum of the sea-quark distributions are already very well determined, as they are directly related to the accurate data on DIS structure functions from muon (electron) and neutrino-scattering experiments. Figure 4 shows the comparison of these distributions from two recent independent distribution sets.

As far as the other parton distributions are concerned, three areas of uncertainty stand out. These are described in the following sections.

7.1 *The Shape of the Gluon Distribution*

Is the gluon distribution shape relatively soft or hard in the large-x region? How does it behave at small values of x?

Some progress has been made in answering these questions in recent years. High-statistics deep inelastic scattering data provide some constraints on the gluon distribution in the region below x of about 0.3, while fixed-target lepton pair and direct photon production experiments probe it in a higher x region, up to 0.6. Figure 5 shows five different recent $\overline{\text{MS}}$ gluon distributions, each evaluated at $Q^2 = 20\ \text{GeV}^2$. The Aurenche et al (ABFOW) distributions were obtained through a joint fit to the BCDMS deep inelastic data and the WA70 direct photon data (74). This analysis was intended primarily to demonstrate the utility of including the direct photon data and is not a true global analysis in the sense discussed here. The Harriman et al (HMRS-B) distributions were obtained by fitting these same data and, in addition, data from neutrino deep inelastic scattering and from lepton pair production (90). The resulting gluon distributions from these two fits are, as would be expected, quite similar, since the data that most strongly constrained the gluon were the same in each case. The third and fourth gluon distributions come from the Morfin & Tung (MT-S1 and MT-S2) global fits to deep inelastic and lepton pair production data, but do not include direct photon data (73). Set MT-S2 has a suppressed strange-quark sea contribution at the initial value of Q_0 (as does HMRS-B), and its gluon distribution agrees well with the above two. The MT-S1 set uses an

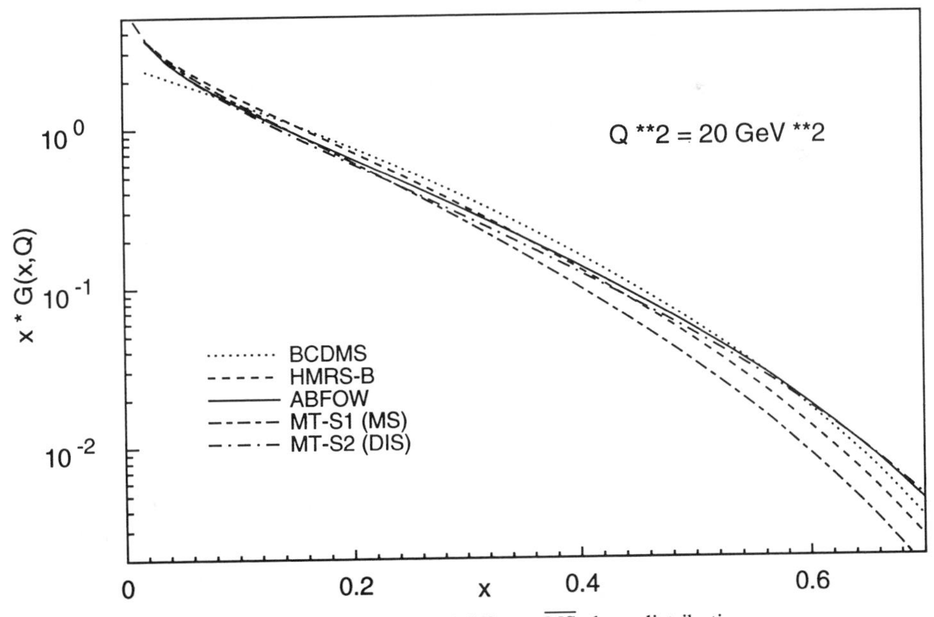

Figure 5 Comparison of several different $\overline{\text{MS}}$ gluon distributions.

SU(3) symmetric sea; the gluon distribution is somewhat softer than the others above x about 0.25, but is quite similar to them below that value. Finally, the fifth curve shows the result from a recent analysis by the BCDMS group (94) using their data and the recently reanalyzed SLAC data. Both target mass corrections and higher-twist terms are included. This new result is very similar to the gluon distributions from the global fits and is not as soft as the results reported earlier by the same group (95). In order to appreciate that some progress has been made in the determination of the gluon distribution, the spread between these curves should be compared with the much larger differences between, for example, the two gluon distributions in (96) (both are of leading order and based on the same data sets). Also note the sharp contrast of Figure 5 to similar plots showing a wide spread of the gluon distributions often seen in the literature where "apples and oranges" are all thrown in the same basket.

As more NLO calculations for various large momentum transfer processes have been completed, and as the data have improved, the very soft and very hard gluon distributions appearing in some older parton distribution sets are clearly no longer viable. Furthermore, with new data becoming available soon from both fixed-target and collider experiments, and with improved theoretical understanding of the uncertainties and newly developed tools to control these (Section 4.3), we can expect continued progress on this front. In addition, since jet cross sections arise primarily from gluon scattering, the increasingly active area of NLO-QCD jet analysis (77) holds the promise of placing useful constraints on the gluon distribution, if not directly contributing to its extraction.

The results in Figure 5 also serve to underline another point touched on previously. The only change between the MT-S1 and MT-S2 fits was in the input boundary condition on the strange sea. However, the shape of the resulting gluon distribution was significantly affected. This reveals the manner in which the input assumptions can affect the output distributions in unexpected ways. Inclusion of the direct photon data in these fits might have reduced the spread, a point currently under investigation.

Another method for studying the gluon distribution is through the longitudinal structure function in deep inelastic scattering, as emphasized by the original papers on QCD. HERA has the potential of finally making this approach feasible (97), especially in the small-x region (30). We can anticipate strong constraints over a large range of x resulting from the simultaneous comparison of data of this type and new direct photon data from both collider and fixed-target experiments.

7.2 *Flavor Differentiation of the Sea-Quark Distributions*

Is the strange-sea distribution the same as the nonstrange ones? Is up-sea the same as the down-sea?

Most LO parton model analyses of existing data on opposite-sign dimuon production in neutrino scattering indicate that the strange-sea content of the nucleon is suppressed compared to the nonstrange sea at a typical Q_0 (98 and references therein), contrary to the naive expectation of an SU(3) symmetric sea. This feature has been imposed on some of the parton distribution sets obtained in global fits, as mentioned above. However, because of the significant mixing between gluons and sea quarks at moderate scale Q (see Section 6), the gluon contribution to this process, so far left out of these analyses, can be rather large, and hence affect this conclusion. The ongoing analysis of currently available data on dimuon production will incorporate the important NLO contributions [R. Brock (FMM) and M. Shavitz (CCFR), private communication], and thus allow a much more meaningful determination of the strange-quark distribution defined in a specific renormalization scheme. The planned next-generation neutrino experiment at Fermilab expects to measure one order of magnitude more dimuon events (99). This will allow a definitive QCD analysis of the charm production process.

The new results from NMC (69) raised doubt whether the $\bar{u}$ and $\bar{d}$ distributions are the same, as is usually assumed. In order to settle this issue, it is desirable to exploit constraints imposed by different sources, such as lepton pair production and forward-backward asymmetry in W production, in addition to deep inelastic scattering. The experimental accuracy for these processes is improving, and careful combined analysis should allow the determination of the limits on the breaking of flavor SU(2) symmetry of the sea.

Significant advances in flavor differentiation of the sea quarks in general can be achieved with a definitive experiment on neutrino scattering that would allow the separate measurement of neutrino and antineutrino structure functions—by way of the y distributions of the respective cross sections.[Currently, because of insufficient statistics, the structure functions are obtained from the sum and difference of the two cross sections, supplemented by assumed relations among the structure functions (26).] Separate measurements would allow the determination of more linear combinations of parton distribution functions, and hence provide the necessary information to extract parton distributions of all flavors unambiguously. Such an experiment has been proposed at Fermilab for the 1990s (99). It should crown the remarkable achievement of the line of DIS experiments that have played such a central role in the QCD parton framework.

7.3 *Small-*x *Behavior of Parton Distribution Functions*

What is the range of possible small-x behavior? Is it possible to calculate the "small-x evolution" of parton distributions?

Predictions on cross sections for standard model and new physics processes at the next-generation colliders depend critically on the behavior of the parton distribution functions at small-x values beyond current measurements. Uncertainty on this front is now rather large. A systematic phenomenological study of the allowed range of small-x behavior, with current data as constraints and using a generalized initial distribution function containing a logarithm factor [$P(x)$ of Equation 18] in addition to the traditional power law, reveals an effective power-law exponent in the range $(0, -0.45)$ at $Q = 2.5$ GeV converging to the much narrower range of $(-0.58, -0.63)$ at $Q = 10^4$ GeV (see Figure 4 of Reference 25). In the immediate future, the DIS experiments NMC (69) and E665 (100) will extend the x range somewhat before the much anticipated results from HERA become available. From hadron collider experiments, we expect low-mass lepton pair and heavy-flavor (especially b-quark) production to be particularly sensitive to small-x behavior of the parton distribution functions (101). If the rapidity range of these measurements can be expanded beyond the central region, x values below 10^{-3} can be effectively probed. Data from the Tevatron on this process will soon be available for analysis.

As briefly discussed in Section 3.3, much theoretical effort is being devoted to understanding the small-x behavior and to developing methods for calculating the parton distributions in this region. Of particular interest are possible generalizations of the familiar Q evolution of parton distributions in other directions on the x-Q plane—"vertically" toward small x or "diagonally" toward large s ($\propto Q^2/x$) based on different resummations in these varied regions. Initial numerical studies of the currently available formalisms indicate remarkable (and unexpected) agreement between the various resummation approaches (102). This agreement lends more credibility to the existing calculations based on conventional twist-2 QCD, but it does not yet provide useful predictive power toward the small-x region. Theoretical progress in this direction in the near future would be most welcome, as comparison with results from HERA should provide tests of QCD in yet another dimension.

8. CONCLUSION

The determination of the universal parton distribution functions in the QCD framework is intimately linked to all aspects of rigorous tests of the standard model: the effort to extract more precise distribution functions from an ever-wider range of processes becomes inseparable from efforts to check the consistency and to push the limit of the standard model in order to discover signs of new physics. At the same time, knowledge of these distribution functions allows us to make predictions on anticipated measurements at much higher energies than achievable at present. As described above, much challenging work remains to be done in both theory and experiment. With regard to global analyses, several near-term goals can be suggested: (*a*) inclusion of additional direct photon data to constrain further the gluon distribution, with attention being paid to how the scale dependence affects the resulting distributions and also the effects of isolation cuts and the bremsstrahlung contribution; (*b*) further study of the flavor dependence of the sea distributions and the inclusion of neutrino-produced dimuon data in the analyses; (*c*) further study of the theory and phenomenology of the small-x regime and the inclusion of results from HERA for both the longitudinal and transverse structure functions; (*d*) continued study of appropriate methods for treating the statistical and systematic errors of the different data sets; and (*e*) systematic investigation of the range of uncertainty of the QCD parameters, including the shape parameters of the uncalculable "initial distributions," and of possible experiments to reduce these uncertainties. With coordinated efforts on all these fronts, QCD analysis and phenomenology are entering an era of unprecedented precision, similar (but of course not yet comparable) to electroweak phenomenology. It is a very rich field, and it will continue to play a central role in high-energy physics in the context described above.

Literature Cited

1. Langacker, P., in *Proc. 24th Int. Conf. on High Energy Physics*, ed. R. Kotthaus, J. H. Kühn. Berlin/Heidelberg: Springer-Verlag (1989), p. 190
2. Altarelli, G., *Annu. Rev. Nucl. Part. Sci.* 39:357 (1989)
3. Geesaman, D. F., et al, eds., *Proc. Workshop on Hadron Structure Functions and Parton Distributions*. Singapore: World Scientific (1990), 500 pp.
4. Collins, J. C., Soper, D. E., *Annu. Rev. Nucl. Part. Sci.* 37:383 (1987)
5. Collins, J. C., Soper, D. E., Sterman, G., in *Perturbative Quantum Chromodynamics*, ed. A. Mueller. Singapore: World Scientific (1989)
6. Owens, J. F., *Rev. Mod. Phys.* 59:465 (1987)
7. Tung, W. K., see Ref. 3, p. 18
8. Appelquist, T., Carazzone, J., *Phys. Rev.*D11:2856 (1975)
9. Collins, J. C., Wilczek, F., Zee, A., *Phys. Rev.* D18:242 (1978); Collins, J. C., Tung, W. K., *Nucl. Phys.* B278:934 (1986); Marciano, W., *Phys. Rev.* D29:580 (1984)

10. Gribov, L. V., Lipatov, L. N., *Sov. J. Nucl. Phys.* 15:78, 1218 (1972)
11. Altarelli, G., Parisi, G., *Nucl. Phys.* B126:298 (1977)
12. Floratos, E. G., Ross, D. A., Sachrajda, C. T., *Nucl. Phys.*B129:66 (1977); erratum, *Nucl. Phys.* B139: 545 (1978); Floratos, E. G., Lacaze, R., Kounnas, C., *Phys. Lett.* 98B:89 (1981)
13. Gonzalez-Arroyo, A., Lopez, C., Yndurain, F. J., *Nucl. Phys.* B153:161 (1979); B166:429 (1980)
14. Curci, G., Furmanski, W., Petronzio, R., *Nucl. Phys.* B175:27 (1980); Furmanski, W., Petronzio, R., *Phys. Lett.* 97B:438 (1980)
15. Herrod, R. T., Wada, S., *Phys. Lett.* 96B:195 (1980); Herrod, R. T., Wada, S., Webber, B. R., *Z. Phys.* C9:351 (1981)
16. Hamberg, R., van Neerven, W. L., Univ. Leiden Rep. INLO-PUB-13/91 (1991)
17. Tung, W. K., *Phys. Rev. Lett.* 35:490 (1975); *Phys. Rev.* D12:3613 (1975)
18. Collins, J. C., Qiu, J., *Phys. Rev.* D39:1398 (1989)
19. Bartel, J., ed., *Proc. Workshop on Small-x Behavior of Parton Distributions.* Amsterdam: North Holland (1991)
20. Lipatov, L. N., *Sov. J. Nucl. Phys.* 23:338 (1976); Kuraev, E. A., Lipatov, L. N., Fadin, V. S., *Sov. Phys. JETP* 45:199 (1978)
21. Collins, J. C., Ellis, K., see Ref. 19, p. 80
22. Gribov, L. V., Levin, E. M., Ryskin, M. G., *Phys. Rep.* 100:1 (1982); Levin, E. M., Ryskin, M. G., *Phys. Rep.* 189:269 (1990); Levin, E. M., Ryskin, M. G., see Ref. 19, p. 92; and Levin, E. M., in *Proc. EP-HEP 91 Conf.,* Geneva, ed. Hegerty. Singapore: World Scientific (1992)
23. Mueller, A. H., Qiu, J., *Nucl. Phys.* B268:427 (1986); Mueller, A. H., see Ref. 19, p. 125
24. Collins, J. C., in *Physics Simulations at High Energy,* ed. V. Barger, T. Gottschalk, F. Halzen. Singapore: World Scientific (1987), p. 265
25. Tung, W. K., see Ref. 19, p. 55
26. Tung, W. K., et al, in *Proc. 1988 Summer Study on High Energy Physics in the 1990s,* ed. S. Jenson. Singapore: World Scientific (1989), p. 305
27. Bardeen, W. A., et al, *Phys. Rev.* D18:3998 (1978)
28. Altarelli, G., Ellis, R. K., Martinelli, G., *Nucl. Phys.* B143:521 (1978)
29. Duke, D. W., et al., *Phys. Rev.* D25:71 (1982); Devoto, A., et al, *Phys. Rev.* D30:541 (1984)
30. Cooper-Sarkar, A. M., et al, *Z. Phys.* C39:281 (1988)
31. Drell, S., Yan, T. M., *Phys. Rev. Lett.* 25:316 (1970)
32. Altarelli, G., Ellis, R. K., Martinelli, G., *Nucl. Phys.* B157:461 (1979)
33. Kubar, J., et al, *Nucl. Phys.* B175:251 (1980)
34. Matsuura, T., Hamberg, R., van Neerven, W. L., *Nucl. Phys.* B345:331 (1990)
35. Sterman, G., *Phys. Lett.* 179B:281 (1986); *Nucl. Phys.* B281:310 (1987); and see Ref. 3, p. 423; see also Catani, S., Trentedue, L., *Nucl. Phys.* B327:323 (1989)
36. Sterman, G., presented at Workshop on Hadron Structure Functions and Parton Distributions, see Ref. 3, p. 423
37. Altarelli, G., et al, *Nucl. Phys.* B246:12 (1984)
38. Collins, J. C., Soper, D., Sterman, G., *Nucl. Phys.* B250:199 (1985)
39. Arnold, P. B., Reno, M. H., *Nucl. Phys.* B319:37 (1989); erratum B330:284 (1990); Arnold, P. B., Kauffman, R. P., *Nucl. Phys.* B349:381 (1991)
40. Aurenche, P., et al, *Phys. Lett.* 140B:87 (1984)
41. Berger, E. L., Braaten, E., Field, R. D., *Nucl. Phys.*B239:52 (1984)
42. Camilleri, L., in *Proc. 19th Symp. on Multiparticle Dynamics,* Arles, France (1988), p. 203
43. Baer, H., Ohnemus, J., Owens, J. F., *Phys. Rev.* D42:61 (1990)
44. Baer, H., Ohnemus, J., Owens, J. F., *Phys. Lett.* B234:127 (1990)
45. Åkesson, T., et al (AFS Collaboration), *Sov. J. Nucl. Phys.* 51:836 (1990)
46. Koller, K., Walsh, T. F., Zerwas, P. M., *Z. Phys.* C2:197 (1979)
47. Berger, E. L., Qiu, J., *Phys. Lett.* B248:371 (1990)
48. Berger, E. L., Qiu, J., *Phys. Rev.* D44:2002 (1991)
49. Aivazis, M. G., Olness, F., Tung, W. K., *Phys. Rev. Lett.*65:2339 (1990)
50. Gottschalk, T., *Phys. Rev.* D23:56 (1981)
51. van der Bij, J. J., van Oldenborgh, G. J., *Z. Phys.* C51:477 (1991)
52. Olness, F. I., Tung, W. K., *Nucl. Phys.* B308:813 (1988)
53. Nason, P., Dawson, S., Ellis, R. K., *Nucl. Phys.* B303:607 (1988); B327:49 (1989); erratum, B335:260 (1990)

54. Beenakker, W., et al, *Phys. Rev.* D40:54 (1989); *Nucl. Phys.* B351:507 (1991); Laenen, E., et al, *Nucl. Phys.*, Stony Brook Rep. ITP-SB-91-14 (1991)
55. Ellis, S. D., Kunszt, Z., Soper, D. E., *Phys. Rev.* D40:2188 (1989); *Phys. Rev. Lett.* 62:726 (1989); 64: 2121 (1990)
56. Aversa, F., et al, *Z. Phys.* C46:253 (1990); *Nucl. Phys.* B237:105 (1989); *Phys. Lett.* 210B:225 (1988); 211B:465 (1988)
57. Mangano, M. L., Parke, S. J., *Phys. Rep.* 200:301 (1991)
58. Giele, W. T., Glover, E. W. N., Kosower, D. A., FNAL Rep. FERMILAB-CONF-91-243-T, FERMILAB-CONF-91-274-T (1991)
59. Bern, Z., Kosower, D., FNAL Rep. FERMILAB-91-111-T (1991)
60. Schuler, G., et al, in *Proc. Snowmass 90 Workshop* (1991)
61. Mishra, S., Sciulli, F., *Annu. Rev. Nucl. Part. Sci.* 39:259–310 (1989)
62. Benvenuti, A. C., et al (BCDMS Collaboration), *Phys. Lett.* B189:483 (1985); B195:91 (1987); B223:485 (1989); B237:592 (1990); B237:599 (1990)
63. Allaby, J. V., et al (CHARM Collaboration), *Z. Phys.* C36:611 (1987)
64. Berge, J. P., et al (CDHSW Collaboration), *Z. Phys.* C49:187 (1991)
65a. Cobau, W. (FMM Collaboration), presented at DPF91 Conf., Vancouver, August (1991)
65b. Mishra, S. R., et al (CCFR Collaboration), Nevis Rep. NEVIS-1466 (1992)
66. Whitlow, L. W., et al (SLAC Experiments), Rep. SLAC-PUB-5442 (1991), *Phys. Lett.* B250:193 (1990)
67. Bazizi, K., Wimpenny, S. J., Sloan, T., in *Proc. 25th Int. Conf. on High Energy Physics,* Singapore (1990)
68a. Mount, R., in *Proc. 24th Int. Conf. on High Energy Physics,* Munich, 1988, ed. R. Kotthaus. Berlin: Springer (1989)
68b. Bodek, A., Milsztajn, A., Wimpenny, S. J., see Ref. 3,; in Proc. 1991 Int. Symp. on Lepton and Photon Interactions at High Energies, Geneva (1992), to be published
69. Amaudruz, P., et al, *Z. Phys.* C51:387 (1991)
70. Roberts, R. G., Whalley, M. R., *J. Phys.* G17:1 (1991)
71. Rutherford, J. P., see Ref. 3, p. 234
72. Brown, C. N., et al, *Phys. Rev. Lett.* 63:2637 (1989)
73. Morfin, J. G., Tung, W. K., *Z. Phys.* C52:13 (1991)
74. Aurenche, P., Baier, R., Fontannaz, M., Owens, J. F., Werlen, M., *Phys. Rev.* D39:3275 (1989)
75. Bonensini, M., et al, *Z. Phys.* C38: 371 (1988)
76. Alverson, G., et al (E706 Collaboration), *Phys. Rev. Lett.* 68:2584 (1992)
77. Huth, J. (CDF Collaboration), Fermilab Rep. Fermilab-Conf-91/223-E (1991)
78. Berger, E. L. (Argonne Collaboration), Halzen, F., Kim, C. S., Willenbrock, S., *Phys. Rev.* D40:83 (1989); erratum, D40:3789 (1989)
79. Ellis, R. K., Furmanski, W., Petronzio, R., *Nucl. Phys.* B207:1 (1982); B212:29 (1983)
80. Qiu, J. W., Sterman, G., *Nucl. Phys.* B353:105, 137 (1991)
81. Milsztajn, A., see Ref. 3, p. 76; Virchaux, M., see Ref. 3, p. 124
82. Devoto, A., et al, *Phys. Rev.* D27:508 (1983)
83. Arneodo, M., et al, *Nucl. Phys.* B331:1 (1990)
84. Martin, A. D., Roberts, R. G., Stirling, W. J., *Phys. Rev.* D43:3648 (1991)
85. Glück, M., Reya, E., Vogt, A., Dortmund Rep. DO-TH 91/07, *Z. Phys.* (1992), in press
86. Reya, E., Dortmund Rep. DO-TH 91/09, in *Proc. Workshop on High Energy Physics Phenomenology II*, Calcutta, Jan. 1991. Singapore: World Scientific (1992), to be published
87. Duke, D. W., Owens, J. F., *Phys. Rev.* D30:49 (1984); Eichten, E., et al, *Rev. Mod. Phys.* 56:579 (1984) and erratum 58:1065 (1986); Glück, M., Reya, E., Hoffmann, E., *Z. Phys.* C13:119 (1982)
88. Diemoz, M., et al, *Z. Phys.* C39:21 (1988)
89. Martin, A. D., Roberts, R. G., Stirling, W. J., *Phys. Rev.* D37:1161 (1988); *Mod. Phys. Lett.* A4:1135 (1989)
90. Harriman, P. N., Martin, A. D., Roberts, R. G., Stirling, W. J., *Phys. Rev.* D42:798 (1990)
91. Kwiecinski, J., et al, *Phys. Lett.* B266:126 (1991)
92. Plothow-Besch, H., "PDFLIB:Structure Functions and α_s Calculation Users's Manual," CERN-PPE 1991.03.21, W5051 PDFLIB (1991)
93. Owens, J. F., *Phys. Lett.* B266:126 (1991)

94. Virchaux, M., Milsztajn, A., *Phys. Lett.* B274:221 (1992)
95. Benvenuti, A. C., et al, *Phys. Lett.* B195:97 (1987)
96. Duke, D. W., Owens, J. F., *Phys. Rev.* D30:49 (1984)
97. Peccei, R. D., ed., *Proc. HERA Workshop,* DESY (1988)
98. Foudas, C., et al, *Phys. Rev. Lett.* 64:1207 (1990)
99. Mishra, S., "A Next Generation Neutrino Experiment at the FNAL Tevatron," Nevis Rep. 1435 (1991)
100. Adams, M. R., et al, *Nucl. Instrum. Methods* A291:533 (1990)
101. Olness, F., Tung, W. K., in *From Colliders to Super Colliders,* ed. V. Barger, F. Halzen. Singapore: World Scientific (1987), p. 523
102. Krawczyk, M., see Ref. 19, p. 64; Badelek, B., et al, DESY Rep. DESY- 91-124 (1991)

Annu. Rev. Nucl. Part. Sci. 1992. 42:333–365

NUCLEONS AS SKYRMIONS

Makoto Oka

Department of Physics, Tokyo Institute of Technology, Meguro, Tokyo 152, Japan

Atsushi Hosaka

TRIUMF, 4004 Wesbrook Mall, Vancouver, British Columbia V6T 2A3, Canada

KEY WORDS: chiral symmetry, soliton, dibaryon, nucleon-nucleon interaction

---

CONTENTS

0163-8998/92/1201-0333$02.00

1. INTRODUCTION

The chiral theory of hadrons has been very successful as an effective theory for quantum chromodynamics (QCD) at low energies (e.g. 1). The theory is based on the chiral $SU(N_f)_L \times SU(N_f)_R$ symmetry, which is preserved (approximately) in QCD with N_f light quarks. This symmetry is not fully reflected in the low-energy hadron spectrum because it is spontaneously broken, leaving the vector flavor symmetry, $SU(N_f)_V$. According to the Goldstone's theorem (2, 3), $N_f^2 - 1$ massless Nambu-Goldstone bosons appear and they are identified with the low-lying pseudoscalar mesons (pions, kaons, etc). The small masses of the mesons indicate that chiral symmetry is explicitly broken as a result of the small masses of quarks. The dynamics of mesons and baryons at low energies can be determined almost uniquely by the chiral symmetry and its realization in the Goldstone mode (e.g. 4). Various low-energy theorems have experimentally been tested with great success, such as relations among the s-wave π-π scattering lengths (5), the Goldberger-Treiman relation (6) and the Adler-Weisberger sum rule for the axial-vector coupling constant g_A and the pion-nucleon interaction (7), and the Kroll-Ruderman relation for pion photoproduction amplitudes (8). In nuclear physics, chiral symmetry explains important roles of the pion in nuclear force and in the electromagnetic properties of nuclei (9). For instance, the forms of the pion-nucleon coupling and the exchange currents are determined almost uniquely by chiral symmetry.

A composite baryon (not as an elementary field) was introduced in the chiral effective theory by Skyrme (10–12). He considered a nonlinear theory of the pion (nonlinear sigma model) and identified a topological soliton solution of the theory as baryon. A conserved topological number plays the role of the baryon number. Thus a soliton with the topological number one corresponds to a baryon and behaves as a particle.[1]

Long after Skyrme's work, Witten realized that this picture of a baryon is indeed sensible in the large N_c limit of QCD (13, 14). Under the assumption of color confinement, it can be proved that large N_c QCD reduces to an effective field theory of weakly interacting mesons. In order to reach a nontrivial limit, the QCD coupling constant g, is set to scale as $1/\sqrt{N_c}$. Then it is shown that an M-point meson vertex scales as $1/N_c^{M/2-1}$. The meson (or $q\bar{q}$) in this limit has a mass of $O(N_c^0)$ and a width of $O(1/N_c)$. Witten argued that the baryon with N_c

[1] The existence of the topological invariant is a manifestation of a nontrivial homotopy group, $\pi_3[SU(2)_V] = Z$.

quarks has a mass of $O(N_c)$ and that the size of the baryon is of $O(N_c^0)$, which is indeed consistent with the solitonic picture (15). In principle, the meson theory equivalent to the large N_c QCD will contain infinitely many meson excitations. It may, however, be justified to truncate mesons of higher masses when we consider a low-energy effective theory for hadrons. Then the Skyrme theory is a very plausible candidate for such a theory.

The Skyrme model is also very attractive from phenomenological points of view. The solitonic baryon is not a point, but is an extended object that has appropriate form factors. This enables us to treat low-energy baryon dynamics consistently with its subnucleonic structure. It should be emphasized that the Skyrme model is most appropriate in long-distance (and/or low-energy) regions, while other subnucleonic models such as the quark model (16, 17) may be useful for short-distance properties of the nucleon and nuclear interaction (18, 19).

In this article, we report the current understanding of the Skyrme model, mostly from phenomenological points of view. We apply the model to meson-baryon, baryon-baryon, and multibaryon systems. In order to make the text self-contained, we first discuss ground-state baryons and explain the basic idea of semiclassical quantization (20) (Sections 2.1–2.4). The pion-nucleon system and excited baryons are treated in Sections 2.6 and 2.7.

The semiclassical method is subject to quantum corrections or, equivalently, to $1/N_c$ corrections in higher orders.[2] Generally speaking, we expect $1/N_c \approx 30\%$ corrections to its predictions and hence the lowest-order predictions to be at best qualitative. [Witten has argued, however, that the expansion parameter could contain a numerical factor like $1/4\pi$, which may reduce the correction by the same factor (14).]. In principle, the $1/N_c$ expansion can be carried out systematically by calculating loop corrections of pions. This method, however, cannot be rigorously applied, since the Skyrme theory is not renormalizable (22). We introduce in Section 2.5 another approach to $1/N_c$ corrections based on the algebraic structure of the Skyrmion rotation (23–25).

A study of baryonic interactions in the Skyrme model was started originally by Skyrme (11). By computing the static energy of a system of two Skyrmions far apart, he pointed out that the long-range force is nothing but a static one-pion-exchange potential, because the Skyrmion can be regarded as a source of pions surrounded by a static pion

[2] The action scales by $F_\pi/e \approx O(N_c)$, where F_π is the pion decay constant and e is the Skyrme parameter (see Section 2.1.). Thus the $\hbar$ expansion based on the functional method (e.g. 21) is equivalent to the $1/N_c$ expansion.

cloud. Shorter-range baryonic interactions are harder to calculate because of the technical complexity of treating the nonlinear nature of the Skyrme model. Yet a great deal of progress has been made during the last several years, and we review the current situation in Section 3. In addition to the two-body system, one can treat the multibaryon system directly in the Skyrme model. Several attempts to apply the model to light nuclei and nuclear matter at high densities are presented in Section 5.

Various generalizations and extensions of the original Skyrme model have been proposed and studied. Most important of all is to include the third flavor, strangeness [the SU(3) Skyrme model]. It brings a new understanding of the topological structure of the Skyrme model (15), in addition to a phenomenological application to strange baryons. We review the SU(3) Skyrme model in Section 4 and present its special solution of the SU(3) Skyrme model, which represents a dibaryon with strangeness (26).

Another important extension is the inclusion of vector mesons, which makes the model more realistic. A systematic method to introduce the ρ, a_1 (the chiral partner of the ρ), and ω mesons has been formulated in the hidden local symmetry approach (27, 28). The vector mesons play a major role in the electromagnetic properties of hadrons through vector meson dominance (29), and some predictions of the model are indeed in better agreement with experiment (30).

In this article, we concentrate on the original Skyrme model, because those generalizations do not change basic qualitative features of the model, and also because it is instructive to apply the simplest model to various phenomena in order to provide a systematic understanding of the model.

2. THE SKYRMION

2.1. *The Lagrangian*

The elementary field of the Skyrme model is that of the Nambu-Goldstone bosons associated with spontaneous breaking of chiral symmetry of QCD. At low energies, it is reasonable to consider only these light mesons as explicit dynamical degrees of freedom, since other heavier mesons are effectively suppressed. In the case of flavor SU(2), they are the pion fields ϕ_a $(a = 1, 2, 3)$. It is convenient to introduce an SU(2) matrix parametrized by the pions:

$$U(x) = \exp(2i\tau^a\phi_a/F_\pi), \qquad 1.$$

where τ^a are the Pauli matrices and the pion decay constant, F_π, is introduced to normalize the pion fields. In the linear sigma model (31), $f_\pi = F_\pi/2$ (= 93 MeV) is the radius of the chiral circle, $\sigma^2 + \boldsymbol{\pi}^2 = f_\pi^2$, and is the vacuum condensate of the sigma field: $F_\pi = 2\langle\sigma\rangle$.

The Skyrmion is a static, finite-energy configuration of U, which has a nontrivial topology. A static field $U(\mathbf{x})$ is regarded as a mapping: $R^3 \rightarrow \mathrm{SU}(2) \sim S^3$. As the restriction to a finite-energy configuration leads to $U(|\mathbf{x}| \rightarrow \infty) \rightarrow 1$, the mapping is topologically equivalent to $S^3 \rightarrow S^3$, implying a nontrivial homotopy $\pi_3(S^3) = Z$. In this case, the topology of the mapping is classified by an integer winding number,

$$B \equiv \int \mathrm{d}^3x B_0. \qquad 2.$$

B_0 is the time component of a conserved topological current

$$B_\mu \equiv \frac{1}{24\pi^2} \epsilon_{\mu\nu\alpha\beta} \operatorname{tr} L^\nu L^\alpha L^\beta, \qquad 3.$$

where $L_\mu \equiv U^\dagger \partial_\mu U$ is the left current. Skyrme proposed to identify the topological invariant B as the baryon number and thus the Skyrmion with the baryon. Although the relation of the topological current (Equation 3) to the baryonic current in QCD, $(1/N_c)\bar{\psi}\gamma_\mu\psi$, is not clear, an adiabatic or $1/N_c$ expansion of the QCD baryonic current gives Equation 3 as the leading term (32–34).

The dynamical contents of the model reflect the symmetries observed in low-energy hadron systems. Chiral symmetry is the most important one, as indicated by the success of the current algebra and low-energy theorems. Skyrme proposed the following chirally symmetric Lagrangian:

$$L = -\frac{F_\pi^2}{16} \operatorname{tr} L_\mu L^\mu + \frac{1}{32e^2} \operatorname{tr}([L_\mu, L_\nu])^2. \qquad 4.$$

It is easy to see that the Lagrangian in Equation 4 is invariant under a global chiral transformation $U \rightarrow g_L U g_R{}^\dagger$, where $g_{L,R}$ are elements of the chiral group $\mathrm{SU}(2)_L \times \mathrm{SU}(2)_R$. The first term is the standard nonlinear sigma model. As a simple scaling argument shows, this term alone cannot support a finite-size soliton solution. Thus Skyrme proposed to add the second term of fourth-order derivatives (the Skyrme term). Physically, the Skyrme term may be interpreted as the kinetic energy term of the ρ meson in the limit $m_\rho \rightarrow \infty$, where the Skyrme parameter e is identified as the coupling constant of the nonlinear interactions of ρ (11). In Equation 4, the mass of the pion has been neglected, as exact chiral symmetry implies. In order to implement a small breaking of chiral symmetry, the pion mass term

$$L_{\text{mass}} = \frac{F_\pi^2 m_\pi^2}{16} \text{tr}[U + U^\dagger - 2] \qquad 5.$$

is added, if necessary.

The model contains two parameters: the pion decay constant F_π, and the dimensionless parameter e. They may be fixed phenomenologically using mesonic processes such as the decay of the pion and the π-π scattering. They may also be regarded as effective parameters for a specific purpose of the model, e.g. to describe the nucleon. Here we take $F_\pi = 129$ MeV and $e = 5.45$ as determined by Adkins et al (20) so as to reproduce the empirical masses of the nucleon and delta (Δ) in the model with the massless pion.

2.2. *The Hedgehog Ansatz*

The $B = 1$ soliton solution is obtained using the spherically symmetric hedgehog ansatz:

$$U_{\text{H}}(\mathbf{x}) = \exp[i\boldsymbol{\tau}\cdot\hat{\mathbf{r}}F(r)], \qquad 6.$$

where the isospin of the pion field is pointing parallel to the radial unit vector $\hat{r}$. With the hedgehog ansatz, the baryon number is simplified to

$$B = \int d^3x B_0 = \frac{1}{\pi}[F(0) - F(\infty)]. \qquad 7.$$

The radial function $F(r)$ is determined by solving the Euler-Lagrange equation, or equivalently by minimizing the static energy

$$E_{\text{H}} = -\frac{F_\pi^2}{16}\int d^3x \, \text{tr}\, L_i L^i - \frac{1}{32e^2}\int d^3x \, \text{tr}([L_i, L_j])^2, \qquad 8.$$

subject to the boundary conditions $F(0) = \pi$ and $F(r \to \infty) = 0$. Note that using the nondimensional length $\tilde{r} = eF_\pi r$, the energy E_{H} is factored by F_π/e, which implies that the Skyrmion mass scales as $O(N_c)$ and its size as $O(1)$, in agreement with large N_c QCD.[3] The exact solution is available only numerically. Calculations show that the minimum energy is $36.5F_\pi/e$. The function $F(r)$ decreases monotonically, and for $r \to \infty$ it falls off as the one-pion tail: $F(r \to \infty) \sim 1/r^2$ for the massless pion.

It is interesting to see that the energy in Equation 8 has a lower

[3] F_π and e scale as $F_\pi \sim O(N_c^{1/2})$ and $e \sim O(N_c^{-1/2})$.

bound for a given B. Using $L_i^a = (-i/2)\ \mathrm{tr}\ \tau^a L_i$, the energy (Equation 8) can be rewritten as

$$E_{\mathrm{H}} = \int d^3x \left[\frac{1}{8}\left(F_\pi L_i^a - \frac{1}{e}\,\epsilon_{abc}\epsilon_{ijk}L_j^b L_k^c\right)^2 + \frac{F_\pi}{4e}\,\epsilon_{abc}\epsilon_{ijk}L_i^a L_j^b L_k^c\right] \geq 3\pi^2\frac{F_\pi}{e}\,|\,B\,| = 29.6\,\frac{F_\pi}{e}. \qquad 9.$$

The equality is attained when the equation $F_\pi L_i^a - (1/e)\epsilon_{abc}\epsilon_{ijk}L_j^b L_k^c = 0$ is satisfied. The Skyrmion solution does not satisfy this condition and the energy E_{H} is 23% higher than the lower bound.

2.3. *Approximate Solutions*

Recently, Atiyah & Manton proposed a simple analytic method for an approximate configuration of $U(\mathbf{x})$ (35). The method is particularly useful for the $B = 2$ system (35, 36). The idea is that a suitable variational ansatz for $U(\mathbf{x})$ is generated by an instanton solution of the (flavor) SU(2) Yang-Mills theory in the four-dimensional Euclidean space (e.g. 37). To be specific, let $A_\mu(x)$ be an instanton with winding number n. A topological consideration tells us that the (nonabelian) holonomy integral

$$U(\mathbf{x}) = T\exp\int_{-\infty}^{+\infty} d\tau A_4(\tau, \mathbf{x}) \qquad 10.$$

yields a static configuration $U(\mathbf{x})$ with $B = n$. In Equation 10, the time ordering T is taken along the line of the time integration. It does not appear that the dynamics of instantons is related to that of the Skyrmion. However, as shown below, Equation 10 provides a remarkably good variational ansatz for the Skyrmion.

In general, the time-ordered integral of Equation 10 is not easily performed. There are, however, a few cases in which the integral can be determined analytically. Let us take an instanton solution given by 't Hooft (38). For an arbitrary winding number $n = B$, it takes the form

$$A_4(x) = \frac{i}{2}\,\boldsymbol{\tau}\cdot\boldsymbol{\nabla}\ln\rho(x), \qquad 11.$$

where the superpotential ρ is given by

$$\rho = 1 + \sum_{i=1}^{B}\frac{\nu_i^2}{(x - X_i)^2}\,. \qquad 12.$$

The pole positions X_i are fixed four-vectors, and the weights ν_i are positive parameters. For $B = 1$, setting $X_1 = 0$ and $\nu_1 \equiv \nu$, one can verify that the time-ordered integral in Equation 10 reduces to an elementary integral, yielding the hedgehog U field in Equation 6 with $F(r)$:

$$F(r) = \pi \left(1 - \frac{r}{\sqrt{r^2 + \nu^2}}\right). \qquad 13.$$

Thus we confirm that the Skyrmion carries the right baryon number $B = n = 1$, since the function $F(r)$ satisfies the boundary condition $F(0) - F(\infty) = \pi$. The scale parameter ν is determined variationally by minimizing the energy. The minimum energy is $36.8F_\pi/e$ at $\nu^2 = 8.44$, which is larger than the exact value $36.5F_\pi/e$ by less than 1%.

2.4. *The Nucleon: Semiclassical Quantization*

The hedgehog ansatz (Equation 6) is not simply a nucleon that has a definite spin J and isospin I. The solution is spherical in the combined space of spin and isospin, since the operator $\mathbf{K} = \mathbf{J}+\mathbf{I}$ leaves the hedgehog configuration invariant. However, it is regarded as a deformed state in an individual space of spin or isospin. In fact, the solution can be rotated without changing the total energy by a constant $A \in SU(2)$:

$$U_H(\mathbf{x}) \rightarrow A U_H A^\dagger. \qquad 14.$$

The hedgehog symmetry makes a rotation in coordinate space equivalent to an inverse rotation in isospin space.

The degeneracy under the rotation in Equation 14 leads to zero energy excitation modes. The presence of the zero modes is due to the rotational symmetry of the Lagrangian and its breaking in the hedgehog solution. In order to recover the rotational symmetry and to construct states with definite spin and isospin, we employ a semiclassical approach (20), assuming that collective motions dominate for the low-lying baryon dynamics. The rotation of the Skyrmion as a rigid body is described by A, regarding A as a time-dependent collective coordinate. One obtains the Lagrangian for a "rotating Skyrmion" (Equation 14) (20):

$$L = \frac{\mathcal{I}}{2}\mathbf{\Omega}^2 - E_H, \qquad 15.$$

where the angular velocity is defined by $\mathbf{\Omega} = -i\,\mathrm{tr}[\boldsymbol{\tau} A^\dagger \dot{A}]$. The spherical symmetry of the Skyrmion leaves the single principal value for the moment of inertia $\mathcal{I}$. Quantizing the rotation, one gets the rotational spectrum $J(J+1)/(2\mathcal{I})$ with equal spin J and isospin I, and the corresponding states, the nucleon ($I = J = 1/2$), the delta isobar ($I = J = 3/2$), and so on. Since the moment of inertia scales as $O(N_c)$, the rotational energies appear as higher order corrections in the order of $O(N_c^{-1})$. Therefore, the semiclassical approach based on the rigid body rotation, or equivalently a slow rotation, is justified in the large N_c region. In the SU(2) Skyrmion, the nucleon can be quantized either as a boson or as a fermion. The ambiguity, however, will be removed in the SU(3) Skyrme model. The Wess-Zumino term introduces a constraint for projection, resulting only in fermion states when N_c is odd (15).

Static properties of the nucleon have been studied using the collective coordinate method (20). In Table 1, we summarize the model predictions for single-nucleon properties and compare them with the empirical values. We also show the results from the Atiyah & Manton method to see how well the method works. Generally speaking, both predictions agree with experimental data to within 30%. Some quantities, such as the moment of inertia and magnetic moments, show relatively large discrepancies, since they are rather sensitive to the Skyrmion profile.

Some comments on these predictions follow: (*a*) We have chosen a small pion decay constant, F_π, which is about 2/3 of the empirical value, but can reproduce the masses of the nucleon and delta. The empirical value $F_\pi = 186$ MeV makes their masses too large (39). The difficulty might be removed by subtracting a spurious zero-point energy associated with the rotation (and translation) in the order of $O(N_c^0)$, which is missing in Equation 15 (22, 40). No field theoretical calculation is

Table 1 Single baryon properties in the Skyrme model from Adkins, Nappi & Witten (ANW) (20) as compared with those in the Atiyah-Manton (AM) method

Quantity	ANW[a]	AM method	Experiment
Pion decay constant, F_π (MeV)	129	129	186
Skyrme parameter, e	5.45	5.45	—
Hedgehog mass, M_H (MeV)	864	871	—
Nucleon mass, M_N (MeV)	939*	927	939
Isobar mass, M_Δ (MeV)	1232*	1152	1232
Moment of inertia, $\mathcal{J}$ (fm)	—	1.32	1.01
Isoscalar r.m.s. radius, $\langle r^2 \rangle_B^{1/2}$ (fm)	0.59	0.61	0.79
Isoscalar g factor, $g_{I=0}$	1.12	0.89	1.76
Isovector g factor, $g_{I=1}$	6.38	8.36	9.40
Axial coupling constant, g_A	0.61	0.48	1.25
Pion-nucleon coupling constant, $g_{\pi NN}$	8.9	9.2	13.5

[a] The asterisks indicate that the nucleon and delta masses are used to fit the parameters F_π and e.

available, however, since the Skyrme theory is not renormalizable. (*b*) Both the axial-vector coupling constant g_A and the isovector magnetic moment are too small. We discuss below the fact that finite N_c corrections modify these values significantly and give better agreement to experiment (23, 24, 39) (Section 2.5). (*c*) The charge radius of the nucleon is also too small.[4] It is known that introduction of the vector mesons (the omega for the isoscalar radius and the rho for the isovector one) is necessary to explain the discrepancy (30).

2.5. *Finite* N_c *Corrections*

One of the difficulties with the semiclassical quantization is that the rotational ($I = J$) spectrum of the baryon contains infinitely many excitations with high spin and isospin. Most of these states, except for the nucleon and the delta isobar, are not observed experimentally, nor are they predicted in a simple quark model. It may be argued that the existence of those exotic states is an artifact of the Skyrme model in the limit $N_c \to \infty$. Indeed, the same exotic states appear in the quark model, if N_c is larger than 3.

Motivated by that observation, Amado et al (23–25) proposed a realization of a cutoff for the baryon spectrum. It is based on a dynamical group symmetry of the Skyrmion rotation. The rotational algebra of spin-isospin $SU(2)_J \times SU(2)_I$ is imbedded into a larger algebra U(4), and the symmetric (N_c) representation of U(4) is taken as the lowest energy band. The nucleon and its rotational excitations belong to this representation, and their spin and isospin are given by $I = J = 1/2, 3/2, \ldots, (N_c - 2)/2, N_c/2$. This simple implementation of finite N_c in the rotational spectrum has many interesting features: (*a*) The rotational spectrum has the upper bound $I = J = N_c/2$, which agrees with the quark model prediction. (*b*) It obeys Witten's rule for quantization of the Skyrmion (15), i.e. if N_c is odd (even), then the spectrum contains only half-odd (integer) spin and isospin. This rule is further confirmed in the SU(3) Skyrmion (Section 4.1). (*c*) One can evaluate a finite N_c correction for some operators associated with the rotation, such as $\sigma_i \tau_a$, which is modified by the factor $1 + 2/N_c$. The axial-vector coupling constant g_A and the isovector magnetic moment of the nucleon are therefore enhanced from the predictions in the semiclassical method by a factor 5/3 at finite $N_c = 3$. Similar finite N_c correction factors modify the pion-baryon coupling constants, the spin-dependent terms of the baryon-baryon interaction, the pion-baryon scattering ampli-

[4] The isovector charge radius, which diverges in the chiral limit, becomes finite when the pion mass is not zero.

tudes, and so on. (*d*) Those correction factors make some of the Skyrmion predictions coincide with those of the SU(4) quark model, such as the ratio of the πNN and $\pi N\Delta$ coupling constants. This may suggest a connection between the symmetry structure of the nonrelativistic quark model and the hedgehog structure of the Skyrmion (41).

The algebraic $1/N_c$ corrections considered here will not exhaust the entire $1/N_c$ effect. Indeed, field theoretical calculations of the quantum corrections to the Skyrmion yield systematic $1/N_c$ expansions of the Skyrme model (22). The algebraic method is analogous to the interacting boson model approach (42) to the collective rotation of a deformed nucleus, while the field theory approach corresponds to a microscopic calculation of the coupling of intrinsic motions to the rotation. We discuss the full field theoretical approach in the following sections.

2.6. *Pion Fluctuations*

Once the ground-state baryons are understood in the Skyrme model, one can go beyond the semiclassical approach and apply the model to pion-baryon system and excited baryon resonances. General quantization of the Skyrmion can be performed in two different ways: (*a*) quantization of field fluctuations around the classical solution under constraints (Dirac's method) (see, for example, 43), and (*b*) quantization of the whole field using sum rules of the field matrix elements [Kerman-Klein method (44, 45)].

The first is based on the semiclassical method as described in the previous section. Field fluctuations around the Skyrmion are introduced by

$$\phi^a(\mathbf{x}, t) = \phi_0^a(\mathbf{x} - \mathbf{R}) + \eta^a(\mathbf{x} - \mathbf{R}, t), \qquad 16.$$

where ϕ^a is the full field, $\phi_0{}^a$ the static Skyrmion, and $\mathbf{R} \equiv \mathbf{R}(t)$ is a collective coordinate for translational motions of the Skyrmion. Here we omit the collective rotations for simplicity. The general field fluctuation η is not totally independent of the collective coordinate, and therefore the canonical quantization is subject to a constraint and is carried out in a manner similar to that used by Dirac (46). When η is expanded in normal modes, the constraint requires the removal of the zero-frequency modes associated with the collective translational (and rotational) motions of the Skyrmion. Those motions are treated separately by the collective coordinates. The derived Hamiltonian describes collective motions and the coupling of nonzero modes to these collective motions. One can see that the Hamiltonian contains the Skyrmion static mass of $O(N_c)$, the pion single-particle Hamiltonian of

$O(N_c^0)$ with a potential produced by the background Skyrmion field, the rotational kinetic energy of $O(1/N_c)$, and the pion-Skyrmion couplings of higher order in $1/N_c$ (47).

Because the Skyrmion solution makes the static energy stationary against pion field fluctuations, the pion-Skyrmion coupling Hamiltonian does not have the linear (Yukawa) coupling that is expected, for instance, from the Goldberger-Treiman relation (48). (A Yukawa-type coupling of higher order in $1/N_c$ arises from the elimination of the zero modes.) It has, however, been pointed out that the π-N scattering amplitude for this Hamiltonian contains Born terms as a background (potential) scattering (49), and also that the N-N interaction has the correct one-pion-exchange potential (11). This puzzling behavior of the Skyrmion-pion Hamiltonian can be understood by distinguishing the Skyrmion from the point nucleon in the π-N field theory. The Skyrmion as a classical solution bears (static) pion clouds around a core source of pions. This is analogous to the Coulomb electric field around a point charge. The extra pion-Skyrmion coupling in the Hamiltonian yields nonstatic (recoil) effects of higher order in $1/N_c$, while the static part is taken care of by the lowest-order $1/N_c$ contributions. In a $1+1$ dimensional model, Ohta (50) showed that in fact the meson-soliton Hamiltonian derived using the collective coordinate method can be transformed into an ordinary meson-baryon Hamiltonian by a unitary transformation (51).

The second approach using the Kerman-Klein method is based on a double expansion in the coupling constant [of $O(1/N_c)$], and in the Skyrmion recoil momentum, P/M. Field matrix elements are calculated order by order using the equations of motion. It has been successfully applied to a $1+1$ dimensional scalar field theory (45). The soliton mass, for instance, is calculated including the recoil effect induced by emitting mesons in the intermediate state. It has also been applied to the full Skyrme model (52).

2.7. *Excited Baryons*

Excited baryon states in the Skyrme model can be studied as resonances in pion-Skyrmion scattering process (53). To the leading order in $1/N_c$ the pion-nucleon scattering is described by a background potential produced by the (classical) Skyrmion field configuration. This potential scattering yields an $O(N_c{}^0)$ amplitude. No bound state is found (53), and the scattering amplitudes show many resonances. Because of the $\mathbf{K} = \mathbf{I}+\mathbf{J}$ invariance of the Skyrmion, the scattering amplitudes are labeled by the quantum numbers, K, L (the initial orbital angular momentum), and L' (the final one). The semiclassical projection of the spin and the isospin of the Skyrmion yields πN (or $\pi\Delta$) scattering amplitudes. This method predicts remarkably simple relations among the πN scattering amplitudes in various channels. For instance, one can prove a relation like

$$T(D_{33}) = \frac{1}{10} T(D_{13}) + \frac{9}{10} T(D_{15}), \qquad 17.$$

for the $L = 2$ partial waves, where $T(L_{2I+1,\,2J+1})$ is the T matrix for the $L = L'$ wave scattering with the total isospin I and spin J. Such a relation can be tested against experiment and most relations agree very well, with a few exceptions: (*a*) For low partial waves, the relations do not hold well because they are derived without treating the translational and rotational zero modes of the Skyrmion correctly; and (*b*) the Δ^*-N^* isospin splittings, which are of $O(1/N_c)$, are not yet included. It is nevertheless surprising that the relations are valid even at scattering energies as high as about 1 GeV, because the Skyrme model is supposed to be valid only at low energies (although we do not know how low that is).

Attempts to calculate the π-N scattering amplitudes directly from the Skyrme Lagrangian have met with little success. Again, the correct elimination of the zero modes is essential for the lower partial waves. A more serious difficulty is that the predicted inelasticity is usually too small. This could again be attributed to an artifact of the large N_c limit of the Skyrme model (54).

3. TWO-BARYON SYSTEMS

3.1. *The Product Ansatz and its Problems*

The principal goal of studying the $B = 2$ sector of the Skyrme model is to apply it to the nucleon-nucleon (NN) interaction. Characteristic features of the nuclear force shared by various semiphenomenological potentials (55–57) are as follows: (*a*) a long-range ($\gtrsim 1.4$ fm) interaction dominated by the one-pion-exchange potential (OPEP), (*b*) a medium-range (~ 1 fm) attraction in the central force, and (*c*) a strong short-range repulsion ($\lesssim 0.7$ fm). The OPEP is well established, while the microscopic origins of the medium- and short-range interactions are less well understood.

A simple study has already been made by Skyrme himself, which surprisingly contains many essential features, at least qualitatively. He proved that a matrix product of two $B = 1$ configurations, $U = U_1 \cdot U_2$, belongs to the $B = 2$ sector. He then proposed the "product ansatz'' for two Skyrmions separated by **R** and individually iso-rotated by A and B:

$$U_{\mathrm{PA}}(x) = AU_{\mathrm{H}}\left(\mathbf{x} + \frac{\mathbf{R}}{2}\right) A^{\dagger}BU_{\mathrm{H}}\left(\mathbf{x} - \frac{\mathbf{R}}{2}\right) B^{\dagger}. \quad 18.$$

The classical energy is calculated using this ansatz, giving

$$E_{B=2} = 2E_{\mathrm{H}} + V(\mathbf{R},C), \quad 19.$$

where the static potential V depends on the relative distance **R** and the relative orientation $C \equiv A^{\dagger}B$. A global rotation of U does not change the total energy. Skyrme pointed out the following two facts. (*a*) For large $R = |\mathbf{R}|$, $V(\mathbf{R}, C)$ approaches the one-pion-exchange potential (OPEP):

$$V(\mathbf{R}, C) \sim T_{ab}\nabla_a\nabla_b\left[\frac{\exp(-m_{\pi}R)}{R}\right] \quad 20.$$

with

$$T_{ab} = \frac{1}{2}\,\mathrm{tr}[\tau_a C\tau_b C^{\dagger}]. \quad 21.$$

(The Yukawa function in Equation 20 is replaced by $1/R$ for the massless pion.) This is natural, since the Skyrmion contains a one-pion cloud in the tail region. (*b*) At $R = 0$ and for $A = B = 1$, the configuration U_{PA} takes the form of the hedgehog with $B = 2$ [$F(0) = 2\pi$ and $F(\infty) = 0$ in Equation 6], and its energy is about 1 GeV higher than the two-Skyrmion threshold. This excess of energy at $R = 0$ suggests a short-range repulsion between two nucleons.

More quantitative analyses using the product ansatz have been performed by several authors (58–60), and the precise form of the potential $V(\mathbf{R}, C)$ was worked out. It is found that $V(\mathbf{R}, C)$ depends strongly on the relative orientation C. For instance, the lowest energy configuration is obtained by choosing $C = i\tau_2$ for the separation **R** parallel to the x axis. This represents a system of a hedgehog and a rotated hedgehog by 180 degrees around the axis perpendicular to the line joining the two.

The NN interaction is then obtained by the semiclassical method from $V(\mathbf{R},C)$. Each Skyrmion is rotated by the collective coordinate A or B, and the relevant spin-isospin component is extracted (Section 3.3). One finds that the NN potential consists of three terms, given by

$$V_{\mathrm{NN}}(\mathbf{R}) = V_0(R) + (\sigma_1\cdot\sigma_2)(\tau_1\cdot\tau_2)V_{\sigma}(R) + (\tau_1\cdot\tau_2)S_{12}(\hat{\mathbf{R}})V_{\mathrm{T}}(R), \quad 22.$$

where S_{12} is the standard tensor operator for nucleons:

$$S_{12}(\hat{\mathbf{R}}) = 3(\boldsymbol{\sigma}_1 \cdot \hat{\mathbf{R}})(\boldsymbol{\sigma}_2 \cdot \hat{\mathbf{R}}) - (\boldsymbol{\sigma}_1 \cdot \boldsymbol{\sigma}_2). \tag{23}$$

The last two terms in Equation 22 tend asymptotically to the OPEP, as expected, while the first term has a shorter range. Unfortunately, it is found that the spin-isospin-independent part of the central force V_0 is strongly repulsive at medium distances (see Figure 3*a*) so that V_{NN} in Equation 22 does not account for the nuclear binding.

This raised many questions concerning the Skyrme model calculation of the nuclear force. Some argue that the Skyrme model may be too simple to account for the details (61, 62). We, however, would like to emphasize that present calculations still contain many shortcomings, mostly related to technical difficulties in dealing with nonlinearity of the model.

First, the validity of the product ansatz has been questioned when the two Skyrmions are close together. It is implausible that the Skyrmions do not deform at all under the interaction (63). It has also been pointed out that the product ansatz violates some of the reflection symmetries that the true solution is expected to preserve (64, 65). For instance, the product ansatz for the two-hedgehog system ($C = 1$) breaks the reflection symmetry in the *x*-*z* plane:

$$\boldsymbol{\tau} \cdot \boldsymbol{\phi}(x, -y, z) = -\tau_2 \boldsymbol{\tau} \cdot \boldsymbol{\phi}(x, y, z) \tau_2. \tag{24}$$

This symmetry, however, holds for infinitely separated hedgehogs. The asymmetry in Equation 24 can be traced back to the asymmetry of the product $U_1 U_2 \neq U_2 U_1$. Indeed, we expect that $U_2 U_1$ represents the same two-Skyrmion system as $U_1 U_2$. Extra components of the pion field, which disappear at $R \rightarrow \infty$, are responsible for this asymmetry, and also for the broken reflection symmetries. This suggests that a suitably symmetrized configuration would have a lower energy.

The second difficulty is that the product ansatz cannot describe the lowest energy configuration with $B = 2$. It is believed, although not rigorously proved, that the lowest energy $B = 2$ configuration is axially symmetric (66). Indeed, numerical studies have shown that it has a torus-shaped distribution of the baryon number density and is tightly bound to about 70 MeV of the two-Skyrmion threshold (66–70). The binding energy is about four times larger than that obtained with the product ansatz, and the size of the torus (about 0.5 fm in radius) is rather small. In this situation, the identity of individual Skyrmions is completely lost. Thus, we need a prescription beyond the product ansatz for short and medium ranges incorporating the torus configuration.

We should also note that in the presence of interactions between Skyrmions accompanied by strong deformations in the field configuration, the choice of relevant collective coordinates becomes somewhat ambiguous (71, 72). According to Manton (71), this is a problem of finding an unstable collective manifold for the two-Skyrmion dynamics at low energies. Major progress has been made in the past few years in finding such a manifold without relying on the product ansatz. Some ambiguity in choosing collective coordinates, particularly at medium and short distances, may be removed by calculating the corresponding inertial masses and by taking the kinetic energy term into account (Section 3.5).

Finally, one has to take quantum effects into account. In the standard approach to the nuclear force, the medium-range attraction comes mainly from a two-pion exchange, e.g. a box diagram with delta (Δ) intermediate states (73). This is an effect that a classical calculation cannot provide. We present some attempts of calculating the quantum effects in Section 3.4.

3.2. *Skyrmion-Skyrmion Interaction*

Classical configurations beyond the product ansatz and the resulting Skyrmion-Skyrmion (SS) interaction were extensively studied numerically. Verbaarschot et al studied the lowest energy configuration at fixed Skyrmion separation (64). Asymptotically, their configuration reduces to that of well-separated Skyrmions interacting through the OPEP (Equation 20) in the most attractive way (corresponding to $C = i\tau_2$ in the product ansatz). As the Skyrmions get closer, the exact solution acquires a greater binding energy than predicted in the product ansatz. The minimum energy then occurs at a rather short distance, where the solution becomes the axially symmetric torus whose symmetry axis is perpendicular to the line joining the two Skyrmions. The channel that interpolates the two well-separated Skyrmions and the strongly bound torus is called the attractive channel.

In order to calculate the NN interaction using the semiclassical method, we need the Skyrmion-Skyrmion interaction with various relative orientations C. Direct numerical calculations of the potential $V(\mathbf{R}, C)$ at an arbitrary C are not available. There are, however, two more cases suited to numerical simulation, in which the U field respects reflection symmetries analogous to Equation 24. The first one corresponds to $C = i\tau_1$, where one of the two Skyrmions is rotated by 180 degrees about the joining axis (x axis). Asymptotically, the two Skyrmions interact through the OPEP in the most repulsive way. Therefore, this channel is called the repulsive channel. The second one is for C

$= 1$, where the two Skyrmions are unrotated hedgehogs: the hedgehog-hedgehog channel. Asymptotically, the OPEP in this channel is weakly repulsive (for the massless pion, the OPEP vanishes).

Walhout et al carried out a direct numerical calculation for these channels (74). The NN interactions were then obtained by the semi-classical projection method (see the next subsection). They have found that the repulsion in the spin-isospin independent part of the central force is significantly reduced compared to results for the product ansatz calculations. A similar result has been obtained by another less numerical approach, solving the quasi-static equation at a fixed R (75, 76).

A full calculation is not yet available for the potential energy V and the inertial masses M and $\mathcal{I}$ in Equation 30 for arbitrary C and $\mathbf{R}$. This is the place where the Atiyah-Manton (AM) method would be most powerful, as the AM ansatz contains a set of parameters that can be regarded as collective coordinates appropriate for the low-energy dynamics of the $B = 2$ system. To be specific, let us take the instanton solution by Jackiw, Nohl & Rebbi (JNR) (77). For winding number $n = B = 2$, the superpotential consists of three pole terms:

$$\rho(x) = \frac{\lambda_1^2}{(x - X_1)^2} + \frac{\lambda_2^2}{(x - X_2)^2} + \frac{\lambda_3^2}{(x - X_3)^2}. \qquad 25.$$

Atiyah & Manton argue that the resulting U field describes various two-Skyrmion configurations such as well-separated Skyrmions with arbitrary orientations, and the torus (35). Indeed, it contains 15 independent parameters, which can play the role of collective coordinates.[5] They account for positions, orientations, and sizes of the two Skyrmions asymptotically, while for the torus they represent the symmetric axis and the size.

The relation between the instanton configurations (Equation 25) and the corresponding Skyrmion fields has been worked out (35, 36). Here we present just one example, which describes the attractive channel, including the torus configuration. Suppose that the three poles in Equa-

[5] Although the function $\rho(x)$ in Equation 25 contains 15 parameters for positions of poles and their weights, a precise counting of the 15 parameters in the U field is slightly intricate. First, we have to add three additional degrees of freedom for global rotations of the A_4 field. Then, we have to subtract the following three redundant degrees: (*a*) one of the weight parameters because the logarithmic derivative in Equation 11 depends only on their relative magnitudes; (*b*) the origin of the time coordinate; and (*c*) a parameter associated with the gauge symmetry of the A_μ field. Because of the scale invariance of the instanton, one of the parameters always represents the overall scale, which can be determined by minimizing the energy of the Skyrmion system.

tion 25 form an isosceles triangle in the x-y ($z = 0$, $t = 0$) plane. We take the pole X_3 on the symmetry axis of the isosceles triangle, chosen as $x = 0$, and choose the weight parameters as $\lambda_1 = \lambda_2 = 1$, $\lambda_3 = \lambda$. Consider a continuous change in λ, accompanied by a deformation of the triangle from an ortho-triangle at $\lambda \to \infty$ to an equilateral triangle at $\lambda = 1$. For $1 < \lambda < \infty$, the forms of the triangles are specified by λ (35, 36). We find that the ortho-triangle ($\lambda \to \infty$) yields two well-separated Skyrmions centered at X_1 and X_2, respectively. On the other hand, the equilateral triangle generates the torus with its symmetry axis z, passing normally through the center of the triangle. In between, the symmetry of the isosceles triangle makes the Skyrmion configuration symmetric with the desired reflection symmetries. In Figure 1, contour plots of the baryon number density are shown for several λ. The deformation occurs so as to interpolate the two separated Skyrmions and the torus.

A calculation of the total energies has been carried out for the attractive (ATR), hedgehog-hedgehog (HH), and repulsive (REP) chan-

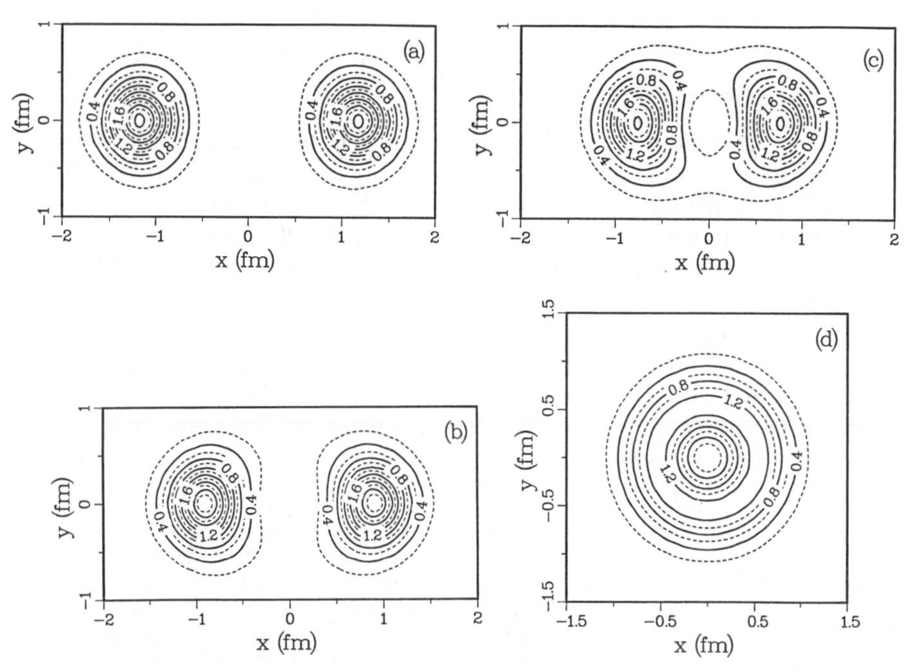

Figure 1 Contour plots for baryon number density (1/fm^3) in the $z = 0$ plane for the most attractive configuration. Skyrmion separations are (*a*) 2.25 fm, (*b*) 1.68 fm, (*c*) 1.35 fm, and (*d*) the torus.

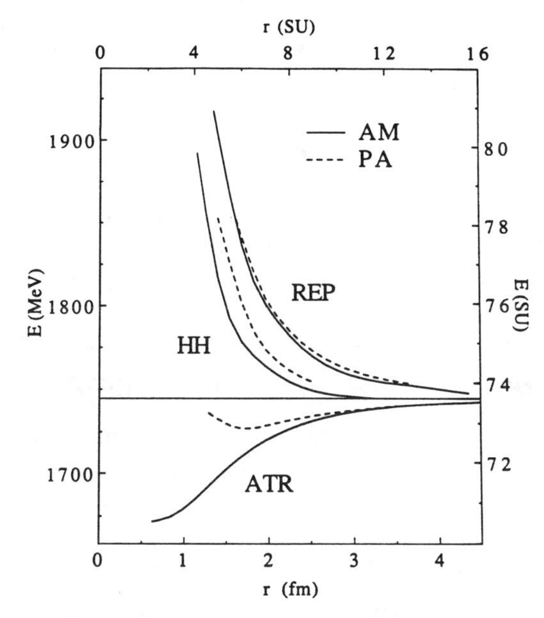

Figure 2 The energy of the $B = 2$ system as a function of Skyrmion separation for various configurations in the Atiyah-Manton (AM) ansatz as compared with the corresponding product ansatz (PA) results. ATR = the attractive channel, REP = the repulsive channel, and HH = the hedgehog-hedgehog channel.

nels using similar pole configurations. The results are plotted in Figure 2, as compared with the product ansatz (PA) calculations. As is expected, the AM and the PA results coincide at large r, and both the potentials approach the OPEP ($\sim 1/r^3$ for the massless pion). As r is decreased, however, the AM configuration yields a lower energy than the PA in all the channels. In the attractive channel, the torus that has the lowest energy cannot be described by the product ansatz, and therefore the potential for the product ansatz deviates markedly at short distances. Symmetrization of the two Skyrmions also lowers the energy significantly, which is seen in the HH channel. The repulsive channel potential does not change very much, since this interaction is dominated by the OPEP.

3.3. *Nucleon-Nucleon Interaction*

Quantization of the two-Skyrmion system is less straightforward than the $B = 1$ case. The relevant collective motions are the individual isorotations and the relative translational motions. When the Skyrmions are far apart and therefore the interaction is small, we may assume that the linear relative momentum is sufficiently small. We cannot, however, assume the same for the angular momenta of the Skyrmions, as they are quantized discretely. We therefore treat them separately, by regarding the rotational motion as a faster motion than the translational one. Then taking the adiabatic limit for the translational motion, the rotation can be quantized at a fixed separation R. In the product ansatz, collective coordinates for the rotational and translational motions are explicitly given. In the AM construction, it is also possible to identify a set of collective coordinates among the 15 instanton parameters.

It is convenient to parametrize the SS potential at a separation $\mathbf{R}$ as

$$V_{SS}(\mathbf{R}) = V_1 + V_2 W + V_3 Z + V_4 W^2 + V_5 WZ + V_6 Z^2 + \ldots, \qquad 26.$$

where the V_i's depend only on $R = |\mathbf{R}|$, and the angular variables are defined by

$$W = T_{ai}(A) T_{ai}(B)$$

$$Z = 3T_{ai}(A) T_{aj}(B) \hat{R}_i \hat{R}_j - T_{ai}(A) T_{ai}(B), \qquad 27.$$

with T_{ai} as in Equation 21, and with A and B representing rotations of the two Skyrmions. The expansion in Equation 26 in principle contains an infinite number of terms. In the product ansatz, however, it is known that the expansion terminates at the first six terms in Equation 26, and further that the latter three (V_4, V_5, and V_6) are much smaller than the first three (24). Although this property may not persist at small R in the AM ansatz, it seems reasonable to assume the dominance of the first three terms in Equation 26. Then the three terms V_1, V_2, and V_3 can be calculated from the SS interactions with three independent orientations C.

In quantizing the Skyrmion rotations, W and Z are regarded as operators acting on nucleon wave functions represented by the collective coordinates A and B. Taking matrix elements between two nucleon states, they are effectively replaced by the Pauli matrices:

$$W \rightarrow \frac{1}{9} (\boldsymbol{\sigma}_1 \cdot \boldsymbol{\sigma}_2)(\boldsymbol{\tau}_1 \cdot \boldsymbol{\tau}_2),$$

$$Z \rightarrow \frac{1}{9} (3\boldsymbol{\sigma}_1 \cdot \hat{\mathbf{R}} \boldsymbol{\sigma}_2 \cdot \hat{\mathbf{R}} - \boldsymbol{\sigma}_1 \cdot \boldsymbol{\sigma}_2)(\boldsymbol{\tau}_1 \cdot \boldsymbol{\tau}_2). \qquad 28.$$

Thus the components in the NN interaction (Equation 22) are related to the components in the SS interaction (Equation 26) by

$$V_0 = V_1, \qquad V_\sigma = \frac{1}{9} V_2, \qquad V_T = \frac{1}{9} V_3. \qquad 29.$$

The factor 1/9 obtained in the semiclassical method is enhanced by the finite N_c correction factor, $(1+2/N_c)^2$, in the algebraic method (Section 2.5). This enhancement is important, especially in the N-Δ transition potential (Section 3.4). Figure 3 shows the NN potential (without the finite N_c correction) obtained both for the product and the AM ansatz, where the central ($V_1 = V_0$), spin-spin ($V_2 = 9V_\sigma$), and tensor ($V_3 = 9V_\tau$) components are separately shown.[6] The change in V_{ATR} and V_{HH} in the SS interaction from the product ansatz to the AM ansatz is reflected almost exclusively in the central channel of the NN interaction, V_1. The repulsive central force in the product ansatz calculation is substantially reduced, although the reduction is not sufficient to give a net attraction, which is necessary for nuclear binding.

We conclude that the semiclassical method for the two-Skyrmion system does not reproduce the medium-range NN attraction. The situation is not improved much when we consider other possible forms of the Skyrme Lagrangian, or change the choices of the parameter values (61, 63, 78). For instance, a symmetric term of fourth-order derivative yields a small attraction for a small value of the coupling constant, while a larger coupling destroys the stability of the single Skyrmion (61). Other models include a scalar sigma meson explicitly with a finite mass, in which case a medium-range attraction is obtained (62). This, however, does not seem realistic because the sigma scalar is a fictitious meson mimicking the multi(dominantly two)-pion-exchange force (57).

3.4. *Quantum Corrections*

Our goal is to go beyond the semiclassical approximation and to explain the nuclear attraction within the original Skyrme model. Quantum effects will emerge as higher $1/N_c$ contributions. There is an effect known to be important in the conventional meson exchange NN interaction, namely the multi-pion exchanges with delta excitations in intermediate states (73). Because the nucleon-delta energy difference originates from the Skyrmion rotation and thus is of order $1/N_c$, the classical calculation cannot incorporate this effect properly.

The N-Δ mass difference is generated only by rotational kinetic energy. The recoil of the Skyrmion due to translations and rotations, as well as quantized pion exchanges, must be considered. This problem is related to the difficulty in describing the pion-nucleon scattering and the excited baryon states. A consistent treatment of the collective mo-

[6] We do not compare the results here with a phenomenological potential, since the AM method generates Skyrmion configurations with a massless pion.

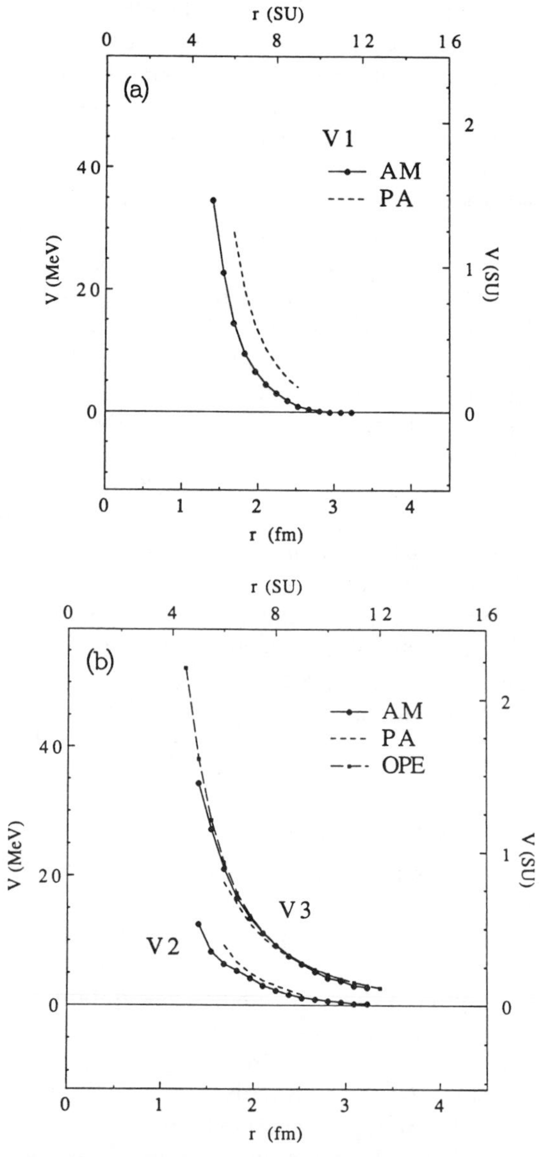

Figure 3 Various components of the nucleon-nucleon interaction in the AM ansatz as functions of nucleon separation: (*a*) the spin- and isospin-independent part of the central force V_1, and (*b*) the spin-spin-dependent part V_2 and tensor part V_3 as compared with the product ansatz results (PA). V_2 and V_3 are also compared with the one-pion-exchange (OPE) potential for the massless pion. The potentials are shown in both physical and Skyrmion units. They are related by $r_{SU} = eF_\pi r_{phys}$ and $V_{SU} = (e/F_\pi)V_{phys}$.

tions of the Skyrmion and quantum pion dynamics is indispensable for further progress.

Although a systematic calculation of the higher order terms in the $1/N_c$ expansion is desirable, a qualitative evaluation of this effect is possible within the adiabatic approach.

Under the Born-Oppenheimer approximation, the separation R is regarded as a slow variable, while the mixing of delta may happen quickly. The SS interaction (Equation 26) mixes $|N\Delta\rangle$ and $|\Delta\Delta\rangle$ components to the lowest energy $|NN\rangle$ state at each R. The problem is then a diagonalization of a 3×3 matrix whose components are $\langle NN|H|NN\rangle$,$\langle N\Delta|H|NN\rangle$, etc. Here the Hamiltonian H consists of the baryon masses (M_N+M_N, M_N+M_Δ, etc.) and the interaction $V(\mathbf{R})$ in Equation 26. Note that H contains the N-Δ mass splitting of order $1/N_c$ besides the static contribution of order N_c. Thus the lowest energy state mixes with NΔ and $\Delta\Delta$ components and the total energy is lowered. This effect was previously computed for the product ansatz and was found not to be so significant (79). However, a recent evaluation by Walet et al (80) uses an improved configuration for the $B = 2$ Skyrmion and exhibits a substantial medium-range attraction that qualitatively agrees with the phenomenological nuclear force. It turns out to be crucial to take into account the algebraic finite N_c correction in the matrix elements of H to obtain sufficiently large attraction.

Thus we find that the Skyrme model seems to account for (the static part of) the nucleon-nucleon interaction at long and medium ranges (~1 fm). It should be emphasized that both the non-product-ansatz configuration and the inclusion of Δ intermediate states with appropriate finite N_c corrections are equally important in reaching this result.

3.5. *Nonadiabatic Interactions*

The theory of interacting Skyrmions is not yet complete. Even if the static part of Skyrmion-Skyrmion interaction is calculated, there is no guarantee that the dynamics is adiabatic. In fact, a fully time-dependent (classical) calculation, which deals with the scattering of two Skyrmions, reveals nonadiabatic effects.[7] Using the collective coordinate approach, a systematic way of introducing the nonadiabatic effect starts by including a velocity-dependent force. The quadratic time derivative term of the Lagrangian gives

$$\mathcal{L}_{\mathrm{kin}} = \frac{1}{2} M(R, C)\dot{R}^2 + \frac{1}{2} \mathcal{I}(R, C)\Omega^2, \qquad 30.$$

[7] Also, radiation of pions occurs at sufficiently high energies ($E_{CM} > 150$ MeV) (81, 82). Here, we consider only the quantization of the collective coordinates at low energies.

where $\dot{R}$ (Ω) is the linear (angular) velocity. (There are no higher order time derivatives in the standard Skyrme Lagrangian.) The inertial mass M for the relative motion (R) and the moment of inertia $\mathcal{I}$ for the relative rotation (C) in general depend on both R and C, and thus the kinetic energy for the relative motion in the Hamiltonian contains

$$K = \frac{1}{2M} P^2 = \frac{1}{2M_0} P^2 + \left(\frac{1}{2M} - \frac{1}{2M_0}\right) P^2. \qquad 31.$$

The last term of K is to be interpreted as a velocity-dependent interaction. The effect of this nonadiabatic interaction can be studied by solving the Schrödinger equation with an R- and C-dependent inertial mass (although one always faces an ordering ambiguity in the velocity-dependent term). This was done for the product ansatz and the effect was found to be attractive but not large (83–85). The inertial mass grows at short distance. There is, however, no complete analysis including both the translational and rotational degrees of freedom even with the product ansatz.

For nonproduct configurations, not much study has been done along this line. In the AM ansatz, the R-dependence of $M(R)$ looks quite different from that in the product ansatz for some channels. That difference is an interesting question for the future.

4. STRANGENESS IN THE SKYRME MODEL

4.1. *The SU(3) Skyrmion*

An important extension of the original Skyrme model is in the direction of increasing the number of flavors involved. For arbitrary flavor N_f, chiral symmetry is $SU(N_f)_L \times SU(N_f)_R$, which is broken into $SU(N_f)_V$ by the $\langle\sigma\rangle$ condensation leading to $(N_f^2 - 1)$ pseudoscalar Nambu-Goldstone bosons. Clearly, $N_f = 3$ barely fits into this picture but $N_f \geq 4$ certainly does not because the chiral symmetry will not hold for the charm and heavier quarks at all. Even for $N_f = 3$ the chiral symmetry is significantly broken by the quark mass term for the strangeness. Nevertheless the study of the SU(3) Skyrme model sheds light on new important qualitative features of the Skyrmion.

The most significant of all is the existence of the Wess-Zumino-Witten (WZW) term (15), which vanishes (unfortunately) for SU(2) in the effective action. The WZW term is required in order to break an extra symmetry that the Skyrme Lagrangian possesses. Indeed the Lagrangian in Equation 4 is invariant separately under

$P_1 : \mathbf{x} \rightarrow -\mathbf{x}$

and

$P_2 : U \rightarrow U^\dagger$,

which are combined into the parity transformation $P = P_1 P_2$. In QCD, however, these symmetries, P_1 and P_2, are broken in processes like $K^+K^- \rightarrow \pi^+\pi^-\pi^0$. The WZW action, which breaks these symmetries but leaves the combined parity P invariant, is given by a five-dimensional integral over a disk Q whose boundary is the four-dimensional space-time,

$$S_{WZW} = -\frac{iN_c}{240\pi^2} \int_Q d^5x \epsilon^{\alpha\beta\gamma\delta\epsilon} \operatorname{tr} L_\alpha L_\beta L_\gamma L_\delta L_\epsilon . \qquad 32.$$

The coefficient in front of the integral, proportional to N_c, is determined by computing electromagnetic processes such as $\pi^0 \rightarrow 2\gamma$ for this action with an additional gauge-invariant coupling to the electromagnetic field and comparing it with the direct QCD calculation.

The explicit N_c dependence of the WZW action leads to various important results for the properties of the Skyrmion. First, it restricts the spin and flavor quantum numbers of the (semiclassically) quantized Skyrmion. Witten showed, by applying an adiabatic rotation to the Skyrmion, that for odd (even) N_c, the quantized Skyrmion has a half-odd (integer) spin and is thus a fermion (boson). It has also been shown that the baryon spectrum consists of the SU(3) representations that contain a state with hypercharge $Y = N_c/3$ (15, 26, 86). For $N_c = 3$ (the real world) these representations agree with the observed baryon spectrum, although the semiclassical quantization predicts a large number of extra baryon excitations with higher spin, isospin, and strangeness. Again the algebraic finite N_c implementation eliminates those unobserved baryons (25).

The importance of the $SU(3)_f$-breaking effect has been noticed by several authors. Two extreme approaches are (*a*) applying a perturbation method in which the symmetry breaking is evaluated by the first-order perturbation (87), and (*b*) using the kaon bound state method in which the strange sector is treated separately from the SU(2) Skyrmion (88). The perturbation fails to reproduce the spectrum of the ground-state baryons, and therefore the SU(3)-breaking effect is considered to be large (25, 89).

The way that the SU(3) is broken is not unique. Not only the mass splitting among the octet pseudoscalar mesons but also the difference in the decay constants f_m contribute to the breaking. Riska & Scoccola showed that the strange baryon spectrum can be better reproduced by including a new SU(3)-breaking term, which originates the breaking by the decay constants (90).

4.2. *The SO(3) Skyrmion*

A special stable solution with $B = 2$ is known for the SU(3) Skyrme model (26). The solution is spherical under a simultaneous rotation in the ordinary three-dimensional space and the space of an SO(3) subgroup of SU(3), generated by λ_7, $-\lambda_5$, and λ_2. This solution is especially interesting because it describes a dibaryon H, with strangeness -2. The dibaryon H is a possible bound state of two Λ's originally predicted in the quark model (91). The one-gluon-exchange interaction among valence quarks, especially its magnetic part, is a major driving force for the bound state. It is extremely interesting to see the same states occur as the lowest energy $B = 2$ state in the Skyrme model. Is this accidental? We do not know the answer yet.

The energy of H has been calculated by various groups (26, 92); and it is found that the H dibaryon appears below the ΛΛ threshold, i.e. it is a bound state that is stable against the strong interaction. This result, however, could be sensitive to the choice of the SU(3)-breaking mechanism.

5. MULTIBARYON SYSTEM

5.1. *Light Nuclei*

In Section 3 we studied the two-Skyrmion system with an emphasis on the NN interaction. One might apply the two-body interaction to multinucleon systems, nuclei, using the conventional method in nuclear physics (e.g. 93). In the Skyrme model, however, a quite different approach is also available, which is based on a solution to the classical field equation. Development along this line owes a lot to the discovery of new classical solutions for $B = 2, \ldots, 6$ by Braaten et al (94). Numerically they found classical solutions with the characteristic geometric shapes of the baryon number density distribution. For $B = 1$, 2, 3, and 4, these are the sphere, the torus, the tetrahedron, and the octahedron, respectively. They also found that the field configurations have discrete symmetries associated with their shapes.

The classical solutions must be quantized to identify them as physical nuclei. For $B > 1$, however, a difficulty arises from the presence of internal motions of individual Skyrmions. A fully consistent treatment including both the exact zero modes of global Skyrmion rotations and the internal motions has not yet been achieved, even for the simplest system, the deuteron. Nevertheless, it is interesting to see that quantization of the exact zero modes alone predicts the right quantum numbers for some light nuclei (69, 95, 96) In this case the symmetries of the field configuration play a crucial role. They restrict the possible spectrum of the nuclei through a constraint according to Finkelstein et al (97). It has then been shown that the lowest states for the $B = 2$ and 3 Skyrmion carry the quantum numbers of the deuteron and ^{3}H.

The quantized objects do not seem quite realistic, however, when physical observables are calculated. For instance the binding energies are too large; the rotating torus and tetrahedron yield binding energies of 158 and 350 MeV respectively, as compared with the empirical values of 2.2 and 8.3 MeV. Also, the sizes of the objects are too small (69, 96). An apparent contradiction arises because the motions of individual Skyrmions are frozen. A simple calculation for the deuteron including the relative translational motion was performed by Walhout (98). An associated zero-point energy is not negligible and must be added to the potential energy given above. The total binding energies are pushed up almost to the threshold. Also, the oscillation of the relative motion causes a large mixing of higher energy configurations into the wave function. This makes the size of the deuteron significantly larger than that of the classical solution. Thus, although the classical solutions for $B > 1$ look quite exotic, it appears that a consistent treatment of both the global rotations and internal motions provides us with a picture of nuclei not too far from the conventional one.

5.2. *Skyrmion Matter*

As the baryon number is increased, the method of the previous section becomes extremely complicated. For an infinite B, however, a great deal of simplification would be possible. Considering an array of Skyrmions with suitable (periodic) boundary conditions, one only has to solve the problem in a unit volume containing a single Skyrmion. So far, most calculations have been restricted to the classical level.[8] Yet interesting properties of Skyrmion matter have been revealed, particularly in high density regions.

In his pioneering work, Klebanov considered a simple cubic lattice (100), where a Skyrmion is placed at each lattice point. The boundary conditions were chosen such that the six pairs in the nearest neighbors were arranged in the most attractive orientations (Section 3.2). He calculated the energy per Skyrmion as a function of the lattice spacing

[8] Walhout (99) quantized a unit cell Skyrmion and discussed the equation of state for neutron matter.

a. At $a \sim (5.5/F_\pi e)$, the minimum energy is $32.0F_\pi/e$, which is only about 8% higher than the topological lower bound in Equation 9.

Although implicit in Klebanov's work, a remarkable fact was realized in the subsequent work by Wüst et al, who explicitly showed contour plots of the baryon number density at various values of a (101). When a is large, nothing special happens: the unit baryon number is localized near a lattice point, which looks like a single Skyrmion. As a is decreased, however, the baryon number density gradually spreads out, and for $a \leq (8.2/F_\pi e)$, the distribution seems to acquire a new symmetry: an array of half Skyrmions in a Wigner-Seitz cell of half volume of the unit cube (102). Arrays of half Skyrmions at high densities (small a) were further studied by several authors (103). It was found that the face-centered array is energetically preferred to the simple cubic lattice and that the minimum energy is only 4% higher than the lower bound, Equation 9.

The transition from a single Skyrmion array to a half Skyrmion array is of second order and seems to be related to the restoration of chiral symmetry. Naively the realization of chiral symmetry is dictated by the radius of the chiral circle of the linear sigma model. Since the Skyrme model is based on the nonlinear sigma model, where the chiral fields are subject to the constraint

$$\sigma^2 + \boldsymbol{\phi}^2 = f_\pi^2, \tag{33}$$

where $f_\pi = F_\pi/2$, the phase transition does not seem to occur in the model unless $f_\pi \to 0$.[9]

A plausible prescription proposed by Forkel et al (105) is to replace the quantities in Equation 33 with their mean field values averaged over the entire space:

$$\langle\sigma\rangle^2 + \langle\boldsymbol{\phi}\rangle^2 = \bar{f}_\pi^{\,2}, \tag{34}$$

and identify $\bar{f}_\pi$ as an effective pion decay constant at finite density. The resulting $\bar{f}_\pi$ indeed vanishes in the half Skyrmion phase, while as the density is decreased, it reduces to the free space value. Similar evidence of the chiral phase transition is observed for the Skyrmion on the hypersphere S^3.

5.3. *The Skyrmion on the Hypersphere*

The Skyrmion on the hypersphere was initially studied in the search for a solution satisfying the topological lower bound in Equation 9 (106). Since $R^3 \sim S^3$, it is natural to consider the mapping $U(\mathbf{x})$ from the

[9] Indeed, the density dependence of the pion decay constant f_π could be introduced in the Skyrme model through the scale invariance of QCD (104).

coordinate space of $S^3(L)$ to the functional space of $S^3(L = 1) \sim SU(2)$, where L is the radius of the sphere S^3. Then R^3 may be regarded as $S(L \to \infty)$.

When L is sufficiently large, the $B = 1$ Skyrmion on $S^3(L)$ has a localized baryon density around a pole of S^3, which is nothing but the Skyrmion in R^3. As L is decreased, the baryon density spreads out, and for $L \leq \sqrt{2}$, it distributes uniformly over the S^3. The field configuration here is the uniform map in which the chiral fields sweeps the S^3 in the same way as the coordinate variables do. The energy of the uniform map scales as

$$E = \frac{3\pi^2}{2}\left(L + \frac{1}{L}\right)\frac{F_\pi}{e}. \qquad 35.$$

The minimum occurs at $L = 1$ with $E = 3\pi^2 F_\pi/e$, saturating the lower bound of Equation 9. For small $L \leq \sqrt{2}$, the uniform map is stable, while for $L \geq \sqrt{2}$, a localized solution is energetically favorable.

The description here sounds quite analogous to the one presented in the previous section, but with L replaced by the lattice spacing a (with an appropriate scale). Indeed there are number of suggestions that the Skyrmion on S^3 could provide a simple model for high density baryonic matter (107). First, if one suitably relates the radius of S^3 to the density of the matter, the energy curve as a function of the density remarkably resembles the one obtained from lattice calculations. Second, the effective pion decay constant $\bar{f}_\pi$, calculated in S^3 as in Equation 34, behaves very much like $\bar{f}_\pi$ from the array calculations. Third, it is possible to calculate explicitly the masses of various symmetry modes. A detailed investigation shows that for $L \leq \sqrt{2}$ there are six massive degenerate modes associated with the chirally symmetric phase, while three of them become massless for $L \geq \sqrt{2}$ (108). These three indeed correspond to the massless pions. Thus at the point $L = \sqrt{2}$, the chiral phase transition occurs. The spread of the baryon density distribution has also been interpreted as deconfinement of the quarks (107, 109).

6. CONCLUSION

The Skyrme model has provided us with a novel picture of the nucleon and other low-lying baryons. The baryon number conservation, conventionally understood as the conservation of the quark baryonic current, is realized as the topological stability of an extended meson field configuration. This new picture fits the historical and current understanding of the low-energy hadron dynamics, supporting chiral symmetry, the low-energy theorems, the meson-exchange potentials, and the exchange currents. Especially from the nuclear physics point of view, the essential long-range dynamics seems to be fairly well represented in the Skyrme model. This is an advantage of the model over models such as the quark model, which supposedly describes the short-distance dynamics efficiently. It is extremely encouraging that we now have a qualitative understanding of the two-nucleon system in the Skyrme model.

A question raised at this point is whether the model should be applied to more than few-nucleon systems. In nuclear physics, the conventional wisdom tells us that many-body forces are not important; the two-body force is usually sufficient in the study of nuclear properties. This conventional wisdom is supported by a consistent p/m expansion in the chiral perturbation theory (110). The Skyrme model predicts, however, interesting shape structures for a few- to many-body Skyrmion systems. Are they real? Or do quantum corrections destroy the structure seen in the classical solutions? Even for the $B = 2$ system, this seems likely because the torus structure does not seem to be directly related to the deuteron (98). We still need a careful evaluation of quantum effects.

Although we already have a theory for the strong interaction, quantum chromodynamics (QCD), it is not yet fully controlled in the low-energy regime. The confinement of color makes a quantitative treatment highly nontrivial, and yet effects of local color inside the hadron do not manifest themselves much in the low-energy hadron physics. Naturally, effective theory approaches, with only color singlet hadrons as dynamical ingredients, have achieved reasonable success, and people are beginning to believe that mesons are all we need to understand low-energy hadron physics. There is, however, a warning raised by Jaffe, who provided an example of an "ineffective effective theory" (111). It is clearly urgent to establish a stronger tie between the Skyrme model and QCD. The argument based on the large N_c limit of QCD must be carefully checked, for example by studying its bosonization process (112). Various fancy (and often complicated) models derived by modification and/or extension from the Skyrmion must also be subjected to the "effectiveness" test.

ACKNOWLEDGMENT

We thank Ralph Amado and Niels Walet for stimulating discussions. This work is supported in part by a grant from the US National Science Foundation. A.H. acknowledges financial support from the Natural Sciences and Engineering Research Council of Canada.

Literature Cited

1. Wilczek, F., *Annu. Rev. Nucl. Part. Sci.* 32:177 (1982)
2. Nambu, Y., Jona-Lasinio, G., *Phys. Rev.* 122:345 (1961)
3. Goldstone, J., *Nuovo Cimento* 19:154 (1961)
4. Bjorken, J. D., Nauenberg, N., *Annu. Rev. Nucl. Sci.* 18:229 (1968); Adler, S. L., Dashen, R. F., *Current Algebra and Applications to Particle Physics*. New York: Benjamin (1968)
5. Weinberg, S., *Phys. Rev. Lett.* 17:616 (1966)
6. Goldberger, M., Treiman, S., *Phys. Rev.* 110:1178 (1958)
7. Adler, S. L., *Phys. Rev. Lett.* 14:1051 (1965); *Phys. Rev.* 143:1144 (1966); Weisberger, W. I., *Phys. Rev. Lett.* 14:1047 (1965); *Phys. Rev.* 143:1302 (1966)
8. Kroll, N., Ruderman, M., *Phys. Rev.* 93:233 (1954)
9. Rho, M., Wilkinson, D. W., eds., *Mesons in Nuclei*. Amsterdam: North-Holland (1979); Rho, M., *Annu. Rev. Nucl. Part. Sci.* 34:531 (1984)
10. Skyrme, T. H. R., *Proc. Roy. Soc. London* A260:127 (1961)
11. Skyrme, T. H. R., *Nucl. Phys.* 31:5 (1962)
12. Zahed, I., Brown, G. E., *Phys. Rep.* 142:1 (1986)
13. 't Hooft, G., *Nucl. Phys.* B72:461 (1974); *Nucl. Phys.* B75:461 (1974)
14. Witten, E., *Nucl. Phys.* B160:57 (1979)
15. Witten, E., *Nucl. Phys.* B223:422,433 (1983)
16. Close, F. E., *An Introduction to Quarks and Partons*. New York: Academic (1979)
17. Thomas, A. W., *Adv. Nucl. Phys.* 13:1 (1983)
18. Oka, M., Yazaki, K., *Phys. Lett.* B90:41 (1980); *Prog. Theor. Phys.* 66:556, 572 (1981)
19. Toki, H., *Z. Phys.* A294:173 (1980)
20. Adkins, G. S., Nappi, C. R., Witten, E., *Nucl. Phys.* B228:552 (1983)
21. Coleman, S., *Aspects of Symmetry*. Cambridge Univ. Press (1985), Ch. 5
22. Zahed, I., Wirzba, A., Meissner, U.-G., *Phys. Rev.* D33:830 (1986)
23. Amado, R. D., Bijker, R., Oka, M., *Phys. Rev. Lett.* 58:654 (1987)
24. Oka, M., et al, *Phys. Rev.* C36:1727 (1987)
25. Oka, M., *Phys. Lett.* B205:1 (1988)
26. Balachandran, A. P., et al, *Phys. Rev. Lett.* 52:887 (1984); *Nucl. Phys.* B256:525 (1985)
27. Bando, M., et al, *Phys. Rev. Lett.* 54:1215 (1985); Bando, M., Kugo, T., Yamawaki, K., *Phys. Rep.* 164:217 (1988)
28. Meissner, U.-G., *Phys. Rep.* 161:213 (1988)
29. Sakurai, J. J., *Currents and Mesons*. Chicago Univ. Press (1969)
30. Meissner, U.-G., Kaiser, N., Weise, W., *Nucl. Phys.* A466:685 (1987)
31. Gell-Mann, M., Levy, M., *Nuovo Cimento* 16:705 (1960)
32. Goldstone, J., Wilczek, F., *Phys. Rev. Lett.* 47:986 (1981)
33. Balachandran, A. P., et al, *Phys. Rev.* D27:1153 (1981)
34. Kahana, S., Ripka, G., *Nucl. Phys.* A429:462 (1984)
35. Atiyah, M. F., Manton, N. S., *Phys. Lett.* B222:438 (1989); Manton, N. S., in *Proc. of the LMS Symposium: Geometry of Low Dimensional Manifolds,* Durham, England, 1989
36. Hosaka, A., et al, *Phys. Lett.* B251:1 (1990); Hosaka, A., Oka, M., Amado, R. D., *Nucl. Phys.* A530:507 (1991)
37. Coleman, S., see Ref. 21, Ch. 7
38. 't Hooft, G., *Phys. Rev. Lett.* 37:8 (1976)
39. Jackson, A. D., Rho, M., *Phys. Rev. Lett.* 51:751 (1983)
40. Urbano, J. N., Goeke, K., *Phys. Rev.* D32:2396 (1985); Hosaka, A., *Phys. Lett.* B207:249 (1988)
41. Gervais, J.-L., Sakita, B., *Phys. Rev. Lett.* 52:87 (1984); Bardakci, K., *Nucl. Phys.* B243:197 (1984); Manohar, A. V., *Nucl. Phys.* B248:19 (1984)
42. Arima, A., Iachello, F., *Adv. Nucl. Phys.* 13:139 (1983)
43. Rajaraman, R., *Solitons and Instantons*. Amsterdam: North-Holland (1982)
44. Kerman, K., Klein, A., *Phys. Rev.* 132:1326 (1963)
45. Goldstone, J., Jackiw, R., *Phys. Rev.* D11:1486 (1975)
46. Dirac, P. A. M., "Lectures in Quantum Mechanics", Belfer Grad. Sch. of Sci., Yeshiva Univ. (1964)
47. Uehara, M., *Prog. Theor. Phys.* 80:768 (1988); Saito, S., *Prog. Theor. Phys.* 78:746 (1987); Holzwarth, G., Hayashi, A., Schwesinger, B., *Phys.*

Lett. B191:27 (1989); Liu, H., Oka, M., *Phys. Rev.* D40:883 (1989)
48. Schnitzer, H. J., *Nucl. Phys.* B261:546 (1985)
49. Kawarabayashi, K., Ohta, K., *Phys. Lett.* B216:205 (1989); Verschelde, H., *Phys. Lett.* B232:15 (1989); Ohta, K., *Nucl. Phys.* A511:620 (1990); Uehara, M., *Prog. Theor. Phys.* 83: 790 (1990); Hayashi, A., Saito, S., Uehara, M., *Phys. Lett.* B246:15 (1990); Holzwarth, G., Pari, G., Jennings, B. K., *Nucl. Phys.* A515:665 (1990); Ikehashi, T., Ohta, K., *Nucl. Phys.* A536:521 (1992)
50. Ohta, K., *Phys. Lett.* B259:404 (1991)
51. Skyrme, T. H. R., *J. Math. Phys.* 12: 1735 (1971); Mandelstam, S., *Phys. Rev.* D11:3026 (1975); Rajeev, S. G., *Phys. Rev.* D29:2944 (1984)
52. Cebula, D., Walet, N. R., Klein, A., *J. Phys.* G18:499 (1992)
53. Breit, J. D., Nappi, C. R., *Phys. Rev. Lett.* 53:889 (1984); Hayashi, A., et al, *Phys. Lett.* B147:5 (1984); Zahed, I., Meissner, U.-G., Kaulfuss, U. B., *Nucl. Phys.* A426:525 (1984); Liu, K. F., Zhang, J. S., Black, G. R. E., *Phys. Rev.* D30:2015 (1984); Mattis, M. P., Peskin, M. E., *Phys. Rev.* D32:58 (1985)
54. Amado, R. D., Oka, M., Mattis, M. P., *Phys. Rev.* D40:3622 (1989)
55. Reid, R. V. Jr., *Ann. Phys. (NY)* 50:411 (1968)
56. Lacombe, M., et al, *Phys. Rev.* C21:861 (1980)
57. Machleidt, R., Holinde, K., Elster, C., *Phys. Rep.* 149:1 (1987)
58. Jackson, A., Jackson, A. D., Pasquier, V., *Nucl. Phys.* A432:567 (1985)
59. Vinh Mau, R., et al, *Phys. Lett.* B150:259 (1985)
60. Yabu, H., Ando, K., *Prog. Theor. Phys.* 74:750 (1985)
61. Vinh Mau, R., et al, *Phys. Lett.* B161:31 (1985)
62. Lacombe, M., et al, *Phys. Lett.* B169:121 (1986)
63. Oka, M., Liu, K. F., Yu, H., *Phys. Rev.* D34:1574 (1986); Oka, M., *Phys. Rev.* C36:720 (1987)
64. Verbaarschot, J. J. M., et al, *Nucl. Phys.* A468:520 (1987)
65. Nyman, E., Riska, D. O., *Phys. Lett.* B203:13 (1988)
66. Manton, N. S., *Phys. Lett.* B192:177 (1987)
67. Kopeliovich, V. B., Stern, B. E., *JETP Lett.* 45:203 (1987)
68. Verbaarschot, J. J. M., *Phys. Lett.* B195:235 (1987)
69. Braaten, E., Carson, L., *Phys. Rev.* D38:3525 (1988)
70. Kurihara, T., et al, *Prog. Theor. Phys.* 81:858 (1989)
71. Manton, N. S., *Phys. Rev. Lett.* 60:1916 (1988)
72. Klein, A., Walet, N. R., Do Dang, G., *Ann. Phys. (NY)* 208:90 (1991)
73. Brown, G. E., Jackson, A. D., *The Nucleon-Nucleon Interaction.* Amsterdam: North-Holand (1976)
74. Walhout, T. S., Wambach, J., *Phys. Rev. Lett.* 67:314 (1991)
75. Amado, R. D., et al, *Phys. Rev. Lett.* 63:852 (1989)
76. Oka, M., *Phys. Rev. Lett.* 66:1019 (1991)
77. Jackiw, R., Nohl, C., Rebbi, C., *Phys. Rev.* D15:1642 (1977)
78. Kalbermann, G., et al, *Phys. Lett.* B179:4 (1986)
79. Depace, A., Muther, H., Feassler, A., *Phys. Lett.* B188:307 (1987); Kalbermann, G., Eisenberg, J. M., *J. Phys.* G15:157 (1989)
80. Walet, N. R., Amado, R. D., Hosaka, A., preprint Univ. Penn. (1992)
81. Verbaarschot, J. J. M., et al, *Nucl. Phys.* A461:603 (1987)
82. Adler, A. E., et al, *Phys. Rev. Lett.* 59:2836 (1987);
83. Oka, M., *Phys. Lett.* B175:15 (1986)
84. Odawara, H., Morimatsu, O., Yazaki, K., *Phys. Lett.* B175:115 (1988)
85. Riska, D. O., Nyman, E. M., *Phys. Lett.* B183:7 (1987).
86. Aitchison, I. J. R., *Acta Phys. Pol.* B18:207 (1987)
87. Guadanini, E., *Nucl. Phys.* B236:35 (1984)
88. Callan, C. G., Klebanov, I., *Nucl. Phys.* B262:365 (1985); Callan, C. G., Hornbostel, K., Klebanov, I., *Phys. Lett.* B202:269 (1988)
89. Yabu, H., Ando, K., *Nucl. Phys.* B301:601 (1988)
90. Riska, D. O., Scoccola, N. N., *Phys. Lett.* B265:188 (1991)
91. Jaffe, R. L., *Phys. Rev. Lett.* 38:195 (1977); Oka, M., Shimizu, K., Yazaki, K., *Phys. Lett.* B130:365 (1983); Takeuchi, S., Oka, M., *Phys. Rev. Lett.* 66:1271 (1991)
92. Jaffe, R. L., Korpa, C. L., *Nucl. Phys.* B258:468 (1985); Lee, H. K., Kim, J. H., *Mod. Phys. Lett.* A5:887 (1990)
93. Fetter, A. L., Walecka, J. D., *Quantum Theory of Many Particle Systems.* New York: McGraw-Hill (1971)
94. Braaten, E., Townsend, S., Carson, L., *Phys. Lett.* B235:147 (1990)

95. Carson, L., *Phys. Rev. Lett.* 66:1406 (1991)
96. Carson, L., *Nucl. Phys.* A535:479 (1991)
97. Finkelstein, D., Rubinstein, J., *J. Math. Phys.* 9:1762 (1968); Williams, J. G., *J. Math. Phys.* 11:2611 (1970)
98. Walhout, T. S., *Nucl. Phys.* A531:596 (1991)
99. Walhout, T. S., *Nucl. Phys.* A484:397 (1988)
100. Klebanov, I., *Nucl. Phys.* B262:133 (1985)
101. Wüst, E., Brown, G. E., Jackson, A. D., *Nucl. Phys.* A468:450 (1987)
102. Goldhaber, A. S., Manton, N. S., *Phys. Lett.* B198:231 (1987)
103. Jackson, A. D., Verbaarschot, J. J. M., *Nucl. Phys.* A484:419 (1988); Castillejo, L., et al, *Nucl. Phys.* A501:801 (1989); Dyakonov, D. I., Mirlin, A. D., *Sov. J. Nucl. Phys.* 47:421 (1988); Kugler, M., Shtrikman, S., *Phys. Lett.* B208:491 (1988)
104. Gomm, H., et al, *Phys. Rev.* D33:801, 3476 (1986)
105. Forkel, H., et al, *Nucl. Phys.* A504:818 (1989)
106. Manton, N. S., Ruback, P. J., *Phys. Lett.* B181:137 (1986)
107. Jackson, A. D., Wirzba, A., Castillejo, L., *Nucl. Phys.* A486:634 (1988)
108. Jackson, A. D., Manton, N. S., Wirzba, A., *Nucl. Phys.* A495:499 (1989)
109. Campbell, B. A., Ellis, J., Olive, K. A., *Nucl. Phys.* B345:57 (1990)
110. Weinberg, S., *Phys. Lett.* B251:288 (1990)
111. Jaffe, R. L., *Phys. Lett.* B245:221 (1990)
112. Dhar, A., Shanker, R., Wadia, S. R., *Phys. Rev.* D31:3256 (1985); Ebert, D., Reinhardt, H., *Nucl. Phys.* B271:188 (1986); Wakamatsu, M., Weise, W., *Z. Phys.* A311:173 (1988); Zaks, A., *Nucl. Phys.* B260:241 (1985)

Annu. Rev. Nucl. Part. Sci. 1992. 42:367–399

HADROPRODUCTION OF CHARM PARTICLES[1]

J. A. Appel

Physics Department, Fermi National Accelerator Laboratory, Batavia, Illinois 60510

KEY WORDS: gluon fusion, structure functions, fragmentation, hadronization, *A* dependence

---

CONTENTS

1. INTRODUCTION

The existence of the charm quark was postulated in 1970 by Glashow, Iliopoulos & Maiani to account for the absence of strangeness-changing neutral currents (1). The first direct evidence for charm quarks came in 1974 from observations in a hadroproduction experiment at Brookhaven National Laboratory (2) and from e^+e^- annihilations at the Stanford Linear Accelerator Center (3). These observations marked the discovery of the J/ψ, a particle understood to be composed of a charm quark and charm antiquark. The observations were rather quickly followed by measurements of higher mass quark-antiquark states and open charm states, states containing a single charm quark or charm antiquark. The J/ψ was immediately linked to the hypothesis of the charm quark. In time, evidence accumulated that the constituents of the J/ψ were indeed strongly interacting quarks. However, in the dozen years after the very first observations of charm, hadroproduction experiments played a limited, often confusing, role.

1.1 *Recent Progress*

In the last five years the situation has improved. Hadroproduction experiments at CERN and Fermilab have resulted in cleaner, less biased data of much higher statistical precision (e.g. Figure 1). These characteristics of recent experiments justify labeling them as a second generation of charm hadroproduction experiments. This second generation is the focus of this review. It is marked by precision measurements of the trajectories of charged particles originating from the beam interaction point and decay vertices of the charm particles. In addition, we can recognize the beginning of a third generation of experiments in which hadroproduction contributes to the detailed study of the decays of charm particles. This development arises as a continuation of the experiments, which have demonstrated the ability to produce, detect, and cleanly reconstruct copious quantities of charm particles.

There has also been progress in theoretical calculations. Calculations of the quantum chromodynamic (QCD) next-to-leading order contribution to charm hadroproduction are available. The agreement of these calculations with the experimental cross-section measurements has increased confidence in the ability to interpret charm results in terms of this accepted theory. The calculations include differential cross sections for inclusive processes and, most recently, next-to-leading order calculations including correlations between the pairs of charm particles.

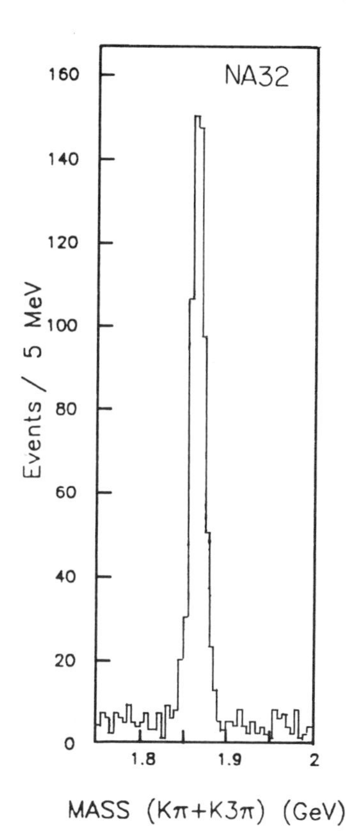

Figure 1 Effective mass spectrum of D^0 candidates from NA32 showing the benefits of precision tracking and decays in free space.

1.2 *The Physics of Charm Hadroproduction*

Three basic elements enter the calculation of the hadroproduction process as understood by QCD. These are (*a*) a basic hard scattering between quarks and gluons of the incident hadrons, (*b*) the distribution of quarks and gluons inside the incident hadrons, and (*c*) the hadronization of the produced charm quarks to charm particles observed in the laboratory.

In the hard-scattering subprocess, one applies the universality of the QCD theory, coupling charm quarks in the same way as lighter quarks. The process is dominated at fixed-target energies by gluon-gluon fusion and thus provides the opportunity to study gluons in a rather well-defined environment. Dependence of the hadroproduction process on the distribution of gluons in the incident hadrons provides a method of studying the gluon structure function in a variety of particles. Beams of pions, kaons, protons, and their antiparticles have been used in experiments, and these particles can have their gluon structure explored in this way.

Finally, as the charm quarks emerge from the hard production subprocess, they must pass through whatever residual hadronic matter there is in the target or projectile. Since the formation time for the charm particle is comparable to nuclear distances, one may view the charm quark as traveling through nuclear matter for much of its "separate" existence. This provides an opportunity to probe yet another area of physics.

The wealth of physics processes involved in the hadroproduction of charm is both a benefit and a problem. The complexity of the full process allows for multiple interpretations of effects. Extracting parameters of the most basic physics quantities requires convolutions and calculations that depend on the quantitative input as well as on the basic theoretical assumptions. On the more purely theoretical level, there are parameters that are less well known than one would like for interpretations of the physics. Furthermore, there is the uncertainty due to the size of the charm quark mass. Is the charm quark heavy enough that perturbative QCD calculations can be used, even including contributions in next-to-leading order, to extract the physics of interest?

1.3 *Relation to Other Processes*

Data on the production of charm quarks at fixed-target energies are a source of copious interactions in which gluons interact in an identifiable way, i.e. the gluon-gluon fusion process. At hadron collider energies, the reaction $g \rightarrow c\bar{c}$ is an increasingly important source of charm quarks.

Processes that involve gluon fusion provide tools for the direct determination of the gluon distribution in the incident hadrons. These structure functions are also relevant in lepton, photon, and neutrino production of heavy quarks. In these processes, the gluons occur only in the target nucleons. The gluon structure function is obtained in deep inelastic scattering as a part of the evolution of the quark structure functions, i.e. as a correction to the dominant quark scattering. In direct photon production, gluons can enter directly, becoming quite large for some parts of the production. Nevertheless, subtractions due to quark sources are necessary. In charm hadroproduction, the quark-antiquark annihilation background is at most only on the order of 10% in leading order.

The hadronization of the charm quarks is among the least explored parts of the hadroproduction of charm. Hadronization occurs in any process in which the charm quarks are created. Fragmentation functions are a measure of this process. The most precise measure of hadronization comes from e^+e^- annihilations producing charm quarks. However, e^+e^- annihilation is different from hadroproduction; it is unaffected by the presence of hadronic matter. Only the effects of the strong color fields between the charm quark and antiquark are relevant.

Hadroproduction is a copious source of particles containing the charm quark. As such, hadroproduction experiments are already beginning to contribute to studies of the physics of heavy flavor decay. This role is likely to continue and expand, but it is not the subject of this review.

2. A BRIEF HISTORY OF MEASUREMENTS

The early years of open charm hadroproduction were limited by the capability of detectors in the face of difficult experimental conditions. Among the difficulties (still faced today) are (*a*) the small fractional charm production cross section (one $c\bar{c}$ pair event per 10^3 interactions, typically), (*b*) the high multiplicity of particles in the charm events, and (*c*) the small branching ratios to specific final states (typically 1–10%). As it turned out, many of the more reliable early measurements were indirect. Among these were the observations of prompt leptons resulting from the semileptonic decays of charm particles. Most of these early experiments had goals other than charm production and decay as their primary motivation. Nevertheless, leptons with intermediate transverse momentum have been interpreted to come from charm decay. Electrons and muons were observed at rates of 10^{-4} to 10^{-3} of the charged pion at fixed-target and collider energies. Muons and neutrinos were also measured in beam dump experiments. The physics results, charm cross sections times average branching ratios, were extrapolated from total observed lepton rates under varied experimental conditions. These observed rates typically included much larger numbers of leptons from photon conversions or decays of particles containing strange, not charm, quarks.

More direct measurements of charm decays resorted to limited regions of phase-space. These were typically in the very forward direction or at high transverse momentum, where the production is better than the one part in a 1000 that is characteristic of the charm cross section relative to the total cross section. Alternately, specialized cuts were chosen to select specific production or decay characteristics. Such techniques ordinarily did result in charm signals, but required very large extrapolations to go from the observation to the physics parameters of interest. One even required multiplication by factors as large as 10^6 to go from observed signals to total cross sections.

The experimental difficulties above led to many strange and controversial results and interpretations. The various results and attempts to understand them are well described in the excellent reviews of these early days by Kernan & VanDalen (4), and by Tavernier (5). The observational difficulties led to discrepancies of factors up to 100 in total cross sections and steep, even unimaginable, energy dependence in going from fixed-target energies ($\sqrt{s}$ in the range 17–39 GeV) to ISR collider energies ($\sqrt{s}$ of 53–62 GeV). Longitudinal momentum distributions suggested very strong leading-particle effects: charm particles having a quark in common with the incident beam particle appeared to dominate in the forward direction. Power-law behaviors for longitudinal momentum distributions appeared to have powers that could vary by experiment and final state from 1 to 11. The average transverse momentum, p_t, of observed charm particles varied from 0.5 to 1 GeV. Some experiments did not see the major signals reported by other experiments with similar detectors. Mass distributions of a given charm particle peaked at different values in different data sets or even appeared to have double peaks. Attempts to explain the plethora of conflicting results were unavailing.

More recent experiments have benefited enormously from the new technology, resulting in sufficiently improved spatial resolution to "see" the decay of charm particles separated from their production point. As described in Section 4, the application of these techniques very much improved signal-to-background ratios. These, in turn, have allowed for charm measurements with full forward acceptance and less restrictive event selection. Additionally, much greater on-line and off-line computing capabilities have permitted data samples that include thousands of decays per experiment instead of hundreds or less. More recent results have shown much greater agreement among experiments, with less controversial and more straightforward interpretations. This review, therefore, describes a more coherent picture of charm hadroproduction, one in which the theoretical ideas are amenable to test and experimental measurements can be used in further theoretical calculations.

3. A BRIEF HISTORY OF THEORY

During the first generation of charm hadroproduction experiments theoretical efforts were focused on explaining the conflicting early data. Quantum chromodynamics calculations, done to leading order at the time, fell short in explaining the measured cross sections and were unable to explain many features of the data. An uncomfortably low charm quark effective mass of 1.2 GeV was required to explain even the lowest of charm cross-section measurements. This left the field open to unconventional explanations for the existing charm data. These possibilities included the existence of intrinsic charm quarks in hadrons (6), excitation of virtual charm quarks (7, 8), diffractive production as analogs of the vector meson dominance model for photoproduction (9), and various enhancements (10) associated with the fragmentation or hadronization process of the partons of the basic interaction.

Coincident with the improved data of the second generation of hadroproduction experiments came QCD calculations complete in next-to-leading order (11). These calculations indicated a nearly constant factor of three increase in charm cross section over the cross sections calculated to leading order in QCD. A more natural charm quark effective mass near 1.5 GeV was adequate to explain the charm cross-section measurements. The success of these calculations has removed the impetus to look for unconventional sources of charm production beyond the basic QCD. More details of the current theory of charm hadroproduction are given in Section 5 of this review.

4. DISCUSSION OF TECHNIQUES

The second generation of charm hadroproduction experiments is distinguished from first-generation experiments by the ability to identify, by tracking alone, those particles that come from the decay of a charm particle. These particles can be separated from all other particles in the event, i.e. particles from the beam interaction point and particles from the decay of the second charm particle in the event. This capability has been provided by high-resolution small bubble chamber, emulsion, and solid-state detectors. These devices have been installed in multiparticle spectrometers that serve to identify the produced leptons and hadrons, measure the momenta of charged particles, and generally select for further investigation those events most likely to contain charm particles. In hadroproduction, so far, these techniques have only been applied in the fixed-target experiments.

4.1 *Precision Tracking*

Two important advantages result from the ability to reconstruct the trajectory of charged particles with high precision and to detect displaced vertices. The first advantage is the selection of events with decays of relatively long-lived particles (a few times 10^{-13} s), and thus events that are highly enriched in charm relative to average events. Another significant advantage is the improvement in signal-to-background ratios due to the reduction of the combinations of particles that must be examined to find a charm decay. Hadronic interactions that produce charm are characterized by higher multiplicity than average events. Selecting only the trajectories of particles that come from a well-defined secondary decay vertex reduces the number of combinations enormously. Overall background reductions by factors of more than a hundred are achieved. Efficiencies for the detection of charm decays with separate decay vertex can be as large as 50%, depending on the spatial resolution and charm particle lifetime.

Table 1 provides a comparison of the charm hadroproduction experiments of the second generation. Experiments prefixed by NA and WA for North Area and West Area are located at CERN, while those prefixed with an E are at Fermilab. A very large increase in the number of recorded events and in the number of reconstructed charm decays is evident in the table. Within the second generation of experiments, the spotlight has shifted from visual techniques associated with bubble chambers and emulsions to purely electronic detectors. These more recent techniques provide the capability of very large statistics. The large statistics, in turn, have reduced the statistical fluctuations and enhanced the ability to adequately study backgrounds. These problems plagued the first-generation experiments and even the early second-generation experiments.

4.2 *Downstream Spectrometers*

In addition to the precision tracking device near the target, all second-generation experiments have multiparticle spectrometers downstream. A typical example, the Tagged Photon Laboratory at Fermilab, is shown in Figure 2. It is characteristic of the other spectrometers in having charged particle tracking through magnetic fields and charged particle identification through two or more threshold Čerenkov counters or a ring imaging Čerenkov counter. Electron, photon, and muon identification is achieved with electromagnetic calorimetry and with absorbers for filtering out hadronic particles.

Precision tracking provides spatial resolutions near the target ranging from 5 to 20 microns. This precision is at least an order of magnitude higher than that available in the proportional wire chambers or drift chambers used in the downstream tracking systems. On the other hand, the physical extent of the precision tracking is only a few centimeters. The wire chamber devices used downstream extend over at least ten meters. This results in comparable angular resolution in the down-

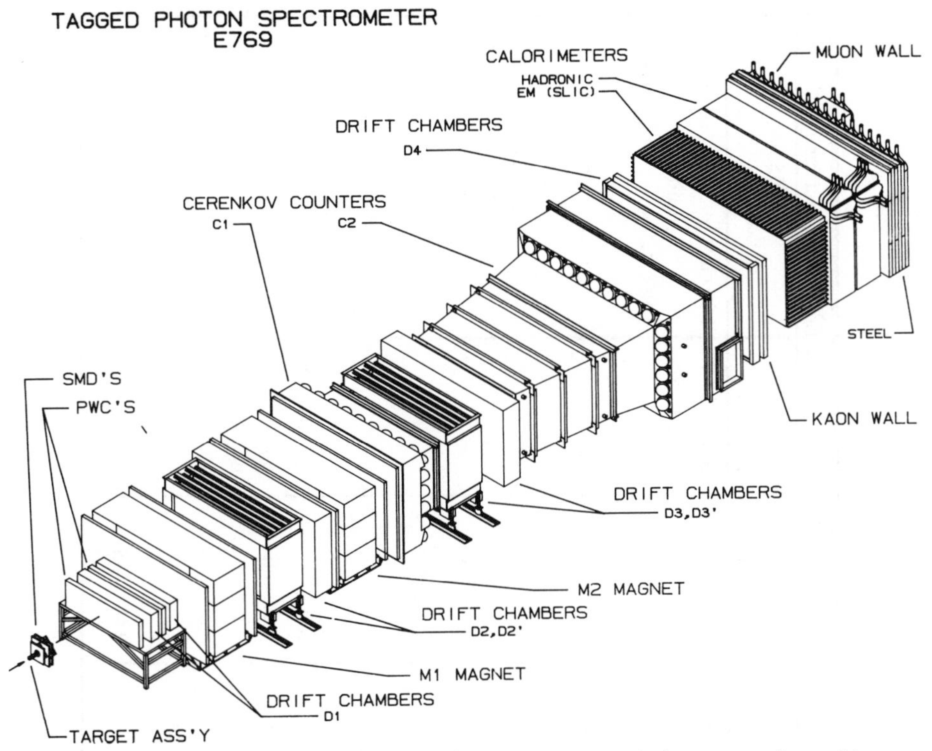

Figure 2 The Fermilab Tagged Photon Spectrometer, a typical apparatus for multi-particle detection in a fixed-target experiment.

stream tracking to that achieved with the higher precision tracking near the target. Wire chamber tracking has poor position resolution compared to solid-state devices, but comparable angular resolution.

The limited spatial extent of the precision trackers limits the size of the useful target. Thus, these experiments all have small luminous regions for targets, even in the direction parallel to the incident beam. Rather than long liquid hydrogen targets, all of the experiments with solid-state detectors use thin (nuclear) targets to maintain large solid angle acceptance. This increases the importance of measurements of the dependence on the target material. One sees a variety of targets in each of these experiments.

4.3 *Selecting Events Containing Charm*

In hadronic interactions, the small probability of producing charm remains a technical problem for experiments. One must sift through enormous numbers of events before finding those that contain the charm

Table 1 Experiment comparisons

Identifier	LEBC-EHS		LEBC-MPS	Hybrid Emulsion Spectrometers		
Experiment Number (reference)	NA27 (12)		E743 (13)	WA75 (14)	E653 (15)	
Beam	360 GeV π^-	400 GeV p	800 GeV p	350 GeV π^-	800 GeV p	600 GeV π^-
Target	H_2	H_2	H_2	Emulsion	Emulsion	
Year data taken	1982	1984	1985		1984-5	1987-8
# Si Planes Beam	2	2	0	6	6	
Downstream	0 (LEBC)	0	0 (LEBC)	10	18	
Resolution	2.5μm	2.5μm	2μm	20μm	5μm	
#DC/PWC planes	66+ISIS	66+ISIS	24	49	55+12	
ΔP_t MeV/c			700	900	336	
Hadronic part. ID	TRD + 2Č	TRD + 2Č	TRD + 2Č		TOF	
(Č or TRD) pion thresholds for Č	0.56, 17	0.56, 17	5.7, 16.7			
EM calorimeter	$10-15\%/\sqrt{E}$	$10-15\%/\sqrt{E}$			$15\%/\sqrt{E}$	
Hadron calorimeter	$150\%/\sqrt{E}$	$150\%/\sqrt{E}$			$80\%/\sqrt{E}$	
μ ID	—	—	not used	2mFe+2mW	5m Fe	
Trigger	interaction $n_{ch} > 2$	interaction $n_{ch} > 2$	interaction $n_{ch} > 2$	muon $n_{ch} > 3 + \mu$	muon	
Data sample size ($\times 10^6$ events)	0.115	1.0	0.5	1.5	5.4	9.6
Reconstructed charm	183†	425†	148†, 33	288†	146	~1000

Table 1 Experiment comparisons (con't)

Identifier	ACCMOR			Omega		TPL	
Experiment Number (reference)	NA11 (16)	NA32 (17, 18)		WA82 (19)		E769 (20)	E791
Beam	175 GeV π^-	200 GeV π^-Kp	230 GeV π^-	340 GeV π^-	370 GeV P	250 GeV h±	500 GeV π^-
Target	Be	Si	Cu	Si, W, Cu,W	Cu, W	Be, Al, Cu, W	C, Pt
Year data taken	1982	1984	1986	'87-88-89	'89-88	1987-88	1991
# Si Planes Beam	6	7		8		2	6
Downstream	6	7	8 + 2 CCDs	13+1		11	17
Resolution	6μm	5-8μm		10μm		20μm	
#DC/PWC planes	48	48		~6		37	
ΔP_t (MeV/c)	270.7 +617.4	270.7 +617.4		2190		212 +320	
Hadronic part. ID (Č or TRD) pion thresholds for Č	3Č 12 GeV	3Č 2.4, 6.1, 12.8	2Č 6.5, 12	RICH in '87-88	—	2Č 6, 10.5	6, 11
EM calorimeter	$18\%/\sqrt{E}$	$18\%/\sqrt{E}$	None	$15\%/\sqrt{E}$		$20\%/\sqrt{E}$	
Hadron calorimeter	None	None		None		$75\%/\sqrt{E}$	
μ ID	None	None		None		2m Fe	
Trigger	Single electron	Interaction	K^+K^- pK^- $\bar{p}K^+$ $\bar{p}p$	Impact parameter		Global E_t $n_{ch} > 5$	$n_{ch} > 4$
Data sample size ($\times 10^6$)	6.3	3.8	17	18 + 30	10	500	20,000
Reconstructed charm	130	170	~1300	~3,000		~4,000	(>100,000)

† topological only
() anticipated

particles. This selection is done by some combination of event-selecting triggers at the time of data taking (on-line) and the more detailed analysis made later (off-line).

The need to sort through large numbers of interactions to find the charm signals has led to a variety of approaches (21). While the bubble chamber has limited its event recording to beam particle interactions in the chamber, all other experiments have used more restrictive triggers. The emulsion experiments demanded a high transverse momentum muon to flag events with semileptonic charm decays. The experiments with solid-state detectors have used triggers based on particle identification in the downstream spectrometer (electrons, muons, and kaons or protons), evidence for secondary vertices or downstream decays and the detection of large transverse energy. Each of these triggers helps to select events with a fraction of charm that is higher than in a random sample of interactions.

As methods to select the charm events have been developed off-line, the methods have been incorporated more and more into the data-taking process. Fast processors with exotic names like ESOP (22), FAMP (23), and MICE (24) have been used in CERN hadroproduction experiments. These specialized processors have calculated electron identification probabilities and evidence for downstream decays, for example. The results enter the decision on whether to record the event on magnetic tape for later analysis. An alternate approach has been pursued at Fermilab, where networks of microprocessors have been assembled for rapid reconstruction of larger data sets off-line. This approach allows looser on-line requirements, but also requires faster event readout time to achieve larger sample sizes. One experiment (E791) recorded 20 billion events. This required faster digitization and readout of front-end electronics and significant parallelism in the data acquisition system. An average-sized event could be read out in 50 microseconds. This is much faster than the time needed by any of the processor systems used so far at CERN to select events. The further event selection is pushed off-line in E791 to the massively parallel network of microprocessors organized in what are called "processor farms" (25). These farms are used not only for selecting events, but also for the full reconstruction of the selected events.

4.4 *Hadron Collider Experiments in the Future*

Precision vertexing is now being added to the CDF hadron collider experiment (26) at Fermilab, motivated by the search for top quarks and the aim of tagging events containing B decays. It is possible that CDF, and eventually other hadron collider experiments, will contribute to charm hadroproduction measurements in the future. Such contributions would be welcome since the kinematic range of production is very different from that in fixed-target experiments. Additional processes such as gluon splitting may be explored as a different probe of QCD. Nevertheless, since the range of fractional momenta x are very low, theoretical interpretations are more difficult. They involve yet higher order processes and more nonperturbative effects. There are also significant technical difficulties arising from the added backgrounds for production so far from threshold (e.g. charm from bottom quark decay). Thus, hadron collider charm production measurements will presumably also begin with measurements in limited regions of phase-space.

Without the precision vertexing and its capability for background rejection, collider experiments have been limited so far to measuring the fractional rate of D^* production. Using the mass difference between the D^* and the D^0, both UA1 (27) and CDF (28) have reported that about 10% of jets contain D^* mesons in the fractional momentum range where the D^* carries at least 10% of the jet energy. As was characteristic of other first-generation charm results, the recent UA1 result differs from the value published by the same group five years earlier (29).

4.5 *Other Attempts and Techniques*

A wide variety of other techniques have been developed for use in charm experiments. The techniques have focused on precision reconstruction of the production and decay region. Holographic bubble chambers (30) and streamer chambers (31) have been developed. However, while these chambers enlarge the volume in which decays can be reconstructed with high resolution, the inherent cycle times of the devices make them uncompetitive today. Inherently faster devices based on bundles of scintillating fibers (32) have been tried at CERN and Fermilab. However, the conversion of light images to digital format for recording and later analysis has added a technical problem, which, coupled to the somewhat marginal light output of the devices, has shifted attention from these techniques. Another technique tried at both CERN and Fermilab, so-called active targets of silicon (33), has failed to achieve notable success. Here, fluctuations in ionization and signals due to nuclear interactions in the target have limited this application. The main goal has been to see a multiplicity jump caused by the decay of charm particles within the target.

Solid-state devices remain the favorite technique for recognizing and measuring trajectories of charged particles from charm decays. We do not review these devices here at any length; we simply note the preponderance of silicon microstrip detectors, arrays with one-dimensional readout. Charged coupled devices (CCDs) have been used by the ACCMOR collaboration at CERN, but their slow readout has prevented their wider use, in spite of the advantages of the two-dimensional readout and higher spatial resolution they provide. Efforts at greater integration of the readout electronics using modern microelectronic technology (34) may change the situation in the future. However, the enormous increase in the number of readout elements, already quite large for the microstrip detectors, provides a challenge, both for their readout and for their use in fast trigger schemes.

5. QCD THEORY OF CHARM HADROPRODUCTION

The leading order QCD diagrams (i.e. schematic representations of the basic processes) for charm hadroproduction are shown in Figure 3. It is assumed that there is a hard scattering of one parton (quarks, antiquarks, or gluons) from each of the incident hadrons. The hard scattering is necessary to produce the relatively high mass of the charm quark pair. The calculations are performed in this parton model starting with Equation 1.

$$\sigma = \sum_{ij} \int dx_1\, dx_2 f_i(x_1, \mu) f_j(x_2, \mu) \hat{\sigma}(x_1 x_2 s, \mu^2). \qquad 1.$$

In Equation 1, f_i and f_j are the distributions of the relevant partons in the incident hadrons. These distributions are called structure functions, the number densities of the light quarks and gluons as a function of their momentum fraction, x_i and x_j, evaluated at a scale μ. The

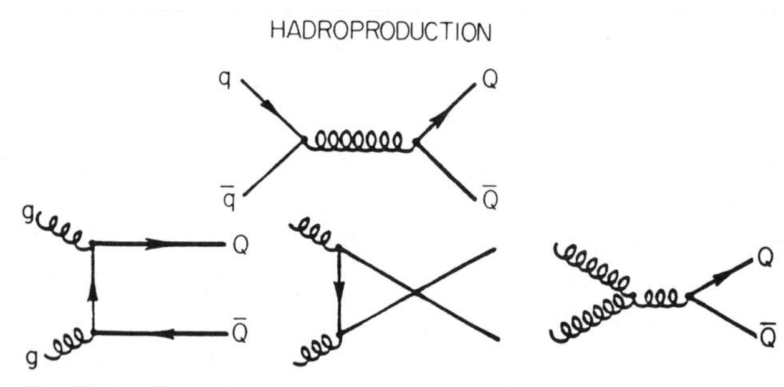

Figure 3 Leading order QCD diagrams for the hadroproduction of charm quarks, Q.

partons from the incident hadrons interact with the cross section, $\hat{\sigma}$, that is a function of the momenta of the incident partons and the scale μ. The total center-of-mass energy squared of the incident hadrons is s. The total center-of-mass energy squared of the partonic subprocess is x_1x_2s.

5.1 *QCD Perturbative Calculations*

The parton level cross section, $\hat{\sigma}$, is normally calculated in a perturbative series in the strong interaction coupling constant $\alpha_s(\mu^2)$. The use of Equation 1 for calculations ignores all of the possible effects discussed in Section 3. Flavor excitation is ignored, as is the role of the spectator (noninteracting) non-charm-producing partons in the incident hadrons. This leading order cross section is proportional to α_s^2. These first-order calculations have been done by a number of authors (35), and the justification of the factorization represented by Equation 1 has been examined on the basis of the lowest order contributions (36).

More recent calculations by Nason, Dawson & Ellis (11, 37) and by Beenakker et al (38) provide a complete next-to-leading order calculation of $\hat{\sigma}$. In this calculation, a great many additional diagrams are included. The cross section varies as the square of the sum of the amplitudes represented by these diagrams, and terms proportional to α_s^3 are kept. As a net result of these calculations, the total cross section increased by a factor of about three relative to the leading order calculation. Agreement with the total cross section data is comfortably achieved with charm quark masses on the order of 1.5 GeV, as seen in Figure 4 (39). Differential cross-section shapes do not change much by the addition of the next-to-leading order diagrams. There is no intrinsic charm quark included in the calculations, but diagrams having the same topology as flavor excitation (Figure 5) (8) are included in these higher order calculations.

Given the large increase in rate provided by the next-to-leading order calculation, one might ask whether the next order terms are negligible. In addition, significant uncertainties remain in these theoretical calculations. These include the precise values of the effective mass of the charm quark, and the renormalization scale. There are also effects associated with the spectator partons, of the order of Λ/m_c (the strong interaction scale over the mass of the charm quark). In spite of these uncertainties, the QCD framework provides a useful tool in understanding the process itself. Furthermore, interactions involving different incident hadrons and outgoing charm particles are expected to have in common basic calculations and high order calculations even if these

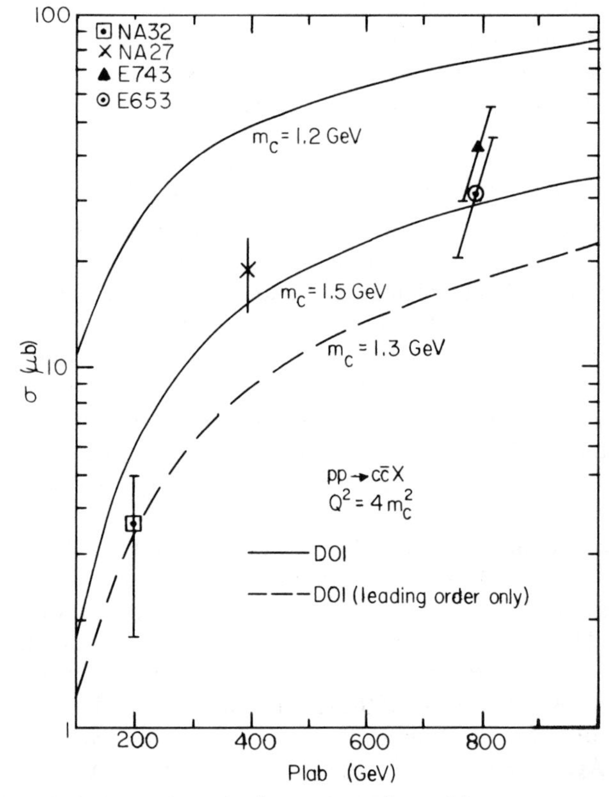

Figure 4 Total charm quark production cross section vs laboratory momentum of the incident proton. The dashed curve represents a calculation to leading order only. The two solid curves represent calculations in next-to-leading order; the charm quark masses are indicated for each curve.

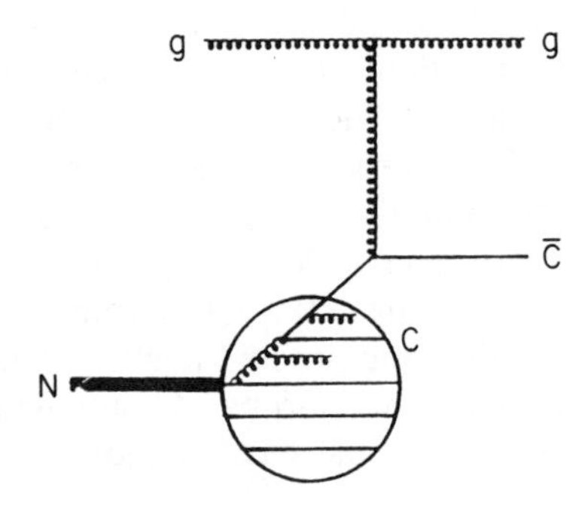

Figure 5 Charm quark excitation diagram from Odorico (8).

cannot be performed explicitly. It may be useful to see how far one can pursue application of the current theory.

5.2 *Structure Functions*

So far, gluon and quark structure functions have been taken as input to calculations of charm production. These inputs come from compilations of data on deep inelastic lepton and neutrino scattering data. This is particularly appropriate for the nucleons' quark structure functions in which lepton data are easier to interpret and more precise. In contrast, the gluon structure function is not well determined. In pions and especially kaons, the gluon structure function is essentially unmeasured. One may anticipate that the best information on these structure functions will come from analysis of the high statistics production data from the second-generation hadroproduction of charm experiments. Especially interesting are the energy dependence of the cross section and the longitudinal momentum dependence of charm particles. Uncertainties associated with the perturbative calculations above and the hadronization process described below will affect the absolute determination of structure functions. However, comparisons of production of charm by pion and kaon beams striking the same target should allow determination of the relative hardness of gluon distributions in pions and kaons. Does the extra mass of the strange quark influence the gluon distribution of the kaon relative to the pion? Similarly, the gluon distributions in the quark-antiquark mesons may be different from the gluon distributions in the three-quark baryon. Such comparisons are of basic interest and provide relevant tests for lattice gauge calculations.

5.3 *Fragmentation/Hadronization*

The calculations described so far relate to the inclusive production of charm quarks. The development of these quarks into charm hadrons requires additional, nonperturbative calculations. These are normally represented by fragmentation functions, which, one might hope, would be the same as those measured in e^+e^- annihilation into charm quarks. However, the direct applicability of these functions is not apparent.

The fragmentation has not been calculated as extensively or analytically as the quark production itself. Monte Carlo calculations are more typical of theoretical efforts. Three models dominate the efforts at this time. These are string fragmentation, cluster fragmentation, and independent fragmentation. The status of these fragmentation models was reviewed recently by Sjöstrand (40). The Monte Carlo calculations are used widely by experimentalists modeling the final-state particles in their detectors. Well-known and widely available Monte Carlo programs such as ISAJET (41), PYTHIA (42), and EUROJET (43) are used. The focus of these programs has been on collider energies and, in fact, none of the event generators of these or other programs have been tuned to match the results of fixed-target charm production. ISAJET uses an independent fragmentation model (44) while JETSET is based on string fragmentation. FRITIOF (45), the newest of the Lund programs, is more relevant for fixed-target experiments. It is aimed at hadron-nucleus collisions and pays more attention to the spectator partons in events.

6. RECENT DATA ON INCLUSIVE CHARM HADROPRODUCTION

The three physical quantities describing the inclusive production of charm particles are the total cross section, the longitudinal momentum distribution, and the transverse momentum distribution. The longitudinal momentum distributions are usually parameterized in terms of the Feynman x, x_F. This is the longitudinal momentum in the center of mass divided by its maximum value:

$$x_F = \frac{p(\text{parallel to the incident beam})}{p(\text{maximum})}. \qquad 2.$$

Data on correlations between the two charm particles produced in an event are discussed in Section 7. The dependence of results on the nuclear target material is discussed in Section 8.

6.1 *Total Charm Cross Section*

Recent data on the total charm cross section for the proton-induced interactions are shown above in Figure 4. The recent data (Table 2) can be easily modeled with the QCD predictions including next-to-leading order contributions and using a charm quark effective mass of about 1.5 GeV. This fits both the absolute cross section and the energy rise from 400 to 800 GeV. There are more experiments on charm production in meson beams, but for cross-section measurements the energy range is still small. More results with a broader incident energy range should be available soon. However, the size of the pion and kaon production cross sections appears to be larger than that for protons at the same incident momentum, consistent with a harder gluon distribution in the mesons.

Table 2 Recent open charm cross section data

Exp't	Particle	# Events	σ or σB $x_F>0$ μb/nuclear	ref.
NA32	D^0	543	$\sigma = 6.3\pm0.3\pm1.2$	(46)
π^-	D^+	249	$\sigma = 3.2\pm0.2\pm0.7$	
230 GeV	All D	792	$\sigma = 9.5\pm0.4\pm1.9$	
	$D_s \to KK\pi$	60	$\sigma B = 0.067\pm.011\pm.010$	
	D*	147	$\sigma = 3.4\pm0.3\pm0.8$	
	$\Lambda_c \to pK\pi$	154	$\sigma B = 0.18\pm.02\pm.03$	(47)
			$R(\Lambda_c/\overline{\Lambda}_c) = 0.99\pm.16$	
	$\Xi_c^0 \to pKK^*(892)$	3	$\sigma B = 0.019\pm.011^{+.066}_{-.009}$	(48)
	$\Xi_c^+ \to \Xi^-\pi^+\pi^+$	3	$\sigma B = 0.13\pm0.08^{+.07}_{-.05}$	(49)
	$\Xi_c^+ \to \Sigma^+K^-\pi^+$	2	$\sigma B = 0.012$	(49)
E769	All D	2283	$8.7\pm0.7\pm0.1$	(50)
π^-	$D_s \to \phi\pi + K^*K$	29	$\sigma B = 0.036\pm0.015$	(51)
250 GeV				
NA32	All D	31	$\sigma = 8.5\pm1.6\pm1.2$	(46)
K^-			$(K^-N \to D/\pi^-N \to D) = 0.9\pm0.2$	(46)
230 GeV				
	$D_s \to KK\pi$	4	$\sigma B = 0.11\pm0.06\pm0.02$	(46)
	Λ_c	7	All $\overline{\Lambda}_c$!	(47)
	Ξ_c^0	1		(48)
	Ξ_c^+	1		(49)
NA32	All D	9	$1.5\pm0.7\pm0.1$	(52)
P				
200 GeV				
E653	$c,\bar{c}$ (from D^0)	146 + 35	$38\pm3\pm13$ μb	(53)
p	$c,\bar{c}$ (from D^+)		$38\pm9\pm14$ μb	
800 GeV			All x_F	

6.2 *Longitudinal Momentum Distributions*

The longitudinal momentum distribution is conventionally fit to the form

$$\frac{d\sigma}{dx_F} \propto (1 - x_F)^n. \qquad 3.$$

This form was motivated historically by kinematic considerations at high x_F, and there are predictions by Gunion (54) for the power n at high x_F. Nevertheless, the form appears to fit the data down to quite low values of x_F. Most of the new experiments have significant acceptance only for $x_F > 0$. Although the theory predicts a maximum at forward x_F for mesons, data are not yet sufficiently precise to verify such an offset. Figure 6 (50) is an example of the scaled longitudinal momentum distribution and a fit to the above form. A single contribution with this functional form seems to fit the data over two orders of magnitude. There is no evidence (for 10–20% of the total cross section) for a second contribution to the x_F distribution as reported earlier (55).

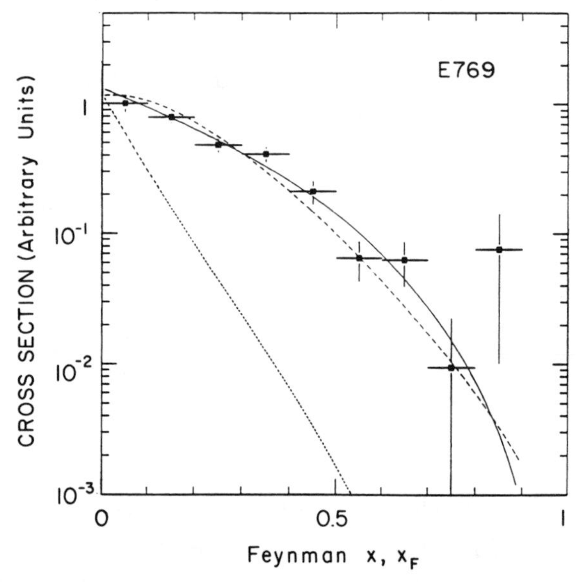

Figure 6 Scaled longitudinal momentum distribution of $D^{\pm}$ mesons. The solid curve is the best fit to the data using Equation 3. The dashed curve is for charm quark production from Nason et al (37). The dotted curve is this distribution convoluted with charm fragmentation functions derived from e^+e^- annihilations (50).

Table 3 Recent x_F and p_t dependence data

Expt	Beam	Particle	n	b(GeV^{-2})	ref.
NA32	200 GeV π^-	All D	$2.5\pm^{+0.4}_{-0.3}$	$1.06^{+0.12}_{-0.11}$	(52)
		Leading D	$2.1\pm^{+0.5}_{-0.4}$	$1.22^{+0.20}_{-0.17}$	
		Non Leading	$3.3\pm^{+0.6}_{-0.5}$	$0.91^{+0.12}_{-0.11}$	
NA32	230 GeV π^-	All D	3.74±0.23±0.37	0.83±0.03±0.02	(46)
		Leading D	$3.23\ ^{+0.30}_{-0.28}$	0.74±0.04	
		Non Leading	$4.34\ ^{+0.36}_{-0.35}$	0.95±0.05	
		Λ_c	$3.52\ ^{+0.51}_{-0.49}$	0.84±0.09	
		D_s	$3.94\ ^{+0.93}_{-0.86}$	0.59±0.10	
E769	250 GeV π^-	$D^{\pm}$	3.21±.24		(50)
		Leading D	2.84±0.31		
		Non Leading D	3.50±0.36		
		D^0	4.2±.5	1.09±.15	(57)
WA82	340 GeV π^-	All D	2.9±0.3	0.78±0.04	(58,59)
NA27	360 GeV π^-	All D	3.8±0.6	1.18±0.17	(55)
		Leading D	$1.8\ ^{+1.6}_{-0.5}$		
		Non Leading	$7.9\ ^{+1.6}_{-1.4}$		
NA32	200 GeV K^-	All D	4.7±0.9	$2.7^{+0.7}_{-0.5}$	(52)
NA32	230 GeV K^-	All D	$3.56\ ^{+1.08}_{-0.99}$	$1.36\pm^{0.32}_{0.26}$	(46)
NA32	200 GeV p	All D	$5.5^{+2.1}_{-1.8}$	$1.4^{+0.6}_{-0.4}$	(52)
WA82	370 GeV p	All D	5.5±0.8	0.79±0.08	(58,59)
NA27	400 GeV p	All c	4.9±0.5	1.0±0.1	(60)
E743	800 GeV p	All c	8.6±2.0	0.8±0.2	(61)
E653	800 GeV p	All c	$6.9\ ^{+1.9}_{-1.8}$	$0.84\pm^{0.10}_{0.08}$	(53)
			$6.8^{+2.1}_{-1.9}$ $x_F > 0$		(53)
			$8.3^{+6.0}_{-5.6}$ $x_F < 0$		(53)

Figure 6 also shows two theoretical curves. The dashed line is taken from the first-order calculation of Nason, Dawson & Ellis (37) and comes quite close to fitting the data, though the normalization is arbitrary. Since the calculation is done for charm quarks, the fragmentation process would have to be such that the charm particle takes all the momentum of the charm quark (a δ function for fragmentation!). Using fragmentation functions from e^+e^- annihilation reduces the average x_F. A convolution of one such fragmentation function with the theoretical prediction for charm quarks is shown by the dotted line in Figure 6. Obviously, this simple convolution of fragmentation functions doesn't reproduce the data from hadroproduction on nuclear targets. It has been speculated that the drag due to color strings attached to the opposite anticharm and residual spectator partons is balanced by a pull of other color strings toward spectator partons going in the same direction as the charm quark. Such processes are included in the Lund fragmentation, but overestimate the leading-particle effect (56).

All the data on charm production by pions and kaons (Table 3) exhibit the power-law behavior of Equation 3, with the power in the range of 3 to 4. Protons at similar energy have a softer distribution, corresponding to a larger n. There has been a suggestion that protons have an increasingly soft distribution with increasing energy. However, such an energy dependence is not well established in recent data nor is it explained by current theoretical calculations.

6.3 *Transverse Momentum Distribution*

The data on transverse momentum distributions (Table 3) now extend out to transverse momenta of 4 GeV. The data from WA82 (59), shown in Figure 7, and from E769 (50) extend to large p_t values and suggest a simple exponential behavior beyond the lowest p_t. These experiments and previous experiments with smaller p_t range have traditionally fit their data with a Gaussian distribution:

$$\frac{d\sigma}{dp_t^2} \propto \exp(-bp_t^2). \qquad 4.$$

The average p_t is measured to be approximately 1 GeV. This is in agreement with QCD calculations, which give average p_t values comparable to the charm quark mass.

At high enough incident energy and p_t, there could be a small contribution from bottom quark production to charm production. However, neither the measurements of bottom production at these energies nor the statistics at the highest p_t values are sufficient to demonstrate such an effect. Both E791, the follow-on experiment to E769 at Fer-

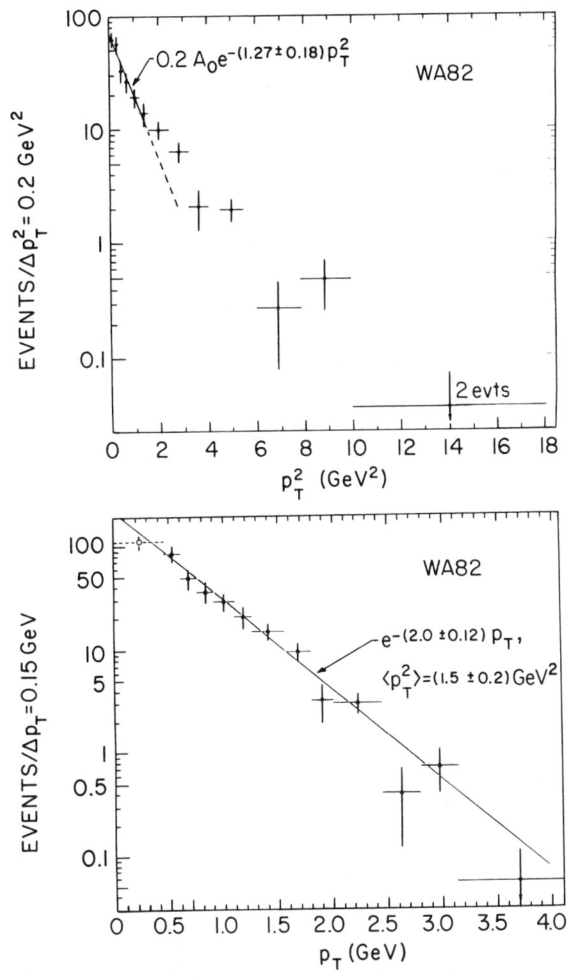

Figure 7 Transverse momentum distributions from WA82 for D mesons indicating the preference for a simple exponential fit at high transverse momentum.

milab, and WA92, the follow-on to WA82 at CERN, may be able to address this possibility.

6.4 *Comparing Structure Functions*

The measurements of the total cross section and x_F dependence data are useful for comparing structure functions among mesons and protons. On the basis of what is known so far, it appears that the pion and kaon have similar gluon structure functions. The mass of the strange quark does not appear to change the gluon distribution dramatically. Baryons have softer gluon distributions (i.e. larger n).

6.5 *Fragmentation to Specific Final States*

In the simplest models with equal partition of energy among possible states, one might anticipate (ignoring mass effects) charm final states produced in proportion to the spin factor, $(2J+1)$. This would argue for equal numbers of directly produced D^+ and D^0, and a factor of three for each of the charged states of the D^*. The lowest mass baryon might have a factor of two larger production relative to the single-charge pseudoscalar mesons. Beyond this, one might anticipate a factor of 10 to 20 reduction for each additional strange quark in the final state. Such a picture is grossly consistent with the data for the lowest mass states shown in Table 2. However, detailed quantitative statements require more data and measurements of the absolute branching ratios of the baryons and charm-strange states. Again, it will be interesting to compare results here with the e^+e^- annihilation and photoproduction.

7. CORRELATIONS IN PRODUCTION OF CHARM PAIRS

Observations of both of the charm particles in a hadron-produced event hold promise for additional information on the production process and for tests of QCD. In addition, if the production dynamics are known, it is possible to determine absolute branching ratios for charm decays from such data.

Observations of pairs of charm particles in events fall into two classes. In the first class, the individual charm decays are identified by topological reconstruction in bubble chamber pictures (62, 63) or emulsion (64, 65). In the second class of observations, one reconstructs fully one or both final-state charm decays using the more complete information from spectrometers. In the first class of observations, the efficiency for observing a second decay, given the first decay, is quite large, approaching 60–80%. This high efficiency is important, of course, since the number of events with at least one charm decay is limited by the number of events in the sample. In the second class of experiments, the number of double charm events is more typically 1% of the number of inclusive charm decays observed. This difference is due not only to the geometrical acceptance of the detectors (typically smaller by at least a factor of two), but more so to the much smaller fraction of charm decays that can be fully reconstructed in the existing detectors. The much larger sample of inclusive charm decays can be used to overcome the branching fraction limitation. In addition, the use of fully reconstructed charm pairs allows the measurement of more precise physical quantities since both longitudinal and transverse momenta are explicitly measured. In the topological measurements, one must basically be satisfied with azimuthal correlations or heavy dependence on models for unfolding other kinematic correlations. Table 4 lists the recent experiments with published results on double charm events. So far, the physics results are scattered over a variety of topics. These include kinematic correlations, meson-meson and meson-baryon production ratios, and qualitative tests of QCD with evidence for next-to-leading order processes.

Angular correlations are the most plentiful results from the various experiments. The charm and anticharm particles are expected to be produced in opposite directions in the transverse plane in the simplest parton models. Only the intrinsic transverse momentum of the incident partons, k_T, modifies this. Indeed, the observed distributions confirm this hypothesis (62–66). However, the width of the correlations is much broader than one predicts from leading order gluon fusion model calculations. This might be interpreted as evidence for the importance of next-to-leading order effects, by which emitted gluons would further modify the nearly back-to-back production of the final charm hadrons. In general, the correlation increases (as shown in Figure 8) with kinematic variables such as the effective mass of the charm pair, the transverse momentum of the charm pair, and the rapidity gap of the charm particles (65). This is as expected in the hardest collisions, tagged by high values of these parameters. On the other hand, E653 observes a polar angle correlation that is more pronounced than predicted by QCD models.

Table 4 Charm correlations

Experiment	Energy	Beam/Target	Number Events	Physics Results	ref.
NA32	230	π^-/Cu	584	Final States	(66)
				Absolute BR's	(66)
				Angular Correlations	(66)
WA75	350	π^-/Emulsion	102	Final States	(64)
				Angular Correlations	(64)
LEBC-EHS	360	π^-/p	53	Angular Correlations	(62)
LEBC-EHS	400	p/p	233	Angular Correlations	(63)
E653	800	p/Emulsion	35	Angular Correlations	(65)

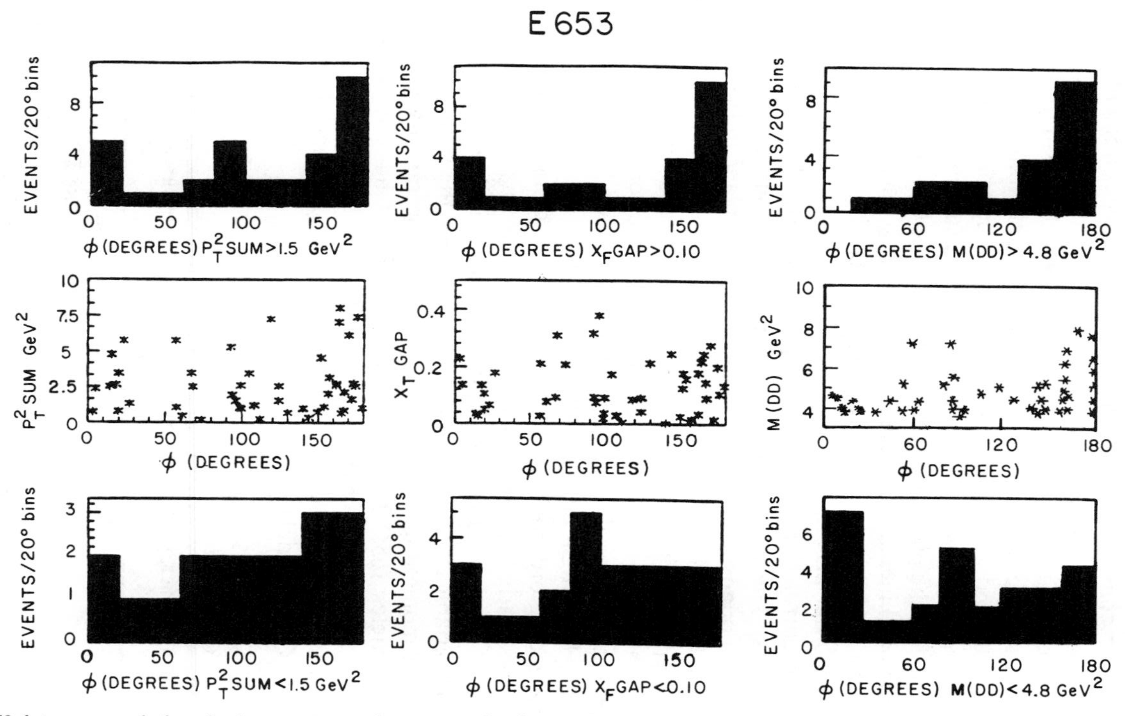

Figure 8 E653 data on correlations in the transverse plane vs production parameters, showing increased correlations with increased transverse momentum squared of the charm pair, the x_F gap between them, and the effective mass of the charm pair.

NA32 has the largest pair sample, with 642 charm pair events (only 584 are identified as charm-anticharm, the others being either cc or $\bar{c}\bar{c}$ in appearance). From the charm-anticharm events, Barlag et al (66) report a dominant fraction of pair-produced charm ($D\bar{D}$ and $\Lambda_c\bar{\Lambda}_c$, etc) relative to associated production ($D\bar{\Lambda}_c$, etc). Among these events, the $D\bar{D}$ seem to be produced back-to-back in the plane perpendicular to the beam. The $D\bar{D}_s$ are not so, and the associated production seems to fall in between. Barlag et al explore the technique of determining absolute branching ratios (67) and obtain (model-dependent) results consistent with the more precise measurements from e^+e^- collisions. The data sample and our general understanding are not adequate yet to determine absolute branching ratios for D_s and charm baryons.

Theoretical work has been limited to leading order QCD calculations so far (68). However, Mangano, Nason & Ridolfi (69) have performed calculations relevant to collider energies and have indicated ongoing work about to be completed that will be relevant for fixed-target experiments.

One should also note that WA75 reported events with four charm particles per event (double associated charm), production at a rate of approxmiately 1% of the total charm pair rate (70).

8. NUCLEAR TARGET DEPENDENCE

The use of small, solid targets is required in order to benefit from solid-state tracking devices. This necessarily raises the issue of the dependence of results on the nuclear material used. Not only is this important for comparing results among experiments with different targets, but also for extrapolating to experiments with hydrogen targets and to theoretical calculations of the basic hadroproduction process. More fundamentally, if the process of producing charm quarks is truly the result of hard parton-parton interactions, one anticipates that the cross section will be proportional to the number of partons and, therefore, to A, the number of nucleons in the target material, with $\alpha = 1$:

$$\sigma = \sigma_0 A^{\alpha}. \qquad 5.$$

This is unlike diffractive production, for example, for which $\alpha = 2/3$. Thus, the A dependence is a measure of the applicability of the QCD calculations. In fact, the early (indirect) measurements by WA78 (71, 72) and E613 (73) indicated a cross section that was proportional to $A^{3/4}$. This contributed to the belief that charm hadroproduction was not a domain in which perturbative QCD calculations could be applied.

There is also the possibility that the structure functions and fragmentation could vary in different nuclei.

The *A* dependence of charm production is best measured by using directly observed charm particles in data taken in a single experiment from a variety of target materials at the same time. Both WA82 at CERN and E769 at Fermilab followed this technique. Many experimental uncertainties cancel in the determination of α from the data. In E769, the beam went through all of the targets at the same time, while in WA82 sections of the beam went through different target materials, resulting in a less complete cancellation of the systematic errors and requiring knowledge of the beam profile and position. Data for D meson production from the four target materials of E769 (56) are shown in Figure 9. The parameterization of Equation 5 represents the data very well. Yet this and other recent results (Table 5) are inconsistent with the earlier indirect measurements.

Neither of the earlier experiments, WA78 and E613, showed significant x_F dependence in their data. However, the E613 collaboration described the result as "suggestive" of an x_F dependence. This may have been motivated by the well-measured nuclear target dependence for lower mass mesons and baryons, summarized by Barton et al (74). This possibility was supported in the review by Tavernier (5). However, the most recent data do not support the suggestion (Figure 10) even though the data cover the same x_F range (75). In trying to explain the differences in the experimental results, it is easiest to imagine changes in experimental conditions during the different target configurations used in the beam dump experiments e.g. the beam halo or other instrumental effects. In any case, the systematic uncertainties are inherently smaller for the more recent data.

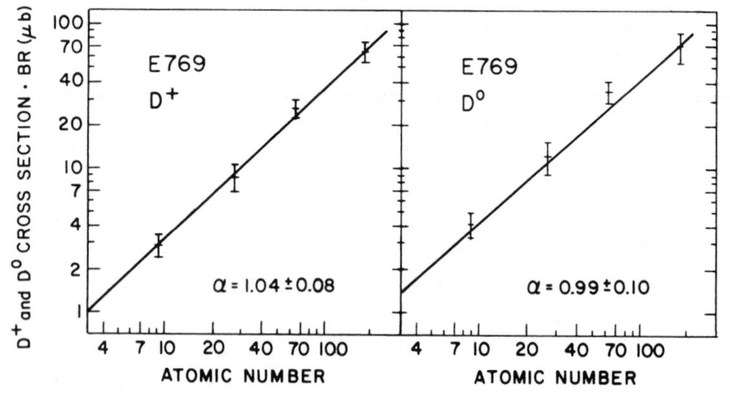

Figure 9 Nuclear target dependence of the D meson cross sections from E769.

Table 5 A-Dependence charm data

Experiment	Process	α	x_F range	ref.
WA82	$pA \rightarrow D$'s	$0.88^{+0.04}_{-0.05}$	0. <x_F> 0.7	(57)
			<x_F>=0.24	
E769	$\pi A \rightarrow D^+$	1.04±0.08	0. < x_F < 0.5	(56)
	$\pi A \rightarrow D^0$	0.99±0.10	0. < x_F < 0.5	(56)
WA78	π-yield of prompt single μ^+	0.76±0.08	x_F > 0.2	(71)
	π-yield of prompt single μ^-	0.83±0.06	x_F > 0.2	(71)
	p yield of prompt single μ^+	0.79±0.12	x_F > 0.1	(72)
	p yield of prompt single μ^-	0.76±0.13	x_F > 0.1	(72)
E613	p prompt ν_e event rates	0.75±.0.05	< x_F > = 0.15	(73)

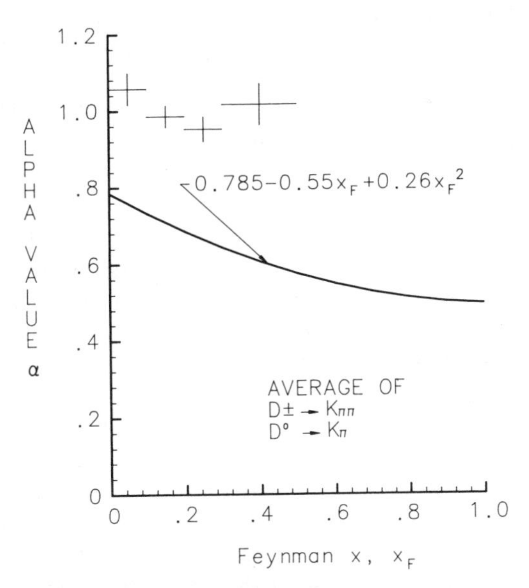

Figure 10 Dependence of the slope parameter α from Figure 9 vs the scaled longitudinal momentum. The solid curve is from Barton et al (74) and summarizes lower mass, non-charm meson and baryon data.

The very high statistics measurements of A dependence for the production of hidden charm in the hadroproduction of the J/ψ and ψ′ (76) gave an α value of 0.920 ± 0.008. One could imagine that the open charm and hidden charm A dependence would be the same. At this point, the data are consistent with this view. Higher statistics will be needed for the charm particles for better A dependence measurements as a function of x_F and p_t. It will be interesting to compare these dependences with those observed for the J/ψ.

The motivation for developing theoretical models to explain an A dependence that is different from $\alpha = 1$ is greatly reduced by the recent data. Nevertheless, small deviations from $\alpha = 1$ may be due to absorption of the incident hadrons (77), changes in the gluon structure functions as a function of the nuclear environment (78), or color screening effects on the outgoing charm quark (79).

9. SUMMARY AND OUTLOOK

Charm hadroproduction is in the midst of its second generation, in both experiment and theory. This is a highly productive period, one with high statistics, open acceptance and good signal/background ratios from experiments, and a consistent theoretical picture based on quantum chromodynamics calculated to next-to-leading order. Progress in experimental technique beyond the first generation has been dependent on precision tracking of charged particles in a very large sample of fixed-target data. Those data are in general, even quantitative, accord with QCD predictions. The predictions are based on the dominance of the gluon-gluon fusion process, but they require detailed and quantitative understanding of the next-to-leading order QCD processes.

There is no recent evidence for the large effects that are not easily explained in QCD models. Recent data do not show steep energy dependence of the cross section, nor evidence for large leading-particle production or intrinsic charm in nucleons.

There is evidence for similarity in the gluon structure functions in pions and kaons, and softer gluon distributions in protons. The fragmentation process for charm quarks appears to be very hard, i.e. the charm particle takes essentially all the energy of the initial charm quark. The hadronization process remains to be fully explained; Monte Carlo simulations have not yet been tuned to match hadronization data.

The most recent measurements of nuclear target dependence also encourage a QCD interpretation of the production process. Earlier indirect measurements have been supplanted by higher statistics direct measurements that indicate a cross section nearly proportional to the number of nucleons in the target, as anticipated for constituent hard-scattering processes.

At the time this review was written, final results from the high statistics experiments that recorded data at the end of the 1980s were just appearing in print. Yet to come are the detailed interpretations of these data and the results of even higher statistics data to be taken in the 1990s. New techniques to further exploit the power of solid-state detectors used for precision tracking are needed for any major increase in capability for hadroproduction of charm experiments. Such improvements would be useful both in the on-line selection of charm events and in the rate of recording of interesting events. Although progress is being made, the technologies are both expensive and time consuming to develop. Such improvements, as well as a new kinematic regime of hadroproduction data, may come from hadron collider experiments, to which the high-energy physics community is devoting much more money and effort. In the meantime, charm hadroproduction is enjoying a period rich in data and providing an opportunity for detailed comparisons with theory, a period that should see increases in both basic understanding and quantitative parameterization.

ACKNOWLEDGMENTS

I gratefully acknowledge my colleagues on the series of charm experiments at the Fermilab Tagged Photon Laboratory for enlightening discussions over the many years of work there, and recently, especially to Paul Karchin, Robert Jedicke, Simon Kwan, and Jeff Spalding. I also acknowledge Leonardo Rossi for discussions of recent CERN data and Keith Ellis for explaining many theoretical points. Thanks go to Sudeshna Banerjee, Tom Carter, Austin Napeir, Jean Slaughter, and Keith Thorne for help with this article, and to Kristen Ford for her extra effort and for expertly preparing this manuscript.

Work on this review was supported by the US Department of Energy under contract No. DE-AC02-76CH03000.

Literature Cited

1. Glashow, S. L., Iliopoulos, J., Maiani, L., *Phys. Rev.* D2:1285 (1970)
2. Aubert, J. J., et al, *Phys. Rev. Lett.* 33:1404 (1974)
3. Augustin, J. E., et al, *Phys. Rev. Lett.* 33:233 (1975)
4. Kernan, A., VanDalen, G., *Phys. Rep.* 106:297 (1984)
5. Tavernier, S. P. K., *Rep. Prog. Phys.* 50:1439(1987)
6. Brodsky, S. J., et al, *Phys. Lett.* 93B:451 (1980); Brodsky, S. J., Peterson, C., Sakai, N., *Phys. Rev.* D23:2745 (1981); Bertsch, G., et al, *Phys. Rev. Lett.* 47:297 (1981)
7. Combridge, B. L., *Nucl. Phys.* B151:429 (1979)
8. Odorico, R., *Nucl. Phys.* B209:77 (1982)
9. Fritzsch, H., Streng, K. H., *Phys. Lett.* 78B:447 (1978); Halzen, F., Keung, W.-Y., Scott, D. M., *Phys. Rev.* D1631:28 (1983)
10. Likhoded, A. K., Slabospitskii, S. R.,

Suslov, M. V., *Sov. J. Nucl. Phys.* 38:433 (1983); Roy, D. P., Desai, B. R., *Z. Phys.* C22:149 (1984); Cooke, M. A. L., *Z. Phys.* C26:71 (1984); Brodsky, S. J., Gunion, J. F., Soper, D. E., *Phys. Rev.* D36:2710 (1987)
11. Nason, P., Dawson, S., Ellis, R. K., *Nucl. Phys.* B303:607 (1988)
12. Aguilar-Benitez, M., et al, *Nucl. Instrum. Methods* A258:26 (1987)
13. Ammar, R., et al, *Phys. Lett.* B183:110 (1986)
14. Aoki, S., et al, *Phys. Lett.* B187:185 (1987)
15. Kodama, K., et al, *Nucl. Instrum. Methods* A289:146 (1990)
16. Bailey, R., et al, *Z. Phys.* C30:51 (1986)
17. Bailey, R., et al, *Z. Phys.* C28:357 (1985)
18. Bailey, R., et al, *Nucl. Instrum. Methods* 213:201 (1983)
19. Adamovich, M., et al, *Nucl. Instrum. Methods* A309:401 (1991)
20. Raab, J. R., et al, *Phys. Rev.* D37:2391 (1988)
21. Appel, J. A., in *Proc. Europhysics Study Conf. on High-Energy Physics,* Erice, Italy, 1981, ed. G. Bellini, S. C. C. Ting. New York/London: Plenum (1984), p. 555
22. Damerell, C., et al, *Comput. Phys. Commun.* 22:349 (1982)
23. Daum, C., et al, *Nucl. Instrum. Methods* 217:361 (1983)
24. Anthonioz-Blanc, J., et al, CERN-DD/80/14 CERN (1980); Adamovich, M., et al, *IEEE Trans. Nucl. Sci.* 37:236(1990)
25. Nash, T., in *Proc. Computing and High Energy Physics Conf.,* Tsukuba, Japan, FERMILAB-CONF-91-87 (1991)
26. Haber, C., et al, *Nucl. Instrum. Methods* A289:388 (1990)
27. Ikeda, M., et al, in *Proc. 8th Topical Workshop on Proton-Antiproton Collider Physics,* Castigilione, Italy, 1989, ed. G. Bellettini, A. Scribano. Singapore: World Scientific (1990), p. 558
28. Abe, F., et al, *Phys. Rev. Lett.* 64:348 (1990)
29. Arnison, G., et al, *Phys. Lett.* 147B:222 (1984)
30. Fisher, C. M., in *Proc. Photonics Applied to Nuclear Physics,* Strasbourg, 1981., CERN 82-01, Geneva (1982), p. 92; Harigel, G. G., *Nucl. Instrum. Methods* A257:614 (1987); Harigel, G. G., *Nucl. Instrum. Methods* A279:249 (1989)
31. Majka, R. D., et al, *IEEE Trans. Nucl. Sci.* 36:63 (1989); Sandweiss, J., et al, *Phys. Rev. Lett.* 44:1104 (1980); Eckardt, V., Wenig, S., *Nucl. Instrum. Methods* 213:217 (1983)
32. Angelini, C., et al, *Nucl. Instrum. Methods* A289:342 (1990); Ruchti, R., et al, *IEEE Trans. Nucl. Sci.* 35:441 (1988)
33. Amendolia, S. R., et al, *Nucl. Instrum. Methods* 226:78 (1984); Coteus, P., et al, *IEEE Trans. Nucl. Sci.* 32:585 (1984); Barlag, S. et al, *Z. Phys.* C37:17 (1987)
34. Shapiro, S., et al, *Nucl. Instrum. Methods* A257:580 (1989); Gaalema, S., et al, in *Proc. Int. Industrial Symp. on the Supercollider,* New Orleans, ed. M. McAshan. NewYork/London: Plenum (1989); Anghinolfi, F., et al, CERN/ECP 91-26 CERN (1991); Parker, S., et al, *Nucl. Instrum. Methods* A275:494 (1989); Heinemann, B., et al, *Nucl. Instrum. Methods* A305:517 (1991)
35. Gluck, M., Owens, J. F., Reya, E., *Phys. Rev.* D17:2324 (1978); Combridge, B. L., *Nucl. Phys.* B151:429 (1979)
36. Collins, J. C., Soper, D. E., Sterman, G.,*Nucl. Phys.* 263B:37 (1986)
37. Nason, P., Dawson, S., Ellis, R. K., *Nucl. Phys.* B327:49 (1989)
38. Beenakker, W., et al, *Phys. Rev.* D40:54 (1989); Beenakker, W., et al, *Nucl. Phys.* B351:507 (1991)
39. Berger, E., in *Physics at Fermilab in the 1990's,* Breckenridge, Colorado, ed. D. Green, H. Lubatti. Singapore: World Scientific (1989), p. 70 (with additional data from Refs. 52 and 53)
40. Sjöstrand, T., *Int. J. Mod. Phys.* A3:751 (1988)
41. Paige, F. E., Protopopescu, S. D., in *Proc. 1986 Summer Study on the Physics of the Superconducting Super Collider,* Snowmass, CO, ed. R. Donaldson, J. Marx (1986), p. 320
42. Bengtsson, H.-U., Sjöstrand, T., *Comput. Phys. Commun.* 46:43 (1987)
43. Ali, A., van Eijk, B., ten Have, I., *Nucl. Phys.* B292:1 (1987)
44. Field, R. D., Feynman, R. P., *Nucl. Phys.* B136:1 (1978)
45. Andersson, B., Gustafson, G., Nilsson-Almqvist, B., *Nucl. Phys.* B281:289 (1987); Nilsson-Almqvist, B., Stenlund, E., *Comput. Phys. Commun.* 43:387 (1987)
46. Barlag, S., et al, *Z. Phys.* C49:555 (1991)

47. Barlag, S., et al, *Phys. Lett.* B247:113 (1990)
48. Barlag, S., et al, *Phys. Lett.* B236:495 (1990)
49. Barlag, S., et al, *Phys. Lett.* B233:522 (1989)
50. Wu, Z., PhD dissertation, Yale Univ., New Haven, CT (unpublished) (1991)
51. Jedicke, R., PhD dissertation, Univ. Toronto, Toronto, Canada (unpublished) (1991)
52. Barlag, S., et al, *Z. Phys.* C39:451 (1988)
53. Kodama, K., et al, *Phys. Lett.* B263:573 (1991)
54. Gunion, J. F., *Phys. Lett.* 88B:150 (1979)
55. Aguilar-Benitez, M., et al, *Phys. Lett.* B161:400 (1985)
56. Gay, C. W., PhD dissertation, Univ. Toronto, Toronto, Canada (unpublished) (1991)
57. de Mello Neto, J. T., PhD dissertation, Centro Brasileiro de Pesquisas Fisicas, Rio de Janeiro, Brazil (unpublished) (1992)
58. Adamovich, M., CERN-PRE-91-095 (1991)
59. Antinori, F., et al, Presented at Aspen Center for Physics, Aspen, CO, January 11, 1991
60. Aguilar-Benitez, M., et al, *Z. Phys.* C40:321 (1988)
61. Ammar, R., et al, *Phys. Rev. Lett.* 61:2185 (1988)
62. Aguilar-Benitez, M., et al, *Phys. Lett.* 164B:404 (1985)
63. Aguilar-Benitez, M., *Z. Phys.* C40:321 (1988)
64. Aoki, S., et al, *Phys. Lett.* B209:113 (1988)
65. Kodama, K., et al, *Phys. Lett.* B263:579 (1991)
66. Barlag, S., et al, *Phys. Lett.* B257:519 (1991)
67. Barlag, S., et al, *Z. Phys. Lett.* C48:29 (1990)
68. Kunszt, Z., Pietarinen, E., *Nucl. Phys.* B164:45 (1980); Kunszt, Z., Gunion, J. F., *Phys. Lett.* 178B:296 (1986); Ellis, R. K., Sexton, J. C., *Nucl. Phys.* B269:445 (1986)
69. Mangano, M., Nason, P., Ridolfi, G., GEF-th-10/1991 and UPRF-91-308 (1991)
70. Aoki, S., et al, *Phys. Lett.* B187:185 (1987)
71. Cobbaert, H., et al, *Phys. Lett.* B191:456 (1987)
72. Cobbaert, H.., et al, *Phys. Lett.* B206:546 (1988)
73. Duffy, M. E., et al, *Phys. Rev. Lett.* 55:1816 (1985)
74. Barton, D. S., et al, *Phys. Rev.* D27:2580 (1983)
75. Alves, G., PhD dissertation, Centro Brasileiro de Pesquisas Fisicas, Rio de Janeiro, Brazil (unpublished) (1992)
76. Alde, D. M., et al, *Phys. Rev. Lett.* 66:133 (1991)
77. Gavin, S., Gyulassy, M., *Phys. Lett.* B214:241 (1988)
78. Close, F. E., Qiu, J., Roberts, R. G., *Phys. Rev.* D40:2820 (1989); Aubert, J. J., et al, *Phys. Lett.* 152B:433 (1985); Sokoloff, M. D., et al, *Phys. Rev. Lett.* 57:3003 (1986)
79. Mueller, A., in *Proc. 17th Recontre de Moriond,* ed. Tran Thanh Van, J., Gif-sur-Yvette, France: Editions Frontieres, (1982) p. 13; Brodsky, S. J., in *Proc. 13th Int. Symp. on Multiparticle Dynamics,* ed. W. Kittel, W. Metzger, A. Stergiou. Singapore: World Scientific (1983), p. 963; Appel, J. A., et al, BNL-45319, Brookhaven, NY; in *Proc. for DPF Summer Study on High Energy Physics in the 1990s,* Snowmass, CO (1990)

Annu. Rev. Nucl. Part. Sci. 1992. 42:401–446

EXCITATION ENERGY DIVISION IN DISSIPATIVE HEAVY-ION COLLISIONS

J. Tõke and W. U. Schröder

Department of Chemistry and Nuclear Structure Research Laboratory, University of Rochester, Rochester, New York 14627

KEY WORDS: damped heavy-ion collisions, energy dissipation, transport phenomena

---

CONTENTS

1. INTRODUCTION

In the early stages of heavy-ion reaction studies, during the 1970s, the presence of a few major interaction modes (1, 2) was established in the total reaction cross section, based on characteristic correlations between the experimental observables. Subsequently, a simple picture of the collision scenario was proposed and adopted; it utilizes the fact the de Broglie wave lengths for moving massive ions are short, and thus permit the use of classical trajectories. Such theoretical trajec-

0163-8998/92/1201-0401$02.00

tories depend on the initial conditions expressed in terms of impact parameter, relative velocity, mass, and atomic number of the colliding nuclei, as well as on the assumptions regarding the macroscopic forces acting between these nuclei. Such macroscopic forces arise from conservative Coulomb and nuclear interactions and from friction-like nonconservative or dissipative interactions, which reflects the presence of couplings between the macroscopic and microscopic degrees of freedom.

A classical view of the interaction, illustrated in Figure 1, distinguishes four qualitatively different types of reaction scenarios. For very large impact parameters associated with distant collisions, a negligible overlap of the mass distributions of the colliding nuclei occurs, and all induced processes are attributable solely to the well-known Coulomb interaction. These processes include elastic scattering and Coulomb excitation.

With decreasing impact parameter, the maximum attained overlap of the mass distributions increases, which first results in "weakly" inelastic processes such as direct reactions. With a further decrease in

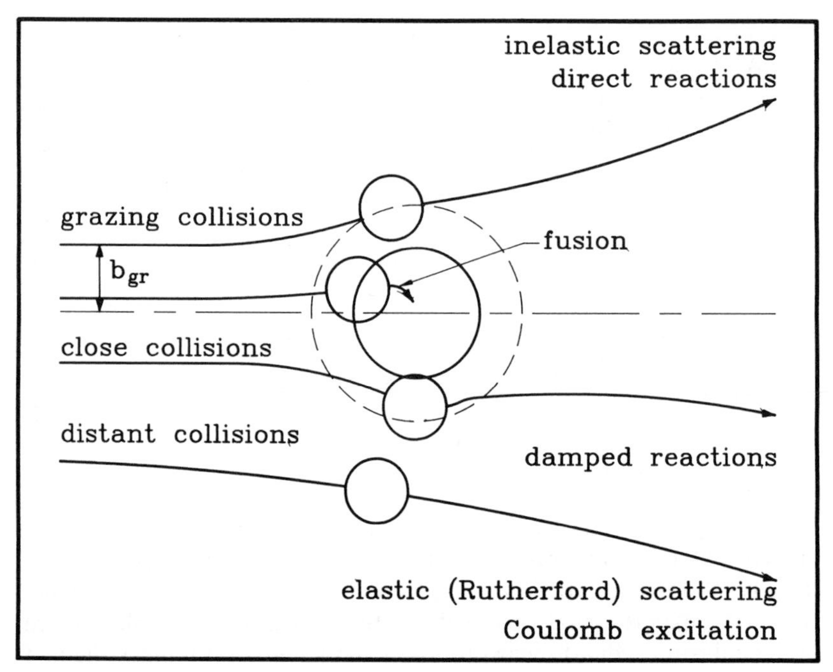

Figure 1 Classes of heavy-ion collisions associated with different values of impact parameters.

impact parameter, the average trajectories of the interacting nuclei begin to show systematic deviations from those predicted for the Coulomb interaction alone. The latter signature is commonly used to define so-called grazing collisions, i.e. those collisions whose trajectories emerge asymptotically at angles at which the elastic scattering cross section drops to one quarter of the value expected for pure Coulomb scattering.

With yet a further decrease in impact parameter, the matter distributions of the interacting nuclei overlap for a sufficiently long time and to a sufficiently large extent to lead to noticeable dissipation of the available kinetic energy into the intrinsic degrees of freedom of the two fragments. When the impact parameter is not too small, the disruptive centrifugal and Coulomb forces eventually overcome the cohesive nuclear forces and two fragments emerge with a relatively good memory of their history, i.e. of their original masses, atomic numbers, and directions of motion. These collisions, termed dissipative or damped collisions, are the subject of the present review. When the impact parameter is reduced further, the two nuclei may fuse and form a long-lived composite nucleus. Products of all the reactions considered above are usually excited and decay via the emission of particles and γ rays, or via fission.

One of the most prominent features of damped collisions is the conversion of a significant part of the initial macroscopic or collective kinetic energy into microscopic or intrinsic energy (heat) of the fragments, in an irreversible dissipative process. Results of the early experiments on isotopic and isobaric fragment distributions were taken as indicative of a completed equilibration of the dissipated kinetic energy in the dinuclear system formed transiently in a damped reaction (3). Generally, dissipation of kinetic energy is reflected in the experimentally observed loss of total kinetic energy E_{loss} (or TKEL) along with the missing energy converted into deformation and rotation energy of the fragments. It is accompanied by other characteristic features such as conversion of relative orbital angular momentum into the intrinsic momenta (spins) of the two interacting fragments and by a broadening and a possible drift of the mass and atomic number distributions with increasing E_{loss}. The general goal of heavy-ion studies in the domain of dissipative collisions is to gain a quantitative understanding of all of the above features in terms of an internally consistent and complete theoretical model. It is highly desirable for such a model to be free of adjustable parameters or ad hoc assumptions that cannot be determined or verified experimentally.

The experimental observable E_{loss} or dissipated energy E_{diss} has traditionally played a key role in heavy-ion reaction studies. It owes such a role to the manifestly evolutionary way in which other observables vary as functions of E_{loss}. For example, observables such as centroids and variances of the atomic or mass number distributions appear to evolve with E_{loss} in a fashion characteristic of diffusion processes, which suggests that E_{loss} can be considered a proper measure of the interaction time. In the energy balance of the reaction, the dissipated kinetic energy appears in the sum of excitation energies of the primary fragments. The way in which this excitation energy is divided between the two primary fragments reflects the properties of heat generation and relaxation mechanisms. Therefore, considerable effort has been focused in the past on an experimental determination of the excitation energy division in damped collisions, as a means of conducting one of the most stringent tests of theoretical models of the interaction of two heavy nuclei and, consequently, of the current understanding of the dynamics of heavy-ion collisions. The present review summarizes the results of theoretical and experimental efforts undertaken to study the excitation energy division in damped collisions. The review is limited to the energy domain below 20 MeV/nucleon, although a brief outlook on studies at higher bombarding energies is given in the concluding section.

2. THEORETICAL MODELS

The dissipation and relaxation of large amounts of kinetic energy in damped nuclear collisions are fundamental time-dependent processes, characteristic of the underlying interaction dynamics, that have attracted theoretical interest since the discovery of this class of reactions. Because of enormous computational or technical difficulties, a full quantum-mechanical microscopic description of these processes has not yet been attempted. Instead, various models have been proposed that describe these complex processes in terms of a weak coupling of the collective motion to a few effective intrinsic degrees of freedom and that predict dissipation of kinetic energy to occur in many small steps, consistent with experimental observations (1, and references therein). The simplest, macroscopic models postulate phenomenological friction forces as responsible for this coupling. More detailed microscopic-macroscopic models derive classical friction forces for the collective motion by considering particular nucleonic degrees of freedom. Other models invoke the excitation of collective or diabatic particle-hole modes as intermediate steps in the dissipation process. All these models are inherently limited. For example, some of them consider only the total excitation energy, and hence make no predictions regarding the excitation energy division between the reaction partners. The following overview emphasizes models that implement coupling mechanisms in a more detailed fashion, such that predictions can be made for the excitation energies of the two interacting fragments separately.

2.1 *Macroscopic Models*

Although purely macroscopic models have historically played an important role in the study of damped collisions, their utility in investigating energy relaxation processes is limited to the low kinetic energy losses associated with peripheral collisions. These macroscopic models link the average system trajectories $\{q_i, \dot{q}_i\}$ in the space of some collective variables q_i to the collective Lagrangian L and the Rayleigh dissipation function F, as given by the set of Lagrange-Rayleigh equations:

$$\left(\frac{d}{dt}\frac{\partial}{\partial \dot{q}_i} - \frac{\partial}{\partial q_i}\right) L = -\frac{\partial}{\partial \dot{q}_i} F. \qquad 1.$$

Equation 1 describes a partial conversion of the initial kinetic energy into thermal and collective energies of the reaction partners. However, only collective energies are treated explicitly, which leaves the distribution of the intrinsic thermal energies outside the scope of the model. The collective types of excitation, dominating at low kinetic energy losses, include rigid rotations as well as static and dynamical deformations and shape oscillations of the dinuclear system and its constituents.

Rigid rotations are induced by tangential friction forces resisting the sliding motion of the reaction partners on each others' surface." In the "rolling limit," the orbital angular momentum is reduced from the initial value of ℓ_i to the final $\ell_f = (5/7)\ell_i$. The dissipated angular momentum is apportioned to the nuclei according to their radii R_i, leading to rotational energies of the fragments inversely proportional to their mass numbers, i.e. to the constant ratio $E_1^*/E_2^* = A_2/A_1$. For higher kinetic energy losses, the "sticking limit" can be reached, where the nuclei attain rotational energies in proportion to their moments of inertia, i.e. $E_1^*/E_2^* = (A_1/A_2)^{5/3}$.

The above "sticking limit" is realized in a thermally equilibrated system of two rotating Fermi gases (4). The concept of such a thermal equilibrium has been generalized by Moretto et al (5) to include a more complete set of angular-momentum-bearing modes, such as tilting, bending, twisting, and wriggling of the dinucleus. Assuming that the system of intrinsic degrees of freedom fulfills approximately the function of an external heat bath of temperature T, one calculates the energy contents of the two nuclei in contact from the canonical partition function (5)

$$Z \approx \int dI \, dI_P \exp[-E(I, I_P, I_T)/\tau], \qquad 2.$$

where I, I_P, and I_T are the total angular momentum of the system and the angular momenta of the two fragments, respectively, and τ is the temperature of the system. Since the thermal excitation of the fragments is not considered explicitly in the above models, their application to the problem of excitation energy division is justified at most for rather peripheral, quasi-elastic collisions leading to intrinsically cold reaction fragments.

A somewhat different mechanism of conversion of kinetic energy of relative motion, considered by Broglia et al (6, 7) and others (1, 8), entails excitation of collective surface vibrations of the constituents of the dinuclear system. Such oscillations could be viewed as the doorway states for a class of dissipative collisions. Theoretical calculations proved that this mechanism can explain shapes of the energy spectra of the reaction fragments (9). Although no detailed model calculations have been reported for the correlations between the excitation energies acquired by the fragments, one expects that these excitation energies will vary smoothly with the surface areas of the nuclei, i.e. the total excitation energy will be divided between the two fragments according to a power law $E_1^*/E_2^* = (A_1/A_2)^{2/3}$.

The effect of the dinuclear shape on the average partition of excitation energy and mass, as well as on the associated fluctuations, has been considered explicitly by Brosa et al (10–13), in a model termed "random neck rupture mechanism." In this model, originally developed for fission, a transient equilibrated dinuclear configuration is assumed to be formed, featuring a massive "neck" between the fragments. This neck then ruptures in a random fashion when its length exceeds a certain critical value associated with the hydrodynamical Rayleigh capillarity instability. The division of the total mass, as well as of the considerable surface energy of the pre-scission configuration, between the two fragments is determined by the location of the rupture point. Furthermore, it is assumed that the thermal excitation energy of the system divides between the fragments in proportion to their masses. During the subsequent acceleration by their mutual Coulomb repulsion, the fragments assume spherical shapes, while their deformation energies are converted into intrinsic excitations. Such a mech-

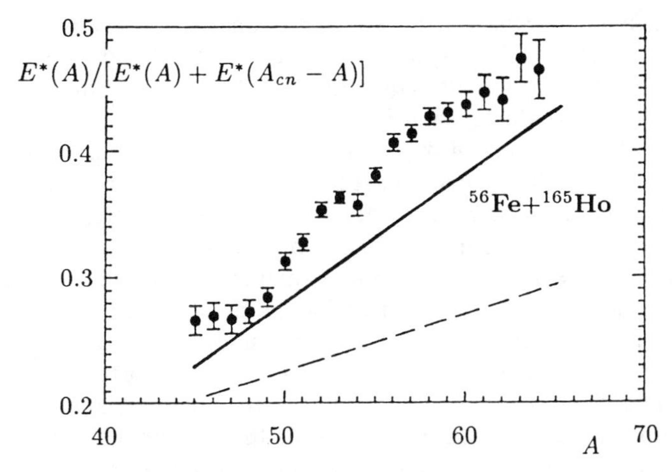

Figure 2 Fraction of the total excitation energy acquired by projectile-like fragments (PLFs) from the reaction ^{165}Ho + ^{56}Fe versus mass number A of the PLF, as predicted by the "random neck rupture model" (*solid curve*). The dashed curve reflects the equilibrium partition of the thermal energy, and the circles represent uncorrected data. $E^*(A)$ and $E^*(A_{cn} - A)$ denote the excitation energies of the PLF and the target-like fragment, respectively, and A_{cn} is the atomic number of the composite system (from 12).

anism induces strong correlations between the final total kinetic energy and the partitions of excitation energy and mass.

Results of model calculations for the reaction ^{165}Ho + ^{56}Fe at E/A = 8.5 MeV are depicted in Figure 2 (12), where the fraction E^*/E_{tot}^* of the total excitation energy E_{tot}^* residing in the lighter, projectile-like fragment is plotted versus fragment mass number A. The solid line represents the model calculations, whereas the dashed curve reflects an expectation based on an equilibrium partition of the thermal excitation energy alone. The solid symbols in Figure 2 represent results of the analysis of experimental data (14), not corrected for the systematical instrumental effects discussed in Section 4.

2.2 *Microscopic-Macroscopic Models*

A more complete explanation of the macroscopic features of damped heavy-ion collisions has to invoke the fundamental microscopic processes of particle-hole excitations and nucleon transfer between the interacting nuclei. It has been shown in several theoretical approaches (1, and references therein) that these modes give rise to transport phenomena when the collective motion is adiabatic, i.e. relatively slow compared to changes on the nucleonic time scale. For example, the conversion of energy of relative nuclear motion into intrinsic excitation energy is a process that can be well approximated (15–18) by the action of classical friction forces, and hence can be described in terms of a classical Rayleigh dissipation function (see Equation 1). Similarly, classical Boltzmann, Master, and Fokker-Planck equations have been derived for the transport of mass and charge in damped collisions (19–21). Since, in such models, a host of experimentally observable, macroscopic phenomena are results of the same underlying transport mechanism, characteristic correlations are predicted to occur between observables. Correlations between mass and energy partition in damped reactions are of particular interest for the present purpose, since the various models proposed in the literature make distinctively different predictions about the presence or absence of such correlations.

Detailed dynamical theories employing the adiabatic approximation for damped reactions assume an intrinsic thermal equilibrium for each individual fragment at any time during the collision, while the dinuclear system overall may retain a nonequilibrium distribution of energy, spin, mass, and charge density. This assumption, leading to Markovian transport processes, appears to be well justified in the regime of low bombarding energies. In the limit of long interaction times, damped reactions are expected to evolve toward a global thermodynamical equilibrium situation characterized by the maximum of the entropy, i.e. of the combined density of states of the system. Using Fermi gas level density formulas with level density parameters a_1 and a_2, respectively, the above maximum entropy condition leads (22) to an average excitation energy partition given by

$$E^*_{\text{ave}} = E^*_{\text{tot}} \frac{a_1}{(a_1 + a_2)}, \qquad 3.$$

with fluctuations represented by the variance

$$\sigma^2_{\text{E}} = 2\tau_{\text{av}} \frac{a_1 a_2}{(a_1 + a_2)}, \qquad 4.$$

where $\tau_{\text{av}} = E_{\text{tot}}{}^*/a_{\text{av}}$ is the average temperature of the dinuclear system. Consequently, in thermal equilibrium, one expects the average excitation energy of a fragment to be proportional to its mass. The magnitude of the fluctuations, on the other hand, is determined by the reduced mass of the system.

A microscopic-macroscopic model of heavy-ion interaction dynamics was proposed by Randrup and collaborators (17, 18, 23–25), in which all transport phenomena are assumed to be mediated by the exchange of independent nucleons between the interacting nuclei. Because of its broad scope, its intuitive clarity, and its self-consistency, this model, termed one-body nucleon exchange (NEM) or transport model, found wide recognition and was further expanded upon in various works (26–30). In this one-body transport model, fundamental drift and diffusion coefficients v_x and D_{xx}, respectively, are derived for all macroscopic observables x of interest, from the differential one-way nucleon fluxes $N'_{ik}(\epsilon)$ between nuclei i and k, at single-particle energy ϵ, and the elementary increments Δx_{ik} associated with individual transfers, according to

$$v_x = \frac{\partial}{\partial t}\bar{x} = \left\langle \int d\epsilon [N'_{12}(\epsilon)\,\Delta x_{12} + N'_{21}(\epsilon)\,\Delta x_{21}] \right\rangle \qquad 5.$$

$$D_{xx} = \frac{1}{2}\frac{\partial}{\partial t}\sigma_x^2 = \left\langle \int d\epsilon\, \frac{1}{2}\,[N'_{12}(\epsilon)(\Delta x_{12})^2 + N'_{21}(\epsilon)(\Delta x_{21})^2] \right\rangle . \qquad 6.$$

In Equations 5 and 6, $\bar{x}$ and $\sigma_x{}^2$ stand for the time-dependent average and the variance, respectively, of the x-distribution that is obtained from the time-dependent joint probability distribution $P(x\ .\ .\ .\ ; t)$, and the angle brackets denote a flux averaging. The time-dependent probability distribution P is assumed to obey a Fokker-Planck equation (1, and references therein). For the transfer of one nucleon, $\Delta A_{12} = 1$, and the elementary amount of energy deposited in the recipient nucleus 2 is calculated as

$$\omega_{12}{}^{(2)} = \epsilon_{F1} - \epsilon_{F2} - \mathbf{p}\cdot\mathbf{u} = F_{12} - \mathbf{p}\cdot\mathbf{u}, \qquad 7.$$

where ϵ_{F1} and ϵ_{F2} are the Fermi energies in the respective nuclei, $\mathbf{p}$ is the momentum of the exchanged nucleon in the donor nucleus, and $\mathbf{u}$ is the relative velocity of the interaction partners. The (hole) excitation energy

$$\omega_{12}{}^{(1)} = \epsilon_F - p^2/2m \qquad 8.$$

generated in the donor nucleus is estimated to be small and comparable in magnitude to the temperature τ of the donor, as most transfers start from single-particle donor levels close to the Fermi surface. The quantity $F_{12} = \epsilon_{F1} - \epsilon_{F2}$ in Equation 7 represents the static driving force for nucleon exchange. Its contribution to the energy loss at finite velocities is usually an order of magnitude smaller than that of the kinetic term $\mathbf{p}\cdot\mathbf{u}$ in this equation. It is worth noting that an evaluation of the transport coefficient v_E for the kinetic energy observable according to Equation 5 results in the classical "window friction formula" (16). In this model, significant amounts of energy are dissipated when large numbers of nucleons are transferred in alternate directions. Therefore, one expects that both nuclei receive on average comparable amounts of excitation energy, in spite of the donor-acceptor asymmetry experienced in a single transfer event. These energies are assumed to be equilibrated in each nucleus independently, such that the nuclear temperature τ is constant over the volume of each of the nuclei. For a very asymmetric system with mass numbers $A_1 < A_2$, an equipartition of the total excitation energy E_{tot}^* between the two nuclei would lead to a strong thermal disparity $\tau_1 > \tau_2$. However, such a temperature disparity will alter the transport characteristics in a way that leads to its decrease in subsequent nucleon exchanges. The strong feedback effect (54) on the temperatures is due to Pauli blocking favoring different net flow directions below and above the Fermi level, as illustrated in Figure 3. It has been suggested (73, 74) that it can also affect the magnitude and direction of the net mass transfer in partially damped collisions when the single-particle barrier between the interacting nuclei is sufficiently high (75). In Figure 3, the occupation probabilities f_1 and f_2 are plotted versus single-particle energy ϵ, for two interacting nuclei at different temperatures, and the transfer probabilities are indicated by arrows. Quantitatively, the average excitation energies per nucleon $E^*/A \propto \tau^2$ develop in time according to

$$\frac{d(\tau_1^2 - \tau_2^2)}{dt} = -\frac{(\tau_1^2 - \tau_2^2)}{t_{rel}} + \frac{(a_2 - a_1)}{(2a_1a_2)}\dot{E}_{loss}. \qquad 9.$$

The t_{rel} in Eq. 9 is a thermal relaxation time, approximated by

$$t_{rel} = \tfrac{2}{3} A_1A_2/(A_1 + A_2) \times [\epsilon_F N'(\epsilon_F)]^{-1}, \qquad 10.$$

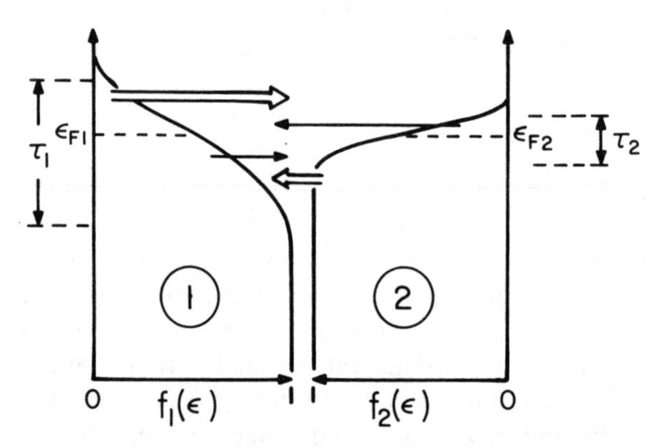

Figure 3 Nucleon transfers between two communicating volumes of Fermi gas with Fermi energies ϵ_{F1} and ϵ_{F2} and temperatures τ_1 τ_2, respectively favored (*open arrows*) and disfavored (*solid arrows*) by the single-particle level occupation probabilities (*solid curves*) in the donor and acceptor volumes.

where ϵ_F is the average Fermi energy. This relaxation time is a dynamical variable, since the differential current $N'(\epsilon)$ depends on the area of the interface between the two nuclei. Typical relaxation times are of the order $t_{rel} \approx 10^{-22}$ to 10^{-21} s, i.e. they are comparable to interaction times. One then expects the thermal equilibrium $\tau_1 = \tau_2$ to be reached only in central collisions associated with relatively long interaction times.

As an example, the history of $^{139}La + ^{40}Ar$ collisions at $E/A = 10$ MeV with $\ell = 135\hbar$ is illustrated in Figure 4, as modeled (31) with a computer code (32) based on the NEM. Of most interest here is the time evolution of the temperatures of the two fragments. The high dissipation rate in the entrance channel causes a temperature differential to build up, with the largest PLF-to-TLF temperature ratios occurring at the beginning of the collision. After passage of the classical turning point at $t = t_{aps}$, only the heavier, colder TLF gains excitation energy. The PLF-to-TLF temperature ratio decreases monotonically, as the PLF transfers approximately as much energy to the colder TLF as it receives in the dissipation process. As a result, the disparity in average temperatures is seen to decrease continuously with time, but it remains finite at the end of the collision. Although significant fluctuations in the fragment temperatures can be expected, no quantitative predictions are available as yet.

As discussed previously in the context of Equations 7 and 8, and disregarding for the sake of simplicity the above heat convection mechanism, one expects an asymmetry in the generation of excitation energy in donor and acceptor nucleus in each nucleon transfer process. Unfortunately, detailed model calculations of the accumulated effects in damped collisions are not yet available. However, it is relatively easy to derive an upper estimate of the magnitude of the effect. Considering two successive nucleon transfers (from nucleus 1 to 2 and back), one calculates from Equations 7 and 8 a rate of change in the excitation energy difference, $\eta = E_1^* - E_2^*$, of

$$d\eta/dt = 2v_A\tau^* = 2D_{AE}, \qquad 11.$$

assuming that the relative velocity $\mathbf{u}$ does not change appreciably in each transfer. In Equation 11, τ^* is an "effective temperature" parameter describing an average energy transferred in a nucleon exchange process, and v_A is the drift coefficient for the mass asymmetry. Equation 11 identifies the static driving force F_{12} (see Equation 7) as responsible for any asymmetry in excitation energy of the interacting fragments that would survive the averaging effect of multiple nucleon exchanges. Since one often finds $v_A \approx 0$ in damped reactions, it follows

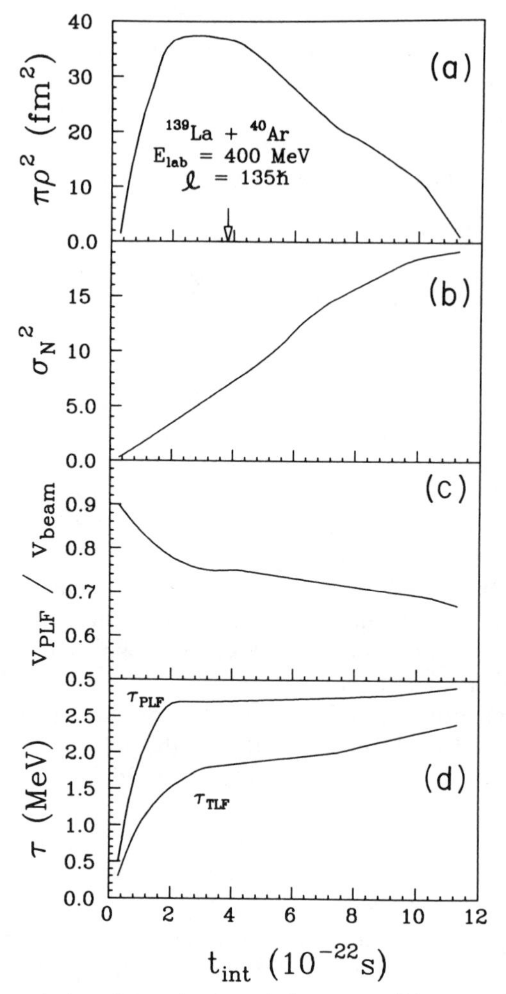

Figure 4 Time evolution of the (circular) window area $\pi\rho^2$ between the projectile-like (PLF) and target-like (TLF) fragments (*a*), the variance of the number of exchanged neutrons σ_N^2 (*b*), the ratio of the PLF and beam velocities v_{PLF}/v_{beam} (*c*), and the fragment temperatures τ_{PLF} and τ_{TLF} (*d*), as predicted by the transport model for the $\ell = 135\hbar$ trajectory for the reaction ^{139}La + ^{40}Ar. The vertical arrow in panel (*a*) indicates the time at which the classical turning point is reached (from 31).

that one would obtain $D_{AE} \approx 0$ also, and no significant asymmetries in the fragment excitation energies are expected to develop over time. A nonzero mixed diffusion coefficient D_{AE} leads to statistical correlations in the joint probability distribution for mass asymmetry and dissipated energy, reflected in a nonzero covariance $\sigma_{A\eta}$ and correlation coefficient $\rho_{A\eta}$. The latter quantity is of the same order of magnitude as the mass-energy correlation coefficient ρ_{AE}, i.e.

$$\rho_{A\eta} \approx \rho_{AE} = \sqrt{v_A F_{12}/v_E}, \qquad 12.$$

where v_E denotes the rate of energy dissipation.

For asymmetric heavy-ion systems, this coefficient assumes values no larger than a few times 10^{-2}, consistent with experimental values. Such small correlation coefficients imply an approximate statistical independence of mass and energy transfer processes. One then expects that, for a given total energy loss, the projectile-like fragments acquire very similar excitation energies, regardless of whether they have suffered a net gain or loss in mass. This is a result characteristic of the diffusion-like nature of the multinucleon exchange processes, in which a given net mass transfer ΔA results from an exchange of a much larger number of nucleons ($N \gg \Delta A$) between the reaction partners. For the rare events associated with the tails of the PLF (or TLF) mass distribution, some net excitation energy difference could result from the donor-acceptor excitation asymmetry; nevertheless the heat convection mechanism discussed earlier is still expected to reduce the effect. Consequently, a correlation between mass asymmetry and energy partition is expected to survive at most in quasi-elastic, few-nucleon transfer reactions.

This one-body transport model has been expanded by Ayik (33) to include two-body effects and particle-hole excitations. Particle-hole excitations are emphasized even more strongly in the model of Gross et al (15, 34), in which they lead to frictional forces that are already strong at internuclear distances of the order of the strong-absorption radius. Since nucleon transfer is less significant at such large distances, this approach leads effectively to a decoupling of nucleon transfer from energy dissipation processes. Neither of these two models has been applied yet in numerical studies of energy division in damped collisions.

For elevated relative velocities of the interacting nuclei, where the adiabatic approximation loses accuracy, Nörenberg (8) predicts new features associated with diabatic interactions to appear. In this approximation, particle-hole states are excited in the two approaching or separating nuclei as a result of diabatic transitions between the interacting single-particle levels of a deformed and changing one-body potential. Although such Landau-Zener transitions are reversible, the ensuing two-body collisions drive the system toward local thermal equilibrium and lead to an irreversible dissipation of kinetic energy. This model of a dissipative process relies strongly on single-particle level schemes in the two-center potential well of the interacting nuclei, and it leads to predictions for the excitation energy division different from those of the NEM. For example, for a heavier nucleus, the density of states at the Fermi surface is higher and these states belong to higher major shells than those of a lighter nucleus. Consequently, at large interfragment separations the heavier fragment is more likely to undergo diabatic single-particle transitions than the lighter one and to acquire on the average more excitation. However, as shown by detailed calculations, the excitation energy per nucleon ϵ^* is predicted to be generally smaller for the heavier reaction partner (35). This behavior results in a nonmonotonic dependence of the energy partition on the total kinetic energy loss that is qualitatively different from that predicted by the nucleon exchange model. In the diabatic model, the ratio $\epsilon_{PLF}^*/\epsilon_{TLF}^*$ is expected to increase moderately with energy loss, from unity to some maximum value, and then decrease with a further loss of kinetic energy (35). So far, energy sharing has been calculated in a simplified diabatic model only for reactions induced by 14.7-MeV per nucleon ^{98}Mo projectiles on ^{147}Sm and ^{238}U targets (35).

3. EXPERIMENTAL METHODS

Many different approaches have been taken to determine the excitation energy division. In the discussion below, they are divided into three classes according to the degree of complexity involved. Experimentally simplest are inclusive or singles types of experiments in which only one massive fragment is detected. More complex measurements include kinematical coincidence experiments, in which both massive projectile-like (PLF) and target-like (TLF) fragments are detected in coincidence, and exclusive studies of the decay products of the primary fragments in coincidence with the secondary massive fragment remnants. Each of the approaches relies to some extent on theoretical simulation calculations. The role of such calculations is discussed in the following section.

3.1 *Theoretical Simulation Calculations*

A determination of the excitation energy division in damped collisions poses significant experimental challenges because of the inherent difficulties of measuring the fragment excitation energies directly. Ideally, one would measure all particles and γ rays associated with the decay of both fragments separately and calculate the two excitation energies of interest in more or less model-independent fashion. Practically, however, one is able to measure only a fraction of the ideal set of data, and extensive simulation calculations are necessary in order to compensate for the missing fraction. The most common and sound strategy consists then in measuring only a limited set of secondary quantities and comparing them to simulation calculations with the aim of identifying the most important properties of the primary products. All primary quantities—such as fragment mass and atomic numbers, total kinetic energy loss E_{loss}, reaction angle, and excitation energy division—enter such simulation calculations as adjustable parameters. Their actual values are determined from a comparison of the results of the simulation calculations to the experimental data.

Because of their complexity and the computational efforts associated with realistic simulation calculations, studies often limit the evaluation of the sensitivity of the available data to a change in the assumptions made in the simulations. Typically, a number of simplified simulation calculations are conducted in the experiment planning stage to verify that certain quantities can be obtained relatively unambiguously from the data to be obtained. A special example of this general strategy, of interest here, is an approach in which primary quantities, such as E_{loss}, total excitation energy E_{tot}^*, excitation energy of the projectile-like fragment E_{PLF}^*, or primary mass number A_{prim}, are reconstructed in an iterative fashion and are used subsequently in comparisons to theoretical model predictions.

A variety of experimental methods have been employed in the past that make different choices of the measured secondary quantities and of the simulation calculations. Such a variety reflects the lack of a single approach that is universally the most accurate for all systems studied, as well as the possibility of significant compromises between the complexity of the experiment and the difficulties of the simulation calculations.

3.2 *Inclusive Experiments*

One of the simplest ways of determining the excitation energy division relies on measuring only the emission angle θ_{lab}, atomic number Z, and kinetic energy E_{lab} of one of the secondary fragments. Such an approach was pursued, for example, by Awes et al (36), whose analysis procedure included the following steps:

1. An effective total kinetic energy loss E_{loss} was reconstructed from the measured E_{lab} and Z assuming binary collision dynamics. No corrections for particle evaporation were made at this stage, and it was assumed that the effective primary mass A_{prim} corresponds to the minimum of the valley of β stability.
2. Simulation calculations were performed for the primary fragments

that used the one-body transport model (17, 18) predictions regarding the fragment atomic and mass number distributions as functions of E_{loss}. These distributions were then used as input for the statistical evaporation code LILITA.

3. In the evaporation simulation calculations, the excitation energy division was varied so as to obtain the agreement between the model predictions and the experimental observations for the centroids of the secondary Z distribution as functions of E_{loss}.

A rather serious deficiency of a simplified experimental approach such as that pursued by Awes et al (36) is its model dependence resulting from an absolute reliance on the predictions of the one-body transport model. Such a model dependence can be significantly reduced by measuring additional experimental observables.

3.3 *Kinematical Coincidence Methods*

One of the possible ways of reducing the reliance on model predictions consists in measuring kinematical quantities associated with both massive fragments in coincidence. Traditionally, such coincidence experiments measure enough observables to solve the binary-kinematics equations. While the inclusive experiments are thought to provide experimental values of secondary, i.e. postevaporative quantities, the kinematical coincidence experiments are commonly thought to provide the values of some of the respective primary quantities as well. This is so because the velocity vectors of the fragments do not change on the average in the course of statistical particle evaporation. Consequently, one can substitute, in the binary-kinematics formulas, velocities of the secondary fragments for those of the primary ones and reconstruct effective values of the primary quantities. It must be kept in mind, however, that such effective values coincide with the true primary values only on average. Event by event, the calculated values not only differ from the true values but may show strong correlations not present in the primary quantities, i.e. correlations of instrumental origin. As pointed out by Tõke et al (37, 38), in order to study correlations between the true primary quantities, it is essential that the simulation calculations account not only for the physical processes of the interaction and subsequent decay of the primary fragments but also for the response of the experimental setup as faithfully as possible.

The minimum requirement for a kinematical coincidence experiment for binary collisions is measurement of three quantities. Such a minimal approach was adopted for example by Benton et al (14) and by Płaneta et al (39), who measured the full velocity vector for the PLF but only the recoil angle for the TLF. A schematic diagram of the experimental setup of Benton et al (14) is shown in Figure 5. As seen in this figure, the velocity v_{PLF} was measured using a time-of-flight method in conjunction with a measurement of the emission angle θ_{PLF}. The recoil angle θ_{TLF} was measured using a position-sensitive detector. Based on the three measured quantities, primary PLF mass and E_{loss} were calculated using binary-kinematics equations.

In the work by Benton et al (14) and Płaneta et al (39), the secondary mass of one fragment was determined independently from the joint measurement of kinetic energy and of time of flight. Subsequently, the fragment excitation energy was deduced from the difference between secondary mass and a primary mass, A_{calc}, calculated event by event using the predictions of the statistical evaporation code PACE (40). As shown by Tõke et al (38), unless a full simulation calculation including the response of the setup is attempted, the calculation of the mass from data obtained in this type of experiment must account for the system-

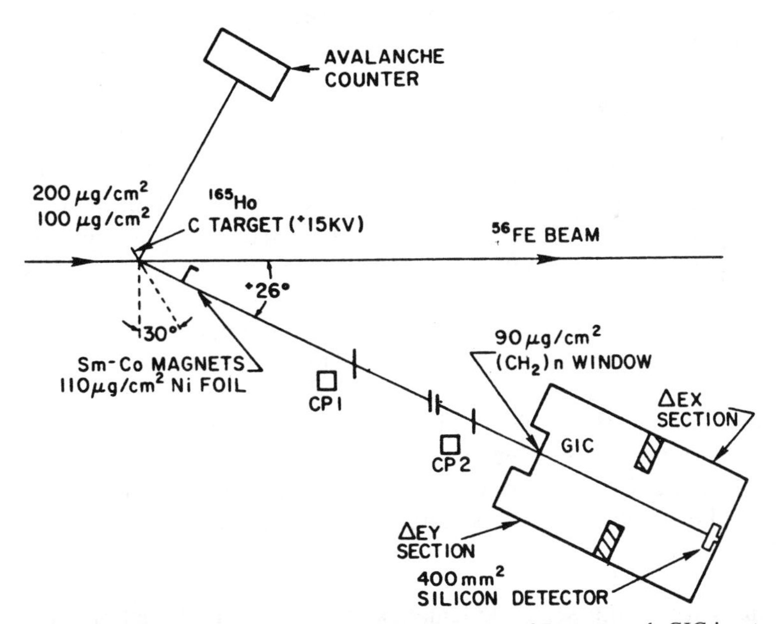

Figure 5 Schematic diagram of the experimental setup of Benton et al. GIC is a two-element x-y position-sensitive gas ionization chamber, measuring energy loss ΔE, energy E, and emission angle θ_{PLF} of the projectile-like fragment, while CP1 and CP2 are microchannel plate detectors used to measure time of flight or velocity v_{PLF} of this fragment. A position-sensitive avalanche counter measures the recoil angle θ_{TLF} of the target-like fragment (from 14).

atical errors induced by finite-resolution effects, i.e.

$$A_{prim} = A_{calc} + (A_{calc} - A_0) \frac{\sigma_A^2}{\sigma_A^2 + \sigma_{res}^2}, \quad 13.$$

where A_0 is the centroid of the primary mass distribution and $\sigma_A{}^2$ and $\sigma_{res}{}^2$ are the variances of the primary mass distribution and of the resolution function respectively. As seen from Equation 13, the difference between the mass calculated event by event and the corrected primary mass is correlated with the mass. Through the essentially linear dependence of the calculated excitation energy on the primary mass, the neglect of such a correction results in nonphysical correlations between the calculated excitation energy and mass of the fragment.

A different choice of kinematical observables was made by Cauvin et al (41), who measured energies and emission angles of both coincident massive fragments, along with the atomic number of the PLF. In this work, the calculated quantities were the total number of evaporated nucleons and the values of the primary masses of the fragments. These calculated quantities were then used as a guide in an iterative procedure of finding the excitation energy division in such a way that full consistency was obtained between the results of the theoretical simulation calculations for the fragment decay and the observed correlations between experimental observables.

A larger set of observables was used by Babinet et al (42); it consisted of the emission angles, energies and times of flight of both PLFs and TLFs, and the atomic number of the PLFs. The two primary masses were then calculated independently from the momentum conservation formulas.

Casini et al (43) showed that one can actually improve the resolution obtained for the primary masses or E_{loss} by measuring the full velocity vectors of all fragments. Such measurements would overdetermine the system in the case of binary and ternary exit channels. The improvement is obtained by using all of the measured velocities v_i^{exp} and by allowing their values to deviate from the measured ones so as to minimize the weighted sum of squares of individual deviations,

$$\Delta^2 = \sum w_i(v_i^{exp} - v_i)^2, \quad 14.$$

while solving the set of kinematical equations. In Equation 14, the summation runs over all measured components of the velocity vectors, the quantities v_i^{exp} and v_i are the experimental and best-fit values of those components, respectively, and w_i are weighting factors reflecting the relative experimental uncertainties of the velocities v_i^{exp}. Subsequently such an improved kinematical coincidence measurement was carried out by Casini et al (44), where it served to reconstruct not only two-body events but also three- and four-body processes associated with sequential fission of primary fragments.

3.4 *Exclusive Light-Particle Measurements*

The excitation energies of the primary fragments are reflected not only in the numbers and properties of the massive remnants of their decay but also in the spectra and angular distributions of their light decay products. Many experiments on the excitation energy division in damped collisions have used this fact and deduced the fragment excitation energies from the quantities related to such light decay products. A vast majority of such experiments utilized neutrons as probes, since neutrons are by far the most abundant decay products of the excited fragments.

Two distinctly different types of exclusive neutron experiments were reported. One measures neutron energy spectra and angular distributions, while the other measures neutron multiplicities in coincidence with one or two massive fragments. In the experiment by Hilscher et al (45), for example, neutron energy spectra were measured at twelve angles, in coincidence with PLFs measured at three different angles. After it had been ascertained that to a sufficient accuracy the observed neutron emission patterns could be understood assuming sequential emission from the fully accelerated primary fragments, simulation calculations were carried out in which the excitation energies of the fragments were allowed to vary so as to produce a reasonably accurate fit to the experimental spectra. It is worth noting that in this type of analysis, both spectral shapes and their absolute magnitudes are explained by the simulation calculations, providing for valuable internal consistency checks of the method.

Another type of exclusive neutron experiment limits itself to the measurement of neutron multiplicities alone. Although several such experiments were carried out using small solid-angle coverage, as was done, for example by Péter et al (46) and Gould et al (47), the hallmark of such studies is the use of large-volume neutron multiplicity meters (NMMs). Because of statistical delays in the response of the NMM to individual neutrons that enter its active volume, one can actually count the reaction neutrons one by one. Simulations show that to obtain a sensitivity of the setup to the excitation energy division one can make either two measurements with one NMM placed at two different positions or one measurement involving two NMMs simultaneously, placed at optimal positions. The latter approach was pursued by Pade et al (48), who operated two NMMs in coincidence with PLF detectors.

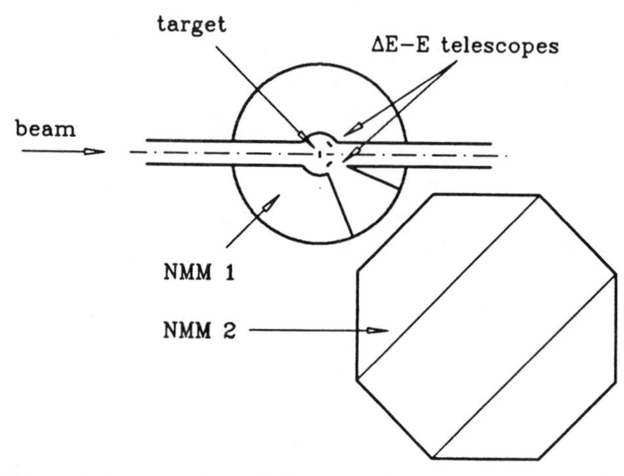

Figure 6 Schematic diagram of a two-NMM experiment of Pade et al. Neutrons emitted isotropically from the slow TLF are counted mostly by the NMM1 surrounding the target; neutrons from the fast PLF, moving along the axis of the conical escape channel, are kinematically focused along this channel and counted mostly by the NMM2 (from 31).

As illustrated in Figure 6, one NMM surrounded the target in an almost 4π geometry, and allowed one to count with high-efficiency neutrons emitted by the slow TLFs. This NMM was equipped with an escape channel in the direction of flight of the fast PLFs through which the kinematically focused neutrons emitted by the PLFs could reach the larger NMM. This larger NMM presented only approximately 4% of 4π coverage for neutrons emitted by the TLF while the effective coverage for the neutrons emitted by the PLF was approximately 30% of 4π. In order to draw conclusions from the correlations between the measured joint neutron multiplicity distributions and the energies and Z values of the coincident PLFs, extensive simulations of the damped collisions, of the statistical decay of the resulting PLFs and TLFs, and of the response of the NMM were necessary (48).

3.5 *Sequential Fission Measurements*

In a few experiments, the sequential fission phenomenon was used to probe the fragment excitation energy. As an example, in an experiment by Vandenbosch et al (49), the TLF sequential fission fragment mass distribution was measured in coincidence with the PLFs. The total excitation energy of the primary fragments was deduced iteratively from the measured energy and atomic number of the PLF. The excitation energy of the TLF was determined from the shape of the fission fragment distribution. The shape of this distribution for the uranium-like fragments involved was known to be of a bimodal character at low excitation energies and to broaden with increasing excitation energy so as to become a Gaussian-like distribution at excitation energies of the order of 100 MeV. More recently, this decay channel was used by Casini et al (44), who deduced the excitation energies of the tin-like fragments from their measured fission probabilities. In order to determine these fission probabilities, a coincidence experiment was carried out in which both binary and sequential-fission reaction channels were kinematically reconstructed.

Exploitation of the sequential-fission phenomenon in the determination of the excitation energy appears to be limited to rather low excitation energies. Furthermore, it clearly suffers from the sensitivity of the fission probability to the atomic number and angular momentum of the fragment, the quantities that cannot be directly determined in a sequential-fission type of experiment.

3.6 *Other Methods*

Relatively few attempts have been made in the past to determine the excitation energy division from an exclusive measurement of light charged particles emitted by the primary fragments. One example of such an experiment is reported by Wilczyński et al (50), who inferred the excitation energies of the PLFs from the charged-particle survival fractions. In order to measure these survival fractions, which represent the probabilities for the fragments to decay solely by neutron and γ emission, the light charged particles emitted by the PLFs were measured with a large solid-angle coverage. Subsequently, simulation calculations utilizing statistical decay codes were used to link the observed charged-particle survival fraction of the PLFs to their excitation energies. In a different type of exclusive light charged-particle experiment of Schmidt et al (51), alpha-particle decay channels were studied to probe the excitation energy of the PLFs emitted in quasi-elastic reactions. In this latter experiment, the excitation energy of the PLF was calculated from the observed kinetic energies of the alpha particle and the PLF, assuming that no other particles were emitted.

4. EXPERIMENTAL RESULTS

Experiments on the excitation energy division in dissipative collisions have addressed two distinctly different types of questions. The first of these questions pertains to whether or not, and on what time scales, an interfragment thermal equilibrium is reached during the relatively short interaction stage of the collision. Quantitatively, such an equilibrium would result in two primary fragments leaving the interaction site with equal temperatures. This type of question is addressed typically in experiments measuring an average excitation energy division. The second type of question concerns the possible difference in excitation energies acquired in the course of a multinucleon transfer reaction by net donor and net acceptor fragments. This latter question is addressed by more complex experiments, in which the excitation energy division is measured as a function of both E_{loss} and mass or atomic number. Results of experiments addressing these two types of questions are discussed below in Sections 4.1 and 4.2, respectively.

4.1 *Thermal Relaxation*

Early experiments on excitation energy division appeared to provide evidence that thermal equilibrium is established between the two interacting fragments on a very short time scale characteristic of collisions leading to low E_{loss} values. The signature of such a thermal equilibrium is attainment of equal temperatures by the fragments, i.e. a division of the total excitation energy between the two fragments approximately in proportion to their masses.

Figure 7 displays the results of measurements made by Gould et al (46) of the multiplicities of neutrons in coincidence with both PLFs and TLFs produced in the reaction $^{197}Au + ^{132}Xe$ at 7.5 MeV/nucleon. With a simultaneous detection of both fragments, this experiment demonstrated the binary character of the reaction. Multiplicities of neutrons associated with PLFs and TLFs were inferred from the neutron spectra measured at a few selected angles. As seen in the upper panel of Figure 7, the ratio of multiplicities of neutrons emitted from TLFs (Au-like) and PLFs (Xe-like) is consistent within the experimental error bars, with a TLF-to-PLF mass ratio of approximately 1.5. This observation is made for any value of the reaction Q value, including the smallest ones measured, associated with short interaction times.

The results of Figure 7 suggest that a thermal equilibrium, reached on a very short time scale persists for the entire duration of the collision. Similar conclusions were reached in several other early experiments involving an experimental determination of neutron multiplicities (46, 52, 53). Figures 8 and 9 show experimental results reported by Tamain et al (52) and by Tserruya et al (53). In Figure 8, the ratio of the neutron multiplicities ν_1 and ν_2, associated with the light and heavy fragment, respectively, from the $^{197}Au + ^{63}Cu$ reaction at 6.3

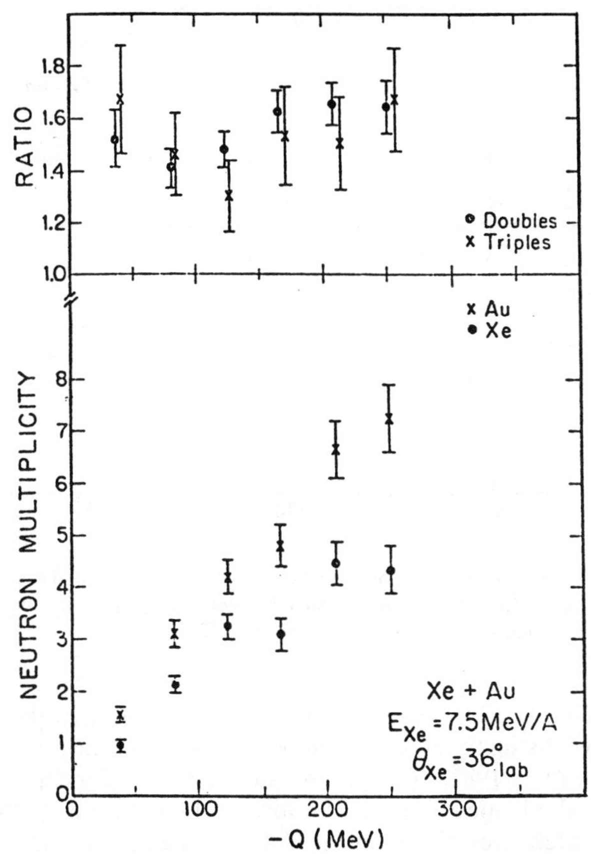

Figure 7 Neutron multiplicities (*lower panel*) associated with TLF and PLF from the reaction $^{197}Au + ^{132}Xe$ at 7.5 MeV/nucleon, and their ratio (*upper panel*) as a function of reaction Q. The two sets of data points in the upper panel refer to double (neutron and PLF) and triple (neutron, PLF, and TLF) coincidence events (from 47).

MeV/nucleon is plotted versus E_{loss}. It is seen that this ratio is approximately equal to the projectile-target mass ratio and stays constant for the full range of E_{loss}. In Figure 9, neutron multiplicities associated with light and heavy fragments from the $^{166}Er + ^{86}Kr$ reaction at 11.9 MeV/nucleon are plotted versus total excitation energy, which is directly related to E_{loss}. Based on these data, shown in Figure 9 for three different mass bins, Tserruya et al concluded that thermal equilibrium is attained virtually immediately in a collision and then sustained throughout the whole collision history (53).

After these early neutron multiplicity studies, experimental evidence accumulated, however, showing that in the initial stages of the damped

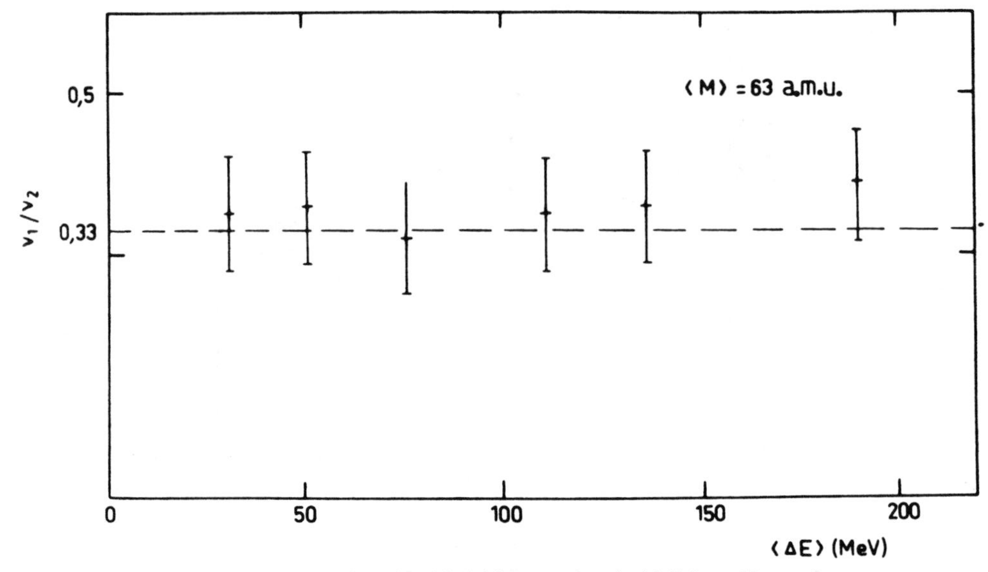

Figure 8 Experimental ratios of multiplicities associated with light and heavy fragments from the $^{197}Au + ^{63}Cu$ reaction at 6.3 MeV/nucleon, compared with the predictions (*dashed line*) assuming a thermal equilibrium (from 52).

collisions, probed by collisions with low values of E_{loss}, the two fragments have distinctly different temperatures if their masses are different. As early as 1982, an experimental study (54) concluded that the neutron multiplicities observed for low total kinetic energy losses, i.e. for short interaction times, in the $^{165}Ho + ^{56}Fe$ reaction at $E/A = 8.5$ MeV (experiment of Reference 45) cannot be reconciled with the assumption of equal temperatures of the fragments, unless unrealistically large fluctuations in the excitation energy division are admitted (see Figure 10). The observed discrepancies were instead suspected to be manifestations of a disparity of the average temperatures of the fragments for small energy losses, as was concluded in most of the subsequent studies. In the same $^{167}Ho + ^{56}Fe$ experiment of Hilscher et al (45), at higher energy losses, $E_{loss} > 50$ MeV, the excitation energy appeared to be divided between the fragments in proportion to the fragment masses, a conclusion based on an invalid assumption of equal charge-to-mass ratios for both fragments. Most of the subsequent studies of the excitation energy division performed using various experimental methods concurred with the conclusion of the lack of thermal equilibrium, at least for low E_{loss} values.

Rather indirect evidence for the lack of thermal equilibrium at short interaction times was obtained by Awes et al (36) in an inclusive or

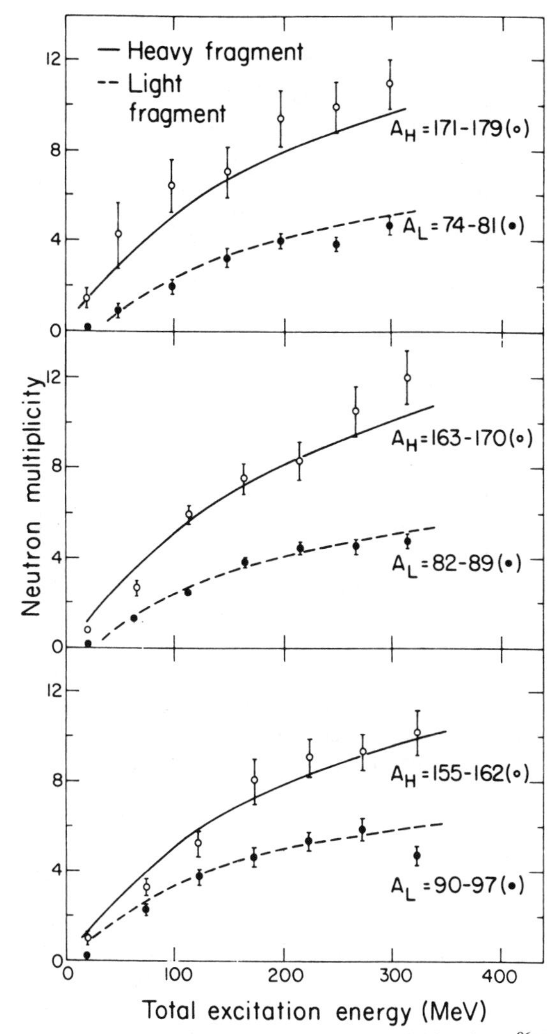

Figure 9 Experimental and theoretical neutron multiplicities from $^{86}Kr + ^{166}Er$ at 11.9 MeV/nucleon for different ranges of mass number for the light (A_L) and heavy (A_H) fragment, as indicated (from 53).

singles experiment measuring the atomic number distributions of the PLFs only. Figure 11 shows the results obtained in a study of the $^{197}Au + ^{58}Ni$ reaction at 15.3 MeV/nucleon (36). In this figure, the experimental centroids of the atomic number distributions appear to agree much better with the results of the statistical model calculations assuming equipartition of total excitation energy than with those assuming an equilibrium division over the whole range of energy losses. Con-

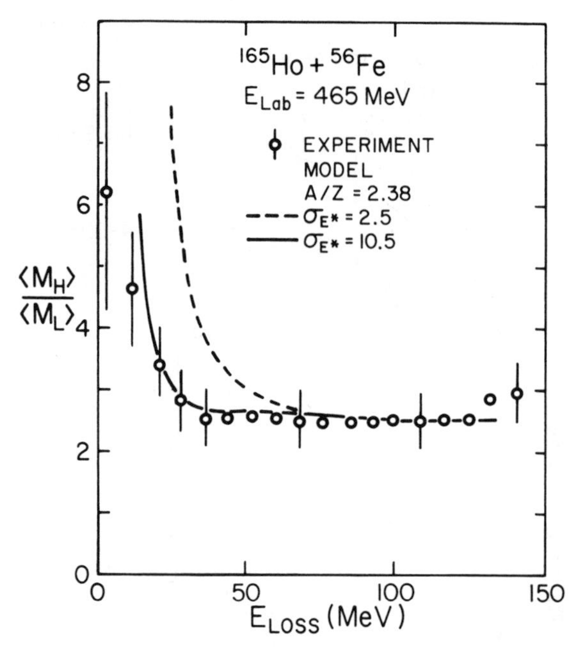

Figure 10 Ratios of average neutron multiplicities associated with light and heavy fragments from the $^{165}Ho + ^{56}Fe$ reaction at E_{lab} = 465 MeV, interpreted in terms of fluctuations in excitation energy division. Open circles represent experimental results (45), while the curves are results of theoretical calculations assuming fluctuations in excitation energy division with standard deviations of 2.5 MeV (*dashed curve*) and 10.5 MeV (*solid curve*) (from 54).

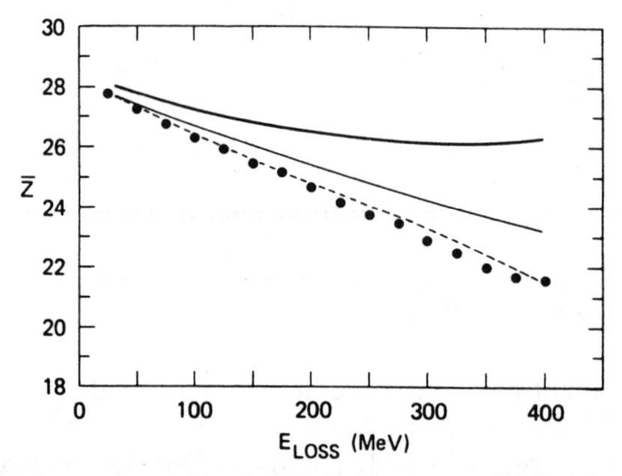

Figure 11 Experimental centroids of the Z distributions (*solid dots*) for $^{58}Ni + ^{197}Au$ at 880 MeV compared with predictions assuming thermal equilibrium (*heavy solid curve*) and excitation energy equipartition with theoretical N/Z ratios (*light solid curve*) and with N/Z ratio equal to that of the combined system (*dashed curve*) (from 36).

clusions from such inclusive measurements unfortunately must rely very strongly on the assumptions regarding the primary N,Z distributions. Similar conclusions were obtained in an exclusive experiment of Sobotka et al studying the $^{174}Yb + ^{80}Kr$ reaction at 8.55 MeV/nucleon (55). In the latter work, it was found that for up to E_{loss} of 160 MeV the excitation energy division corresponds to the nonequilibrium equipartition limit. As in other work, these conclusions relied strongly on assumptions regarding the primary isotope distributions that were not directly verified.

A unique experiment reported by Vandenbosch et al (49) also relies to some extent on the assumptions regarding the primary mass and Z distributions. In this work, sequential-fission fragment mass distributions from the $^{238}U + ^{56}Fe$ reaction at 8.5 MeV/nucleon were studied as a function of the total kinetic energy loss. Figure 12 shows the strong

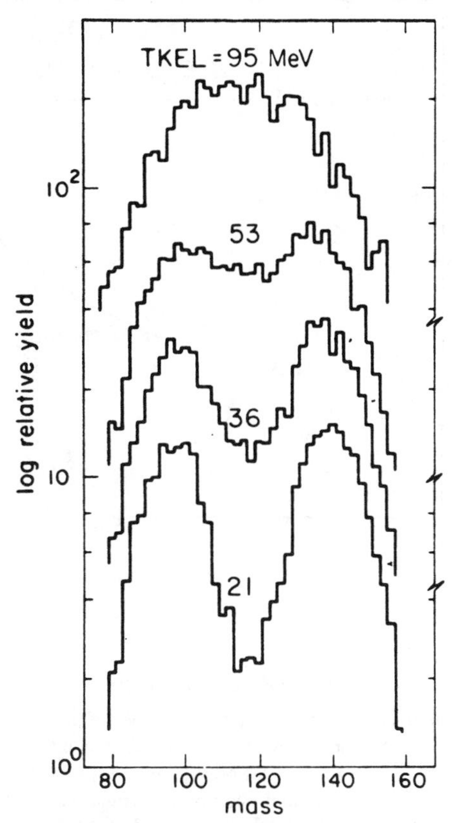

Figure 12 Samples of fission fragment mass distributions from $^{238}U + ^{56}Fe$ at 476 MeV for different kinetic energy losses (TKEL) (from 49).

dependence of the shape of such a distribution on total excitation energy deduced from the kinematics of the measured PLF. From a comparison of the shapes of such distributions measured for different E_{loss} with the empirical shapes, known from low-energy fission studies (56) as a function of excitation energy, the excitation energy of the primary uranium-like fragment was deduced. Results of this analysis are shown in Figure 13; a clear deviation of the experimental excitation energy division from that corresponding to interfragment thermal equilibrium is observed. At low energy losses, to which this unique method appears to be confined, the excitation energy is divided in almost equal portions between the two damped reaction fragments of largely differing masses.

A virtually full range of kinetic energy losses was covered in recent kinematical coincidence experiments (14, 39, 57, 58). As shown in ear-

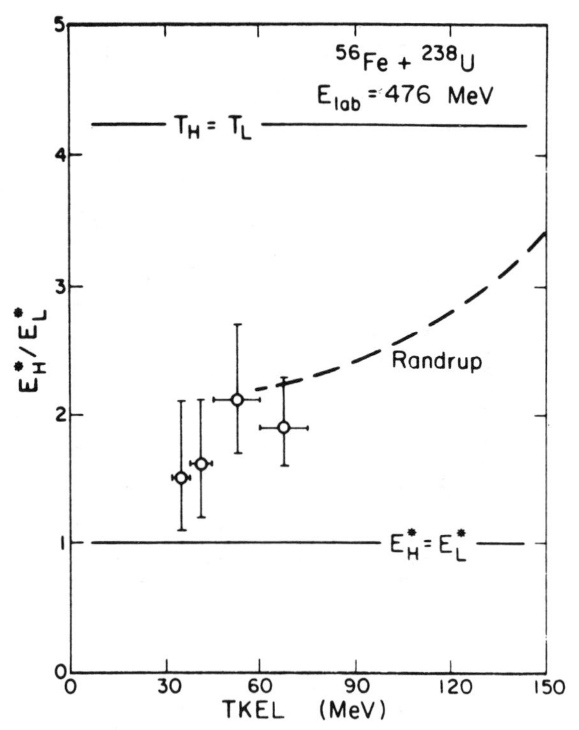

Figure 13 Experimental ratio of the excitation energies of the light and heavy fragments from the $^{238}U + ^{56}Fe$ reaction at 476 MeV (*open circles*) as a function of kinetic energy loss (TKEL), compared with theoretical predictions. Solid horizontal lines labeled $T_H = T_L$ and $E_H^* + E_L^*$ illustrate thermal equilibrium (equal temperatures) and excitation energy equipartition, respectively, while the dashed curve is a prediction of the one-body transport model of Randrup (from 49).

lier studies (41, 42) limited to fully relaxed events, the kinematical coincidence method largely avoids assumptions regarding the primary masses. Typical results are shown in Figure 14 (39) for the reaction ^{74}Ge + ^{165}Ho at 8.5 MeV/nucleon. A clear transition from an equipartition to the limit of thermal equilibrium division of the excitation energy is observed with increasing E_{loss}. Full thermal equilibrium, however, appears not to be reached in this reaction.

A similar evolution of the excitation energy division with E_{loss} is observed in more recent inclusive studies of neutron emission from the two fragments of damped collisions (31, 48, 59, 60). For example, Figure 15 (31) shows the double-differential neutron multiplicities from the very asymmetric ^{139}La + ^{40}Ar reaction at 10 MeV/nucleon, whereas Figure 16 shows the ratios of the spectral slope parameters (*a*) and of the angle-integrated neutron multiplicities (*b*) associated with PLFs and TLFs from this reaction, as plotted versus E_{loss}. As seen in the latter figure, at small kinetic energy losses the experimental data are consistent with the assumption of an equipartition of the excitation energy between PLF and TLF, while at larger E_{loss} values, they are almost consistent with the assumption of interfragment thermal equilibrium. Figure 17 shows the result of a unique experiment (48), in which two high-efficiency neutron multiplicity meters (see Figure 6) were used simultaneously to measure the joint multiplicity distribution of neutrons

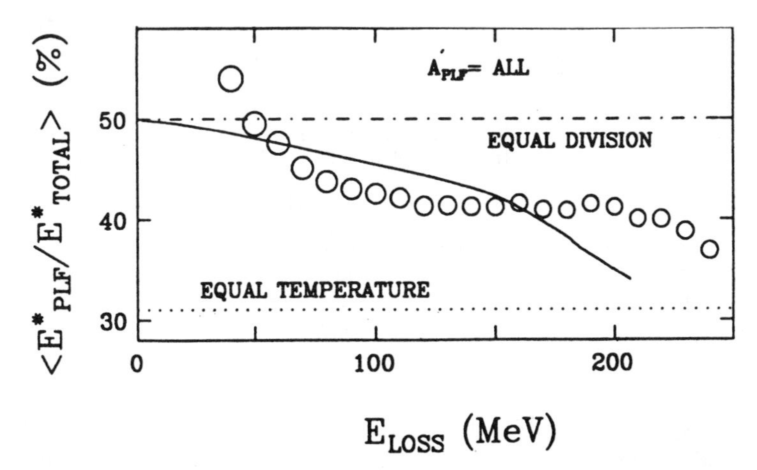

Figure 14 Experimental (*open circles*) average excitation energy division for ^{74}Ge + ^{165}Ho at 8.5 MeV/nucleon as a function of E_{loss}, compared with the predictions of the one-body nucleon exchange model of Randrup (*solid curve*). Dot-dashed and dotted lines illustrate excitation energy equipartition and interfragment thermal equilibrium (equal temperatures), respectively (from 39).

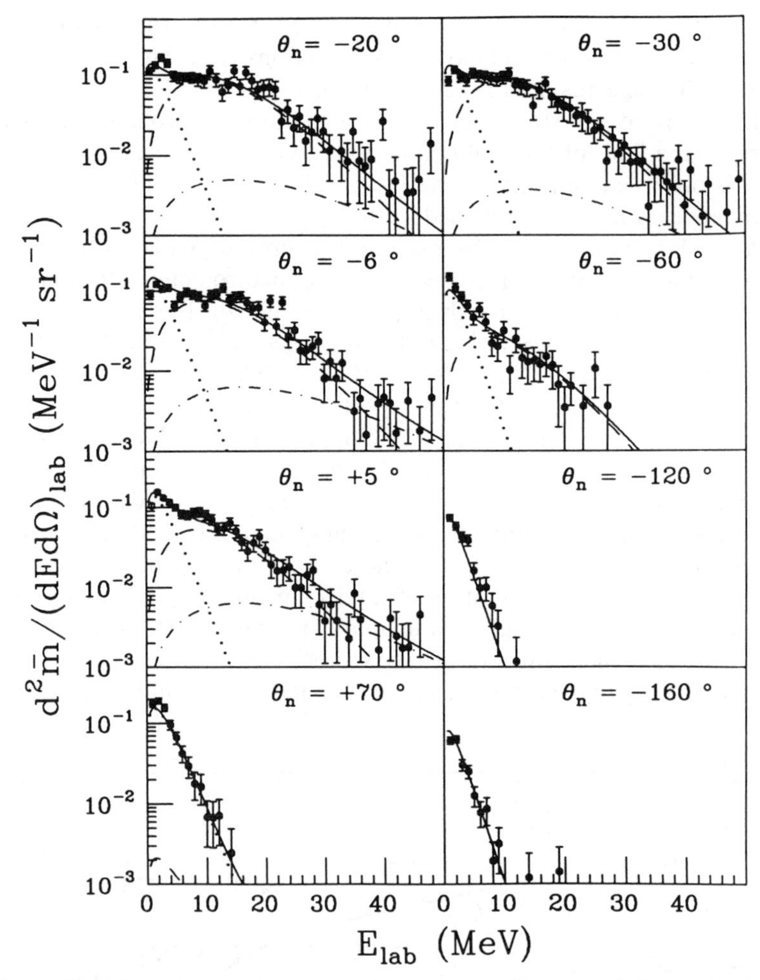

Figure 15 Laboratory energy spectra of neutrons at different emission angles from $^{139}La + ^{40}Ar$ at 400 MeV, normalized to a single reaction event. The solid curves represent a three-source model fit accounting for contributions from PLF (*dashed curves*), TLF (*dotted curves*), and nonequilibrium (*dot-dashed curves*) sources (from 31).

associated with PLFs and TLFs. Although the neutron detection efficiencies in this experiment proved insufficient for extracting useful information on the variances of the neutron multiplicity distributions [as suggested by Morrissey et al (22)], the ratio of the average multiplicities associated with the fragments was measured with good accuracy. The experimental data from this study are seen to follow clearly the same trends established in earlier experiments.

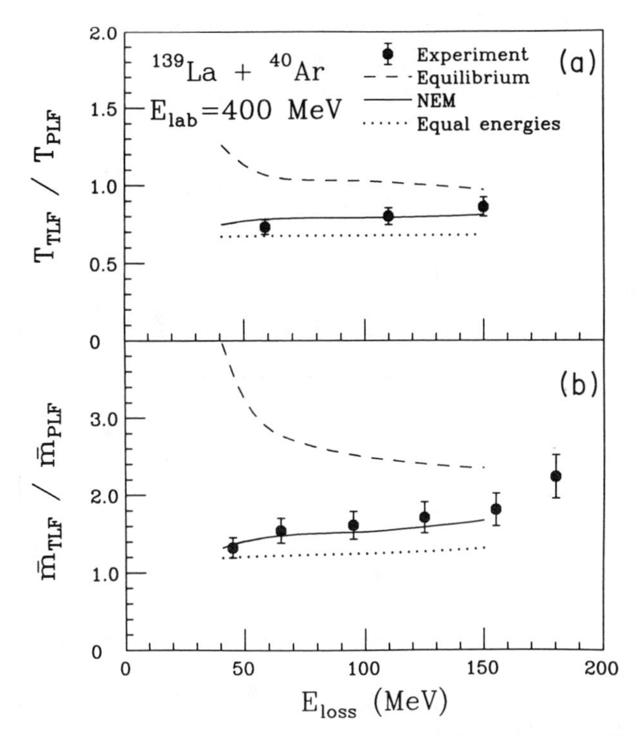

Figure 16 Ratios of the logarithmic slope parameters T_{TLF} and T_{PLF} of the energy spectra (*a*) and of the respective multiplicities $\bar{m}_{\mathrm{TLF}}$ and $\bar{m}_{\mathrm{PLF}}$ (*b*) of neutrons associated with PLFs and TLFs from ^{139}La + ^{40}Ar at 400 MeV. The solid curves represent predictions of Randrup's one-body nucleon exchange model (NEM), while the dashed and dotted curves illustrate interfragment thermal equilibrium and excitation energy equipartition, respectively (from 31).

In several experiments, the excitation energy division was studied for special cases. Such cases include reactions leading to full relaxation of energy (and angular momentum), reactions at energies very near the interaction barrier, and peripheral collisions. Generally, conclusions from these studies are consistent with those discussed above for a wider range of interaction times, kinetic energy losses, or impact parameters.

Figure 18 shows the results of early kinematical coincidence studies of the reaction ^{nat}Ag + ^{40}Ar at 8.5 MeV/nucleon (41) confined to the fully relaxed events. Similar results are obtained for ^{58}Ni + ^{40}Ar at 7 MeV/nucleon (42). In both of these experiments, the excitation energies of the fragments were deduced from the measured evaporated masses. As seen in the figure, the measured evaporated masses appear to be well explained by assuming an excitation energy division in proportion

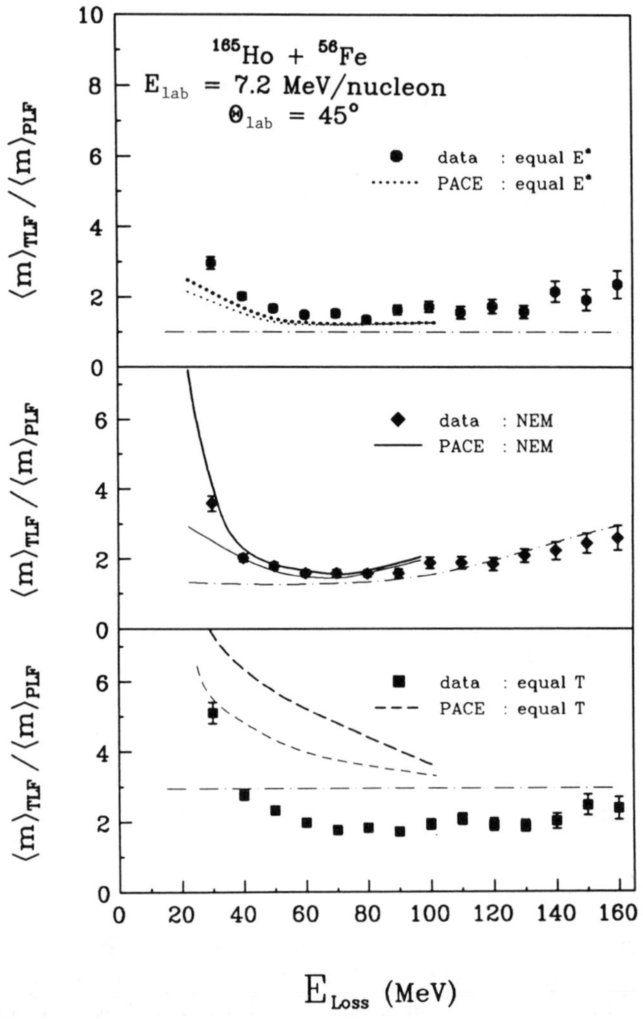

Figure 17 Ratios of the average multiplicities $\langle m\rangle_{TLF}$ and $\langle m\rangle_{PLF}$ of neutrons associated with PLF and TLF from $^{165}Ho + ^{56}Fe$ at 7.2 MeV/nucleon as a function of kinetic energy loss E_{loss}. The solid symbols represent "experimental" values obtained by correcting the measured multiplicities for the efficiencies of the neutron multiplicity meters, calculated under three different assumptions regarding excitation energy division as illustrated by the dot-dashed lines in the panels: equipartition (*top panel*), as predicted by the one-body nucleon exchange model (NEM) (*middle panel*), and as corresponding to equal temperatures of the fragments (*bottom panel*). Results of the theoretical simulations using the statistical model code PACE in conjunction with the above assumptions are shown in the respective panels in sets of dotted, solid, and dashed curves obtained with (*light curves*) and without (*heavy curves*) taking into account the fragment spins.

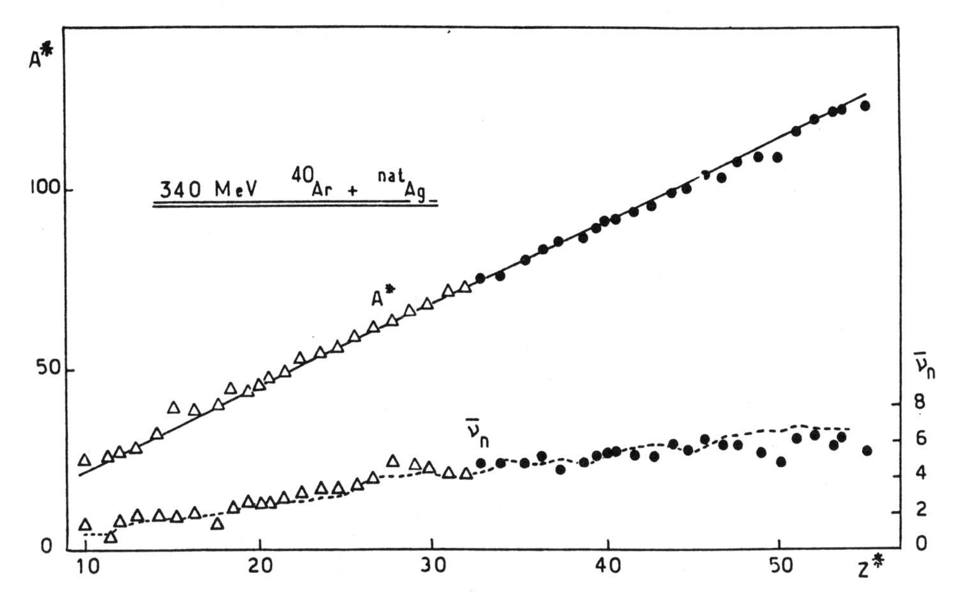

Figure 18 Primary masses (A^*) and number of evaporated nucleons ($\bar{\nu}_n$) as functions of secondary Z for light (*triangles*) and heavy (*dots*) fragments from the reaction $^{nat}Ag + {}^{40}Ar$ at 8.5 MeV/nucleon (from 41).

to the fragment masses. In the absence of significant shell effects and, for example, in the level density parameter a, such an excitation energy division would indicate the attainment of thermal equilibrium. Figure 19 shows the results of exclusive studies of Eyal et al (61) of multiplicities of neutrons in coincidence with both fragments of fully damped $^{166}Er + {}^{86}Kr$ collisions at 7 MeV/nucleon. The ratio of the neutron multiplicities attributed to the PLFs and TLFs reflects well the ratio of the PLF and TLF masses, as it should when the fragments acquire excitation energies in proportion to their masses.

Studies of the peripheral collisions or quasi-elastic reactions generally find a nonequilibrium division of the excitation energy between the fragments, with a strong dependence on the structure of the nuclei involved (62–65) or direction of the transfer. Most of these experiments focus on the net transfer effects, as is elaborated upon in the next section.

From the experimental results discussed above, a coherent picture emerges of the excitation energy generation along the collision trajectory. According to this picture, at very small interaction times associated with quasi-elastic or peripheral collisions, excitation energy is generated as a result of inelastic scattering and of transfer of individual

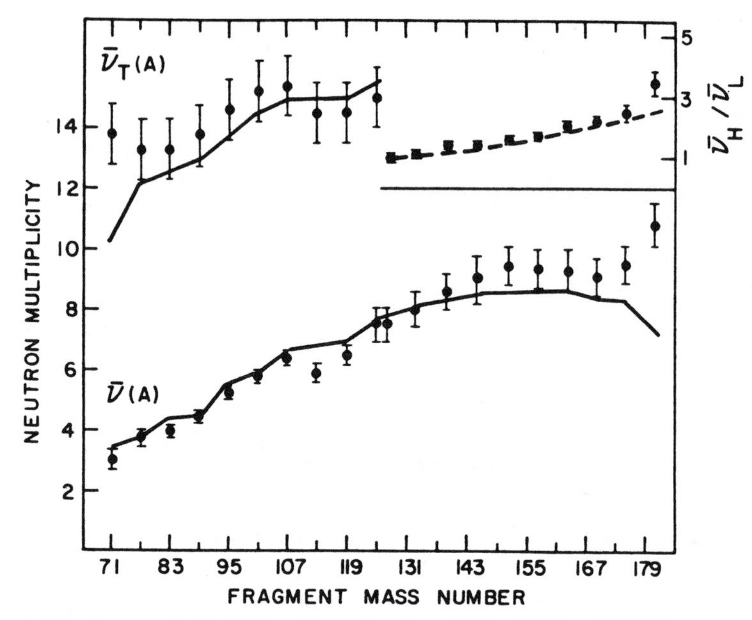

Figure 19 Average multiplicities of neutrons from $^{166}Er + ^{86}Kr$ at 602 MeV, associated with individual fragments, $\bar{\nu}(A)$, and with the system as a whole, $\bar{\nu}_T(A)$. The upper right panel shows the ratio of average multiplicities associated with heavy ($\bar{\nu}_H$) and light ($\bar{\nu}_L$) fragments as a function of the mass number of the heavy fragment. The solid curves are results of evaporation calculations, while the dashed curve represents the average fragment mass ratio (from 61).

nucleons or clusters of nucleons. The division of excitation energy is then governed by nuclear structure or nucleon transfer probability effects. The stochastic nucleon exchange model is generally inapplicable in these cases, and the data can be understood either through quantum-mechanical DWBA calculations (62, 64) or through some approximate geometrical models (66). At somewhat higher kinetic energy losses, the generation of excitation energy is quite well understood within the model of stochastic nucleon exchanges. The excitation energy is seen to be divided in equal portions between the two fragments, as expected within the framework of Randrup's one-body transport model based on the exchange of independent nucleons across the interfragment neck (see Section 2.2). At even higher energy losses, associated with longer interaction times, the excitation energy equilibration mechanism (through nucleon exchange) becomes more effective and eventually succeeds in reducing the temperature gradient faster than the successive nucleon exchanges can feed the thermal disequilibrium. As a result, the temperatures of the fragments approach each other; however, a full parity has not yet been observed.

4.2 *Correlations Between Mass and Energy Flow*

The role of net transfer of mass in the process of excitation energy generation has attracted significant experimental attention in recent years. Some of the theoretical implications were summarized in Section 2.4. There are three distinct contexts in which the correlations between net mass transfer and energy flow are important. They are discussed in the following sections.

4.2.1 PERIPHERAL COLLISIONS Peripheral reactions are characterized by small losses of total kinetic energy. Their dominating mechanism is inelastic scattering or the transfer of a few nucleons. Consequently, the transfer of energy and the resulting excitation energy division is expected to be governed by the quantum-mechanical effects of transfer and of intrinsic nuclear structure, which can be difficult to model in semiclassical macroscopic models without recourse to ad hoc assumptions. As an example, Figure 20 shows the results of studies of Schmidt et al (51) of the $^{197}Au + ^{20}Ne$ reaction at 11 MeV/nucleon, in which projectile-like fragments were measured in coincidence with light charged particles. Clearly, the PLFs are seen to acquire excitation energy as they acquire mass. The trends seen in Figure 20 can be explained by assuming that in the course of the transfer of a nucleon, the acceptor fragment acquires all of the excitation energy that results both from the static effects of change in macroscopic binding energies (Q_{gg}) and from the dynamic effects of transfer of linear momentum. This assumption appears justified for peripheral reactions, in which the quantum-mechanical reaction amplitudes favor a transfer between the loosely bound outer orbits. Such transfers leave the donor nucleus in a state with a weakly excited hole, and they produce an acceptor nucleus in a state with a highly excited particle.

Similar conclusions were drawn in the magnetic spectrometer studies of Sohlbach et al of the $^{208}Pb + ^{86}Kr$ reaction at 10, 13, and 18.2 MeV/nucleon (62) and of the $^{208}Pb + ^{197}Au$ reaction at 10 and 18.2 MeV/nucleon (64). Results of the latter studies are summarized in Figure 21 displaying the trends in excitation energy division as a function of the identified quasi-elastic channel. The authors conclude that there is a strong channel dependence of the excitation energy division, in general with the net acceptor fragment significantly more excited than the donor.

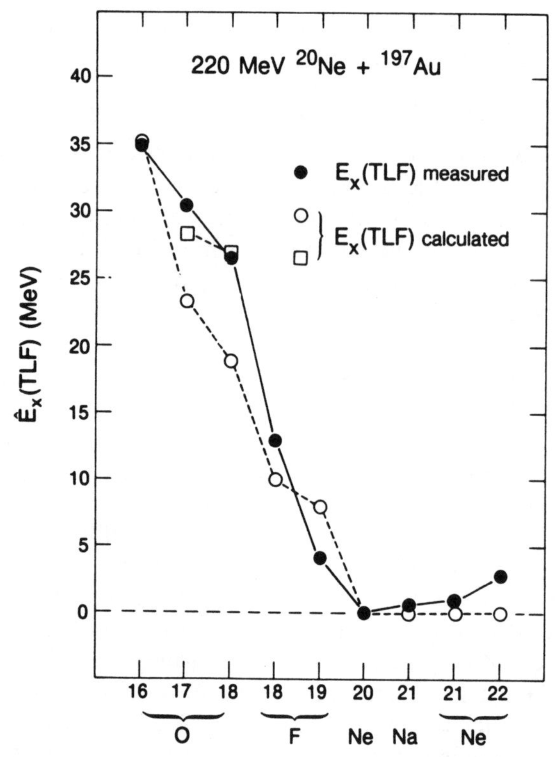

Figure 20 Comparison of experimental excitation energies of TLFs from the $^{197}Au + ^{20}Ne$ reaction at 11 MeV/nucleon as a function of fragment mass number with calculations based on optimal Q value assuming unidirectional (*open circles*) and bidirectional (*boxes*) mass transfer (from 51).

4.2.2 DAMPED COLLISIONS Studies of energy relaxation have demonstrated that the total kinetic energy loss in a damped collision is divided between the fragments approximately in proportion to their masses, at least when some averaging over the primary mass distributions is carried out. More detailed experimental studies (14, 39, 57, 58) were undertaken to determine whether, for a given E_{loss}, there is a dependence of the excitation energy division on the primary mass ratio and if so how strong such a correlation is. Benton et al (14, 57) studied the reaction $^{165}Ho + ^{56}Fe$ at 9 MeV/nucleon, using a kinematical coincidence method, in conjunction with the direct measurement of the secondary PLF mass through a time-of-flight method (see Figure 5). Differences between the calculated primary and measured secondary masses were taken in this work to demonstrate the existence of significant correlations between the excitation energy division and pri-

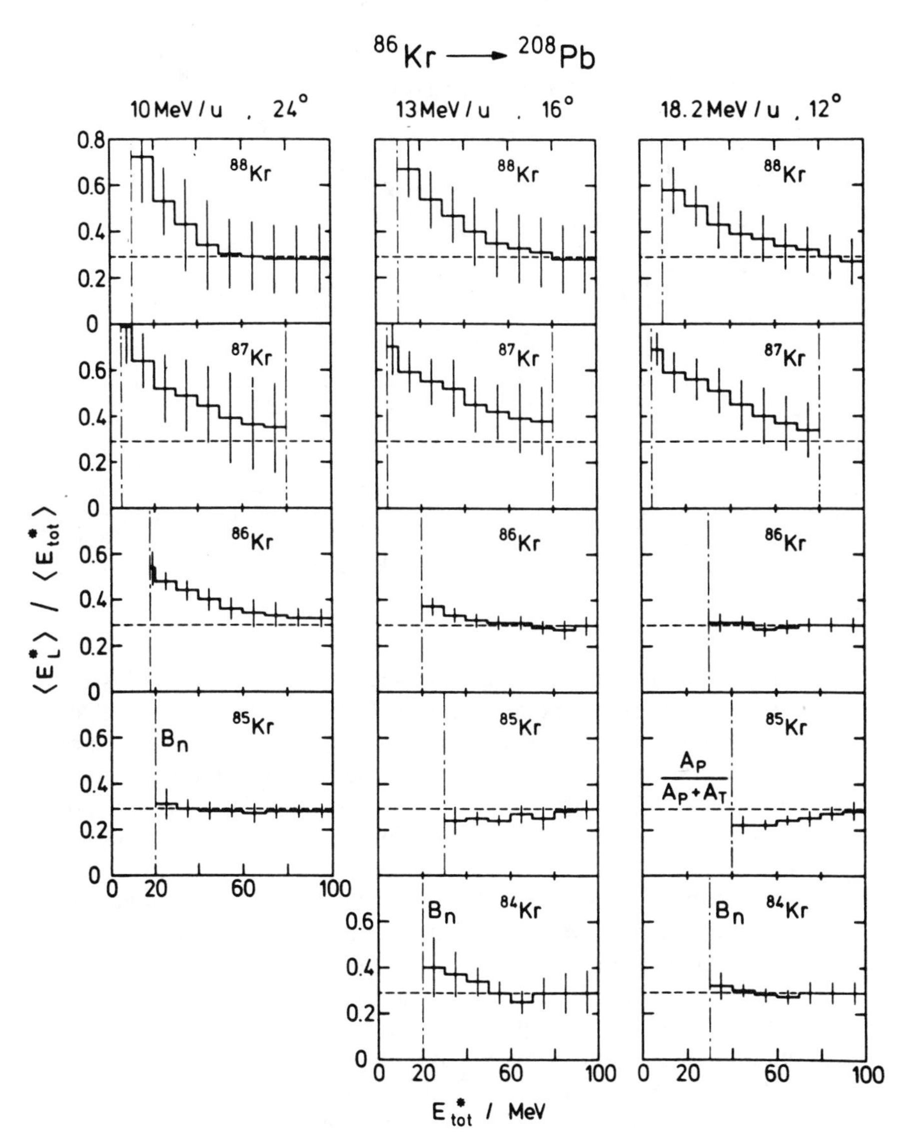

Figure 21 The fraction $\langle E_L{}^* \rangle / \langle E_{tot}{}^* \rangle$ of the total excitation energy residing in the lighter fragment (PLF) from the quasi-elastic reaction $^{208}Pb + {}^{86}Kr$ at $E/A + 10$, 13, and 18.2 MeV as a function of the total excitation energy, measured at three spectrometer angles. The vertical dot-dashed lines mark the limits of the total excitation energy range accessed in the experiment, which depend on the neutron binding energy B_n (from 64).

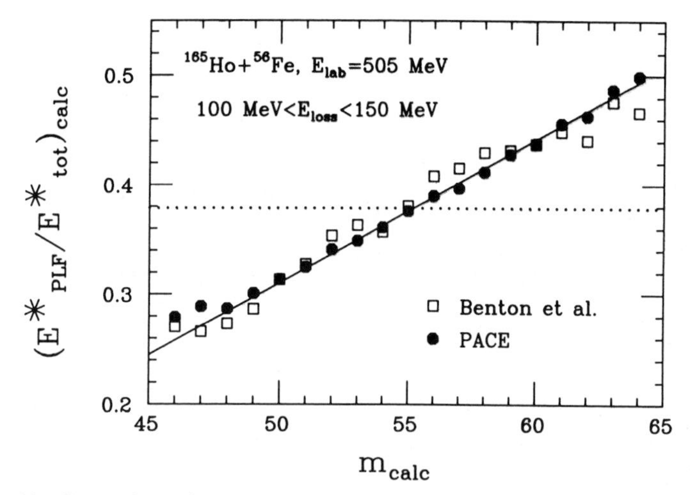

Figure 22 Comparison of the correlations between the excitation energy division and the calculated PLF mass number m_{calc} extracted from a kinematical coincidence experiment of Benton et al (14) with the results of a simulation based on a statistical decay code PACE and assuming a constant mass-independent energy division (*solid dots*) (from 37).

mary fragment masses over the full range of kinetic energy losses. However, subsequent reexamination by Tõke et al (37) showed that, at least for higher energy losses, the observed correlations may be of an instrumental character and arise from the finite resolution inherent in the kinematical coincidence method. The relevant data (14) are shown in Figure 22, along with simulation calculations (37), in which a constant, mass-independent excitation energy division was assumed.

The reaction ^{165}Ho + ^{74}Ge at 8.5 MeV/nucleon was later studied by Płaneta et al (39) and Kwiatkowski et al (58), using experimental techniques and analysis procedures similar to the ones employed by Benton et al (14, 57). Figure 23 illustrates typical results of the original analysis by Płaneta et al (39), showing strong correlations between the division of excitation energy and primary masses at small energy losses. These correlations become weaker as E_{loss} increases.

A reanalysis of the data from the experiment of Płaneta et al (39) was subsequently carried out by Tõke et al (67). In the latter work, a new method (38) was employed that, unlike the original analysis, did not rely heavily on a precise knowledge of the experimental resolutions. The reanalysis confirmed the presence of some correlations between the excitation energy and the net mass transfer. However, these cor-

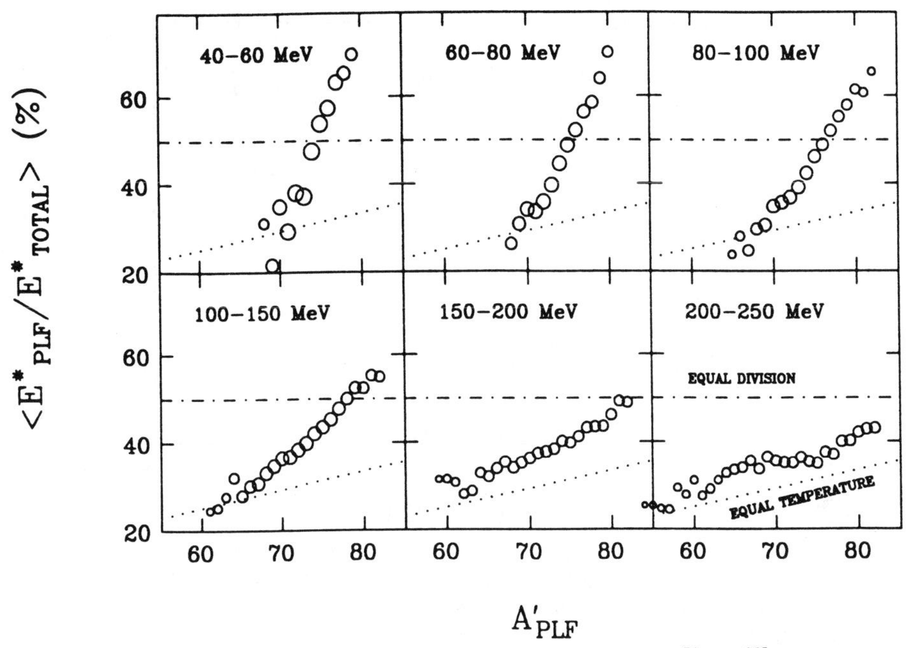

Figure 23 Average ratio of PLF to total excitation energies from $^{74}Ge + ^{165}Ho$ at 8.5 MeV/nucleon, plotted versus the calculated primary mass for the different kinetic energy loss bins as indicated. The dot-dashed and dotted lines illustrate excitation energy equipartition and interfragment thermal equilibrium, respectively (from 39).

relations were found to be significantly weaker than claimed in the original analysis (39). The main result of the work by Tõke et al (67) is summarized in Figure 24, in which the strength R of the correlations is plotted versus the total kinetic energy loss. The quantity R is defined via the equation:

$$E_{PLF}^* = C + RE_{tot}^*(A_{prim} - A_0), \qquad 15.$$

where E_{PLF}^* and E_{tot}^* are the PLF and total excitation energies, respectively, A_{prim} is the primary mass, A_0 is the centroid of the primary mass distribution, and C and R are two constants. Additionally, Tõke et al showed that the correlations between net mass transfer and excitation energy do not depend noticeably on the species of the transferred nucleons, i.e. that they are similar in strength for net proton and net neutron transfers (67).

The evolutionary scenario of a damped collision raises the following

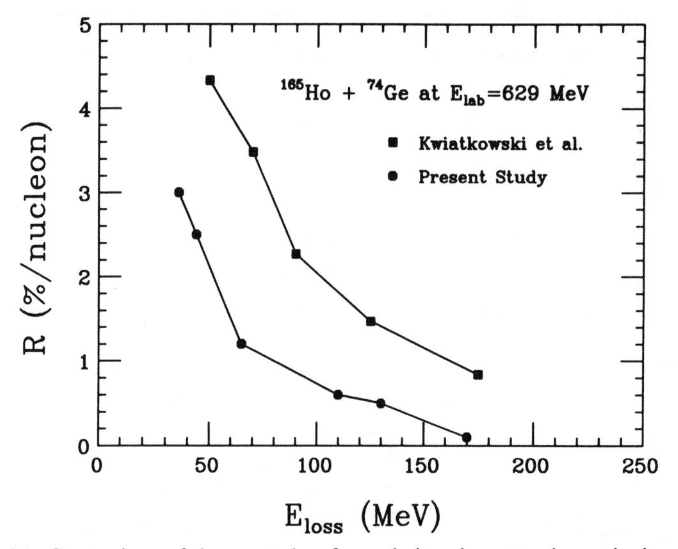

Figure 24 Comparison of the strengths of correlations between the excitation energy and fragment mass as a function of kinetic energy loss obtained with limited (*boxes*) and more complete (*dots*) accounts of finite resolution effects (from 67). See Equation 15 for the definition of R.

question: to what extent can the correlations between net mass transfer and excitation energy division, observed at higher energy losses, be attributed to the early stage of the collision, which is known to generate different amounts of heat in donor and acceptor fragments? To address this question, a quantity η is introduced (67), describing the asymmetry between the excitation energies e_a and e_d generated in the acceptor and donor fragments, respectively, in the course of a single nucleon transfer. This parameter has a differential character and can be linked via the equation

$$\eta = \frac{e_a - e_d}{e_a + e_d} = \frac{d(E^*_{tot} R \sigma^2_A)}{dE_{loss}} \bigg/ \frac{d\langle E^*_{PLF}\rangle}{dE_{loss}} . \qquad 16.$$

to the experimentally observable correlation strength R of Equation 15 and the variance $\sigma_A{}^2$ of the mass distribution. Surprisingly, for the data of Kwiatkowski et al (58), the quantity η turned out to be almost constant over the large range of E_{loss}, always favoring the acceptor fragment. If confirmed by other studies, this constancy of the asymmetry η poses a challenge to the models of heat generation in damped collisions.

Correlations between net transfer and excitation energy division were also studied in unique experiments by Wilczyński et al (50) on the $^{197}Au + ^{40}Ar$ reaction at 11.25 MeV/nucleon and by Casini et al (44) on the $^{100}Mo + ^{120}Sn$ reaction at 19.1 MeV/nucleon. In the former experiment (50), light charged particles were measured in coincidence with massive fragments, in order to determine the charged-particle emission "survival fractions." The experimental survival fractions, representing the probabilities for the fragments to decay solely via neutron and γ-ray emission, were then compared with the results of simulation calculations based on a statistical decay model. From these comparisons, it was concluded that the net acceptor is strongly favored in the division of the total excitation energy. In appreciating this work, one realizes that the analysis of any experiment of this type relies much more strongly on simulation calculations and assumptions regarding the primary mass distributions than does the analysis of the kinematical coincidence experiments. In the experiment by Casini et al (44), the excitation energy division was determined from the fission probabilities measured for the two massive primary fragments. Again a strong dependence of the excitation energy division on net mass transfer was concluded.

4.2.3 COLD FRAGMENT PRODUCTION IN NEAR-BARRIER COLLISIONS The magnitude of the asymmetry in the division of the excitation energy is important for the production of transactinide and superheavy nuclei. In this context, near-barrier energies and small total kinetic energy losses are of main interest, and the hallmark of pertinent studies is a search for a cold-donor fragment production. The reaction $^{248}Cm + ^{48}Ca$ was studied, using radiochemical methods, by Gäggeler et al (68) at an energy equivalent to 1.04 to 1.1 times the interaction barrier. Based on fission survival rates for the reconstructed primary reaction fragments, they concluded that surviving reaction products are produced at very low excitation energies. Similarly, radiochemical studies by Keller et al (69) of the $^{207,208}Pb + ^{50}Ti$, ^{54}Cr, ^{58}Fe reactions at near-barrier energies conclude that the net donor fragments leave the interaction site cold.

The probability for cold-donor fragment production was studied at somewhat higher bombarding energies by Chatterjee et al (70). They measured exclusive energy spectra of neutrons from the $^{58}Ni + ^{208}Pb$ reaction at 6.65 MeV/nucleon in coincidence with Ni-like fragments. In contrast to the conclusions drawn in the radiochemical work discussed above, the observed neutron spectra were found to be compatible with the assumption that the heavy Pb-like fragment receives most of the total excitation energy, independent of the direction and the magnitude of the net mass transfer. Results of this study (70) are summarized in Figure 25, in which the experimental neutron multiplicities are compared with the results of theoretical calculations assuming various ratios for the excitation energies of PLF and TLF. Although in the subset of data covered by Figure 25, the Pb-like fragment is a net donor, it is always seen to receive most of the excitation energy. These observations cannot be reconciled with those of the radiochemical studies of Keller et al (69), but they appear consistent with the trends established at higher bombarding energies. They are also consistent with the results obtained (71) in a kinematical coincidence measurement of fission probabilities for heavy fragments from the $^{110,124}Sn + ^{238}U$ reaction at 6.17 MeV/nucleon.

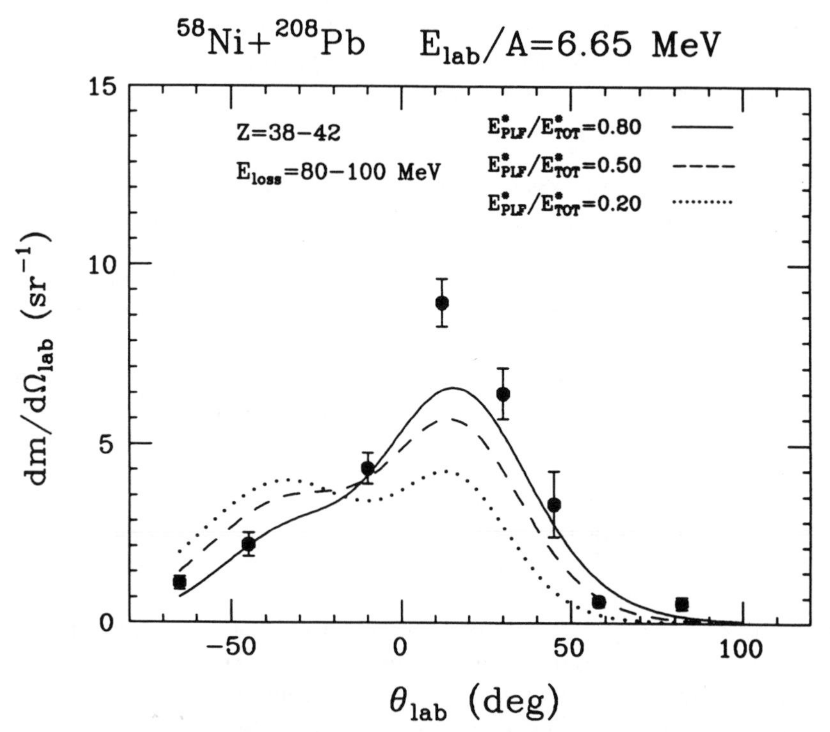

Figure 25 Neutron multiplicities per steradian associated with events in which the heavy fragment in the reaction ^{58}Ni + ^{208}Pb at 6.65 MeV/nucleon is a net mass donor, as a function of neutron laboratory angle, compared with model calculations assuming different excitation energy division (from 70).

5. SUMMARY AND OUTLOOK

The present review is limited to the bombarding energy domain from the near-barrier energies to approximately 20 MeV/nucleon. Such a limitation in scope is natural and justified for several reasons. Firstly, in this energy domain, the dominance of damped or dissipative collision dynamics is well established. This type of collision has distinct signatures that allow one to separate it experimentally from competing reaction scenarios. Secondly, most of the conclusive experimental work on the excitation energy division was performed in this energy domain—if not for other, then for historical reasons. Thirdly, in this energy domain, all of the characteristic features of the damped collisions can be understood at least qualitatively within the mean field approach, as expressed, for example, in the stochastic nucleon exchange model. The underlying assumption of this model of a dominantly one-body interaction is expected to become less justified at higher bombarding energies.

The experimental task of determining the excitation energy division in dissipative or damped collisions is quite difficult, as it requires not only advanced measuring techniques but also complex analysis methods. The latter include consistent simulation calculations of both the physical processes of primary interaction and the subsequent decay, taking account of the response of the experimental setup to the occurrence of such processes. Quantitative discrepancies among the conclusions of several of the reviewed experiments can be understood mostly as resulting from the simplifying assumptions made in the course of the analysis rather than from experimental errors. Among such critical assumptions are, for example, one regarding the rate of the N/Z equilibration process and one concerning the primary masses or spin-dependences of the fission probabilities.

In spite of some quantitative disagreement between the reviewed experiments, however, a quite solid overall picture of the excitation energy generation phenomenon has emerged that is consistent with the stochastic nucleon exchange model. In this picture, it is only at very small kinetic energy losses that the conversion of the kinetic energy into intrinsic excitation energy of the two fragments is governed by nuclear structure or transfer probability effects. With increased interaction time, and hence higher kinetic energy losses and overall excitation energies, the amounts of excitation generated in both fragments begin reflecting the stochastic character of exchanges of nucleons across the neck opening between the fragments. At small energy losses, or at the beginning of a collision, an equipartition of the excitation energy results from the fact that nucleon transfer between equally cold fragments generates on the average equal amounts of energy in both fragments. Equal excitation energies lead to a thermal disparity between the fragments of unequal masses, which gives rise to a feedback effect such that in subsequent nucleon exchanges this disparity is diminished and the system is driven to thermal equilibrium. In agreement with the stochastic nucleon exchange model, however, a full interfragment thermal equilibrium is not attained even for the highest energy losses observed.

There appears to be sound evidence that, in an elementary nucleon exchange, more excitation energy is generated in the acceptor nucleus than in the donor. It is not clear as yet whether the observed magnitude of this kind of asymmetry can be explained by the present version of the stochastic nucleon exchange model. Alternative models of interfragment interaction have been considered that provide a framework for interpreting the observed phenomena of a mass-asymmetry dependence of the excitation energy division. Such models point out the possible importance of effects that are not treated realistically or in enough detail in the stochastic nucleon exchange model, for example the effects related to the neck and deformation degrees of freedom.

There are clearly many questions regarding the excitation energy generation in dissipative collisions that still await unambiguous and conclusive answers. One of these questions refers to the asymmetry between the energies generated in the acceptor and donor. Other questions concern, for example, the possible importance of shell and deformation effects. Experiments proposed in the literature (22), which would measure the joint excitation energy distribution, event by event, for the two fragments, seem difficult to carry out (48) but may eventually prove very useful in shedding additional light on the subject. It also seems worthwhile to perform additional exclusive neutron and/or light charged-particle experiments, in conjunction with the high-precision kinematical coincidence measurements, of the type reported by several groups (14, 39, 44). One could, for example, consider mass-symmetrical systems for which the net transfer effects can be studied unobstructed by the systematic drifts in atomic and mass numbers.

The dominance of the binary reaction scenario has been proven at significantly higher bombarding energies than those considered in the present review (72). In these experiments, leading in some cases to a complete disassembly of at least one of the original fragments, large numbers of neutrons and charged products are seen that presumably originate from these fragments of the primary dissipative collision. It will be interesting to extend the study of excitation energy division to these higher bombarding energies where the dissipative mechanisms are expected to reflect more of the two-body interactions. In the domain of intermediate energies, such studies may provide insights into the transition from dynamics dominated by mean field effects to dynamics dominated by two-body interactions.

Literature Cited

1. Schröder, W. U., Huizenga, J. R., in *Treatise in Heavy-Ion Science,* ed. D. A. Bromley. New York: Plenum (1984), 2:115–734
2. Gobbi, A., Nöenberg, W., in *Heavy-Ion Collisions,* ed. R. Bock. Amsterdam: North Holland (1980), 2: 128–273
3. Volkov, V. V., *Phys. Rep.* 44:95–157 (1978)
4. Friedman, W. A., *Phys. Lett.* 98B:21–25 (1981)
5. Moretto, L. G., Blau, S. K., Pacheco, A. J., *Nucl. Phys.* A364:125–43 (1981)
6. Broglia, R. A., Dasso, C. H., Winther, A., *Phys. Lett.* 53B:301–5 (1974)
7. Broglia, R. A., Winther, A., *Heavy-Ion Reactions.* New York: Addison-Wesley (1991)
8. Nöenberg, W., in *New Vistas in Nuclear Physics,* ed. P. J. Brussard. New York: Plenum (1986), p. 91
9. Esbensen, H., Winther, A., Broglia, R. A., *Phys. Rev. Lett.* 41:296–98 (1978)
10. Brosa, U., Grossmann, S., Müller, A., *Z. Phys.* A310:177–87 (1983)
11. Grossmann, S., Brosa, U., *Nucl. Phys.* A481:340–54 (1988)
12. Brosa, U., et al, *Nucl. Phys.* A502:423c–42c (1989)
13. Brosa, U., Grossmann, S., Müller, A., *Phys. Rep.* 197:167–262 (1990)
14. Benton, D. R., et al, *Phys. Lett.* B185:326–30 (1987)
15. Gross, D. R. E., Kalinowski, H., *Phys. Rep.* 45:175–210 (1978)
16. Blocki, J., et al, *Ann. Phys. (NY)* 113:330–86 (1978)
17. Randrup, J., *Nucl. Phys.* A307:319–48 (1978)
18. Randrup, J., *Nucl. Phys.* A327:490–516 (1979)
19. Nöenberg, W., *Z. Phys.* A274:241 (1975)
20. Hofmann, H., Siemens, P. J., *Nucl. Phys.* A257:165–88 (1976)
21. Weidenmüller, H. A., *Prog. Part. Nucl. Phys.* 4:273–81 (1980)
22. Morrissey, D. J., Moretto, L. G., *Phys. Rev.* C23:1835–38 (1981)
23. Randrup, J., *Nucl. Phys.* A383:468–508 (1982)
24. Døssing, T., Randrup, J., *Nucl. Phys.* A433:215–79 (1985)
25. Døssing, T., Randrup. J., *Nucl. Phys.* A433:280–350 (1985)
26. Feldmeier, H., Spangenberger, H., *Nucl. Phys.* A428:223c–38c (1984)
27. Feldmeier, H., *Rep. Prog. Phys.* 50:1 (1984)
28. Sammaddar, S. K., De, J. N., Krishan, K., *Phys. Rev.* C31:R1053–56 (1985)
29. Tassan-Got, L., Orsay Rep. IPNO-T-89–02; PhD thesis (1988)
30. Borderie, B., et al, *Ann. Phys. (France)* 15:287 (1990)
31. Wile, J. L., et al, *Phys. Rev.* C39:1845–55 (1989)
32. Schröder, W. U., et al, *Nucl. Sci. Res. Conf. Ser.*. Chur/London/Paris/New York: Harwood Academic (1987), 11:255–92
33. Ayik, S., *Phys. Rev.* C35:2086–89 (1987)
34. Gross, D. H. B., Hartmann, K. M., *Phys. Rev.* C24:2526–44 (1981)
35. He, Z., et al, *Nucl. Phys.* A473:342–52 (1987)
36. Awes, T. C., et al, *Phys. Rev. Lett.* 25:251–54 (1984)
37. Tõke, J., Schröder, W. U., Huizenga, J. R., *Phys. Rev.* C40:R1577–80 (1989)
38. Tõke, J., Schröder, W. U., Huizenga, J. R., *Nucl. Instrum. Methods* A288:406–12 (1990)
39. Płaneta, R., et al, *Phys. Rev.* C39:R1197–1200 (1989)
40. Gavron, A., *Phys. Rev.* C21:230–36 (1980)
41. Cauvin, B., et al, *Nucl. Phys.* A301:511–32 (1978)
42. Babinet, R., et al, *Nucl. Phys.* A296:160–76 (1978)
43. Casini, G., et al, *Nucl. Instrum. Methods* A277:445–57 (1989)
44. Casini, G., et al, *Phys. Rev. Lett.* 67:3364–67 (1991)45. Hilscher, D., et al, *Phys. Rev.* C20:576–91 (1979)
46. Péter, J., et al, *Z. Phys.* A283:413–14 (1977)

47. Gould, C. R., et al, *Z. Phys.* A284: 353–54 (1978)
48. Pade, D., et al, *Phys. Rev.* C43:1288–97 (1991)
49. Vandenbosch, R., et al, *Phys. Rev. Lett.* 52:1964–66 (1984)
50. Wilczyński, J., et al, *Phys. Lett.* B220: 497–501 (1989)
51. Schmidt, H. R., et al, *Phys. Lett.* B180:9–13 (1986)
52. Tamain, B., et al, *Nucl. Phys.* A330: 253–68 (1979)
53. Tserruya, I., et al, *Phys. Rev.* C26: 2509–24 (1982)
54. Huizenga, J. R., et al, *Nucl. Phys.* A387:257c–82c (1982)
55. Sobotka, L. G., et al, *Phys. Lett.* B175:27–31 (1986)
56. Vandenbosch, R., Huizenga, J. R., *Nuclear Fission*. New York: Academic (1973), 422 pp.
57. Benton, D. R., et al, *Phys. Rev.* C38: 1207–24 (1988)
58. Kwiatkowski, K., et al, Phys, Rev. C41:958–72 (1990)
59. Petit, G. A., et al, *Phys. Rev.* C40:692–705 (1989)
60. Fiore, L., et al, *Phys. Rev.* C41:R419–21 (1990)
61. Eyal, Y., et al, *Phys. Rev. Lett.* 41: 625–28 (1978)
62. Sohlbach, H., et al, *Phys. Lett.* 153B: 386–91 (1985)
63. Van Engelen, et al, *Nucl. Phys.* A457: 375–400 (1986)
64. Sohlbach, H., et al, *Z. Phys.* A328: 205–17 (1987)
65. Sohlbach, H., et al, *Nucl. Phys.* A467: 340–64 (1987)
66. Wilczyński, J., Wilschut, H. W., *Phys. Rev.* C39:2475–76 (1989)
67. Tõke, J., et al, *Phys. Rev.* C44:390–97 (1991)
68. Gäggeler, H., et al, *Phys. Rev.* C33: 1983–87 (1986)
69. Keller, H., et al, *Z. Phys.* A328:255–56 (1987)
70. Chatterjee, M. B., et al, *Phys. Rev.* C44:R2249–52 (1992)
71. Beier, G., et al, *Z. Phys.* A336:217–22 (1990)
72. Lott, B., et al, Rep. *Phys. Rev. Lett.* 68:3141–44 (1992)
73. Moretto, L. G., *Z. Phys.* A310:61–63 (1983); *Nucl. Phys.* A409:115c–34c (1983)
74. Moretto, L. G., Lanza, E. G., *Nucl. Phys.* A428:137c–44c (1984)
75. Feldmeier, H., Spangenberger, H., *Nucl. Phys.* A428:223c–38c (1984)

Annu. Rev. Nucl. Part. Sci. 1992. 42:447–481

ANGULAR MOMENTUM DISTRIBUTIONS IN SUBBARRIER FUSION REACTIONS[1]

Robert Vandenbosch

Department of Chemistry and Nuclear Physics Laboratory, University of Washington, Seattle, Washington 98195

KEY WORDS: heavy ion, spin distributions

CONTENTS

1. INTRODUCTION

1.1 *General Remarks*

Interest in subbarrier heavy-ion fusion was stimulated by the recognition some years ago that subbarrier fusion cross sections were enhanced by many orders of magnitude over what would be expected from quantum mechanical one-dimensional barrier penetration. This enhancement is now understood to arise from the coupling of other degrees of freedom to the relative-motion degree of freedom. Nuclear subbarrier fusion is especially interesting because of the rich interplay between the dynamics of the reaction and the nuclear structure of the participating species. It is interesting to note that quantal effects on the relative contribution of different angular momenta in the D+D fusion reaction were recognized many decades ago (1). Tunneling plays a crucial role in many other areas of physics and chemistry, such as condensed matter physics and atom-surface or molecule-surface reactions, and it is the basis of the scanning tunneling electron microscope. The roles of other degrees of freedom are particularly evident in chemical reactions. Some of the important theoretical approaches that we discuss in connection with subbarrier heavy-ion fusion are also employed in understanding chemical reactions. These include coupled-channels calculations (2) and multiple reaction path calculations on multidimensional potential energy surfaces (3).

This review focuses on the angular momentum (spin) distributions in heavy-ion fusion reactions. This distribution is given by the partial wave contributions of $\sigma_f(l)$ where l is the orbital angular momentum. Classically, l is the product of the impact parameter and the linear momentum. The fusion cross section is given by the zeroth moment of the spin distributions,

$$\sigma_f = \sum_l l^0 \sigma_f(l). \qquad 1.$$

It is often useful to characterize the spin distribution by its first and second moments, which correspond to the mean l and the mean square l. As discussed below, these higher moments provide additional constraints on the theoretical models for subbarrier fusion.

The general subject of heavy-ion subbarrier fusion was reviewed by Beckerman in 1985 (4) and by Steadman & Rhoades-Brown in this series in 1986 (5). There have also been a number of topical conferences or workshops on this and related subjects, and proceedings of these meetings have been published (6, 7). The text of several review talks or lectures with particular emphasis on angular momentum distributions in subbarrier fusion are available (8, 9).

This review is organized as follows. I conclude this introduction with an example of how coupling another degree of freedom to the relative-motion degree of freedom both enhances the fusion cross section and broadens the angular momentum (spin) distributions. The energy dependences of these observables predicted by a simple model are presented as an orientation for what is to follow. In the second section I outline different experimental approaches for measuring various moments of the compound nuclear spin distribution. In the third section I discuss various theoretical approaches for obtaining spin distributions in heavy-ion fusion reactions.

1.2 *Deformation: An Example of the Effect of Another Degree of Freedom*

As a particularly simple and physically transparent example of the effect of coupling another degree of freedom to the relative motion in the entrance channel, consider the nuclear shape degree of freedom. If one of the reaction partners is deformed, the interaction barrier will depend on the orientation of the deformed nucleus relative to the direction of approach of the reaction partner. If the approach is along the axis of symmetry of a prolately deformed nucleus, the potential barrier is lowered and fusion is enhanced. If the approach is along the direction perpendicular to the symmetry axis, the barrier is increased and fusion is reduced. The net effect on the barrier transmission at subbarrier energies is to enhance the fusion cross section, as is illustrated schematically in Figure 1. One gains more from the lower barrier than one loses from the higher barrier, because the cross section depends exponentially on energy. There is also an increase in the mean value and the width of the spin distribution.

Several interesting results can be obtained if one assumes that the different partial waves have parabolic barriers at the same radius R_B and are displaced by the rotational energy $l(l+1)\hbar^2/2\mu R_B{}^2$. The barrier curvature is often characterized by $\hbar\omega$, which can be estimated from a realistic nucleus-nucleus potential $V(r)$ from $\hbar\omega = [\mu^{-1}|d^2V(r)/dr^2|]^{1/2}$. In these relations, μ is the reduced mass in the entrance channel. For energies well below the barrier $[(E - V_B) \ll \hbar\omega/2\pi]$, the partial cross sections are given by

$$\sigma_l = \pi\lambda\!\!\!^{-2}(2l + 1) \exp\left\{-\frac{2\pi}{\hbar\omega}\left[V_B - E + \frac{l(l + 1)\hbar^2}{2\mu R_B^2}\right]\right\}. \qquad 2.$$

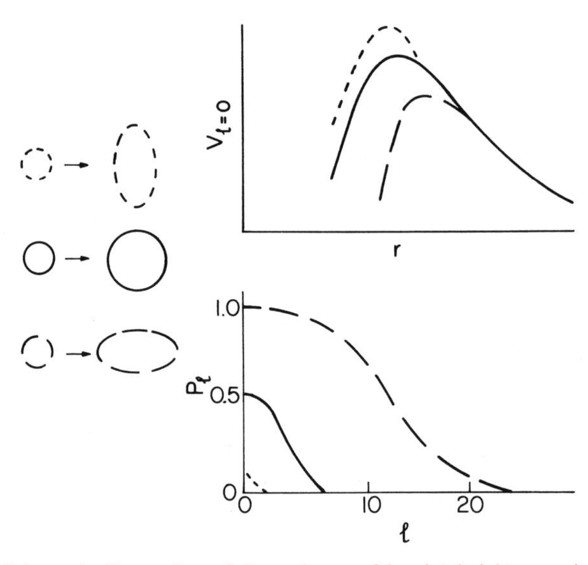

Figure 1 Schematic illustration of dependence of barrier height on orientation of a deformed target nucleus. The lower part of the figure indicates the dependence of the penetration coefficients on angular momentum l for the different orientations and for a bombarding energy equal to the spherical $l = 0$ barrier height.

Thus at low energies the shape of the distribution no longer changes and $\langle l \rangle$ saturates at the fixed value $(\mu R_B{}^2\hbar\omega/4\hbar^2)^{1/2}$ as the energy is lowered. The total fusion cross section at low energies is given by

$$\sigma = \frac{R_B^2\hbar\omega}{2E}\exp\{-[2\pi(V_B - E)/\hbar\omega]\}. \qquad 3.$$

From this latter expression one sees that the asymptotic slope at low energy of ln $E\sigma(E)$ vs E is given by $2\pi/\hbar\omega$. These asymptotic dependences for a single $l = 0$ barrier (no deformation of reacting nuclei) are shown by the solid curves in Figure 2. The behavior at higher energies is also illustrated. Now consider a deformed nucleus, in which a distribution of $l = 0$ barriers results from the different orientations of the deformed nucleus with respect to the direction of approach of the reaction partner (see inset to Figure 2). After taking the appropriate average over the distribution of orientations of the randomly oriented nuclei (10, 11), one finds the leading edge of the fusion excitation function is displaced to lower bombarding energies but exhibits the same asymptotic slope (dashed curve in Figure 2). The mean l value is increased significantly at energies near the barrier, as expected from the

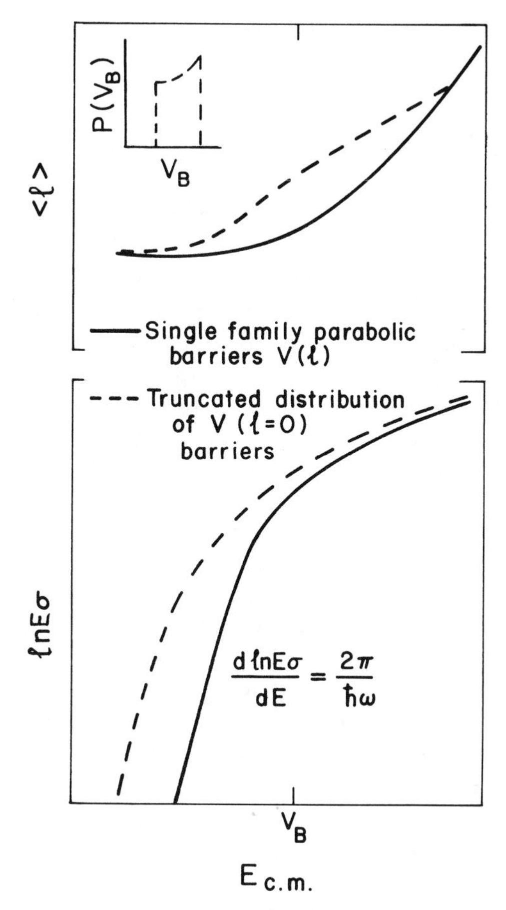

Figure 2 Schematic illustration of the dependence of the mean spin value and the product of the energy times the cross section as a function of bombarding energy. The solid curve is for a single $l = 0$ barrier, and the dashed curve includes averaging over a distribution of $l = 0$ barriers. This distribution is truncated at a barrier energy higher than the lowest bombarding energy shown. The inset shows the probability distribution for barriers of height V_B.

schematic illustration in Figure 1 and illustrated by the dashed line in Figure 2. At energies well below or far above the barrier, the mean l value approaches that of a single barrier (no deformation). The maximum increase in the mean l value occurs at an energy just below the barrier, giving rise to what I call here the "barrier bump" in the dependence of the mean l on bombarding energy.

2. EXPERIMENTAL PROBES

Before discussing specific experimental probes for deducing information about spin distributions, it is important to recognize that in all cases one is not directly measuring angular momentum and that one must convert the observable to angular momentum. This conversion is often somewhat model dependent. The reliability of the conversion of an observable to angular momentum can be checked by performing a calibration reaction with lighter ions. It is a helpful fact that lighter ions on the appropriate target nuclei will form compound nuclei with comparable angular momenta at above-barrier bombarding energies when compound nuclear excitation energies are matched. Since the angular momentum distributions can be reliably predicted for light-ion reactions at above-barrier energies, measurement of the observables of interest in such reactions provides a valuable calibration of the conversion of these observables to angular momentum (12).

2.1 *Gamma-Ray Multiplicities*

The methods discussed in this section rely on the observation that most of the angular momentum acquired by a compound nucleus is carried away by gamma rays at the latest stage of the deexcitation process. Thus a measurement of the average gamma-ray multiplicity or the gamma-ray multiplicity distribution provides information about the initial compound nuclear angular momentum distribution.

Since at subbarrier energies the fusion cross section is less than the inelastic and transfer cross sections, it is necessary to have a fusion "tag" to define when a fusion event has occurred. Often the detection of an evaporation residue provides such a tag. Electrostatic deflectors (12, 13) and velocity filters can be used to separate the evaporation residues from beam or projectile-like fragments. Less inclusive techniques involve the detection of characteristic gamma rays from specific residual nuclei (14). In such cases one must be careful about exit-channel biases (15).

2.1.1 AVERAGE MULTIPLICITIES

The simplest gamma-ray multiplicity measurement is its average. This is defined by

$$M_{\gamma} = \frac{N_{\text{coin}}}{N_{\text{tag}}\epsilon_{\gamma}}, \qquad 4.$$

where N_{coin} is the number of coincidences between the fusion tag detector and the gamma-ray detector, N_{tag} is this number of single events in the fusion tag detector, and ϵ_{γ} is the efficiency of the gamma-ray detector. A convenient gamma-ray detector is a 3-inch by 3-inch NaI(Tl) detector with a graded Cu-Pb absorber. This detector exhibits a flat response for gamma rays over a broad energy range, a fact that has been exploited in the simple expression above. In order to avoid summing corrections, the gamma-ray detector should not subtend too large a solid angle. Placement of the gamma-ray detectors at angles corresponding to zeros of the Legendre polynomial $P_2(\cos\theta)$ minimizes any corrections due to the angular distribution of the gamma rays. An important feature of this technique is that the efficiency of the tag detector cancels out in the derivation of the multiplicity.

2.1.2 MULTIPLICITY DISTRIBUTIONS Measurement of the distribution of gamma-ray multiplicities from event to event can in principle provide information on higher moments of the angular momentum distribution. In addition to a tag detector, a multi-element gamma-ray detector is required. Efficiency and multiplicity resolution corrections are large unless the number of elements and the solid-angle coverage is large. As discussed in the next section, care must be taken in the conversion of the observed multiplicity distributions to angular momentum distributions. Quantitative interpretation of the limited amount of experimental data available from multi-element detectors has been compromised by the lack of calibration data from fusion reactions of known angular momentum content.

2.1.3 CONVERSION OF MULTIPLICITY TO ANGULAR MOMENTUM Compound nuclei produced in near-barrier fusion reactions deexcite first by particle emission and finally by gamma-ray emission. For not-too-light compound nuclei the evaporated particles carry off only small amounts of angular momentum and most of the angular momentum is carried off by gamma rays. For heavy collective nuclei, much of the angular momentum is carried off by "stretched" transitions down the ground-state rotational band. For even-even nuclei these will be quadrupole (E2) gamma rays; for odd-A nuclei a mixture of M1 and E2. Thus the average angular momentum carried off by "stretched" gamma rays is between 1 and 2 units. Empirical observations of gamma-ray multiplicities for systems of known angular momentum suggest that the average angular momentum is related to the average gamma-ray multiplicity by a relation such as

$$\langle l \rangle = 1.6(M_\gamma - 3). \qquad 5.$$

The constant 3 is identified as the average number of nonstretched gamma rays that do not carry off a significant amount of angular momentum, and the coefficient 1.6 reflects the average mixture of dipole and quadrupole stretched transitions. Although relations of this form are useful for approximate deductions of angular momenta, quantitative deduction of $\langle l \rangle$ values requires a more detailed approach. The angular momentum carried off by neutrons and by statistical gamma rays, as well as the differences in the decay pattern of even-even as compared to odd-A and odd-odd nuclei, needs to be taken into account. Halbert et al (16) have described and used a procedure for converting multiplicity to angular momentum that relies on a statistical model code. This method has also been used by Gil et al (12) and by Wuosmaa et al (13). Gil et al (12) have also performed light-ion calibration reactions to establish parameters and confirm the validity of the method. The gamma-ray multiplicity method has to be used with special care for subbarrier reactions with light projectiles in which the mean angular momentum for the fusion reaction approaches the mean spin of the compound nuclear level density spin distribution. As indicated in the introduction, the mean angular momentum is expected to reach a minimum (saturation) value of $(\mu R_B{}^2\hbar\omega/4\hbar^2)^{1/2}$, whereas the mean spin of a Fermi gas level density distribution is given by $(IT/\hbar^2)^{1/2}$, where I is the moment of inertia and T is the nuclear temperature. Typical empirical values of $(IT/\hbar^2)^{1/2}$ at lower excitation energies are between 3 and 5.

2.2 *Isomer Ratios*

The relative population of ground and isomeric states of different spin reflects the initial angular momentum distribution. The advantage of this method is its high sensitivity. When suitable half lives are involved, characteristic gamma rays can be observed after irradiation. The disadvantages of the method include the sensitivity to statistical model parameters, particularly the spin dependence of the nuclear level density, and to the location of specific levels above the isomeric state, which determines the pathways that selectively feed or bypass the isomeric state. Another disadvantage of this method is the fact that there is usually only a single exit channel with a suitable isomeric state, so that the method can only be reliably used over a limited bombarding energy range.

Isomer ratios have been successfully used for several reactions leading to the $I = 11/2$ isomeric state and $I = 3/2$ ground state in ^{137}Ce (17, 18). Calibration reactions with ^{3}He and ^{4}He projectiles confirm the statistical model analysis used to extract quantitative information about the first moment of the compound nuclear spin distribution. An attempt to extend the method to an isomeric state in ^{195}Hg was less successful (18), probably reflecting a less fortuitous location of specific levels above the isomeric state.

2.3 *Rotational State Populations*

Using rotational state populations is similar in principle to using isomer ratios. Information about the original spin distribution is inferred from the population pattern of members of the ground-state rotational band. The method has the advantage of determining the population of a large number of levels rather than just two, and it therefore contains more differential information about the spin distribution. It suffers the same disadvantages of the isomer ratio method, as it is only practical for specific residue channels (in this case when an even-even nucleus is the product) and is sensitive to the level structure and feeding pattern of higher-lying states. Calibration reactions need to be studied in order to extract quantitative information. The rotational band deexcitation gammas have to be measured on-line, and measurements far below the barrier are difficult if the target can be easily Coulomb-excited. This method is currently being applied to the oft-studied $^{16}O + {}^{154}Sm$ reaction (19).

2.4 *Fission Fragment Angular Distributions*

The idea behind this approach is that the direction of emission of fission fragments relative to their angular momentum vector reflects the difference in rotational energy for angular momentum parallel as compared with perpendicular to the fission axis. The projection of the initial angular momentum on the fission axis (designated K) is presumed to get frozen in at some stage during the fission process and to determine the final fission fragment angular distribution. Although from time to time it has been proposed that this stage is close to the final scission stage, the strong dependence of the effective moment of inertia $I_{\text{eff}} = I_\perp I_\parallel/(I_\perp - I_\parallel)$ on Z^2/A is strong evidence for the general belief that the angle between the fission axis and the angular momentum vector is frozen in at an earlier stage closer to the saddle configuration (20). The statistical theory of fission fragment angular distributions is well developed, and the anisotropy can be reasonably well approximated by $W(180°)/W(90°) = 1 + \langle l^2 \rangle / 4 I_{\text{eff}} T$ for not too large anisotropies. This formula shows that one must either take I_{eff} and T from fission and statistical models, or determine the product experimentally using calibration reactions. Fortunately, there is a large body of information on alpha-induced reactions at energies roughly twice the barrier and resulting in compound nuclear excitation energies and spins comparable to those of interest in subbarrier fusion reactions (21). These anisotropies can be reasonably well reproduced by diffuse-surface-liquid-drop model calculations of saddle-point moments of inertia and by conventional level density parameters, although discrepancies of about 20% are not uncommon. The advantage of using $I_{eff}T$ values directly from experiment is that the appropriate average over the contributions from fission following neutron emission is obtained automatically. Unfortunately, as discussed below, the reactions most suitable for studying heavy-ion fusion spin distributions lead to compound nuclei that cannot be made in alpha-induced "calibration" reactions.

The early hope (22) that fission fragment angular distribution could be used as a general probe of compound nuclear spin distributions has not been realized because, for the more fissionable targets at subbarrier energies, there is appreciable contamination from transfer-induced fission (23). The direction of this effect on perturbing the anisotropy for fusion fission is rather subtle, with the lower magnitude and the broader spatial distribution of the angular momentum vectors for transfer reactions resulting in a lower anisotropy, but with the lower average excitation energy tending to increase the anisotropy. In fact, Lestone et al (24) used coincidence techniques to show that for the $^{16}O + ^{232}Th$ reaction at 86 MeV the fusion-fission and transfer-fission anisotropies happen to be nearly identical. This happy accident is not expected to persist at all bombarding energies, and the coincidence measurements required to separate the contributions are too time-consuming to be applied generally. Back et al (25) suggested an alternative approach based on measuring the masses of the fission fragments at different angles. This approach works at energies where the overlapping mass distributions of fusion fission and transfer fission can be reliably decomposed.

2.5 *Alpha Angular Distributions*

The angular distributions of evaporated alpha particles are also sensitive to the second moment of the compound nuclear spin distribution. The anisotropies are generally lower, and one must include spin fractionation when extracting quantitative information. Spin fractionation arises because alpha emission competes more favorably with neutron and proton emission as the compound nuclear spin increases. Borges et al (26) attempted to deduce information about the spin distribution in a series of related reactions of sulfur isotopes with medium mass nuclei.

2.6 *Evaporation Residue* x*N Distributions*

The basis of this method is the fact that excitation energy used for rotation is not readily available for nucleon evaporation. Thus the partial cross section excitation function for a particular evaporation residue following emission of x nucleons (N) is displaced to a higher excitation energy with increasing compound nuclear angular momentum. To put it a different way, the ratio of two channels with $x+1$ as compared to x nucleons evaporated decreases with increasing $\langle l \rangle$ at a given excitation energy. The sensitivity of the $(x+1)$N to xN residue ratio on the spin distribution has been recognized for some time (27; S. Gil, private communication). A quantitative deduction of the mean angular momentum from partial cross sections is sensitive to statistical model parameters and the location of low-lying high-spin levels (the yrast band).

Das Gupta et al (29) exploited this technique to deduce the mean l for three fairly symmetric entrance-channel reactions leading to the ^{96}Ru compound nucleus. A study of light-ion calibration reactions would be valuable.

3. THEORETICAL CONSIDERATIONS

3.1 *Relationship Between Fusion Excitation Function and Spin Distribution*

It has been pointed out by several workers (30–32) that if certain approximations are made in a particular class of models there is an intimate connection between the spin distribution and the fusion excitation function. The method assumes that the different partial wave barriers have the same shape and penetrability as the $l = 0$ barrier and that the barriers are displaced from the $l = 0$ barrier by a fixed, energy-independent amount. With these assumptions, one can deduce the partial wave distribution at energy E from the slopes of the fusion cross section at energies below and up to the energy E. The method requires accurate cross-section data at energies well below that for which one is interested in the spin distribution. For example (31),

$$\langle l^2 \rangle = \{(2\mu R_B^2)/[\sigma_f(E)E\hbar^2]\} \int_{-\infty}^{E} dE' E' \sigma_f(E'). \qquad 6.$$

This procedure is equivalent to fitting a fusion excitation function with a model that incorporates the above assumptions and then calculating the spin distribution from this model. A direct experimental measurement of $\langle l^2 \rangle$ that disagrees with the expectation based on the fusion excitation function would indicate a failure of one of the model assumptions. These in turn can be due either to a failure of approximations, such as the constancy of the shape and position of the barriers for different partial waves, or to more fundamental assumptions, such as the independence of the inertial mass on internuclear separation or on angular momentum. The first approximations are not made in quantum mechanical models that numerically solve the Schrödinger equation for realistic barriers whose shape and position depend on l. The shape and position dependence in more realistic models arises from the different radial dependences of the nuclear, Coulomb, and centrifugal potentials.

3.2 *Static Deformations*

We presented in our introduction a qualitative discussion of the role of static deformations in broadening the spin distribution. More quantitative treatments are possible. Wong (10) has presented closed-form analytic expressions assuming parabolic barriers and frozen shapes and taking the proper average over the distribution of orientations of randomly oriented nuclei. Approximations made in Wong's treatment lead to an overestimate of the effect of deformation, or to an underestimation of the deformation parameter β to achieve a given effect (A. W. Charlop, J. R. Leigh, private communications, 1992). Esbensen (11) has presented the results of numerical calculations for permanently deformed nuclei using a nuclear potential based on that given by Christensen & Winther (33). The code CCDEF (34) includes an option for averaging over the random orientation of a static (permanently) deformed nucleus. Nagarajan et al (35) showed that the classical barrier distributions obtained from the different possible orientations are equivalent to the barrier distribution from a sudden-limit coupled-channels calculation (see below) in which all members of the rotational band are included but their excitation energies are neglected.

Most studies of static deformations have concentrated on quadrupole deformations. The expected role of hexadecapole deformations has been explored systematically by Fernandez-Niello & Dasso (36). They find that positive hexadecapole deformations always enhance the fusion cross section, whereas negative hexadecapole deformations may lead to a small decrease in the fusion cross section for nuclei with a sizeable positive prolate deformation. The effect on the spin distribution has not been explored.

3.3 *Vibrational Zero-Point Motion*

If the vibrational motion is slow compared to the collision time, each individual collision samples a particular nuclear shape whose probability of occurrence is given by the square of the amplitude for finding a particular deformation. The standard deviation of the zero-point motion deformation is proportional to the $B(E\lambda)$ value connecting the ground and relevant vibrational state,

$$\sigma_\lambda = \frac{R}{Z(\lambda + 3)} \left[(2\lambda + 1) \frac{B(E\lambda)}{B_W(E\lambda)} \right]^{1/2}. \qquad 7.$$

$B_W(E\lambda)$ is the Weisskopf single-particle estimate for a transition of multipolarity λ. Thus even if the equilibrium shape is spherical, the zero-point vibrations lead to an ensemble of instantaneous shapes and a manifold of effective barriers. The shape of the barrier probability distribution for vibration about a spherical equilibrium shape is approximately Gaussian, rather than the truncated concave distribution (see Figure 2 inset) for a static deformation (37).

If a vibrational state has a high energy and hence high vibrational frequency, the condition $\omega_{vib} \ll \omega_{collision}$ is no longer satisfied and the frozen-shapes approximation is no longer valid. Esbensen, Wu & Bertsch (38) examined the attenuation of the subbarrier enhancement by comparing results of frozen-shapes calculations with coupled-channels calculations. They find that if $E = \hbar\omega_{vib}$ is of the order of a few MeV, the enhancement is appreciably reduced. This can be thought of in terms of an averaging over the fluctuations in shape for each collision, so that the effect of the shape fluctuations is absorbed into a single effective potential and does not give rise to barrier fluctuations. This partly explains why the effect of the projectile shape degrees of freedom are relatively unimportant for ^{12}C and ^{16}O projectiles with high-energy first-excited states. (Another reason that shape degrees of freedom are less important for light nuclei is that the barrier lowering depends on the product βR of the deformation β and the nuclear radius R, not on β alone).

3.4 *Coupled-Channels Approach*

A more quantitative treatment of multidimensional effects on barrier penetration uses the coupled-channels formalism (39). In this formalism one expands the total wave function Ψ in terms of channel states, Φ_i:

$$\Psi = \sum_i \frac{\chi_i(r)}{r} \Phi_i. \qquad 8.$$

The $\chi_i(r)$ are the radial distorted waves in channel i, while Φ_i contains the internal structure of the projectile and target. The channels that are coupled can be either inelastic channels, such as those arising from quadrupole or octupole vibrations, or they can be transfer channels. The total wave function has to satisfy a Schrödinger equation, which leads to coupled equations of the form

$$\frac{d^2\chi_i}{dr^2} + \frac{2\mu_i}{\hbar^2}[E_i - V_i^{\text{eff}}(r)]\chi_i = \frac{2\mu_i}{\hbar}\sum_{j\neq i}\langle\Phi_i \mid V \mid \Phi_j\rangle. \qquad 9.$$

Broglia et al (40) suggested that the inelastic coupling strengths near the top of the barrier can be estimated from the expression

$$\langle\Phi_i \mid V \mid \Phi_j\rangle \approx \frac{1}{4\pi}[-\beta R(\mathrm{d}V/\mathrm{d}r)_{R_\mathrm{B}} + F^{\mathrm{c}}(R_\mathrm{B})], \qquad 10.$$

where the second term is the form factor for Coulomb excitation. The coupling strength for a transfer channel in the same approximation is given by

$$\langle\Phi_i \mid V \mid \Phi_j\rangle \approx \frac{1}{4\pi}(3\ \text{MeV})\exp[-(R_\mathrm{B} - R_1 - R_2)/a]. \qquad 11.$$

This gives a strength typically 1/5 of that for an inelastic excitation. These expressions are useful for orientation purposes, but realistic calculations require taking into account the full radial dependence of the coupling. Dasso & Landowne (41) used a series expansion about the barrier radius to incorporate finite range effects in a simplified coupled-channels code (42). To make use of these couplings for a fusion cross-section calculation, one has to make an assumption as to what defines fusion. Usually, one assumes that the flux that reaches the nuclear interior (inside the classical turning point) is associated with fusion. This can be accomplished by using an incoming wave boundary condition or an absorptive potential localized inside the barrier ($r_0 \lesssim 1.0$ fm). This code has been used extensively in the interpretation of experimental data.

One instructive result that emerges from a simple two-level coupling model (43) is the realization that any coupling interaction will split the barrier into a lower and a higher barrier, so that at subbarrier energies there is always an enhancement from coupling effects.

3.5 *Multidimensional Tunneling*

Coupled-channels calculations have generally used as basis states those corresponding to the collective or single-particle (transfer) excitations of the target and projectile nuclei. This may not be an efficient basis set for describing more complicated degrees of freedom such as neck formation of the dinuclear system. Alternatively, one can construct a multidimensional potential energy surface based on a more macroscopic approach in which radial separation and neck formation (or some other macroscopic shape variable) are selected as the most important degrees of freedom. In this approach, one must first define both the multidimensional potential energy surface and the inertial parameters accompanying this representation. One then finds the quantum mechanical probability for tunneling through the potential barrier. This method has been applied to the problem of spontaneous nuclear fission, and more recently to heavy-ion fusion.

Krappe et al (44) performed one of the early two-dimensional quantum mechanical tunneling calculations for subbarrier fusion. The internuclear separation was taken as one coordinate, and a second unspecified coordinate (of harmonic oscillator form) was parameterized by fitting excitation function data. The parameters of this coordinate were determined for a variety of systems, and were shown to correlate with the distance between half-density surfaces at the barrier. They suggest that such a trend is compatible with neck formation. Both Krappe et al and Aguiar et al (45) suggested that empirical extraction of the energy shift associated with subbarrier fusion enhancements is comprised of two parts: one associated with specific reactant degrees of freedom that vary from system to system because of nuclear structure effects, and a second part associated with the neck degree of freedom, which varies more smoothly with system size.

Aguiar et al (46) also constructed two-dimensional potential energy surfaces from liquid drop model considerations. They have not, however, done two-dimensional quantal calculations. At subbarrier energies they assume that the approach to the classical turning point is only along the radial coordinate. They then assume a subbarrier trajectory at right angles on this potential energy surface and calculate tunneling along the neck degree of freedom using a WKB approximation and an inertial parameter based on irrotational fluid flow. They have calculated the cross section and spin distributions for the $^{80}Se + ^{80}Se$. They get reasonable agreement with the mean spin at all but the lowest energies, but at the expense of seriously overestimating the fusion cross section. Iwamoto & Harada (47) also performed schematic two-dimensional calculations and compared them to data with an empirical fitting parameter.

Schneider & Wolter (48) constructed a potential energy surface based on a parameterization of Krappe, Nix & Sierk (49). They use some approximations for the inertial parameters. They describe the motion of the system on this surface by the time-dependent propagation of a wave packet, which has the advantage of providing a very intuitive picture of the fusion process in two dimensions. One important result obtained from a sample calculation near the barrier is that the wave packet does not follow the adiabatic path but rather spreads out along the barrier ridge line with many paths effectively contributing to fusion. As might be expected from this observation, the calculated spin distribution is rather broad.

Semirealistic calculations of this sort are still in their infancy but offer the possibility of a priori identification of the most important degrees of freedom in fusion. It is generally believed that the neck degree of freedom will be more important for heavier systems, in which the density overlap at the separation corresponding to the top of the barrier is largest.

3.6 *Phenomenological Approaches*

3.6.1 ABSORPTION UNDER THE BARRIER Udagawa and collaborators (50) proposed that fusion can occur under the barrier, i.e. before reaching the inner classical turning point. They utilize an optical model, identifying that part of the imaginary (absorbing) potential inside a "fusion radius" R_f with fusion, and that part beyond with flux going into direct reaction channels. This imaginary potential can then be inserted into either an optical model or a coupled-channels code, depending on whether the coupling to specific reaction channels is weak or strong. Most analyses have, however, used only an optical model.

The optical potential is somewhat constrained by comparison with elastic scattering data, and the fusion radius R_f is determined by comparison with fusion data. Typical values of R_f are given by $r_f = R_f/(A_1^{1/3}+A_2^{1/3})$ of 1.4 to 1.5. This is quite large, and leads to larger mean spin values than do optical model calculations, which identify flux reaching the inner turning point (or equivalently a fusion potential with $r_f \sim 1.0$) as contributing to fusion. Mean spins for several reactions have been calculated and compared with experimental results (51, 52). This model does not seem to predict a "barrier bump" in the mean spin. No fundamental justification for identifying absorption at large distances (under the barrier) with fusion has been identified (53).

3.6.2 SURFACE FRICTION MODEL For systems with large Z_1Z_2 values, there may be appreciable probability for damping of the entrance channel kinetic energy into rather complicated internal degrees of freedom before getting inside the barrier. Wolfs (53b) has found significant strength for deeply inelastic scattering at subbarrier energies in the reactions of ^{58}Ni with ^{112}Sn and ^{124}Sn. Coupled channels calculations are impractical in such a situation, and Fröbich & Richert (53c) have suggested using a surface friction model originally developed to model reactions at higher energies. A first attempt to extend this approach to low energies has been reported. The Langevin method—with an empirically modified Einstein relation and an empirically modified conservative potential—has been used to try to reproduce fusion cross sections and mean spin values for the $^{64}Ni + ^{100}Mo$ reaction (53c). The extension of this method to subbarrier energies, in which tunneling is important, is not straightforward.

3.6.3 ENERGY-DEPENDENT OPTICAL MODEL PARAMETERIZATION It is now well established that the real part of the nucleus-nucleus potential increases rapidly as the bombarding energy approaches the barrier (54, 55). This is sometimes referred to as the threshold anomaly. Empirical analyses show that the imaginary potential weakens as the energy drops down to the barrier. This is to be expected since, as the energy is lowered, reaction channels become closed by energetic limitations. Nagarajan et al (56) were able to understand the energy variation of the real potential as a consequence of a dispersion relation relating the real and imaginary parts of the optical potential.

Subbarrier fusion enhancement can now be expected as a consequence of the increase of the attractive real potential and the associated lowering of the fusion barrier. If one assumes that the geometry of the potential does not change, one can use the energy dependence of the real potential determined from scattering data (which determines the potential only at the strong absorption radius) and a barrier penetration model to calculate the fusion cross section. This has been done for the relatively well-studied $^{16}O + ^{208}Pb$ system, and the subbarrier enhancement is reproduced quite well (55). Another consequence of the barrier lowering is an increase in the mean square spin of the spin distribution. This has also been calculated from the empirical energy-dependent potential and compared with experiment. Although the agreement is better than for an energy-independent potential that fits scattering at higher energies, the mean square spin is still significantly underestimated at subbarrier energies.

Dasso, Landowne & Pollarolo (57) pointed out that the underestimation of the width of the spin distribution is a consequence of comparing the results from an energy-dependent potential parameterization of a given fusion cross section with that from a fully dynamical coupled-channels calculation. This can be understood in terms of the fact that an energy-dependent barrier parameterization leads to a single barrier for each l, whereas dynamical effects as calculated in the coupled-channels formalism give rise to a distribution of barriers for each l.

4. COMPARISON OF EXPERIMENTAL RESULTS WITH MODEL CALCULATIONS

In this section I review the available experimental results and compare them with model calculations. I emphasize particularly those results that illustrate the characteristic dependence of mean angular momentum on bombarding energy or that indicate failure of model assumptions. When appropriate model comparisons are not available in the literature, I compare experimental results with calculations performed with the CCFUS/CCDEF code (34, 42) using literature values of $\beta_2 R$ and $\beta_3 R$ (58, 59). In performing these calculations, I first adjust the barrier height to reproduce the fusion cross section, and then with those parameters calculate spin distributions. Failure to reproduce measured moments of the spin distribution then indicates a failure of one or more of the model assumptions. I conclude this section by developing a systematics of the ratio of the experimental to theoretical mean spin.

I organize my review of experimental results roughly by the mass asymmetry of the entrance channel, starting with results for the lightest projectiles ^{12}C, ^{16}O, and ^{19}F.

4.1 *Very Asymmetric Entrance Channels*

The $^{12}C + ^{128}Te$ reaction has been studied by the isomer ratio technique (18). The results for this system are illustrated in Figure 3. They provide the clearest demonstration of the saturation of the mean angular momentum with decreasing energy at energies well below the Coulomb barrier. The small reduced mass of this system, together with the small collectivity of the target, results in a vanishingly small barrier bump in the dependence of $\langle l \rangle$ on energy. The experimental results are in excellent agreement with expectations.

The $^{16}O + ^{152}Sm$ and $^{16}O + ^{154}Sm$ reactions have been studied by two different groups. The multiplicity distributions for the former reaction was studied by Wuosmaa et al (13) and the latter reaction by Gil et al (12). The results for both systems are shown in Figure 4. The barriers for the two systems differ by less than 1 MeV, and occur at $E_{lab} = 65$ MeV. The mean angular momentum for the former system seems to agree with expectations based on a coupled-channels calculation (11) taking into account the shape degrees of freedom of the target, whereas the mean angular momentum for the ^{154}Sm target exceeds expectations based on a similar calculation. The reason for this quantitative discrepancy is not clear, but it may reflect differences in corrections for unobserved transitions and in the conversion of multiplicity to angular momentum. The discrepancy is largest at bombarding energies for

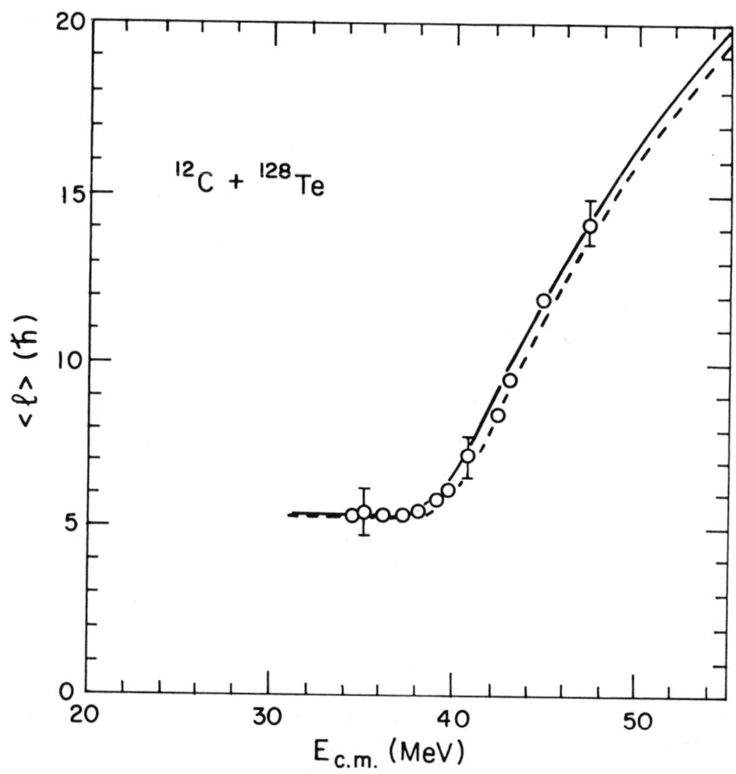

Figure 3 Mean spin deduced from isomer ratios for the 3n evaporation product. The curves are predictions from a coupled-channels calculation with (*solid*) and without (*dashed*) coupling included. After (18).

which the 3n channel is dominant. In the ^{152}Sm study no correction was made for the ground-state spin of the evaporation residue. The conversion of multiplicity to angular momentum for the ^{154}Sm experiment was calibrated using light-ion data (12). The dominance of the quadrupole deformation of ^{154}Sm in enhancing the subbarrier fusion cross section has been demonstrated by Wei et al (60). They performed high-precision measurements of the fusion excitation function and differentiated $E\sigma$ twice to obtain the distribution of barriers. The deduced barrier distributions exhibited the asymmetric shape for a nucleus with a permanent quadrupole deformation.

The angular momentum distributions inferred from the multiplicity distribution data for the $^{16}O + ^{152}Sm$ reaction are compared with coupled-channels calculations in Figure 5. The shapes of the observed distributions are in quite good agreement with calculations.

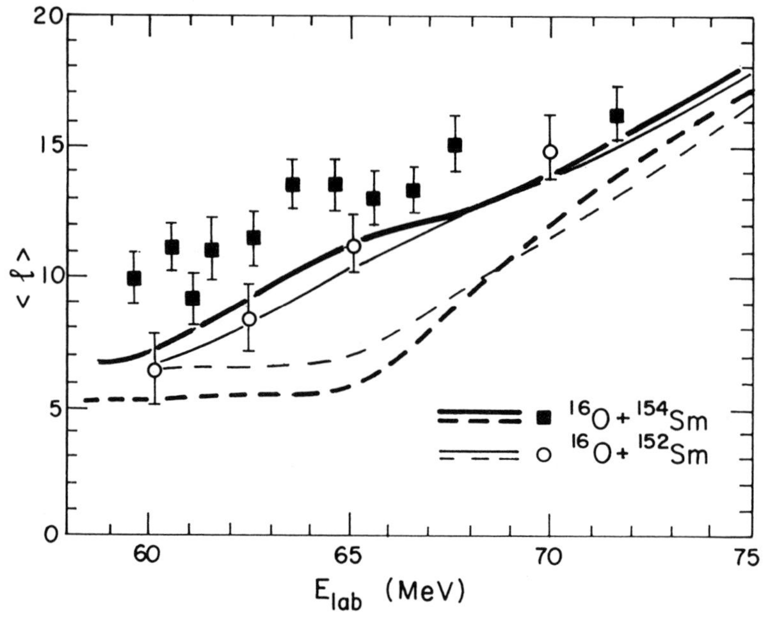

Figure 4 Mean angular momentum deduced from gamma-ray multiplicity measurements. The $^{16}O + ^{152}Sm$ results are from Wuosmaa et al (13) and are compared with no-coupling (*dashed curves*) and coupled-channels (*solid curves*) calculations based on the Esbensen model (11). The $^{16}O + ^{154}Sm$ results are from Gil et al (12) and several of the data points are averages for the two tagging techniques employed. The solid curves show the average of two calculations taking into account the shape degrees of freedom (10, 34) and the dashed curves represent a no-coupling calculation (12).

The ^{16}O and $^{19}F + ^{208}Pb$ reactions and the $^{16}O + ^{232}Th$ reactions have been studied by the fission fragment angular distribution technique. For the latter reaction, we only consider results from experiments that discriminate against fission following inelastic or transfer reactions. Because of the complication of transfer-induced fission, we concentrate our attention on fission of the less fissionable target ^{208}Pb, where the fission barrier of the transfer products is sufficiently high to suppress fission following transfer. The $^{16}O + ^{208}Pb$ reaction is of particular interest, since the elastic and transfer channels have also been studied. This system has also been the subject of extensive theoretical work (61, 62). Because of the lack of suitable targets, the ^{224}Th compound nucleus produced in this reaction cannot be reached in alpha-induced reactions. One must therefore rely on model calculations of $I_{eff}T$. These calculations must take into account fission following neutron emission. The evaporation residue yields of Vulgaris et al (63) provide an im-

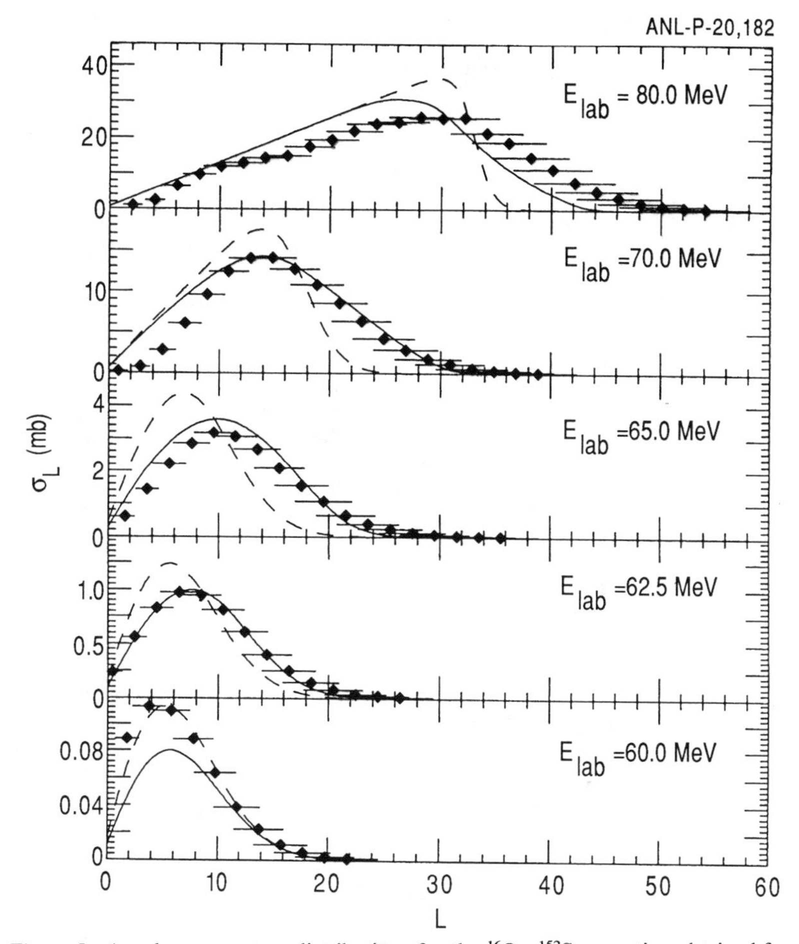

Figure 5 Angular momentum distributions for the $^{16}O + ^{152}Sm$ reaction obtained from γ-ray multiplicity distribution data. The solid curves are the results of the coupled-channels plus deformation calculation, and the dashed curves are from an adjusted barrier penetration model. The partial wave distributions have been normalized to the experimental total fusion cross section. From Wuosmaa et al (13).

portant constraint on the statistical model parameters used. The effective moment of inertia was taken from the diffuse surface model of Sierk (64). The mean square spin deduced from the measured (63, 65) anisotropies generally exceeded theoretical model predictions, as shown in Figure 6. Rossner et al (65b) recently claimed that mean square spin values deduced from fission fragment anisotropies for $^{16}O + ^{208}Pb$ are in good agreement with expectations for near-barrier

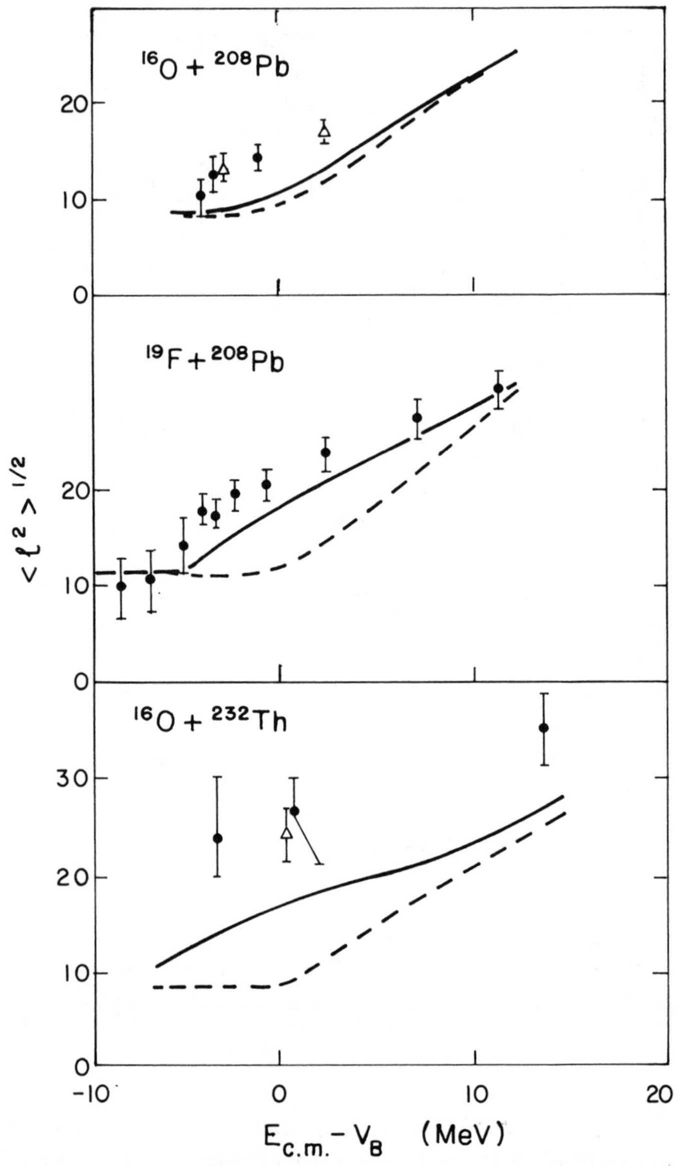

Figure 6 Root-mean-square spins derived from fission anisotropy data. The solid curve is from coupled-channels calculations and the dashed curve is without coupling (V_B = 75.7 MeV). The experimental data for $^{16}O + ^{208}Pb$ are from Murakami et al (65) (*circles*) and Vulgaris et al (63) (*triangles*). The $^{19}F + ^{208}Pb$ data and calculations are from Zhang et al (66, 67). The $^{16}O + ^{232}Th$ data are from Back et al (25) (*circles*) and from Lestone et al (24) (*triangle*). The solid curve is a CCDEF calculation with V_B = 81 MeV.

bombarding energies and that there is no evidence for anomalous anisotropies if appropriate corrections for pre-fission neutrons are made. In fact, this claim is not supported by the quantitative results appearing in their Figure 7, as the discrepancy between experiment and expectation in this figure is comparable to or even larger than that reported by Murakami et al (65) (and shown in Figure 6 of this review) at sub-barrier energies.

Zhang et al (66, 67) measured fission cross sections and fission fragment anisotropies for the $^{19}F + ^{208}Pb$ reaction. $I_{eff}T$ values estimated from systematics and from the Sierk model (64) are said to be in reasonable agreement. Figure 6 shows that the mean square spin values deduced from the anisotropies are qualitatively consistent with but somewhat larger than theoretical expectations. Comparison with model calculations that reproduce the fission cross section excitation function suggests that the plateau region may have been reached at the lowest bombarding energy, although the experimental errors are too large to confirm this. The barrier bump in the mean square spin is expected to be larger for $^{19}F + ^{208}Pb$ than for $^{16}O + ^{208}Pb$ because of lower-lying collective states in ^{19}F than in ^{16}O.

The $^{16}O + ^{232}Th$ fission angular distributions were originally studied in an inclusive measurement that did not suppress contributions from transfer fission (22). These results led to anomalously large mean square spin values. Lestone et al (24) used coincidence techniques to show that at 86-MeV bombarding energy the anisotropy from fusion fission was even a little higher (by 5%) than the inclusive value. Back et al (25) used mass distribution data to isolate fusion-fission angular distributions at lower bombarding energies. Although the anisotropies are somewhat lower than the inclusive values, they still lead to root-mean-square spin values almost twice the expected values at the lowest energy.

DiGregorio & Stokstad (69) compared available fusion and mean angular momentum data with CCFUS calculations. From comparisons of the experimental to calculated mean angular momentum ratios as derived from different experimental probes, they suggest that the fission angular distribution method is suspect. They included inclusive fission data for heavy systems, however, where fission following transfer is an important contaminant. When these results are removed from their comparison, the largest deviations in their systematics no longer arise from fission angular distribution data. Their concern is well taken that some of the largest deviations from expectations do arise from measurements based on fission angular distributions (discussed in Section 4.3). DiGregorio & Stokstad suggest that the apparent failure of the fission anisotropy method may arise from a failure of the assumption as to when the projection of the angular momentum on the nuclear symmetry axis is frozen in. This particular suggestion seems untenable to this reviewer, as calibration reactions with light projectiles at above-barrier energies (21) support the assumptions made. Also, Back et al (70, 71) examined a large body of fission anisotropy data and concluded that for projectiles lighter than Mg anisotropies are consistent with a saddle-point statistical model.

Ramamurthy et al (72) challenge the interpretation of the fission fragment anisotropies from a different point of view. They suggest that the anomalous fission fragment anisotropies may result from a form of pre-equilibrium fission in which the nucleus undergoes fission before ever equilibrating the K degree of freedom. They further suggest that the occurrence of this kind of pre-equilibrium fission depends on the entrance-channel mass asymmetry $a = (A_T - A_P)/(A_P + A_T)$. When this asymmetry is less than some critical value (the "Businaro-Gallone critical asymmetry"), the evolution of the system from contact to reseparation may occur before equilibration of the orientation of the projection K of the angular momentum on the dinuclear symmetry axis. For values of the asymmetry parameter larger than the critical value, estimated on the basis of charged liquid drop model considerations to be about 0.9, the driving forces in the entrance channel might drive the system faster toward a true compound nucleus, in which the K degree of freedom is equilibrated.

Ramamurthy et al measured fission anisotropies for a number of systems with entrance-channel asymmetries varying from 0.85 to 0.92, and found evidence that systems with smaller asymmetry values (heavier projectiles) exhibit the most anomalous anisotropies. Unfortunately, their measurements were inclusive and were made on the fissionable targets ^{232}Th and ^{237}Np so that some fission following transfer as well as fusion fission is to be expected. They also did not control for bombarding energy relative to the barrier, and we have found that the data can alternatively be correlated with energy relative to the barrier (73). The ^{19}F + ^{208}Pb system has an asymmetry value of 0.83 and, therefore, would be expected by the criterion of Ramamurthy et al to exhibit anomalous anisotropies. The results for this system, shown in the middle panel of Figure 6, do not exhibit anomalous anisotropies once the bombarding energy exceeds the barrier by about 10%. Furthermore, Toke et al (74) have measured fragment mass-angle correlations and find no evidence for pre-equilibrium fission for ^{16}O + ^{238}U, a possible hint for ^{27}Al, and clear evidence for ^{48}Ca. Back et al (75) measured gamma-ray multiplicity-fragment mass correlations and found no evidence for pre-equilibrium fission for $^{16}O + ^{238}U$, a hint for ^{26}Mg, and clear evidence for ^{27}Al. On the basis of these observations one would not expect a pre-equilibrium contribution for the systems used to probe subbarrier fusion spin distributions. Thus the importance of pre-equilibrium fission in the systems used for spin distribution studies remains an open question. Mean spin values deduced from fission fragment angular distributions may be questionable, but there seems to be no compelling evidence to dismiss them at the present time.

4.2 *More Symmetric Systems*

The $^{28}Si + ^{154}Sm$ system has been studied carefully by the gamma-ray multiplicity method (76). The conversion of multiplicity to angular momentum has been calibrated using the $^{16}O + ^{166}Er$ reaction to form the same compound nucleus. The results for the $^{28}Si + ^{154}Sm$ system, illustrated in Figure 7, provide the most dramatic evidence to date for the barrier bump in the dependence of mean spin on bombarding energy. This is a consequence of the large target deformation and of the larger barrier and reduced mass in the entrance channel compared to lighter systems. The Wong model, incorporating only the quadrupole deformation of the target, accounts for most of the increase in $\langle l \rangle$. A coupled-channels calculation including projectile excitation and a few transfer channels (relatively unimportant) accounts reasonably well for the data.

It is instructive to plot the ratio of the mean l obtained experimentally or from a coupled-channels calculation to that from a no-coupling calculation. This is done in Figure 7*d*, where one sees that there is a pronounced bump in this ratio centered at an energy just below the barrier. In contrast to this behavior for the mean l, the ratio of the experimental or coupled-channels fusion cross section to the no-coupling expectation continues to increase as the bombarding energy decreases, only approaching a saturation value at an energy corresponding to the point where the mean l is approaching its plateau value. Neglecting penetration effects, this energy can be identified as the lowest barrier for the most favorable orientation of the colliding ^{28}Si and ^{154}Sm.

The $^{32}S + ^{100}Mo$ and $^{36}S + ^{96}Mo$ reactions leading to the same ^{132}Ce compound nucleus have been studied by Hennrich et al (77). Pengo et al (78) had shown earlier that the subbarrier fusion enhancement is larger in the former system than in the latter system. Hennrich et al were able to fit the latter system excitation function with a simple coupled-channels calculation by increasing the literature β_i values by 50%, but still were unable to reproduce the enhancement for the

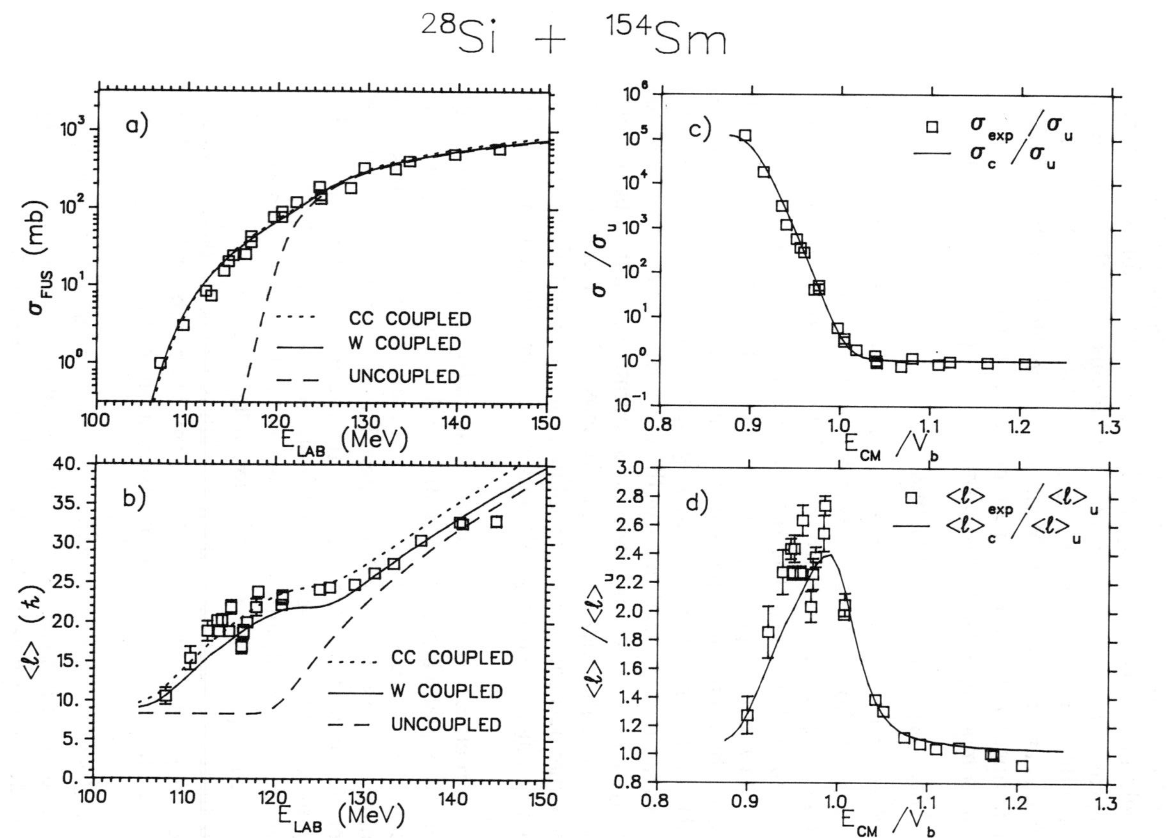

Figure 7 Fusion cross section (*a*) and average angular momentum (*b*) versus bombarding energy deduced from evaporation residue gamma multiplicities. The solid and dotted curves are from the Wong model and coupled-channels calculations, respectively. The dashed curve is obtained when deformation effects are suppressed. Energy dependence of the ratio of the experimental and of the theoretical cross sections to the expected cross section without couplings (*c*) and of the ratio of the experimental and of the theoretical mean spin to the expected mean spin without couplings (*d*). V_b = 101.5. After Gil et al (76).

$^{32}S + ^{100}Mo$ system. The use of deformation parameters 50% larger than literature values is troubling, and we searched for an alternative description. We found that it is possible to reproduce the $^{36}S + ^{96}Mo$ excitation function without increasing the literature β_i values if one makes a modest adjustment (1%) of the Bass model barrier. The inclusion of a single-neutron transfer channel with $Q = -3.1$ MeV has virtually no effect. Adjustment of the barrier in the $^{32}S + ^{100}Mo$ system, however, does not lead to satisfactory agreement with the excitation functions, as expected from the conclusions of Hennrich et al that the enhancement in this channel cannot be understood from the inelastic couplings alone. An attempt by Hennrich et al to include the very positive Q-value 2n transfer channel in a CCFUS calculation failed as a result of the unphysical results obtained with this simple implementation of coupled channels (41; S. Landowne private communication).

The $^{32}S + ^{100}Mo$ system seems to exhibit some similarities to the $^{58}Ni + ^{64}Ni$ system, in which many different couplings were required to reproduce the observed enhancement (80, 81). In particular, multiphonon vibrational excitations were shown to be important. Our best fits to the Pengo et al excitation function data with literature deformation parameters and including single-nucleon transfer channels are shown in the top part of Figure 8. At the lowest energies, the transfer contribution dominates for $^{32}S + ^{100}Mo$, while inelastic excitations dominate for $^{36}S + ^{96}Mo$. The lower panels compare the mean angular momentum values with the experimental values of Hennrich et al. The agreement with the $^{36}S + ^{96}Mo$ system is satisfactory, but the mean angular momentum is underestimated for the $^{32}S + ^{100}Mo$ system, as concluded by Hennrich et al.

Another study of several reactions leading to the same compound nucleus is that of Das Gupta et al (29). Only for the $^{28}Si + ^{68}Zn$ reaction is there possibly significant data below the barrier. We find the mean $\langle l \rangle$ to be reasonably well reproduced by CCFUS calculations fitting the fusion excitation function.

The $^{64}Ni + ^{100}Mo$ reaction has been studied with the spin spectrometer at the Oak Ridge National Laboratory in Tennessee and has been the subject of considerable analysis (16). The authors conclude that the excitation function cannot be reproduced with a priori coupling strengths, and that even when the coupling strengths are increased to 1.5 times their expected strength the mean angular momenta are underestimated. We found that the necessity to use larger than expected coupling strengths depends rather strongly on the demand to fit the highest energy point on the fusion cross section. We recalculated the fusion cross sections and mean angular momentum values using ex-

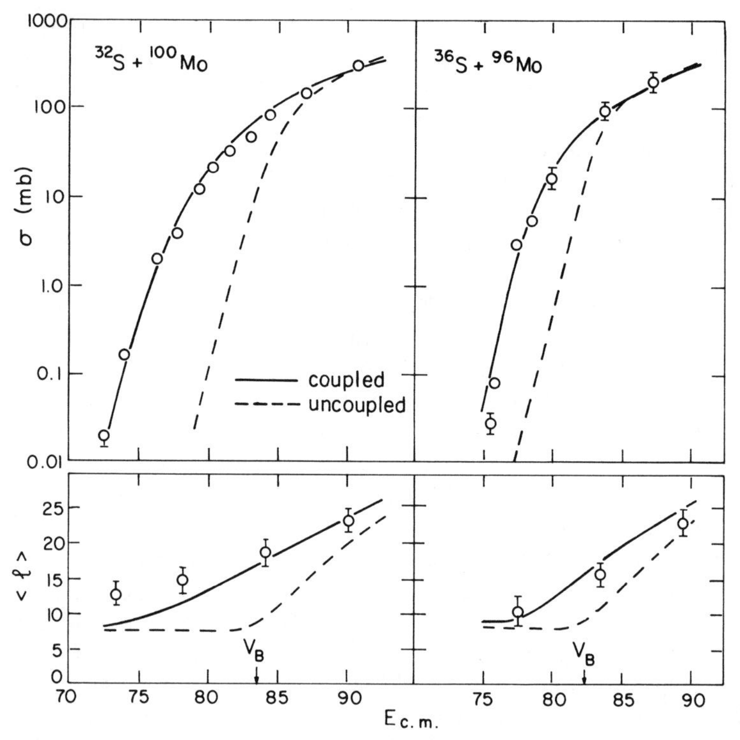

Figure 8 The upper panels show coupled-channels fits to the excitation functions of Pengo et al (78). The lower panels compare the mean spin from the coupled-channels calculations with the values deduced from multiplicity data (77).

pected values of coupling strengths. Our fit to the cross section is shown in the upper part of Figure 9. This calculation fits the lower points very well, but overestimates the fusion cross section at the highest energy point. In contrast, Halbert et al (16) required their calculations to reproduce this point. The two sets of calculations also differ somewhat in the diffuseness of the real potential used; Halbert et al used a folding model potential while the CCFUS (34) calculations presented here used a Saxon-Woods parameterization of the Christensen-Winther potential (33). But the primary difference is in the weighting given to the highest energy point. It would be desirable to have more extensive fusion cross-section data for this system; measurements of the spin distributions and fusion cross section with a thinner target at low energies and an extension of the cross-section measurement to higher energies would be particularly valuable.

The mean angular momenta from the new calculations are compared

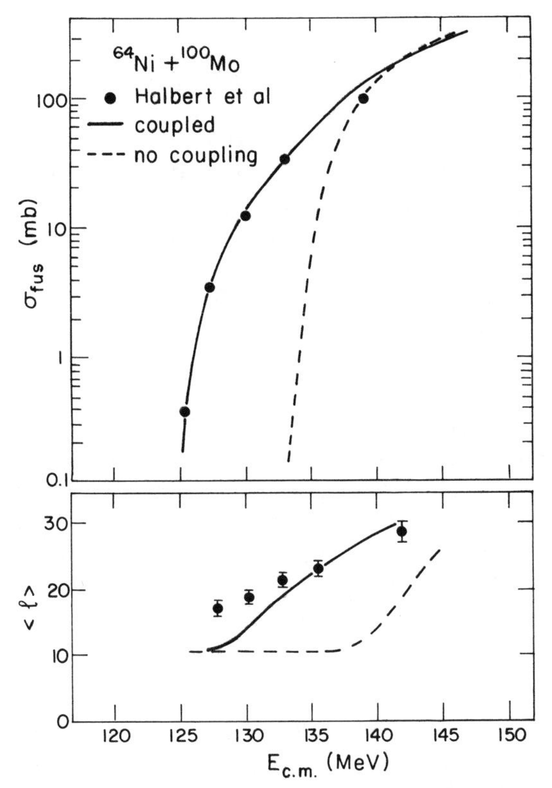

Figure 9 The upper panel shows a coupled-channels fit to the cross-section data of Halbert et al (16). The lower panel compares their mean spin values with the coupled-channels values.

with experiment in the lower part of the figure. The calculations still considerably underestimate the mean angular momenta at the lower energies, which supports the main conclusion of Halbert et al. Pioneering studies of near-barrier spin distributions for symmetric entrance channels were performed by Nolan et al (82) and by Haas et al (27). Haas et al studied four entrance channels at near-barrier energies leading to the ^{160}Er compound nucleus, including the ^{80}Se + ^{80}Se reaction. The latter reaction was studied by Nolan et al, who extended the measurements to subbarrier energies. These results have been the subject of considerable analysis (84) and comparison with model calculations (27, 46, 83). The calculations generally agree well with the data at above-barrier energies, but underestimate the mean angular momentum at subbarrier energies.

4.3 *Systematics of Results*

We present in this section an updated summary of experimental results on the mean spin (or in the case of fission data on root-mean-square spin) values. In order to compare different systems it is helpful to compare the ratio of the mean spin derived from experiment to that from a model that incorporates the expected effects from coupling of shape and nucleon transfer degrees of freedom. The latter have generally been taken from previous calculations, but in several cases we have performed new calculations using the CCFUS or CCDEF codes. We also scale the bombarding energy to that of the fusion barrier in the absence of coupling effects. (See 15, 85 for summaries of the early data.)

The results of our updated survey are plotted in Figure 10. The experimental results are generally in agreement with theoretical expectations at energies above the barrier, where coupling effects are weak. The situation at energies a little below the barrier, where coupling effects are expected to be strongest, is less clear. There is considerable evidence that the coupling of simple degrees of freedom to the entrance channel is inadequate to explain the observations fully. This is apparent

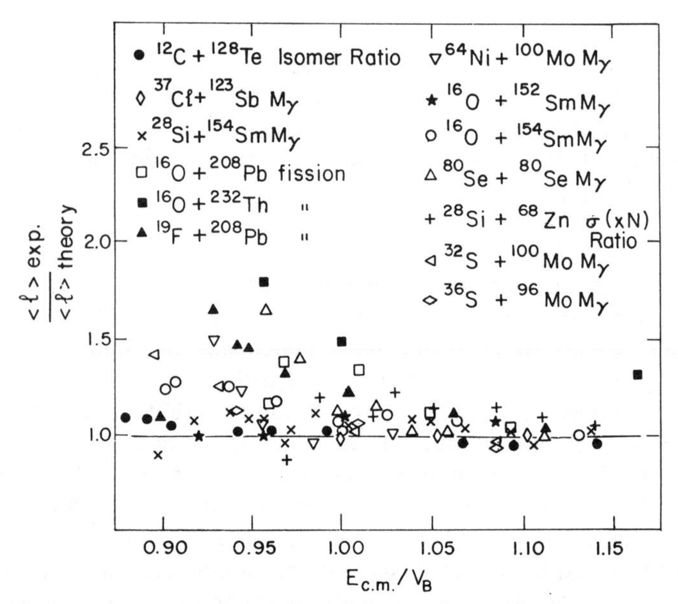

Figure 10 A comparison of the ratio of experimental to theoretical mean spin values (or ratio of root-mean-square values in cases of fission anisotropy probes) as a function of the ratio of bombarding energy to barrier energy.

for values obtained from both gamma-ray multiplicity data and from fission anisotropy data. We have not shown error bars on the ratios in Figure 10. They were omitted because the errors in general are not dominated by statistical or other known experimental uncertainties, but rather by the uncertainty in conversion of the observable to the angular momentum. This uncertainty is typically of the order of 20%, unless careful calibration reactions have been performed. The $^{16}O + ^{232}Th$ fission data (24, 25) require special comment. The errors on the data derived from the anisotropies of Back et al (25) are rather large (about 25%) because of the decomposition into fusion and transfer components. The results obtained, however, are in good agreement with that obtained at one energy by Lestone et al (24) using a very different approach. The failure to reproduce the expected value well above the barrier is disturbing. It may reflect the fact that the barrier has become so low for this very fissile and high-spin compound nucleus that statistical considerations fail. Or it may be a consequence of a nonequilibrium entrance-channel effect as suggested by Ramamurthy et al (72) and mentioned above.

Apart from this special case, there is considerable evidence from a variety of probes and from different laboratories for larger than expected mean spin values. The origin of the underestimation of the mean angular momentum is not clear. We previously suggested that it may arise from lack of convergence of the coupled-channels approach (85). This approach, using as basis states the asymptotic two-body channels, accounts for the dominant effects of shape and transfer degrees of freedom. These basis states, however, lead to slow convergence in accounting for dinuclear degrees of freedom such as neck formation and density saturation. If not treated explicitly, these effects may lead to a departure of the inertial mass from the reduced mass in the entrance channel. It is interesting to note that there is some evidence that the largest deviation from expectations based on only reactant degrees of freedom occurs for systems with large Z_1Z_2 products (85). It is these systems that have the largest density overlap at the internuclear separation corresponding to the top of the barrier. This might also explain why fission probe measurements, which require systems with large Z_1+Z_2, are among those with the largest deviations.

There are also more empirical approaches to accounting for spin distributions. Broader spin distributions are obtained in the phenomenological optical model parameterization of Udagawa et al (50) and Kim et al (51), which allows fusion absorption under the barrier. Broader spin distributions are also sometimes obtained (29, 86) from empirical barrier distributions, obtained by fitting the near-barrier region of the fusion cross-section excitation function as proposed by Stelson et al (86). This is not a unique result, as there are many examples in which a satisfactory reproduction of the fusion cross-section excitation function is obtained yet the mean spins are not reproduced (Figure 8; 12, 65–67).

5. OPEN PROBLEMS

At the end of the preceding section I focused on residual discrepancies between experimental deductions and theoretical expectations, having divided out the expected effects of coupling certain simple degrees of freedom to the entrance channel. What are the successes in our understanding of subbarrier fusion? For weakly coupled systems such as $^{12}C+^{128}Te$ or $^{16}O+^{208}Pb$, the enhancement of the subbarrier fusion cross section is modest and only a small increase in the broadening of the spin distribution is expected. The quantal expectation that the mean spin should saturate at a nonzero value as the bombarding energy decreases has been experimentally verified for the $^{12}C+^{128}Te$ system (Figure 3). For systems in which one of the reactants has a large static deformation, the coupling of the shape degree of freedom alone increases the cross section by many orders of magnitude and produces a pronounced bump in the mean spin just below the barrier. This behavior has been clearly demonstrated in the $^{28}Si+^{154}Sm$ reaction (Figure 7). It is interesting to note that these effects can be predicted with a priori nuclear potentials and nuclear deformations. Only slight ($\sim 1\%$) adjustments of the barrier height are required to reproduce quantitatively the fusion excitation function, and the expectation for the mean spin is essentially unaffected by this fine-tuning. This ability to perform an absolute calculation of the quantum tunneling is not possible in some areas of physics, such as the Josephson effect in condensed matter physics.

What are the open problems in subbarrier fusion, particularly in regard to spin distributions? On the experimental side, the unsatisfactory scatter in the empirical observations on the mean spin below the barrier (Figure 10) needs to be understood. The absolute determination of the spin distribution is very difficult when the mean spin of the compound nucleus becomes comparable to the most probable spin of the compound nuclear level density distribution. The observable signals become very weak and their translation into spin information becomes more model dependent. Calibrations of this translation need to be performed. These problems become somewhat more tractable for systems with larger barriers and entrance-channel reduced masses as the mean spins involved are larger. The range of applicability of the fission fragment angular distribution probe needs to be explored further with respect to entrance-channel mass asymmetry, bombarding energy relative to the fusion barrier, and temperature of the compound nucleus relative to the fission barrier.

On the theoretical side, one needs to find some representation in which both the degrees of freedom associated with excitation and transfer of the reactants and at least one degree of freedom of the overlapping fragments (such as neck formation) can be treated simultaneously. A step in this direction might be taken by using studies of multidimensional tunneling (such as that of Schneider & Wolter) to define an effective path and its associated inertia. That path and its inertia might be used to replace the pure radial coordinate in coupled-channels calculations, where the known collective and transfer degrees of freedom of the reactants are treated explicitly.

ACKNOWLEDGMENTS

I thank A. W. Charlop, S. Gil, S. Kailas, and S. Landowne for helpful discussions. This work was supported in part by the US Department of Energy.

Literature Cited

1. Konopinski, E. J., Teller, E., *Phys. Rev.* 73:822 (1948)
2. Yeager, D. L., Miller, W. H., *J. Phys. Chem.* 95:8212 (1991)
3. Marcus, R. A., *J. Phys. Chem.* 95:8236 (1991)
4. Beckerman, M., *Phys. Rep.* 129:145 (1985)
5. Steadman, S. G., Rhoades-Brown, M. J., *Annu. Rev. Nucl. Part. Sci.* 36:649 (1986)
6. Signorini, C., et al, eds., *Proc. Symp. on Heavy Ion Interaction Around the Coulomb Barrier,* Legnaro, Italy, in *Lecture Notes in Physics,* Vol. 317. Berlin: Springer-Verlag (1988)
7. Nagarajan, M. A., ed., Heavy ion collisions at energies near the coulomb barrier, in *Proc. Workshop,* Daresbury, England, Inst. Phys. Conf. Ser. No. 110. Bristol: Inst. Phys.
8. Vandenbosch, R., Heavy ions in nuclear and atomic physics, in *Proc. 20th Mikolajki Summer School,* Mikolajki, Poland. Bristol: Adam Hilger (1989), pp. 44–70
9. Betts, R. R., in *Proc. Int. Symp. Towards a Unified Picture of Nuclear Dynamics,* Nikko, Japan, AIP Conf. Proc. 250. New York: Am. Inst. Phys. (1991)
10. Wong, C. Y., *Phys. Rev. Lett.* 31:766 (1973)
11. Esbensen, H., *Nucl. Phys.* A352:147 (1981)
12. Gil, S., et al, *Phys. Rev.* C43:701 (1991)
13. Wuosmaa, A. H., et al *Phys. Lett.* B263:23 (1991)
14. Gil, S., et al, *Phys. Rev.* C31:1752 (1985)
15. Charlop, A., et al, see Ref. 6, p. 157
16. Halbert, M. L., et al, *Phys. Rev.* C40:2558 (1989)
17. Stokstad, R. G., et al, *Phys. Rev. Lett.* 62:399 (1989)
18. DiGregorio, D. E., et al, *Phys. Rev.* C42:2108 (1990)
19. Bierman, J. D., et al, *Annu. Rep. Nucl. Phys. Lab.,* Univ. Washington (1991), p. 22; (1992), p. 22
20. Vandenbosch, R., Huizenga, J. R., *Nuclear Fission.* New York: Academic (1973)
21. Reising, R. F., Bate, G. L., Huizenga, J. R., *Phys. Rev.* 141:1161 (1966)
22. Vandenbosch, R., et al, *Phys. Rev. Lett.* 56:1234 (1986); 57:1499 (1987)

23. Cheifitz, E., Britt, H. C., Wilhelmy, J. B., *Phys. Rev.* C24:519 (1981)
24. Lestone, J. P., et al, *Nucl. Phys.* A509:178 (1991)
25. Back, B. B., et al, in *Proc. 6th Winter Workshop on Nuclear Dynamics,* Jackson Hole, Wyoming, Lawrence Berkeley Lab. Rep., LBL 28709 (1990), p. 9
26. Borges, A. M., et al, in *Proc. Symp. on Many Facets of Heavy Ion Fusion Reactions,* Argonne Natl. Lab., ANL-PHY-86-1, Argonne, IL (1986), p. 441
27. Haas, B., et al., *Phys. Rev. Lett.* 54:398 (1985)
28. Deleted in proof
29. Das Gupta, M., et al, *Phys. Rev. Lett.* 66:1414 (1991)
30. Reisdorf, W., et al, *Nucl. Phys.* A438:212 (1985)
31. Dasso, C. H., Esbensen, H., Landowne, S., *Phys. Rev. Lett.* 57:1498 (1986)
32. Balantekin, A. B., Reimer, P. E., *Phys. Rev.* C33:379 (1986)
33. Christensen, P. R., Winther, A., *Phys. Lett.* 65B:19 (1976)
34. Fernandez-Niello, J., Dasso, C. H., Landowne, S., *Comput. Phys. Commun.* 54:409 (1989)
35. Nagarajan, M. A., Balantekin, A. B., Takigawa, N., *Phys. Rev.* C34:897 (1986)
36. Fernandez-Niello, J., Dasso, C. H., *Phys. Rev.* C39:2069 (1989)
37. Back, B. B., et al., *Phys. Rev.* C32:195 (1985), C33:385 (1986)
38. Esbensen, H., Wu, J.-Q., Bertsch, G. F., *Nucl. Phys.* A411:275 (1983)
39. Austern, N., *Direct Nuclear Reaction Theories*. New York: Wiley (1970)
40. Broglia, R. A., et al, *Phys. Lett.* 133B:34 (1983)
41. Dasso, C. H., Landowne, S., *Phys. Lett.* B183:141 (1987)
42. Dasso, C. H., Landowne, S., *Comput. Phys. Commun.* 46:187 (1987)
43. Dasso, C. H., Landowne, S., Winther, A., *Nucl. Phys.* A405:381 (1983)
44. Krappe, H. J., et al, *Z. Phys.* A314:23 (1983)
45. Aguiar, C. E., et al, *Nucl. Phys.* A326:201 (1987)
46. Aguiar, C. E., et al, *Nucl. Phys.* A479:571 (1987)
47. Iwamoto, A., Harada, K., *Z. Phys.* A326:201 (1987)
48. Schneider, J., Wolter, H. H., *Z. Phys.* A339:177 (1991)
49. Krappe, H. J., Nix, J. R., Sierk, A. J., *Phys. Rev.* C20:992 (1979)
50. Udagawa, T., Kim, B. T., Tamura, T., *Phys. Rev.* C32:124 (1985)
51. Kim, B. T., Udagawa, T., Tamura, T., *Phys. Rev.* C33:370 (1986)
52. Kubo, K.-I., Manyum, P., Hodgon, P. E., *Nucl. Phys.* A534:393 (1991)
53. Kim, B. T., Udagawa, T., *Phys. Lett.* B273:37 (1991)
53b. Wolfs, F. L. H., Phys. Rev. C36:1379 (1987)
53c. Fröbich, P., Richert, J., Phys. Lett. 237B:328 (1990)
54. Lilley, J. S., et al, *Phys. Lett.* 151B:181 (1985)
55. Fulton, B. R., see Ref. 7, p. 15
56. Nagarajan, M. A., Mahaux, C., Satchler, G. R., *Phys. Rev. Lett.* 54:1136 (1985)
57. Dasso, C. H., Landowne, S., Pollarolo, G., *Phys. Lett.* B217:25 (1989)
58. Raman, S., et al, *At. Data Nucl. Data Tables* 36:1 (1987)
59. Spear, R. H., *At. Data Nucl. Data Tables* 42:55 (1989)
60. Wei, J. X., et al, *Phys. Rev. Lett.* 67:3368 (1991)
61. Thompson, I. J., et al, *Phys. Lett.* 157B:250 (1985)
62. Pieper, S. C., Rhoades-Brown, M. J., Landowne, S., *Phys. Lett.* 162B:43 (1985)
63. Vulgaris, E., et al, *Phys. Rev.* C33:2017 (1986)
64. Sierk, A., *Phys. Rev.* C33:2039 (1986)
65. Murakami, T., et al *Phys. Rev.* C34:1353 (1986)
65b. Rossner, H., et al, *Phys. Rev.* C45:719 (1992)
66. Zhang, H., et al *Nucl. Phys.* A512:531 (1990)
67. Liu, Z., et al, *Phys. Rev.* C42:2752 (1990)
68. Deleted in proof
69. DiGregorio, D. E., Stokstad, R. G., *Phys. Rev.* C43:265 (1991)
70. Back, B. B., et al, *Phys. Rev.* C32:195 (1985)
71. Back, B. B., *Phys. Rev.* C31:2104 (1985)
72. Ramamurthy, V. S., et al, *Phys. Rev. Lett.* 65:25 (1990)
73. Vandenbosch, R., in *Proc. 8th Winter Workshop on Nuclear Dynamics,* Jackson Hole, Wyoming, ed. W. Bauer, B. Back. Singapore: World Scientific (1992)
74. Toke, J., et al, *Nucl. Phys.* A440:327 (1985)
75. Back, B. B., et al, *Phys. Rev.* C41:1495 (1990)
76. Gil, S., et al, *Phys. Rev. Lett.* 65:3100 (1990)

77. Hennrich, H.-J., et al, *Phys. Lett.* B258:275 (1991)
78. Pengo, R., et al, *Nucl. Phys.* A411:255 (1983)
79. Deleted in proof
80. Esbensen, H., Landowne, S., *Phys. Rev.* C35:2090 (1987)
81. Esbensen, H., Landowne, S., *Nucl. Phys.* A492:473 (1989)
82. Nolan, P. J., et al, *Phys. Rev. Lett.* 161B:36 (1985)
83. Dasso, C. H., Landowne, S., *Phys. Rev.* C32:1094 (1985)
84. Vandenbosch, R., see Ref. 26, p. 155
85. Vandenbosch, R., see Ref. 7, p. 269
86. Stelson, P. H., et al, *Phys. Rev.* C41:1584 (1990)

Annu. Rev. Nucl. Part. Sci. 1992. 42:483–536

NUCLEAR STRUCTURE AT HIGH EXCITATION ENERGY STUDIED WITH GIANT RESONANCES

Jens Jørgen Gaardhøje

Niels Bohr Institute, University of Copenhagen, Blegdamsvej 15-17, Copenhagen 2100, Denmark

KEY WORDS: giant dipole resonance, hot nuclei, high-energy photons, collectivity, shapes and fluctuations, pre-fission emission, fission time scales

CONTENTS

0163-8998/92/1201-0483$02.00

1. INTRODUCTION

Studies of the properties of atomic nuclei have traditionally focused either on the reaction mechanism in nuclear collisions or on the structure of the fragments from such collisions, after most of the energy has been dissipated. Between these widely different regions it has, for many years, been difficult to study in detail the properties of atomic nuclei with large internal energy (temperature) and high angular momentum. This is because the spectra of the particles that are emitted in the course of the decay of the hot nucleus carry only limited information on the structural properties of the emitting system. In recent years, however, a specific experimental probe has become available that can provide new information on diverse properties of nuclei at finite temperature: for example, on the shapes and fluctuations of hot nuclei, on the mechanisms underlying the damping of collective excitations in hot nuclear matter, and on dissipative effects in hot-fission-unstable nuclei.

These developments are based on the measurement of the high-energy gamma rays (with energies typically in the range 10–20 MeV) emitted when isovector giant dipole resonances (GDRs), linear oscillations of the protons and neutrons in the hot nucleus, are damped. The so-called excited-state resonances are useful for several reasons. First, the damping of the vibrations and the associated gamma-ray emission occurs on a time scale that is sufficiently short that the GDR photons can compete with other modes of nuclear decay, such as particle emission or fission, although with a rather modest relative probability (approximately 10^{-3} high-energy photons are emitted per nuclear decay). The GDR emission occurs early in the decay of excited nuclei and probes the conditions prevailing at that time. Second, the peaked resonance shape (typical effective GDR widths are in the range 5–10 MeV) results in a rather well-localized strength in continuum gamma-ray spectra. Third, the resonances couple strongly to other nuclear degrees of freedom, such as the shape degrees of freedom, and thus measurements of the spectral distribution of the photons from GDR decay can provide information on the nuclear ensemble at finite temperature and fast rotation.

Historically, the first quantitative analysis of excited-state giant dipole resonance spectra was carried out for the photons seen in the spontaneous fission of ^{252}Cf (1). The major interest arose, however, when the GDR decay from hot and rotating nuclei formed in heavy-ion-induced fusion reactions was observed in 1981 (2), an observation that promised to open up for gamma-ray spectroscopy the totality of the phase-space available for the decay of hot nuclei. The main efforts of the first years of research of this new field were directed at establishing the universality of GDR decay in nuclei, formed mostly with moderate angular momentum and excitation energy ($E^* < 100$ MeV) and to determine the overall features. For earlier overviews, we refer the reader to Snover's 1986 article in this series (3), and to the proceedings of the 1987 Legnaro conference (4). For reviews of the properties of giant resonances in cold nuclei, see, for example, the experimental and theoretical reviews (5–10).

In the last few years the field has developed rapidly, and significant progress in the experimental techniques and methods of analysis has resulted in data of much higher quality and reliability. The development of more powerful experimental gamma-ray detection systems makes possible exclusive measurements in which the GDR properties can be correlated with the type of reaction (fusion or quasi-fusion, peripheral, fission, etc), the angular momentum of the gamma-ray-emitting nucleus, and specific properties of the (cold) reaction fragments. These advances have also made it possible to extend such studies into the intermediate-energy domain. Recently, investigations of heavily charged nuclei in which the GDR decay precedes fission have also been initiated. In a symbiotic relationship with the experimental advances, new and detailed theoretical models at nonzero temperature have been proposed.

Figure 1 displays schematically the phase-space in the variables excitation energy, E^*, and angular momentum, I, available to a decaying nucleus. Indicated are also the three main regions being investigated by excited-state GDR techniques. The subject matter of this review is also subdivided according to these regions.

At the lowest excitation energies ($E^* < 100$ MeV), covering angular momenta up to the fission limit ($I < 65$–70), the main thrust of GDR research is aimed at following the evolution from the region close to the nuclear yrast line, where quantum physics dominates, up to the region where the specific features of the shell structure are expected to disappear, and where nuclei can mostly be described by concepts familiar from classical and statistical physics. A diagnostic for such changes is provided by the shapes of heated nuclei, which change as the excitation energy and the angular momentum increase. This type of study also provides new information on fluctuation phenomena in finite quantal systems. This region is still the most extensively studied. In Section 3, we briefly review the GDR observables and bulk properties. Shape and fluctuation effects are discussed in Section 4.

At higher bombarding energies, nuclei can be produced with thermal energies (temperatures) high enough that the individual nucleons ul-

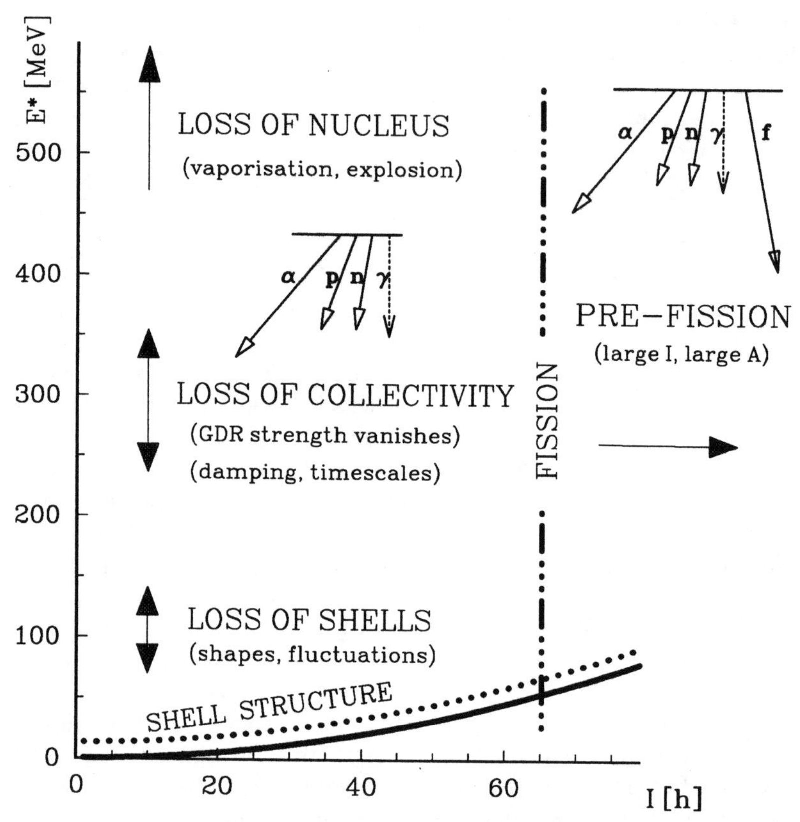

Figure 1 Schematic representation of the phase-space available for the decay of a hot nucleus. Gamma rays from the decay of excited giant dipole resonances (GDR) compete with other decay modes at high excitation energy. Areas of current research interest are indicated.

timately become unbound or the system fragments into a large number of smaller pieces. Initially it was hoped that in this domain the disappearance of the GDR with temperature would characterize the critical temperature of nuclear matter, and thus bring a new point of view to bear on the equation of state. Instead it was found that the observable collective strength in the resonance region vanishes well before the nucleus disappears. A detailed understanding of this phenomenon may provide new information on the damping mechanism of collective excitations and on the time scales for equilibration and decay in hot nuclear matter, as discussed in Section 5.

Finally, GDR gamma-ray spectroscopy promises to become an important tool for the study of nuclei that are unstable with respect to fission, because of either their high rotation or their high charge (Section 6). Indeed, recent work shows that GDR gamma rays compete favorably also with the fission process, which provides the possibility of studying nuclei under extreme conditions prior to dissociation. An understanding of the relative time scales involved in these decay modes can shed new light on dissipative processes in nuclei.

2. EXPERIMENTAL METHODS

2.1 *Detectors for High-Energy Photons*

Detectors for high-energy photon spectroscopy must contain a large fraction of the electromagnetic shower with good efficiency. Scintillators are a natural choice since they can be produced in large volumes and at reasonable cost. For in-beam reactions at energies in the range 5–30 MeV/u, charged particles and neutrons of high energies are copiously produced. These deposit energy in the detectors in the energy range of the GDR. Such events can be identified and excluded by a measurement of the time of flight of the particles from the target to the detector. The evolution toward complicated coincidence experiments simultaneously requires large solid angles for the photon detection, and thus small target-to-detector distances.

These criteria have led to a growing usage of scintillators of bariumdifluoride (BaF_2) for high-energy photon measurements. This material combines sub-nanosecond time resolution with a low cross section for neutron absorption and a discrimination capability against charged particles. While the stopping power is similar (the radiation length is 2.3 cm in BaF_2) to that for NaI, the energy resolution is somewhat inferior (typically 5% at 15 MeV). Several detection systems based on BaF_2 are currently in use. The Karlsruhe 4π array (11) is used for studies of neutron capture reactions of astrophysical relevance. The HECTOR array (12), a Danish-Italian collaboration, consists of eight single large crystals, 145 mm in diameter and 175 mm long, operated with a solid angle of approximately 10%. The German-Dutch-French system TAPS is a multidetector array (13) consisting of several hundred hexagonal detectors (59 mm in diameter and 275 mm long). These detectors can be used in closely packed groups adding back to a central detector the energy scattered into neighboring detectors. The Italian MEDEA system (14), consisting of about 300 detectors, is a multidetector completely mounted in vacuum that makes possible the simultaneous measurement of photons and low-energy charged particles. For experiments at lower bombarding energies, large shielded NaI detectors are operated by different groups [Stony Brook (15), Seattle (16)]. The time resolution is worse than for BaF_2, but superior energy resolution can be obtained, in some cases better than $\Delta E_\gamma / E_\gamma = 2\%$ at $E_\gamma = 20$ MeV.

Solid-state germanium detectors have also been used for high-energy photon studies. The main advantages of such detectors are their intrinsic gain stability and their high resolution ($\Delta E_\gamma / E_\gamma = 0.2\%$). Since germanium crystals can only be manufactured with limited volumes, they have been combined with the BGO anticoincidence shields normally used for high-resolution, discrete line, gamma-ray spectroscopy in add-back mode (17). With such systems, a combined resolution of about 1.2% at $E_\gamma = 15$ MeV can be obtained, in spite of the poor intrinsic resolution of BGO, by exploiting the fact that most of the radiation escaping from the germanium crystal has low energy since it originates from annihilation of e^+e^- pairs. A limiting factor is the poor time resolution of germanium detectors ($\Delta t \approx 10$ ns). In the near future, clusters of closely packed germanium detectors will become available in connection with the Euroball project (18), a large European collaboration. With such systems the energy resolution limit will most likely be set by the Doppler broadening resulting from the finite opening angle of each detector segment. Energy resolutions as good as 0.4% may be expected.

2.2 *Energy and Response Calibration*

A particular problem is associated with the energy calibration of detectors. Indeed, no radioactive sources producing gamma rays with energies in the GDR region exist. Commonly used is the $Pu^{13}C$ source, producing gamma rays with $E_\gamma = 6.13$ MeV by the reaction $\alpha + {}^{13}C \rightarrow n + {}^{12}C + \gamma$. The reaction $p + {}^{11}B$ populates the GDR in ^{12}C (located around 23 MeV), which can decay to the ground state by emission of a single gamma ray with energy $E_\gamma = E_{cm} + Q$ MeV. Several reactions populate the 15.1-MeV state in ^{12}C, which decays to the ground state by a strong M1 transition. Among these we mention the reaction $^2H(^{11}B,n\gamma)^{12}C$ at 19 MeV, and the $^{12}C(p,p'\gamma)^{12}C$ reaction (19) at $E_p \approx 27$ MeV. The latter reaction is particularly efficient. At higher energies cosmic rays (ultrarelativistic muons) deposit energy in the crystal proportionately to the distance traversed, resulting in a wide peak.

The efficiency and response function of detectors to high-energy photons can be established by exposing them to monochromatic hard photon beams from annihilation in flight or from tagged-photon facilities. The response function can be calculated by Monte Carlo simulations of the propagation of the electromagnetic shower in the detector (active and passive material). Such simulations, based for example on the EGS (20) or GEANT (21) computer programs, are able to reproduce the measured line shapes and efficiencies (22–24) to a high degree of accuracy, enabling the calculation of detector response functions on a fine grid and over a wide energy range.

Finally, we mention that the energy stabilization of scintillator assemblies often causes problems in experiments. These effects, if not controlled, are particularly important for experiments on excited-state GDR, in which the number of emitted photons typically decreases by an order of magnitude for an increase in transition energy of $E_\gamma = 5$ MeV. Thus, gain shifts of less than 1% can change the amplitude of angular distributions measured in the GDR region by about 10%. Typical schemes include the monitoring of a reference light signal injected into the scintillator crystal from a laser or a temperature-stabilized LED.

3. GDR OBSERVABLES AND METHODS OF ANALYSIS: BULK PROPERTIES

The spectrum of the photons emitted from a hot nucleus is entirely dominated by the density of the final states that can be reached in the decay. Since the nuclear level density varies exponentially with excitation energy, the number of emitted photons decreases exponentially with transition energy. Three distinct regions can be identified in such spectra, as illustrated in Figure 2. At the highest energies ($E_\gamma > 30$ MeV), the dominant contribution is from photons produced in the initial stages of the collision, prior to equilibration, and associated with nucleon-nucleon bremsstrahlung. This part of the spectrum varies to good accuracy exponentially with E_γ. At lower transition energies, the gamma rays originate from the hot system after thermalization. In the range $E_\gamma = 10$–20 MeV, gamma rays from the GDR give rise to a characteristic bump in the spectrum. Below $E_\gamma = 8$–10 MeV (approximately the particle separation energy, S_{part}), the gamma rays are mostly emitted from the low-temperature residual nuclei at $E^* < S_{\text{part}}$, as recognized by the steeper slope. Information on the GDR from the analysis of such continuum gamma-ray spectra stems today from two sources: the analysis of the spectrum shape with the statistical model and the measurement of the angular distribution of the emitted photons.

3.1 *The Spectrum: Statistical Decay*

A hot nucleus formed in a heavy-ion collision is an isolated system. Its energy is, after thermalization, evenly distributed among the nucleons, although not all the energy is stored in random thermal motion.

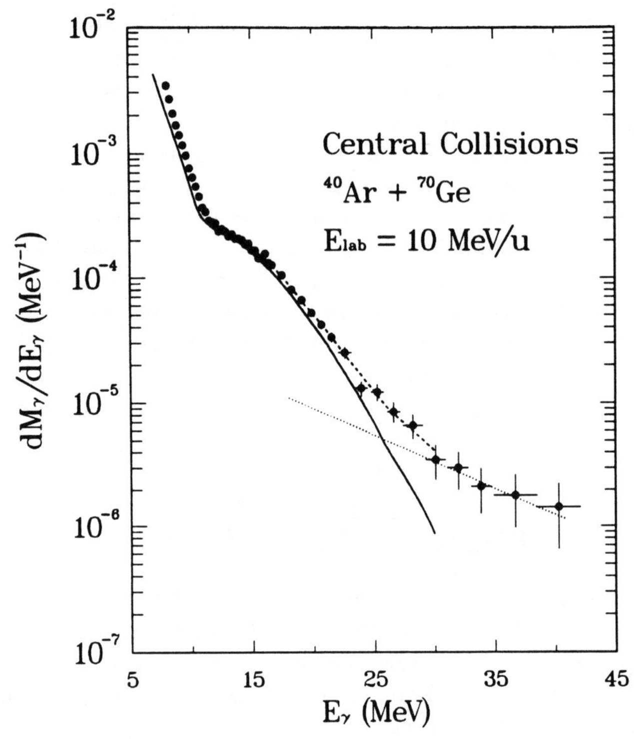

Figure 2 Measured gamma-ray spectrum from the reaction $^{40}Ar + ^{70}Ge \rightarrow ^{110}Sn$ at 10 MeV/u. The ordinate shows the measured multiplicity of high-energy gamma rays. The dotted line shows the contribution from pre-equilibrium gamma rays due to nucleon-nucleon collisions. Gamma rays from the decay of the GDR in the excited nucleus are concentrated in the range E_γ = 10–25 MeV. The solid line shows a statistical model calculation of the type described in the text. The spectrum is well described by the sum of these two contributions (*dashed line*). From (66).

Some of the energy may be found in collective degrees of freedom such as the large-amplitude quadrupole shape oscillations and the small-amplitude giant resonances, like the GDR. The damping and decay of the GDR may be thought of as a two-step process. (*a*) the GDR state exists as a collective state built on another state characterized by the remaining energy, angular momentum, etc; and (*b*) the GDR state couples, either directly by particle emission to the continuum or to the dense background of complicated compound states, modifying these and acquiring a width and subsequently gamma decaying. One may think of the latter process as a thermalization of the GDR occurring on a time scale given by its width (typically 5 MeV, which implies $\tau \approx 1 \times 10^{-22}$ sec). In the original formulation by Brink (25), in connection with a study of neutron resonance widths, the GDR was assumed not to depend on the detailed properties of the state on which it was built. However, the present interest in excited-state GDR is based on the fact that the GDR does depend on the characteristics of the mother state, although the general features appear to be remarkably stable.

The expectation is that the gamma-ray decay from the GDR can be treated as a statistical decay. This is in fact borne out by the now massive experimental evidence available in the range $E^* \approx 30–100$ MeV. At higher E^*, new elements have to be considered. We discuss these in Section 5. In the following we briefly recall the main aspects of the statistical analysis of the GDR spectra (for details see, for example, 26).

In the statistical model, the gamma-ray decay rate from an initial state (E_i,I_i) to a final state (E_f,I_f) can, in analogy with the particle decay rate, be written

$$R_\gamma \, dE_\gamma = \frac{\rho(E_f,I_f)}{\hbar\rho(E_i,I_i)} f_{GDR}(E_\gamma) \, dE_\gamma, \quad 1.$$

where ρ is the density of levels at the given spin and excitation energy (30). The function f_{GDR} (E_γ) is the giant dipole resonance strength function. Assuming that only collective E1 decay is important, it may be written

$$f_{GDR}(E_\gamma) = \frac{4e^2}{3\pi\hbar cmc^2} S_{GDR} \frac{NZ}{A} \frac{\Gamma_{GDR} E_\gamma^4}{(E_\gamma^2 - E_{GDR}^2)^2 + \Gamma_{GDR}^2 E_\gamma^2}, \quad 2.$$

where S_{GDR} is the strength (in terms of the classical E1 sum rule), and E_{GDR} and Γ_{GDR} the centroid energy and width of the GDR, respectively. This expression assumes that the line shape of the GDR can be represented by a Lorentzian function, as is the case for the GDR in cold nuclei measured in photonuclear reactions. Higher multipolarities, such as the giant quadrupole resonance (isoscalar, $E_{ISGQR} \approx 65A^{-1/3}$ MeV, and isovector, $E_{IVGQR} \approx 130A^{-1/3}$ MeV) have not been clearly identified built on excited states, but they can be included in the statistical analysis (27–29) in analogy with Equation 2. The quadrupole radiation strength is in general about two orders of magnitude lower than the dipole strength in the relevant energy range. The normalization to the E1 sum rule, and consequently to the ground-state photoabsorption cross section (30), is based on the principle of detailed balance. These issues, and GDR sum rules at finite T have been discussed by several authors in more detail (3, 31).

The now standard analysis method of hot GDR spectra is based on the comparison of measured spectra to gamma-ray spectra calculated with a statistical model code, usually a modified version of the CASCADE computer program (32). The parameters S_{GDR}, E_{GDR}, and Γ_{GDR} are varied, preferably in a χ^2 minimization routine until the measured spectra are reproduced. This procedure involves calculating the gamma-ray contribution from all the nuclei in the decay cascade, and can be heavily time consuming, particularly in light nuclei and for high E^*. The comparison to experiment also requires the convolution of the calculated experimental spectra with the measured or simulated (energy-dependent) detector response function. Examples of analyses of experimental spectra with the statistical model are shown in Figures 2, 3, 15, 18, and 19. The calculations are able to reproduce the spectra well over many orders of magnitude. It is often convenient to display the GDR region in a linearized form. This may be done either by multiplying the measured gamma-ray spectra by a factor $\exp(E_\gamma/T_{eff})$, in order to cancel in an average way the effect of the level density (where

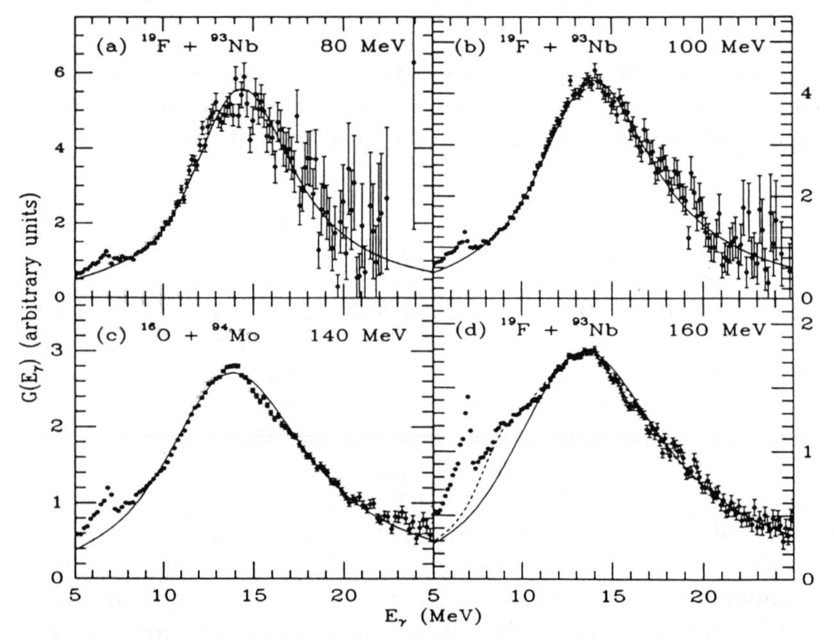

Figure 3 Measured GDR spectra from the decay of the GDR in ^{110,112}Sn isotopes at different excitation energies. The lines show statistical model calculations in which the GDR width was varied. Spectra and calculations are presented on a linear scale after division with a statistical model calculation using constant E1 matrix elements. From (23).

T_{eff} represents an average temperature for the region of states the GDR is built on), by dividing the spectra by a calculation using constant matrix elements, or by calculating the ratio of the fitted spectrum shape and response function. Examples of the latter procedures are exhibited in Figures 3 and 11.

The applicability of the statistical picture rests on the assumption of equilibration of the nucleus and of the GDR. This assumption is now well established in this energy range and for a large range of masses. In Figure 4 we display a compilation of the existing systematics for the strength, centroid energy, and width of excited-state GDRs. The data are collected from the available literature. The left-hand side of the figure shows well-known similar information for cold nuclei, obtained from photonuclear reactions (33). The main features of this comparison are the following. The strengths deduced for excited-state resonances are generally consistent with the full energy-weighted sum rule, although in quite a number of cases this quantity has not been measured directly from the fusion cross sections or from the multiplicity of gamma rays per reaction. For the points shown without error bars, S_{GDR} has been either fixed to the sum rule value or determined as a free parameter in the best fit to the spectrum shape. The centroid energies, although showing larger scatter, are in overall agreement with the systematics from ground-state GDRs (left), which indicates that the main physical quantities responsible for E_{GDR}, namely the symmetry energy and the volume, have not changed appreciably. Nevertheless, we propose the parametrization $E_{GDR} = 17A^{-1/3} + 25A^{-1/6}$ (MeV) for the centroid energy of the GDR built on excited states. This suggest that surface effects may play a larger role at higher temperature, although it must be kept in mind that only the isospin $T_<$ component of the GDR is populated in fusion reactions (this plays a role in lighter nuclei due to isospin splitting).

The width of excited-state GDR are normally considerably larger than in cold nuclei, and they do not exhibit the strong correlation with the shell structure seen for ground-state resonances. The reasons for this are several: weakening of the shell structure, thermal fluctuations, and changes of deformation.

In the analysis of deformed nuclei, f_{GDR} may be written as a superposition of Lorentzians, to account for the fact that the GDR will split up into components corresponding to vibrations along the principal axes, with frequencies inversely proportional to the length of these axes. When deformation effects are considered, the existing analyses have normally been restricted to using two Lorentzian components, implicitly assuming axial symmetry. For nonrotating nuclei, the ratio of the strength in the upper component to that in the lower component should be equal to 2 for prolate nuclei, and equal to 0.5 for oblate nuclei (the ratio S_2/S_1 is shown in the bottom row of Figure 4). From such an analysis, effective widths and an effective deformation can be obtained. The deformation is often estimated from $\beta = 1.05(d - 1)d^{-1/3}$. Here $d = b/a$ is the ratio of the axes lengths, related to the ratio of the vibrational energies along the symmetry axes (b) and the perpendicular axes (a) by $E_b/E_a = 0.911d+0.089$.

This procedure obviously has a limited applicability insofar as nuclei with changing and fluctuating shapes are concerned. Nevertheless it establishes the occurrence of nuclear deformation at finite temperature as a common phenomenon (see Figure 4, fourth panel on the right).

3.2 *The Angular Distribution*

The measurement of the angular distribution of the GDR photons provides a complementary method for studying the GDR in hot nuclei that have an alignment due to rotation. Measuring angular distribution has several advantages over the more standard analysis of the spectrum shape. Indeed, experimental angular distributions are to a large extent free from possible systematic errors originating from the statistical analysis of the spectrum, for example errors arising from assumptions about the nuclear level density at finite temperature. Furthermore, the angular distribution depends markedly on the orientation of the nucleus with respect to the direction of the total angular momentum vector, I_{tot}.

The angular distribution depends on whether a given vibration is along an axis parallel or perpendicular to the direction of I_{tot}. For example, in a prolate nucleus rotating collectively, I_{tot} is perpendicular to the symmetry axis. The angular momentum associated with the low GDR component, which corresponds to a vibration along the symmetry axis, is therefore parallel to I_{tot} and couples to I_{tot}. The associated transition is stretched ($|\Delta I| = 1$). The two degenerate high-energy GDR components, associated with vibrations along the short axes, that are parallel and perpendicular to I_{tot}, respectively, correspond to $\Delta I = 0$ (unstretched) and $|\Delta I| = 1$. The angular distribution of these compo-

Figure 4 Systematic comparison of available data on giant dipole resonances in cold (*left*) and excited (*right*) nuclei as a function of mass number. From top to bottom: Strength in units of the classical sum rule, centroid energy (shown are usual parametrizations of the ground-state systematics and a proposed parametrization for finite-temperature GDRs), total width, quadrupole deformation parameter β, and strength ratio of the upper to lower components.

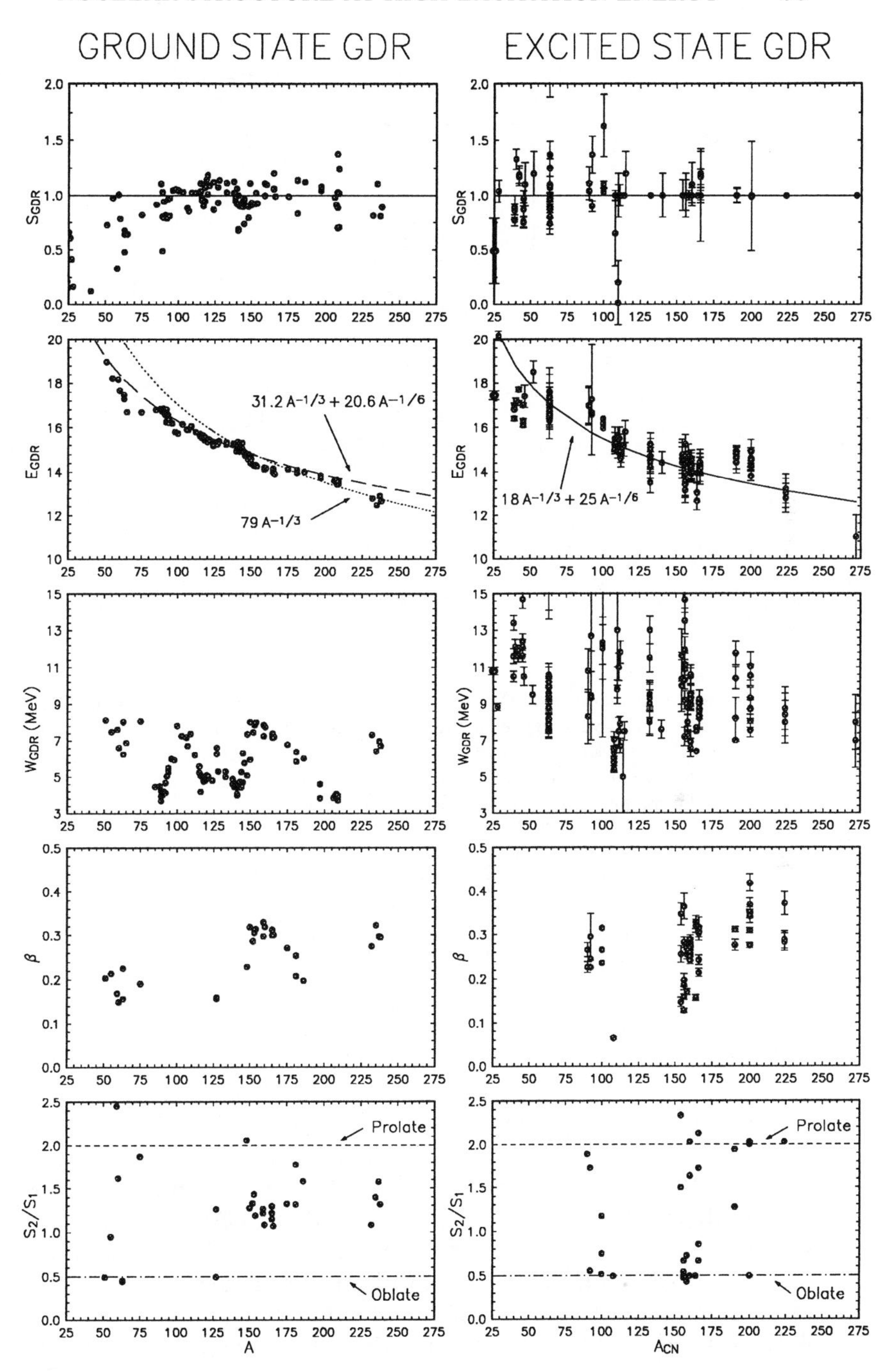
GROUND STATE GDR
EXCITED STATE GDR
SGDR
EGDR
31.2 A−1/3 + 20.6 A−1/6
79 A−1/3
18 A−1/3 + 25 A−1/6
WGDR (MeV)
β
S2/S1
Prolate
Oblate
A
ACN

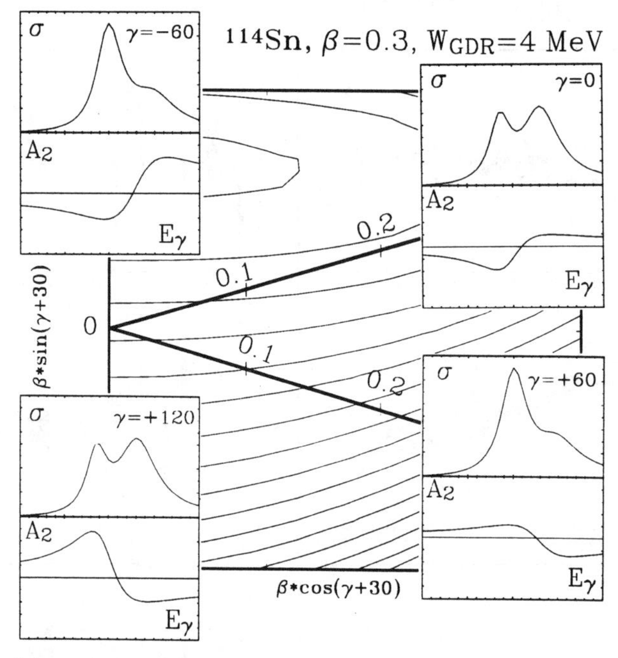

Figure 5 Schematic representation of the GDR strength function and angular distribution coefficients relative to the beam direction for different shapes of the nucleus ^{114}Sn. The convention for the quadrupole shape parameters β and γ used here is that $\gamma = 0$ corresponds to a prolate nucleus rotating collectively and $\gamma = -60$ to an oblate nucleus with the angular momentum aligned along the symmetry axis.

nents in the laboratory frame of reference is given by

$$W^{|\Delta I|=1}(\theta,E_\gamma) = W_0(E_\gamma)\left[1 - \frac{1}{4}P_2(\cos\theta)\right] \qquad 3.$$

and

$$W^{\Delta I=0}(\theta,E_\gamma) = W_0(E_\gamma)\left[1 + \frac{1}{2}P_2(\cos\theta)\right], \qquad 4.$$

where P_2 is a Legendre polynomial in the polar angle θ between the direction of emission of the gamma ray and the beam axis. In an oblate nucleus, rotating noncollectively, the two longer axes give rise to two degenerate components at lower energy, both with $|\Delta I| = 1$, while the shorter symmetry axis, being parallel to the direction of I_{tot} corresponds to a vibrational component with $\Delta I = 0$. In experimental analyses, data are normally fitted to the function $W(\theta,E_\gamma) = W_0[1 + a_2(E_\gamma)P_2\cos(\theta)]$. The resulting pattern of the $a_2(E_\gamma)$ coefficients in the GDR energy region is therefore sensitive both to the magnitude of the deformation (the splitting of components with different angular distributions) and to the shape and orientation of the density distribution.

This is illustrated in Figure 5, which shows calculated angular distribution patterns for four different shapes of the nucleus ^{114}Sn. It is seen that nuclei with prolate shapes rotating collectively and oblate nuclei rotating noncollectively have similar angular distribution patterns, although the amplitude is reduced for the prolate case. For prolate nuclei rotating noncollectively and oblate nuclei rotating collectively, the sign of the $a_2(E_\gamma)$ pattern is reversed, as is the amplitude relation, because the axes are inverted with respect to I_{tot}. All measurements of the angular distribution pattern in the GDR region made to date are consistent with nuclear shapes in the sector corresponding to collective prolate to oblate noncollective rotation. In situations in which shape fluctuations are important, the amplitude of the $a_2(E_\gamma)$ can be significantly attenuated (34) because of the changing overlap of the various vibrational components. Likewise, if I_{tot} is not perpendicular to one of the major axes, the angular distribution will be attenuated because of the loss of alignment (26, 53).

4. THE SHAPES AND FLUCTUATIONS OF HOT NUCLEI

The splitting of the GDR in cold deformed nuclei is well known, as exhibited by the spectacular spectra from (γ,n) reactions on stable targets. Most of the early excitement, in the beginning of the 1980s, about excited-state GDR was indeed tied to the expectation that similar deformation effects would be observable in nuclei produced in nuclear reactions with high angular momentum and internal excitation energy. A determination of the shapes of such nuclei would carry nuclear structure studies far from the yrast line and afford a view of the breakdown of the nuclear shell structure. It was therefore a significant step forward in the new field when around 1984 to 1985 it became possible to observe deformation effects (35, 36) in the spectra of gamma rays from heavy nuclei ($A = 108$ and $A = 160$–166) populated at excitation energies up to $E^* = 65$ MeV and angular momenta up to $I = 35$. It rapidly became clear, however, that at finite temperature the GDR strength function was much more featureless than in cold nuclei, which made the determination of nuclear deformations and shapes a complicated task. Indeed, with the exception of the lowest excitation energies, the individual components of the GDR were no longer apparent in the spectra. It was early realized that shape fluctuations play an important role in determining the effective GDR strength function (37). In spite of this, the GDR in hot nuclei exhibits a strong dependence on the angular momentum and the temperature of the nucleus in which it is excited, as may be seen in Figure 6. The figure shows the strong observed increase of the width of the GDR, up to $E^* = 130$ MeV, in Sn isotopes with $A = 108–112$, as a function of the E^* (and implicitly also the angular momentum) transferred to the compound nuclei in the reaction.

4.1 *The Shapes of Hot and Rotating Nuclei*

Hot and rotating atomic nuclei are expected to exhibit a rich variety of different shapes. At excitation energies close to the yrast line the shapes are determined by the shell structure. At higher temperature the finite occupation probability of orbitals above and below the Fermi surface limits the influence of specific valence orbitals. The temperature at which the shell effects should be substantially weakened can be estimated (30) to approximately $T = 1.5–2$ MeV, from the expected decrease of shell energies as a function of T. Once shell effects are gone, nuclei are expected to follow the shapes predicted for a charged liquid drop (38), i.e spherical shapes at zero rotation that develop into oblate shapes of increasing deformation with increasing angular momentum. At very high angular momentum, nuclei should develop large prolate deformations prior to fission.

Studies of shape probability distributions at high excitation energy have normally been done in the canonical ensemble characterized by the variables temperature (T) and rotational frequency (ω) in terms of the free energy in the rotating frame $F = E - TS - \boldsymbol{I} \cdot \boldsymbol{\omega}$, where S is the entropy. F can be calculated using standard methods in for example the Nilsson-Strutinsky approach in which F is written as a sum of liquid drop terms and a shell correction term.

Particularly useful has been the development of a formalism based on the Landau theory of phase transitions (39), which in a simple form reveals the global features of the shape landscapes. In this approach, the free energy at a given T and I can be parametrized in terms of the

→

Figure 6 From top to bottom: systematics of the strength, centroid energy, and total width of the GDR as a function of the excitation energy of the $^{108-112}$Sn compound nuclei. The lowest panel shows the angular momentum transferred to the nucleus in the reactions. Dashed lines show average values for the GDR in cold Sn nuclei. The dash-dotted line is a parametrization of the width increase seen at lower E^*.

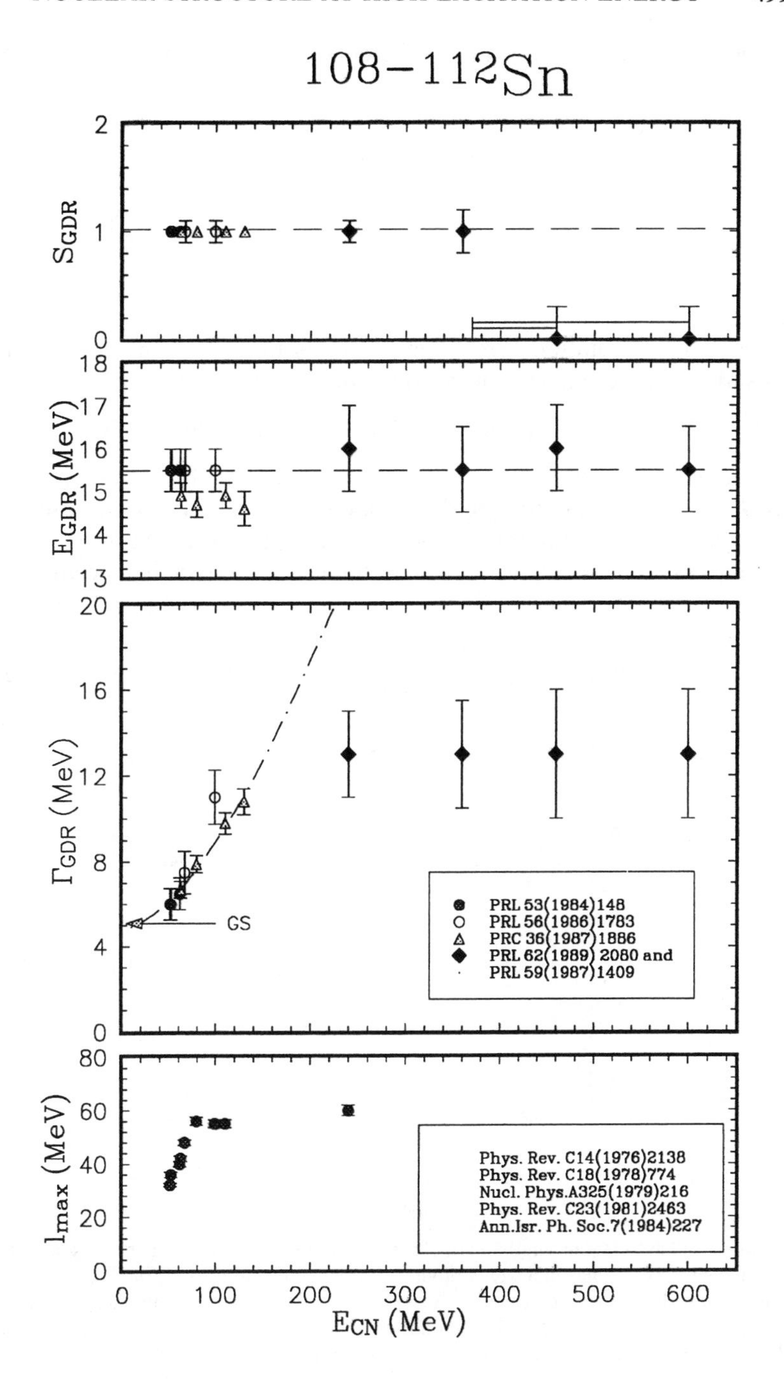
108−112Sn
S_GDR
E_GDR (MeV)
Γ_GDR (MeV)
GS
PRL 53(1984)148
PRL 56(1986)1783
PRC 36(1987)1886
PRL 62(1989) 2080 and
PRL 59(1987)1409
l_max (MeV)
Phys. Rev. C14(1976)2138
Phys. Rev. C18(1978)774
Nucl. Phys.A325(1979)216
Phys. Rev. C23(1981)2463
Ann.Isr. Ph. Soc.7(1984)227
E_CN (MeV)

deformation parameters with coefficients that vary smoothly with temperature,

$$F(T,\omega,\beta,\gamma) = F(T,\omega = 0,\beta,\gamma) - \frac{1}{2}J_z\omega^2$$

$$= F_0(T) + A(T)\beta^2 - B(T)\beta^3 \cos(3\gamma) + C(T)\beta^4 - \frac{1}{2}J_z\omega^2, \quad 5.$$

where ω is the rotational frequency (here assumed to coincide with the z axis). J_z is the moment of inertia about the rotation axis,

$$J_z = J_0(T) - 2R(T)\beta \cos \gamma + 2J_1(T)\beta^2 + 2D(T)\beta^2 \sin^2 \gamma. \quad 6.$$

For actual cases the temperature-dependent coefficients F_0, A, B, C, J_0, R, J_1, and D must be determined by a fit to a full microscopic calculation. The equilibrium shape at a given (T,ω) is then the one that minimizes F. The transition to the variables (E^*, I) can then be effected

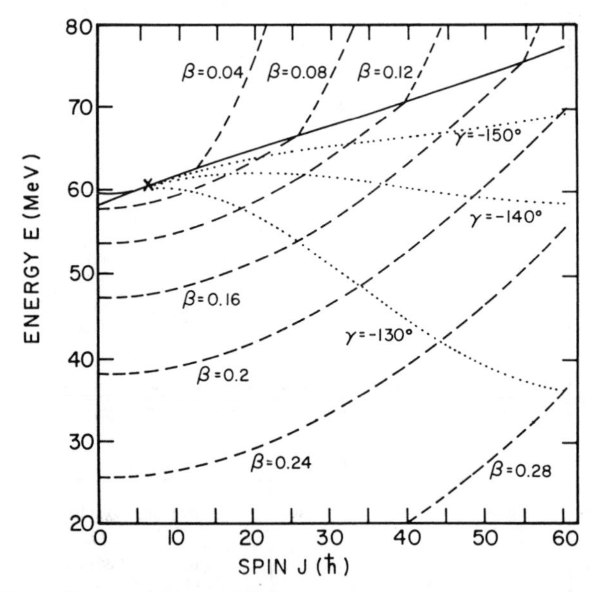

Figure 7 Phase diagram, in the variables excitation energy and angular momentum for the nucleus ^{166}Er. Lines indicate contours of equal elongation β and triaxiality γ for the equilibrium shape. Note that here $\gamma = -120$ corresponds to a prolate nucleus rotating collectively, while $\gamma = -180$ corresponds to noncollective rotation of an oblate shape. The solid line shows the location of the shape phase transition. From (39).

using the expression $T \approx (E^*/a)^{1/2}$, where a is the level density parameter, and assuming an effective moment of inertia.

Figure 7 shows the calculated phase diagram in the variables E^* and I for the nucleus ^{166}Er (39). Note that the convention used in this figure and in the above formulas is as follows: $\gamma = -180$ corresponds to collective rotation of a prolate shape, and $\gamma = -120$ corresponds to an oblate shape with the angular momentum vector aligned along the symmetry axis. The so-called Lund convention, which is extensively used in nuclear spectroscopy and employed in Figures 5 and 9, assumes rotation about the x axis and defines $\gamma_L = \gamma + 120$. The features of this diagram are generic, namely the existence of a phase transition (predicted to be of second order) from prolate-triaxial shapes to oblate shapes at a characteristic excitation energy that depends on the rotation and the considered nucleus. The evolution of the critical excitation energy E^*_c as a function of the neutron number (40) is shown in Figure 8. Mid-shell nuclei (around $N = 104$ and $Z = 66$) have the highest E^*_c, which is of the order of 65 MeV. The behavior with angular momentum can be widely different. For example, in transitional nuclei (such as ^{162}Yb, which has $N = 90$), the phase transition boundary is much more down-curved so that oblate shapes should be reached at $I > 40$ even at yrast energies. Such shape transitions have been observed in discrete line spectra from cold rotating nuclei. These features of the shape landscapes have also been studied with similar results (41) in finite-temperature cranked Hartree-Fock-Bogoliubov (HFB) calculations, for the nuclei ^{166}Er and ^{158}Yb, including a treatment of the pairing phase transition (pairing effects are predicted to vanish at $T \approx 0.5$ MeV).

4.2 *The GDR in Rotating Nuclei*

In nonrotating deformed nuclei the GDR splits up in vibrations with frequencies roughly inversely proportional to the length of the axes. The frequencies can be estimated from the often used Hill-Wheeler (30) formula

$$\omega_k = \omega_{GDR} \exp\left[-\sqrt{\frac{5}{4\pi}}\,\beta\cos\left(\gamma - \frac{2\pi k}{3}\right)\right], \qquad 7.$$

where $k = 1, 2, 3$ labels the principal axes in the intrinsic frame and ω_{GDR} is the average frequency of the dipole vibration.

In the case of rotation, the problem has been treated by many authors (37, 42–45). In general, Coriolis forces cause a further splitting of the components perpendicular to the rotation axis of order ω ($\approx$ 1 MeV).

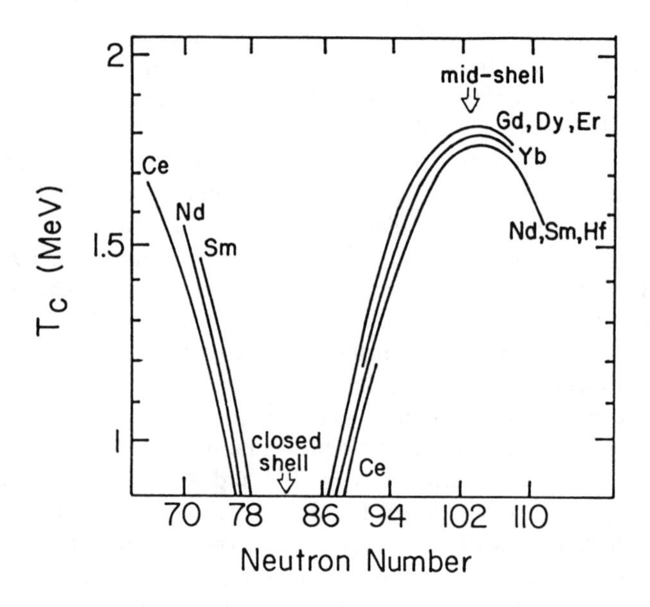

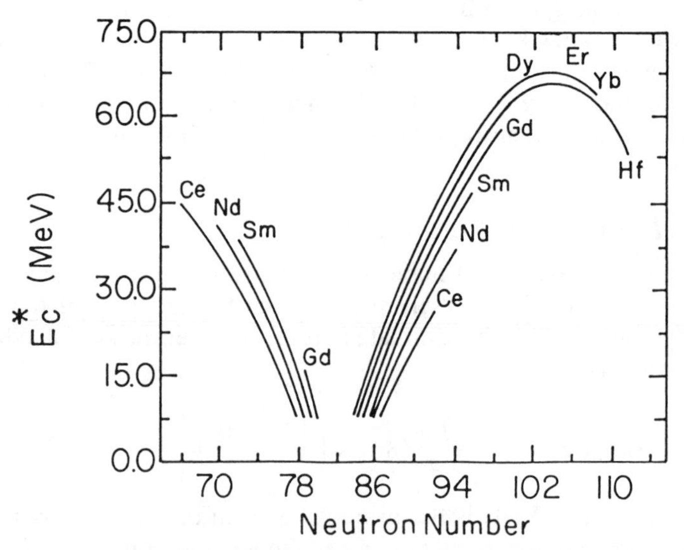

Figure 8 The calculated critical temperature T_c and the corresponding critical excitation energy E_c^* of the shape phase transition as a function of the neutron number for even-even rare earth nuclei. From (40).

When rotation is around the symmetry axis, the frequencies in the laboratory frame are identical to the nonrotating frequencies.

4.3 *Fluctuations: Theoretical Approaches*

The above discussion of nuclear shape landscapes has so far concentrated on the equilibrium shape, the one that minimizes the free energy F. While this shape is the relevant one for discussing the structure of the GDR in cold nuclei, the free energy distributions as a function of the shape variables no longer show a deep minimum at finite temperature. On the contrary, the free energy landscapes broaden considerably, which suggests that fluctuations of the shape degrees of freedom will be important even at temperatures $T \geq 1$ MeV. This is exhibited in Figure 9, which shows free energy distributions calculated for the nucleus ^{162}Yb at $\omega = 0.5$ MeV (corresponding to $I \approx 35$) and $T = 0.1$, 0.5, and 1.0 MeV. The shape transition for the equilibrium deformation is apparent. As far as the measured effective GDR strength functions are concerned, they will reflect the extent to which the collective vibration couples to the various shapes and consequently carry information on the relative time scales for shape rearrangements.

4.3.1 ADIABATIC MODELS The approach follows from standard statistical mechanics. For a system at constant temperature, the probability that the nucleus has a deformation and shape (β,γ) in terms of the probability of having the equilibrium shape (β_0,γ_0) is given by

$$\frac{P(T,\omega,\beta,\gamma)}{P(T,\omega,\beta_0,\gamma_0)} = \exp\left\{\frac{-[F(T,\omega,\beta,\gamma) - F(T,\omega,\beta_0,\gamma_0)]}{T}\right\} = \exp\left(\frac{-\Delta F}{T}\right). \qquad 8.$$

Such shape probability distributions for the nucleus ^{162}Yb are displayed in the right-hand side of Figure 9. They demonstrate that even at moderate temperature, shape fluctuations are very important and that they may to a large extent obscure the shape phase transition predicted for the most probable deformation. These issues have been explored quantitatively by several authors (46–48).

For the GDR, the simplest approach consists in assuming that, during the time it takes for a GDR to damp, the nuclear shape does not change. This is the adiabatic approximation. In this picture, the effective GDR strength distribution, measured in experiment, is a sum of the strength distributions corresponding to vibrations built on each of the shapes that the nucleus can explore, weighted with the probability that the

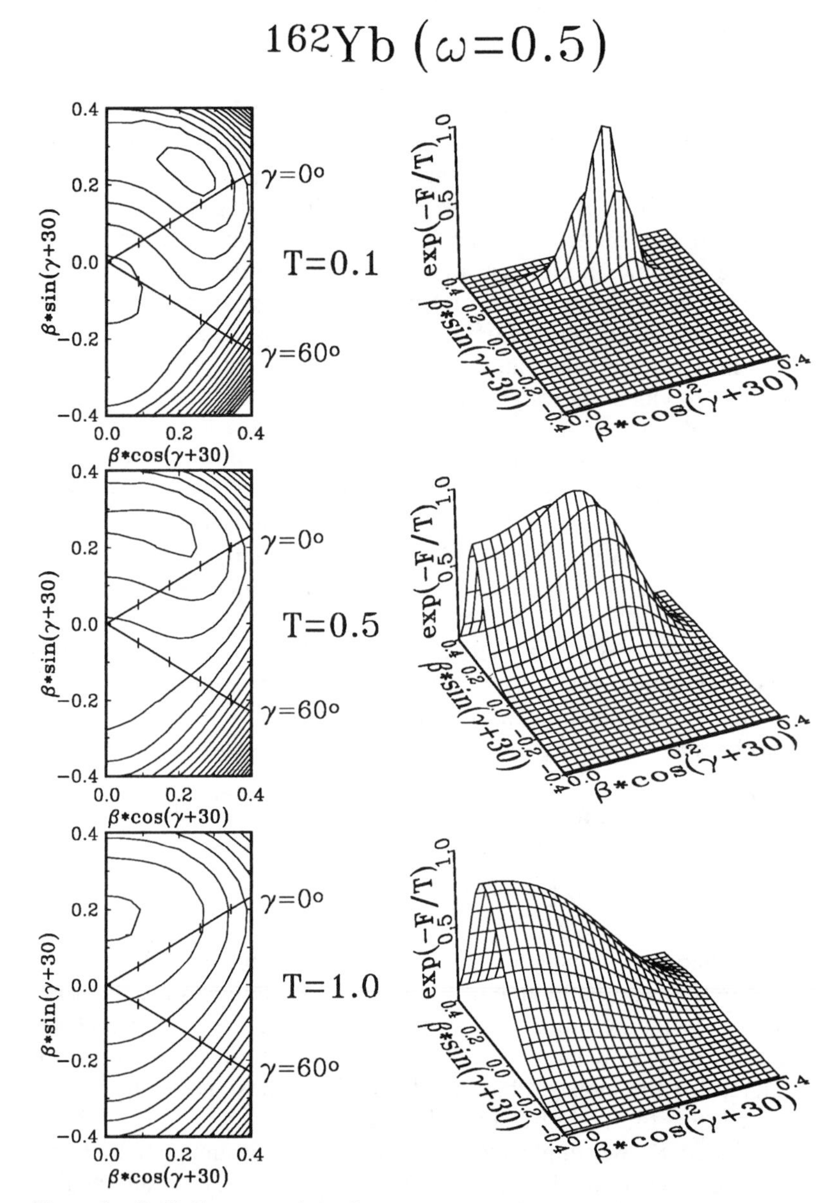

Figure 9 (*Left*) Contours of the free energy as a function of the quadrupole shape variables β and γ for the nucleus ^{162}Yb at fixed rotational frequency for three temperatures. (*Right*) The associated shape probability distributions as given by the Boltzmann factor $\exp(-F/T)$.

nucleus has that shape,

$$\langle f_{\mathrm{GDR}}(E_\gamma,T,\omega)\rangle = \frac{\int f_{\mathrm{GDR}}(E_\gamma,\omega,\beta,\gamma)\exp[-F(T,\omega,\beta,\gamma)/T]\,\mathrm{d}\tau}{\int \exp[-F(T,\omega,\beta,\gamma)/T]}. \qquad 9.$$

Here f_{GDR}, denotes the GDR strength function calculated either from the harmonic oscillator or in some realistic model. The term $\mathrm{d}\tau$ is the volume element associated with the parameters β and γ. The choice of the proper metric has been a subject of quite some controversy in later years. Some researchers (37, 47–49) use a volume element in the variables (β,γ),

$$\mathrm{d}\tau = \beta\,\mathrm{d}\beta\mathrm{d}\gamma, \qquad 10.$$

Others (50, 51) use the metric

$$\mathrm{d}\tau = \beta^4|\sin(3\gamma)|\sin\theta\,\mathrm{d}\beta\mathrm{d}\gamma\,\mathrm{d}\theta\mathrm{d}\phi. \qquad 11.$$

The latter volume element follows from the Jacobian associated with the transformation of the five quadrupole coordinates $(\beta, \gamma, \theta, \phi, \psi)$, thus also including orientation degrees of freedon (30), as discussed below. In both cases the averaging should be done over an interval in gamma of 60 degrees, covering shapes ranging from prolate to oblate, in order not to include an implicit averaging over the orientation degrees of freedom (see below). The differences between these metrics can be quite important when performing an average over shapes. For example, the latter metric will enhance the contribution of larger deformations because of the higher power of β and of triaxial shapes. Such an enhancement will result in a larger spread of GDR frequencies and large $a_2(E_\gamma)$ coefficients, as compared in Figure 10.

As in the treatment of shape fluctuations, fluctuations in the orientation degrees of freedom can be considered (52, 53) in terms of the Euler angles, θ and ϕ, describing the orientation of the angular frequency vector relative to that of the density distribution. Equation 8 is extended by including the orientation angles explicitly in the free energy. Equation 5 is modified to

$$F(T,\omega,\beta,\gamma,\theta,\phi) = F(T,\omega = 0,\beta,\gamma) - \frac{1}{2}(J_x\sin^2\theta\cos^2\phi + J_y\sin^2\theta\sin^2\phi + J_z\cos^2\theta)\omega^2, \qquad 12.$$

where J_x, J_y, and J_z denote the intrinsic moments of inertia about the principal axes. This formulation assumes that the GDR couples adia-

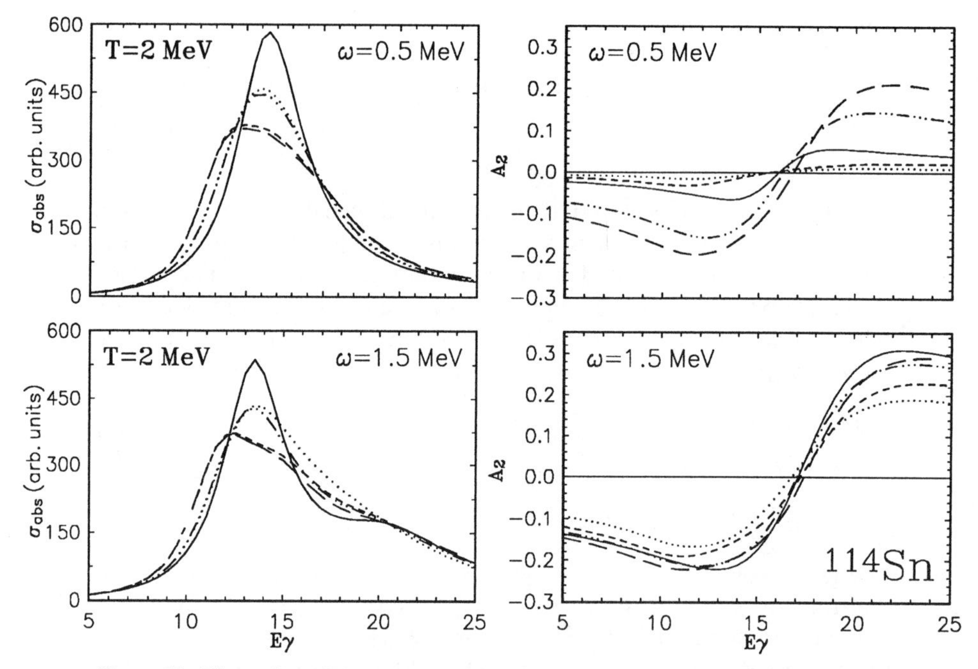

Figure 10 Illustration of the effect of fluctuations on the GDR strength function and angular distribution, as obtained in various adiabatic approaches: (*solid lines*) equilibrium shape, (*long-dashed and short-dashed lines*) adiabatic (using Equations 10 and 11, respectively), (*dotted and dot-dashed lines*) also including averaging over orientations.

batically to the orientation degrees of freedom. The averaging volume element must be extended by

$$d\tau' = \sin\theta \, d\theta d\phi. \qquad 13.$$

As mentioned in Section 3, the angular distribution is also affected by shape and thermal fluctuations. Neglecting the effect of the Coriolis splitting on the GDR components (see Section 4.2), one can derive an analytical expression for the angular distribution (53)

$$a_2(E_\gamma) = -\frac{1}{2}\frac{[f_x(E_\gamma) + f_y(E_\gamma)]/2 - f_z}{f_x + f_y + f_z}\frac{3\cos^2\theta - 1}{2} + \frac{3}{8}\frac{f_x - f_y}{f_x + f_y + f_z}\sin^2\theta\cos(2\phi), \qquad 14.$$

where f_x, f_y, and f_z are the absorption cross sections for the GDR along the principal axes. The discussion above was carried out in terms of the canonical variables T and ω, not in the physical variables E^* and I. The differences arising in the treatment of shape fluctuations from conserving T constant and not E^* were explored by Goodman (54). Since the available energy is constantly repartitioned between the thermal energy and the deformation energy, deformations away from equilibrium are less populated because the thermal energy is reduced. In contrast, all shapes are populated with the constant temperature constraint. The constant energy constraint can be treated by expressing the shape probability not in terms of the free energy as in Equation 8, but in terms of the entropy [replacing $\exp(-\Delta F/T)$ by $\exp(\Delta S)$]. The differences have been explored quantitatively for the nucleus ^{166}Er. It is found that the shape probability distributions are not altered significantly for $E^* > 30$ MeV, although the T constraint in general gives broader shape probability distributions.

The differences between the constant rotational frequency constraint and the physical constant angular momentum constraint have also recently been explored (55) in microscopic calculations, based on a two-dimensional cranking model for a single j shell ($j = 13/2$). For each orientation, the rotational frequency is varied in order to maintain the same average I. An issue here is whether, at finite T, the moment of inertia about the principal axes are affected by orientation fluctuations. It is found that the constant ω constraint may significantly overestimate the orientation fluctuations at finite temperature. However, these issues have not yet been explored in terms of the observable effects on the GDR angular distribution.

4.3.2 DYNAMICAL MODELS In the previous section we discussed the adiabatic situation, disregarding the fact that the coupling of the GDR to the shape degrees of freedom may depend on the relative time scales of the collective shape and vibrational motions and therefore may require a dynamical treatment. The idea (56) is the following. With increasing temperature, the time spent by the excited nucleus in a configuration characterized by a given deformation and orientation decreases. Hence, the nucleus may not spend enough time in a given point in deformation space for the GDR to adjust its frequency to the shape. Rather, the jumping between different shapes implies that the GDR never explores the extreme deformations but feels only the average shape. Such a mechanism would substantially reduce the effect of shape and orientation fluctuations.

The proposed effect, called motional narrowing, is a nuclear analogue to that responsible for the narrowing of the line shape in the nuclear magnetic resonance (NMR). It occurs whenever a periodic resonant effect undergoes a random time-dependent perturbation occurring on a time scale shorter than the time needed for the system to adjust its frequency to that perturbation. In NMR, the relevant time scales are those needed to change the magnetic field at a given lattice point in a crystal (as a result of the temperature-dependent mobility of its neighboring atoms) compared to the time needed to adjust to a change of the field. In nuclei, the relevant parameters are the inverse of the hopping time between different shapes ($\Gamma = \hbar/\tau$) and the quantity $\delta\omega$ measuring the spread in dipole frequencies corresponding to different nuclear shapes. If $\delta\omega \ll \Gamma$, motional narrowing will occur. For fluctuating nuclei, $\delta\omega$ is typically 1–2 MeV, corresponding to $\sim 5 \times 10^{-22}$ sec. Hence, an experimental determination of the validity of the motional narrowing theory could provide new information on the characteristic time scales for nucleonic rearrangements in hot nuclei.

These issues, and the consequences for the GDR were recently explored quantitatively by several authors. Alhassid & Bush (57) described the time evolution of the quadrupole shape parameters α in terms of a Langevin equation

$$\frac{\partial\alpha}{\partial t} = -\frac{1}{\chi}\frac{\partial F}{\partial\alpha} + f(t), \qquad 15.$$

where f is a random force that causes statistical fluctuations in the shape parameters and makes the process stochastic. χ is a parameter that scales the driving force for α and is a measure of the degree of adiabaticity; it is proportional to the average relaxation time of the quadrupole motion.

Ormand et al (51) described the jumping in terms of a Kubo-Anderson process. In this model the conditional probability of having a given deformation α at time t after having been at α_0 at time t_0 is written

$$P(\alpha,t \mid \alpha_0,t_0) = \exp[-\Gamma(t - t_0)]\delta(\alpha - \alpha_0)$$
$$+ (1 - \exp[-\Gamma(t - t_0)] \exp(-F/T) \left(\int \exp(-F/T)\, d\tau\right)^{-1}, \qquad 16.$$

where Γ is the mean shape jumping rate. In both cases the quadrupole coordinates α include both shape and orientation degrees of freedom. In an extended formulation the shape and orientation coordinates have been treated separately, which allows for the possibility that the relaxation time for the shape and orientation degrees of freedom may be different (58).

4.4 *Summary of Experimental Data and Comparison with Theory*

4.4.1 REVIEW OF EXPERIMENTAL DATA AND ANALYSES In this section we briefly review the experimental information on GDR properties in hot rotating nuclei. Until recently this information was primarily based on the analysis of the spectrum shape in terms of the simplified one- or two-Lorentzian analysis described in Section 3.3, although realistic analyses of the angular distribution are becoming available. In Figure 4 we summarized the effective deformation parameters derived from the statistical model analysis. From the strength ratios of the two fitted components information on the shape (prolate or oblate) can be obtained, although it must be stressed that the shape fluctuations and the resulting departure from axial symmetry limit the conclusions that may be drawn, particularly at higher E^*. In general, strong experimental evidence for changes and fluctuations of the shape is supplied by the consistent picture of increased widths for the GDR in nuclei covering a span of masses from $A = 40$ to 200 and excitation energies from $E^* = 30$ to 130 MeV.

A = 39–45 In Seattle experiments (59), these nuclei were produced at roughly the same temperature ($T = 1.7$–1.8 MeV) but average spins varied between 8 and 18.5. A systematic broadening of the GDR, between 10.5 and 14.5 MeV, is observed and ascribed to spin effects. The overall larger widths are ascribed to thermal fluctuations of the shapes. For comparison, the ground-state width of ^{45}Sc is 5 MeV. In these light nuclei, isospin effects (60) are important and must be considered in the analysis. Nonzero $a_2(E_\gamma)$ coefficients (between -0.09 and -0.12) in the low-energy side of the GDR are measured, consistent with shape effects.

A = 63 The ^{63}Cu isotopes have been studied at $E^* = 22$–77 and $\langle I \rangle = 2$–23 using different light-ion-induced reactions (61–63). GDR widths from $\Gamma = 7.5$ to 10.6 MeV are determined, growing with increasing E^* and I. The systematics of the Cu isotopes are well reproduced by adiabatic calculations (62).

A = 90–100 In inclusive studies of ^{92}Mo and ^{100}Mo in Seattle, deformations of $\beta \approx 0.22$ were found by analyzing the spectra with a two-Lorentzian strength function (64). These deformations are in excess of the rotating liquid drop predictions in this mass region. In these nuclei, populated at $E^* = 48$–67 MeV and $\langle I \rangle = 9$–24, the widths of the individual GDR components ($\Gamma_1 = 5.5$–7.1 MeV, $\Gamma_2 = 6.8$–11.5 MeV) and of entire GDR strength function ($\Gamma = 7.6$–10 MeV) are larger than the corresponding ground-state widths (5.5 MeV). A difference be-

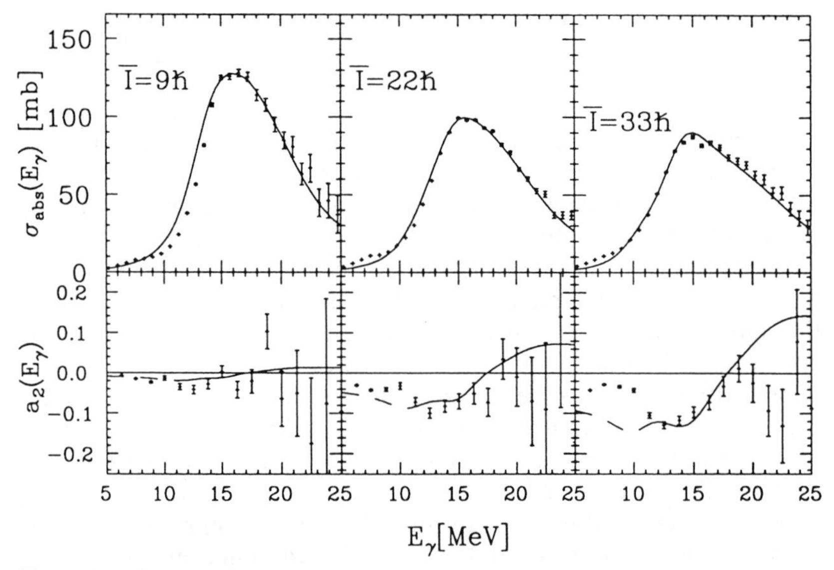

Figure 11 Measured GDR absorption cross sections and angular distribution coefficients, for excited Zr and Mo nuclei with $A = 90$–92. The average angular momentum increases as indicated. The data are compared to adiabatic calculations using the metric in Equation 11 and including the effect of orientation fluctuations. From (65).

tween the total widths of the two nuclei of about 1.5–2.0 MeV is observed. This difference, somewhat smaller than the width difference seen in the cold nuclei, is interpreted as a persistence of shell effects in spite of the role obviously played by thermal fluctuations. In ^{90}Zr nuclei with $\langle I \rangle = 9$–33 and $E^* = 54$–75 MeV, the inclusive GDR width increases from 8.2 to 9.7 MeV (65). These data are interpreted in terms of a spin-dependent change of an oblate deformation, based on the observed increase of the measured $a_2(E_\gamma)$ and comparisons to theory (see Figure 11 and Section 4.4.2).

A = 108–114 Systematic information on the GDR over a broad range of E^* and I is available for the $^{108-112}$Sn isotopes (23, 27, 35, 66–68), which have been studied from $E^* \approx 50$ MeV up to $E^* \approx 600$ MeV (see Figure 6). In all the experiments contributing to the shown Sn systematics (from Copenhagen, Berkeley, Stony Brook, and Grenoble), the high-energy gamma rays have been correlated with emission by a fused system, identified either by the multiplicity of the evaporation residues or by the detection of the recoiling residues. A width increase from 5.8 to 10.8 MeV, in the range $E^* = 50$–130 MeV, is observed (Figure 6). This can be parametrized (23) in terms of E^* (but implicitly also as a function of I_{max}) by the expression $\Gamma = 4.8 + 0.0026 E^{*1.6}$. Beyond

$E^* = 130$ MeV, the width increase is very small. At $E^* \geq 230$ MeV a width of 13 MeV is found. This is interpreted as evidence for the saturation of a strong spin dependence of the GDR, correlated with the limitation by fission of the angular momentum of the compound nucleus.

The role played by thermal fluctuations and shape changes in the Sn isotopes has been the object of several theoretical studies. A strong increase of the magnitude of the $a_2(E_\gamma)$ (from -0.10 to -0.21), as a function of growing I, is observed (Figure 12) (84, 85), which supports

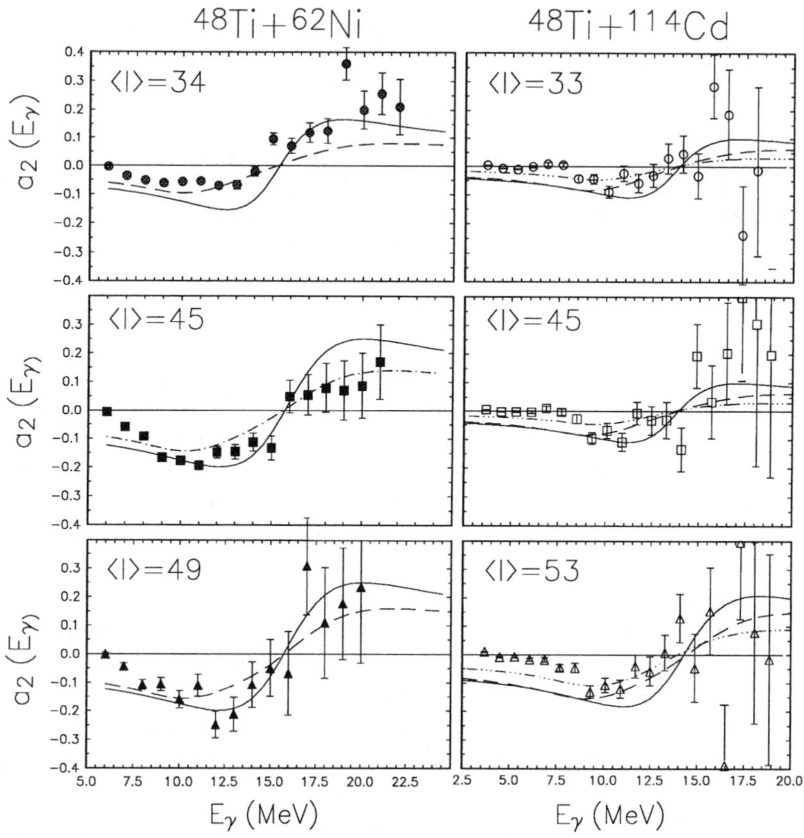

Figure 12 Measured GDR angular distributions for ^{110}Sn and ^{162}Yb nuclei excited to $E^* = 90$ and 75 MeV, respectively. The angular momentum increases from top to bottom as determined by gating on different intervals of the multiplicity of low-energy gamma rays. Also shown are adiabatic calculations as discussed in Section 4, including the effect due to orientation fluctuations: (*solid lines*) equilibrium shape, (*long-dashed lines*) adiabatic with Equation 11, (*dot-dashed lines*) adiabatic with Equation 10. Data from (84, 86).

the role played by deformation effects and indicates the reduced role of orientation fluctuations at higher I. A comparison between ^{109}Sn and ^{110}Sn for the same angular momentum gates, but for total E^* differing by approximately 12 MeV, results in nearly identical $a_2(E_\gamma)$ distributions, an indication that T effects are not dominant.

A = 132 Data on ^{132}Ce compound nuclei at $E^* = 80$–125 are available from recent experiments at JAERI (69, 70), correlating the GDR properties with the angular momentum by measuring the multiplicity of low-energy gamma rays in the residual nuclei. An increase of the width from 8 to 13 MeV is observed. This was interpreted as an E^* effect, also on the basis of data at higher E^* (see the discussion in Section 5). The width increase is parametrized as $\Gamma = 5 + 0.074E^* + 1.7 \times 10^{-8} E^{*4}$.

A = 152–166 For nuclei in the rare earth and actinide regions, experimental spectra, as a rule, require an analysis in terms of several Lorentzian components. Indeed it was in this region that the first clear evidence for deformation effects at high E^* was found.

The gamma decay from ^{156}Dy has been studied independently by two groups with similar results, using differential techniques. In a Copenhagen-Daresbury experiment (17), utilizing an array of germanium-BGO spectrometers and a 4π BGO multiplicity filter, a marked dependence of the spectrum shape on I (defined to three bins with $\langle I \rangle$ between 20 and 40) was found. Measured $a_2(E_\gamma)$ coefficients were roughly reproduced by calculations assuming prolate, triaxial, and oblate shapes for the three studied angular momentum regions and using the effective widths and shapes deduced from the statistical model analysis. In an Amsterdam-Groningen experiment (71), a single large shielded NaI detector was used for the high-energy gamma-ray detection and a sum spectrometer for the angular momentum determination. The statistical model analysis is consistent with a transition from prolate shapes to oblate shapes. In both experiments the widths of the GDR components were in the range $\Gamma_{GDR} = 8$–12 MeV, largely in excess of the $T = 0$ widths, which points to substantial shape fluctuations. A similar experiment (72) on ^{164}Yb ($N = 94$) with the Heidelberg Crystal Ball, an array of 162 NaI detectors, was analyzed with similar conclusions. Using the measured angular distributions the spectrum was further decomposed into the contributions from $|\Delta I| = 1$ and $\Delta I = 0$ radiation. The analysis suggested a reduction of the GDR centroid, and in particular of the $|\Delta I| = 1$ component, with increasing spin. This effect, however, is not systematically seen in other experiments (17, 34). In a study with the Chalk River 8π detector of excited ^{154}Gd nuclei as a function of I, deformations increasing from $\beta = 0.08$ to 0.35 were extracted from the data (73). In this case, however, the authors concluded prolate shapes up to the highest spins.

In an inclusive study of ^{160}Er compound nuclei, with $E^* = 75$–90 MeV and $I_{max} \approx 60$, no sensitivity to the nuclear shape was found, in the sense that fits to the spectra assuming either prolate or oblate shapes were found to be equally good (74). This result may be taken as evidence that, for nuclei formed with high E^* and over a broad range of I, the shape can no longer be interpreted in terms of a single axial symmetric shape. It also emphasizes the fact that more sophisticated methods of analysis and more exclusive data sets providing data integrated over smaller regions of the phase-space are required.

In Seattle experiments (36) at moderate I in reactions forming ^{160}Er and ^{166}Er nuclei at $E^* \approx 43$–47 MeV, prolate shapes similar to the known ground-state shapes of these nuclei were found, which suggests a persistence of shell effects up to $T \approx 1$ MeV. The widths of the two fitted Lorentzian components were found to be rather similar to widths measured for the ground state, in agreement with the reduced effect of thermal fluctuations at low T. At higher energy, the Copenhagen experiment (35) on ^{166}Er at $E^* = 60$ MeV and $I_{max} = 35$ concluded a large proportion of oblate shapes from the ratio of the strengths of the two fitted components and larger fluctuations. In a subsequent study at Copenhagen (75), correlating the properties of the GDR with the angular momentum of the compound nucleus using a sum-spectrometer technique, a transition from prolate to triaxial shapes was concluded at $E^* < 60$ MeV and $\langle I \rangle \approx 20$–25 for ^{166}Er, while the analysis indicated triaxial-oblate shapes already at $E^* < 47$ MeV for the soft ^{158}Er ($N = 90$) nucleus. In this region effective deformations $\beta = 0.20$–0.30 are found.

Deformations at higher E^* have also been deduced from the analysis of the angular distribution of alpha particles relative to the direction of the angular momentum vector (76). Small a_2 amplitudes have been observed in an inclusive study of ^{166}Er (77) and recent experiments (84) on ^{167,166,165}Er and ^{176,175}Hf nuclei at moderate spin ($I_{max} < 35$). In a differential experiment using ^{48}Ti projectiles to form ^{162}Yb nuclei at high angular momentum (see Figure 12), a_2 coefficients were found to increase with increasing angular momentum input from -0.05 to about -0.12. For ^{162}Yb the angular distribution of gamma rays defined to a narrower interval of excitation energy has also been measured (86). The idea is based on the subtraction (78) of gamma-ray spectra from reactions forming nuclei differing by one particle (in this case a neutron) and produced at excitation energies differing by the energy removed by that particle. If the angular momentum input is the same in the two reactions, the difference of the two gamma-ray spectra, normalized to the same number of low-energy statistical transitions in the region of $E_\gamma = 5$ MeV, yields the gamma rays emitted solely by the compound nucleus. This restricts the spread of E^* of the final states to the typical width of the GDR, rather than sampling over all steps in the nuclear decay. In the case of an imperfect spin matching and changing deformation, the magnitude of the observed a_2 distribution is enhanced, but can be corrected. The idea underlying this procedure is illustrated in Figure 13.

A = 190–224 In the actinide region, nuclear deformation has also been identified (79), both from an analysis of the spectrum shape and from a_2 coefficients measured relative to the fission plane. In experiments populating neutron-deficient Pb nuclei (nearly spherical in their ground state) at $E^* = 65$–102 MeV, deformations with $\beta \approx 0.32$ were deduced. From a comparison to the ratio of the widths of the GDR ground-state components, a prolate shape was preferred. In the analysis, a contribution from the fission fragments was calculated and subtracted, since fission becomes important in this region. In an experiment leading to the formation of excited ^{190}Hg compound nuclei using the Daresbury recoil mass separator, the GDR decay in the 2n channel (leading to ^{188}Hg) was studied using a Monte Carlo version of the CASCADE code to simulate the experimental gating (80). Deformations of 0.28 were deduced, although the data could equally well be described by a single Lorentzian GDR with a width of 7 MeV.

4.4.2 COMPARISON TO MODEL CALCULATIONS For nuclei in mainly two mass regions, namely the region of the nearly spherical $A = 90$–110 nuclei (at $T = 0$) and the region of well-deformed prolate rotors around $A = 166$, more detailed comparisons have been made, based on calculations of the GDR strength function. We note that some of the earlier comparisons have, however, not always taken proper care to average the calculations over the (E^*, I) regions populated in experiment.

The width increase in the Sn isotopes was early on analyzed by Gallardo et al (37) within the adiabatic model using the metric given in Equation 10. The absorption cross section for the GDR was calculated at each point in deformation space within the framework of the random-phase approximation, a procedure that does not predict the absolute magnitude of the width. Nevertheless, the calculations were able to reproduce quite well the relative increase of the observed GDR width. The main effect underlying this increase was found to be a significant change of the equilibrium deformation (from spherical to oblate) of the Sn nuclei with increasing spin, while the effect of thermal

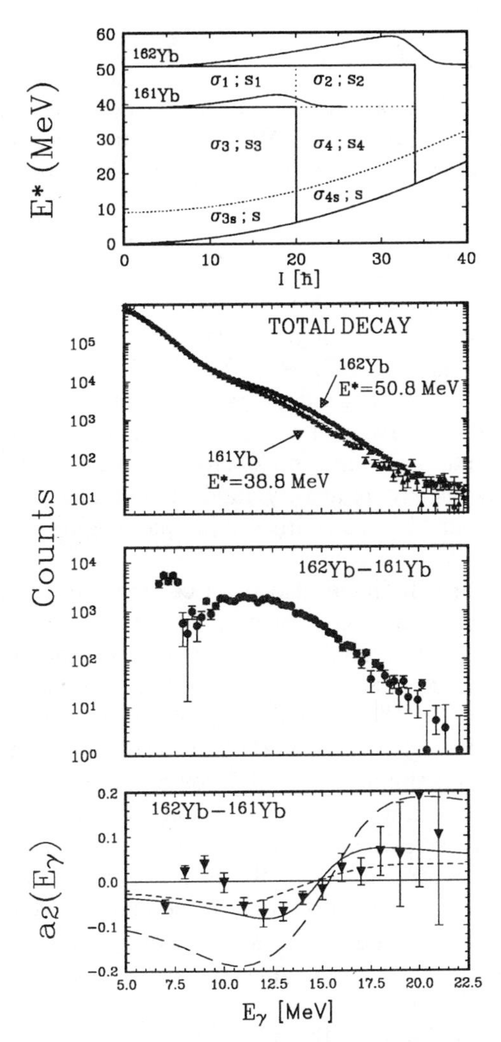

Figure 13 Illustration of a method to isolate the first-chance GDR decay, applied to the nuclei ^{162}Yb and ^{161}Yb. In the absence of a complete matching of the spins in the two reactions, the angular distribution of the difference spectrum can be enhanced if the GDR changes with spin. A correction factor can be simply estimated from the cross sections of the reaction (σ_{1-4}, in top panel). The bottom panel shows measured corrected $a_2(E_\gamma)$ of the difference compared to adiabatic calculations with Equation 11, with and without orientation fluctuations (*long-dashed* and *short-dashed lines*, respectively). The solid line shows a calculation for the equilibrium deformation. From (85, 86).

fluctuations was estimated to contribute to the width due to a spread in frequencies

$$\delta\omega \sim 1.3T^{1/2}\ \text{(MeV)}. \qquad 17.$$

Subsequent studies have taken as a starting point the parametrization of the width of the individual GDR components according to the phenomenological relation established for ground-state resonances (81)

$$\Gamma_{\text{GDR}} = \Gamma_0 \left(\frac{E_{\text{GDR}}}{E_0}\right)^{\delta}, \qquad 18.$$

in tune with the expectation that the large-amplitude shape fluctuations are indeed the dominant element in the description of the GDR in nuclei up to excitation energies of $E^* < 100$ MeV. Such descriptions (47, 50) have been able to give a good overall account of the GDR observed in ^{166}Er at low temperature and spin, although it must be remarked that the angular momentum and excitation energy range covered in these comparisons are relatively modest and that different metrics (Equations 10 and 11) and width scaling parameters ($\delta = 1.6$ or 1.9) have been used.

Similar comparisons have also been carried out for the Sn nuclei. The comparison between the measured effective strength functions and those calculated in the adiabatic model suggest that the role of fluctuations is overestimated. Ormand et al (51) found that the GDR strength function for ^{110}Sn could only be reproduced by calculations averaged over the (E^*, I) region relevant for the experiment, assuming a time-dependent coupling of the GDR to the shape degrees of freedom, as discussed in Section 4.3. With the value $\Gamma = 6\delta\omega$ deduced in that work, motional narrowing effects would be expected to play a role even at nuclear temperatures of the order of $T \approx 1$ MeV. Estimates may be made for the relaxation time of the quadrupole degree of freedom in various limits. Assuming that the configurations of the excited nucleus are uncorrelated multi-quasi-particle states, Γ was related to the single-particle width (82), $\Gamma \approx 0.34\ E^* \approx 0.034AT^2$, leading to estimates $\Gamma \approx 1$–100 MeV. In the other extreme, where the nuclear configurations are coherent surface states, information on the relaxation time may be obtained from the slow dynamical fission process, leading to a much smaller value of $\Gamma \approx 0.01$ MeV (see the discussion in Section 6). A similar result was found in the work of Alhassid & Bush for ^{112}Sn requiring a value of χ intermediate between the fully motional narrowed and adiabatic limits, although the comparison was made not for a distribution of temperatures and angular momenta, but for values reasonably representative for the average of these quantities.

Additional constraints on these theories are provided by comparisons to measured angular distributions. In two recent works on ^{90}Zr nuclei (65, 83) the GDR strength function and angular distribution were calculated in the adiabatic approximation using the metric in Equation 11, including orientation fluctuations, and averaging over the (E^*, I) ensemble relevant for the experiment. The calculations give a good account of the effective strength function and of the magnitude of the $a_2(E_\gamma)$ in the low-energy part of the GDR (see Figure 11), although they fail to reproduce the high-energy part of the GDR distribution. The latter effect may be understood by noting that the effect of the density of final states on the decay probabilities of the various ΔI components of the GDR has so far been neglected in the analysis of the $a_2(E_\gamma)$ distributions and of the strength functions. Indeed, this can be quite important (Figure 14) in lighter mass nuclei, where the low moments of inertia result in large rotational frequencies, in large Coriolis splitting of the GDR components, and simultaneously in a steep yrast line. This work (65), which is the first to attempt a simultaneous description of the spectrum shape and of the a_2 distribution in terms of a theory including shape and orientation fluctuations, also provides a confidence limit for the parameters extracted from the two-Lorentzian statistical model analyses of the spectrum shape. It is suggested that the effective deformations extracted from such analyses lie between that of the equilibrium deformation and that corresponding to the most probable deformation.

Figure 12 shows angular distributions as a function of the angular momentum measured (84, 85) for ^{110}Sn and ^{162}Yb. Also shown are adiabatic calculations as described above, and calculations for the equilibrium shape. The latter may be taken as representative of the fully motional narrowed situation. As mentioned earlier the data, like other

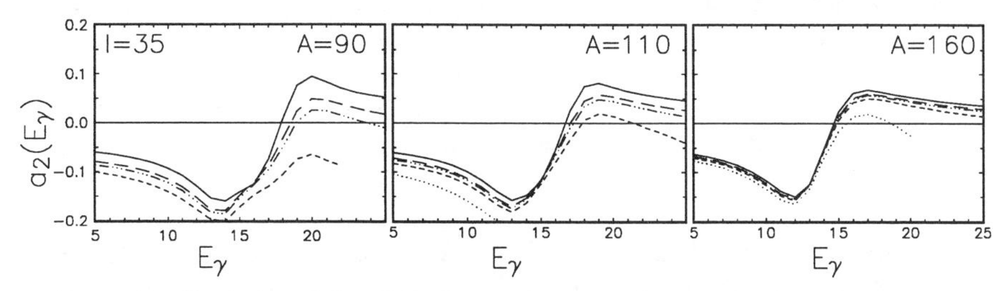

Figure 14 Illustration of the effect of including the appropriate density of final states for the different ΔI components of the GDR in the simulation of angular distribution coefficients for a prolate shape. The effect is more pronounced in light nuclei and at high rotation: (*solid lines*) effect not included, (*long-dashed lines*) $E^* = 75$ MeV, (*dot-dashed lines*) $E^* = 60$ MeV, (*short-dashed lines*) $E^* = 45$ MeV, (*dotted lines*) $E^* = 30$ MeV.

spin-gated data in heavier nuclei, show a strong increase of the $a_2(E_\gamma)$ with increasing angular momentum. The main reasons for this effect that may be inferred from the calculations are that (*a*) the nuclei readily become oblate at finite temperature because of the weakening of the shell structure and acquiring larger deformations as a function of angular momentum; and (*b*) the effect of orientation fluctuations diminishes with increasing rotation. The reduction in a_2 observed from ^{110}Sn to ^{162}Yb reflects the increased importance of orientation fluctuations in the heavier nucleus (due to the lower rotational frequency) and the increase in thermal fluctuations close to the critical temperature. While in this data set a preference between the various model calculations does not appear reasonable, the exclusive measurements (86) shown in Figure 13 seem to be better reproduced by the calculation for the equilibrium shape. Shown in Figure 13 are $a_2(E_\gamma)$ values calculated without including orientation fluctuations (short-dashed line). It is seen that orientation fluctuations (assumed uniform and adiabatic, short-dashed line) play a strong role counteracting the effect of thermal shape fluctuations that, with this metric, tend to favor large deformations and therefore increase the a_2.

We conclude this discussion by remarking that, while a general understanding of the shapes and fluctuations of hot nuclei in this energy domain seems to be at hand, more specific questions such as the relevance of the motional narrowing picture for hot atomic nuclei and the details in the averaging procedure must wait for a larger body of more specific and systematic experimental information and comparisons to theory. We anticipate that experiments in which the measured GDR gamma rays are defined to originate from narrow cells in (E^*, I) will provide such detailed information in the near future.

5. LIMITS OF COLLECTIVITY IN HOT NUCLEAR MATTER

With the development of heavy-ion beams of energies up to 100 MeV/u, it has become possible to form atomic nuclei with thermal energies per particle exceeding the binding energy of the individual nucleons. In this situation the system will vaporize or explode. In recent years a considerable effort has been devoted to the investigation of the reaction mechanism involved in such energetic collisions and trying to determine a limiting temperature for the existence of the nucleus [see for example the overviews by Suraud et al (87) and Guerreau (88)]. Such studies have been based primarily on measurements of the particles and fragments that result from such energetic collisions, although high-energy photons from nucleon-nucleon bremsstrahlung have also been intensively studied, providing information on the initial stages of nuclear reactions [for a recent review, see Nifenecker & Pinston (89)].

Measurements of the gamma-ray decay of the GDR excited in nuclei produced with very high internal energies constitute a new and specific tool to study hot nuclei on a longer time scale and to explore the conditions prevailing during and after thermalization of the nucleus.

5.1 *Saturation of the GDR Strength*

The initial expectation was that the GDR decay could serve as an indicator for the disappearance of the nucleus with increasing excitation energy. The first study of GDR decay in fusion-like reactions at high E^* was from a Grenoble-Milano-Copenhagen collaboration that used ^{40}Ar beams at 15 and 24 MeV/u from the SARA accelerator to produce compound nuclei around ^{110}Sn, excited up to $E^* \approx 600$ MeV (90). The gamma rays were measured in coincidence with two parallel plate avalanche counters (PPACs) positioned at forward angles, and were used to identify the recoiling fused residues. The linear momentum transfer in the reaction was estimated from the time of flight of the particles. Figure 15 shows the measured gamma-ray spectra for the two energies, both for central collisions (large momentum transfer) and for peripheral collisions (panel at right). The figure shows the important contribution from pre-equilibrium photons associated with bremsstrahlung from neutron-proton collisions, which must be subtracted from the spectra. A statistical model calculation for the reaction at 15 MeV/u spectrum accounts well for the observed spectrum shape and intensity. The calculation assumes that the decaying nuclei were formed in almost complete fusion reactions with excitation energies $E^* \approx 320$ MeV.

By increasing E^* further, the number of gamma rays emitted in the GDR region per fusion reaction should increase, reflecting the additional number of chances to decay en route to the ground state. Surprisingly, this is not the case. The yield of GDR gamma rays was found to be practically the same for the reaction at 24 MeV/u as for the reaction at almost half the energy, more than a factor of two lower than the statistical model prediction. Subsequent studies of the same reaction at 19 MeV/u support this observation (66, 67). In contrast, the observed strength at 10 MeV/u ($E^* \approx 230$ MeV) is consistent with the full energy-weighted sum rule. Although, the onset of significant incomplete fusion around 15 MeV/u limits the transferred E^*, a comparison to available systematics on the transfer of linear momentum indicates that such an effect cannot explain the observations (91). Neither can a decrease of the nuclear level density parameter with E^*,

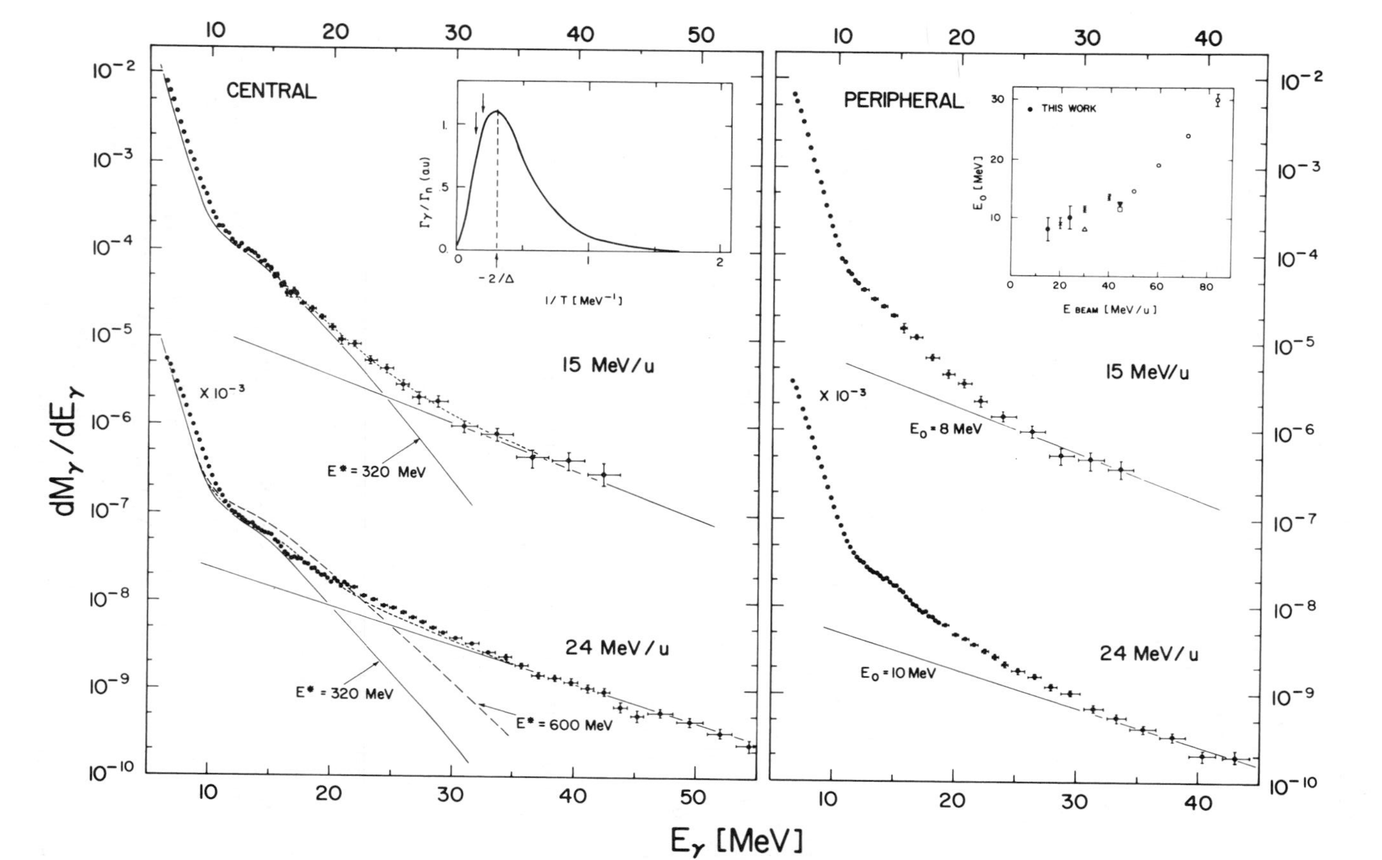

Figure 15 Measured gamma-ray spectra for the reaction $^{40}Ar + ^{70}Ge \rightarrow ^{110}Sn$ at $E(^{40}Ar) = 15$ and 24 MeV/u for central collisions (*left*) and peripheral collisions (*right*). Straight lines indicate the contribution from nucleon-nucleon bremsstrahlung. Solid curves are statistical model calculations assuming $E^* = 320$ MeV. The short-dashed curve indicates the sum of the statistical and pre-equilibrium components. The long-dashed curve, which assumes $E^* = 600$ MeV (close to the expected E^* transfer), overpredicts the measured spectra at the higher energy. From (90).

although it reduces the discrepancy. Such a change from $a = A/8$ to $a = A/13$ at $T = 3.5$–4 MeV has been suggested based on differential studies of the spectra of evaporated charged particles (92, 93). Hence, this experiment suggests a limiting temperature for the GDR of $T \approx 4.5$ MeV, although the disappearance of the GDR is in all likelihood not a sudden phenomenon.

A similar study has been carried out at the Riken and JAERI facilities in Japan for the reactions $^{40}Ar + ^{92}Mo$ at 21 and 26 MeV/u, leading to nuclei with $A \approx 130$, and for the $^{40}Ar + Ni$ and $^{40}Ar + ^{122}Sn$ reactions at 26 MeV. A similar particle detection method was used and an array of BaF_2 detectors. Again a saturation of the observed experimental gamma-ray multiplicity between $E_\gamma = 12$ and 20 MeV was observed with increasing E^*, as shown in Figure 16. Limiting temperatures for the GDR between 3.5 and 4.5 MeV, decreasing with increasing mass of the composite system, were deduced. The values depend on the method of analysis, as discussed in the next section. For these reactions, the neutron multiplicities and spectrum shapes were also measured in coincidence with the same particle gates. While the gamma multiplicity saturates with increasing E^*, the neutron multiplicity continues to grow in a manner consistent with the expected E^* transfer (94). From the analysis of the neutron spectra initial temperatures up to $T \approx 6$ MeV were deduced. In a Giessen-Ganil study of the reaction $^{90}Mo + ^{90}Mo$ at a bombarding energy of 19 MeV/u, a similar saturation of the gamma-ray multiplicity above 15 MeV was observed for $T \approx 3.5$ MeV (95).

Although not directly connected with this discussion, we mention that GDR gamma-ray decay has also been used to study the transfers of E^* to target- and projectile-like fragments in grazing heavy-ion collisions at intermediate energies (96, 97).

5.2 *Damping of the GDR at High Excitation Energies*

The observation of a limiting temperature for the GDR, which according to available evidence is not coincident with the limiting temperature for the nucleus, indicates that excited nuclei cool substantially by particle emission before the gamma-ray emission from the GDR can occur. The question is whether the collective vibration exists at all during the time of this cooling, for example if the nucleus is not thermalized or is too chaotic for a collective vibration to exist, or whether the gamma decay of the collective mode is prevented by the competition with faster modes of decay. Such diverse effects should, however, have different consequences for the width of the GDR as observed in experiment.

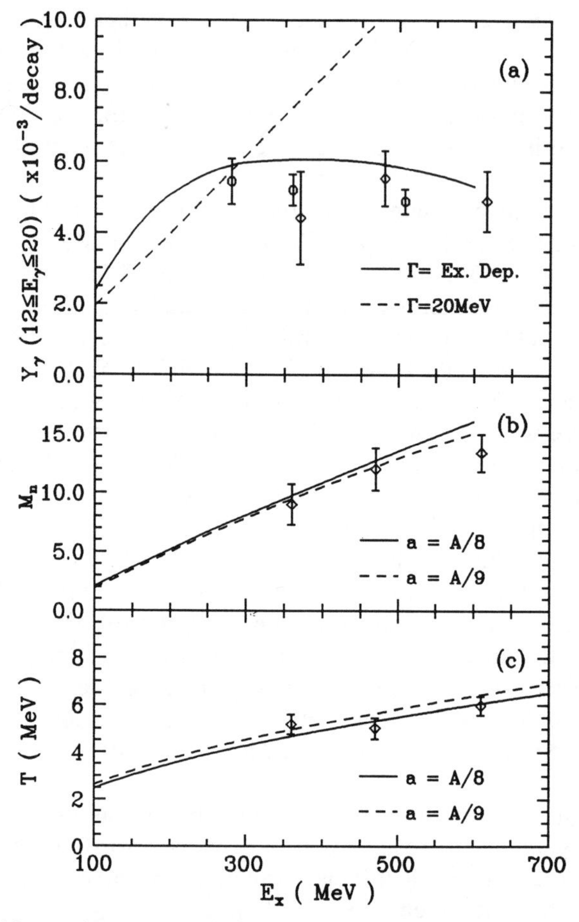

Figure 16 Measured gamma-ray (*a*) and neutron (*b*) multiplicities as a function of the maximum E^* of the compound nucleus for central collisions in the reaction $^{40}Ar + ^{92}Mo$ at 21 and 26 MeV/u. Statistical model predictions using a fixed GDR width (*dashed line*) and a GDR width increasing with excitation energy (*solid line*) are indicated. The bottom panel (*c*) shows the estimated temperature of the emitting system assuming different values of the level density parameter, *a*. From (94).

In the analysis of the 10-, 15-, and 19-MeV/u reactions leading to the formation of excited Sn isotopes (see Figures 6 and 15), the gamma-ray spectra were reasonably well reproduced by statistical model calculations in which the GDR width was fixed to $\Gamma_{GDR} = 13$ MeV. Even for the 24-MeV/u reaction, it appears that the width is less than 15 MeV. These values are consistent with a modest influence on the GDR arising from temperature effects (see Equation 17). The low value of the width, similar to the width at $E^* \approx 130$ MeV, was interpreted by Bracco et al (66) as being consistent with a limitation of the angular momentum that can be transferred to the hot compound system if it is to survive fission. A saturation of the GDR width is also seen by G. Enders et al (*Phys. Rev. Lett.* 69:249, 1992). The latter argument seems substantiated by the recent measurements of the angular distribution of the Zr and Sn isotopes as a function of angular momentum. These suggest, by comparison to model calculations, that spin and shape effects are the dominant cause of the width increase seen at lower bombarding energies (see the discussion in Section 4).

Theoretical estimates of the temperature dependence of the spreading width, $\Gamma^{\downarrow}$, of the GDR, have been made that support this interpretation. The coupling between the GDR and the low-lying 2p-2h states has been studied and found to be roughly independent of temperature (98, 99). The influence of nucleon-nucleon collisions on the GDR width has also been explored in a phase-space model (82). A weak temperature dependence, $\Gamma_{coll} = 0.35T^{1.6}$ was found.

In contrast to this approach, the saturation of the GDR width can be phenomenologically explained by assuming that the GDR width grows rapidly with temperature, and thereby spreads strength outside the region of transition energies $E_{\gamma} = 10$–20 MeV. A problem is that strength outside the main GDR region is difficult to identify experimentally. On the low-energy side of the GDR, the neutron threshold (around 10 MeV) limits gamma emission dramatically. On the high-energy side, the bremsstrahlung contribution complicates a precise analysis. A parametrization of the width in terms of the thermal energy, $\Gamma_{GDR}(E_x) = \Gamma_0 + 0.036E_x + 1.6 \times 10^{-8} E_x^4$ was suggested and used in the statistical model analysis of the experimental spectra (69, 94). With this procedure, the observed gamma-ray multiplicities can be reproduced, as well as the GDR widths measured in $A = 130$ nuclei up to $E^* = 120$ MeV. However, this parametrization implies a very strong growth of the intrinsic GDR width at higher excitation energies. For example, $\Gamma(E^* = 130) \approx 14$ MeV and $\Gamma(E^* = 250) \approx 75$ MeV, which implies strongly overdamped vibrations at values as low as $E^* \approx 150$ MeV. Defining the limiting E^* for the GDR as the excitation energy at which $\Gamma_{GDR} = 30$ MeV, limiting excitation energies of 140, 180, and 180 MeV were deduced for GDR in Cd, Sm, and Yb. The corresponding temperatures are $T \approx 3.9$, 3.6, and 3.3. Analyzing the data in terms of a fixed width ($\Gamma_{GDR} \approx 15$–20 MeV), limiting temperatures of 4.6, 4.1, and 3.8 MeV were deduced. Since the width for direct neutron emission is expected to be small ($\Gamma^{\uparrow} < 1$ MeV), such an increase would have to come about from a strong increase of the spreading width. A recent calculation (100) based on the Vlasov equation, with a relaxation time approximation using semiclassical methods, predicts, however, a very strong increase of the damping width as a function of increasing T, due to nucleon-nucleon collisions.

Based on time-dependent Hartree Fock calculations (101) describing collisions in a relaxation time approximation, a vaporization of the surface of ^{40}Ca has been predicted at $E^*/A \approx 3$ MeV. Such an effect, resulting in a strong increase in the particle emission widths in the pre-equilibrium phase, could deplete the available E^* and degrade the ordered motion of the GDR. However, currently available experimental data does not indicate a strong pre-equilibrium particle component. Other investigations predict a reduction of the GDR absorption cross section at $T \approx 6$ MeV. These predictions are based on thermal random-phase approximation (RPA) and cranked Hartree-Fock-Bogoliubov (HFB) calculations. A dynamical strength function is derived, which is somewhat temperature dependent. In a recent work, the angular distribution at high temperature is calculated (102).

Recently, general arguments have been advanced for a constancy of the GDR width as a function of temperature. These arguments are based on the expectation that, although equilibrium of the single-particle degrees of freedom is attained rapidly ($\tau \approx 10^{-23}$ sec) after only few collisions among the nucleons, the collective modes (like the GDR) only develop after some time. This time delay can be thought of as an equilibration time for the GDR, determined by the width of the coupling of the GDR to the compound nucleus states. The temperature at which this width becomes comparable to the particle emission width defines a limiting temperature for the GDR decay. This problem has been investigated, as illustrated in Figure 17, in a model comprising two classes of states, the compound nucleus states, C, and the GDR states, D (103). To each C state corresponds a state D shifted up by the energy of the GDR. The transition rates between these classes are denoted λ and μ. Both classes of states can decay by particle emission, while only the dipole states can decay by high-energy gamma-ray emission. By considering the time-dependent probabilities of being in either class of states, the probability for gamma decay, $P\gamma$, can be evaluated to yield

$$P_\gamma = \frac{\gamma_\gamma}{\gamma_{ev}} \left(\frac{\lambda}{\gamma_{ev} + \lambda + \mu} \right) \simeq \frac{\gamma_\gamma}{\gamma_{ev}} \left(\frac{\lambda}{\gamma_{ev} + \mu} \right), \quad 19.$$

where γ_γ and γ_{ev} are the decay rates for gamma rays and particles, and $\lambda \ll \mu$ because of the very different level densities. In the limit where $\gamma_{ev} \gg \mu$, the gamma-ray emission probability will decrease strongly with increasing particle emission rate, i.e. with increasing T.

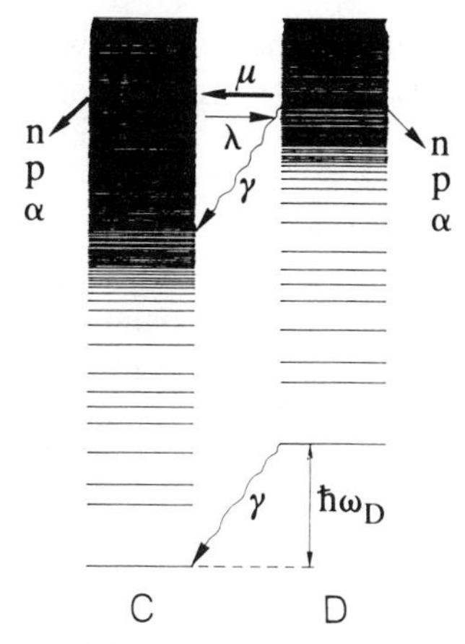

Figure 17 Schematic representation of nuclear energy levels. Levels C are ordinary states. D states are GDR states built on C states, shifted upward by the average GDR energy. Each type of state can decay by particle emission. Only D states can decay by high-energy gamma-ray emission. Transition rates between C and D states are denoted by λ and μ. From (103).

By equating the transition rate from the dipole states to the compound states with the spreading width ($\Gamma^{\downarrow} \approx 5$ MeV), and then evaluating the temperature dependence of the particle widths, Bortignon et al (103) argue for a limiting excitation energy of the order of $E^* \approx 250$ MeV.

We end this section by addressing again the question of motional narrowing, introduced in Section 4. If one assumes a temperature dependence for the relaxation time for the quadrupole degrees of freedom, motional narrowing effects should become more prominent at higher temperature. In nuclear reactions at very high energy, such dynamical effects should manifest themselves as a reduction of thermal fluctuations, which would in turn reduce the effective GDR width. In nonspherical nuclei, a strong reduction of thermal fluctuations could lead to strength functions resembling those observed in cold nuclei. In a few instances, narrower structures in high-temperature GDR spectra have been reported. In ^{110}Sn, produced with ^{40}Ar beams at 24 MeV/u, two narrow components were identified in the GDR region (90). These structures were not analyzed quantitatively and are not understood. In a study of the reaction ^{16}O + ^{159}Tb at 18.7 and 25 MeV/u, using a similar experimental technique, narrow structures were also observed (104). Although contributions from background reactions cannot be ruled out (for example a contamination from the 15.1-MeV line in ^{12}C, excited in inelastic scattering), we note that both experiments were coincidence experiments in which the high-energy gamma rays were measured in coincidence with slow, heavy residues. Widths of the order of 2.4–5.5 MeV were deduced, although only after assuming a large unexplained gamma-ray contribution around 10–12 MeV (105, 106).

It must be concluded that there is no firm evidence yet for motional narrowing mechanisms in hot nuclei. Such effects are probably best sought at elevated temperatures, although the inhibition of the gamma-ray strength apparently sets an upper limit of about T = 4–5 MeV to the temperature range that can be explored. A better understanding of the damping of the GDR must come from differential experiments, where gamma rays emitted from rather narrow regions in E^* are isolated. A simultaneous determination of the magnitude of the angular distribution coefficients in the GDR region at elevated temperature should be able to answer the questions regarding the GDR width and contribute new information on the time scales involved in the damping of collective excitations in hot nuclear matter.

6. PRE-FISSION GAMMA RAYS FROM GIANT RESONANCES

Nuclear fission has been a subject of continuous interest since the late 1930s (107). Many basic questions, regarding dissipative processes in nuclei and the time scale of the fission process are still open, however. Some of these questions may be addressed by investigations of giant resonances that gamma decay before the system fissions. Such techniques may also provide a unique way of studying nuclei formed with extreme angular momenta, which fission under the influence of strong centrifugal forces and which otherwise are inaccessible to gamma-ray spectroscopic studies. Likewise, pre-fission GDR studies with beams in the intermediate-energy domain may bring a renaissance to the study of nuclei with charge numbers in excess of Z = 100, the so-called superheavy nuclei.

6.1 *Pre-fission Gamma Emission in Hot Nuclei*

Strong evidence has been collected in the last 7–8 years for a significant slowing down of the fission process as compared to expectations based on the statistical model for nuclear decay. Experimentally, this evidence comes from a study of the multiplicities of neutrons emitted before and after scission (108–117). Similar, but less extensive, information is available for charged particles (118–121). The number of prescission particles considerably exceeds the number predicted by the statistical model, and the number of emitted post-scission particles no longer grows above excitation energies $E^* = 100$–200 MeV. This pattern indicates that fission is a cold process occurring late in the decay of excited nuclei. The observations have been attributed to a slow, large-scale mass diffusion process governed by the nuclear viscosity (122–126).

Pre-fission gamma rays associated with the decay of GDR modes in hot ^{224}Th nuclei prior to fission were observed by Thoenessen et al (127), in support of these ideas. In a pioneering study of the $^{16}O + ^{208}Pb$ reaction at bombarding energies of 100, 120, and 140 MeV, at Stony Brook, highly fissile Th nuclei were produced at $E^* = 44$, 64, and 82 MeV with average angular momenta of 25, 37, and 43, respectively. The measured gamma-ray spectra were analyzed with the statistical model in terms of a pre-scission and a post-scission gamma-ray component. These two contributions are generically different and contribute differently to the total gamma-ray spectrum.

The component corresponding to post-fission emission contains essentially all the gamma rays with energies $E_\gamma < 7$–8 MeV, since these are emitted in the particle-bound region after particle evaporation from the excited fragments. Furthermore, the high-energy gamma rays reflect the properties of the GDR excited in nuclei with approximately half the mass of the compound system ($E_{GDR} \approx 15$–16 MeV). The excitation energy of each fragment can be calculated from $E^*_{fis} = \frac{1}{2}(E^*_{CN} + Q_{fis} - E_{kin}) - E_{rot} - E_{def}$, where E^*_{CN} is the excitation energy of the compound nucleus, Q_{fis} the Q value for the fission process (for example assuming symmetric fission), E_{kin} the total kinetic energy released [obtained from systematics (128)], E_{rot} the energy due to the rotation of the fragments, and E_{def} the energy bound in deformation of the fragments (129). The angular momentum of the fragments may be calculated from $I_{fis} = 1/2[(2/7)I_{CN} + S(I)]$, where $S(I) = 18.0 - 0.17I$ describes the deviation from a rigid rotation of two touching spheres (130).

In contrast, the spectrum associated with emission prior to fission emission contains essentially no low-energy gamma rays. The spectrum shape is characterized by the lower resonance energy of the GDR in the heavy system and by the higher temperature of the compound nucleus.

Thoenessen et al (127) began their analysis by calculating the gamma-ray spectra using the standard statistical model, thus neglecting all dynamical effects. In this approach the fission rate is simply determined by the level densities of the final states at the saddle point, where the available energy is the energy of the compound system corrected for the height of the fission barrier, the rotational energy, and the kinetic energy (131). Such calculations, although able to reproduce the low and high parts of the spectrum, could not account for the observed gamma-ray strength around $E_\gamma = 11$–12 MeV, the region of the compound nucleus GDR. A satisfactory description of the spectrum shape could only be obtained by reducing the fission probabilities in the initial steps of the decay, thus enhancing the contribution of the pre-fission gamma-ray emission to the total spectrum.

As further support of the idea of a strong pre-fission GDR component, the anisotropy of the gamma rays was also measured in coincidence with the fission fragments detected in four small solid-state detectors located in the plane perpendicular to the beam direction and spaced by azimuthal angles differing by 90 degrees. The observed large anisotropies at $E_\gamma \approx 10$–11 MeV, measured relative to the fission plane, are consistent with those expected for the low-energy component of a GDR in a deformed heavy nucleus, although we remark that simulations of the anisotropies including the effect of thermal fluctuations have not yet been done for such systems.

Butsch et al (132) analyzed these issues more quantitatively in terms of the nuclear friction constant, γ (see Figure 18). A time-dependent fission width

$$\Gamma_{\text{fis}}(t) = \Gamma^{\text{K}}_{\text{fis}}[1 - \exp(t/\tau_{\text{f}})] \qquad 20.$$

was used, where $\Gamma^{\text{K}}_{\text{fis}}$ is the Kramers width, reduced as compared to the Bohr-Wheeler width due to friction

$$\Gamma^{\text{K}}_{\text{fis}} = \Gamma^{\text{BW}}_{\text{fis}}(\sqrt{1 + \gamma^2} - \gamma). \qquad 21.$$

In the above expressions the time constant τ_{f} is associated with the evolution of the nucleus toward the saddle point. The time from the saddle point to scission was also included as $\tau_{\text{ssc}} = \tau^0_{ssc}[(1+\gamma^2)^{1/2}+\gamma]$, assuming $\tau^0_{ssc} = 3 \times 10^{-21}$ sec. Finally, τ_{f} was related to γ by assuming an overdamped motion, $\tau_{\text{f}} = (\gamma/\omega_1) \ln(10B_{\text{f}}/T)$, in terms of the fission barrier height, the temperature, and the knocking frequency. The spectrum of the gamma-ray decay of ^{244}Th, at $E^*_{\text{CN}} = 90$ MeV was analyzed by varying the viscosity parameter (Figure 18). The best fits to the spectrum shape were obtained with a value of $\gamma = 5$–10, which suggests a strong overdamping of the collective mass flow. These values imply average fission times $\tau_{\text{f}} \approx 3.4$–$6.4 \times 10^{-19}$ sec. We remark that these results are consistent with the neutron multiplicity measurements that were analyzed in a similar approach, as well as with recent studies of fission times deduced from an analysis of measured fission probabilities from target residues in peripheral collisions of ^{40}Ar$+^{232}$Th (133). If these times are associated with the time for the relaxation of the shape

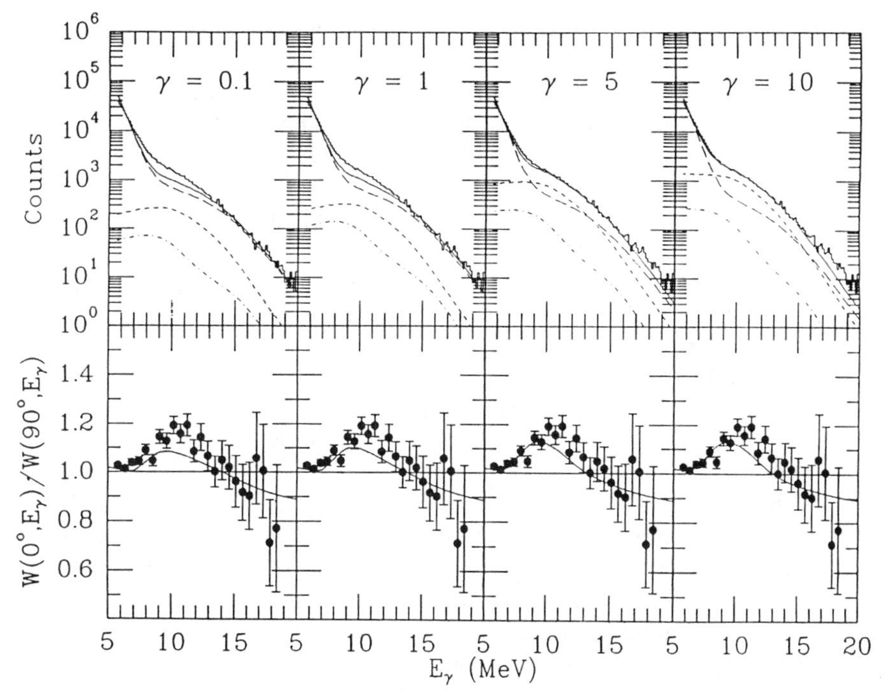

Figure 18 Gamma-ray spectra (*top*) and anisotropy spectra for the reaction $^{16}O + ^{208}Pb$ at 140 MeV measured in coincidence with fission fragments. The fits correspond to different values of the friction coefficient, discussed in Section 6. The indicated contributions are presaddle (*short-dashed lines*), postsaddle (*dot-dashed lines*), and post-scission (*long-dashed lines*). The total contribution from the various effects is shown with the solid line. From (132).

degrees of freedom, then motional narrowing effects would not be relevant for nuclear conditions.

6.2 *Pre-fission GDR Emission and Superheavy Nuclei*

A fascinating aspect of pre-fission gamma-ray studies is the prospect of obtaining new information on superheavy nuclei. As is well known, a variety of calculations predict an island of stable nuclei around Z = 114 and N = 184 [see for example the review by Kumar (134)]. Among the candidates proposed to be the most stable are the elements $^{294}[110]$, and $^{300}[116]$, the latter with a predicted total half life of about 103 years. Extensive experimental efforts have been devoted to trying to synthesize such heavy nuclei using heavy-ion-induced fusion reactions, but with little success. The heaviest element identified so far in such reactions is Z = 109. Most of these experiments have used beams with energies very close to the barrier, to produce nuclei with as low thermal excitation energy as possible in order to minimize the chances for fission and to exploit stabilizing effects related to the shell structure. GDR decay offers the possibility of studying such heavy systems prior to fission by the methods discussed in the previous section. This type of analysis requires, however, that the mass and initial excitation energy of the conglomerate system be known and that the system have time to equilibrate. A strong indicator of compound formation is fission of the system into fragments of nearly equal mass and that the angular distribution of the fragments is isotropic in the center-of-mass system around the spin direction.

The production of heavy fused elements with $Z > 110$ requires the projectiles of Ar or Ca, with masses around $A = 40$–50. With increasing mass of the projectile and in reactions close to the barrier, however, symmetric fission becomes less favored. This can be identified by the observation of a substantial spreading of the mass distributions and of characteristic anisotropies in the emission of the fragments. Such effects have been suggested to correspond to a situation in which the system is trapped in an intermediate state behind the conditional saddle. The system subsequently fissions on a shorter time scale than for fusion-fission. This process is called quasi-fission or fast fission. The proportion of symmetric and quasi-fission varies according to the projectile-target combination (135, 136). In the intermediate situation it has been predicted that the system may be in a highly deformed state (137), called the mononucleus. It is possible that the nucleus may also have time to establish a GDR in this state.

Under these conditions the excitation energy and time available for GDR decay may depend significantly on the chosen reaction, affecting the observed GDR strength. This effect has been explored by comparing inclusive gamma-ray spectra from the reactions $^{16}O + ^{208}Pb$ at 120 MeV, $^{24}Mg + ^{196}Pt$ at 150 MeV, and $^{32}S + ^{nat}W$ at 185 MeV. These reactions all lead to the formation of Th nuclei at $E^* = 64$, 70, and 72 MeV respectively, and with similar angular momenta. A systematic decrease of the gamma-ray yield in the GDR region is observed as a function of increasing projectile mass, which is not reproduced by the statistical model calculations. This effect is attributed to the reaction mechanism, since the change in fissility ($\chi \propto Z^2/A$) caused by the change in neutron number is small. The strong influence of quasi-fission appears to limit the usefulness of pre-fission GDR in studying very heavy systems at bombarding energies around the Coulomb barrier. However, there is strong experimental evidence (138) that at higher bombarding energies, the influence of quasi-fission on the total cross sections is reduced.

In an experiment with the SARA cyclotron in Grenoble (139), the gamma-ray emission from the reaction $^{40}Ar + ^{232}Th$ at bombarding energies of 6.8 and 10.5 MeV/u was studied, producing compound nuclei with $A = 272$, $Z = 108$. The choice of bombarding energies was motivated by the desire to reduce the contribution from quasi-fission and by the expectation that, at sufficiently high excitation energies of the compound nucleus, fission may no longer compete efficiently with gamma-ray emission. Such a situation appears to be realized at $E^* = 150$–250 MeV, as deduced from the observed saturation of the post-scission neutron multiplicity discussed in the previous section. In such a case one would expect the absolute strength of the post-scission GDR emission to become independent of the initial excitation energy. In contrast, the strength of the pre-fission component should increase with E^*_{CN}. A subtraction of the measured spectra would then isolate the pre-fission contribution.

Photons with $E_\gamma > 6$ MeV were detected in four large BaF_2 scintillators by the HECTOR collaboration in coincidence with the detection of both fission fragments in two PPACs. From the angular correlation and flight times of the fragments, events corresponding to symmetric fission were identified and used for gating the gamma-ray spectra. The measured gamma-ray spectra for the two energies are displayed in Figure 19. The maximum compound energies were $E^* \approx 105$ and 230 MeV respectively. In both cases, a statistical model calculation using GDR parameters for the fission fragments cannot alone reproduce the measured spectra. In order to describe the observed spectral shapes, a pre-fission GDR component centered at around $E_{GDR} = 11.5$ MeV must be included. This value agrees reasonably well with an extrapolation of the ground-state systematics (Figure 4).

The fact that the multiplicity of GDR rays increases, both in the pre-fission and in the post-fission regions, indicates that fission still competes with gamma-ray emission at energies above 100 MeV. From the observed pre-fission GDR multiplicity it can be concluded that the lifetime of the composite system is of the order of 10^{-19} to 10^{-20} sec. In a very recent experiment (140), the same reaction was studied at bombarding energies of 10.5 and 15 MeV/u in an experimental arrangement also permitting the measurement of the angular distribution of the GDR relative to the fission plane. A problem with increasing the bombarding energies may arise because of the limitation of the high-energy GDR emission, discussed in Section 5. However, it is likely

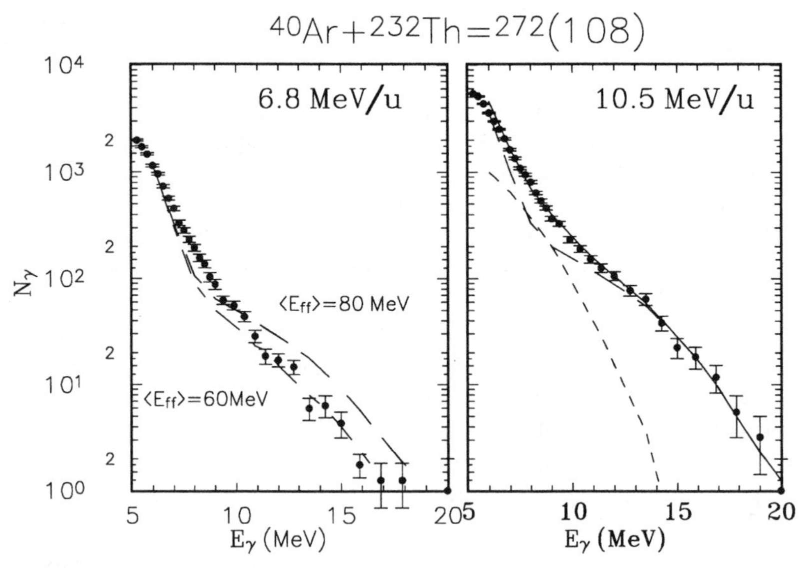

Figure 19 Measured gamma-ray spectra for the reaction $^{40}Ar + ^{232}Th$ at 6.8 and 10.5 MeV/u. (*Left*) Statistical model calculations assuming different excitation energies for the fission fragments. (*Right*) Post-fission component (*long-dashed lines*), pre-fission component (*short-dashed lines*), and total contribution (*solid lines*). From (139).

that the appropriate parameter is the excitation energy per nucleon, which in the present case is only around 1 MeV/u, while it is close to 3 MeV in the $A = 110$ region.

6.3 *Pre-fission Gamma Emission in Cold-Fission-Unstable Nuclei*

We conclude our discussion of pre-fission GDR emission by returning to the high-energy gamma-ray emission from nuclei that undergo spontaneous fission. Because of the large amount of energy available, of the order of 185 MeV for ^{252}Cf, this process has lately prompted various searches for bremsstrahlung emission (141, 142), neutral pions (143, 144), and charged pions and various exotic radioactivities (145). Also, the gamma-ray decay from the GDR decay is being revisited after a long rest. The interest in this reaction lies in the fact that the entire energy available for GDR decay must be produced by dissipative processes in the nucleus on the way to scission, since, at the start, the nucleus only has the zero-point energy. In a recent experiment in Groningen (146), the spectrum and angular distribution relative to the fission plane of the gamma rays with $E_\gamma = 2$–40 MeV were measured employing an experimental technique similar to that used for the in-beam experiments described above. An analysis of the spectrum in terms of a deformed pre-scission GDR component and a small deformation post-scission component with GDR parameters following the systematics reproduces the observed spectrum shape. Although a pronounced anisotropy in the region $E_\gamma = 8\text{–}20$ MeV, consistent with GDR emission from a heavy deformed system prior to fission, was first reported (146), this has since been retracted based on a subsequent experiment. The observed anisotropies appear consistent with emission from the moving fission fragments. It is thus not clear whether GDR emission occurs on the way to scission. Although the time available for GDR decay during the evolution toward scission must be taken into account, the limiting factor for the gamma-ray probability is presumably set by the level density of final states. This would imply that any pre-fission GDR emission must occur rather close to scission. In fact, the window plus wall dissipation model (147, 148), predicts only dissipated energies of the order of 17 MeV and long time scales (10^{-20} sec). More detailed and accurate studies of the gamma-ray emission from spontaneous fission reactions, correlating, for example, the gamma emission with the kinetic energy of the fission fragments may, however, provide new information on nuclear dissipation at low temperature.

7. SUMMARY AND OUTLOOK

In this article we reviewed the current understanding of collective dipole vibrations in hot atomic nuclei. Such nuclei are produced in energetic nuclear collisions using heavy ions. The experimental data, now covering a large span of masses, excitation energies, angular momenta, and reaction types, establish the universality of the GDR in excited nuclei. The systematics of the measured strengths, resonance energies, and widths show that the resonances are stable features, that the gamma-ray decay is statistical, and that the resonances couple strongly to the shape degrees of freedom of excited nuclei. The latter aspect provides a unique possibility for learning about the detailed properties of hot rotating atomic nuclei.

A unifying theme in today's research is the understanding of the time scales relevant for the compound nucleus and its decay: the time of equilibration, the time scale for the gamma-ray and particle decay, the time for the damping of elastic excitations, such as the GDR, and the time for large-scale (plastic) mass and surface rearrangements. The balance among these various times, exhibited in the spectra of the high-energy photons from the GDR, promises to shed new light on the microscopic and macroscopic properties of the nuclear many-body system at finite temperature.

ACKNOWLEDGEMENT

The author gratefully acknowledges the support of the Danish Carlsberg Foundation.

Literature Cited

1. Dietrich, F. S., et al, *Phys. Rev.* C10:795 (1974)
2. Newton, J. O., et al, *Phys. Rev. Lett.* 46:1383 (1981)
3. Snover, K. A., *Annu. Rev. Nucl. Part. Sci.* 36:545–603 (1986) and references therein
4. Bortignon, P. F., Gaardhøje, J. J., Di Toro, M., eds., in *Proc. First Topical Meet.,* Legnaro; *Nucl. Phys.* A482 (1988)
5. Berman, B. L., Fultz, S. C., *Rev. Mod. Phys.* 47:713 (1975)
6. Bertrand, F. E., *Annu. Rev. Nucl. Part. Sci.* 26:457–509 (1976); *Nucl. Phys.* A354:129c (1981)
7. Speth, J., van der Woude, A., *Rep. Prog. Phys.* 44:719–86 (1981)
8. van der Woude, A., *Prog. Part. Nucl. Phys.* 18:217 (1986)
9. Goeke, K., Speth, J. *Annu. Rev. Nucl. Part. Sci.* 32:65 (1982)
10. Bertsch, G., Bortignon, P. F., Broglia, R. A., *Rev. Mod. Phys.* 55:287–314 (1983)
11. Wisshak, K., Käppeler, F., Müller, H., *Nucl. Instrum. Methods* A251: 101 (1986)
12. Gaardhøje, J. J., et al, *Annu. Rep.,* Niels Bohr Inst. (1991)
13. Schwalb, O., et al, *Nucl. Instrum. Methods* A295:191–98 (1990)
14. Minheco, E., The MEDEA array, LNS, Catania, preprint (1991)
15. Sen, S., et al, *Nucl. Instrum. Methods* A264:407 (1988)
16. Hasinoff, M., *Nucl. Instrum. Methods* 117:375 (1974)
17. Bruce, A. M., et al, *Phys. Lett.* 215:237–41 (1988)
18. Gerl, J., Lieder, R. M., eds., *Upgrading to Euroball,* GSI preprint (1991)
19. Howell, R. H., Dietrich, F. S., Petrovich, F. *Phys. Rev.* C21:1158–61 (1980)
20. Ford, R. L., Nelson, W. R., SLAC Rep. 210 (1978) (unpublished)
21. Brun, R., et al, CERN DD/EE/84-1 (1986)
22. Michel, C., et al, *Nucl. Instrum. Methods* A251:119 (1986)
23. Chakrabarty, D. R., *Phys. Rev.* C36:1886–95 (1987)
24. Hingmann, R., Thesis, Univ. Giessen (1987)
25. Brink, D. M., Dr. Phil. Thesis, Oxford Univ. (1955)
26. Gaardhøje, J. J., in *Frontiers of Nuclear Dynamics,* ed. R. A. Broglia, C. H. Dasso. New York: Plenum (1985), pp. 133–70
27. Gaardhøje, J. J., et al, *Phys. Rev. Lett.* 56:1783–86 (1986)
28. Harakeh, M. N., et al, *Phys. Lett.* B176:297–301 (1986)
29. Kicinska-Habior, M., et al, *Phys. Rev.* C36:612–27 (1987)
30. Bohr, Aa., Mottelson, B. R., *Nuclear Structure.* New York/Amsterdam: Benjamin (1969), Vols. 1–2
31. Brink, D. M., see Ref. 4, pp. 3–12
32. Pühlhofer, F., *Nucl. Phys.* A280:267 (1977)
33. Berman, B. L., *Nucl. Data Tables* 15: 321 (1975)
34. Gallardo, M., Luis, F. J., Broglia, R. A., *Phys. Lett.* B191, 222–226 (1987)
35. Gaardhøje, J. J., et al, *Phys. Rev. Lett.* 53:148–51 (1984)
36. Gossett, C. A., et al, *Phys. Rev. Lett.* 54:1486–90 (1985)
37. Gallardo, M., et al, *Nucl. Phys.* A443:415–34 (1985)
38. Cohen, S., Plasil, F., Swiatecki, W., J., *Ann. Phy. (NY)* 82:557 (1974)
39. Alhassid, Y., Levit, S., Zingman, J., *Phys. Rev. Lett.* 57:539 (1986); *Nucl. Phys.* A482:205 (1987)
40. Alhassid, Y., Manoyan, J. M., Levit, S. *Phys. Rev. Lett.* 63:31–34 (1989)
41. Goodman, A. L., *Phys. Rev.* C35:2338–40 (1987)
42. Neergaard, K., *Phys. Lett.* B110:7 (1982)
43. Hilton, R. R., *Z. Phys.* 309:233 (1983)
44. Ring, P., see Ref. 4, pp. 27c–44c, and references therein
45. Szymanski, Z., presented at 14th Masurian Summer School on Nucl. Phys., Mikolajki (1981) (unpublished)
46. Egido, J. L., et al, *Phys. Lett.* B178:139 (1987)
47. Pacheco, J., Yannouleas, C., Broglia, R. A., *Phys. Rev. Lett.* 61:294–97 (1988)
48. Goodman, A. L., *Phys. Rev.* C37:2162–69 (1988)

49. Bertsch, G. F., *Phys. Lett.* B95:157 (1980)
50. Alhassid, Y., Bush, B., Levit, S., *Phys. Rev. Lett.* 61:1926–29 (1988)
51. Ormand, W. E., et al, *Phys. Rev. Lett.* 64:2254–57 (1990)
52. Døssing, T., Neergaard, K., in *Proc. 4th Nordic Meet.*, Fugløs, Denmark (1982)
53. Alhassid, Y., Bush, B., *Phys. Rev. Lett.* 65:2527–30 (1990)
54. Goodman, A. L., *Nucl. Phys.* A528: 348–80 (1991)
55. Goodman, A. L. in *Proc. Symp. on Future Directions in Nucl. Phys.*, Strasbourg (1991) and to be published
56. Lauritzen, B., et al, *Phys. Lett.* B207: 238–42 (1988)
57. Alhassid, Y., Bush, B., *Phys. Rev. Lett.* 63:2452–55 (1989); *Nucl. Phys.* A514:434–60 (1990)
58. Ormand, W. E., et al, to be published
59. Kicinska-Habior, M., et al, *Phys. Rev.* C41:2075 (1990)
60. Harakeh, M. N., et al, *Phys. Lett.* 176B:297 (1986)
61. Kicinska-Habior, M., et al, *Phys. Rev.* C36:612 (1987)
62. Viesti, G., et al, *Phys. Rev.* C40: R1570 (1989)
63. Fornal, B., et al, *Z. Phys.* 340:59–62 (1990)
64. Kicinska-Habior, M., et al, *Phys. Rev.* C45:569–75 (1992)
65. Gundlach, J. H., et al, *Phys. Rev. Lett.* 65:2523–26 (1990)
66. Bracco, A., et al, *Phys. Rev. Lett.* 62: 2080–83 (1989)
67. Bracco, A., et al, *Nucl. Phys.* A519: 47c (1990)
68. Stolk, A., et al, *Nucl. Phys.* A505: 241–56 (1989); *Phys. Lett.* B200:13 (1988)
69. Kasagi, J., et al, *Nucl. Phys.* A538: 585–92 (1992)
70. Kasagi, J., in *Proc. Int. Symp. Towards a Unified Picture of Dynamics,* Nikko, Japan (1991)
71. Stolk, A., et al, *Phys. Rev.* C40: R2454-R2457 (1989)
72. Thirrolf, P., et al, see Ref. 4, pp. 93c–119c
73. Flibotte, S., et al, *Nucl. Phys.* A, to be published (1992)
74. Chakrabarty, D. R., et al, *Phys. Rev.* C37:1437–41 (1988)
75. Gaardhøje, J. J., Bruce, A. M., Herskind, B., see Ref. 4, pp. 121c–39c
76. Nicolis, N. G., et al, in *Variety of Nuclear Shapes,* ed. Kalfas, Ganett, et al. Singapore: World Scientific (1987), pp. 526–43
77. Snover, K. A., see Ref. 4, pp. 13c–26c
78. Gaardhøje, J. J., et al, *Phys. Lett.* 139B:273–75 (1984)
79. Chakrabarty, D. R., et al, *Phys. Rev. Lett.* 58:1092–95 (1987)
80. Atac, A., et al, *Phys. Lett.* 252B:545–49 (1990)
81. Carlos, P., et al, *Nucl. Phys.* A219: 61 (1974)
82. Bortignon, P. F., et al, *Nucl. Phys.* A495:155c (1989)
83. Alhassid, Y., Bush, B., *Phys. Rev. Lett.* 65:2527 (1990)
84. Camera, F., et al, *Phys. Lett.* B (1992). In press
85. Gaardhøje, J. J., et al, *Nucl. Phys.* A538:573–84 (1992)
86. Maj, A., et al, *Phys. Lett.* B (1992). In press
87. Suraud, E., Gregoire., C., Tamain, B., *Prog. Nucl. Part. Sci.* 23:357 (1989)
88. Guerreau, D., *Nuclear Matter and Heavy Ion Collisions*. New York: Plenum (1989), pp. 187–230
89. Nifenecker, H., Pinston, J. A., *Prog. Part. Nucl. Phys.* 23:271 (1989)
90. Gaardhøje, J. J., et al, *Phys. Rev. Lett.* 59:1409–12 (1987)
91. Bertholet, R., et al. *Nucl. Phys.* A280:267 (1987)
92. Nebbia, G., et al, *Phys. Lett.* B176 20 (1986)
93. Natowitz, J. B., see Ref. 4, pp. 171–86
94. Yoshida, K., et al, *Phys. Lett.* 245B: 7 (1990)
95. Herrmann, N., et al, *Phys. Rev. Lett.* 60:1630 (1988)
96. Hingman, R., et al, *Phys. Rev. Lett.* 58:759 (1987)
97. Thoennessen, M., et al, *Phys. Rev.* C43:R12–R43 (1991)
98. Bortignon, P. F., et al, *Nucl. Phys.* A460:149 (1986)
99. Dinh Dang, N., *Nucl. Phys.* A504:171 (1989)
100. Smerzi, A., Bonasera, A., Di Toro, M., *Phys. Rev.* C44:1713–16 (1991)
101. Okolowicz, J., et al, *Nucl. Phys.* A501:289 (1989)
102. Sugawara-Tanabe, K., Tanabe, K., see Ref. 69; *Phys. Lett.* B192:268 (1987), *Prog. Theor. Phys.* 76:356 (1986)
103. Bortignon, P. F., et al, *Phys. Rev. Lett.* 67:3360 (1991)
104. Morsch, H. P., et al, *Phys. Rev. Lett.* 64:1999 (1990)
105. Thoennessen, M., *Phys. Rev. Lett.* 66:1640 (comment) (1991)
106. Morsch, H. P., et al, *Phys. Rev. Lett.*

66:1641 (reply) (1991)
107. Bohr, N., Wheeler, J. A., *Phys. Rev.* 56:426 (1939)
108. Hinde, D. J., et al, *Phys. Rev. Lett.* 52:986 (1984)
109. Zank, W. P., et al, *Phys. Rev.* C33: 519 (1986)
110. Hinde, D. J., et al, *Nucl. Phys.* A452: 318 (1987)
111. Gavron, A., et al, *Phys. Rev.* C35:579 (1987)
112. Hinde, D. J., et al, *Nucl. Phys.* A472: 318 (1987)
113. Newton, J. O., et al, *Nucl. Phys.* A483:126 (1988)
114. Hinde, D. J., et al, *Phys. Rev.* C37: 2923 (1988)
115. Hilscher, D., et al, *Phys. Rev. Lett.* 62:1099 (1989)
116. Hinde, D. J., Hilscher, D., Rossner, H., *Nucl. Phys.* A502:497c (1989)
117. Hinde, D. J., et al, *Phys. Rev.* C39: 2268 (1989)
118. Vaz, L. C., et al, *Z. Phys.* A315:169 (1984)
119. Ajitanand, N. N., et al, *Z. Phys.* A316:169 (1984)
120. Scahad, L., et al, *Z. Phys.* A318:179 (1984)
121. Lestone, J. P., et al, *Phys. Rev. Lett.* 67:1078–81 (1991)
122. Kramers, H. A., *Physica* 7:284 (1940)
123. Grange, P., Weidenmüller, H. A., *Phys. Lett.* 96B:26 (1980)
124. Grange, P., et al, *Phys. Rev.* C34:209 (1986)
125. Bhatt, K. H., Grange, P., Hiller, B., *Phys. Rev.* C33:954–68 (1986)
126. Delagrange, H., et al, *Z. Phys.* A323: 437 (1986)
127. Thoennessen, M., et al, *Phys. Rev. Lett.* 59:2860–63 (1987)
128. Viola, V. E., Kwiatkowski, K., Walker, M., *Phys. Rev.* C31:1550 (1985)
129. Keller, J. G., et al, *Phys. Rev.* C36: 1364 (1987)
130. Schmitt, R. P., Mouchaty, G., Haenni, D. R., *Nucl. Phys.* A427:614 (1984)
131. Vandenbosch, R., Huizenga, J. R., *Nuclear Fission*. New York/London: Academic (1973), p. 228
132. Butsch, R., et al, *Phys. Rev.* C44: 15151 (1991)
133. Eckert, E. M., et al, *Phys. Rev. Lett.* 64:2483–86 (1990)
134. Kumar, K., *Superheavy Elements,* Bristol/New York: Adam Hilger (1989)
135. Back, B. B., et al, *Phys. Rev.* C32: 195–213 (1985)
136. Toke, J., et al, *Nucl. Phys.* A440: 327–65 (1985)
137. Swiatecki, W. J., *Phys. Scr.* 24:113 (1981)
138. Shen, W. Q., et al, *Phys. Rev.* C36: 115 (1987)
139. Gaardhøje, J. J., Maj, A., *Nucl. Phys.* A520:575c–94c (1990)
140. Ramsøy, T., et al, to be published
141. Kasagi, J., et al, in *Proc. Fifth Int. Conf. on Clustering Aspects in Nuclear and Subnuclear Systems,* Kyoto (1988); *J. Phys. Soc. Jpn. Suppl.* 58:620–25 (1989)
142. Luke, S. J., Gossett, C. A., Vandenbosch, R., *Phys. Rev.* C44:1548–54 (1991)
143. Beene, J. R., Bemis, C. E., Halbert, M. L., *Phys. Rev.* C38:569 (1988)
144. Cerruti, C., et al, *Z. Phys.* A329:283 (1988)
145. Ion, D. B., Ivascu, M., Ion-Mihai, R., *Ann. Phys. (NY)* 171:237 (1986)
146. van der Ploeg, H., et al, *Phys. Rev. Lett.* 68:3145–47 (1992)
147. Swiatecki, W. J., *Prog. Part. Nucl. Phys.* 4:383 (1980)
148. Nix, J. R., in *Proc. Conf. on 50 Years with Nuclear Fission,* Guithersburg, Maryland, preprint LA-UR-89-1500 (1989)

Annu. Rev. Nucl. Part. Sci. 1992. 42:537–597

RELATIVISTIC HEAVY ION PHYSICS AT CERN AND BNL[1]

Johanna Stachel

Physics Department, State University of New York, Stony Brook, New York 11794

Glenn R. Young

Physics Division, Oak Ridge National Laboratory, Oak Ridge, Tennessee 37831

KEY WORDS: QCD, quark-gluon plasma, deconfined matter, restoration of chiral symmetry, relativistic heavy ion collisions, AGS, SPS, RHIC

CONTENTS

1. INTRODUCTION

Collisions between relativistic nuclei are studied in order to explore nuclear matter in the entirely uncharted regions of high energy and baryon density. One is interested in the space-time development of the hadronic and partonic interactions in nuclear volumes under such conditions. Quantum chromodynamics (QCD) predicts the existence of a new phase of strongly interacting systems at high energy density. Whereas in the normal hadronic environment, quarks and gluons are confined inside of bags (hadrons) sustained by the pressure of the physical vacuum surrounding them, the new phase would consist of extended regions of the perturbative vacuum (previously found inside hadrons only) with unconfined quarks and gluons. Since the constituents carry color charge, we can consider this perturbative vacuum state a plasma, the quark-gluon plasma (QGP). We furthermore expect from the QCD property of asymptotic freedom that chiral symmetry, which is spontaneously broken in hadronic matter, will be restored in the deconfined phase at high energy density.

QCD has supplied the theoretical basis for this two-phase picture. The equation of state of the QGP can be evaluated perturbatively in the high density limit (1). However, the entire density range of strong interaction thermodynamics is accessible only via nonperturbative calculations, as provided for example by the lattice formulation of QCD (see 2 for an earlier review in this series). Lattice QCD simulations have indeed confirmed the expectations for a deconfinement phase transition and restoration of chiral symmetry (2, and updates in 3–7). The exact conditions and the nature of the phase transition depend somewhat on the number of flavors included as well as on the quark masses. This question and the current best estimates are discussed in Section 5. We are looking for energy densities at least an order of magnitude larger than the energy density of a nucleus in its ground state (0.15 GeV/fm^3).

Collisions between relativistic heavy ions offer the best chance to create and observe the QGP in the laboratory. For a given accelerator energy and thus center-of-momentum (cm) energy, nuclei that are as heavy as possible produce the highest energy density. To enhance the possibility of observing the QGP, we want high energy density over a sufficiently large volume, with dimensions larger than the scattering length of the constituents, so that equilibrium thermodynamics may be applicable. Once created, the system must exist long enough to equilibrate so that a phase in the thermodynamic sense can be established. These considerations have led to the present experimental programs at the Brookhaven Alternating Gradient Synchrotron (AGS) and the

CERN Super Proton Synchrotron (SPS). At each of these facilities one expects to create a QGP in a baryon-rich environment. The quest for a QGP in a region with no (or a very low) net baryon density, but significantly higher energy density, led to the decision to build the Relativistic Heavy Ion Collider (RHIC), currently under construction at Brookhaven.

An initial concern was whether the collision between two relativistic nuclei would actually transfer enough energy into an equilibrated system. This question arose from the knowledge that as bombarding energies in hadron-hadron collisions increase the absorption of energy into a nuclear target becomes less complete. Such increasing nuclear "transparency" might severely limit the usefulness of heavier projectiles and targets. Therefore the primary objective of the first round of experiments is to show how efficiently high energy nuclei slow each other and deposit energy in the midrapidity region. Experimental evidence for this comes from distributions of leading baryons, discussed in Section 4.1, and from the global energy and particle distributions discussed in Section 3.

The second objective is to find indications for (or against) thermalization of the deposited energy. This question is addressed by the global observables as well as the spectral distributions of constituent nucleons and created particles and their relative abundances, discussed in Sections 4.1 and 4.2. Conclusions on the nuclear stopping power and the energy densities achieved in current experiments are presented in Section 5.

The third objective is the search for special signals indicating the presence of a QGP. We do not list here all the signals proposed for experimental investigation. Instead, the reader is referred to the article by Kajantie & McLerran (8) in this series and the proceedings of a number of "Quark Matter" conferences, which give an extensive overview (3–7). We limit our discussion to quantities that have already been studied in current experiments, and we examine to what extent the existing data indicate a need for new physics. Section 4.3 addresses strangeness production, Section 4.4 discusses data on high mass vector mesons, Section 4.5 thermal radiation, and Section 4.6 turns to the question of correlations. Finally, Section 6 summarizes the status and addresses the future of the field.

2. EXPERIMENTAL FACILITIES

Plans to inject heavy ions into the Brookhaven AGS or CERN PS or ISR existed as early as 1974 (9). They took concrete form in the early 1980s after QCD provided a firm basis for the current experimental programs.

At present there are two experimental facilities. The Super Proton Synchrotron (SPS) at CERN accelerates beams of ^{16}O and ^{32}S to momenta per nucleon of $400(Z/A)$ GeV (here and in the following we use $c = 1$). The Brookhaven Alternating Gradient Synchrotron (AGS) has accelerated ^{16}O and ^{28}Si ions to momenta per nucleon of $29(Z/A)$ GeV. Brookhaven has just brought into operation a 1-GeV booster synchrotron that injects into the AGS, making beams as heavy as ^{197}Au available. At CERN an upgrade has started with construction of a Pb injector, and Pb beams are expected for the experimental program in 1994.

The experimental relativistic heavy ion program started in the fall of 1986 both at CERN and at Brookhaven. Results are available from several large-scale electronic detector experiments: NA34 (HELIOS), NA35, NA36, NA38, WA80, and WA85 at CERN, and E802, E810, and E814 at Brookhaven. In addition, over two dozen emulsion experiments were performed at both accelerators. Several new experiments are under construction or were just completed at both facilities. An overview of the experiments can be found in the summary of the Berkeley Particle Data Group and the CERN compilation of current experiments (10). A detailed overview of the detectors and experimental techniques used in relativistic heavy ion research is given by Braun-Munzinger et al (11). The reader is also referred to the proceedings of Quark Matter Conferences (3–7), where the individual experiments are described in detail. In the following, we concentrate principally on the experimental results now available.

3. GLOBAL OBSERVABLES

In relativistic heavy ion physics the basic global observables are the number of particles emitted after a collision and the energy carried by them. The term global is used to distinguish these observables from the spectroscopy of identified particles. The spatial distribution of global observables is typically measured over a large solid angle but without identifying individual particles or the energy any one of them carries. These observables are indicators of the global reaction dynamics and kinematics. They are very important tools for event characterization and event selection according, for example, to centrality. Beyond that, the study of the systematic dependence of the observables on beam energy, target and projectile mass, and centrality of the collision provides our basic understanding of the reaction mechanism. Comparison to model predictions constitutes one possible method to obtain information on the nuclear stopping power and on the energy density achieved in the early stages of the collision. On the other hand, some of these observables are also sensitive to rescattering effects and therefore probe the time scale later in the collision, after secondaries have been produced.

3.1 *Simple Geometric Model*

To categorize particles emitted from a heavy ion collision at high energies, one frequently employs a very simple geometric model. As the two nuclei collide, nucleons in the overlap region, known as the "participants," collide with each other and thereby slow down and create new particles, while the projectile and target remnants outside the overlap volume, known as "spectators," continue to move at nearly projectile and target rapidity. The number of participants and spectators is determined by the collision geometry, i.e. the impact parameter. A good approximation is obtained by treating the colliding nuclei as intersecting spheres with a sharp radius $R = 1.2A^{1/3}$ fm. For central collisions of a small projectile (A_p) with a large target nucleus (A_t), there is an even simpler approximation for the number of target participants: $A_f = 1.5A_t^{1/3}A_p^{2/3}$, which leaves $A_{ts} = A_p - A_f$ target spectator nucleons and zero projectile spectators A_{ps}.

The overlap region of A_p and A_f projectile and target participants has a total energy in the laboratory frame of $E_{part} = (\gamma_p A_p + A_f)m_n$, where m_n is the nucleon mass. Assuming a totally inelastic collision between the participants, i.e. the limit of complete stopping of the participants in the center-of-momentum (cm) frame, we find that the participants move with a velocity of $\beta_{cm} = \gamma_p A_p/(\gamma_p A_p + A_f)$. This system has, in the cm frame, an energy of $E_{cm} = m_n(A_p^2 + A_f^2 + 2\gamma_p A_p A_f)^{1/2}$ and an excitation energy $E^* = E_{cm} - m_n(A_p + A_f)$. For large γ_p, the excitation energy scales as $E^* \propto A_p^{1/2}A_f^{1/2}$; for the small projectile–large target combination mentioned above therefore $E^* \propto A_p^{5/6}A_t^{1/6}$. Table 1 lists a few typical values for excitation energy and rapidity of the overlap region in central collisions at the

Table 1 Excitation energy (in GeV) and rapidity of the target-projectile overlap region for central collisions and assuming full stopping for three different projectile energies per nucleon

			14.6 GeV		60 GeV		200 GeV	
System	A_p	A_f	E^*	y_{cm}	E^*	y_{cm}	E^*	y_{cm}
O + Al	16	24	69	1.5	172	2.2	343	2.8
O + Au	16	55	98	1.2	253	1.8	510	2.4
S + S	32	32	113	1.7	281	2.4	561	3.0
S + Au	32	88	179	1.3	458	1.9	920	2.5
Au + Au	197	197	695	1.7	1736	2.4	3455	3.0

three beam energies available so far. The numbers are based on intersecting spheres. The corresponding three cm energies per colliding nucleon-nucleon pair are $s^{1/2} = 5.4$, 10.7, and 19.4 GeV, increasing roughly by factors of two from one to the next. It should be noted that for each target-projectile combination E^* increases in proportion to $s^{1/2}$. Therefore any scaling observed in $s^{1/2}$ will also hold in E^*.

Obviously, this model only contains the basic collision geometry and the related kinematics. Even if the participants were to stop each other completely, the treatment of the spectators is very rough in that we have not accounted for the surface diffuseness nor any kind of energy transfer to the spectators. We use it in the following only as a baseline for comparison, especially if scaling with energy or masses is considered. We also see that the basic shapes of the spectra of global observables are dominated by geometry. In that sense this simple model gives some insight that might be hidden in a more complex model based, for example, on the individual collisions.

Using this simple geometric model one immediately obtains the excitation energy E^* as function of the impact parameter b and, weighting with $b\,db$, the cross section as function of the excitation energy. Both quantities are plotted on the left side of Figure 1 for S + Au collisions and the three beam energies. It can be seen that for $b < 3$ fm the projectile dives completely into the target. From that point on, the excitation energy grows only very slowly. Also shown in Figure 1 is the cross section as a function of the energy carried by projectile spectators, E_{zd}, normalized to the projectile energy E. This quantity is complementary to E^*.

3.2. *Global Variables*

One frequently used global observable is the energy emitted into a narrow cone about the beam axis, i.e. at "zero degrees," E_{zd}. The cone angle is chosen such that projectile spectators or fragments with momenta in the projectile rest frame of the order of the Fermi momentum are accepted. At AGS energies a 20-mr opening is sufficient, while at SPS energies a factor of four to ten less is required. Because of the small solid angle, contributions from particles from the interaction zone are small. This observable measures to first order the number of projectile spectator nucleons: $E_{zd} \approx A_{ps}E_{lab}/A_p$. It therefore is most relevant for event characterization for events with incomplete overlap of target and projectile.

The complementary global observables are particle and energy flow from the interaction zone. They are measured as a function of the polar

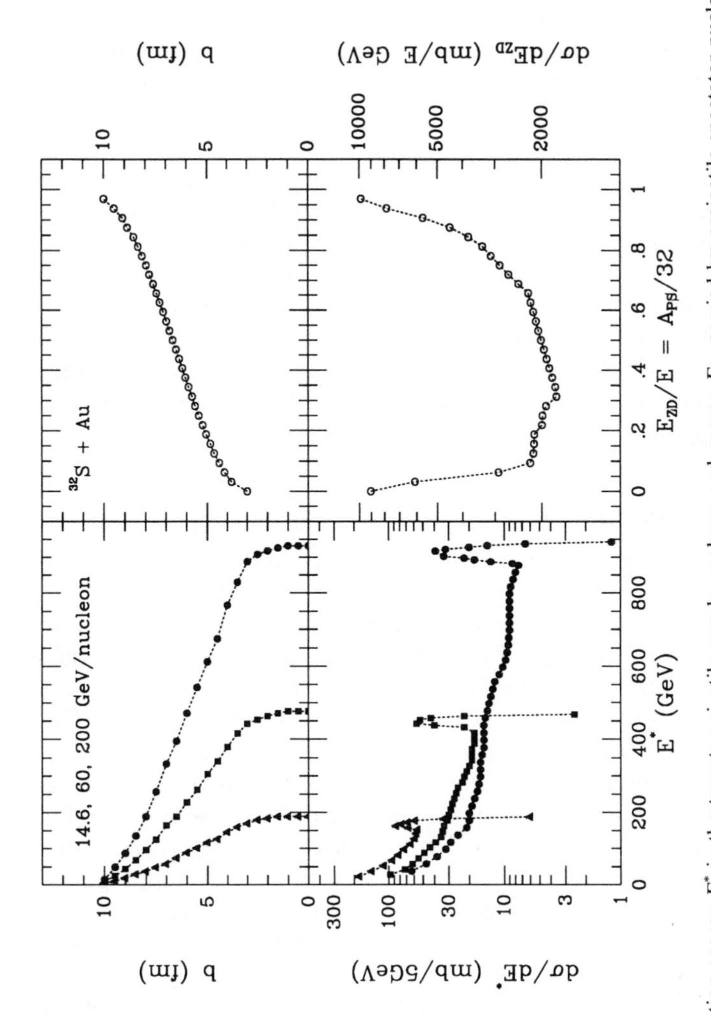

Figure 1 Excitation energy E^* in the target-projectile overlap volume and energy E_{zd} carried by projectile spectator nucleons, normalized to the incident beam energy E in the geometric model. E^* is shown for S + Au collisions at three different beam energies. E_{zd}/E is determined by geometry only, and the values shown here are for sharp spheres with equivalent radius $R = (1.2)A^{1/3}$ fm. (*Top*) Impact parameter dependence. (*Bottom*) Differential cross section, integrating to the geometric cross section.

angle in the laboratory, θ, defined relative to the beam direction, or the pseudorapidity $\eta = -\ln[\tan(\theta/2)]$. For experimental reasons, the measurement of the number of particles is usually restricted to the number of charged particles, and the corresponding global observable is the charged-particle pseudorapidity density $dN_c/d\eta$. The degradation and possible thermalization of initial longitudinal energy of the participating projectile and target nucleons, considered in the participant cm frame, leads to the production of transverse energy E_t. To evaluate this quantity the energy e_i emitted into a given solid angle Ω_i is weighted with the sine of the polar angle before summing over the whole detector: $E_t = \sum[e_i \sin(\theta_i)]$. Since particles are not identified, masses are not included in this energy with two exceptions: neutral mesons decaying into photons, and the decay momenta of charged particles decaying within the time window (typically a few hundred nanoseconds) when the detector is active. Transverse energy and charged-particle multiplicity provide the best event characterization in rather central collisions. In order not to let fluctuations dominate the measured observable, it is essential that a substantial fraction of the solid angle (ideally all of it) be covered.

3.3 *Experimental Results*

3.3.1 TRANSVERSE ENERGY AND CHARGED-PARTICLE MULTIPLICITY SPECTRA Figure 2 shows an example of transverse energy and charged-particle multiplicity distributions and their correlation for Si + Pb collisions from E814 at the AGS (12, 13). The E_t distribution covers angles backwards of midrapidity ($\eta < 0.8$), while N_c is measured for $0.9 < \eta < 4.0$. Grazing collisions produce small E_t and N_c. They have the largest probability because of the geometrical factor $b\, db$, the cause of the rise of $d\sigma/dE^*$ at low E^* in Figure 1. The rather flat part of the distribution corresponds to collisions of increasing centrality. The edge or shoulder of the distribution corresponds to "typical" central (near $b = 0$) collisions. As can be seen from inspecting Figure 1, a shoulder is produced if a decrease of b does not lead to an increase in excitation energy and therefore cross sections for a range of $b =$ 0–3 fm pile up at a fixed E_t and N_c. This shoulder will always be present for very asymmetric systems when measurements are made over the full solid angle. However, a finite experimental acceptance will smear the shoulder because the participant cm frame shifts event by event relative to the laboratory frame. How to select operationally a well-defined degree of centrality using these distributions is discussed below in connection with the mass and energy dependence of global variables.

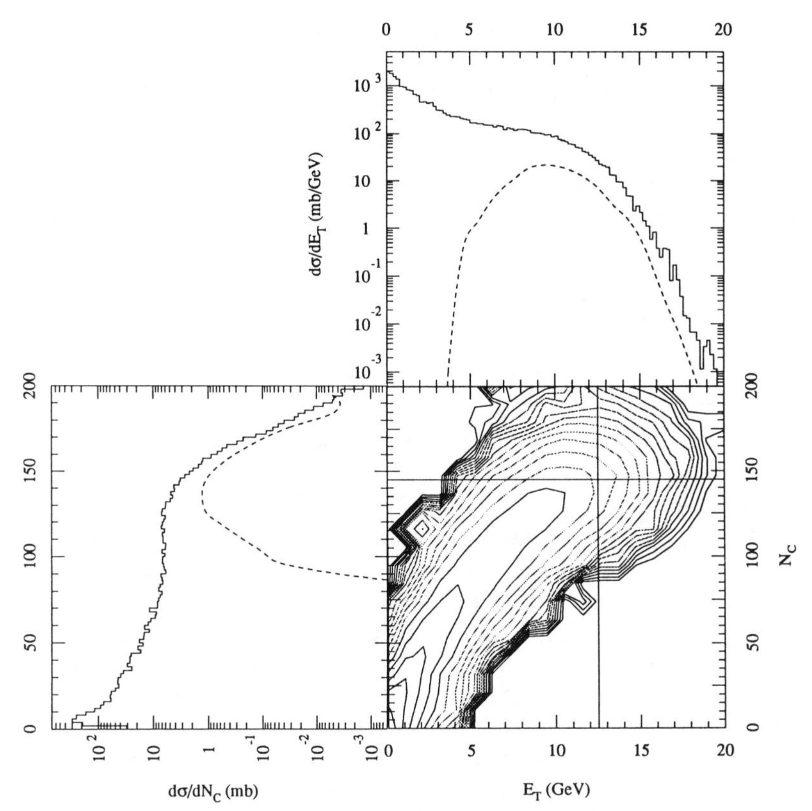

Figure 2 Contour plot of the double differential cross section $d^2\sigma/dE_t dN_c$ for minimum-bias Si + Pb collisions from E814 (12, 13) at 14.6 GeV/nucleon (see text for the acceptance). Three contour lines correspond to a drop of one order of magnitude in cross section. Left and Top: Projections on the N_c and E_t axes for all values of the other variable (*solid lines*) as well as for cuts at E_t = 12.5 GeV or N_c = 145 (*dashed lines*) corresponding to the most central 30 mb of the cross section.

The key difference in shape between the spectra shown in Figure 2 and the schematic behavior indicated in Figure 1 is the much smoother distribution of the data because of fluctuations. E_t and N_c fluctuate even for a fixed collision geometry because of the way the many emitted particles sample the phase space available. There are also fluctuations due to the limited experimental coverage that are of basically the same origin. Finally, there may be fluctuations in the initial energy density achieved for a given geometry, as indicated by fluctuations in the number of participants for a given overlap, or interesting nonstatistical effects in the multiparticle emission. It has been shown that even the simple geometric model described in Section 3.1 can account for the shape of the experimentally observed distributions by including only the first two of the effects mentioned (14). To illustrate the extent of that type of fluctuations we also show in Figure 2 the measured E_t and N_c distributions for fixed values of N_c (145) and E_t (12.5 GeV), respectively. These cuts in one variable lead to nearly Gaussian distributions in the other variable, with widths of $\sigma(E_t) = 1.8$ GeV and $\sigma(N_c) = 15$. Since the two variables are measured with a complementary acceptance, the fluctuations in either variable with respect to the other one are due to fluctuations in E_t per charged particle as well as to fluctuations in the spatial distribution of the emitted particles. If the target nucleus is not spherical but rather is deformed, the relative orientation between target and projectile is an additional source of fluctuations, as shown by the HELIOS collaboration for S+U (15).

Recently the ratio of electromagnetic to hadronic transverse energy was measured by NA35 (16) and WA80 (17). The ratio depends on centrality only weakly and not visibly on target or projectile mass. At 200 GeV/nucleon, the two experiments report values of 0.40 and 0.35 for central collisions, the difference being within the quoted systematic uncertainties. For a pion gas of mean isospin zero, a value of 0.50 would be obtained. The event generator Fritiof (18), which is based on the Lund string model, gives a value of 0.45. A smaller value in the data could indicate a larger abundance of heavy hadrons. Fluctuations of this ratio are found to be very small (5–10% standard deviation) as compared again to Fritiof (15%). This difference, however, is due partly to the way electromagnetic and hadronic transverse energy are disentangled experimentally, a process that involves averaging as well as considering the response of calorimeters to various particles (16, 17).

The transverse energy per charged particle E_t/N_c has been measured at all three beam energies. The results for central collisions are summarized in Table 2. The measured value at 14.6 GeV is only two thirds of the value at the higher beam energy. This can be attributed to the larger ratio of nucleons to pions at AGS energies. There the numbers of positive pions and protons are similar, while at 200 GeV the relative abundance is 6:1. For p_t spectra corresponding to the same temperature, e.g. 0.15 GeV, a pion contributes on average 0.38 GeV to E_t, while the number for a nucleon is 0.12 GeV. At midrapidity the value of E_t/N_c does not change significantly with increasing centrality. The results for 60 and 200 GeV are very close. A very interesting effect has been found by the HELIOS Emulsion experiment in that E_t/N_c rises markedly towards larger angles (see Table 2). This rise depends on centrality and the colliding system. It is currently not explained.

Table 2 Transverse energy (in GeV) per charged particle for different projectiles and targets and three different beam energies

System	E_{lab}/A	E_t/N_c	η range	Experiment	Ref.
Si + Pb	14.6	0.45	1.0	E814	19
O + W	60	0.65	−0.1–2.9	HELIOS	20
O + Au	60	0.53	2.0–4.2	WA80	21
O + Ag	200	0.65	2.0–3.0	HELIOS Emulsion	22
O + Ag	200	0.92	1.0	HELIOS Emulsion	22
O + W	200	0.68	−0.1–2.9	HELIOS	20
O + Au	200	0.57	2.0–4.2	WA80	21
S + Ag	200	0.68	2.0–3.0	HELIOS Emulsion	22
S + Ag	200	0.94	1.0	HELIOS Emulsion	22
S + W	200	0.68	2.0–3.0	HELIOS Emulsion	22
S + W	200	0.75	1.0	HELIOS Emulsion	22

Using event generators, one can convert E_t/N_c into the mean transverse momentum $\langle p_t \rangle$. Using the Lund model one finds the conversion factor at 200 GeV to be 0.59 (22), which leads to values of $\langle p_t \rangle$ = 0.3–0.4 GeV at midrapidity.

Charged-particle and transverse energy distributions have been measured at all three beam energies for various systems and covering large fractions of 4π (for a review see 23). Systematic trends as functions of energy and masses are discussed below. From comparison to models, information can be extracted on the extent of stopping in the nuclear collision. One model to which data have frequently been compared for this purpose is the isotropic fireball model. It is based on the geometric model discussed in Section 3.1 and assumes isotropic emission of particles. The transverse energy corresponding to this fireball is close to $E_t(\text{iso}) = (\pi/4)E^*$. The amount of stopping one obtains for the three beam energies and the heaviest target is $S \propto 100$, 75, and 50% for 14.6, 60, and 200 GeV. Sometimes $E_t(\text{iso})$ is interpreted as the maximal transverse energy attainable with full stopping. In fact the kinematic limit given by energy and momentum conservation is higher (24). There are, however, indications that the event shape, in particular at higher energy, is not isotropic but rather elongated along the beam axis. Taking this into account and using the Landau hydrodynamic model of Stachel & Braun-Munzinger (14), one obtains the stopping probabilities for the three beam energies as $S \propto 100$, 90, and 80. Other approaches to extracting the amount of stopping achieved in the collisions considered are discussed below.

At all three beam energies there has been an indication of rescattering effects involving created particles and (mostly spectator) nucleons (25,

26). The effect was most pronounced for the heaviest target nuclei, where A_{ts} is largest, and at angles backwards of midrapidity. It is nearly negligible for symmetric systems.

3.3.2 ZERO DEGREE ENERGY Typical forward energy spectra from WA80 are shown in Figure 3 for various systems at 60 and 200 GeV/nucleon (17). The basic shape of each spectrum is again determined by geometry. The cross section is peaked at the beam energy and falls off towards lower values of E_{zd} because of the geometry and phase space, as can be seen in Figure 1 for S + Au. The decrease in the cross section at the highest energies is determined by the specific minimum-bias interaction trigger used in the experiment. For symmetric systems the probability for full overlap is zero, and consequently the cross section falls monotonically with decreasing energy for light targets. For heavy targets the peak at low E_{zd} is due to collisions with full overlap. Its location gives another estimate for the amount of stopping achieved in the collision. For full stopping the peak should be at zero (see Figure 1), while in actual data it is seen at small but finite values. The data of NA35 and WA80 (16, 17) agree well and one finds the peak at 3, 8,

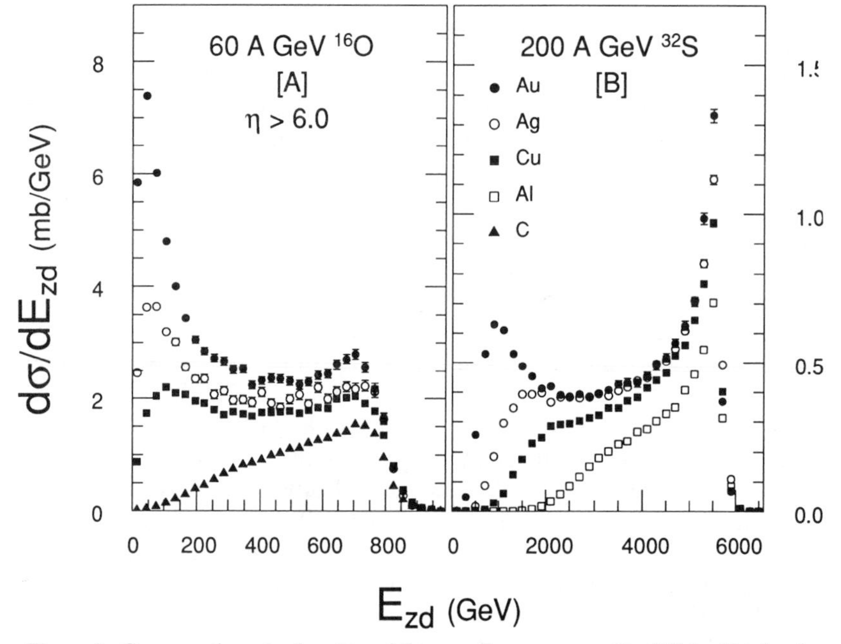

Figure 3 Cross section as a function of the zero degree energy E_{zd} (17) for O + A collisions at 60 GeV/nucleon and S + A collisions at 200 GeV/nucleon measured by WA80.

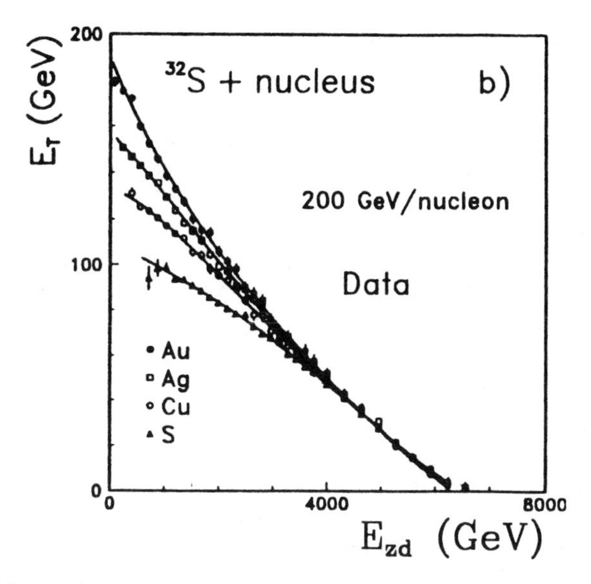

Figure 4 Correlation between zero degree energy and transverse energy E_t for S+A collisions at 200 GeV/nucleon from NA35 (16). Curves correspond (from the top down) to target nuclei Au, Ag, Cu, and S. (Figure from 16)

and 11% of the beam energy for 60- and 200-GeV O and for 200-GeV S beams on Au targets, respectively. These correspond to fractions of projectile energy of 97, 92, and 89%. These fractions suggest values for the stopping power, but they can only provide an estimate for two reasons: (*a*) E_{zd} receives a contribution from participants, although the contribution is suppressed because of the small solid angle. (*b*) E_{zd} only counts energy removed from the beam direction, irrespective of where and how it reappears. One needs to combine the estimates based on E_t and E_{zd} to obtain a consistent picture.

As one would expect from the fact that E_t and E_{zd} spectra are dominated by geometry, there is a close correlation between the two quantities. An example from NA35 is shown in Figure 4. For S+S the correlation is nearly linear. One finds that 7.8 GeV of E_t is produced (into the full solid angle) per interacting projectile nucleon, which corresponds to 45% of the energy available in the cm frame. For heavier targets and $E_{zd} > 0.5\ E_{beam}$ the data follow the same curve, while for more central collisions the E_t production increases more rapidly with decreasing E_{zd}. For 60-GeV/nucleon O projectiles and $E_{zd} > 0.5\ E_{beam}$ the same fraction of the available energy, 45%, appears as E_t per interacting projectile nucleon (16). At 14.6 GeV/nucleon for Si projectiles and otherwise similar conditions, this number is estimated to be 56% from the E814 data (13, 19). The fluctuations in the correlation between E_t and E_{zd} are relatively small. NA35 (16) finds for central O+Au collisions at 200 GeV a standard deviation of about 10%, dominated presumably by the fluctuations in E_t (see above).

3.3.3 DISTRIBUTIONS IN PSEUDORAPIDITY Distributions $dN_c/d\eta$ and $dE_t/d\eta$ have been measured for various systems at all three beam energies and yield good systematic information (for review see 23). The shapes of both types of distributions are found to be close to Gaussian and therefore can be described in terms of three parameters and their dependence on centrality and on target and projectile masses. A typical example of the dependence on centrality (27) from HELIOS is shown in Figure 5.

The centroid of the distribution is close to half the beam rapidity for symmetric systems. For asymmetric systems it generally shifts to lower values with increasing centrality, as expected from the overlap geometry (see Table 1). For all systems studied the centroids are close to the value expected from the participant kinematics (Section 3.1). Nevertheless, one should keep in mind that the distributions are for a mixture of particles with different and nonzero rest masses. If the true rapidity distribution is peaked at y_{cm}, then the pseudorapidity distribution will be peaked more forwards and will be broader even if the rapidity distribution dN/dy is identical for different particles, simply because of the different relation between energy and momentum for different particle species.

The width is generally found to decrease with increasing centrality for both N_c and E_t distributions and for all three beam energies (13, 17, 22, 27) as can also be seen in Figure 5. This decrease indicates that particles and E_t are concentrated more at midrapidity for increasing centrality. The decrease in width is smaller in emulsion experiments (28–30), which are mostly sensitive to pions. This leads to the conclusion that the narrowing is due to the nucleon distribution that shifts toward midrapidity with increasing centrality. However, no clear A dependence of the width has been observed for a given centrality. The width of the distributions does increase significantly as a function of beam energy. A growth as $\sigma_\eta = 0.25 \ln s$ has been established for the N_c distribution (23). This is in clear contradiction to expectations for a system emitting particles isotropically. The distribution for that case is of the type $1/\cosh^2(y_{cm})$ and can be approximated by a Gaussian with $\sigma = 0.88$, independent of the beam energy (31). Measured distributions are found to be wider by 10% already at 14.6 GeV/nucleon; this gap

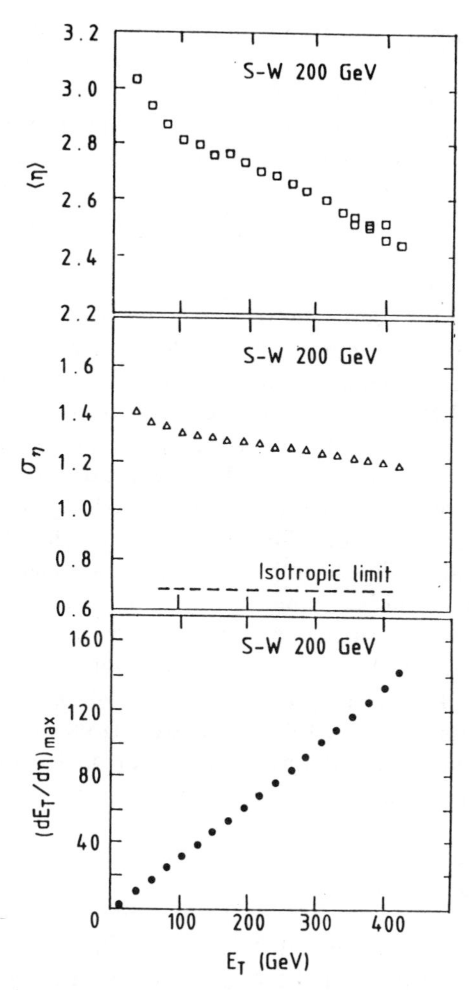

Figure 5 Characteristics of $dE_t/d\eta$ as a function of centrality from HELIOS for S + W collisions at 200 GeV/nucleon (27). (*Top*) Centroid of the distribution. (*Middle*) Width of the distribution given by the standard deviation of a Gaussian fit. (*Bottom*) Peak value of the distribution (in GeV). (Figure from 27)

grows to about 70% at 200 GeV/nucleon. These results clearly rule out an isotropic fireball as the description of the experimental data.

The width of $dE_t/d\eta$ is more difficult to determine experimentally because one has to deal with the spread of electromagnetic and hadronic showers in calorimeters and with the finite granularity of the calorimeters. There is some spread in the experimental data even at comparable centrality (17, 27). Still, the general trend emerges that the

E_t distribution is narrower in η than is the N_c distribution. This is due in part to the fact that the mean transverse momentum is observed to decrease away from midrapidity.

The value of the distributions at the maximum as a function of centrality does not contain much additional information because the measure of centrality is either the same or a strongly correlated variable. It must grow somewhat more steeply than linearly because of the decreasing width with increasing centrality.

3.3.4 SCALING OF E_t AND N_c WITH NUCLEAR MASSES AND ENERGY We have seen that the distributions in pseudorapidity are Gaussians, with width and centroids changing as a function of beam energy, target mass, and centrality. Effects due to these changes must be excluded if one wishes to make a statement about the systematic behavior of E_t and N_c. In particular, it can be very misleading to extract an A dependence by comparing values measured at a fixed angle (η) or in a fixed experimental acceptance (see 17 for an example of the variation).

The A dependence has been parametrized in terms of A^α. To extract values for α one can either compare the maximum of the pseudorapidity densities or the integral of the Gaussians (extrapolating outside the coverage of a specific experiment). An additional difficulty lies in defining similar centrality as a function of A. One method (I) is to use the same fraction X of the total geometric cross section for a target-projectile combination to define the cut in a global variable, E_t(cut) or N_c(cut), corresponding to similar centrality:

$$X = \int_{E_t(\text{cut})}^{\infty} (\mathrm{d}\sigma/\mathrm{d}E_t)\mathrm{d}E_t/\sigma_{\text{geo}}. \qquad 1.$$

Another method (II) aims at finding the same impact parameter. To find the appropriate cut in a global variable, one could integrate from the high (most central) end of the spectrum of this variable down until a certain cross section is reached, which one would associate with πb^2. This method is most reliable if E_{zd} is considered and it is limited to impact parameters $b \geq R_t - R_p$. For smaller impact parameters, fluctuations in E_t and N_c are too big and one must resort to a more complicated method, involving comparison to an event generator, in order to take the fluctuation into account properly. The prescription is as follows:

1. Select b, e.g. $b = 0$.
2. Calculate, for this b, $\mathrm{d}\sigma/\mathrm{d}E_t$ ($\mathrm{d}\sigma/\mathrm{d}N_c$) using the event generator.
3. Find the peak of this distribution (it will typically look like the dashed curves in Figure 2), e.g. $E_t(0)_{\text{th}}$, where the subscript "th"

stands for the model. One thus obtains the mean E_t (N_c) for this choice of b.

4. Evaluate, using the event generator, for minimum-bias collisions:

$$\sigma_{\mathrm{trig}} = \int_{E_t(0)_{\mathrm{th}}}^{\infty} (\mathrm{d}\sigma/\mathrm{d}E_t)_{\mathrm{th}}\, \mathrm{d}E_t. \qquad 2.$$

5. Find the equivalent E_t, e.g. $E_t(0)$, in the data by requiring

$$\sigma_{\mathrm{trig}} = \int_{E_t(0)_{\mathrm{ex}}}^{\infty} (\mathrm{d}\sigma/\mathrm{d}E_t)_{\mathrm{ex}}\, \mathrm{d}E_t. \qquad 3.$$

By using cuts on $E_t(0)_{\mathrm{ex}}$, one can select "typical" $b = 0$ collisions, albeit with an unavoidable admixture of larger impact parameters. This must be taken into account when comparing to those theoretical calculations that employ just a cut in b. Such a simple cut in b is not possible experimentally. Therefore the model calculations must use a centrality measure of the same type as used in the experiment.

Methods I and II give similar results. Many other approaches have been chosen in the literature, which has led to a proliferation of scaling laws and values of α.

Using method I and compiling all available data on charged-particle multiplicity as a function of A_t, values of $\alpha \approx 0.40$ and 0.45 have been obtained for O and Si/S projectiles (23). For transverse energy and S projectiles $\alpha = 0.40$ was found. Models based on individual nucleon-nucleon collisions show A_t^{α} scaling and give $\alpha = 0.33$. The excitation energy in the participant frame E^* does not scale like A_t^{α} but rather grows more slowly with ever increasing A_t. Recent data from NA35 (16) selecting $b = 0$ collisions using method II indeed show this flattening, but the effect is only about half of what one expects from scaling with E^*. This can be understood by an increasing amount of stopping with increasing A_t and thus increasing target thickness.

So far, only two projectile masses can be compared, $A_p = 32$ and 16. Choosing again typical $b = 0$ collisions, a ratio of 1.7–1.8 is obtained for E_t produced into 4π using the two projectiles (16, 22) corresponding to $\alpha \approx 0.8$. This is very close to the $A_p^{5/6}$ dependence expected in the simple geometric model.

There are relatively few data available to study the systematic beam energy dependence of global variables since one would like to study the same system at the same degree of centrality. Because the widths of both $\mathrm{d}E_t/\mathrm{d}\eta$ and $\mathrm{d}N_c/\mathrm{d}\eta$ are dependent on beam energy and because different experiments cover different regions of rapidity, we compare E_t and N_c into the full solid angle. A Gaussian shape has been used to extrapolate beyond the experimental coverage. For E_t the only system comparable for all beam energies is Si + Pb (S + W). Figure 6 shows the extrapolated data points for central collisions as a function of $s^{1/2}$. The rise of E_t is linear with beam energy. Since the total excitation energy E^* is proportional to $s^{1/2}$, E_t is also proportional to E^*, and one finds that 18.5 GeV of E_t are produced per GeV in $s^{1/2}$, corresponding to 0.34 GeV per GeV in E^*. Both the width and the maximum of the Gaussian describing the charged-particle multiplicity were found to scale in proportion to ln $s^{1/2}$ (23); so does the integral N_c. This is also shown in Figure 6 for central Si + Pb (S + W) collisions and O + AgBr collisions at various degrees of centrality. The coefficients of proportionality, for either ln s or ln E^*, are also given in the figure. Note that the different scaling of E_t and N_c is due to the increase of E_t per charged particle with increasing $s^{1/2}$. This result was attributed (see above) mostly to the changing nucleon-to-pion ratio with $s^{1/2}$ and to the fact that a calorimeter measures a nucleon's kinetic energy only.

The dependence on mass and energy thus behaves in many ways as expected from the participant excitation energy E^* but with modifications due to increased stopping with increasing target mass.

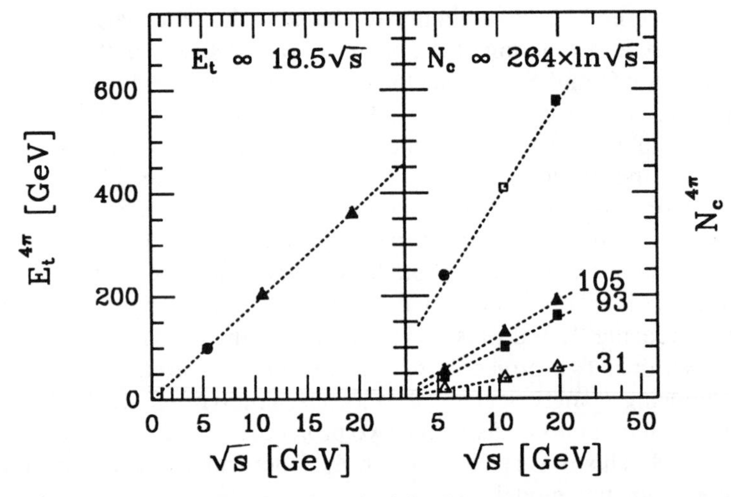

Figure 6 Dependence of E_t and N_c into the full solid angle on cm energy per nucleon-nucleon pair $s^{1/2}$. (*Left*) Typical $b = 0$ collisions of Si(S) + Pb(Au) (12, 16, 19); for the middle point, O + Au data were extrapolated using the $A_p^{5/6}$ scaling law. (*Right*) The top curve corresponds to Si(S) + Pb(W) collisions (13, 22) at $\sigma_{trig}/\sigma_{geo} = 7\%$ (middle point treated as in left panel). The curve second from the top represents O + Emulsion collisions (28) at comparable centrality. The curve second from the bottom represents O + AgBr at 31% of the geometric cross section (30). The bottom curve corresponds to minimum-bias O + Emulsion collisions (28, 29). All curves are labeled by the constant of proportionality for the ln $s^{1/2}$ scaling as displayed by the dashed lines.

4. DISTRIBUTIONS OF IDENTIFIED PARTICLES

4.1 *Baryon Distributions*

Critical information on the issue of nuclear stopping power is obtained by studying the distributions of baryons. The baryons observed at all three energies are predominantly the constituent nucleons of projectile and target. Their rapidity distributions directly reflect the slowing down of projectile nucleons and the acceleration of target nucleons. In this section we concentrate on the rapidity distributions. The p_t distributions are covered, together with the spectra of produced particles, in Section 4.2.

4.1.1 PUNCH-THROUGH NUCLEONS At the AGS beam rapidity, protons and neutrons have been studied by E814 as a function of centrality and for different target-projectile combinations (12, 32). The event characterization allows us to define collisions with full overlap, leading to the term punch-through nucleons. Their measurement gives information complementary to nuclear stopping, which refers to the extent to which the projectile kinetic energy is deposited in the collision zone. These punch-through nucleons represent instead the energy carried through the interaction zone. In a central collision with full target-projectile overlap, the multiplicity of nucleons at beam rapidity $\langle M \rangle_{yb}$ defines a transparency $T = \langle M \rangle_{yb}/A_p$. The experimental conditions for this measurement in E814 were such that only nucleons with very small transverse momenta, less than 0.3 GeV at beam rapidity, were accepted. This strongly suppresses nucleons that have undergone inelastic collisions, and therefore the transparency T as defined above has been identified with the probability $P(0)$ that a projectile nucleon underwent no inelastic collisions. For central Si + Pb collisions, only 0.092 $\pm$ 0.010 nucleons out of 28, or 1 in 300, are transmitted without collisions. The absence of punch-through nucleons is a necessary but not a sufficient condition for full stopping.

If the individual nucleon-nucleon collisions are considered to be independent, we can use Poisson statistics to determine the mean number of inelastic collisions $\langle n \rangle$ using $P(0) = \exp(-\langle n \rangle)$. For a known geometry and the Glauber theory assumption of collinear trajectories, this leads to the in-medium inelastic nucleon-nucleon cross section. Typical $b = 0$ collisions have been selected as described in Section 3.3; the inelastic nucleon-nucleon cross section was obtained from the measured data (32) as $\sigma_{NN^*} = 28.5 \pm 0.5$ mb for Si + Pb. Here the N^* in the subscript refers to the fact that, on average, every target nucleon is hit twice. That means that half of the transmitted projectile nucleons "see" on their path target nucleons that have already been struck. The measured value is very close to the free proton-proton cross section of 28 mb at this energy. This suggests that, on the time scale of the traversal time, the cross section for a second collision is not altered as compared to the first. This argument is frame and density independent. Note, however, that because of the way $b = 0$ is selected experimentally, there is always an admixture of larger impact parameter collisions, and therefore the cross section given is a lower limit.

4.1.2 RAPIDITY DENSITIES Proton distributions for central Si + Au collisions at midrapidity at the AGS have recently given rise to speculations about considerable transparency at 14.6 GeV per nucleon (33). This seemed surprising in light of the data on global observables (Section 3.3). We therefore compile the available experimental evidence.

At the AGS there is information from three experiments concerning the rapidity distribution of protons and it covers a very large rapidity range. Experiment E802 has published proton data in the interval of laboratory rapidity, $y = 0.7$–1.9 for central Si + Au collisions ($\sigma_{trig} = 7\% \sigma_{geo}$) (34, 35). They are displayed as solid squares in Figure 7. Using

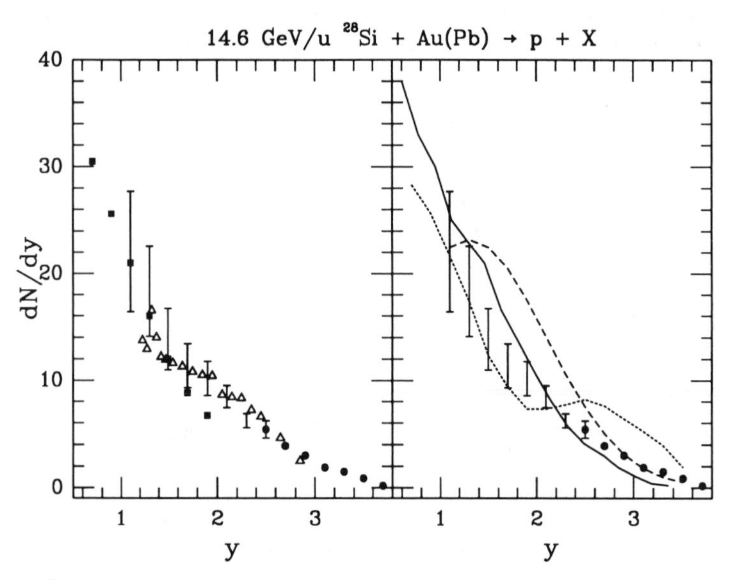

Figure 7 Proton laboratory rapidity distribution for central collisions of Si + Au and Si + Pb at 14.6 GeV/nucleon (see text). (*Left*) Experimental data; solid squares are from E802 (34, 35), open triangles (positive-negative difference) are from E810 (36), solid dots and bars (extrapolations to high p_t) are from E814 (38). (*Right*) Comparison to model predictions. The solid line represents the Lund model as well as RQMD (see 33, 39). The dotted line represents the transparent firestreak model (33). The dashed line represents the Landau fireball (14).

a time projection chamber, experiment E810 obtains the difference between tracks of positive and negative charge for a similar degree of centrality. This difference should be close to the number of protons. Differences in the 10–20% range can be expected in a system with a large neutron abundance, as there should be more negative than positive pions. This is in part compensated by the K^+-K^- difference (see below). Preliminary E810 data are displayed as open triangles in the left panel of Figure 7 (36). They are somewhat above the E802 points at $y = 1.9$ and they cross at $y = 1.5$. One should keep in mind, however, that the E810 acceptance corrections increase rapidly for lower rapidities (37). E814 recently enlarged the opening of the forward spectrometer to increase the proton acceptance. Preliminary data for central Si + Pb collisions ($\sigma_{trig} = 2\% \ \sigma_{geo}$) are also displayed in Figure 7 (38). The solid circles correspond to the rapidity range where a large enough p_t bin has been measured to integrate reliably to obtain dN/dy. In that range the E814 and E810 data agree well. At $y \leq 2.3$, only the low p_t part of the spectrum has been measured in E814, so a slope constant has been assumed in order to extrapolate to infinite p_t. The bars shown in Figure 7 correspond to a range of different extrapolations (38) but do not include statistical errors, which are given as dashed error bars in the right panel. Using the lowest extrapolation considered by E814, the lower end of the statistical error bar comes close to the E802 data. Note that the off-line cut used by E814 was on more central collisions than that used by E802, which could lead to a higher proton rapidity density at midrapidity.

The data are compared to predictions from different models in the right panel of Figure 7. The solid line corresponds to the prediction from the Lund model (18) in the ATTILA version (33). Comparison to the E802 data prompted the suggestion by Chapman & Gyulassy (33) of a multicomponent firestreak model with a long stopping length of $L_s \approx 26$ fm, based on conservation of momentum and baryon number. The corresponding results are displayed as the dotted line in the figure. Also shown is the result for the Landau hydrodynamic model (14) for $y > 1.3$. Since spectator nucleons were not considered in this scenario, it should not be compared to proton data at rapidities backwards of midrapidity. Predictions (39) from the relativistic quantum molecular dynamics (RQMD) approach (40), a microscopic model for the nucleus-nucleus dynamics in phase space, coincide with the solid line. At forward rapidity, the model with a long stopping length (33) strongly overpredicts the data, which are much more in line with the RQMD, Lund, or Landau fireball pictures.

While it may be difficult to account for the low measured values of baryon number at midrapidity, the solution does not appear to lie in increased transparency of nuclear matter. One factor that may add to the apparent discrepancy between data and calculations at midrapidity is that the calculations were performed with a cut on impact parameter, and thus neglected the impact parameter fluctuations toward larger b present in the data (see discussion in Section 3.3).

Protons in the target fragmentation region at rapidities $y = 0.5$–1.0 have been measured (41) for 200-GeV/nucleon p, O, and S beams. A backward rise was observed, similar to that seen at 14.6 GeV (Figure 7). Model comparisons show that such a rise is a strong indication of rescattering effects. At even more backward angles, corresponding to rapidities of $-1.5 < y < 0.5$, protons have been measured by WA80 using the Plastic Ball (42) and compared to the Multi-Chain Fragmentation model by Ranft (43). Although this model includes cascading, or secondary collisions, and in general reproduces the observed rapidity dependence, the momentum spectrum of protons is much harder than predicted by this model.

For O + Au collisions at 60 GeV, the difference between tracks of positive and negative charge has been measured by NA35 using a streamer chamber (44). A continuous rise toward lower rapidities is observed, similar to the observation at AGS energies and in contrast to the Fritiof prediction.

The same method has been employed by NA35 to extract a "quasi-proton" distribution for S + S collisions at 200 GeV (45, 46). The result is shown in Figure 8. The point at $y = 0$ is indirect information. The trigger condition implies a certain number of target and projectile spectator nucleons (45). The dashed line connecting these data can be reflected about midrapidity, because in a symmetric system the rapidity distribution has to be symmetric about midrapidity. This distribution can be compared to a preliminary result from E814 at the AGS (38) for a similar nearly symmetric system, displayed in the top panel of Figure 8. The open circles involve the same type of extrapolation as the data represented in Figure 7. The solid line again employs reflection symmetry about midrapidity. Except for the differences at target and beam rapidity, which are due to the different trigger condition, the distributions at the two energies are not dramatically different. There is, however, a clear indication of target- and projectile-like peaks at the higher energy, while the 14.6-GeV distribution is rather flat between $y = 0.4$ and 3. As already concluded from the systematics of the global variables, some transparency is present at 200 GeV/nucleon.

The information displayed in Figure 8 has been used to make a quantitative estimate of the nuclear stopping power for this reaction (45, 46). One also needs to consider, for this argument, the distribution of protons that have changed into hyperons. This is given by the difference

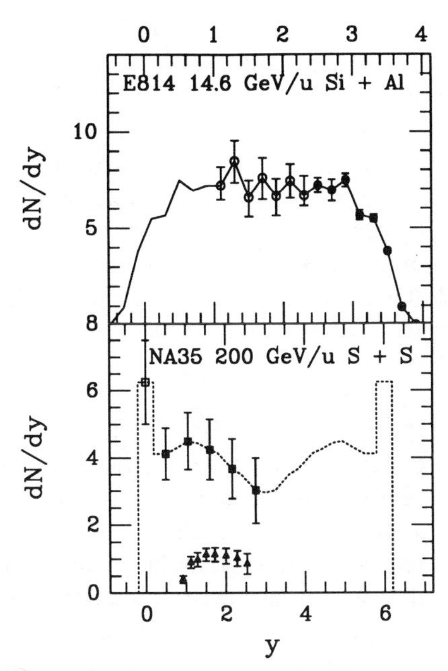

Figure 8 Proton laboratory rapidity distribution for central collisions of Si + Al at the AGS (38) and S + S at the SPS (45, 46). The solid and dashed lines connect data points and are reflected about midrapidity. The solid triangles show the difference between Λ and $\overline{\Lambda}$ rapidity densities.

between Λ and $\overline{\Lambda}$ rapidity densities displayed as triangles in Figure 8 (47) for $p_t > 0.5$ GeV. On average a proton is shifted 1.5 ± 0.1 units in rapidity. From this and considering the mean transverse momentum of a proton, the mean energy loss of a proton is calculated (45, 46) as $\langle dE \rangle_{cm} = 5.8 \pm 0.3$ GeV out of $s^{1/2}/2 = 9.7$ GeV available for a nucleon in the cm frame. This corresponds to a stopping power of 60%. Using this energy loss one can define a nuclear stopping length $L = -t \ln(1 - 2\langle dE \rangle/s^{1/2})$, where t is the average thickness of a sulfur nucleus. The value obtained is $L \approx 5$ fm, in good agreement with results from p + A collisions (48) but in marked contrast to the large value proposed for AGS energies in (33). As already indicated by the 60-GeV data, the Fritiof model predicts more pronounced target and projectile fragmentation peaks corresponding to less stopping. The RQMD and VENUS model predictions come closer to the data (45).

An analysis of stopping power based on the mean rapidity shift may be an underestimate, particularly if the distribution is flat, peaked at midrapidity, or rising backwards, as for example it is for Si(O) + Au at 14.6 and 60 GeV/nucleon. If a proton is detected at a certain rapidity y_x, it is not distinguishable whether it was slowed down from beam rapidity by $dy_1 = y_{beam} - y_x$, accelerated from target rapidity by $dy_2 = y_x$, or emitted from a thermalized source at midrapidity ($dy_3 = y_{cm}$ or $y_{beam} - y_{cm}$) with a momentum such that its final rapidity is y_x. Evidence for the last view is available from measurements made by E814 in Si+Pb collisions of the proton:neutron ratio at forward rapidities, just below the projectile fragmentation peak. A ratio of 0.75 is found (32) at low p_t, very close to the ratio of protons to neutrons in the participant volume (46:63). This agreement indicates that nucleons at rapidities $y = 2$–3 are not just projectile nucleons slowed down only moderately but are nucleons that have lost their identity as either target or projectile nucleons.

4.2 *Spectra of Baryons and Mesons*

One would like to check whether the spectral distributions of emitted particles are consistent with those expected from a thermalized system radiating particles as part of a specific search for creation of a quark-gluon plasma. For a thermalized system the particle spectra in p_t reflect the temperature of the systems at the time when the particles decouple, i.e. the freeze-out temperature. It was suggested by Van Hove (49) that the latent heat associated with the phase transition is visible in a plot of temperature versus entropy; during the phase transition, the temperature remains essentially constant while the change in entropy can be large, depending on the latent heat. Experimentally, temperature would be obtained from the p_t spectra, while the multiplicity of (charged) particles is a measure of the entropy. Should there be collective motion in the system such as hydrodynamic flow, it would influence the shape of the transverse momentum distribution as well. Effects of restoration of chiral symmetry and associated mass changes might manifest themselves in the spectra of particles, in particular at low p_t. Many particles are not created directly in the first stages of the collision but rather by decay of heavier resonances, which will also influence the p_t spectra.

Distributions in transverse momentum p_t have been studied extensively both at the AGS and SPS. A detailed review of transverse momentum distributions is given by Schukraft (50). In this discussion we summarize the systematic behavior that has emerged as a function of beam energy, colliding system, rapidity, and particle species. A few typical examples of p_t spectra are shown in Figures 9 and 10, where results from different experiments are overlaid.

→

Figure 10 Transverse momentum spectra $(1/p_t)dn/dp_t$ versus p_t for negative hadrons and neutral pions for central O+Au(W) collisions at the SPS using data of (52, 54, 55). (Figure from 50)

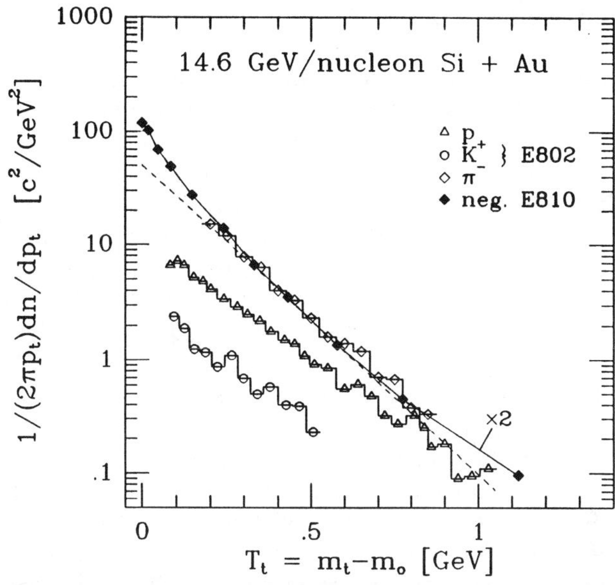

Figure 9 Transverse momentum spectra $(1/2\pi p_t)dn/dp_t$ for p, K^+, π^-, and negative particles as functions of transverse kinetic energy for Si + Au collisions at the AGS (figure from 51). The data from E810 are multiplied by an arbitrary factor of two to take into account the difference in rapidity interval and centrality (see 51) and are connected by a solid line. The dashed line corresponds to the exponential fit to the pion spectra in the interval m_t = 0.3–0.6 GeV.

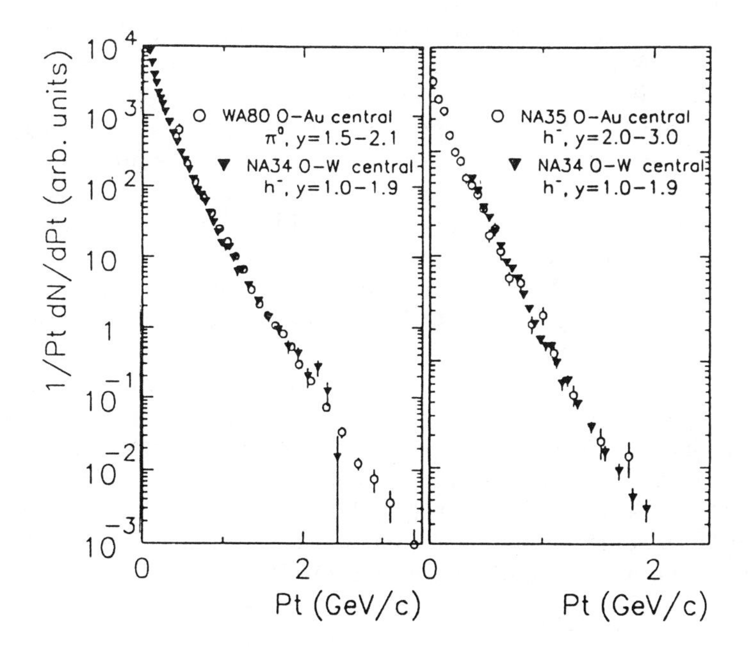

4.2.1 SLOPE CONSTANT OR TEMPERATURE The spectra are typically characterized by the inverse slope constant (hereafter called the slope) of an exponential function. However, different variables and functions have been used. The invariant cross section $\sigma_{inv} = E\ d^3\sigma/dp_t^3 = 1/(2\pi p_t)d^2\sigma/dp_t dy = 1/(2\pi m_t)d^2\sigma/dm_t dy$, where $m_t = \sqrt{p_t^2 + m^2}$, has been parametrized as an exponential with respect to either p_t or m_t, i.e. $\sigma_{inv} \propto \exp(-p_t/T_p)$ or $\sigma_{inv} \propto \exp(-m_t/T_0)$. An isotropic thermal spectrum is given by $\sigma_{inv} \propto E \exp(-E/T)$, where E is the total energy of the particle. Using $E = m_t \cosh(y - y_{cm})$ leads to $\sigma_{inv} \propto m_t \exp(-m_t/T_B)$ in a small rapidity bin (for larger y intervals, see 56, 57). T_B may depend on y; specifically, if isotropy is assumed, $T_B = T/\cosh(y - y_{cm})$.

As can be seen from Figures 9 and 10, spectra of pions and negative particles cannot be fitted over the entire p_t range by any of these simple functional forms. There are clear deviations at low and high p_t values, discussed below. Furthermore, because of the small pion mass, at intermediate p_t all parametrizations mentioned above give equally satisfactory fits to the data. For heavier particles, for which the difference between m_t and p_t is larger, a general preference for parametrizations in m_t has emerged. But very low p_t data are needed to distinguish between an exponential spectrum in m_t and a thermal spectrum; this question has not been answered definitively.

In order to compare results from the various experiments that have used these different parametrizations, one needs a "translation" between the inverse slope constants or temperatures T_p, T_0, and T_B. This translation depends on the interval in p_t or m_t and the particle mass. For the interval $p_t = 0.5\text{–}1.0$ GeV, which is covered by the majority of the experiments, the appropriate translation is given in Figure 11 for protons, kaons, and pions. Values for lambdas are close to the proton lines. The general trend $T_p > T_0 > T_B$ emerges for fitting the same spectrum and the same p_t interval with the different functional forms. The differences among the various slope values are considerable. In Figure 12 results from different experiments and for different particle species are compiled based on the parametrization using T_B. The values shown are obtained fitting all data in the 0.5–1.0 GeV p_t interval. While over this p_t range all data are consistent with that type of parametrization, it is by no means proven that the p_t spectra are in fact thermal distributions. We adopt it as a convenient way to combine data from different experiments. Most of the data points shown are for central collisions of S (Si) projectiles and a heavy target nucleus (W, Au, Pb). Some points from minimum-bias collisions (labeled with "c") and from other colliding systems as O+W,Au ("a") and S+S ("b")

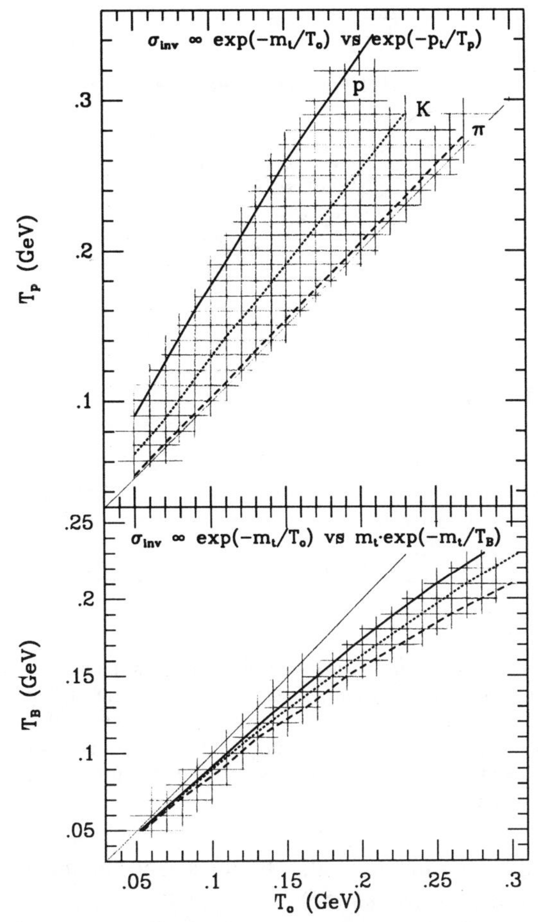

Figure 11 Equivalent slope constants for (*top panel*) an exponential spectrum in p_t (T_p) as compared to an exponential spectrum in m_t (T_0) and (*bottom panel*) for the latter as compared to a thermal spectrum (T_B) at fixed rapidity obtained by fitting the interval p_t = 0.5–1.0 GeV. Solid lines show protons, dotted lines show kaons, and dashed lines show pions.

have been added for a more complete picture. The following trends emerge:

1. The rapidity dependence of T_B is rather weak for rapidities between 20 and 80% of the beam rapidity. The values fall off for smaller and larger rapidities, as would be expected for kinematic reasons. The y dependence is, however, significantly weaker than the $T \propto 1/\cosh(y - y_{cm})$ behavior expected for an isotropic thermal source. This was

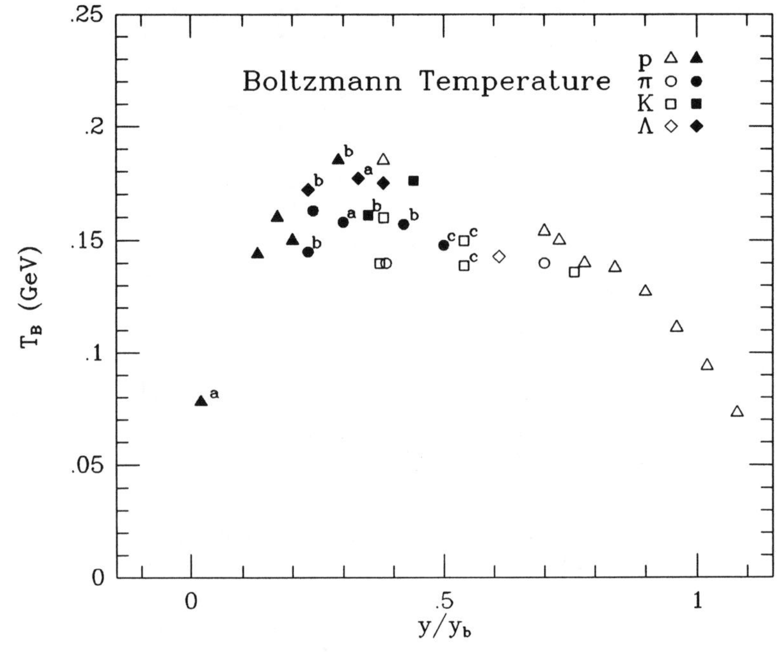

Figure 12 Slope constants of thermal spectra for protons, pions, kaons, and lambdas for central Si(S) + Pb(Au,W) collisions. Open symbols represent 14.6 GeV/nucleon, solid symbols represent 200 GeV/nucleon. To compare data for the two different beam energies, the rapidity is normalized to the beam rapidity y_b. Data points marked with "a" for O projectiles, with "b" for S + S collisions, and with "c" for minimum-bias collisions are shown for comparison.

already noted by Stroebele et al, who discussed the mean p_t as a function of y for negative tracks (55).

2. Comparing the temperature constants T_B measured at the AGS and SPS reveals no strong beam energy dependence. Typical values are 150 and 170 MeV at $s^{1/2}$ = 5.5 and 19.4 GeV, but there are insufficient points in Figure 12 where only $s^{1/2}$ is different to clearly establish a difference in temperature. The similar values are not so surprising since the measured slopes probably reflect the freeze-out temperature and certainly not the initial temperature of the system.

3. At the same rapidity and for the same system, T_B is largest for protons. Values for lambdas and for positive and neutral kaons are similar and somewhat smaller. Pions show smaller slopes still. At the AGS the K^- slopes have been found to be close to the pion values and smaller than the K^+ values (34, 35, 60). A similar effect for K^+ and K^- is seen at 200-GeV/nucleon at $y/y_b \approx 0.18$ (62).

4. There is no clear dependence of the slope constants for produced particles on centrality of the collision or the colliding system, but systematic information is scarce. For proton slopes, a clear trend has been established that slope constants are larger for more central collisions and for Si + Au as compared to Si + Al for midrapidity and less (63). For $y/y_b > 0.7$ the same slopes are found for the two targets (38).

5. Data for central p + A collisions indicate slopes for π, K^0, and Λ spectra in p + A collisions that are about 10 MeV smaller than those for heavy ion collisions (35). There is a striking difference for p spectra, which are found to have 40–50 MeV smaller slopes for p + A collisions. The change in K^+ spectra (35) is intermediate.

6. Inclusive particle spectra from minimum-bias pp collisions and the same values of $s^{1/2}$ give Boltzmann temperatures of 0.125–0.130 GeV at midrapidity (64–66).

The different slope constants for different particle species (trend 3 above) have been interpreted in terms of collective flow (67). In this scenario, for a given flow velocity, the mean (transverse) momentum grows with the mass of the particle. There are some observations that contradict this conjecture, however. The antiproton p_t spectrum is found to be significantly steeper than the proton spectrum (68). Both the differences among K^+, K^0, and K^- spectra and the fact that the Λ spectrum is steeper than the proton spectrum are inconsistent with the interpretation of collective flow. Another possible explanation for these different slope constants is the influence of strong rescattering effects (39, 69). Indeed, at AGS energies calculations within the RQMD model account for the experimental observations and have been associated with rescattering. The fact that the proton slope constant grows with the target mass for $y/y_b < 0.5$ (trend 4 above) also hints at rescattering effects strongly influencing the slope constant for a heavy target (with many target spectator nucleons).

4.2.2 LOW p_t ENHANCEMENT We use the terms low and high p_t enhancement in the following to indicate an excess in σ_{inv} at low and high p_t over the exponential fitted to the data in the intermediate p_t region of 0.5–1.0 GeV. Spectra of negative particles (dominantly pions) show a pronounced enhancement below $p_t \approx 0.25$ GeV, both at AGS (36, 37) and SPS (50) energies. At present this finding is still somewhat controversial. While some of the data may contain a contamination due to electrons, which pile up at low p_t if not identified, this should not be a problem for the E810 data, where a very thin target was used, and the NA34 data, which have been explicitly corrected for electrons. Furthermore, the effect is also seen in π^0 spectra (52). On the other hand, the data of EMU05 show only a very weak enhancement if any (70). For the E810 data, the enhancement over an exponential spectrum contains 25–30% of the pions (36, 37). At SPS energies for the HELIOS O + W data, a number of 40% was estimated (50, 54). A strong enhancement is already seen in central p + W collisions (54).

The observation of this low p_t enhancement triggered a large number of theoretical papers attributing the effect to abundant production and decay of baryonic and mesonic resonances, to a pion collective potential, to a positive pion chemical potential, to restoration of chiral symmetry, and to other phenomena. The scope of this article does not allow a thorough discussion of the theoretical literature, and the reader is referred to a recent paper by Shuryak (71) giving a critique of the various suggestions (as well as references to other theoretical work). While there is not yet an established explanation of the effect, it is clear that more experimental data are needed to confirm or rule out the different scenarios suggested. In particular, the low p_t spectra of other particle species, systematics of the rapidity dependence of the effect, and data from heavier colliding systems (Au + Au or Pb + Pb) are awaited.

4.2.3 BEHAVIOR AT HIGH p_t As can be seen from Figure 10, there is also an excess of particles at high p_t compared to the extrapolated exponential behavior at intermediate p_t. Since such an effect has been seen in pp and pA collisions we briefly summarize the situation before discussing the heavy ion data. Spectra of pions from pp collisions at $s^{1/2} = 23$ GeV (72) show a clear excess at high p_t over an $\exp(-p_t/T_p)$ fit at intermediate p_t. The enhancement becomes much more pronounced with increasing beam energy (73). But at a given beam energy it is also stronger in dd and even stronger in $\alpha\alpha$ data (72). Furthermore, the enhancement increases with increasing centrality.

Spectra at high transverse momenta (1–7 GeV) from p-nucleus collisions were first studied by Cronin et al (74) at $s^{1/2} = 19.4$–27.4 GeV. If the target mass dependence of the pion spectra is parametrized as $A^{\alpha(p_t)}$, $\alpha(p_t)$ is found to increase with increasing p_t in the 2–4 GeV region and to saturate for $p_t = 4$–6 GeV. This effect occurs at values of p_t above 2 GeV, yielding little overlap with current heavy ion experiments. Comparison of measurements made in the same experiments with proton beams and with heavy ion beams will be more relevant since many of the experimental variables, including trigger conditions, are in common.

The enhancements at high p_t have been demonstrated by comparing to pp parametrizations or by plotting ratios with respect to pp or pA spectra. This has not helped to clarify the picture and can, in fact, be rather misleading. The data so compared are frequently at a different rapidity, and we have seen that p_t spectra depend on rapidity. Even if this is taken into account and data are compared at the same rapidity with respect to the beam rapidity, the kinematics depends on the mass numbers of the colliding nuclei and is symmetric about midrapidity only for symmetric colliding systems. Furthermore, a comparison of semi-inclusive spectra from central heavy ion collisions to minimum-bias pp results introduces some bias since the spectral shape has been found to depend on multiplicity also in pp collisions. As was discussed in the previous section, the general trend has emerged that spectra from pp collisions show a lower temperature than those from pA, which in turn have a smaller slope than spectra from AA collisions. Therefore, ratios of AA/pA and AA/pp spectra will primarily reflect this difference in overall slope constant rather than isolate a high p_t enhancement.

We therefore prefer to quote, as a measure for the high p_t enhancement, the measured enhancement over an exponential parametrization at intermediate p_t in order to compare pp, pA, and AA data. We use as a benchmark an exponential function in p_t fitted to the 0.5–1.0 GeV region and we quote the excess $X_2 = (\sigma_{obs} - \sigma_{fit})/\sigma_{fit}$ with respect to this exponential at 2 GeV, which is usually the highest p_t for which all groups have published heavy ion data. For minimum-bias pp collisions at $s^{1/2} = 23$ GeV, this yields an enhancement of $X_2 = 4$–5. Data for neutral pions from minimum-bias p + Au, O + Au, and central O + Au collisions give $X_2 = 1.8$, 2.0, and 2.6 (53, 54). Spectra of negative particles from central p + W, O + W, and S + W collisions yield $X_2 =$ 3.5, 3.0, and 3.2 (54). It appears that under the same conditions the effect is similar in p + A, O + A, and S + A collisions. Furthermore, as compared to pp data, pion spectra from pA and AA collisions show less of a high p_t enhancement in the 1–2 GeV region, i.e. they are closer to exponential. It should be noted again that the Cronin effect for pA collisions refers to significantly higher values of p_t (3–5 GeV).

4.2.4 MEAN p_t AS A FUNCTION OF MULTIPLICITY DENSITY A significant increase followed by gradual saturation of the mean transverse momentum, $\langle p_t \rangle$, as a function of dN/dy was first established in pp collisions at $s^{1/2} = 540$ MeV (73) and subsequently also found at ISR energies for $s^{1/2} = 23$–62 GeV and pp, dd, and $\alpha\alpha$ collisions (72). One possible explanation for this increase is the production of mini-jets, a possibility supported by the strong increase of the effect with $s^{1/2}$. An intriguing and (for current heavy ion collisions) more relevant suggestion was made by Van Hove in terms of a thermodynamical model (49).

In general, the p_t spectrum reflects the temperature of the system and its transverse expansion. The multiplicity is determined by the entropy of the system. In a geometric picture, smaller impact parameters create larger energy density ϵ and entropy density σ (the contribution from pressure is small) and hence create larger dN/dy, larger T, and thus larger $\langle p_t \rangle$ (see Section 5 for more quantitative relations). But in a system that undergoes a phase transition, e.g. from the quark-gluon plasma to hadronic matter, there will be a plateau at a constant temperature T_c where the number of degrees of freedom and σ change. Hadrons are emitted at or below the critical temperature. This leads to a saturation of $\langle p_t \rangle$ as a function of entropy or even a slight decrease. For very high initial energy densities, hydrodynamic calculations including a phase transition (75) show that there will be a second rise in $\langle p_t \rangle$. Because it is a flow effect, it is predicted to depend strongly on the particle mass.

Heavy ion data show that $\langle p_t \rangle$ for pions grows with increasing overlap of target and projectile and stays constant once there is complete overlap (50). This observation is not in disagreement with expectations for a phase transition considering that the energy densities reached in current experiments are moderate (see Section 5). The second rise due to hydrodynamic flow should occur at multiplicities or entropy densities about a factor of three higher than achieved to date. A hydrodynamic calculation for a hadronic gas without a phase transition (75) is, on the other hand, at variance with the observed saturation of $\langle p_t \rangle$.

4.3 *Strangeness Production*

In the current model of high energy collisions, secondary particles are created via hadronization of $q\bar{q}$ pairs created by the breaking of color flux tubes or strings due to the separating color fields of the collision partners. The creation of heavy flavors is suppressed by their mass; for values of $s^{1/2} \leq 60$ GeV, it is sufficient to consider $u\bar{u}$, $d\bar{d}$, and $s\bar{s}$ pairs. The suppression of strange quark pairs has been studied extensively in pp collisions and is quantified by the strangeness suppression factor λ, a quantity that reflects the relative probability of creating $q\bar{q}$ pairs, i.e. $u\bar{u}:d\bar{d}:s\bar{s} = 1:1:\lambda$. We follow here the procedure introduced by Wroblewski (76) to count strange and nonstrange quark pairs at the initial time of hadronization. This procedure explicitly takes out the multiplication of u and d quarks due to resonance decay. In Figure 13 some pp results and the band of pp systematics established by Wroblewski's analysis are displayed as a function of $s^{1/2}$ and of the multiplicity of negative hadrons. One notices a gradual rise with $s^{1/2}$ or mul-

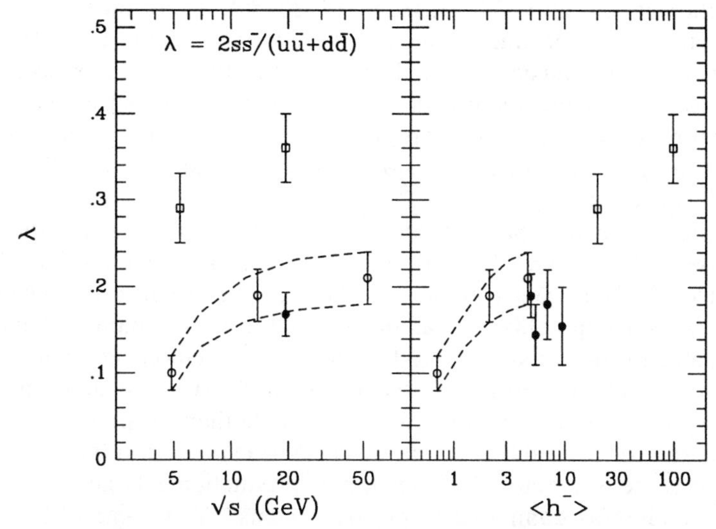

Figure 13 Strangeness suppression factor λ as a function of $s^{1/2}$ (*left*) and the multiplicity of negative hadrons (*right*) for pp collisions [open circles and dashed line, from systematics of Wroblewski (76)], pA collisions [solid dots, from Seyboth (80)], and Si + Si (S + S) collisions (*open squares*) at the AGS [using data of E810 (86), E814 (13), and SPS (80)].

tiplicity from 0.1, with the indication of a saturation at $\lambda = 0.2$ for these pp data.

Increased production of mesons and baryons containing strange quarks has long been predicted as one of the signatures for the formation of a quark-gluon plasma (77; reviewed in 78). Gluon-gluon collisions could be a fast way to establish thermal and perhaps chemical equilibrium, leading to similar number of strange, up, and down quarks. In a plasma the required energy would be given just by the strange quark current mass, i.e. about 300 MeV per pair, as compared to the ΛK associated-production threshold of 700 MeV. This is a large decrease, given that the plasma temperature T is only of the order of 200–300 MeV. The lower Fermi energy in a baryon-rich environment favors strange quarks and in part compensates for their larger mass. The very large strangeness density expected in a plasma as compared to a hadronic gas could give rise to a dramatic increase in hadrons with single and even multiple s or $\bar{s}$ quarks (79).

Many heavy ion experiments both at the AGS and SPS have therefore studied strangeness production. Data have been published for production of charged and neutral kaons, lambdas, antilambdas, φ's, cascades, and even anticascades for various colliding systems and as a function of centrality (34, 47, 58, 59, 61, 80–86). It was found uniformly that strangeness production indeed exceeds expectations in central nucleus-nucleus collisions. We discuss in the following this so-called strangeness enhancement with respect to extrapolations from nucleon-nucleon collisions, production of nonstrange hadrons, model predictions for nucleus-nucleus collisions, and relative abundance of various strange hadrons.

Considering that particle spectra are rather similar in pp and AA collisions (see Figure 12) and that stopping is large even at 200 GeV/nucleon, a first-order estimate for the increase in particle production in AA versus pp collisions can be obtained from the amount of energy available in the cm system. Extending the calculations given in Table 1, we see that for central Si + Au collisions at 14.6 GeV/nucleon (system "a") there is 30 times more energy available than for pp ($s^{1/2} = 5.4$); similarly, for central S + S collisions at 200 GeV/nucleon (system "b") the increase is a factor 29. Indeed, pion production is larger by factors of 37 and 36 as compared to pp for systems "a" (35) and "b" (46). Strangeness production, however, grows much more. For system "a," E802 finds an increase (35) in K^+ and K^- production by factors of 79 and 60 and E810 obtains a factor of 80 for Λ production (58). NA35 obtains for system "b" an enhancement in Λ, K_s^0, and $\overline{\Lambda}$ production by factors of 86, 63, and 115 (47). One measure for the increase in strangeness production is the isospin-averaged ratio $(K^+ + K^-)/(\pi^+ + \pi^-) = 2K_s^0/(\pi^+ + \pi^-)$. This ratio roughly doubles for these two heavy ion systems relative to pp.

An even stronger increase is found for the meson containing two strange quarks; NA38 reports for central S + W collisions an increase in the ratio $\phi/(\rho + \omega)$ by a factor of three relative to pW, for which the ratio was found to be consistent with pp (81). Figure 14 shows that the

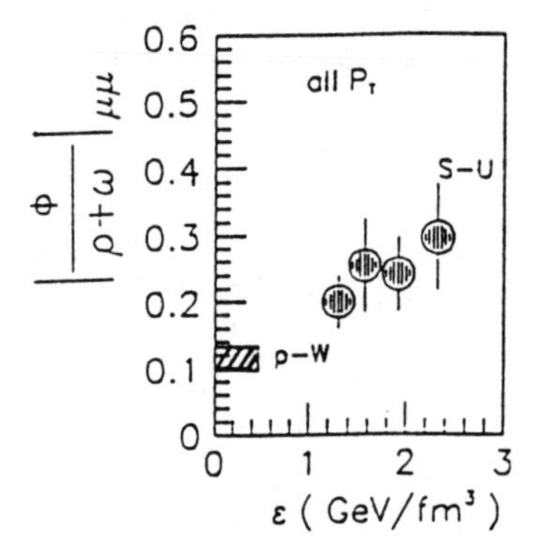

Figure 14 Production of ϕ mesons in S + U collisions at 200 GeV/nucleon from NA38 as compared to the $(\rho + \omega)$ yield as a function of the energy density $\epsilon = 3\langle E_t^0\rangle/(\Delta\eta \cdot \tau S)$, using the neutral E_t measured over $\Delta\eta = 2.4$, $\tau = 1$ fm, and S as the overlap area for a given centrality calculated (104) in a manner similar to method II in Section 3.3. The hatched band indicates the result for p + W collisions. (Figure from 81)

ϕ enhancement grows with increasing centrality. Recent measurements from HELIOS 3 extend to even lower values of p_t and larger rapidity (82). They confirm the ϕ enhancement found by NA38 and show that at low p_t it may be even somewhat larger. A spectacular result on anticascade production at $p_t > 1$ GeV was recently reported by the WA85 collaboration (83). For central S + W collisions a ratio $\overline{\Xi}/\overline{\Lambda}$ = 0.33(11) was found as compared to a value of 0.06(2) for pp collisions (95).

The centrality dependence seen for ϕ production in Figure 14 has been mapped out in more detail for K and Λ production. The combined data from NA35, NA36, and WA85 indicate that the K/π and Λ/π ratios grow with increasing centrality and level off at the enhancement factors quoted above when full target-projectile overlap is approached. In contrast, the Λ enhancement seen by E810 at AGS energies in Si + Au forward of midrapidity is not found to depend on centrality.

For central Si + Si collisions at 14.6 GeV (86) and central S + S collisions at 200 GeV (47, 80), the strangeness suppression factor λ can be computed with Wroblewski's technique (76). Values about a factor of two larger than for pp collisions at the same value of $s^{1/2}$ are obtained. These results are shown in Figure 13 as a function of $s^{1/2}$ and of the multiplicity of negative particles in the event. Also shown are points for pA collisions at 200 GeV as evaluated by the NA35 collaboration (80). They more closely reflect the leveling off observed in the pp data than the rise seen in the heavy ion data. The question remains open whether the increase in λ observed for heavy ion collisions with respect to pp at a given value of $s^{1/2}$ reflects only a scaling with overall multiplicity or whether the enhancement is a genuine heavy ion feature. Estimates for λ when a plasma has been formed can be derived from lattice QCD calculations, in which the energy density is found to be shared by $u\bar{u} + d\bar{d}$, $s\bar{s}$, and gluons in the ratio 4:1:7 (87). Depending on how the gluons hadronize (see also 88), one obtains values of λ = 0.37–0.50. These are close to current heavy ion data; it will be very interesting to see whether results for Pb + Pb collisions level off at the QCD value or keep growing with multiplicity.

Comparison of the data with models based on a superposition of pp collisions also suggests a strangeness enhancement in the data. At AGS energies, K^0 and Λ yields predicted by the HIJET simulation fall a factor of two below the data, even when a maximal amount of rescattering is included (58, 86). The event generator Fritiof significantly underpredicts charged kaons measured by E802 (40). At SPS energies K and Λ production using S beams is underpredicted by the event generators Fritiof and VENUS by factors larger than two (47, 59, 61).

On the other hand, RQMD calculations (89) can account for the K^+ and K^- yield observed at 14.6 GeV; the enhancement has been attributed to secondary interactions. This is somewhat surprising since the dominant channels for strangeness production in secondary interactions do not produce K^- (see below). The same calculation (89) reproduces K^0 and Λ production for 200-GeV O + Au at backward rapidities, but cannot account for the Λ and, in particular, the $\bar{\Lambda}$ yield observed for 200-GeV S + S.

These model comparisons point to the fact that interactions of secondary particles in baryon-rich matter, also called "cascading," do play a role in strangeness production, in particular for very asymmetric target-projectile combinations and at rapidities close to that of the heavy collision partner. For instance, the E802 K^+ enhancement is not peaked at midrapidity as is the pion production but rises backwards. The chief reactions that need to be considered are $NN \rightarrow N\Lambda K$, $\pi N \rightarrow \Lambda K$, and $\bar{K}N \rightarrow \Lambda$. However, these three produce only Λ, K^+, and K^0, and the last one in fact depletes K^- and $\bar{K}^0$. This would lead to comparable enhancements for Λ and K^+, about half the enhancement or less for $K_s{}^0$ (since it is a superposition of K^0 and $\bar{K}^0$) and no enhancement for K^- and $\bar{\Lambda}$.

Strangeness production in pA collisions has also been interpreted in terms of cascading. Nikolaev (90) argues that pXe data from NA5 on Λ and K^0 production are indicative of $\pi N \rightarrow \Lambda K$ and $\bar{K}N \rightarrow \Lambda$ contributing about equally, since the K^0 yield is not enhanced. A recent publication of the FNAL E565/E570 collaboration shows that the K^0 yield is also increased but less than the Λ yield, which points more towards $\pi N \rightarrow \Lambda K$ being the dominant process (91). Data from E802 for pAu collisions show a clear backward enhancement of the K^+ yield, in agreement with a cascading scenario, but also some K^- enhancement at midrapidity (92).

Returning to heavy ion collisions, E810 observes the same enhancement for K^0 and Λ in Si + Si collisions, in contrast to the pA data claimed to be explained in terms of cascading. NA35 observed in SS collisions the same enhancement for Λ, $K_s{}^0$, and $\bar{\Lambda}$. WA85 and NA36 also find, for S + W(Pb) collisions, a ratio of $\bar{\Lambda}/\Lambda$ consistent with the NA35 result, if compared at the same rapidity. All these results make it unlikely that cascading is the dominant source of strangeness production in central SS and SA collisions. Clearly, in this respect heavy ion collisions are not an extension of pA physics. The Λ production in NA35 is peaked at midrapidity (47); this feature does not support a cascading picture. Furthermore, the increase in the strangeness suppression factor λ discussed above is also not in line with a cascading picture. As can be seen from Figure 13, λ is even slightly less for pA collisions as compared to pp, which means that the increased strangeness production is accompanied by an even stronger or at least comparable increase in nonstrange mesons. A novel approach to reinteractions has been chosen by Aichelin & Werner (93). Overlapping hadrons form "parton clusters" of several tens of GeV total energy, which decay (hadronize) statistically. This approach reproduces the $\overline{\Lambda}$ enhancement seen experimentally. One is tempted, however, to view these parton clusters as plasma bubbles in a mixed phase. (This interpretation has interesting consequences for two-particle correlations.)

An interesting alternative to strangeness enhancement through a quark-gluon plasma was proposed by Brown et al (94). They consider the kaon mass in hot and dense matter and suggest a considerable reduction as the phase transition temperature T_c and/or densities of a few times normal nuclear matter density are approached. Using this variation of mass, one can indeed reproduce the E802 kaon enhancement. This mass shift reflects the effects of the onset of restoration of chiral symmetry. It does not contradict a plasma scenario, but rather points toward the gradual changes that occur before a plasma is actually formed.

In summary, strangeness production is enhanced in central AA collisions. While for very asymmetric systems, in particular at 14.6 GeV, a cascading picture can account for some of the experimental data, this is not true for symmetric systems or for S+A collisions at 200 GeV. These data require strangeness to be in or close to chemical equilibrium with the light flavors, whether through gradual restoration of chiral symmetry, the formation of parton clusters or a mixed phase, or through the formation of an actual quark-gluon plasma. This hypothesis is supported by the large effect seen for ϕ production and the observed strong anticascade enhancement, as predicted for example to arise from the sudden hadronization from a quark-gluon plasma near equilibrium (79).

4.4 *High Mass Vector Mesons*

It is conjectured that a quark-gluon plasma will modify properties of the vector mesons created and will change the propagation of certain particles, such as charmed hadrons. Rho mesons should exhibit shifts in mass and width due to the restoration of chiral symmetry (96, 97). Phi mesons were discussed in the previous section. Charmed hadrons, such as D mesons, would exhibit enhanced energy loss in traversing a mixed phase of hadron gas and plasma because of the different manner in which they propagate in the two phases (98). The best studied of such conjectures, by Matsui & Satz (99), is that the analog of the Debye-screening radius in a quark-gluon plasma would at high energy density become smaller than the radius of vector mesons formed of heavy quarks, such as c or b quarks. This would result in a "suppression" of the formation probability for such mesons, in particular for the J/ψ meson, a 1s $c\bar{c}$ pair, because the final-state interaction responsible for the formation of the bound state is weakened by the screening of the intervening partons as the $c\bar{c}$ pair separates to the asymptotic size of a J/ψ. Several authors have elaborated upon this idea (100–102); in particular it was noted that the probability for a J/ψ to escape a plasma is significantly increased for large p_t. The pair's M/p_t ratio is a measure of the proper time required for it to leave the plasma volume. High p_t pairs are more likely to escape before expanding to the size of J/ψ and thus would be less suppressed. This should lead to an effective modification of the p_t spectrum of J/ψ formed in heavy ion collisions.

Measurements have been made of J/ψ cross sections at the CERN SPS by the NA38 collaboration. The J/ψ measurements were made using the NA10 muon pair spectrometer combined with an electromagnetic calorimeter for event characterization (103). The total transverse energy E_t measured by the calorimeter is corrected for hadronic energy and compared to expectations from a geometric model to give an estimate for the energy density for a given event (104). Cross sections for J/ψ mesons are extracted from the opposite-sign muon yield in the mass range 2.7–3.5 GeV. At the J/ψ mass the spectrometer mass resolution is approximately 0.15 GeV, dominated by multiple scattering in the muon filter. Cross sections for continuum pairs are taken from the mass ranges 1.7–2.7 and >3.5 GeV; the former region has 95% of the continuum yield so defined. The relative behavior of these two cross sections, $R = \sigma(J/\psi)/\sigma(\text{cont})$, is displayed in Figure 15 and shows a steady decrease with increasing E_t, or centrality (103, 104). This behavior is seen for all heavy ion systems measured (O + Cu, O + U, S + U), and the ratio for the heavy ion reactions is consistently below that measured for p + Cu and p + U reactions (103, 104). Assuming the continuum yield has no, or weak, dependence on E_t, one could interpret the behavior thus measured as evidence for increasing "suppression," as proposed by Matsui & Satz (99), with increasing energy density. The energy densities quoted are in fact in the range thought to be needed for deconfinement.

The ratio of the J/ψ yield to the continuum yield has been shown to increase with increasing p_t for the heavy ion reactions (105, 106), as can be seen in Figure 7 of the first reference in (106) and in Figure 4 of the second reference in (106). The values of $\langle p_t \rangle$ and $\langle p_t^2 \rangle$ are also

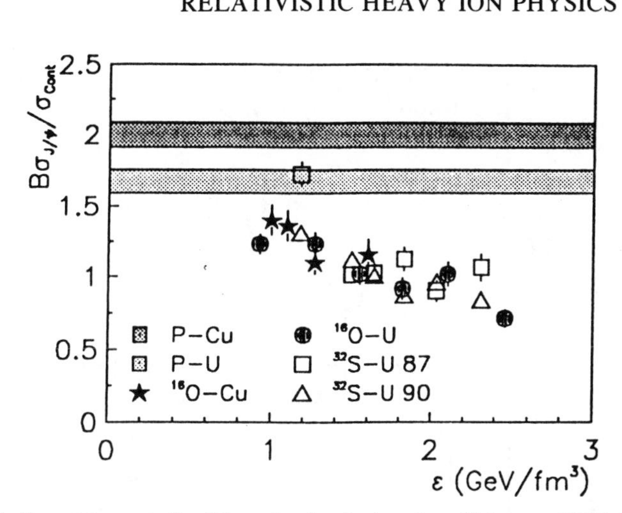

Figure 15 Branching ratio for J/ψ production in $A_1 + A_2$ collisions at 200 GeV/nucleon as compared to the continuum as a function of energy density ϵ from NA38. See Figure 14 for the definition of ϵ. (Figure from 105)

found to increase with increasing reaction centrality; an increase in $\langle p_t \rangle$ of about 20% is found for the most central events compared to the $\langle p_t \rangle$ found for peripheral events. Baglin et al showed that this results from an increase in the inverse slope of the spectrum, as expected if low p_t J/ψ were less likely to be produced in events with large E_t (106). The authors point out, however, that the increase in $\langle p_t^2 \rangle$ with increasing centrality seen for example in Figure 16 may be explained by initial-state effects, as discussed below (107).

Various and conflicting interpretations of these results have been offered. The suppression behavior expected for a quark-gluon plasma has been discussed, for example by Satz (108) and by Matsui (109). They show that it is possible to reproduce the observed p_t dependence of R for various systems as well as the E_t dependence. It has been calculated that the suppression as a function of p_t observed by NA38 is consistent with a plasma proper lifetime of 1.5 fm (110). Measurement of the p_t dependence of cross sections beyond 5 GeV is necessary (111) to distinguish a QGP picture from a picture involving reabsorption in a hadron resonance gas. Reabsorption will occur at any p_t and will lead to a suppression that decreases continuously with increasing p_t. Conversely, in a plasma a "maximum" p_t exists, above which there is no suppression. This occurs because at the "maximum" p_t, the $c\bar{c}$ pair, traveling faster than the expansion front of the plasma, leaves in a short enough proper time that it is outside the plasma before the pair has separated to the point where the effects of screening can influence its

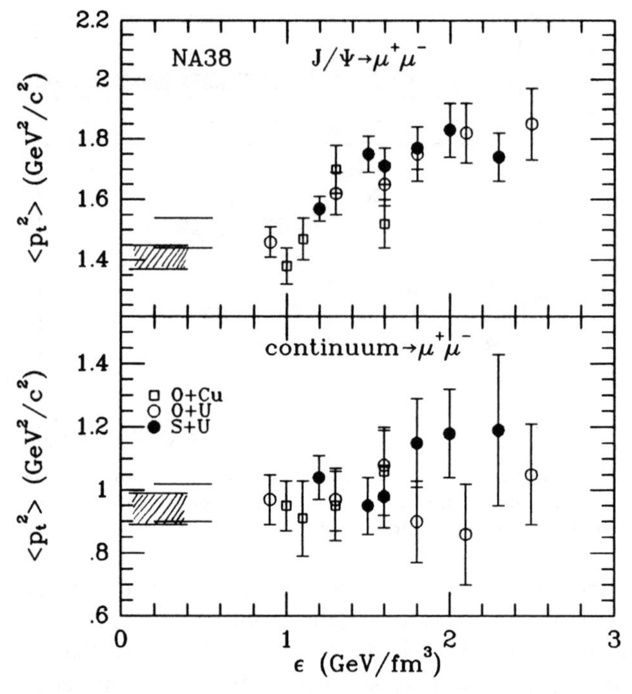

Figure 16 Dependence of the mean squared transverse momentum of J/ψ and continuum pairs as a function of energy density ε from NA38 (106). See Figure 14 for the definition of ε. The hatched band is for p + Cu; the open band is for p + U.

binding. The exact value of this maximum p_t depends on energy density and details of the expansion of the plasma. Measurements of several other $c\bar{c}$ states, such as the ψ′ and χ_c states, or $b\bar{b}$ states such as the Υ and χ_b states, are needed to be sure that the correct dependence of suppression upon energy density for vector mesons of several different radii is observed before a plasma interpretation can be established (111).

The absorption cross section for J/ψ interacting with nucleons is estimated to be of the order of a few mb by using vector dominance models to extract the absorption cross section from measurements of the photoproduction cross section of J/ψ from nucleons. Only about 1–2% of the J/ψ absorption cross section, about 80 μb (112), leads to a J/ψ present in the final channel. Measurements show that the J/ψ-N absorption cross section increases rapidly near the $D\bar{D}$ threshold (113), which indicates that the absorption cross section arises mainly from

$J/\psi + N \rightarrow D\bar{D} + X$. These observations have led to numerous theoretical attempts to reproduce the observed behavior in heavy ion reactions by modeling the final-state absorption of any J/ψ created. Calculations have been made of the probability for a J/ψ to survive as it traverses an expanding hadron resonance gas (114–117). Expressions of the form $P(p_t) = (\tau_f/\tau_i)^\beta$ result, where $\beta = \langle\sigma\rangle \cdot v_{rel} \cdot dN/dy/(\pi R_p^2)$; τ_i is the Lorentz-dilated J/ψ formation time, τ_f is the last time the J/ψ can interact, $\langle\sigma\rangle$ is the mean absorption cross section, v_{rel} is the mean relative speed between the J/ψ and the hadrons in the hadron resonance gas, R_p is the projectile radius, and dN/dy is the rapidity density of the hadron gas. The resulting p_t dependence tends to be much flatter than observed in the experiment (see 108). Dense assemblies of hadrons of a few per fm^3 are implicit in these models. A careful consideration of final-state interactions, reabsorption on dense matter, and scattering from co-moving hadrons is able to reproduce the trend for R as a function of $E_t/A^{2/3}$ in the NA38 data (118, 119).

It was also pointed out that regeneration of J/ψ by decays from parents whose mean lives are much longer than expected plasma lifetimes (e.g. χ_c, η_c, and B mesons) can contribute significantly to the J/ψ yield (117, 120). For example, WA11 found 39% of J/ψ resulted from decays of the χ_{c1}, χ_{c2}, and ψ' states (121). The E806 collaboration found 47 ± 8% of J/ψ result from radiative decays (122). Such effects are expected to dominate the high p_t part of the J/ψ spectrum for heavy ion collisions at Large Hadron Collider (LHC) energies (120).

The weak p_t dependence of models considering only reabsorption effects leads to the question of whether or not initial-state scattering of the incoming partons contributes to the measured increase of $\langle p_t^2 \rangle$ with E_t. It is known from pA studies of J/ψ production that $\langle p_t^2 \rangle$ increases with target mass (123–125). A similar although smaller effect is also measured in Drell-Yan production in pA reactions (124). Here any A-dependent effects are ascribed to initial-state effects because the Drell-Yan process is purely electromagnetic. The total cross section for Drell-Yan events in pA reactions is observed to scale as A times that for pp events (126), which means the initial q and $\bar{q}$ producing the Drell-Yan event underwent only elastic interactions in the entrance channel.

Similar A scaling for continuum events in AA collisions is reported by NA38 (127). Extending these arguments to the gluon fusion producing J/ψ's leads to the following interpretation: the increase of $\langle p_t^2 \rangle$ with increasing E_t (and thus with increasing number of nucleon-nucleon collisions) for J/ψ in AA collisions is another manifestation of gluon multiple scattering (128–130). A relation between $\langle p_t^2 \rangle$ for the J/ψ in pp and pA reactions is given as

$$\langle p_t^2 \rangle_{pA}^{J/\Psi} = \langle p_t^2 \rangle_{pp}^{J/\Psi} + \rho_0 \sigma_{gN} \langle p_t^2 \rangle_{gN} L_A, \quad 4.$$

where ρ_0 is the matter density, σ_{gN} is the gluon-nucleon cross section, $\langle p_t^2 \rangle_{gN}$ is the mean p_t^2 generated in gluon-nucleon collisions, and $L_A = (3/4) r_0 A^{1/3}$ is the impact-parameter-averaged target thickness. From the NA3 and NA10 results, a value for $\sigma_{gN} \cdot \langle p_t^2 \rangle_{gN} = 0.39 \pm 0.06$ fm^{2}GeV^2 is obtained. A similar relation holds for A-B nucleus-nucleus collisions:

$$\langle p_t^2 \rangle_{BA}^{J/\Psi} = \langle p_t^2 \rangle_{pp}^{J/\Psi} + \rho_0 \sigma_{gN} \langle p_t^2 \rangle_{gN} (L_A + L_B). \quad 5.$$

Such an expression, using the value for $\sigma_{gN} \cdot \langle p_t^2 \rangle_{gN}$ taken from pA data, can reproduce the trend of R with p_t seen by NA38 as shown in (105, 107, 108). A similar relation holds for continuum events:

$$\langle p_t^2 \rangle_{BA}^{J/\Psi} = \langle p_t^2 \rangle_{pp}^{J/\Psi} + \rho_0 \sigma_{qN} \langle p_t^2 \rangle_{qN} (L_A + L_B), \quad 6.$$

where now the quark-nucleon cross section enters, with a value $\sigma_{qN} \cdot \langle p_t^2 \rangle_{qN} = 0.12 \pm 0.08$ fm^{2}GeV^2, notably smaller than the gluon-nucleon value, as is expected from QCD. The data for the continuum in fact show only a small increase in $\langle p_t^2 \rangle$ with E_t (107).

Such a treatment needs an overall scale that would be set by reabsorption effects. This has been discussed by Gerschel & Huefner (131), who find they can use a "universal" J/ψ-N cross section of 6.9 ± 1.0 mb combined with the above parton multiple-scattering considerations to reproduce the overall scales and trends of suppression factors. It was noted, however, by Satz (108) that only single gluon exchange is likely, which leads to a saturation of the effects of initial-state gluon scattering on the J/ψ and continuum $\langle p_t^2 \rangle$ with increasing projectile and target mass number and not at all to the linear increase with $(L_A + L_B)$ proposed in (128–130). Thus in this view gluon multiple scattering could only partially explain the experimental growth of $\langle p_t^2 \rangle$.

Further progress can be made in several areas. Better statistics for the continuum pairs will determine how strong their $\langle p_t^2 \rangle$ dependence on centrality is, which should directly address the question of initial-state parton multiple scattering. Measurements made at significantly larger p_t for J/ψ and the continuum will help resolve whether or not there is a limit to the suppression at sufficiently high p_t, as expected in plasma models. Measurements for a range of projectile and target masses will resolve different A-dependent behaviors. Perhaps most importantly, measurements at much larger values of $s^{1/2}$, as will be available at RHIC and perhaps LHC, will open a much greater range of energy densities and plasma initial conditions and lifetimes, thus requiring any suggested interpretation to apply over a large range of conditions.

4.5 *Thermal Radiation*

The system formed during relativistic heavy ion collisions consists of many charged objects moving in close proximity and thus emitting electromagnetic radiation in the form of real and virtual photons. Such radiation has a mean free path orders of magnitude larger than the size of a nucleus and escapes with negligible probability for interaction. It carries with it direct information about conditions at its time of creation, unlike hadrons, which will suffer collisions as the final-state matter expands. The real and virtual photon spectrum has been calculated for a quark-gluon plasma by making assumptions about the initial conditions and the expansion and cooling of the system (132). The most important characteristics are a scaling of the rate with $(dN/dy)^2$, since quark-antiquark or quark-gluon collisions produce the photons, and a T^3 temperature scaling that weights heavily the initial temperature. A recent theoretical update with references to previous work can be found in a publication by Ruuskanen (133). In addition to the thermal radiation, initial-state hard scattering is important at high p_t or mass (134), and bremsstrahlung will dominate at the low p_t end of the spectrum (135).

Photon emission in heavy ion reactions has been measured by WA80 and HELIOS. WA80 measured (one-photon, two-photon, etc) inclusive spectra at 60 and 200 GeV/nucleon using a highly segmented Pb glass calorimeter (136) and reconstructed the two-photon invariant mass and thereby the spectra of π^0 and η mesons (137). The photon decays of these two mesons account for most of the secondary photons in the inclusive spectrum. Using the measured π^0 and η spectra, one can then calculate the contribution of their decays to the inclusive single-photon spectrum and subtract it from the data. Additionally, a few percent of the total yield is expected to arise from radiative decays ($\Sigma^0 \rightarrow \Lambda + \gamma$, $\omega \rightarrow \pi^0 + \gamma$) and hadrons with small two-photon decay branches (e.g. $\eta' \rightarrow \gamma + \gamma$) whose yields must be estimated using known systematics. Any remaining excess can be ascribed to direct thermal and hard-scattering photon emission. Considerable care is required, however, in order to quantify the reconstruction efficiency for photons in the high-multiplicity environment of heavy ion reactions, as this quantity directly enters into the calculation of the fraction of the inclusive photon spectrum arising from hadron decays.

The results for the direct photon component determined by WA80

(138) for several colliding systems are shown in Figure 17. In this figure the measured inclusive γ/π^0 ratios are presented as functions of p_t together with the calculated contributions coming from nondirect sources such as hadron decays; this ratio has advantages in canceling systematic errors in the measurement. The results of WA80 (138) place an upper limit of 15% of the inclusive photon yield on any direct thermal photon component emitted with $0.8 < p_t < 2.4$ GeV in heavy ion reactions at CERN energies. This limit holds for all systems examined and also for all reaction centralities, and thus energy densities, attainable to date.

The inclusive photon spectrum was measured using the conversion technique by HELIOS for p,O,S + W,Pt at 200 GeV/nucleon (139). An iron converter of thickness $t = 5.7\%$ of a radiation length was followed by a tracking spectrometer in order to measure the momenta of the electron and positron from the pair conversion. The use of a thin converter has two consequences. The counting statistics are smaller by at least a factor of $1/t$ (for a 100% efficient spectrometer) for a given integrated luminosity as compared to the calorimetric method used by WA80. The probability of converting both decay photons from a π^0 is $t^2 = 3.2 \times 10^{-3}$, which means it is not feasible with this technique to reconstruct π^0 and η mesons, even putting aside acceptance considerations. A thin converter has an advantage in that it determines the photon direction better than a calorimeter with moderate segmentation.

The yield of photons as a function of p_t was compared with the yield estimated from a Monte Carlo code that used as its input the cross sections for π^0 (taken as being equal to the π^- cross section measured in the same acceptance) together with production cross sections for η, η', and ω mesons normalized to π^0 yields by using ratios determined from ISR p + p data. It should be noted that there are many more neutrons than protons in the entrance channel, and hence the yield of π^0 will not necessarily be equal to the π^- number, but will likely be smaller, depending on the local baryon number. The results reported for the p_t integrated (over $p_t = 0.1$–1.5 GeV) value of γ/π^0 agree within statistical (4–11%) and systematic (9%) errors with the Monte Carlo result for secondary photon decays for a range of centralities (139). This integral result is dominated by the part of the spectrum below $p_t = 0.6$ GeV; statistical errors alone increase to 50% or more for the range 0.6–1.5 GeV. It should be noted that both the integral limit placed by HELIOS and the p_t-dependent limit by WA80 are significantly above the value expected for thermal radiation using the initial conditions reached in present experiments (see Section 5). Recent calculations by Braun-Munzinger, using Ruuskanen's formalism, show

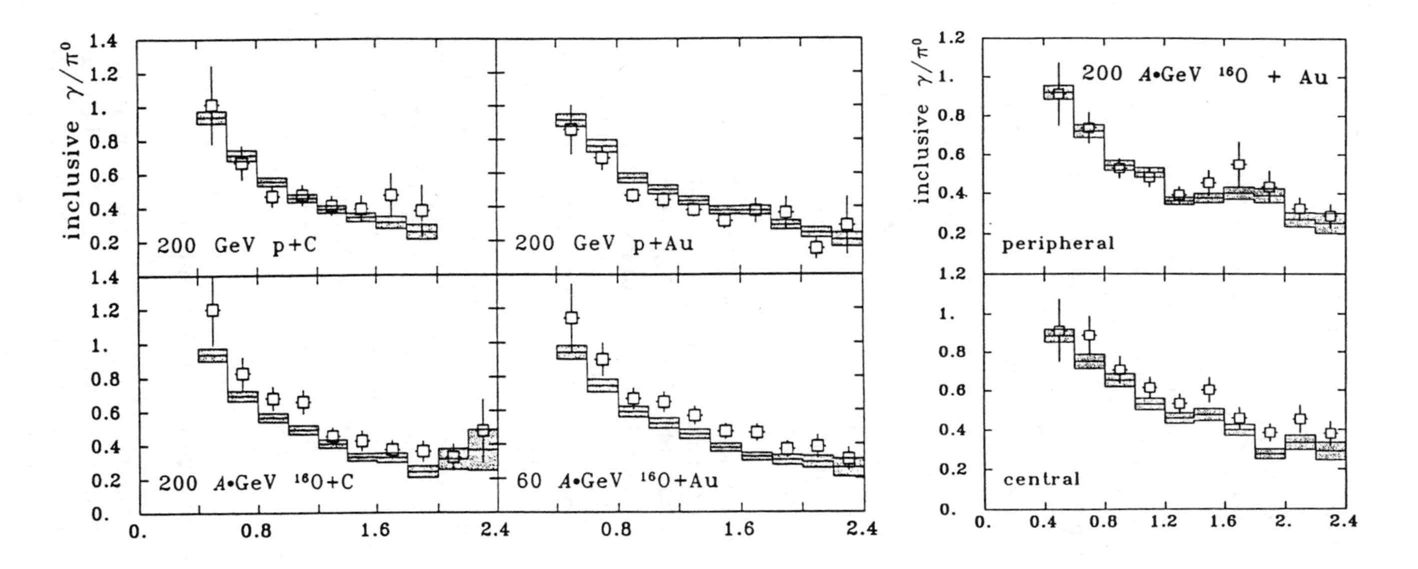

Figure 17 Ratios of inclusive photon to π^0 cross sections (*squares*) as a function of p_t for various systems as measured by WA80 (138). Ratios for photons originating from hadron decay as determined from a Monte Carlo simulation based on measured π^0 and η spectra (see text) are shown as the shaded boxes. If not noted otherwise, the data are for minimum-bias events. (Figure from 138)

that a contribution to γ/π^0 at $p_t = 1$ GeV of 1% or smaller may be expected (140).

The very soft, $p_t < 100$ MeV, component in the photon spectrum has been measured for p+Be collisions by HELIOS (141, 142) and E855 (C. Woody, private communication) following earlier reports of an anomalously large soft-photon component in 70-GeV K^++p reactions measured in the BEBC bubble chamber (143) [an earlier similar measurement for 10.5-GeV π^++p did not show any such excess (144)]. The resulting photon p_t spectra are reproduced within the errors of the experiment by calculations assuming the soft photons are bremsstrahlung emitted by the incident nucleons (141, 142; C. Woody, private communication).

The low-mass ($m < m_\rho$) lepton pair spectrum has been studied by several groups in pA collisions (for a review and references, see 141, 142, 145). An excess was claimed for low-mass pairs beyond the yield due to Dalitz decays of low-mass mesons (such as π^0, η, ρ, ω, η', ϕ); scaling as the square of the associated charged-particle multiplicity was also hypothesized, as is characteristic for the thermal pairs discussed above.

However, a recent investigation by the HELIOS Collaboration has pointed out the need for a careful measurement of the yields of mesons with Dalitz branches (141, 142). They used the HELIOS spectrometer in its open geometry mode (HELIOS/1) to measure e^+e^- pairs over $0.25 < y_{cm} < 1.25$ and $\mu^+\mu^-$ pairs over $0.25 < y_{cm} < 1.5$ simultaneously. A BGO matrix provided photon identification and energy measurement. Dalitz decays such as $\eta \rightarrow e^+e^-\gamma$ and $\eta \rightarrow \mu^+\mu^-\gamma$ could be fully reconstructed. The results show that the yields for these low-mass mesons are significantly above estimates used previously based on scaling from the observed $\rho+\omega$ yield (141, 142); the measured ratio $R_\eta = \sigma_\eta/(\sigma_\rho+\sigma_\omega)$ is 0.55 ± 0.13 and $0.52 \pm 0.06 \pm 0.10$ for e^+e^- and $\mu^+\mu^-$ pairs, respectively, as compared to a value of 0.17 measured at 16 GeV (146). From these results HELIOS concluded that, within systematic errors of 20–25%, there is no evidence for an excess of low-mass pairs in p+Be. See Figure 18. For p+W and S+W collisions the low-mass part of the $\mu^+\mu^-$ spectrum is found to be consistent within errors with a spectrum calculated in the same manner as for p+Be, after scaling of the distributions with A_p and A_t is taken into account (141, 142).

The NA45 Collaboration has brought into operation at the CERN SPS a novel hadron-blind e^+e^- pair spectrometer based on measuring the directions of e^+ and e^- with a pair of azimuthally symmetric RICH counters placed before and after a magnetic field. Taking advantage

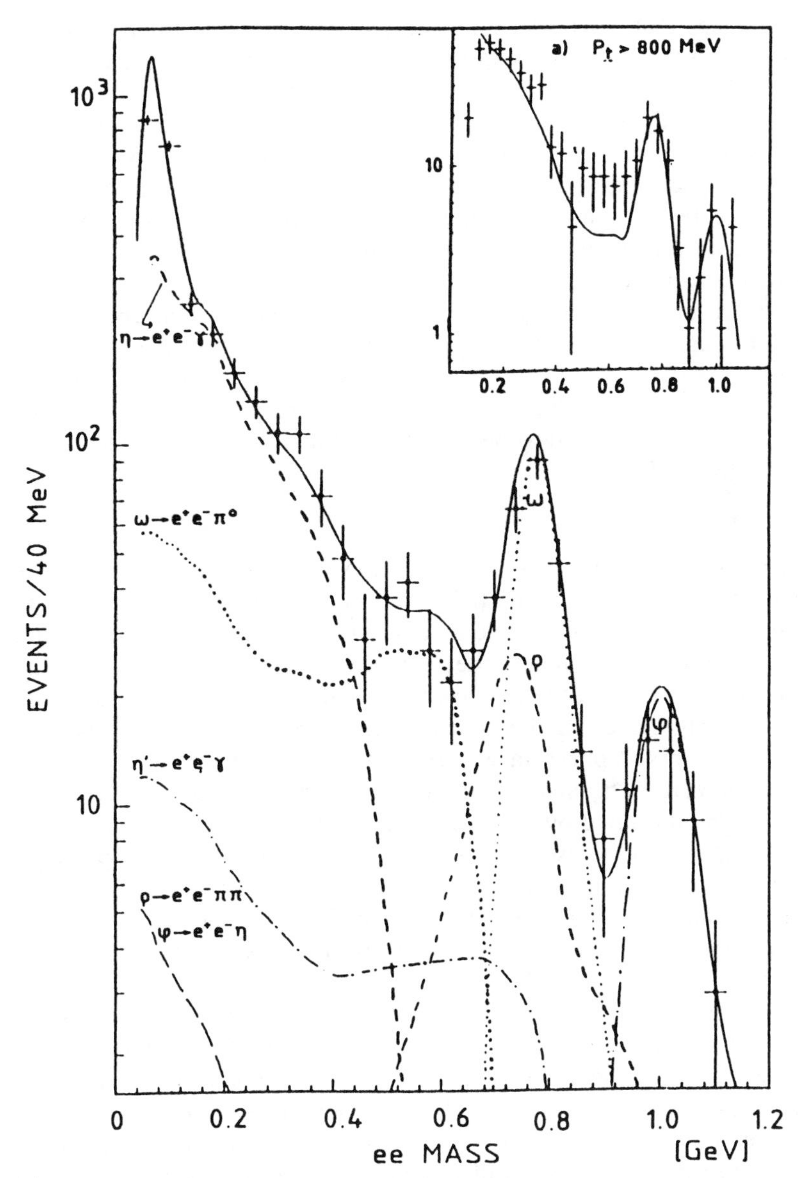

Figure 18 Invariant mass spectrum (*points*) of e^+e^- pairs summed over p_t as measured by HELIOS (141) for p + Be collisions at 200 GeV. Also shown as broken and dashed lines are contributions from Dalitz decays, based on the measured cross sections in the same setup. The contributions from e^+e^- decays of ρ, ω, and φ are also shown, as well as the calculated contribution from $\rho \rightarrow e^+e^-\pi\pi$ and $\phi \rightarrow e^+e^-\eta$ decays. *Solid line*: Sum of all calculated contributions. *Inset*: measured (*points*) and calculated (*solid line*) mass spectra for pairs with $p_t > 800$ MeV. (Figure from 141)

of the fact that a RICH counter measures the direction cosines of a particle, comparison of the displacement of ring centers measured after the field to those measured before yields the particle's momentum. Data from this experiment are expected in the near future.

The lepton pair spectrum at masses $m > m_{\phi}$ was discussed in Section 4.4 as the continuum background for the J/ψ resonance. The data were found to be consistent in terms of mass dependence and A_p and A_t scaling with the expectations for the Drell-Yan process, i.e. initial-state hard scattering of partons.

At present, there is no positive evidence for thermal emission of electromagnetic radiation from heavy ion collisions. In the light of recent estimates for the rate of such radiation (133, 140), it is probably detectable only with the significantly larger energy densities and initial temperatures expected to be encountered in collider experiments, for example at RHIC.

4.6 *Correlations*

There are strong motivations for developing methods to determine the space-time extent of the particle-emitting source created in heavy ion collisions. If the volume and/or time at which freeze-out occurs are determined experimentally, they can be related to the initial size of the participant zone and thereby to energy and number densities achieved in the early stages of the collision. A hadronizing plasma will pass through a mixed phase in which the velocity of sound may become small or zero. The amount of time the system spends in the mixed phase is determined by the latent heat of the phase transition: a large latent heat should result in an expansion lasting a large proper time and thus result in a large size. A large ordered transverse flow—or a high initial temperature, which would lead to a large expansion in size before the system cools enough to hadronize—would also result in a large emitting system (147–149 and references therein).

Two-particle correlations have been studied in heavy ion collisions both at relativistic and at lower energies (150, 151 and references therein). The connections to the geometric and dynamical source properties are outlined in (152, 153). The method is based on the fact, pointed out by Hanbury-Brown and Twiss, that an interference arises between the matter waves of two identical particles if there are two possible paths for the particles from the emission point to the coincidence detectors. If the amplitudes for this process are summed and squared, the result is sensitive to the separation of the emission regions even for incoherent emission amplitudes. The exact shape of the correlation and the resulting spatial and temporal extent of the source require careful consideration of the correlation scales, Lorentz frame in which the analysis is performed, and density of particles in phase space (147, 150).

Different correlation functions have been proposed, using different variables. Fits to experimental correlation functions have been made either using a static Gaussian distribution model (152) in which source points are assumed fixed in space-time and described by Gaussians that are separate in space and time, or using a model incorporating collision dynamics and freeze-out (148). In the former, the Gaussians are taken to be functions of the components of the relative four-momentum of the pair, $Q = p_1 - p_2$, and its conjugate R, the source size four-vector. Q and R can be used in a one-dimensional analysis in terms of Q_{inv}, where $Q_{\text{inv}}^2 = -(p_1 - p_2)^2$, and an associated size R_{inv}, which, however, is not invariant under changes of rest frame (see below). The correlation function is in this case given by:

$$C_2(Q_{\text{inv}}) = 1 + \lambda \exp[-(Q_{\text{inv}}R_{\text{inv}})^2]. \qquad 7.$$

Alternatively, the analysis can be decomposed in longitudinal and transverse components of Q and R, leading to a correlation function

$$C_2(Q_{\text{T}}, Q_{\text{L}}, Q_0) = 1 + \lambda \exp[-(Q_{\text{T}}R_{\text{T}})^2/2 - (Q_{\text{L}}R_{\text{L}})^2/2 - (Q_0\tau)^2/2], \qquad 8.$$

which can be used to study the shape of the source in directions parallel and perpendicular to the beam axis (154). If statistics in the data permits, one can distinguish between two transverse directions, outwards and sidewards, where "outwards" points in the direction of the pair's p_{t} and "sidewards" is perpendicular to the beam and the pair's p_{t} (154–156). This leads to variables $Q_{\text{T}}(\text{out})/R_{\text{T}}(\text{out})$ and $Q_{\text{T}}(\text{side})/R_{\text{T}}(\text{side})$ for the two transverse components. $R_{\text{T}}(\text{out})$ measures the effective depth of the source or the duration of particle emission, and $R_{\text{T}}(\text{side})$ measures the effective transverse source size.

The chaoticity parameter λ can vary from 1 for a totally chaotic source to 0 for a coherent one. The correlation and particularly λ can also be modified greatly by the decay of long-lived resonances (149 and references therein).

The NA35 Collaboration has published results for transverse and longitudinal source radii R_{T} and R_{L} for central O+Au collisions at 200 GeV/nucleon from a two-particle correlation analysis for negative particles, which are dominantly π^- (157). Analyzing the data for several disjoint windows in rapidity, they found R_{T} to reach values as large as 8 fm at midrapidity, decreasing to values of 3–4 fm away from midrapidity. The largest values for λ were also observed at midrapidity, which led to the conjecture that a partly thermalized large source was created at rest in the cm frame. Further analysis, using a sliding rapidity window of width $\Delta y = 1$ and including 200-GeV/nucleon S+S,Ag,Au data, showed that the values of R_T, R_L, and λ display a correlated increase in the window $y_{cm} \pm 0.5$ (80, 158). At larger and smaller rapidities, R_L, evaluated in a fixed observer frame with $y_{obs} = 2.5$, was found to decrease as expected for a longitudinally expanding source with a boost-invariant velocity profile (147). The results for O+Au and S+S,Ag,Au are all near $R_L = 5$ fm, which suggests a freeze-out time of the same order because the initial coherence length in rapidity is thought to be only of the order of $\Delta y = 1$ (147). The transverse radius R_T was found to grow with target and projectile radius, but the actual values are a factor of 1.5–2 larger than the radius of the participant zone (based on the density profile of the intersecting nuclei). The large values of R_L and R_T are an indication that pions are emitted only after the system has expanded considerably, both longitudinally and transversely. The results of NA35 indicate a growth in volume by a factor 10–20. Distinguishing in the analysis between the two transverse dimensions R_T(side) and R_T(out), one finds the resulting values to be equal within errors. This signifies that pions are emitted after considerable expansion but over a time span that is relatively short.

E802/E859 measured two-particle correlations for π^+, π^-, K^+, and p in Si+Al and Si+Au collisions at 14.6 GeV/nucleon (159, 160): they used a single-arm spectrometer and a two-particle trigger to enhance the data sample, an enhancement that is particularly necessary for kaons. The data have been analyzed using the one-dimensional parametrization in terms of Q_{inv} and R_{inv} (see above) and assuming $R = \tau$. The two-pion correlation results are consistent with values for the equivalent rms radius $r_{rms} = \sqrt{3}R = 3.5 \pm 0.3$ fm, evaluated in the rest frame of the colliding nuclei. There is no significant dependence on target or charge of the pair. The pion source size was studied as function of centrality or target-projectile overlap geometry, and an increase of about 30% was found in going from peripheral to central collisions. The two-proton results used the fermion correlation analysis developed by Pratt (154). The data are well reproduced by that functional form. For Si+Al, a value of $r_{rms} = 3.56 \pm 0.16$ fm was found, which is very similar to the result for pions, while that for Si+Au was significantly larger, 4.75 ± 0.14 fm.

This experiment also obtained the first results for two-K^+ correlations for central Si+Au collisions. It yielded a value of $R_{inv} = 2.47$

fm in the two-K^+ rest frame, to be compared with $R_{inv} = 4.19$ fm for two-π^+ data. However, as discussed by Zajc et al, R_{inv} is frame dependent (159). R_{inv} measures the distribution of separations in the source of the two observed particles, as determined in the rest frame of the pair. Kaons will have a smaller Lorentz factor with respect to the source than will pions. Transforming into the rest frame of the two colliding nuclei (147, 159) results in $r_{rms} = 3.4$ fm for kaons and 3.5 fm for pions, as quoted above. Thus at AGS energies it appears that the source radii measured via pions or kaons are only slightly larger than the initial geometric distribution of the nucleons. The proton radii, on the other hand, show a 30–40% increase for heavy targets, corresponding to a growth in volume of a factor 2.5. Most likely this is due to target spectator protons emitted from secondary collisions.

Correlations for 200-GeV/nucleon S+Pb were presented by NA44, also from a single-arm spectrometer (161). An analysis using Equation 7 yielded values of $R_{inv} = 6.2$–6.4 fm. The preliminary analysis does not yet indicate a clear trend with centrality or pair p_t. The values for two-Gaussian fits were $R_T = 6.0$–6.2 fm and $R_L = 6.0 \pm 0.7$ fm for central and 7.2 ± 0.8 fm for peripheral reactions.

There is much interest in the scaling behavior of source sizes. The E802/E859 Collaboration (159) notes their rms radii are consistent with $A_p^{1/3}$ scaling, as found earlier (162) in work done at the Bevalac and at Dubna. However, the CERN results (80, 157, 161) lie well above such a curve, which in combination with data from ISR and Sp$\bar{p}$S data suggests that a scaling of r_{rms} with $(dN_c/dy)^{1/2}$ might be more accurate (150, 157, 162, 163). This behavior of source sizes was proposed as evidence of isotropic (chaotic) expansion of a system in which pions freeze-out when the system reaches dimensions close to the pion mean free path (163), which in turn leads to the $(dN_c/dy)^{1/2}$ scaling. Improved statistics now indicate that a scaling according to $(dN_c/dy)^{1/3}$ gives better agreement with the data. This suggests rather an isentropic expansion with ordered flow: a particle interacts only with its local environment, and therefore freeze-out occurs at a density such that the average distance to the nearest neighbor is equal to the mean free path, which results in a $(dN_c/dy)^{1/3}$ dependence of the radii (164), as shown in Figure 19.

The results to date for two-particle correlations of mesons are consistent with emission (*a*) from a source with a transverse size of the order of the projectile size for reactions at beam energies up to those at the AGS, and (*b*) from a source that has expanded in radius by less than a factor of two above the projectile radius at SPS energies. The scaling with particle multiplicity is suggestive of an isentropic expan-

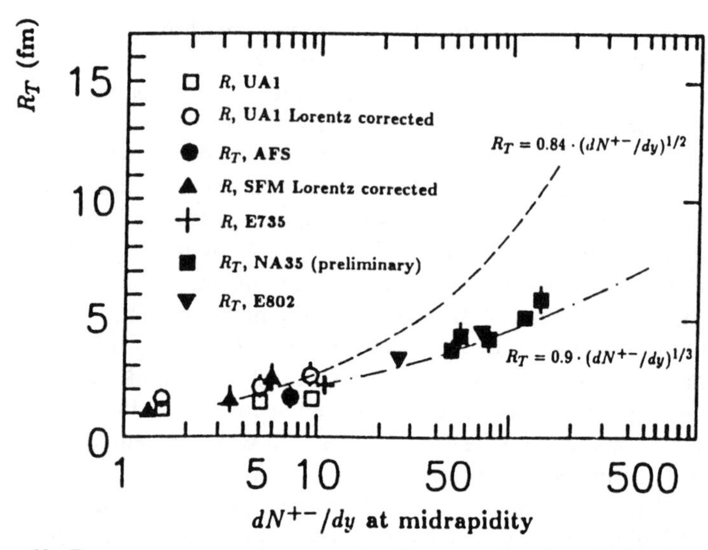

Figure 19 Transverse source size parameter R_T versus the charged pion density at midrapidity. A model with freeze-out at $R_T - \Lambda(\pi\pi)$, the pion mean free path, yields a scaling as $(dN^{+-}/dy)^{1/2}$ (*dashed line*), while one with freeze-out at pion density $\rho_\pi = 0.5\ fm^{-3}$ yields a scaling as $(dN^{+-}/dy)^{1/3}$ (*dash-dotted line*). (Figure from 164)

sion, which has implications for the energy densities attained (see below). Future measurements with high statistics for identified particles are planned in order to perform a full separation of the different source dimensions. It remains to study correlations as a function of p_t in order to understand the possible effects of resonance decay (149, 150).

Nonstatistical multiplicity fluctuations in small intervals of rapidity could be a sign of a deconfined phase (165) leading to intermittent behavior. Bialas & Peschanski (166) have proposed an analysis of the rapidity density distributions in terms of normalized factorial moments to study intermittency. This was taken up by experimental and theoretical groups (see references in 4–7). However, the experimental situation is unclear and conflicting results have been reported. On the theoretical side Carruthers & Sarcevic (167) cautioned that the intermittency phenomenon in hadron-hadron collisions could be understood in terms of conventional short-range correlations and is completely determined by two-particle correlations (not correlations of identical particles as discussed above, but long-range correlations over typically $\Delta y = 1.0$). This suggests that it is prudent to study as a first step the two-particle correlation functions obtained in heavy ion experiments carefully considering such effects as resonance decays, pair conversion, residues of Hanbury-Brown & Twiss correlations, and Coulomb effects.

5. NUCLEAR STOPPING POWER AND ENERGY DENSITIES

The amount of stopping S in a heavy ion collision can be deduced from experimental data in different ways, as outlined in Section 3. In particular, we have seen that comparison of measured transverse energy distributions with models can lead to quite a range of possible answers. To select among them one should require that the stopping fraction so deduced agrees with other, less model-dependent measures and that the model used to deduce S reproduces the data for a particular reaction overall. As can be seen from the discussion in Sections 3 and 4, the isotropic fireball model does not meet either of these criteria and thus cannot be used to obtain reliable information about the extent of stopping in a given reaction. Table 3 summarizes the best values for S emerging from an overall evaluation of the information given in the previous sections. The corresponding values for the thermalized energy E_{ex} in the participant volume are obtained as SE^* (see Table 1).

A first estimate for the initial energy density ϵ_0 can be obtained by dividing the energy E_{ex} by the volume of the participants calculated as $V_p = A_t d/\gamma_{cm}$, where A_t is the transverse area of the smaller of target or projectile, and d is the diameter of the larger nucleus. This is the volume at collision time and it is, of course, not clear that the energy E_{ex} is thermalized at that time. In fact there is information from studies within the RQMD model that thermalization takes several fm/c longer (168) and that ϵ_0 at collision time cannot be associated with a thermodynamic phase in equilibrium.

Shuryak (169) and Bjorken (170) have used longitudinal growth and

Table 3 Stopping S achieved in typical $b = 0$ collisions on heavy targets

System	E_{lab}/A (GeV)	S (%)	E_{ex} (GeV)	Volume (fm^3)	ϵ_0 (GeV/fm^3)	Ref
Si + Pb	14.6	100	179	296	0.6	12, 13
O + Au	60	>90	227	129	1.8	16, 17
S + Au	200	>80	737	104	7.1	16, 17
S + S	200	60	337	45[a]	7.5	16, 17, 45

[a] d_{min} = 1 fm used.

proposed a scenario of free hydrodynamical flow. In this scenario the energy density is calculated as

$$\epsilon_{sb} = \langle m_t \rangle \, dN/dy/(A_t \tau), \quad 9.$$

where $\langle m_t \rangle$ is the average transverse mass of the particles emitted, A_t has the same meaning as above, and τ is the formation time, usually taken to be 1 fm. Instead of $\langle m_t \rangle dN/dy$, sometimes dE_t/dy is used and one should note that nucleons contribute quite differently in that case. Quite analogously the entropy density can be calculated and, using $dS/dy = 3.7\, dN/dy$ from Landau theory for multiparticle production (the exact number depends on the particle statistics), one obtains

$$\sigma_{sb} = 3.7\, dN/dy/(A_t \tau). \quad 10.$$

For an ideal gas we have $\epsilon = (\pi^2/30)\kappa T^4$, $\sigma = 4/3(\pi^2/30)\kappa T^3$ with the degeneracy of the gas κ. For a pion gas, $\kappa = 3$; for a gas of gluons and quarks of N_f flavors, $\kappa = 2[16+(21/2)N_f]$. Using these values, one can calculate the energy density for the case of isentropic expansion as proposed by Gyulassy & Matsui (171). For an ideal gas of quarks and gluons, one finds

$$\epsilon_{qg} = 0.33[dN/dy/(A_t \tau)]^{4/3}. \quad 11.$$

The key difference from the Shuryak-Bjorken estimate is that, as the system expands and cools, m_t is decreasing also. Note that ϵ is rising faster than linearly with dN/dy for this isentropic expansion. While in hydrodynamics the total entropy per unit rapidity is a constant of motion, both heating due to viscosity and a phase transition add entropy to the system. Gyulassy & Matsui calculated an upper bound for the entropy generated and obtain, using this upper bound, ϵ_{sb} as a lower bound for the energy density. Both estimates, ϵ_{sb} and ϵ_{qg}, are displayed in Figure 20, together with estimates based on experimental data for Si(S)+Pb(Au) and $\sigma/\sigma_{geo} = 5$–7% corresponding to typical $b = 0$ collisions, as shown in Figure 6. Depending on the estimate chosen, energy densities in the range of 1.5–3 GeV/fm^3 are obtained in current heavy ion reactions using the heaviest systems available.

The biggest uncertainty is introduced by the assumption made for the value of τ. It could be larger or smaller by factors of two. In particular, it has been argued (172) that the initial temperature and the formation time are linked by the uncertainty principle and should both have a characteristic A dependence:

$$T_0 \approx 200 A^{1/6} \text{ MeV} \quad \text{and} \quad \tau_0 \approx A^{-1/6} \text{ fm}. \quad 12.$$

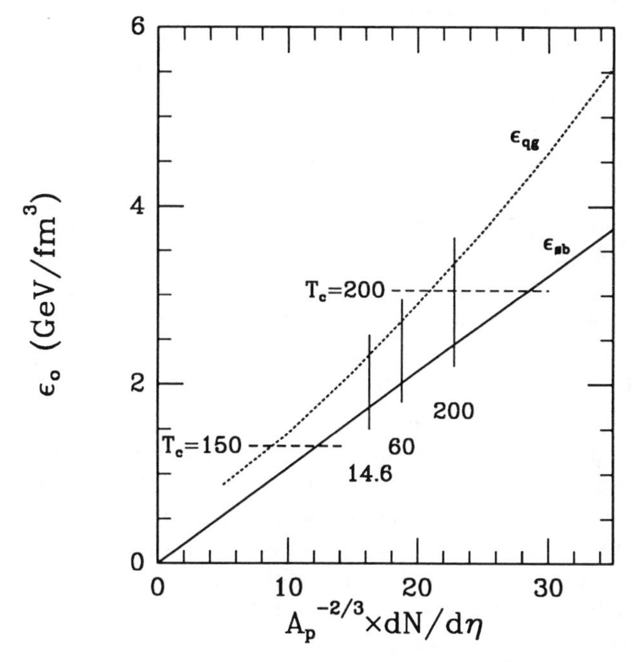

Figure 20 Initial energy density as a function of the multiplicity density. The vertical lines correspond to estimates based on the measured charged-particle multiplicity at midrapidity for the three indicated beam energies using data from E814 (13) and HELIOS (22). Curves represent the theoretical estimates for the Shuryak-Bjorken scenario (ϵ_{sb}) (*solid lines*) and for isentropic plasma expansion (ϵ_{qg}) (*dotted lines*). Critical energy densities for a quark-gluon plasma at two values of T_c are also shown (*dashed horizontal lines*) (see text).

This estimate is relevant at very high energies and would lead, for an ideal gas, to initial energies densities (172),

$$\epsilon_0 \propto T^4 \approx A^{2/3}\ \text{GeV/fm}^3. \qquad 13.$$

Energy densities also have been evaluated within several models that are otherwise in good agreement with the overall characteristics of the experimental data. A calculation within the RQMD model gives for central Si + Au collisions at 14.6 GeV a maximum energy density of 1.7 GeV/fm^3 (168). Within the string model VENUS, energy densities have been evaluated for collisions of symmetric systems and values significantly higher than the Shuryak-Bjorken estimate have been found (173). For central S + S collisions at 200 GeV, an energy density of about 6 GeV/fm^3 has been obtained, close to the simple estimate given in Table 3.

The question of the critical temperature for the phase transition to the quark-gluon plasma and the energy density connected with it has been addressed by lattice QCD simulations. The answer for the pure gauge transition, i.e. a system of gluons only, has been fairly well established for a few years (174). The critical temperature is 200 ± 20 MeV. The energy density was found to rise quickly to the Stefan Boltzmann value, which would be $\epsilon_c = 1.1 \pm 0.4$ GeV/fm^3. Calculations including several flavors of quarks still show finite size effects and only an estimate for T_c can be given. It could be as low as 150 MeV, but there are arguments (174) that it could also approach the pure gauge value of 200 MeV, leading, for $N_f = 2$, to $\epsilon_c = (0.8 - 2.6) + B$, where B is the bag constant. In Figure 20, values are indicated for the two critical temperatures (as dashed lines) using $B = 0.5$ GeV (175). The above QCD results are for baryon-free systems, i.e. $\mu = 0$. This is obviously not the situation in current heavy ion experiments, where the central region has a substantial baryon density at all three bombarding energies. It is expected that the critical temperature will decrease from its $\mu = 0$ value with increasing baryon chemical potential. A study has been made by the Bielefeld group (176) to see the effect of a finite baryon chemical potential on their lattice QCD results. They see a clear shift of the deconfinement point towards lower temperature as expected and they estimate that for $\mu = 400$ MeV T_c has dropped about 25%.

Considering all of this, it could well be that current relativistic heavy ion experiments actually reach the critical temperature and energy density or come very close to it. For the heavier systems available in the near future both at the AGS and at CERN the expectations are that energy densities will increase significantly and that the systems will remain at a high energy density for a longer time. The RQMD model predicts (168) for Pb + Pb collisions at 10 GeV a maximum energy density of 3 GeV/fm^3, nearly a factor of two higher than that for Si + Au collisions at 14.6 GeV. A similar increase is found in the VENUS model at 200 GeV (173).

6. FUTURE PROSPECTS

The discussion in the previous sections showed that energy and particle densities both increase with the use of heavier projectiles and larger cm energies. The initial temperature reached, the length of time a system will stay above the critical energy density, and the strength of many signals of restoration of chiral symmetry and deconfinement, particularly the penetrating ones, are all increased by the same measures. Attainment of the conditions needed to create, identify, and diagnose the quark-gluon plasma is thus most likely if the heaviest nuclei possible and the highest collision energies available are used. These observations have led to the current efforts to accelerate Au nuclei in the BNL AGS and Pb nuclei in the CERN SPS. The newly operational AGS Booster makes it possible to produce Au beams of 11.7 GeV/nucleon in addition to the O and Si beams available so far. Acceleration of Pb ions in the CERN Booster/PS/SPS complex to the top energy of 160 GeV/nucleon is expected by 1994. Upgrades to existing experiments and construction of new ones are proceeding apace, with several technical improvements being made to handle the very large final-state multiplicities, which are expected to reach the order of a few thousand charged particles per central collision. If created at these fixed-target facilities, the QGP would have a large baryochemical potential.

The desire to achieve significantly higher energy densities and temperatures and to create a QGP with low net baryon density led to the decision to build the Relativistic Heavy Ion Collider (RHIC) at BNL and to discussions concerning acceleration of heavy ions in the proposed Large Hadron Collider (LHC) at CERN. RHIC will attain cm energies per colliding nucleon-nucleon pair of $s^{1/2} = 250$ GeV for light nuclei and 200 GeV for Au-Au collisions, while LHC might reach 3.8 TeV for Pb-Pb collisions. The construction of RHIC was begun in 1991, with the expectation of colliding beams operating by 1997. RHIC in particular offers operation as a dedicated facility for such research, an important consideration for experiment planning, for observation of several interesting signals of deconfinement (such as vector meson suppression and/or modification), and for modification of parton propagation, all of which have small cross sections and thus must be studied at large integrated luminosity. Present theoretical estimates indicate that only at these colliding beam facilities will the temperature significantly exceed the critical temperature, a likely requirement for the emission of observable amounts of thermal radiation.

ACKNOWLEDGMENTS

It is our pleasure to thank Drs. Terry Awes, Peter Braun-Munzinger, Ole Hansen, Vladislav Manko, Vesa Ruuskanen, Rainer Santo, Peter Sonderegger, Reinhard Stock, Edward Shuryak, and William Zajc for enlightening discussions and comments. This work was supported in part by the NSF and the US DOE.

Literature Cited

1. Shuryak, E. V., *Phys. Rep.* 61:71 (1980)
2. Satz, H., *Annu. Rev. Nucl. Part. Sci.* 35:245 (1985)
3. Ludlam, T. W., Wegner, H. E., eds., *Proc. Quark Matter 83, Nucl. Phys.*, Vol. A418 (1984)
4. Schroeder, L. S., Gyulassy, M., eds., *Proc. Quark Matter 86, Nucl. Phys.*, Vol. A461 (1987)
5. Satz, H., Stock, R., Specht, H. J., eds., *Proc. Quark Matter 87, Z. Phys.*, Vol. C38 (1988)
6. Baym, G., Braun-Munzinger, P., Nagamiya, S., eds., *Proc. Quark Matter 88, Nucl. Phys.*, Vol. A498 (1989)
7. Blaizot, J. P., et al, eds., *Proc. Quark Matter 90, Nucl. Phys.*, Vol. A525 (1991)
8. Kajantie, K., McLerran, L., *Annu. Rev. Nucl. Part. Sci.* 37:293 (1987)
9. Prelec, K., van Steenbergen, A. in Proc. Workshop on BeV per Nucleon Collisions of Heavy Ions, New York, 1974, Columbia preprint
10. Wohl, C. G., et al, "Current Experiments in Elementary Particle Physics," Berkeley Particle Data Group, Rep. LBL-91, revised UC-414, September 1989; update presently in preparation (1992); Experiments at CERN 1991. CERN report
11. Braun-Munzinger, P., Cleland, W., Young, G. R., Detectors for Relativistic Heavy-Ion Experiments," Report prepared for the NSAC Instrumentation Subcommittee, Rep. ORNL/TM-11179 (1989)
12. Barrette, J., et al (E814 Collaboration), *Phys. Rev. Lett.* 64:1219 (1990)
13. Barrette, J., et al (E814 Collaboration), *Phys. Rev.* C46 (1992), in print
14. Stachel, J., Braun-Munzinger, P., *Phys. Lett.* 216B:1 (1989); *Nucl. Phys.* A498:577c (1989)
15. Akesson, T., et al (HELIOS Collaboration), *Phys. Lett.* 214B:295 (1988)
16. Baechler, J., et al (NA35 Collaboration), *Z. Phys.* C52:239 (1991)
17. Albrecht, R., et al (WA80 Collaboration), *Phys. Rev.* C44:2736 (1992)
18. Anderson, B., *Phys. Rep.* 97:31 (1983); Anderson, B., Gustafson, G., Nilsson-Almqvist, B., *Nucl. Phys.* B281:289 (1987); Nilsson-Almqvist, B., Stenlund, E., *Comput. Phys. Commun.* 43:387 (1987)
19. Waters, L., PhD thesis, SUNY, Stony Brook, Energy Flow in Relativistic Heavy Ion Collisions at AGS Experiment 814. (1990)
20. Akesson, T., et al (HELIOS Collaboration), *Nucl. Phys.* B333:48 (1990)
21. Albrecht, R., et al (WA80 Collaboration), *Phys. Lett.* 202B:596 (1988)
22. Akesson, T., et al (HELIOS Emulsion Collaboration), *Nucl. Phys.* B342:279 (1990)
23. Stachel, J., *Nucl. Phys.* A525:23c (1991)
24. Braun-Munzinger, P., Stachel, J., *Nucl. Phys.* A498:33c (1989)
25. Shor, A., Longacre, R., *Phys. Lett.* 218B:100 (1989); Ludlam, T., Pfoh, A., Shor, A., in *Proc. RHIC Workshop I*, Brookhaven Natl. Lab., ed. P. Haustein, C. Woody, Rep. BNL 51921. Upton, NY: Brookhaven Natl. Lab. (1985)
26. Werner, K., Koch, P., *Phys. Lett.* 242B:251 (1990)
27. Akesson, T., et al (HELIOS Collaboration), *Nucl. Phys.* B353:1 (1991); Akesson, T., et al (HELIOS Collaboration), *Nucl. Phys.* B357:208 (1991), erratum
28. Adamovich, M. I., et al (EMU01 Collaboration), *Phys. Rev. Lett.* 62:2801 (1989)
29. Jain, P. L., et al (EMU08 Collaboration), *Phys. Lett.* 235B:351 (1990)
30. Barbier, L. M., et al (KLM Collaboration), *Phys. Rev. Lett.* 60:405 (1988)
31. von Gersdorff, H., PhD thesis, Univ. Minnesota, Minneapolis (1989)
32. Barrette, J., et al (E814 Collaboration), *Phys. Rev.* C45:819 (1992)
33. Chapman, S., Gyulassy, M., *Phys. Rev. Lett.* 67:1210 (1991)
34. Abbott, T., et al (E802 Collaboration), *Phys. Rev. Lett.* 64:847 (1990)
35. Abbott, T., et al (E802 Collaboration), *Phys. Rev. Lett.* 66:1567 (1991)
36. Love, W. A., et al (E810 Collaboration), *Nucl. Phys.* A525:601c (1991); Love, W. A., et al (E810), in *Proc. Workshop on Heavy Ion Physics at the AGS*, Brookhaven Natl. Lab., ed. O. Hansen, Rep. BNL 44911. Upton, NY: Brookhaven Natl. Lab. (1990), p. 27
37. Ahmad, S. et al (E810 Collaboration), *Phys. Lett.* B281:29 (1992)
38. Braun-Munzinger, P., et al (E814 Collaboration), in *Proc. Quark Matter 91, Nucl. Phys.* A544 (1992), in print
39. Sorge, H., et al, *Phys. Rev. Lett.* 68:286 (1992)
40. Sorge, H., Stoecker, H., Greiner, W., *Ann. Phys.* 192:266 (1989); Mattiello, R., et al, *Phys. Rev. Lett.* 63:1459 (1989) *Nucl. Phys. (Proc.*

Suppl.) 24B:221 (1991)
41. Akesson, T., et al (HELIOS Collaboration), preprint CERN-PPE/91-77 (1991); submitted to *Z. Phys.* (1992)
42. Albrecht, R., et al (WA80 Collaboration), *Z. Phys.* C45:529 (1989)
43. Ranft, J., *Z. Phys.* C43:439 (1989)
44. Tonse, S. R., et al (NA35 Collaboration), *Nucl. Phys.* A525:689c (1991)
45. Wenig, S., PhD thesis, Goethe Univ., Frankfurt, Proton Spektren in zentralen ^{32}S-^{32}S Reaktionen bei 200 GeV/Nukleon, CERN Experiment NA35 (1990)
46. Stroebele, H., et al (NA35 Collaboration), *Nucl. Phys.* A525:59c (1991)
47. Bartke, J., et al (NA35 Collaboration), *Z. Phys.* C48:191 (1990)
48. Wong, C. Y., *Phys. Rev.* D30:972 (1984); Huefner, J., Klar, A., *Phys. Lett.* 145B:167 (1984); Csernai, L. P., Kapusta, J. I., *Phys. Rev.* D32:619 (1985); Bowlin, J. B., Goldhaber, A. S., *Phys. Rev.* D34:778 (1986)
49. Van Hove, L., *Phys. Lett.* 118B:138 (1982)
50. Schukraft, J., in *Proc. Int. Workshop on QGP Signatures*, Strasbourg, 1990, ed. V. Bernard et al. Gif-sur-Yvette: Edition Frontieres (1991) p. 127
51. Stachel, J., *Nucl. Phys.* A527:167c (1991)
52. Albrecht, R., et al (WA80 Collaboration), *Z. Phys.* C47:367 (1990)
53. Albrecht, R., et al (WA80 Collaboration), *Z. Phys.* C38:97 (1988)
54. Akesson, T., et al (HELIOS Collaboration), *Z. Phys.* C46:361 (1990)
55. Stroebele, H., et al (NA35 Collaboration), *Z. Phys.* C38:89 (1988)
56. Hagedorn, R., *Riv. Nuevo Cimento* 6:1 (1983)
57. Chasman, R., Hansen, O., Wegner, H., in *Proc. 9th Int. Seminar on High Energy Physics Problems, Relativistic Nuclear Physics and Quantum Chromodynamics*, JINR, Dubna, USSR, June 1988, preprint BNL-41534 (1988)
58. Eisemann, S. E., et al (E810 Collaboration), *Phys. Lett.* 248B:254 (1990)
59. Greiner, D., et al (NA36 Collaboration), in *Proc. Quark Matter 91, Nucl. Phys.* A544 (1992), in print; Andersen, E., et al (NA36 Collaboration), *Phys. Rev.* C (1992)
60. Cole, B., et al (E802 Collaboration), in *Proc. Quark Matter 91, Nucl. Phys.* A544 (1992), in print
61. Bamberger, A., et al, *Z. Phys.* C43: 25 (1989)
62. van Hecke, H., et al (HELIOS Collaboration), *Nucl. Phys.* A525:227c (1991)
63. Bloomer, M. A., PhD thesis, MIT, Boston, Energy and Baryon Densities in ^{28}Si + A Collisions at 14.6 A GeV/c, AGS experiment 802 (1990)
64. Alper, B., et al, *Nucl. Phys.* B100:237 (1975)
65. Boeggild, H., et al, *Nucl. Phys.* B57: 77 (1973)
66. Becker, U., et al, *Phys. Rev. Lett.* 37: 1733 (1976)
67. Lee, K., Heinz, U., Schnedermann, E., *Z. Phys.* C48:525 (1990)
68. Abbott, T., et al (E802 Collaboration), preprint BNL-46545 *Phys. Lett.* B (1991)
69. Werner, K., *Nucl. Phys.* A525:501c (1991)
70. Takahashi, Y., et al (EMU05 Collaboration), *Nucl. Phys.* A525:591c (1991)
71. Shuryak, E. V., *Nucl. Phys.* A533: 761 (1991)
72. Breakstone, A., et al, *Z. Phys.* C33: 333 (1987); Bell, W., et al, *Z. Phys.* C27:191 (1985); Breakstone, A., et al, *Phys. Lett.* 183B:227 (1987)
73. Arnison, G., et al (UA1 Collaboration), *Phys. Lett.* 118B:167 (1982)
74. Cronin, J. W., et al, *Phys. Rev.* D11: 3105 (1975)
75. Kataja, M., et al, *Phys. Rev.* D34: 2755 (1986)
76. Wroblewski, A., *Acta Phys. Pol.* B16:379 (1985)
77. Rafelski, J., Mueller, B., *Phys. Rev. Lett.* 48:1066 (1982)
78. Koch, P., Mueller, B., Rafelski, J., *Phys. Rep.* 142:167 (1986)
79. Jacob, M., Rafelski, J., *Phys. Lett.* 190B:173 (1987)
80. Seyboth, P. (NA35 Collaboration), in *Proc. Quark Matter 91, Nucl. Phys.* A544 (1992), in print
81. Ferreira, R. (NA38 Collaboration), in *Proc. Quark Matter 91, Nucl. Phys.* A544 (1992), in print
82. Mazzoni, M. A. (HELIOS 3 Collaboration), in *Proc. Quark Matter 91, Nucl. Phys.* A544 (1992), in print
83. Abatzis, S., et al (WA85 Collaboration), *Phys. Lett.* 270B:123 (1991)
84. Abatzis, S., et al (WA85 Collaboration), *Phys. Lett.* 259B:508 (1991)
85. Abatzis, S., et al (WA85 Collaboration), *Phys. Lett.* 244B:130 (1990)
86. Foley, K. (E810 Collaboration), in *Proc. Quark Matter 91, Nucl. Phys.* A544 (1992), in print
87. Kogut, J., Sinclair, D., *Phys. Rev. Lett.* 60:1250 (1988)

88. Barz, H. W., et al, *Nucl. Phys.* A486: 661 (1988)
89. Mattiello, R., et al, *Nucl. Phys. (Proc. Suppl.)* B24:221 (1991)
90. Nikolaev, N. N., *Z. Phys.* C44:645 (1989)
91. Brick, D. H., et al (E570 Collaboration), *Phys. Rev.* D45:734 (1992)
92. Abbott, T., et al (E802 Collaboration), preprint BNL-45018, *Phys. Rev.* D (1992), in print
93. Aichelin, J., Werner, K., Univ. Heidelberg preprint HD-TVP-91-18 (1991)
94. Brown, G. E., et al, *Phys. Rev.* C43: 1881 (1991)
95. Akesson, T., et al (AFS Collaboration), *Nucl. Phys.* B246:1 (1984)
96. Chin, S. A., *Phys. Lett.* 119B:51 (1982)
97. Pisarski, R., *Phys. Lett.* 110B:155 (1982)
98. Svetitsky, B., *Phys. Lett.* B227:450 (1989)
99. Matsui, T., Satz, H., *Phys. Lett.* B178:416 (1986)
100. Karsch, P., Petronzio, R., *Phys. Lett.* B212:255 (1988)
101. Chu, M. C., Matsui, T., *Phys. Rev.* D37:1851 (1988)
102. Blaizot, J. P., Ollitrault, J. Y., *Phys. Rev.* D39:232 (1989)
103. Baglin, C., et al (NA38 Collaboration), *Phys. Lett.* B220:471 (1989); B255:459 (1991)
104. Baglin, C., et al (NA38 Collaboration), *Phys. Lett.* B251:472 (1990)
105. Abreu, M., et al (NA38 Collaboration), in *Proc. Quark Matter 91, Nucl. Phys.* A544 (1992), in print
106. Baglin, C., et al (NA38 Collaboration), *Phys. Lett.* B251:465 (1990); B262:362 (1991)
107. Baglin, C., et al (NA38 Collaboration), *Phys. Lett.* B268:453 (1991)
108. Satz, H., *Phys. Lett.* B242:107 (1990); *Nucl. Phys.* A525:473c (1991); "Color Screening and Quark Deconfinement in Heavy Ion Collisions, in *Quark-Gluon Plasma*, ed. R. C. Hwa. Singapore: World Scientific (1990), p. 593
109. Matsui, T., *Ann. Phys.* 196:182 (1989)
110. Karsch, F., Petronzio, R., *Z. Phys.* C38:627 (1988)
111. Karsch, F., Satz, H., Z. Physik C51: 209 (1991); see also Cleymans, J., Redlich, K., Satz, H., CERN preprint CERN-TH. 6079/91 (1991)
112. Aubert, J. J., et al (EMC Collaboration), *Nucl. Phys.* B213:1 (1983)
113. Anderson, R. L., et al, SLAC-Pub 1741 (1976)
114. Ftacnik, J., Lichard, P., Psuit, J., *Phys. Lett.* B207:194 (1988)
115. Huefner, J., Gerschel, C., *Phys. Lett.* B207:253 (1988)
116. Gavin, S., Gyulassy, M., Jackson, A., *Phys. Lett.* B207:257 (1988)
117. Vogt, R., Prakash, M., Koch, P., Hanson, T. H., *Phys. Lett.* B207:263 (1988)
118. Gavin, S., Vogt, R., *Nucl. Phys.* B345:104 (1990)
119. Vogt, R., Brodsky, S. J., Hoyer, P., *Nucl. Phys.* B360:67 (1991)
120. Sonderegger, P., in *Proc. Large Hadron Collider Workshop*, Aachen, October 1990, CERN 90-101/ECFA 90-133, Geneva: CERN Vol. II, pp. 1252-10 (1990)
121. Lemoigne, Y., et al (WA11 Collaboration), *Phys. Lett.* B11:3509 (1982)
122. Cobb, J. H., et al (R806 Collaboration), *Phys. Lett.* B72:497 (1978); Kourkemelis, C., et al (R806 Collaboration), *Phys. Lett.* B81:405 (1979)
123. Badier, J., et al (NA3 Collaboration), *Z. Phys.* C20:101 (1983)
124. Bordalo, P., et al (NA10 Collaboration), *Phys. Lett.* B193:373 (1987)
125. Katsanevas, S., et al (E537 Collaboration), *Phys. Rev. Lett.* 60:2121 (1988)
126. Bordalo, P., et al (NA10 Collaboration), *Phys. Lett.* B193:368 (1987)
127. Baglin, C., et al (NA38 Collaboration), *Phys. Lett.* B270:105 (1991)
128. Huefner, J., Kurihara, Y., Pirner, H. J., *Phys. Lett.* B215:218 (1988)
129. Gavin, S., Gyulassy, M., *Phys. Lett.* B214:241 (1988)
130. Blaizot, J. P., Ollitrault, J. Y., *Phys. Lett.* B217:392 (1989)
131. Gerschel, C., Huefner, J., in *Proc. Quark Matter 91, Nucl. Phys.* A544 (1992), in print
132. Shuryak, E., *Phys. Lett.* 78B:150 (1978); McLerran, L., Toimela, T., *Phys. Rev.* D31:545 (1985); Hwa, R., Kajantie, K., *Phys. Rev.* D32:1109 (1985); Kajantie, K., et al *Phys. Rev.* D34:2746 (1986)
133. Ruuskanen, P. V., in *Proc. Quark Matter 91, Nucl. Phys.* A544 (1992), in print
134. Ferbel, T., Molzon, W. R., *Rev. Mod. Phys.* 56:181 (1984)
135. Bjorken, J. D., McLerran, L., *Phys. Rev.* D31:63 (1985)
136. Albrecht, R., et al (WA80 Collaboration), *Phys. Lett.* B201:390 (1988)
137. Albrecht, R., et al (WA80 Collaboration), *Z. Phys.* C47:367 (1990)
138. Albrecht, R., et al (WA80 Collaboration), *Z. Phys.* C51:1 (1991)

139. Akesson, T., et al (HELIOS Collaboration), *Z. Phys.* C46:369 (1990)
140. Braun-Munzinger, P., David, G., in *Proc. 21st Int. Workshop on Gross Properties of Nuclei and Nuclear Excitations*, ed. H. Feldmeier, preprint GSI Darmstadt and TH Darmstadt, ISSN 0720-8715:8 (1992); and Braun-Munzinger, P., private communication.
141. Goerlach, U. (HELIOS Collaboration), in *Proc. Quark Matter 91, Nucl. Phys.* A544 (1992), in print
142. Erd, Ch., Diploma thesis, Technical University, Vienna (1991)
143. Chliapnikov, P. V., et al, *Phys. Lett.* B141:276 (1981)
144. Goshaw, A. T., et al, *Phys. Rev. Lett.* 43:1065 (1979)
145. Willis, W. J., in *Proc. Int. Symposium on Lepton and Photon Interactions at High Energies*, Geneva 1991, in print
146. Bartke, J., et al, *Nucl. Phys.* B118: 360 (1977)
147. Sinukov, Y. M., *Nucl. Phys.* A498: 151 (1989)
148. Kolehmainen, K., Gyulassy, M., *Phys. Lett.* B180:203 (1986)
149. Gyulassy, M., Padula, S., *Phys. Lett.* B217:181 (1988)
150. Zajc, W. A., in *Hadronic Matter in Collision*, ed. P. Carruthers. Singapore: World Scientific (1988), p. 43; Zajc, W. A., *Nucl. Phys.* A525:315c (1991) and references therein
151. Boal, D. H., Gelbke, C. K., Jennings, B. K., *Rev. Mod. Phys.* 62:553 (1990)
152. Yano, F. B., Koonin, S. E., *Phys. Lett.* 78B:556 (1978)
153. Gyulassy, M., Kaufmann, S. K., Wilson, L. W., *Phys. Rev.* C20:2267 (1979)
154. Pratt, S., *Phys. Rev. Lett.* 53:1219 (1984): *Phys. Rev.* D33:1314 (1986)
155. Bertsch, G. F., *Nucl. Phys.* A498: 173c (1989)
156. Bertsch, G. F., Tohyama, M., Gong, M., *Phys. Rev.* C37:1896 (1988)
157. Bamberger, A., et al (NA35 Collaboration), *Phys. Lett.* B203:320 (1989)
158. Lahanas, M., et al (NA35 Collaboration), in *Proc. Quark Matter 90, Nucl. Phys.* A525:327c (1991)
159. Zajc, W. A., et al (E802/E859 Collaboration), in *Proc. Quark Matter 91, Nucl. Phys.* A544 (1992), in print
160. Morse, R. A., PhD thesis, MIT, Cambridge, Bose-Einstein Correlation Measurements, in 14.6 A GeV/c Nucleus-Nucleus Collision, AGS experiment 802 (1991)
161. Sarabura, M. et al (NA44 Collaboration), in *Proc. Quark Matter 91, Nucl. Phys.* A544 (1992), in print
162. Bartke, J., Kowalski, M., *Phys. Rev.* C30:1341 (1984)
163. Stock, R., *Ann. Phys.* 48:195 (1991)
164. Stock, R., in *Proc. Quark Matter 91, Nucl. Phys.* A544 (1992), in print
165. Gyulassy, M., in *Proc. Quark Matter 83*, ed. T. W. Ludlam, H. W. Wegner, *Nucl. Phys.* Vol. A418 (1984); van Hove, L., *Z. Phys.* C21:93 (1984)
166. Bialas, A., Peschanski, R., *Nucl. Phys.* B273:477 (1984)
167. Carruthers, P., Sarcevic, I., *Phys. Rev. Lett.* 63:1562 (1989)
168. Sorge, H., et al, *Phys. Lett.* 243B:7 (1990)
169. Shuryak, E. V., *Phys. Lett.* 78B:150 (1978) and references therein
170. Bjorken, J. D., *Phys. Rev.* D27:140 (1983)
171. Gyulassy, M., Matsui, T., *Phys. Rev.* D29:419 (1984)
172. von Gersdorff, H., et al, *Phys. Rev.* D34:794 (1986)
173. Werner, K., Cerello, P., Giubellino, P., in *Proc. ECFA Large Hadron Collider Workshop*, Aachen 1990, preprint HD-TVP-91-1, December (1990)
174. Petersson, B., *Nucl. Phys.* A525:237c (1991)
175. Shuryak, E. V., *Phys. Lett.* 79B:135 (1978)
176. Berg, B., et al, *Z. Phys.* C31:167 (1986)

CUMULATIVE INDEXES

CONTRIBUTING AUTHORS, VOLUMES 33–42

CHAPTER TITLES, VOLUMES 33–42

PARTICLE SPECTROSCOPY

PARTICLE THEORY

GOSHEN COLLEGE - GOOD LIBRARY
3 9310 01078807 1